5th International Conference on Information Technology and Digital Applications (ICITDA 2020)

IOP Conference Series: Materials Science and Engineering
Volume 1077

Yogyakarta, Indonesia
13 - 14 November 2020

Editors:

Taufiq Hidayat
Ridho Rahmadi

ISBN: 978-1-7138-4582-9
ISSN: 1757-8981

Printed from e-media with permission by:

Curran Associates, Inc.
57 Morehouse Lane
Red Hook, NY 12571

Some format issues inherent in the e-media version may also appear in this print version.

Institute of Physics
Dirac House, Temple Back
Bristol BS1 6BE UK

Phone: 44 1 17 929 7481
Fax: 44 1 17 920 0979

techtracking@iop.org

Additional copies of this publication are available from:

Curran Associates, Inc.
57 Morehouse Lane
Red Hook, NY 12571 USA
Phone: 845-758-0400
Fax: 845-758-2633
Email: curran@proceedings.com
Web: www.proceedings.com

TABLE OF CONTENTS

IOP Conf. Series: Materials Science and Engineering 1077 (2021) 011001 doi:10.1088/1757-899X/1077/1/011001

PREFACE

Welcome to the fifth ICITDA, November 2020. This year, the pandemic of Covid-19 is happening. That said, we all agreed that the learning process should not stop here. We witness that the pandemic yields some obstacles but also believe that it leads to other opportunities. We, therefore, with the very spirit of ICITDA, continue the endeavour by holding this year's ICITDA virtually.

We would take this opportunity to thank all of the people who made this conference possible, insya Allah, and the enthusiasm they bring to the atmosphere. This year we are happy to have 6 keynote speakers and 6 workshop speakers from 6 countries, to receive more than 180 paper submissions (among them 71 are accepted & presented) and to admit hundreds of participants, coming from 8 countries. We also thank the University and Department of Informatics for all of the support. Last but not least, I personally want to express my thanks to all of my teammates of the fifth ICITDA, for the hard--and sincere--work.

Finally, once again thank you for joining this conference, and we are looking forward to seeing you in the next ICITDA.

Best regards,

Ridho Rahmadi
On behalf of the 5th ICITDA's Committee

IOP Conf. Series: Materials Science and Engineering 1077 (2021) 011001 doi:10.1088/1757-899X/1077/1/011001

CHAIRS

Host

- Department of Informatics, Faculty of Industrial Technology, Universitas Islam Indonesia

Editors

- Taufiq Hidayat
- Ridho Rahmadi

Organizing Committee

- Ridho Rahmadi (Chairman of The Committee)
- Hendrik
- Ahmad M.R. Pratama
- Taufiq Hidayat
- Septia Rani
- Arrie Kurniawardhani
- Rian Adam Rajagede

Program Committee

- Prof. Fathul Wahid (UII, Indonesia)
- Izzati Muhimmah, Ph.D. (UII, Indonesia)
- Dr. Mukhammad Andri Setiawan (UII, Indonesia)
- Dr. Raden Teduh Dirgahayu (UII, Indonesia)
- Dr. Sri Kusumadewi (UII, Indonesia)
- Dhomas Hatta Fudholi, Ph.D (UII, Indonesia)
- Dr. Syarif Hidayat (UII, Indonesia)
- Dr. Jefferson Tan (IBM Research)
- Dr. Agfianto Eko Putra (UGM, Indonesia)
- Ir. Kridanto Suhendro, Ph.D. (ITB, Indonesia)
- Dr. Taufiq Asyhari (Cranfield University, UK)
- Prof. Reyer Zwiggelaar (Aberystwyth University, UK)
- Dr. Uraiwan Inyaem (RMUTT, Thailand)
- Dr. Shofwatul 'Uyun (UIN Yogyakarta, Indonesia)
- Prof. Wei-Chung Teng (NTUST, Taiwan)

- Dr. Dini Oktarina Dwi Handayani (Taylor's University, Malaysia)
- Dr. Noor Zaman Jhanjhi (Taylor's University, Malaysia)
- Dr. Raja Kumar Murugesan (Taylor's University, Malaysia)
- Marini Othman, Ph.D. (IIUM, Malaysia)
- Dr. Hamwira Sakti Bin Yaacob (IIUM, Malaysia)
- Dr. Adamu Abubakar Ibrahim (IIUM, Malaysia)
- Dr. Madihah Sheikh Abdul Aziz (IIUM, Malaysia)
- Dr. Murni Mahmud (IIUM, Malaysia)
- Dr. Normi Sham Awang Abu Bakar (IIUM, Malaysia)
- Dr. Noor Azura Zakaria (IIUM, Malaysia)
- Dr. Abdul Qayoom Hamal (Help University, Malaysia)
- Dr. Mustafa Ali Abuzaraida (UUM, Malaysia)

IOP Conf. Series: Materials Science and Engineering 1077 (2021) 011001 doi:10.1088/1757-899X/1077/1/011001

Keynote Speeches

LEARNING CAUSAL REPRESENTATIONS AND USING THEM

Dr. Kun Zhang

Carnegie Mellon University, USA

RISE OF THE DIGITAL TWIN IN INDUSTRIAL IOT AND INDUSTRY 4.0: OPPORTUNITIES AND FUTURE ASPECTS

Dr. Noor Zaman Jhanjhi

Taylor's University, Malaysia

STRENGTHENING INFORMATION TECHNOLOGY ECOSYSTEMS IN HIGHER EDUCATION: LESSONS FROM THE FIELD

Prof. Fathul Wahid, PhD

Universitas Islam Indonesia, Indonesia

SYNERGISTIC COMBINATIONS FROM HIGH-THROUGHPUT DRUG SCREENING

Dr. Tjeerd Djikstra

Max Planck Institute for Developmental Biology, Tubingen, Germany

THE TRANSITION TO POST-QUANTUM CRYPTOGRAPHY: CHALLENGE AND CHANCE

Prof. Peter Schwabe

Radboud University, Nijmegen, the Netherlands

HOW TO MAKE MACHINE LEARNING WORKS IN INDUSTRY?

Elena Sokolva, Ph.D

Amazon

IOP Conf. Series: Materials Science and Engineering 1077 (2021) 011001 doi:10.1088/1757-899X/1077/1/011001

Hands-on Workshop

IDENTITY FEDERATION FOR RESEARCH AND EDUCATION NETWORK

Dr. Mukhammad Andri Setiawan

Board of Information System, UII

EEG DATA ANALYSIS

Dr. Hamwira Sakti Bin Yaacob

International Islamic University Malaysia

MOBILE APPLICATION BASED ON DEEP LEARNING FOR IMAGE RECOGNITION

Dhomas Hatta Fudholi, Ph.D

Center of Data Science, UII

DIGITAL FORENSIC

Dr. Yudi Prayudi

Center of Digital Forensics Studies, UII

DATA VISUALIZATION WITH TABLEAU

Dr. Dini Oktarina Dwi Handayani

Taylor's University

SMART CONTRACTS IN BLOCKCHAIN DECENTRALIZED APPLICATIONS

Dr. R. Teduh Dirgahayu

Centre for Enterprise Information Systems, UII

Peer review declaration

All papers published in this volume of *IOP Conference Series: Materials Science and Engineering* have been peer reviewed through processes administered by the Editors. Reviews were conducted by expert referees to the professional and scientific standards expected of a proceedings journal published by IOP Publishing.

- **Type of peer review: Double-blind**
- **Conference submission management system: Easychair**
- **Number of submissions received: 185**
- **Number of submissions sent for review: 160**
- **Number of submissions accepted: 74**
- **Acceptance Rate (Number of Submissions Accepted / Number of Submissions Received X 100): 40%**
- **Average number of reviews per paper: 2.23**
- **Total number of reviewers involved: 44**
- **Any additional info on review process: Each submitted paper is reviewed by the editor for plagiarism checking. If the paper passes the plagiarism checking, it is sent to 2 or 3 reviewers. Each reviewer gives recommendation of a paper's acceptance along with confidence and argumentations. The reviewer also suggests revision if any. The editor will decide a paper's acceptance based on recommendation of all reviewers. A paper is sent to the author(s) for revision if the decision is accepted. The author(s) should consider all suggestions from all reviewers. Final acceptance of a paper is decided by the editor based on the revised paper.**
- **Contact person for queries: septia.rani@uii.ac.id**

IOP Conf. Series: Materials Science and Engineering 1077 (2021) 012001 doi:10.1088/1757-899X/1077/1/012001

Sentiment Analysis on Twitter using Neural Network: Indonesian Presidential Election 2019 Dataset

Ahmad Fathan Hidayatullah[1], Siwi Cahyaningtyas[2], Anisa Miladya Hakim[3]

Department of Informatics, Universitas Islam Indonesia
Jalan Kaliurang KM 14,5 Sleman Yogyakarta Indonesia

E-mail: [1]fathan@uii.ac.id, [2]siwictyas@gmail.com, [3]anisanisa.nh@gmail.com

Abstract. Due to the utilization of Twitter by Indonesian politicians ahead of the Indonesian presidential election 2019, Indonesian people have given diverse responses and sentiments to the politicians. This study aims to classify sentiment on Indonesian presidential election 2019 tweet data by using neural network algorithms and obtain the best algorithm. In our study, we train our dataset using some variants of deep neural network algorithms, including Convolutional Neural Network (CNN), Long short-term memory (LSTM), CNN-LSTM, Gated Recurrent Unit (GRU) -LSTM and Bidirectional LSTM. Moreover, as a comparison with our deep learning model, we also train our dataset using other traditional machine learning algorithms, namely Support Vector Machine (SVM), Logistic Regression (LR) and Multinomial Naïve Bayes (MNB). Our experiments showed that Bidirectional LSTM achieved the best performance with the accuracy of 84.60%.

1. Background

Currently, with the popularity of social networking media which has become a powerful tool to influence people and share opinion to the public. For example, Twitter is one of social media that often be used by politician for political campaign in Indonesia [1,2]. Among several social networking sites, Twitter is one of the most well-known tools used by politicians [2]. Twitter has been used by many Indonesian politicians to attract public sympathy and increase popularity ahead of the election [3]. Through Twitter, politicians post their opinion, thoughts and activities to attract and influence people to vote them in the general election. In addition, Twitter has been utilized as a data source to analyze public sentiment in the election in some countries such as Indonesia, India, Germany, United States, UK and Bulgaria [1,4–8].

The Indonesian presidential which held on April 17, 2019, is getting more interesting considering the two presidential candidates in this year are the same name in the previous presidential election in 2014. In the 2019 Indonesian presidential election, the battle of public figures and supporters of presidential candidates on Twitter is very exciting to discuss. It can be seen from several Twitter hashtags from two different sides such as #2019GantiPresiden (means: 2019 replace the president) and #Jokowi2Periode (means: Jokowi 2 periods). The tweets posted by those politicians will certainly get diverse responses and sentiments from netizens. Therefore, knowing the sentiment towards the tweet posted is necessary. The goal of sentiment analysis is to identify the sentiment polarity of a given sentence or statement into three polarity categories, namely positive, negative, or neutral [9,10].

Knowing the sentiment polarity of tweets posted by politician can illustrate the public response toward a specific politician.

However, sentiment analysis task in for short texts like Twitter is quite challenging because of the limitation of contextual information in the texts [11]. Therefore, several algorithms have been continuously developed to obtain the best sentiment analysis model result. Several previous studies have proven that traditional machine learning method such as Support Vector Machine (SVM) and Naïve Bayes Classifier (NBC) have shown great performance for the sentiment analysis task. Almatrafi et al [4] proposed a location-based sentiment analysis system using Twitter dataset to discover several topic trends during Indian general elections in 2014 using Naïve Bayes Classifier. Hidayatullah and Sn [3] applied both Support Vector Machine and Naïve Bayes Classifier to conduct sentiment analysis and category classification on Indonesian politicians using Twitter dataset. The study revealed that Support Vector Machine using term frequency as a feature obtained better accuracy than Naïve Bayes. Hasan et al [12] also employed both SVM and Naïve Bayes algorithms for Twitter sentiment analysis task. They combined the two algorithms with sentiment lexicon. In addition, the sentiment analysis task was conducted by comparing several sentiment lexicons with three libraries, including SentiWordNet, W-WSD and TextBlob.

Currently, with the increasing numbers and size of data, researchers have been developing a deep neural networks approach to be applied in the sentiment analysis task. Santos and Gatti [13] conducted perform sentiment analysis by applying deep convolutional neural networks architecture joined with three different level representations, such as character-level, word-level and sentence-level representations. The proposed method has been applied on two different corpora, namely Stanford Sentiment Treebank and Stanford Twitter Sentiment corpus. Ali, et al [14] carried out sentiment analysis using movie reviews dataset by utilizing four deep learning algorithms, namely Multilayer Perceptron (MLP), Convolutional Neural Network (CNN), Long short-term memory (LSTM) and CNN_LSTM. The experiments have shown that CNN_LSTM outperformed other deep learning methods. Jianqiang, et al [10] proposed a method by combining Glove (Global Vectors) as word representation and Deep Convolutional Neural Networks (DCNN) to conduct sentiment analysis utilizing Twitter dataset. Their study revealed that the Glove-DCNN method performed better than bag-of-words model with Support Vector Machine (SVM) classifier. Wang et al [11] performed sentiment analysis by applying two Deep Learning algorithms, Convolutional Neural Networks (CNNs) and Recurrent Neural Networks (RNNs). The study revealed that the model from the combination between CNNs and RNNs achieved higher accuracy than other existing models.

This study applies several different deep neural network algorithms to classify the sentiment on Indonesian Presidential Election 2019 Twitter dataset. In addition, this work aims to classify the sentiment of the dataset into positive and negative. To obtain the best model, we train our dataset using some different deep neural network algorithms, including Convolutional Neural Network (CNN), Long short-term memory (LSTM), CNN-LSTM, Gated Recurrent Unit (GRU)-LSTM and Bidirectional LSTM. As a comparison, we also train our dataset using other well-known traditional machine learning algorithms, namely Support Vector Machine (SVM), Logistic Regression (LR) and Multinomial Naïve Bayes (MNB).

The remainder of this paper is organized as follows: Section 2 describes deep learning algorithms; in Section 3, we explain our research methodology; Section 4 describes result and discussion of our study; In Section 5, the conclusion of this study is presented.

2. Deep Learning Algorithm

2.1. Convolutional Neural Network (CNN)

Convolutional Neural Network (CNN) is one of deep learning method that is developed from multilayer perceptron (MLP). CNN consists of three main layers such as convolution, pooling and fully connected layer. The convolution layer is the primary CNN building block which consists of a set independent filters. Pooling layer aims to reduce the complexity of convolved layer representation. The vector results from several convolutions and pooling operations on the multilayer perceptron are known as fully-connected layers that are used to do a particular task such as classification.

2.2. Long-short Term Memory (LSTM)
LSTM is an improved model of RNN that is developed to overcome long time-dependencies problem. An LSTM cell consists of three gates: input gate, forget gate and output gate. These gates are produced by a sigmoid function over the input sequences and the previous hidden state [15]. These gates receive information across LSTM through cell state. The fundamental component of LSTM is cell state as a transport information from memory from previous block to memory from current block.

2.3. Gated Recurrent Unit (GRU)
GRU (Gated Recurrent Unit) is similar to LSTM that is able to deal with long term dependencies, but they have fewer parameters [16]. GRU has gating units like LSTM, except memory cells. GRU aims to overcome the vanishing gradient problem by using update and reset gate. The update gate assists the model to identify the past information from previous steps should be passed along to the future. The reset gate is utilized from the model to choose the amount of the past information to forget.

2.4. Bidirectional LSTM
Bidirectional LSTM (BLSTM) is an improved model of LSTM that connects two hidden layers to the same output. The objectives of BLSTM are to enhance performance of the model on sequential problem. Different from LSTM that using input are only from the past, BLSTM using inputs into two ways, past to future and future to past.

3. Methodology

3.1. Data Collection
Dataset collected by two different periods of time, before and after the presidential election. As for the period before the presidential election, we crawled the tweet five times, in January, February, March and April 2019. The data were crawled the day after the debates of presidential candidates and vice presidential candidates. We used tweepy as a library provided by Python obtained 115,931 raw tweets.

3.2. Data Labelling
Manually labelling data is expensive and time-consuming. To reduce process on labelling data, we can use pseudo-labelling to labelling our data. Pseudo-labelling is a semi-supervised technique to labelling data with still requires labelled data. Pseudo-labelling implements machine learning algorithms for building models. The model then used to generate a pseudo-label as a class of unlabelled data. In this work, we used Multinomial Naïve Bayes as a model. First, we used 1369 raw Twitter sentiment dataset that have been labelled manually as a training set [3]. Training set that was labelled used to train our model using Multinomial Naïve Bayes. Model that build by training set then used to predict labels (pseudo-label) on unlabelled data. Finally, we concatenate the training set with the pseudo-labelled set and use both datasets to retrain our model. The process is schematically shown in Figure 1.

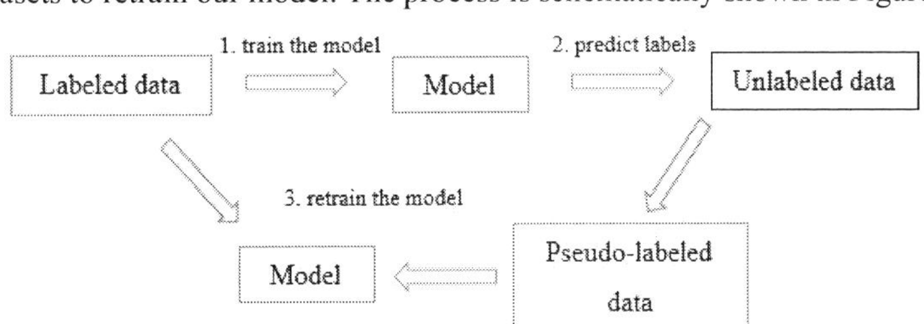

Figure 1. Pseudo-labelling.

3.3. Pre-processing
We applied some pre-processing techniques to our dataset. We omitted several characters from the Twitter data, including ASCII characters; username, hashtags, URLs, retweet; punctuations; duplicate characters in a word; excessive space; stop words; and tweet duplication. In addition, we also performed other pre-processing tasks such as case folding; word normalization; and stemming.

3.4. Sentiment Analysis Model Variations
In this work, we created our sentiment analysis model by performing several experimental model variations. We built the first model variations using three traditional machine learning algorithms, such as SVM, Logistic Regression and Multinomial Naïve Bayes. We also compared the accuracy result by applying Term Frequency-Inversed Document Frequency (TF-IDF) as a feature and not applying TF-IDF on those three algorithms. The second sentiment analysis model variations were built using several deep neural network algorithms, including Convolutional Neural Network (CNN), Long short-term memory (LSTM), CNN+LSTM, Gated Recurrent Unit (GRU)+LSTM and Bidirectional LSTM.

4. Result and Discussion
In this study, we used Holdout Method to split the data into 2 parts, 80% for training and 20% for testing. In addition, we also added validation data which obtained from 10% of the training data. Table 1 shows the accuracy results from traditional machine learning algorithms. Based on the accuracy results, SVM outperformed Multinomial Naïve Bayes and Logistic Regression with the accuracy of 84.04%. In addition, application of TF-IDF can improve the accuracy of all the methods.

Table 1. Results of traditional machine learning algorithms

Method	TF-IDF Feature	Accuracy (%)
SVM	Yes	**84.04**
	No	83.23
Logistic Regression	Yes	83.15
	No	82.74
Multinomial Naïve Bayes	Yes	82.13
	No	82.08

Table 2 shows the accuracy results from deep learning algorithms. We performed training using five different deep neural network algorithms for our dataset with 5 epochs and batch size of 128. In the experiments, we split our dataset into training and testing data with the ratio of 80:20. In addition, we took 10% of our training set for validation set. Based on the accuracy results, Bidirectional LSTM obtained the best result of 84.60%. As we can see, the accuracy of Bidirectional LSTM is higher 0.56% compared to SVM. This achievement shows that Bidirectional LSTM has better performance than the SVM algorithm.

Table 2. Results of deep learning algorithms

Method	Accuracy(%)
LSTM	84.20
CNN	84.05
CNN+LSTM	84.30
GRU+LSTM	84.50
Bidirectional LSTM	**84.60**

Figure 2 shows the layer architecture of Bidirectional LSTM as the best model. Firstly, we put the embedding layer as our input layer which receive three arguments, input dimension, output dimension and input length. Input dimension is the size of vocabulary in the dataset. Output dimension defines the size of the output vector for respective words from this layer. Input length assigns the length of input sequences. The next layer is a dropout layer which aims to overcome overfitting of the model. Dropout layer regularizes a deep neural network by randomly removing inputs to a layer using probabilistic calculation. After that, we added two stack Bidirectional LSTM layers with 256 hidden nodes. In addition, we added a recurrent dropout parameter argument into both Bidirectional LSTM layers to drop the linear transformation of the current state. Moreover, we put dropout layer after the two Bidirectional LSTM layer. Finally, we put the dense layer as the output layer.

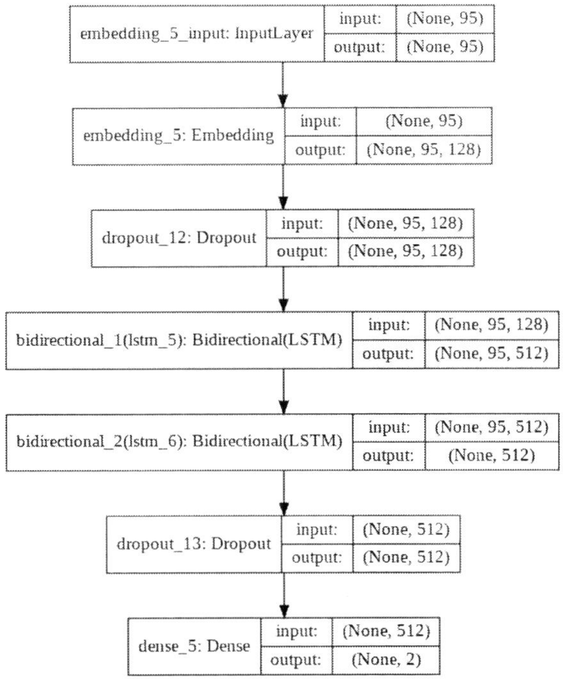

Figure 2. Bidirectional LSTM Architecture

5. Conclusion

In this work, we have presented sentiment analysis on Indonesian presidential election 2019 Twitter dataset using deep neural network algorithms. Overall, our experiments showed that deep neural networks algorithms had better accuracy compared with other three traditional machine learning algorithms (SVM, Logistic Regression and Multinomial Naïve Bayes). In addition, as for deep learning approach, we proposed five different methods, including LSTM, CNN, CNN+LSTM, GRU+LSTM and Bidirectional LSTM. According to our experiments, Bidirectional LSTM outperformed other deep learning methods with the accuracy of 84.60%.

Acknowledgement

This study is conducted in the collaborative research scheme of lectures and students which is funded and supported by the Department of Informatics, Universitas Islam Indonesia.

References

[1] Alashri S, Parriott E, Kandala S S, Awazu Y, Bajaj V and Desouza K C 2018 The 2016 US Presidential Election on Facebook: An Exploratory Analysis of Sentiments *Proc. of the 51st Hawaii Int. Conf. on System Sciences* pp 1771–80
[2] Bansal B and Srivastava S 2018 On predicting elections with hybrid topic based sentiment

analysis of tweets *Procedia Computer Sci.* **135** pp 346–53
[3] Hidayatullah A F and Sn A 2014 Analisis sentimen dan klasifikasi kategori terhadap tokoh publik pada Twitter *Seminar Nasional Informatika 2014 (semnasIF 2014)* pp 115–22
[4] Almatrafi O, Parack S and Chavan B 2015 Application of location-based sentiment analysis using Twitter for identifying trends towards Indian general elections 2014 *Proc. of the 9th Int. Conf. on Ubiquitous Information Management and Communication - IMCOM '15* (Bali, Indonesia: ACM Press) pp 1–5
[5] Budiharto W and Meiliana M 2018 Prediction and analysis of Indonesia Presidential election from Twitter using sentiment analysis *J. of Big Data* **5**
[6] Heredia B, Prusa J D and Khoshgoftaar T M 2018 Location-Based Twitter sentiment Analysis for Predicting the U.S. 2016 Presidential Election *The Thirty-First Int. Florida Artificial Intelligence Research Society Conf. (FLAIRS-31)* pp 265–70
[7] Burnap P, Gibson R, Sloan L, Southern R and Williams M 2016 140 characters to victory?: Using Twitter to predict the UK 2015 General Election *Electoral Studies* **41** pp 230–3
[8] Smailović J, Kranjc J, Grčar M, Žnidaršič M and Mozetič I 2015 Monitoring the Twitter sentiment during the Bulgarian elections *2015 IEEE Int. Conf. on Data Sci. and Advanced Analytics (DSAA)* pp 1–10
[9] Tang D, Wei F, Qin B, Liu T and Zhou M 2014 Coooolll: A Deep Learning System for Twitter sentiment classification *Proc. of the 8th Int. Workshop on Semantic Evaluation (SemEval 2014)* (Dublin, Ireland: Association for Computational Linguistics) pp 208–12
[10] Jianqiang Z, Xiaolin G and Xuejun Z 2018 Deep Convolution Neural Networks for Twitter sentiment analysis *IEEE Access* **6** pp 23253–60
[11] Wang X, Jiang W and Luo Z 2016 Combination of Convolutional and Recurrent Neural Network for sentiment analysis of short texts *Proc. of COLING 2016 The 26th Int. Conf. on computational linguistics: Technical papers* pp 2428–37
[12] Hasan A, Moin S, Karim A and Shamshirband S 2018 Machine learning-based sentiment analysis for Twitter accounts *Mathematical and Computational Applications* **23** p 11
[13] Dos Santos C N and Gatti M 2014 Deep Convolutional Neural Networks form *Proceedings of COLING 2014, the 25th Int. Conf. on Computational Linguistics: Technical Papers* (Dublin, Ireland: Dublin City University and Association for Computational Linguistics) pp 69–78
[14] Ali N M, El Hamid M M A and Youssif A 2019 Sentiment analysis for movies reviews dataset using deep learning models *Int. J. of Data Mining & Knowledge Management Process* **09** pp 19–27
[15] Yin W, Kann K, Yu M and Schütze H 2017 Comparative Study of CNN and RNN for Natural Language Processing *Preprint arXiv:1702.01923 [cs]*
[16] Cho K, van Merrienboer B, Gulcehre C, Bahdanau D, Bougares F, Schwenk H and Bengio Y 2014 Learning Phrase Representations using RNN Encoder-Decoder for Statistical Machine Translation *Preprint arXiv:1406.1078 [cs, stat]*

IOP Conf. Series: Materials Science and Engineering 1077 (2021) 012002 doi:10.1088/1757-899X/1077/1/012002

A Method of Simplifying the Asset Dependency Cycle in Security Risk Analysis

Edri Yunizal[1], Judhi Santoso[2], Kridanto Surendro[3]

[1,2,3]Program Studi Doktor Teknik Elektro dan Informatika, Institut Teknologi Bandung
[1]Manajemen Informatika, IAIN Batusangkar

E-mail: [1]edriyunizal@iainbatusangkar.ac.id, [2]judhi@informatika.org, [3]surendro@gmail.com

Abstract. In recent years, one obstacle to information security risk analysis is the complexity of asset dependency. Developing research tends to produce models with great complexity and some that do not involve dependency cycles, make the model difficult to implement. This research developed the idea of compound nodes to produce a modular risk analysis that simplify the asset dependency cycle. This study uses the following assumptions: aimed at quantitative assessments, values based on likelihood, and based on graph basic models. The stages of the method consist of identification of cyclic in the initial model, reconstructing cyclic graph into acyclic graph, and reconstruction of cyclic sub graph into acyclic sub graph. Through an illustrative example, this research produces a method that helps facilitate understanding the complexity of asset dependencies in risk analysis.

1. Introduction

Large amounts of data on systems connected to the internet certainly need a protection mechanism. The method commonly used to obtain adequate protection is to carry out an information security risk analysis (ISRA), a process to identify and evaluate factors that can interfere with the success of the project or the achievement of objectives [1]. ISRA has an important role in the organization's ability to survive [1] – [3].

One of the main limitations of ISRA is asset dependency complexity [2]. The failure of one information asset in an organization can affect other assets and result in greater system failure. For example, a web has high security, but installed on a server with poor security. The calculation of web risk must also consider the condition of the server, because it is relevant. This makes the risk model assuming asset dependency stronger than not [4]. Unfortunately, traditional risk analysis only partially studied about it [5].

[5] – [15] is research that offers solutions to asset dependency constraints. Unfortunately, no one takes into account the asset dependency cycle. Cyclic dependence in risk analysis can originate from the process of collecting data [16] or requirements analysis [17]. Cyclic data collection may be due to weaknesses of the learning algorithm, with hidden variables. Example of a requirements analysis cycle is a client server application consisting of three components: database, application interface, and backup. Unauthorized access to the application interface will continue to allow unauthorized access to the backup. Unauthorized backup access will allow the attacker to obtain a password from the database, which again provides unauthorized access to the interface. The relationship between the database, application interface, and back up is a form of cycle.

Muller [17]–[19] attempted to solve the limitations of cyclic dependency by introducing the Dependency-Aware Root Cause (DARC) model in the Critical Infrastructure (CI) domain. Unfortunately, CI focuses on a series of important assets with a small size, while in other domains; the number of assets can be very large. In his own perspective, Muller showed that complexity is one of the future jobs. ISRA requires models that can take into account their cyclic dependencies, but in simpler complexity.

2. Related Works

The existence of asset dependencies in ISRA is generally new. Sendi et al. [2] in ISRA's taxonomy identified asset dependency as one of the fundamental problems that current risk tools and methods cannot address.

Suh and Han [6] initiated the asset dependency of ISRA. The model developed to determine the value of assets based on the relationship between the asset and the business function. Assets are valued based on the relationship with other assets that use maximum functions. Rahmad [8] added confidentiality and integrity in the form of threat scenarios corrected the weaknesses of [6], which only focused on availability.

Tatar & Karabacak [10] also observed the weaknesses in terms of availability in [6], and chose the hierarchical valuation of assets considering the CIA (confidentiality, integrity and availability) of these assets; this connectivity then presented in the form of associated tables. Breier [11] considered that the proposed asset dependency analysis [6] was too straightforward and could not describe it in detail because it based only on the maximum value of the related assets. Next, Breier presented the relationship between assets in the form of related graphs and provided asset values based on connection considerations, a risk matrix, physical connection patterns, and logic. The resulting value was combined with the initial value of the asset independently.

Alpcan & Bambos [5] examined the problem of interdependence in organizational risk based on three main factors of risk management: business units, threats/ vulnerabilities and humans. The proposed framework examined dependencies within and between business units, security/vulnerability threats and people using graph theory. Then, the model registered dynamic changes in the risk due to the interactions. Finally, they classified the risks in the organizations by using risk algorithms.

Khanmohammadi & Houmb [13] introduced a new concept in risk assessment focusing on business objectives, ignoring the assets and process which supporting it. The process identified and assessed the risks at the business process level. The risk was also grouped based on the critical level, role and level of importance of the organization as a whole. Schmidt & Albayrak [14] have developed a quantitative framework for risk assessment vertically and horizontally. Horizontal connectivity is between assets, while vertical connectivity brings connectivity into three levels: process, service, and application. At the application level, the hardware gives virtual asset values. Hierarchical virtual asset values is combined with district time algorithms to obtain risk values. This approach evaluates business processes and thanks to infrastructure dependencies, these values are broken down into each component of the computer system.

Lololei et al [15] divided the values of the assets on independent and dependent values. Independent values obtained from two parameters: cost (initial costs, maintenance costs, replacement costs), and the level of importance of the asset in the business process. The dependence value of the asset dependency presented in the form of CIA parameters, focused on availability. Level of dependency classified as partial or total, as indicated by the dependency percentage.

Muller [19] has created a DARC model, a quantitative research that offers convenience because the organization only needs to determine the value of assets ranked as root and attribute the probability of dependence. This search has improved by positioning the assets in layers by himself [18].

Based on the basic description of several previous studies. The theoretically elaborated models can overcome the limits of asset dependence. However, in practice, Rahmad [7]–[9] must use a very large list of threat scenarios [19]. Breier's research [11] showed the connectivity of 10 assets in a building, it will be very difficult to apply it to the real world with all assets. Similarly, with the development of

IOP Conf. Series: Materials Science and Engineering 1077 (2021) 012002 doi:10.1088/1757-899X/1077/1/012002

Muller [19], the connectivity of each asset will be determined in a special graph that makes the mapping very complex, and will be difficult for organization to implement. Muller stated that the fact that graphics dependencies can be large is a problem that must be solved [19]. This clearly shown in [19] of how to apply the proposed DARC model in the real world (Figure 6). This shows how the loss of readability of the graph. Even for the supply chain sector, the development of initial risk has a very large graph, see [20].

The second obstacle is the coverage of the types of risk. The investigation conducted by Loloei and Shahriari [15], Suh and Han [6] is example that admitted only the assets availability risk. The dimensions of the information security objectives only transmitted as separate values sometimes used as the average for each asset, such as research [6]–[9].

Qualitative approaches are indeed more widely used in the field of information security [3]. Indeed, this approach presents the subjectivity of the evaluator [2] which further contributes to the creation of weaknesses in expert dependence.

Ouyang told CI that one of the recommended and advanced methods for overcoming dependencies is high level architecture (HLA). HLA uses the concept of modularization to overcome difficulties in the legibility of the risk analysis model that takes into account asset dependencies [21]. With the similarity of the existing parameters, IS can take advantage of these developments, just like Muller [19], who has developed several CI studies for IS.

Differences in modelling and simulation approaches sometimes lead to conflict, so an integrated approach is needed to differentiate each responsibility into a uniform framework [22]. Modularization of Eusgeld's risk analysis at the highest level of the system of system (SoS), which in turn consists of several subsystems at the critical infrastructure level, and with the lowest, is at the system level. Unfortunately, this remains at the conceptual level with several challenges in its application [22]. Which later became the basis of this article.

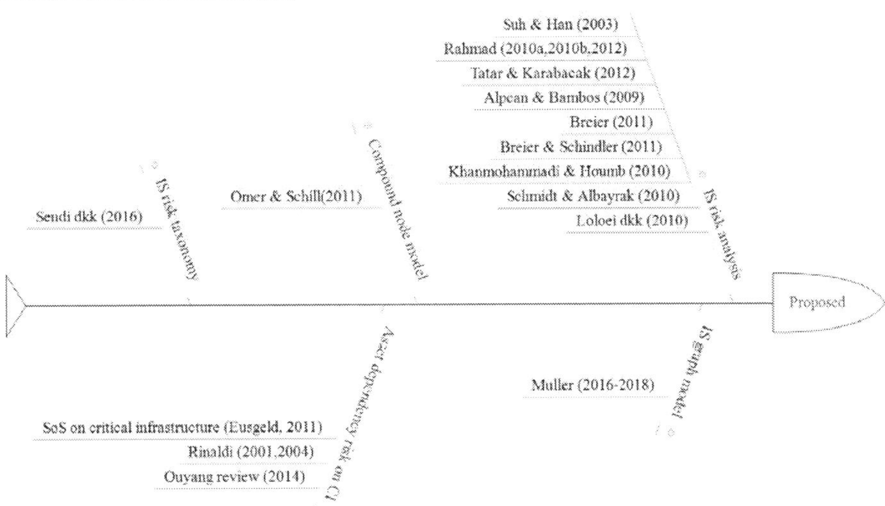

Figure 1. Study of literature

To apply the Eusgeld conceptual model [21] in the risk analysis, the approach with fault tree analysis (FTA), event tree analysis (ETA) and the attack tree cannot be used because it does not support cycle dependencies. The only possible approach is to use the attack graph. A dependency model involving dependency cycles is the connectivity model of the Omer & Schill's [23]. In their study, the cycle was correlated with dependencies for the case of web services.

Based on the study of the literature (Figure 1), this paper proposed a model of risk analysis based on dependence. The idea came from the development of the dependence of the IS risk analysis [5]–[9], [13]–[15], [15] resulting from the Sendi's taxonomy [2]. Based on the faster development of CI, the adaptation of the conceptual model developed by the literature [21], [22], [24], [25]. In the final stage, the conceptual model was applied in the operational model using a combination of Muller risk analysis proposals [17]–[19] followed by a compound node proposed by Omer and Schill [23].

3. Methods

3.1. Assumption

3.1.1. Quantitative. The emphasis on the conceptual model makes HLA a challenge in terms of application. However, the idea of combining several models at the same time will be very difficult to be applied in a qualitative research that emphasizes the subjectivity and experience of the evaluation team. This makes it difficult to give an overview of the risks involved. Only quantitative research makes it possible to observe all the risk scenarios and compare them to each other [17].

3.1.2. Likelihood. The risk calculated based on a combination of threat probability estimates for each asset and the impact that affects the assets successfully [26]. The non-consideration vulnerability used Fenz's statement in [9]: the node's probability of exploiting a vulnerability due to the actual threat is only an intermediate node to facilitate the mapping of the threat. After all, the targeted steps are risk analysis, to demonstrate why risk management needs to be implemented [1], which requires simplicity in the components used. Because risk defined as a product of probability and non-compliance, preventive safety measures aim to reduce the likelihood, while mitigation measures applied to reduce the consequences [27]. The first priority is of course to reduce the likelihood.

The likelihood connection expressed in the formula [26]: Risk (Threat, Asset) = Likelihood (Threat) $\otimes$ Impact (Threat, Asset)

3.1.3. Graph based model. This research adapted the model from [17], because it has considered cyclic in information security. In [17], where for V set as a security incident, statement $V \subseteq A \times S$ can be used, A is the set of all the assets and S is the set of security properties. These security properties known as Security Triads: Confidentiality, Integrity, and Availability (CIA) [17], so S consists of {C, I, A}. Asset confidentiality written as a.S instead of $(a, s) \in A \times S$. So that the hard disk (HDD) availability can be written HDD.A. If incident a have impact on incident b, it can be written a→b instead of (α, β), so that if hard disk availability affects data availability, it can be written HDD.A→DATA.A.

3.2. Construction

Based on the previous assumptions, the simplification process consisted of the following steps:

1. Identification and extraction of cyclic sub graphs.
   Cyclic identification and extraction process used Tarjan algorithm [28], as done by [23] to test the measurement of cyclic dependence on web services. Tarjan's algorithm has the ability to identify elementary circuit, an elementary path (a path contains no node twice) with the exception that its first and last node are identical. Tarjan algorithm requires initial graph (G1) as basic input. Tarjan algorithm uses G1 to identify cyclic sub graphs, and then extracts them into array of cyclic sub graph (G2)
2. Reconstruct initial cyclic graph into acyclic graph.
   The next step is to transform each G2 elements in G1 into a compound node. In the process of transformation, there is a change in the situation at the incoming and outgoing edges. Edges connected to the compound node replaces all the edges connected to the G2 element. The incoming edge of the compound node is from the incoming edge of the G2 elements, the same thing applies to the outgoing edge. So that it can be interpreted that every node that depends on

the G2 elements will depend on the compound node that replaces it. This also applies vice versa; the compound node generated during the reconstruction will remain dependent on the dependent G2 elements.

3. Reconstruct cyclic sub graph into acyclic ones.
Based on the model definition [17], node execution was not performed several times, only once in one experiment. Therefore, every G2 element structure must be acyclic.
Restructuring process used Wang algorithm [29]. The use of Wang's algorithm is for cyclic restructuring. This algorithm used the Bayesian Network (BN) due to the acyclic requirements. Based on Wang's algorithm, a node that is not an entry point of a cycle will be more efficient to calculate the cumulative score without considering the cycle. For each node in the cycle that has only one incoming edge, the cumulative score can be calculated (after the calculation of the point entry is complete) as the node is not in the cycle.
The results of the reconstructions of the cyclic G2 elements are stored in acyclic G4 elements (also in array).

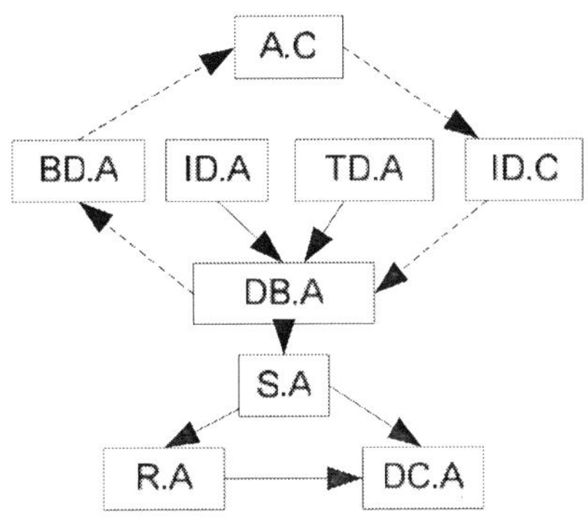

Figure 2. G1 Initial Graph

3.3. Illustration

A data security case study of a client-server application will demonstrate the implementation capabilities of the proposed method.

In the client-server application, a data center requires availability of routers and servers. Server availability requires the availability of the database. Database availability requires the availability of internal data A and transaction data A. For security, database availability requires the availability of database backup. Availability database backup requires the confidentiality of the application obtained from confidentiality of internal data C.

Based on the model requirements, the case presented in the form of the initial graph G1 (Figure 2). G1 shows the connection between:

1. Availability of data center (DC.A)
2. Availability of routers (R.A)
3. Availability of server (S.A)
4. Availability of database (DB.A)
5. Availability of internal data A (ID.A)
6. Availability of transaction data A (TD.A)
7. Availability of database backup (BD.A)
8. Confidentiality of application (A.C), and
9. Confidentiality of data internal (ID.C)

Based on first step, the proposed method identified and extracted cyclic sub graphs from G1. For the sake of simplicity, in this case there was only one cyclic graph: DB.A → BD.A → A.C → ID.C → DB.A, with DB.A as entry point. Result of this step is the extraction of sub graph cyclic G2 (Figure 3).

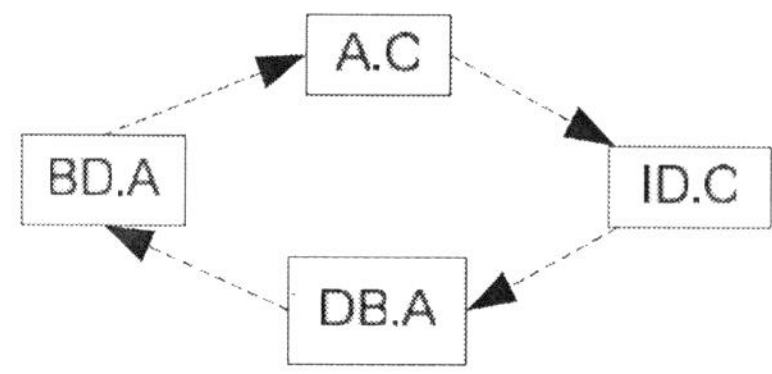

Figure 3. G2 Sub Graph Cyclic

In the second step, the proposed method is reconstructing G1 into a compound node (CN.DB.A). In G2, there are 2 incoming edges: ID.A→DB.A, and TD.A→DB.A, both of which then becsme the incoming edge for CN.DB.A. G2 has one outgoing edge DB.A→S.A, similar to the incoming edge; this also became the outgoing edge of CN.DB.A. G3 stored te results of this reconstruction (Figure 4). The initial form of G1, which originally had a cyclic structure, was fully acyclic in its new form.

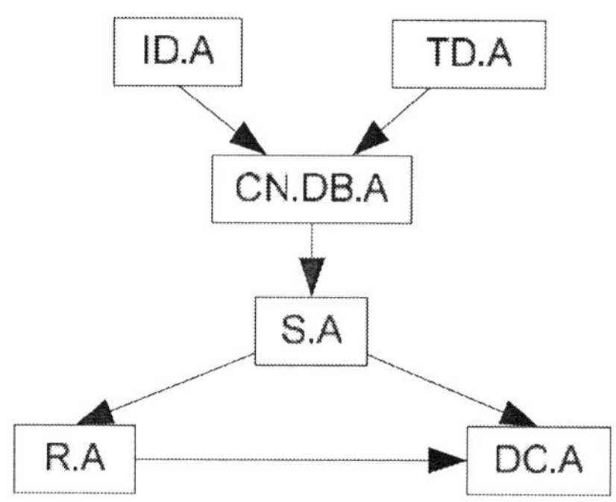

Figure 4. G3 Graph acyclic

Final step is to reconstruct it into the acyclic form because G2 is still cyclic. This stage is different from the G3 reconstruction process, simplifying DB.A as an entry point through one execution with placement as the initial node. The graph became acyclic by removing the edge ID.C→DB.A, then extracted into G4 (see Figure 5).

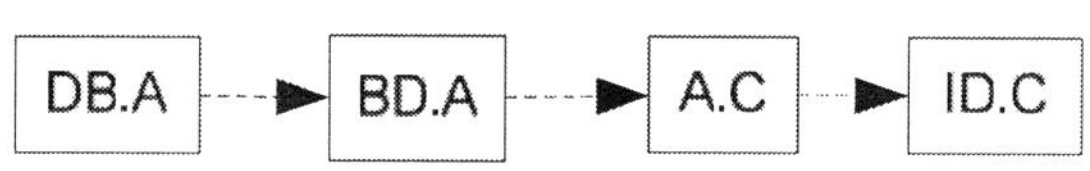

Figure 5. G4 Sub graph acyclic

The use of compound nodes simplifies the attack graphics. The complexity that initially tends to be complicated can br modulated quantitatively. We can see the separate discussion of the composite nodes in acyclic structure.

4. Discussion

Several studies [5]–[15] have offered asset dependency solutions in information security risk taking into account a small number of important assets. In practice, the actual amount of assets in an organization is very large, although the proposed solution might be to model the system in general; the complexity still lies in their future work [1], [19]. The complexity of the solutions offered hinders organizational understanding of the importance of the problems they face, limited funds and staff [3] which force them to focus on problems they considered more important.

Organizational failure to understand the important problem of information security risks is a form of failure to understand its complexity. This minimized the opportunities to implement the solutions offered.

Proposed method has several advantages over some previous approaches. If [[17]–[19] offered more complexity to group assets of the same type, proposed method offered separates the parts of the cycle in a modular way. Complexity can be seen more clearly than [17]. Compound node CN.DB.A represented the cycle.

Not only exceeded cycles, but dependencies also will be simplified. Using the threat scenario concept of Rahmad [7]–[9], assets can have the same set of security incidents. If [17]–[19] focused only on exceeding the dependency cycle, the proposed model is much more complete because it simplified it. For example, the process carried out by Muller [17] was to evaluate risk analysis includes about 50 different assets with three or four threats; this process produces hundreds of node relationships (Figure 6). Proposed method can add the ability to separate existing connections into multiple layers of node dependency. Thus, the number of nodes in the proposed method first layer was less than the number of Muller nodes, which certainly simplified the structure.

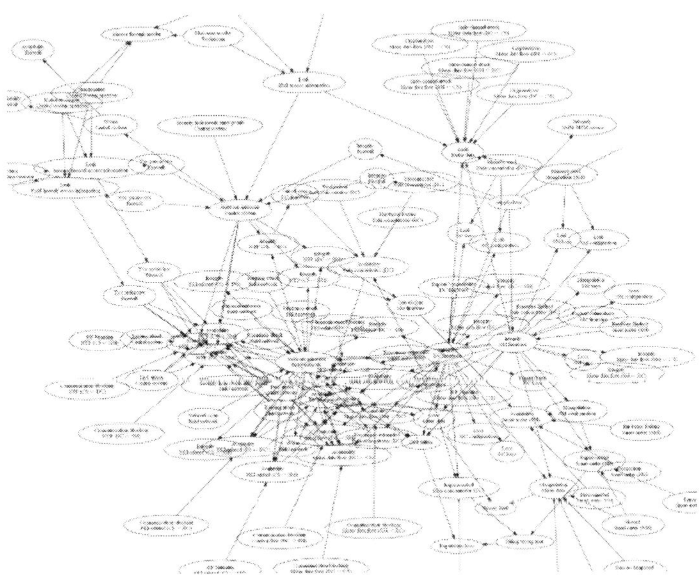

Figure 6. Complexity as Future Work on Muller [19]

5. Conclusion and Future Work

The biggest challenge in analysing information security risks analysis is to involve asset dependencies. It is always difficult to find achievements that can serve as an organizational basis for identifying risks. The popularity of qualitative solutions will depend largely on the availability of experts for the organization, which will generally have increased costs and subjectivity. Developing a quantitative model that can be cost-benefit oriented and easily understood by managers will be a major step towards meeting these needs.

The attack graph as a form of probability is very promising in the analysis of information security risks because of its flexibility and the support of the dependency cycle. It is just that the complexity of some previous studies is difficult to be implemented in the real world. The proposed method makes it possible to analyze the complexity of the risk analysis.

Future work will focus on an evaluation of the proposed method, which may not be presented due to incomplete data. This study will compare several risk analysis models to compare process speed and readability (if possible).

References

[1] Peltier T R 2010 *Information Security Risk Analysis* (Florida: Auerbach publications)
[2] Shameli-Sendi A, Aghababaei-Barzegar R and Cheriet M 2016 Taxonomy of information security risk assessment (ISRA) *Computers & security* **57** pp 14–30
[3] Landoll D J and Landoll D 2005 *The Security Risk Assessment Handbook: A Complete Guide for Performing Security Risk Assessments* (Florida: CRC Press)
[4] Haimes Y Y 2018 Risk modeling of interdependent complex systems of systems theory and practice *Risk Analysis* **38(1)** pp 84–98
[5] Alpcan T and Bambos N 2009 Modeling dependencies in security risk management *2009 Fourth Int. Conf. on Risks and Security of Internet and Systems (CRiSIS 2009)* pp 113–6
[6] Suh B and Han I 2003 The IS risk analysis based on a business model *Information & Management* **41(2)** pp 149–58
[7] Rahmad B, Supangkat S H, Sembiring J and Surendro K 2010 Threat scenario dependency-based model of information security risk analysis *IJCSNS* **10(8)** pp 93–102
[8] Rahmad B, Supangkat S H, Sembiring J and Surendro K 2012 Modeling asset dependency for security risk analysis using threat-scenario dependency *Int. J. of Computer Science and Information Security* **10(4)** pp 103–11
[9] Rahmad B 2010 *Analisa Risiko Keamanan Informasi dengan Mempertimbangkan Dependensi Skenario Threat dan Kontrol sebagai Pereduksi Likelihood dan Impact* Dissertation (Bandung: Institut Teknologi Bandung)
[10] Tatar U and Karabacak B 2012 An hierarchical asset valuation method for information security risk analysis *Int. Conf. on Information Society (i-Society 2012)* pp 286–91
[11] Breier J 2014 Asset valuation method for dependent entities *J. of Internet Services and Information Security* **4(3)** pp 72–81
[12] Breier J and Schindler F 2014 Assets dependencies model in information security risk management *Information and Communication Technol.* (Heidelberg: Springer) pp 405–12
[13] Khanmohammadi K and Houmb S H 2010 Business process-based information security risk assessment *2010 Fourth Int. Conf. on Network and System Security* pp 199–206
[14] Schmidt S and Albayrak S 2010 A quantitative framework for dependency-aware organizational IT Risk Management *2010 10th Int. Conf. on Intelligent Systems Design and Applications* pp 1207–12
[15] Loloei I, Shahriari H R and Sadeghi A 2012 A model for asset valuation in security risk analysis regarding assets dependencies *20th Iranian Conf. on Electrical Eng. (ICEE2012)* pp 763–8
[16] Li K, Ren Y, Fan D, Liu L, Wang Z and Ma Z 2018 Enhance GO methodology for reliability analysis of the closed-loop system using Cyclic Bayesian Networks *Mechanical Systems and Signal Processing* **113** pp 237–52
[17] Muller S, Harpes C, Le Traon Y, Gombault S and and Bonnin J M 2017 Efficiently computing the likelihoods of cyclically interdependent risk scenarios *Computers & Security* **64** pp 59–68
[18] Muller S, Harpes C, Le Traon Y, Gombault S, Bonnin J M and Hoffmann P 2016 Dynamic risk analyses and dependency-aware root cause model for critical infrastructures *Int. Conf. on Critical Information Infrastructures Security* pp 163–75
[19] Muller S 2018 *Risk Monitoring and Intrusion Detection for Industrial Control Systems* PhD Thesis (Luxembourg: University of Luxembourg)
[20] Badurdeen F et al 2014 Quantitative modeling and analysis of supply chain risks using Bayesian theory *J. of Manufacturing Technol. Management* **25(5)** pp 631–54

[21] Eusgeld I, Nan C and Dietz S 2011 System-of-systems: approach for interdependent critical infrastructures *Reliability Eng. & System Safety* **96(6)** pp 679–86
[22] Ouyang M 2014 Review on modeling and simulation of interdependent critical infrastructure systems *Reliability engineering & System safety* **121** pp 43–60
[23] Omer A M and Schill A 2011 Automatic management of cyclic dependency among web services *2011 14th IEEE Int. Conf. on Computational Sci. and Eng.* pp 44–51
[24] Rinaldi S M, Peerenboom J P and Kelly T K 2001 Identifying, understanding, and analyzing critical infrastructure interdependencies *IEEE Control Systems Magazine* **21(6)** pp 11–25
[25] Rinaldi S M 2004 Modeling and simulating critical infrastructures and their interdependencies *37th Annual Hawaii Int. Conf. on System Sciences* pp 1–8
[26] Ionita D 2013 *Current Established Risk Assessment Methodologies and Tools* Master's Thesis (Enschede: University of Twente)
[27] Khakzad N, Khan F and Amyotte P 2013 Dynamic safety analysis of process systems by mapping bow-tie into Bayesian network *Process Safety and Environmental Protection* **91(1–2)** pp 46–53
[28] Tarjan R 1973 Enumeration of the elementary circuits of a directed graph *SIAM J. on Computing* **2(3)** pp 211–6
[29] Wang L, Islam T, Long T, Singhal A and Jajodia S 2008 An attack graph-based probabilistic security metric *IFIP Annual Conf. on Data and Applications Security and Privacy* pp 283–96

Measurement of Information Security Awareness Level: A Case Study of Digital Wallet Users

A L Fadhilah[1], Y Ruldeviyani[2], R Prakoso[3] and K F Arisya[4]

[1,3,4]Magister of Information Technology, Faculty of Computer Science, Universitas Indonesia
[2]Department of Information Technology, Faculty of Computer Science, Universitas Indonesia, Jl. Kenari 2 Kota Jakarta Pusat, Daerah Khusus Ibukota Jakarta, Indonesia

E-mail: amanda.lailatul@ui.ac.id

Abstract. Awareness of a person is an important factor in maintaining information security and reducing the fraud gap. The existence of various types of digital wallet application fraud requires someone to have an awareness of information security. This study was conducted to measure the information security awareness of digital wallet users. The research was conducted by distributing questionnaires to 156 respondents of digital wallet users. These respondents represented users of digital wallet applications in Indonesia. Information security was measured based on the user's knowledge-attitude-behavior. The assessment of information security awareness level was calculated using the Analytic Hierarchy Process (AHP) method. The results showed that the score was at the awareness level of good (80.78%). However, several focus areas still showed score at the average level which was one of the reasons why there are still many information security breaches against users of the digital wallet.

1. Introduction

The development of information technology has led to various innovations in various sectors. These innovations have brought significant changes to the way of living, working and communicating. No wonder these changes have led to a digital lifestyle. The technology-based financial sector or fintech also contributes to changes to a digital lifestyle. Nowadays, the digital wallet is one of the trending technology.

The popularity of digital wallet among the public is also high, this is showed by Dailysocial research that digital wallets (82.7%) were the most popular fintech product category according to respondents, followed by investment (62.4%), pay later (56.7%), and p2p lending that accommodated personal needs (40%). The same source also stated that 81.6% of the considerations for using digital wallets were based on respondents' trust in the product itself. 72.2% of respondents also tend to use it because of its suitability to their needs. Besides, 72.9% of respondents chose that digital wallets have the appropriate functionality. Then, 66.2% of respondents also stated that digital wallets were easy to use and 32.8% of using digital wallets could save time [1].

The continued growth of digital wallets and high popularity among the public has led to the rise of various types of fraud and deception mode on behalf of these service providers. This is evidenced by the Police report which stated that the high number of cases received in 2019 was online fraud. However, the police did not specifically mention the number of online deception figures, especially deception that occurred on digital wallet service. The mode of deception that is carried out can be in the form of suspicious messages, phone calls and so on. From these messages or call, deceiver usually asks for OTP from customers, ask for personal data even ask for a balance transfer. Apart from deception, there are also cybercrimes in the form of digital wallet burglary. These cybercriminals usually perform these 3 types: impersonate and take over SIM cards (SIM Swab), Man-in-the-middle (MiTM) attacks and phishing and also malware attack [2].

The rise of various types of fraud and cybercrime modes related to digital wallet service shows the importance of public awareness of security in the cyberspace. In the research conducted by Ogutcu, et al., stated that security issues are not just a technological problem, but also human attitudes and behavior towards the use of information technology [3]. Therefore, public awareness of information security needs to be properly assessed and evaluated. This is useful to find out more about the scope of information security which still needs to be improved as a preventive measure and self-protection towards various types of fraud and data theft.

Kruger and Kearney have developed a model for measuring information security awareness, namely the Knowledge-Attitude-Behavior (KAB) model. This model is the most commonly used as a basic model for measuring information security [4]. Several studies have adopted the KAB model as a basis for measuring information security awareness, such as a study written by Parsons, et al. [5]. However, there is no specific research that discusses measuring the level of information security awareness for digital wallet users.

The purpose of this research is to find out the security awareness of Indonesians in cyberspace by testing their security level when using a digital wallet. The population of digital wallet users in this study represented the majority of Indonesian users.

2. Literature Review

2.1. Information Security

Information security is defined as the protection of information and information systems from unauthorized access, use, disclosure, interference modification, or destruction [6]. The concept of information security according to Andress is known as the CIA Triad, namely Confidentiality, Integrity, and Availability. Confidentiality is an important component of privacy and refers to an organization's ability to protect the data of unauthorized persons. Integrity refers to the ability to prevent to not change illegally. Availability refers to the ability to access data when needed.

Meanwhile, according to Whitman and Mattord, Information Security is an organization that must have several layers of security to protect its operations. The security in question is as follows [7]:

- Physical security generally protects physical items or objects and areas from illegal access and abuse
- Personnel security generally protects an individual or group of individuals who have the authority to access the organization and its operation
- Operation security generally protects the details of a particular operation or series of activities.
- Communication security generally protects communication media, technology and content.
- Network Security generally protects network component, connection, and their contents.
- Information security generally protects confidentiality, integrity and availability of information assets, whether in terms of storage, processing or transmission. This is achieved through the application of policies, education, training and awareness, as well as technology.

2.2. Information Security Awareness

Information security awareness is characterized by understanding security policy, procedure and practice so that when security problems occur, the public can make appropriate judgments. Improve information security awareness is like implementing a security policy, taking preventive measures, improving user security behavior patterns, or changing routines so that they can implement habits related to information security [8].

Building a level of information security awareness is important because human behavior is considered to be the biggest threat of information security. Various types of information security threats arise due to human behavior such as carelessness and negligence in safeguarding information, unwise mobile use, access unsafe sites and handle the data that does not comply with security policies. This behavior can be risky internally and externally. It caused cybercrime and deception to steal confidential data [9].

2.3. Measurement of Information Security Awareness

Measuring information security awareness is a dynamic process. Risk and threat are constantly changing so that the measurement of information awareness needs to be continuously measured and updated to reflect changes in the risk profile.

The most popular information security awareness measurement model used is the model proposed by Kruger and Kearney. 3 dimensions serve as benchmarks for assessing information security awareness, namely Knowledge, Attitude, Behavior. Knowledge to measure what users know about information security. Having insights related to information security is an initial step in creating information security awareness. Attitude is to measure how users feel or think about the known risk of information security. Behavior is to measure what users do or actions regarding information security [4]. Table 1 is a scale of the level of awareness that is used to help define the level of awareness of the object studied.

Table 1. Level of Awareness

Level	Result
Good	80-100
Average	60-79
Poor	<59

In the dimensions of Knowledge, Attitude and Behavior, it is focused on the variable focus area which refers to the variable developed by Parsons, namely Human Aspects of Information Security (HAIS-Q) [5]. Each dimension variable is measured to the 7 HAIS-Q variables, namely password management, email usage, use of the internet, use of social media, mobile device, information handling, and incident reporting.

2.4. Analytic Hierarchy Process Method

In this study, 3 area dimension variables and 7 focus area variables are used so that need a method that can calculate the validity up to the tolerance level for inconsistencies of various criteria and alternatives chosen by decision-makers that is Analytic Hierarchy Process (AHP). AHP is decision support using a basic approach. This approach is used to select the best number of alternatives that are evaluated based on several criteria [10]. It describes a complex multi-factor or multi-criteria problem into a hierarchy. Hierarchy is defined as a representation of a complex problem in a multilevel structure where the first level is the goal, then the factor level, criteria, sub-criteria, and so on until the last level of alternatives [11]. Data quality from respondents takes precedence over quantity data when applying the AHP method. The sample size in the AHP method does not have a specific formula or standard [12], there is a general consensus that the AHP does not require a very large sample [13]. Figure 1 shows the hierarchy of AHP.

3. Methodology

This section discussed the research stages, research instruments, data collection, hierarchical structure design, and data analysis techniques. The detail of each point explained below.

3.1. Research Stages

This section shows the stages taken in this study. Each stage explained below:

- Determine the problem. The first thing to do was to determine the problem that was researched. In this case, it also looked for facts about the problem and determined the research objectives.
- Literature study. At this stage, looking for various things or information that was relevant in the research, including searching for several previous studies that have similar topics or were related to this research.
- Make a research design. The research design made was compiling a hierarchical structure for the implementation of the AHP. In this hierarchical structure, the KAB (knowledge, attitude and behavior) model acted as a criterion and the HAIS-Q (human aspect information security questionnaire) acted as an alternative.
- Create questions and distribute questionnaires. After determining the research design, the next step was to create a questionnaire question. The questionnaire refers to two variables, namely the dimensional variable and the focus area variable. Furthermore, the questionnaire was distributed online by having predetermined criteria for respondents.
- Processing data and analysis. The next stage was processing the data from the questionnaire and analysis. The data processing technique used was AHP. After getting the research results, the results were analyzed.
- Conclude. In this stage, conclusions can be drawn based on the data that has been collected and analyzed. At this stage, several recommendations were also given that could be made to improve information security awareness.

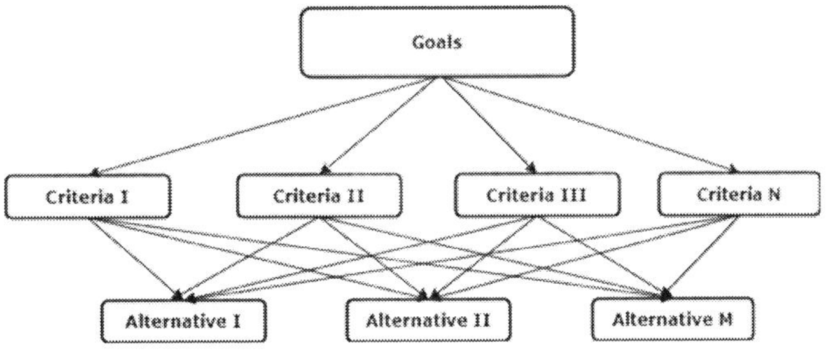

Figure 1. Hierarchy of AHP

Table 2. Likert Scale

Scale	Category
1	Strongly disagree / very negative / very never
2	Disagree / negative / almost never
3	Neutral / hesitant / sometimes
4	Agree / positive / syringe
5	Strongly agree / very positive / always

3.2. Research Instrument
The questionnaire was created using Google form and consisted of 55 questions consisting of 51 questions regarding information security and 4 questions which represent the criteria for respondents such as age, latest education, occupation and gender. The 51 questions regarding information security relating to 3-dimensional variables and 7 focus area variables. The questionnaire was translated into the Indonesian to ensure broader coverage. Google form is also set to only accept one response to prevent duplicate respondents. Respondents were asked to fill out the questionnaire using a Likert Scale with a

scale of 1 to 5. The Likert Scale was designed to measure attitudes in a scientifically accepted way and validated in 1932 [14]. Table 2 is the Likert Scale table used.

3.3. *Data Collection*

The questionnaires were distributed randomly online using random sampling techniques and met the requirements of using a digital wallet application in Indonesia. According to Arif Rahman, random sampling has several assumptions that the population is known as a whole, the sample is drawn randomly from the population, and each member of the population has an equal chance of being sampled [14]. The respondent's data were then processed using the Analytic Hierarchy Process and the results could be drawn a conclusion and analysis.

3.4. *Hierarchical Structure Design*

AHP has a hierarchical structure which is one of the stages to use in this method. The top hierarchical structure is the purpose of the decisions taken, then the next level is the criteria and the lowest level is the alternative [4]. Figure 2 is the hierarchical structure used in this study.

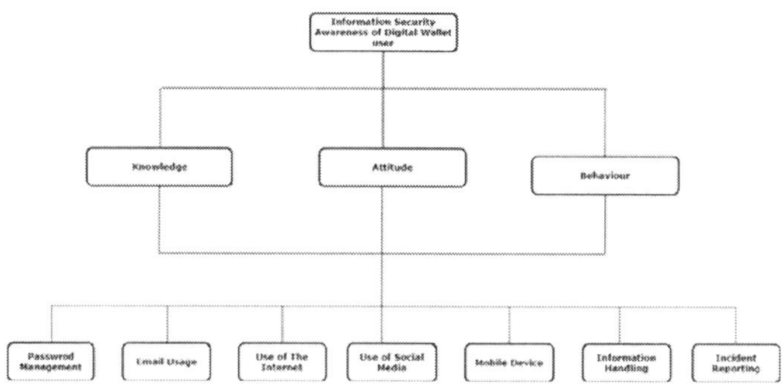

Figure 2. AHP Hierarchy Structure

The highest level of the structure is knowing the level of information security, the next level is the criteria that is filled with 3-dimensional variables (knowledge, attitude, behavior) and the lowest level is 7 focus area variables (password management, email usage, use of the internet, use of social media, mobile device, information handling, and incident reporting).

3.5. *Data Analysis Technique*

After the respondent's data collected, Kruger and Kearney measure the weight of the dimensions and focus areas using AHP. The value of the dimensional variable weight that is used in this study refers to the weight that has been made by Kruger and Kearney [4]. Table 3 represents the weight that was used in the dimensional variable.

Table 3. Dimensional Weight

Dimension	Weight (%)
Knowledge	30
Attitude	20
Behavior	50

Determining the weight of the focus area variable requires expert consideration to fill the pairwise of the comparison matrix [10]. An expert in information security from the University of Indonesia was interviewed to get a priority scale between the focus area variables. Table 4 is the weight of the focus areas used in this study.

The score of each focus area per dimension was calculated and normalized to the sum of one. The total score, v (a), was determined using the formula below [4]:

$$V(a) = \sum_{i=1}^{n} v_i(a)\, w_i \quad (1)$$

Table 4. Focus Area Weight

Focus Area	Weight (%)
Password Management	17.36%
Email Usage	7.69%
Use of Internet	5.22%
Information Handling	25.30%
Mobile Device	15.40%
Use of Social Media	15.40%
Incident Reporting	13.62%

After obtained the value from the awareness level of the digital wallet application users, the value was grouped into 3 categories of information security awareness level [4]. Table 5 represents the value categories that are used and the colors used in the results table. The function of giving color to the table is to make it easier to read the table and provide detailed information so that it can make it easier to take action for each dimension and focus area.

Table 5. Criteria of Information Security Awareness Level

Score (%)	Information	Colour
80-100	Satisfactory, does not require action/improvement	
60-79	Monitor, potentially requiring action/improvement	
0-59	Not satisfactory, requires action/improvement	

4. Result and Discussion

The questionnaire was distributed online to the general public but the respondent has used a digital wallet application. The number of respondents was 156, representing users of digital wallet applications in Indonesia.

The results of measuring the average information security awareness of digital wallet application users can be seen in Table 6. The total score is 80.78%. The result of the measurement is categorized into the good category according to the metering of the level of information security awareness used by Kruger and Kearney [4]. This showed that the information security awareness of digital wallet application users was satisfactory so that no action is needed to improve information security awareness.

Although the overall showed the good category, several focus areas need attention. The focus area of "use of social media" had the lowest average level of awareness of 73.35%. Some respondents still considered that displaying a phone number on social media was an action that would not interfere with their information security. Although this action is quite risky. When other people know someone's phone number, this will be a fraud gap to be able to log in by getting an OTP number from the victim.

The focus area of "use of internet" and "information handling" also still needs attention because each dimension is categorized into an average level and has the potential to take action to improve information awareness. The lowest value in the focus area of "use of internet" was on the knowledge dimension, which was 70.67%. It may some respondents were not aware that there could be a fake application that acts on behalf of the digital wallet service so that they provide their data to the fake application. Many respondents also might access digital wallet applications with a free VPN from the application store. Besides, respondents were also not aware that entering personal data on online sites that lured digital wallet balance vouchers that could compromise their personal data. Meanwhile, the lowest value in the focus area of "information handling" was on the behavior dimension, which was 75.32%. Many respondents still stored important information about digital wallet accounts such as PINs on their phones without providing protection. This is quite risky when someone accesses their phone and accidentally opens information about a digital wallet account.

Table 6. Measurement Results of Information Security Awareness

Focus Area	Dimension			Total Awareness
	K	A	B	
Password Management	89.26%	87.98%	76.92%	82.84%
Email Usage	71.15%	77.56%	85.58%	79.65%
Use of Internet	70.67%	76.92%	78.85%	76.01%
Information Handling	77.24%	79.17%	75.32%	76.67%
Mobile Device	92.31%	90.06%	86.22%	88.81%
Use of Social Media	76.37%	79.12%	69.23%	73.35%
Incident Reporting	90.97%	90.38%	83.33%	87.03%
Total	82.69%	83.75%	78.44%	80.78%

The last focus area that needs to be considered was "email usage", because the score of the knowledge and attitude dimensions still showed the category of needing supervision, 71.15% for the knowledge dimension and 77.56% for the attitude dimension. There were still many respondents who deliberately entered their data because they were lured with reward through fake email notifications on behalf of the digital wallet service. They also might don't understand the threat of email containing malware.

Although none of the dimensions and focus areas showed the poor category, digital wallet service providers need to make some improvements in several focus areas with the average category by the following recommendations:

- Based on focus area of information handling and email usage that has an average score and needs some improvements, so that digital wallet service providers need to provide education to users periodically. This is important to do because unconsciously users tend to do things that can threaten the security of their digital wallets. This education can be in the form of an appeal not to put a phone number on social media, then not to store important information about digital wallets on phone, and an appeal to only trust the official website or email of a digital wallet service provider. The hopes that digital wallet users always remember to keep their data security.
- Based on focus area use of internet that the score still on average and needs an action such as the digital wallet service providers may add policies for digital wallet users. One of the policies that need is to prohibit users from accessing digital wallet applications using a free VPN from the application store. Cybersecurity observers said that using a free VPN can be risky. The types of risks are also varied, such as VPN traffic can easily spread advertisements or malware, the other risk is robbery and data leaks [16]. So that digital wallet providers need to strictly prohibit their users from accessing digital wallets using a free VPN.

5. Conclusion

The measurement of information security awareness to digital wallet application users using the KAB model has been successfully delivered. The KAB model assessed 3 dimensions that are Knowledge,

Attitude, Behavior and 7 focus areas that are password management, email usage, use of the internet, and use of social media, mobile device, information handling, and incident reporting. The results showed that the total score of the respondents' category of awareness was at the good. However, focus areas such as "use of social media", "use of internet", "information handling", and "email usage" still showed average scores or require supervision.

Although there are no dimensions and focus areas that indicate a poor category, these actions are still needed to increase the level of user awareness in several focus areas with an average score. Such as an action to provide education to users regarding appeal not to carelessly display a phone number and reminding them to properly store valuable information, then not easily trusting fake sites and emails. It can also create policies that strictly prohibit the use of free VPN when opening digital wallet applications.

Acknowledgments

This research was supported by PUTI Q2 grant "Measurement of Information Security Awareness Level: A Case Study of Digital Wallet Users" (NKB-850/UN2.RST/HKP.05. 00/ 2020). We would express our gratitude to the Faculty of Computer Science and Directorate of Research and Community Engagement, Universitas Indonesia.

References

[1] DSresearch 2019 *Moving towards a New Era in Indonesia's Financial Industry (Electronic Materials)* (Jakarta: Daily Social)
[2] Nugroho A 2020 *Three Attacks Used to Break Digital Wallets (Electronic Materials)* (Jakarta: cyberthreat.id) [in Indonesian]
[3] Öugütçü G, Testik Ö Z and Chouseinoglou O 2016 Analysis of personal information security behavior and awareness *Comput. Secur.* **56** pp 83–93
[4] Kruger H A and Kearney D W 2006 A prototype for assessing information security awareness *Computers & Security* pp 289–96
[5] Parsons K, McCormac A, Butavicius M, Pattinson M and Jerram C 2014 Determining employee awareness using the Human Aspects of Information Security Questionnaire (HAIS-Q) *Comput. Secur.* **42** pp 165–76
[6] Andress J 2011 *the Basics of information security: understanding the fundamentals of infosec in theory and practice* (Amsterdam: Elsevier)
[7] Whitman M E and Mattord H J 2012 *Principle of Information Security 4th Edition* (Boston: Thomson Course Technology)
[8] Fakeh S K W, Zulhemay M N, Shahibi M S, Ali J and Zaini M K 2012 Information security awareness amongst academic librarians *J. of Applied Sciences Research* **8(3)** pp 1723–35
[9] Bauer S, Bernroider E W N, and Chudzikowski, K. 2017 Prevention is better than cure! Designing information security awareness programs to overcome users' non-compliance with information security policies in banks *Computers and Security* **68** pp 145–59
[10] Saaty T L and Vargas L G 2012 *Models, Methods, Concepts & Applications of the Analytic Hierarchy Process 2nd ed* (New York: Springer)
[11] Saaty T L 2008 Decision Making with the Analytic Hierarchy Process *Int. J. Services Sciences* pp 83–98
[12] Saaty T L 2012 *Decision Making for Leaders: The Analytic Hierarchy Process for Decisions in A Complex World 5th ed* (Pittsburgh: RWS Publication)
[13] Schmidt K, Aumann I, Hollander I, Damm K and Von Der Schulenburg J M G 2015 Applying the Analytic Hierarchy Process in healthcare research: A systematic literature review and evaluation of reporting *BMC Med. Inform. Decis. Mak.* **15** p 112
[14] Edmondson D R 2005 Likert scale: a history *Conf. on Historical Analysis and Research in Marketing* pp 127–33

[15] Rahman A 2013 Randomized random sampling with an inverse-transform random variate generator approach based on the hypergeometric distribution *Seminar Nasional TEKNOIN* **4** pp 106–11 [In Indonesian]
[16] Jakartapost 2019 *Risks of using VPN and why it's difficult to stop (Electronic Materials)* (Jakarta: thejakartapost.com)

The Performance of Indonesia's President: A Sentiment Analysis in Social Media

Agus Mansur[1], Zuhdi Allamsyah, Putri Amalia

Industrial Engineering Department, Universitas Islam Indonesia.

E-mail: [1]agusmansur@uii.ac.id

Abstract. The role of a leader is critical, especially when a person leads a country. Indonesia, as the biggest archipelago country, led by Jokowi since 2014. From his first period in 2014 until his second period started in 2019, many opinions, suggestions, and criticisms addressed to the president, Jokowi. This research aims to measure the performance of Jokowi as Indonesia's president by collecting public opinions on social media, peculiarly in an interview video published on YouTube. All comments are scrapped and processed using machine learning tools, which are Python and TextBlob. The results of this research are positive and negative sentiments related to Jokowi. From the research conducted, the most influential positive sentiments of Jokowi are his personality. The negative sentiments gained in the video are about the poor quality of translation and the English skill of Jokowi.

1. Introduction

The president's duty Indonesia is not just as a head of state, but also as a head of government. Since then, there are many responsibilities that are carried out by the president of Indonesia. The president is also a leader, for both in state organization and for civil society. As in [1], a leader should have many skills such as giving motivational speeches, communicating, and other soft skills. Those competences will be useful for a leader to direct people for accomplishing the organizational goals. Hence, to be a good leader, the president of Indonesia should have those intelligences.

After the internal skills of the leader, another important aspect of a leadership is to be in line with people who are being led. The leader and his subordinates should have the same vision and mission to achieve organizational goals. Those things should also be implemented for the president because no wonder how excellent the strategy proposed and used by a president, if the public does not respond well, it might be unsuccessful to carry out. So, every strategic movement or policy should be conforming to public or civil society.

In order to know how the public responses are, a calculation from public perception when the leader delivers his ideas should be done. However, the method to measure public perception is sumptuous and takes a long time to accomplish. Social media developments make people able to criticize, give opinions, and suggestions to everything shown online. By the enhancement of it, many contents are available online from TV shows until the public press releases. These kinds of media can be used by people to show their expressions towards the material, and their judgments about the information delivered.

This research aims to measure the public's perspective on their leader by a fast and less-expensive method from social media by using natural language processing tool (NLP) method using Python. The outcome of this research is to know how is the public sentiment based on the data. The result of this research can be used by the decision-makers to improve their performance and evaluate the aspects of their leadership which are not preferred by the public.

2. Literature Review

This paper will try to know the performance of Indonesia's president. Many similar research has been done in scope of leadership assessment, one of them using the Delphi approach in [2]. The evaluation was through several steps: literature review, refine the leadership domain area, create survey, grouping, and wording. Leadership itself has many factors that influence the performance of them. Based on [3], four factors may give significances of the leader's performances: culture, constraints, follower, and hierarchical levels.

In order to process the data gained, the researchers will use sentiment analysis. As in [4], sentiment analysis is a popular tool to know the data about the writer's emotion in text. A sentiment analysis research was done based on the user responses on a YouTube video. The research completed by [5], who gained the data from 50 popular YouTube channels about pro and anti-pro anorexia. One reason behind the use of YouTube for the data because YouTube is a universal video social media platform and provides comments for everybody who watched it.

YouTube is a top-rated site to gain data, as social media has reached by most people in the world, sentiment analysis can be done to calculate the probability of a leader becoming elected. Similar research has been done by [6] that collected a thousand hundreds of posts under several related keywords and crawled them with Python as a machine learning tool. The data then went to several processes until the sentiment detection using the Lexicon-based and machine learning approach. The result was a comparison between the first and second candidates, with their own positive and negative sentiments. Public sentiment analysis is useful for many cases, not only to examine the possibility of candidates in an election but also can be used to oversee the quality and consistency of an elected leader.

A study in [7] which also implement sentiment analysis of a president candidate of US, Donald Trump after his primary debates. They used Twitter as a platform to collect the public's opinion. The result of that research is knowing the polarities of the tweets, to know if the public support the candidate or not. Therefore, this research is done to know the president of Indonesia's performance based on its public sentiments from a specific YouTube video by combining the previous researches methods.

3. Data Collection and Analysis

The case used in this research is a video taken from Bloomberg Market News and Analytics on YouTube. The video contains an interview of Indonesia's President, Joko Widodo, regarding his plans for the country's infrastructure, economy, and resources. From the video that was already seen by 162,053 viewers, an enormous number of comments also lied in the comment section. There are more than a thousand comments in total, both from Indonesians and from other countries. However, this research only focuses on the comments related to the video, which are about 300 comments.

There are six steps in conducting this research: scrapping, data cleansing, sentiment analysis, positive sentiment clustering, negative sentiment clustering, and visualizing output as described in the fig. 1.

3.1. YouTube Comments Scrapping

To collect the desired data from YouTube, the researchers used a web scraping tool. Web scraping is employed to get all data from targeted web pages. There are many libraries that are able to do web scraping especially which support Python programming language. The library used in this research is Selenium Library. Generally, Selenium Library is utilized for browser automation tools, but then there are many features that can be done with the library. ChromeDriver is also equipped to conduct comments scraping together with the Selenium Library. The ChromeDriver is a standalone server to implement W3C webdriver. WebDriver here is an open source tool to do testing automation with several capabilities such as navigation, user input, javascript, and many more. The result of the scraping process is a dataset from the specific YouTube content and it is stored in CSV file format.

IOP Conf. Series: Materials Science and Engineering 1077 (2021) 012004 doi:10.1088/1757-899X/1077/1/012004

3.2. Data Processing

The data gathered from machine learning then processed to sort the raw data into data set that is ready to use. This step is conducted by several actions, such as:

a. Deleting the line breaks in each cell and replace it with spaces.
b. Word processing which aimed to change several abbreviations (table 1)

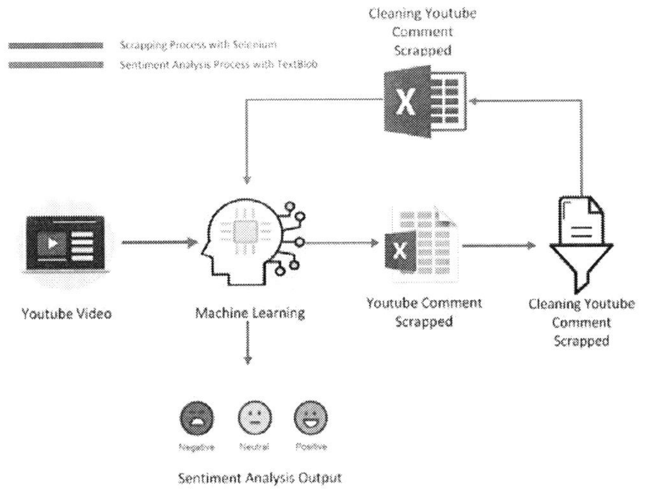

Figure 1. Research Flow

Table 1. Abbreviation Revamping

Before	After
Don't	Do not
Don't	Do not
Dont	Do not
I'm	I am

3.3. Sentiment Analysis

Sentiment analysis is a prevalent natural language processing tool that is easy and convenient to do opinion mining. NLP has a subfield that focuses on text analysis and can recognize whether the data are having positive, negative, or neutral sentiments. As an unsupervised learning tool, the work done does not require cheat sheet or labeled output. The researchers only write the commands in Python without input the correct example data as the model.

Sentiment analysis is done to know in which polarity value the comments are into and tagging them as negative, neutral, or positive sentiment. The tool used is TextBlob. The utilization of TextBlob is to perform NLP in Python. As a Python library, the researchers used TextBlob to calculate the polarity and subjectivity in conducting sentiment analysis. The result of the calculation is in the range of -1 to 1 for the polarity, and 0 to 1 for the subjectivity. The sample data that has been processed shown in Table 2. The criteria utilized to determine the sentiment are:

a. If the polarity value <0, then the data considered as negative sentiment.
b. If the polarity value =0, then the data considered as neutral sentiment.
c. If the polarity value >0 and <1, then the data considered as positive sentiment.

IOP Conf. Series: Materials Science and Engineering 1077 (2021) 012004 doi:10.1088/1757-899X/1077/1/012004

Table 2. Sentiment Analysis Result

Review	Polarity	Subjectivity	Sentiment
I am not an Indonesian but I love this man.	0.5	0.6	Positive
Proud of you My President for Indonesia Jaya raya	0.8	1	Positive
I am not proud of you my president	0.8	1	Positive
his english is not good	-0.175	0.3	Negative
Fake news in action	-0.2	0.55	Negative
Why He do not speak english?	-0.1	0	Negative
The style is very simple.	0	0.4642857143	Neutral
Can we get a version without the voice over dub?	0	0	Neutral
Unity in diversity	0	0	Neutral

Table 3. Leadership Aspects Classification

No	Keyword	Aspects
1	Like	Personality
2	Humble	
3	Attitude	
4	Smart	
5	Patience	
6	Love	
7	Prosperity	
8	Creatively	
9	Outstandingly	
10	Good	
11	Hard work	Hard worker and visionary
12	Hardwork	
13	Vision	
14	Amazing	
15	Good job	
16	Politician	Good Leader
17	Leader	
18	Good leader	
19	Politician	

3.4. Collecting Positive Sentiment and Classifying into Leadership Aspects

After the previous step is done, the researchers got three set of data: positive sentiment, neutral sentiment, and negative sentiment. The positive sentiment then refined and grouped based on the leadership aspects. The aspects and keywords are determined based on the word similarity that appeared in the positive sentiment comment data. There are three leadership aspects and 19 keywords. Table 3 shows the details.

After decided the keywords for each aspect, the data of sentiment analysis result then grouped into those three aspects. The example of aspects classification describes in table 4.

Table 4. Classified Positive Sentiment in Leadership Aspects

Review	Sentiment	Leadership Aspects
For me Indonesia is a sleeping giant slowly awakening from its slumber. Sooner this country will be prosperous and powerfull! Love from Philippines	Positive	Personality
President with his sneaker shoes. Good job sir	Positive	Hard worker and visionary
true, that is how his leadership is no less no more.	Positive	Leadership
Proud of using Indonesian ... even though English is good, sir	Positive	Personality
Our beloved Mr President? proud of him, proud to be Indonesian	Positive	Personality
Good job Mr. President	Positive	Hard worker and visionary

Table 5. Negative Sentiment Aspects Classification based on Words Similarity

Review	Sentiment	Aspects
Subtitle please, the dubbed translation is awful	Negative	Wrong translation
So awful! Please stop to have interview with this 'president'	Negative	Jokowi's English skill
Are there more Palace protocol teams that regulate the President's appearance. Is it appropriate to receive state guests wearing casual shoes with mini socks. Sorry if I think that I do not appreciate the state guests and give the impression that the Republic of Indonesia is still underdeveloped in international relations.	Negative	Jokowi's fashion is not appropriate
Cuih cuih cuih huueeek he cheat the election result he is not president the real one is prabowo subianto, he might seem humble but what kind of president who careless about 600 people dead on 2019 election.	Negative	Disagree with Jokowi's policy
shame on you Mr. President	Negative	Disagree with Jokowi's policy
I am Sorry our president can not speak English	Negative	Jokowi's English skill

3.5. Negative Sentiment Classification
The negative sentiment dataset is processed by screening the data and arranging the data based on word similarity and sentence expression. The researchers made four main aspects that mainly appeared on the data, which are about wrong translation, English skill, inappropriate fashion, and the disagreement of the policy. The author put "other" aspects to cover data that are not belong to those four aspects.

3.6. Visualizing Output
The output of this research is to map the reaction of the video reviewer related to the leadership aspects of President Jokowi and the negative sentiment that emerged. The sentiment analysis result is drawn in charts.

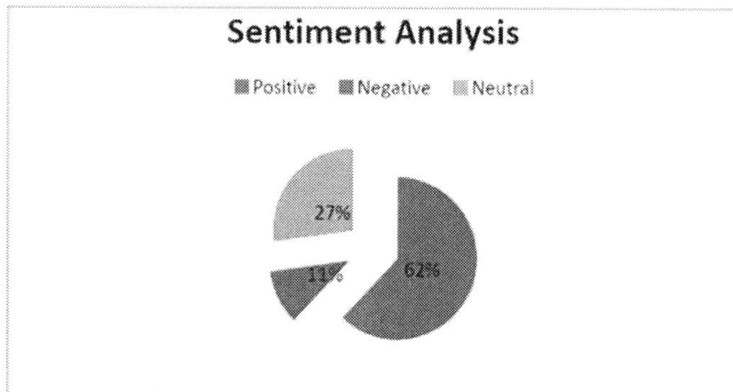

Figure 2. The proportion result of each sentiments

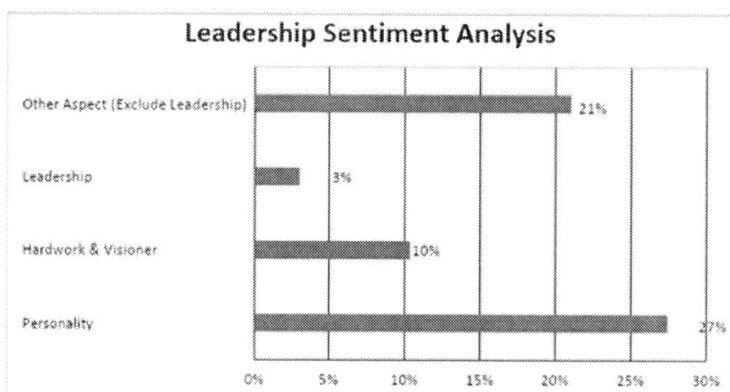

Figure 3. The sentiment analysis result of leadership aspects

There are more than half of the reviewers give positive sentiments towards Jokowi's leadership aspects as shown in Figure 2. This result is in line with the survey of Arus Survey Indonesia (ASI), a public survey institution that examined the satisfaction of Indonesia's people under Jokowi's period during the first year of the second term on his leadership. As in [8], 68,3% of respondents are contented enough with the performance of Jokowi as Indonesia's president. Although the people who voted satisfied are more than 50%, this period of Jokowi's leadership is assumed to be lower than his first period [9]. According to the survey in [10], 68,6% of the public are satisfied enough for the performance of Jokowi as Indonesia's president in the current 3,5 years during his first period of leadership. So, the result of the analysis that has been done in this research draws that the positive sentiment is reduced by 4,2% from the first period of Jokowi's leadership.

Figure 3 describes that personality is the most significant concern of people in supporting Jokowi. It is strengthened by [11] that the leadership success factor is affected by the leader's personality. In this research, it can be reckoned that the highness rate of people who love the personality of Jokowi may lead to the increased success rate of Jokowi in his second period of leadership.

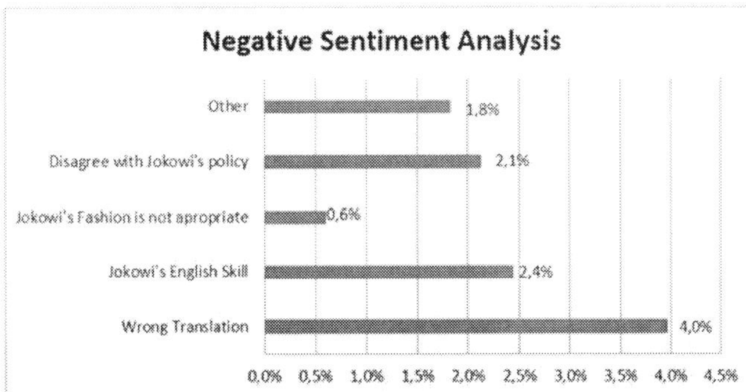

Figure 4. The negative sentiment analysis result

Figure 4 collected many negative sentiments and grouped them based on their similarity. In the video where the researchers gathered the data, the wrong translation is the most common comment in the comment section. Many viewers of the video criticize the translation since the video delivered in English, whereby it is translated from Bahasa, and people noticed there are many mistranslated words. The mistranslated words may lead to misconceptions and misunderstandings for people who watched the video, especially those not the Indonesian natives. The next concerns from the viewers are about the English skill of Jokowi since the video used dubbed audio to cover the speech from Jokowi in Bahasa. Many people asked why Jokowi did not talk in English and wondered about the incompetence of Jokowi in communicating with English.

4. Conclusion

This research successfully measured the public's perceptions towards the Jokowi's leadership skill as the Indonesia's president. The analysis result shows that the popularity, which in this research is positive sentiment, is slightly decreased compared to the previous period of Jokowi's leadership. Aspects that people love from Jokowi is his personality, since personality gained 27% from all sentiments. Many comments related to the Jokowi's personality, such as his appearance and style as a president that different with another typical country's leader. Although this research can describe the public opinion towards their leader, but there are many deficiencies that should be improved in the next research, which is screening for the profile of people who added the comments such as age, gender, job, and others. The future research will be more interested if can review the backgrounds of respondents, whether comes from the opposition or stronghold. This information can be utilized to gage the rational and irrational supporters based on comments submitted.

References

[1] Wijono S 2018 *Kepemimpinan Dalam Perspektif Organisasi* (Jakarta: Prenadamedia Group)

[2] Vaughan R, Kost R G and Brassil 2020 D *Development of a Leadership Assessment Scale in Translational Science 2020*

[3] Wong A and Chan A 2010 Understanding the leadership perceptions of staff in China's hotel industry: Integrating the macro- and micro-aspects of leadership contexts *Int. J. Hosp. Manag.* **29** pp 437–47 DOI: 10.1016/j.ijhm.2010.01.003

[4] Wojcik R 2019 Unsupervised Sentiment Analysis: How to extract sentiment from the data...

Towards Data Science Retrieved on 16 October 2020 from https://towardsdatascience.com/unsupervised-sentiment-analysis-a38bf1906483

[5] Oksanen A *et al* 2015 Pro-anorexia and anti-pro-anorexia videos on YouTube: Sentiment analysis of user responses *J. Med. Internet Res.* **17** pp 1–10 DOI: 10.2196/jmir.5007

[6] Oyebode O and Orji R 2019 Social Media and Sentiment Analysis: The Nigeria Presidential Election 2019 *2019 IEEE 10th Annual Inf. Technol. Electron. Mob. Commun. Conf. IEMCON 2019* pp 140–6 DOI: 10.1109/IEMCON.2019.8936139

[7] Abdullah M and Hadzikadic M 2918 Sentiment analysis of twitter data: Emotions revealed regarding Donald Trump during the 2015-16 primary debates *Proc. Int. Conf. Tools with Artif. Intell. ICTAI 2018* pp 760–4 DOI: 10.1109/ICTAI.2017.00120

[8] Rif'an A 2020 Evaluasi publik jelang 1 tahun pemerintahan Jokowi-Ma'ruf Amin *Arus Survei Indonesia n.d.* Retrieved on 11 August 2020 from https://www.arussurvei.com/evaluasi-publik-jelang-1-tahun-pemerintahan-jokowi-maruf-amin/

[9] Ariant T 2020 *Bedah Survei Persepsi Publik, Jokowi Disarankan Evaluasi Kinerja Menteri | Monitor n.d.* Retrieved on from 11 August 2020 from https://monitor.co.id/2020/06/25/bedah-survei-persepsi-publik-jokowi-disarankan-evaluasi-kinerja-menteri/

[10] DH A 2018 Tingkat Kepuasan Publik Terhadap Kinerja Jokowi-JK 65,1 Persen *Tirto.ID* Retrieved on 12 August 2020 from https://tirto.id/tingkat-kepuasan-publik-terhadap-kinerja-jokowi-jk-651-persen-cKXm

[11] Do M H and Minbashian A 2020 Higher-order personality factors and leadership outcomes: A meta-analysis *Personality and Individual Differences* **63** 110058 DOI: 10.1016/j.paid.2020.110058.

Analyzing Public's Reaction towards Black Lives Matter Campaign using Machine learning-based Approach through Spark

Rio Rizki Aryanto[1], Dhomas Hatta Fudholi[2]

[1]Master Program in Informatics, Universitas Islam Indonesia, Yogyakarta, Indonesia
[2]Department of Informatics, Universitas Islam Indonesia, Yogyakarta, Indonesia

E-mail: [1]rio.aryanto@students.uii.ac.id, [2]hatta.fudholi@uii.ac.id

Abstract. Since the end of May 2020, there is a massive wave of protestant in the United States addressed to the government regarding the case of police violence towards black people. On May 25, 2020, George Floyd, a 46-years-old black American man was killed during an arrest for allegedly using a counterfeit bill in Minneapolis. This study analyzes the public's reactions towards the Black Lives Matter campaign using a supervised machine learning-based approach. The proposed model uses logistic regression with word vectors as its feature. The model classifies the public's reaction represented by tweets and crawls using Spark Streaming into sentiment class, i.e. positive, negative and neutral. In addition, named entity recognition analysis was also conducted in this study. The aim is to find who else besides George Floyd whose rights have been fought for by the public. SparkNLP is used to build the logistic regressions model, sentiment analysis and named entity recognition. This study finds that most of the public tweets had a negative tone addressed to the Floyd incident specifically and to the violence towards black people in general. Another finding is that the campaign not only fought for George Floyd, but also fought for the other victims like Rayshard Brooks, Dominique Fells and Eric Garner.

Keywords: social media analytics, Spark, machine learning, sentiment analysis, black live matter

1. Introduction

Incident of George Floyd's death, a 46-years-old black American man on May 25, 2020 sparked a huge wave of protests in the United States. Even today, protest is still often found and reported in the media as the Black Lives Matter campaign. On May 25, 2020 George Floyds was killed during an arrest for allegedly using a counterfeit bill in Minneapolis. A white police officer named Derek Chauvin was knelt on Floyd's neck for almost 8 minutes. In the end, Floyd's life could not be saved anymore. The day after the incident, the videos made by witnesses and footage from security cameras became public. As a result, anger emerged from the public regarding police violence towards black people and triggered a move known as Black Lives Matter campaign.

The incident is one of many cases that show the inequality received by black people, especially African-American in the United States. This inequality has indeed become an old case and has even become public information for the people of the United States. In the past, there have been countless cases related to police violence towards African-American men. As today, the term white privilege was born, the privileges received by white people in the United States.

According to FBI and US Census Bureau data for 2018 about drug abuse arrest, black people had a higher arrest rate compared to its population rate. The opposite happened to white people and other ethnic where these two had a lower arrest rate compared to its population rate. The data said about 750 for every 100,000 African-Americans were detained for substance abuse reasons. While for the white people, it is just about 350 for every 100,000 people. The United States' national survey of drug use also stated that there are equal numbers of use among white people and African-Americans, however, blacks are more likely to be arrested. This is also the evidence of the inequality received by black people.

African-American are also detained 5 times more than white people and 2 times more than Hispanic-American. In 2018, 13% of the United States' population were African-American, but one-third of prison inmates in the United States' prison were also African-Americans. The majority of the United States' population were white people with about 60%. These people only contribute 30% of prison inmates. That means, for every 100,000 African-American citizens, 1,000 people were imprisoned. By the same token, only 200 white people were imprisoned. All of these statistics were reported by the US Census Bureau in 2018.

The other famous black people inequality was about shootings by police officers. Still according to the US Census Bureau and supported by Statista.com data in 2019, of the 1,004 fatal shooting cases that occurred in the United States, 36.8% happened to the white people, 23.4% African-Americans, 15.7% Hispanic and the rest for the other ethnic. African-American's fatal shooting percentage was higher than its population percentage in the United States. This only happened for black people whereas in others it happened the opposite. Fatal shooting itself is a shooting by a police officer that leads to death of the victim or suspect.

Above data just shows us that there is inequality received by black people in the United States which this inequality usually came from the police department. Recent cases that happened to George Floyd arouse the enthusiasm of African-American in the United States in particular and black people around the world in general together advocate for equality of black people in society. This study analyzed the public reaction towards Black Lives Matter campaign in the United States that was conveyed through social media Twitter. As today, the protest did not only happen on the street, but also through social media. The public reactions would be classified into sentiment classes while for deeper analysis, named entity recognition would be used as an additional.

2. Related Works

Machine learning-based approaches for analysis sentiment have been done many times in the past. Before machine learning-based approaches, the lexicon-based approach was known and widely used in many researches. Lexicon-based approaches make use of dictionaries or corpus to detect the sentiment. The quality of sentiment resulting from lexicon-based approach is very dependent on its dictionary or corpus. If the dictionary or corpus used is complete enough and is able to cover the scope of the problem raised, it is likely that the results of the sentiment classification will also be good.

In the previous work, one example of lexicon-based approach was done using an emoticon corpus [2]. Besides emoticon corpus, the researchers who used this approach also usually used N-Gram or hashtags as done on [3]. This study uses them to label the training data. Lexicon-based approach might feel old, but it could be used for problems or topics that are more general. Something that could be a challenge if using a machine learning-based approach. The other benefit using lexicon-based approach is it is easier to implement compared to machine learning-based approaches. On the other side, lexicon-based approaches also came with some drawbacks. This approach cannot provide the correct sentiment label for different subjects and usually is unable to capture the rich relational structure of the text. Therefore, this approach is mostly used as a complement or preprocessing stage before being complimented by a machine learning-based-based approach.

Lexicon-based approach's challenges could be handled using a machine learning-based approach. Basically, a machine learning-based approach would learn the data and its relationship so it is able to address data that may not have been studied before. Something that could not be done using lexicon based approach. To capture the rich relational structure of the lexicon, researchers used the word vector in their sentiment classifier [4]. Some studies also used supervised machine learning algorithms for its sentiment classifier like logistic regressions [5]. In their study, [5] use logistic regressions as base learner and majority vote as ensemble method. They also integrated their sentiment classifier with Hadoop.

A machine learning-based approach is indeed very closely related to very large amounts of data, where the application requires a lot of effort. The greater the data used, the heavier the workload of the machine which can result in the longer the processing time. Implementation of big data platforms could be the solution to this problem. One of the most popular big data analysis platforms is Spark. Spark is widely used by many researchers due to its ability in fault tolerance and real-time data processing. [6] use Spark for their real-time sentiment analysis for specific advertising targets. They used social media Twitter data for the study. [7] also conducted a study about sentiment analysis using social media Twitter data. The attempt of the study was to make financial decisions such as stock market predictions.

Spark implementations in big data analysis especially social media data were quite popular. [10] uses Spark streaming to stream Twitter data about job vacancies and then classified it into job categories. Spark was used to facilitate the effective job search in their system. Their work succeeded in implementing real-time analysis of filtering job vacancies among million streaming tweets which led to the improvement of the job search's effectiveness. Spark is also capable of conducting sentiment analysis like the one done in [11]. Their study implements Spark in a comparison analysis of sentiment analysis classifiers such as naive bayes, logistic regressions and decision tree. Another Spark implementation is real-time sentiment analysis using Spark streaming by [12], who in the study also conducted fake-account detection analysis. His work found that the classification performances of the system for real-time modes is about 80.93%. The proposed model in this study uses Spark streaming to retrieve the Twitter streaming data and analyze it using SparkNLP. Instead of fake-account detection analysis similar to [12], this study adds named recognition analysis to extract deeper insight from the tweets.

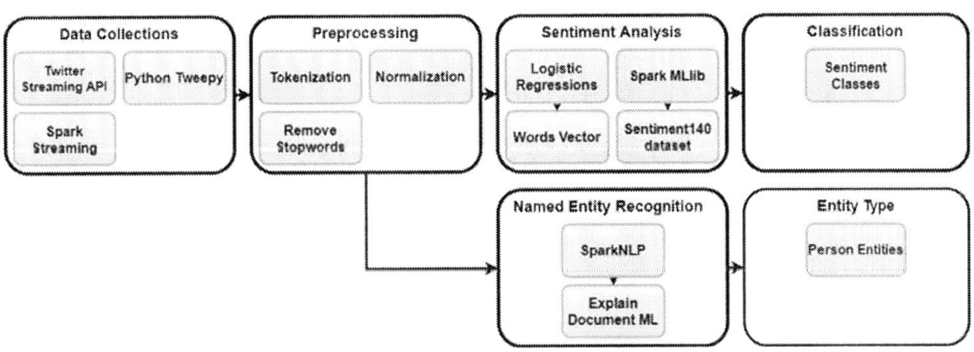

Figure 1. Workflow of the Proposed System

3. Methods

Focus of this study is to analyze the public reactions towards Black Live Matter campaign that happened in the United States. Similar to [6] and [7], this study would make use of real-time features in Spark as its main architecture. Real-time process happened in a crawling process which uses Spark streaming and Twitter streaming API to collect tweet data related to predefined keywords. While for sentiment and named entity recognition analysis, it was done every 15 minutes which the schedule was run using the help of crontab.

This study aims to analyze the public reaction that was conveyed through social media Twitter. Public reactions would be classified into sentiment classes using logistic regression method as a base learner for the classifier. The implementation of several analyses in this study were also done using Spark models such as Spark MLlib for sentiment analysis classification and SparkNLP for its named entity recognition. Figure 1 shows how the workflow of the proposed system which contains data collections, preprocessing, sentiment analysis and named entity recognition. The detail of each component is explained below.

3.1. Data collection

Data were obtained from Twitter that have been crawled in the period between June 4 to June 26, 2020. To note that, in this study there is a small missing time series data gap on June 7 due to technical problems. The system started with a crawling process using Twitter streaming API accessed from Python Tweepy and Spark streaming features. In order to obtain tweets that could represent public reaction toward Black Lives Matter campaign, the keywords filter features in Twitter streaming API is used. Tweets that would be taken are tweets that contain one of these keywords i.e George Floyd, Black Lives Matter or Racism. Using Spark streaming features, real-time tweets that have been collected will be saved in CSV files every 5 seconds. This CSV file will later be used in the sentiment analysis process and named entity recognition.

Instead of using historical data, this study uses stream data that is collected from 4th of June 2020. The data collecting starting date is just a few days after George Floyd's murder happened (on May 25). The decision to use stream data has been triggered by events that happened in early June such as Trump's threat to deploy military on protestant on June 1 and the charge of three former Minneapolis police officers on June 3 following Chauvin's charge on May 29. This study was interested in analyzing public responses toward Black Live Matter campaigns after some movements made by the United States' government. Was the public anger subsided or it did not affect anything at all? Using streaming data, we are able to capture every public response through Twitter after that time and also cover a longer period of time monitoring.

Spark streaming itself gives some benefits on its implementation. First, it recovers fast from failures as Spark streaming is popular as a fault-tolerance streaming system. Spark streaming also could handle heavy computation with its load balancing schema, which it would balance all the jobs across its workers. Interestingly, Spark streaming is also capable of doing interactive queries on streaming data on its worker memory. Moreover, the streaming data are able to be built into a machine learning model using rich libraries such as Spark MLlib.

As mentioned before, the tweets are expected to be able to represent the current condition in the United States. To accommodate this purpose, additional filters will be made before the tweets stored in CSV files. Tweets to be used must be in English and contain one of the following 3 keywords, namely "George Floyds", "Black Lives Matter" or "blacklivematter". In the end, a total of 1,299,181 tweets have been collected during the period of monitoring.

3.2. Preprocessing

Before analysis, the tweets must be preprocessed first. In this step, there are 3 processes done i.e tokenization, removing stop words and word vector formations. Spark was also used in the preprocessing stages. Tokenization is the process of separating single unit text into words. After that, the words which are categorized as stop words such as "this", "that", "was", "were", "during", etc would be removed. Lastly, word vector formation was done. Word vectorization itself is a process to convert filtered words into numerical data. Spark's Word2Vec model was used to create the word vector which in its implementation the vectors were built using the skip-gram model.

It is generally known that logistic regression only accepts numeric input as its feature, while our input text has been converted into a matrix of word vectors. However, we could still use logistic regression as our model since in Spark MLlib, the logistic regression model would automatically convert the matrix of word vectors into a vector by averaging all the values in the sequence. So, instead of using a matrix of word vectors, Spark's logistic regression used a vector representation of it as its feature.

3.3. *Sentiment analysis*

Our logistic regressions classifier was built and trained using Sentiment140's public dataset obtained from [8], which contains two different data i.e data-train and data-test. Our classifier was trained using the data-test, because it has three different sentiment classes and its suit best for this study's case. Actually, Sentiment140's data-train had more records than the data-test one, however, it only has two different sentiment classes.

Basically, logistic regression was used for binary classification problems. In Spark MLlib, binary logistic regression can be generalized into multinomial logistic regression and predict multiclass classification. In binary classification, to predict whether an input is positive or negative, the default threshold of probability i.e 0.5 was used. Probability above 0.5 would be classified as positive or negative otherwise. On the other side, there are two options that could be used for multiclass classification problems in Spark, whether using thresholds or depending only on each class probability returned from the multinomial logistic regression model. Using thresholds, Spark would return a class which has both of the highest probability and its value is above the threshold. Otherwise, Spark would return a class with the highest probability as a predicted class and not consider any thresholds. In this study, we did not define any thresholds for the model.

One-vs-one approach was used in multiclass classification implementation on Spark. Technically, Spark would regress each class against the pivot class using binary logistic regression. By default, Spark would choose the first class with label 0 as a pivot class. For example, for K possible outcomes or class, Spark would have a multinomial logistic regression model, which contains K-1 binary logistic regression. The class with the highest probability from the multinomial logistic regression model would be chosen as the predicted class.

In Figure 1, we can also see that beside sentiment analysis, this study also adds additional analysis using named entity recognition. Only one entity type was considered in this study i.e person entity. This additional analysis was done to obtain deeper insight about Black Lives Matter campaign.

4. Result and discussion

Most of the public tweets found between June 4 to June 26, 2020 had a negative tone, as seen in Figure 2. The gap with other sentiments is also quite high with almost 5 times higher. A total of 73% of the tweets had negative sentiment while for both of positive and neutral sentiment, it has similar percentage with each 14% and 12.8%.

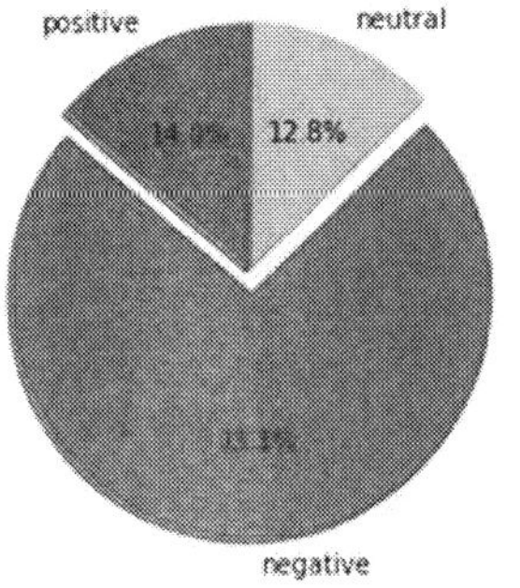

Figure 2. Proportion Sentiment of Public Tweets

During the period of monitoring, the number of negative tweets always surpasses the number of both positive and neutral tweets (see Figure 3). It seems there is a negative conversation happening on Twitter related to Black Lives Matter campaign. This might indicate that the public's rage had happened for a long period of time. They massively expressed their opinion through the Twitter platform with a negative tone.

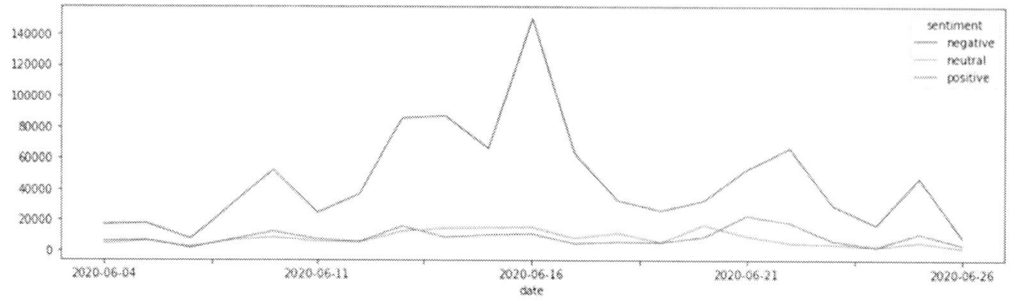

Figure 3. Time Series of Sentiment Classed during Period of Monitoring

Peak of negative tweets happened between June 13 to June 16, 2020. The blue line seen in Figure 3 had a slope up around that time. The peak might have been triggered by the incident that happened to Rayshard Brook, a 27-year-old African American man who was fatally shot by Garret Rolfe, an Atlanta Police Department (APD) officer on the night of June 12, 2020. This incident further exacerbated the situation and increased the public's anger level. After June 16, 2020 the number of negative tweets started to decrease back to before Brook's incident. On the other side, both positive and neutral sentiment had quite a stable number of tweets almost all the time.

Since most of the tweets had a negative sentiment, it could be misleading about the carried message, whether their anger addressed either to Floyd's incident or Black Lives Matter campaign. There is a possibility that several segments of society do not support the Black Lives Matter campaign. This crucial question could be answered using wordcloud analysis. The wordcloud in Figure 4 was formed using a collection of words in negative tweets.

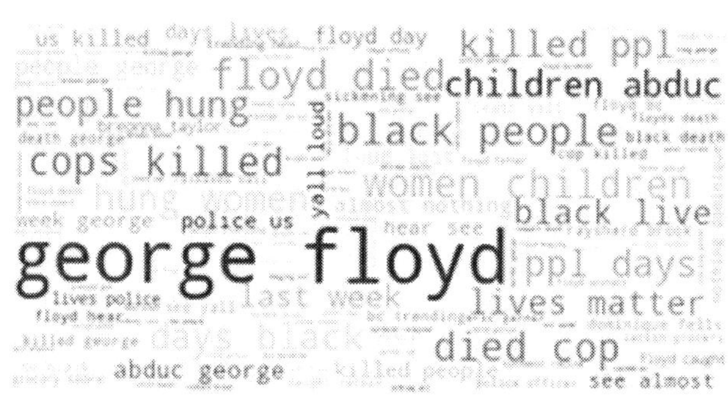

Figure 4. Wordcloud of Negative Tweets

Words like killed, hung, died, cop, police and etc, that are highly related to Floyd's incident have been found. It seems that the public's anger has been addressed to Floyd's incident rather than the message of Black Lives Matter campaign. The anger kept rising as the similar incident happened in close proximity to different people. Since the message of Black Lives Matter campaign was not clearly found in negative tweets, it might be carried on the positive tweets. On the positive tweets' wordcloud as seen in Figure 5, the words like police reform and justice act were found. Here, the message of Black Lives Matter started to be seen. There is an expectation from the public that there would be changes especially in the United States' police department. They demand reform within the police department and expect there would be equality in the society for black people.

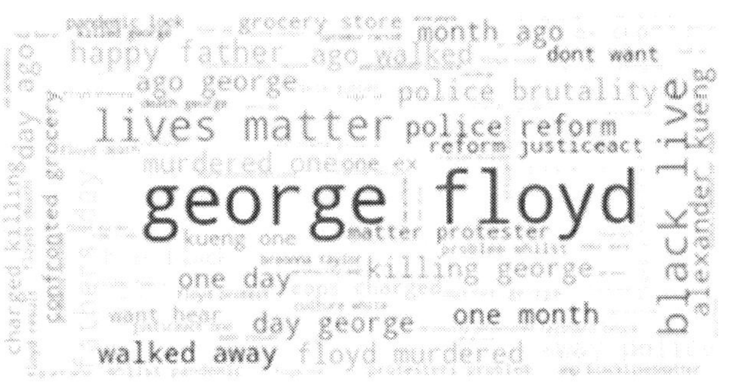

Figure 5. Wordcloud of Positive Tweets

This study also conducted named entity recognition analysis which aims to find deeper insight related to Black Lives Matter campaign. We only consider one entity in this analysis i.e person entity. Black Lives Matter campaign might have been triggered by Floyd's incident, however, our previous analysis found that the rage of the public just rose in the middle of June due to the Rayshard Brook's incident. There is possibility of other similar cases which are also being fought for through this campaign. Please bear in mind that the word coloring both in negative (Figure 4) and positive wordcloud (Figure 5) did not represent classes or anything, it was just for beautification.

Besides George Floyd and Rayshard Brooks, the result of named entity recognition in Figure 6 also shows the other names like Dominique Fells and Eric Garner. These two were also the victims of the violence towards black people in the United States. Dominique Fells, a Philadelphia transgender woman, have been murdered and the body was found on the banks of the Schuykull River in the second week of June 2020. According to CBS Philadelphia, the authorities believe that 36-years-old Akhenaton Jones is responsible for Fells' murder. This case might not relate specifically to police violence however, it is still related to the violence towards black people. It also happened around the same time to Brooks' incident. Fair to say, that the peak of negative sentiment found during the period was also triggered by Fells' murder.

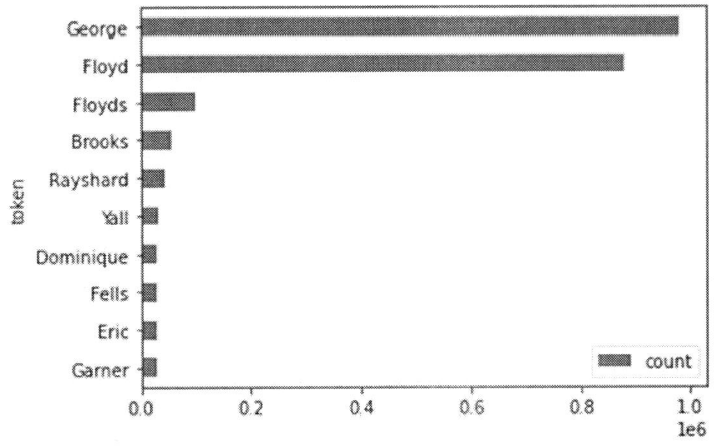

Figure 6. Top 10 Person Entity in Named Entity Recognition Analysis

The other name found was Eric Garner, who on July 17, 2014 died due to prohibited chokehold by a New York Police Department (NYPD) Daniel Pantaleo while arresting him. The incident took place in the New York City borough of Staten Island. Although it happened about 6 years ago, the public still remembered about the incident since there were no significant changes happening in the police department afterwards. Even on Jan 15, 2015, it concluded that Pantaleo's chokehold was administered and recommended charges, according to court documents. The public is back to fight for what they have fought before 6 years ago.

In the end, Black Live Matter campaign was not only about George Floyd but also the other victims such as Rayshard Brook, Dominique Fells and even Eric Garner. It is not only limited to cases about police violence, but about justice and strives for degree equality of black people among the society. The public tries to address this issue to the government of the United Stated and demand a reformation within the police department as their first step.

5. Conclusions

Based on the results of the study, we found that there is a public's rage emerging due to Floyd's incident which triggered a huge movement called Black Lives Matter campaign. About 73.1% of public tweets contain a negative tone, with most of it being addressed to Floyd's death incident. Another finding is about Black Lives Matter campaign's message. It could be found in the collection of positive tweets which some words such as police reform and justice act exist. Public expects a reformation within the United States' police department and hopes there would be equality for black people among the society.

The proposed model used Spark as its main architecture. Spark streaming was used together with Twitter Streaming API to crawl the data while Spark MLlib and SparkNLP were used for sentiment classification and named entity recognition analysis. Even though the system uses a simple algorithm such as logistic regression, the result of the analysis gives us quite an understanding of the public's reaction towards Black Lives Matter campaign. Named entity recognition also gives a deeper insight about what happened and what is being fought for through the campaign.

References

[1] Prabhat A and Khullar V 2017 Sentiment classification on big data using Naïve bayes and logistic regression *Int. Conf. on Computer Communication and Informatics (ICCCI)* (India: Coimbatore) pp 1–5

[2] Read J 2005 Using emoticons to reduce dependency in machine learning techniques for sentiment classification *Proc. of the ACL Student Research Workshop (ACLstudent '05)* (USA: Association for Computational Linguistics) pp 43–8

[3] Zhang L, Ghosh R, Dekhil M, Hsu M and Liu B 2011 *Combining Lexicon-based and Learning-based Methods for Twitter Sentiment Analysis* (California: HP Laboratories)

[4] Mass A L, Daly R E, Pham P T, Huang D, Ng A Y and Potts C 2011 Learning Word Vectors for Sentiment Analysis *Proc. of the 49th Annual Meeting of the Association for Computational Linguistics* (Portland: Oregon) pp 142–50

[5] Lin J and Kolcz K 2012 Large-scale machine learning at twitter *Proc. of the 2012 ACM SIGMOD Int. Conf. on Management of Data (SIGMOD '12)* (New York: Association for Computing Machinery) pp 793–804

[6] Nair L R, Shetty S D and Shetty S D 2017 Streaming Big Data Analysis for Real-Time Sentiment based Targeted Advertising *Int. J. of Electrical and Computer Engineering* **7** pp 402–7

[7] Das S, Behera R K, Kumar M and Rath S K 2018 Real-Time Sentiment Analysis of Twitter Streaming data for Stock Prediction *Procedia Computer Science* **132** pp 956–64

[8] Go A, Bhayani R and Huang L 2012 Twitter Sentiment Classification using Distant Supervision Retrieved from http://help.sentiment140.com/for-students

[9] F.F da silva N, Hruschka E R and Hruschka Jr E R Tweet sentiment analysis with classifier ensembles *Decision Support System* **66** pp 170–9

[10] Nair L R and Shetty DR S D 2015 Streaming Twitter Data Analysis Using Spark for Effective Job Search *J. of Theoretical and Applied Information Technology* **80** pp 349–53

[11] Baltas A, Kanavos A and Tsakalidis A K 2017 An Apache Spark Implementation for Sentiment

Analysis on Twitter Data *Algorithmic Aspects of Cloud Computing ALGOCLOUD 2016 Lecture Notes in Computer Sci.* 10230 ed Sellis T and Oikonomou K pp 15–25

[12] Kılınç D 2019 A Spark-based Big Data Analysis Framework for Real-time Sentiment Prediction on Streaming Data *J. of Software: Practice and Experience* **49** pp 1352–64

[13] BBC News 2020 George Floyd: Tiga 'perlakuan tidak adil' yang dialami warga kulit hitam di Amerika Serikat *BBC News* Retrieved from https://www.bbc.com/indonesia/dunia-52902071

[14] BBC News 2020 Kematian George Floyd: Bagaimana rasanya menjadi warga kulit hitam di AS *BBC News* Retrieved from https://www.bbc.com/indonesia/dunia-52917132

Graph Extraction of Batik Image Using Region Adjacency Graph Representation

Akmal[1], R Munir[2], and J Santoso[3]

[123]School of Electrical Engineering and Informatics, Institut Teknologi Bandung

E-mail: [1]akmal@unpad.ac.id

Abstract. Image feature extraction done in image analysis based on the traditional pixel is not extracted effectively. This is due to the extraction that only represents the content. A very promising approach and the challenge are to extract a graph from an image that represents the content and their relationship. The results of graph extraction are obtained from the process of image segmentation. The selection of an appropriate segmentation method from many graph-based image segmentation methods is interesting to be reviewed. The method of super-pixel segmentation is one way to divide the image into regions. The regions obtained in segmentation are represented by vertices and edges represent connections between adjacent regions. This representation is called Region Adjacency Graph (RAG). In this paper, graph extraction of "batik" images has been successfully performed where its representation uses RAG and segmentation uses the Minimum Spanning Tree method. The results of graph extraction have been tested in a simple image retrieval process using the VF2 graph matching method.

1. Introduction

Recently, graphs have been increasingly used as a representation way for image processing and analysis [1]. Graphs are used to represent information consisting of entities that are described as vertices and relationships between entities that are described as edges. In the field of pattern recognition, these representations have been used since the late 1970s to recognize objects. In addition to images, 2D / 3D shapes, documents, symbols, and characters, these representations are also used for chemical or biological structures, semantic web and web, social and economic networks, and many others.[2].

There are several reasons why image processing and analysis is carried out based on graphs. The first reason is its simple representation which is suitable for efficient method development. The second is that different types of images can be represented flexibly. The third is the reuse and redevelopment of graph theory algorithms and theorems in the field of image analysis and the fourth is the addition of high-level information to low-level representations (related to the semantic gap problem)[3].

Image segmentation is the initial stage in image graph extraction. It will form a homogeneous set of regions. Each region has a set of connected pixels and all pixels in a region have one feature set[4]. Extracting a very comprehensive image feature that can reflect image content is still a challenging problem[5].

In this paper, the extraction of the batik image graph is performed. It is expected to become the basis for analyzing the image of batik based on the graph method. The resulting graph extraction will be tested using graph matching through a simple image retrieval example using the VF2 algorithm. This can contribute to maintaining one of Indonesia's cultures by providing a suitable batik graph dataset. The contribution of this research is to become a model for forming a graph dataset of batik images. This dataset can be used to quickly and accurately classify and recognize various types of Indonesian batik.

In the Great Dictionary of Indonesian Language, batik has the meaning of illustrated fabric that is made specifically by writing "Malam" on the cloth, then processing it in a certain way. Batik is a familiar thing for Indonesian people. Batik is one of the unique archipelago heritage that must be preserved and has also become a world cultural heritage since 2009 which was confirmed by UNESCO. The uniqueness of batik is shown in various kinds of motives that have their own meaning.

The paper will be organized into several sections. The first part is an introduction that explains graph extraction. The second part contains the related work of research on this topic. The third part is the theoretical basis that underlies the extraction of image graphs such as graph theory terminology, segmentation, region adjacency graph, and VF2 algorithm. The fourth part is the results and discussion which is followed by the testing phase on image retrieval. The last section is a conclusion and suggestions for future work.

2. Related Works

There are several advantages of graph-based data structures that can be used to find relationships between objects [6] and can be applied to statistical methods and machine learning in graph theory [7]. Promising research in this area is interactive visual data mining together with graph-based data analysis [8]

Many methods have been proposed to obtain graph representations of objects. It depends on the purpose of the application. The difference between all these methods is in the scale of operation, but generally, in all these methods, the parts of an object in the image are represented by one node[2]. Holzinger extracts the graph structure from the image using HTML canvas and Javascript which is run on a Web browser [5]. Sharma determined the similarity of the histological image from the breast biopsy image using inexact graph matching where the results of the retrieval image made were superior to the histogram-based technique.[9]

There are methods that use a different node for each image pixel. These methods form a graph with a large number of vertices where the edges have a regular structure. An example is the Boykov and Kolmogorov algorithms which perform graph-based segmentation[10]. On the other hand, there are nodes that represent the whole object, or important parts of the object, which are generated from several levels of segmentation. Nodes represent a homogeneous region (e.g. Graph Adjacency Region Tremau and Colantoni) [11].

3. Definition and Terminology

There are a lot of literature and works that have been done related to segmentation based on the graph method. In this section, we briefly discuss the method for this work: some basic things about graph theory, segmentation whose region is used as a node for the Region Adjacency Graph representation, Minimum Spanning Tree algorithm, and graph matching using the VF2 algorithm

3.1. Graph Terminology

A graph is a set of elements and relation pairs between two elements. Nodes represent elements and edges represent relations. Graph G is the set $G = (V, E)$ with $E \in V x V$. The vertex i denoted by $v_i \in V$, and the edge i with $e_{ij} \in$ E, and also denoted $e_{ij} = \{v_i, v_j\}$[1].

There are several graph terminology as the basis for processing and analyzing images with graphs Here are some basic definitions of graph theory such as adjacent, incidence, subgraph, and graph matching [1]:

1. Adjacency: two vertices (v_i and v_j) connected by an edge are adjacent if $\exists e_{ij} \in E$ or $\exists e_{ji} \in E$.
2. Incidence: an edge e_{ij} is incident to a vertex v_i that is one of its endpoints.
3. Subgraph: G'= (V', E ') is a subgraph of G = (V, E) if V' $\subseteq$ V and E' $\subseteq$ E where E' = {$e_{ij} \in$ E | $v_i \in$ V' and $v_j \in$ V' }.

4. Graph matching is a comparison between two graphs with certain criteria to match the nodes and edges of both graphs. There are several basic concepts related to graph matching, such as graph isomorphism and subgraph isomorphism
   a. Graph isomorphism definition : let $G1 = (V_1, E_1)$ and $G2 = (V_2, E_2)$ are two non-directed graphs. A function $f: V_1 \rightarrow V_2$ from G_1 to G_2 is called an isomorphism if $(v_i, v_j) \in E_1$ then $(f(v_i), f(v_j)) \in E_2$
   b. Subgraph isomorphism definition: a subgraph isomorphism from graph G_1 to G_2 is an isomorphism from graph G_2 to subgraph G_3 from graph G_1 $(G_3 \subseteq G_1)$.

The graph formation can use 4-neighbor or 8-neighbor pixels. The Euclidean distance formula is used to find the RGB weight between two pixels [12] :

$$d(p,q) = \sqrt{(R_p - R_q)^2 + (G_p - G_q)^2 + (B_p - B_q)^2} \quad (1)$$

3.2. Segmentation and Region Adjacency Graph

Segmentation and RAG are included as ways to reduce the number of graph nodes. Segmentation will divide the image into several homogeneous regions which are interconnected. The partition graph is the basic principle of the graph-based segmentation method. This means that a search for the set of subgraphs (SG_1, SG_2, ..., SG_n) of graph G will be performed [4].

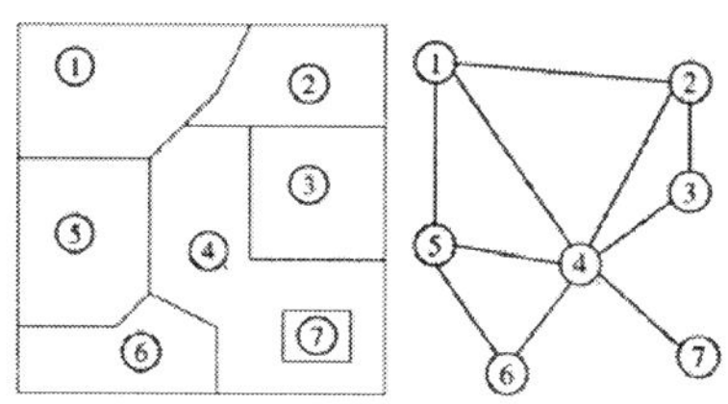

Figure 1. a. Region partition; b. Region Adjacency graph

The formation of a graph with a region adjacency graph representation is done by taking the centroid region as a node and the relation between two neighboring regions as an edge as seen in Figure 1. In this way, the number of vertices is reduced because the pixels in a region are represented by only one node.

The region that consists of pixels that have proximity or similarity is also called super-pixel. Super-pixel is a meaningful pixel grouping and is used as a primitive for further computing or processing [6]. Image processing and analysis using super-pixel bases can increase the processing speed compared to using pixel-based due to the number of vertices that have dropped dramatically.

3.3. Minimum Spanning Tree

Minimum Spanning Tree (MST) is a spanning tree with the smallest weight value compared to all spanning trees on the graph. [13]. Kruskal and Boruvka are MST algorithms that are widely used in the graph-based segmentation method. The complexity of the MST algorithm is polynomial and the algorithm is greedy. Basically, the MST algorithm MST is easy to understand and fast in processing. It is a graph-based clustering where the graph is an undirected graph. The edge weight is the color distance of two related vertices. The grouping is done by removing the edges graph with certain criteria to form a subgraph. The important criteria are gestalt principles such as similarity and proximity[14].

The efficient graph-based image segmentation method proposed by Felzenszwalb and Huttenlocher is MST based on the Kruskal algorithm [14]. This method uses the global differences between the two neighboring subgraphs and the local differences within each subgraph. Meanwhile, the segmentation using the Boruvka algorithm and the Felzenszwalb criteria was done by Ali Saglam. The method developed is to detect objects based on color and cluster size.[12] as seen in Figure 2.

Equation (2) is Felzenszwalb's criteria for grouping similar pixels in one region [12]:

$$Dif\ f(X,Y) < \min\left(Int(X) + \frac{k}{|X|}, Int\ (Y) + \frac{k}{|Y|}\right) \tag{2}$$

Internal Difference (*Int (X)*) is the difference between the edges with the highest weight on subtree *X*. *Diff (X, Y)* is the difference in the smallest weight of the two neighboring subtrees *X* and *Y*. $|X|$ is the size of the subtree X or the total number of vertices and k is a constant. Two subtrees will be combined if these criteria are met.

The following is an algorithm for merging regions in MST:

1. Sort the weights of all edges e ∈ E ascending from smallest to largest
2. For all sides (*e* ∈*E*):
   a. Check that the pixels connected to the edge *e* are in one region. If it is connected then continue processing to the next edge.
   b. If it is not connected then check if there is a boundary between the two observed regions. (Find *Int(X), Int(Y)* and *Dif f(X, Y)*) There is a limit if the minimum edge weight connected to the region is greater than the minimum of the maximum internal weight of the MST edges.
   c. If there is a limit, then go to the next edge.
   d. If not, then combine the region into a new region.

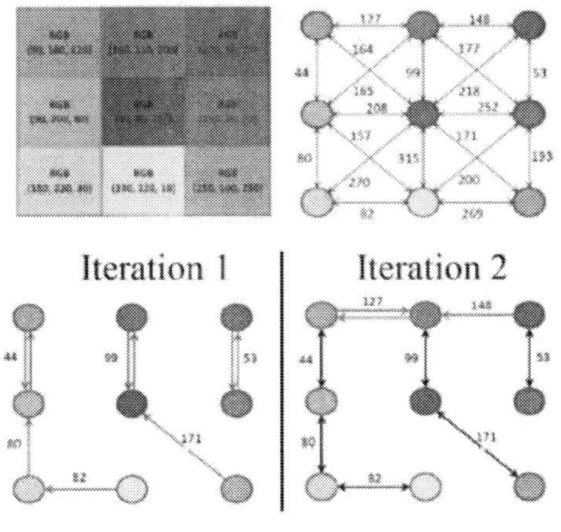

Figure 2. Minimum spanning tree (MST) on RGB 3x3 image [12]

3.4. Graph Matching using VF2 Algorithm

Finding the exact correspondence between two graphs can use the exact matching method. Structurally, ensuring that the mapping between the vertices of the two graphs must maintain the incident edges between these vertices. This means that if two vertices in one graph are connected by an edge when it is mapped to two vertices on another graph, it is also connected by an edge.

Figure 3. An example of 4 classes of the synthetic image dataset

The basic objective of exact matching is to determine whether the structures and labels of two graphs or subgraphs are identical. Two examples of isomorphism subgraph algorithms that are often used are

Ullmann's algorithm [15] and the VF2 algorithm [16]. Ullmann's algorithm is the basic algorithm of the tree search class of algorithms for subgraph isomorphism. Ullmann is the first depth-based search algorithm commonly used with complex heuristics to reduce the number of states visited. The main problem is the space complexity of O (n^3) which becomes impractical when dealing with large sparse graphs.

The Ullman algorithm is developed into the VF2 algorithm which can be used for graph isomorphism. VF2 is an algorithm commonly used in exact matching and is designed to solve pattern matching problems. In more than a decade, VF2 has been the best algorithm and is widely used because of its space efficiency and simplicity. Status-space is the representation used in VF2 and several conditions in each state will be checked to reduce the search tree.

4. Results and Discussion

It will show the process of graph extraction of the batik image using the Region Adjacency Graph representation model, and the testing of the results in image retrieval.

4.1. Data Preparation

The data prepared are synthetic images and real images. Synthetic data consists of 200 digital images with a size of 64 x 64 pixels. There are 4 different data classes, each totaling 50 images per class. This synthetic data will be tested in the graph extraction stages so that the details of the RAG formation can be known. Figure 3 is an example of synthetic image data that represents each image class.

Synthetic data was created using Adobe Photoshop CS6 64 bit. Some shapes with specific colors are drawn manually on page 64 x 64.

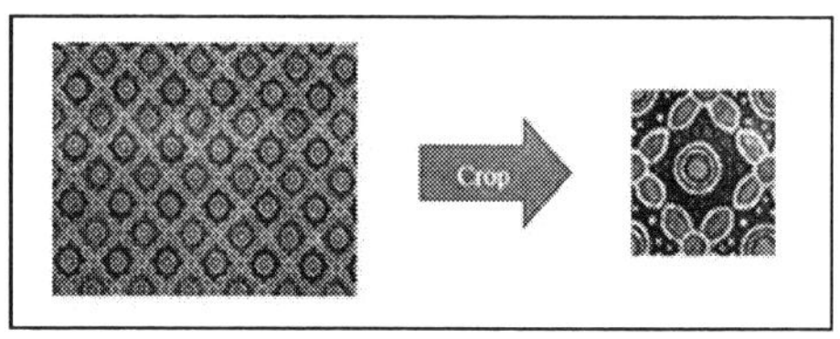

Figure 4. An example of a batik core motif

Batik Grompol, Batik Parang, and Batik Kawung are 3 variants of the core batik motifs used in this research. Each variant contains 60 images. The whole batik cropping technique is used using Adobe Photoshop. Figure 4 is an example of the cropping result of the batik core motif

4.2. Development of Region Adjacency Graph (RAG)

Region Adjacency Graph (RAG) is created on each segmented image. Each region of an image is given a label and its important features are searched. The centroid of the region is calculated and so is the RGB mean value. Then the centroids of the neighboring regions are connected and their weight is searched with the euclidean distance. Figure 5 is an example of the stage of development of RAG

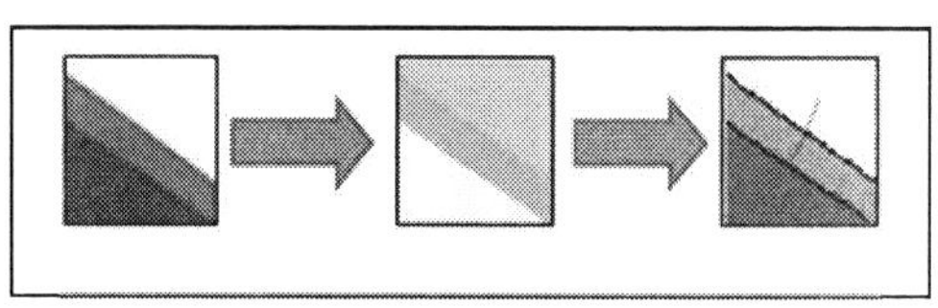

Figure 5. Stage of development of RAG

The algorithm of MST has been tested for real images and artificial images. The running time for this algorithm runs linearly with the number of edges. This method can measure regional boundaries by comparing two quantities: based on differences in the intensity of the regional boundaries that are opposite, and based on the intensity between neighboring pixels in each region.

Table 1. Examples of Image Segmentation and RAG Results for Each Class

File Name	Original Image	Image Segmentation	RAG
img030.jpg	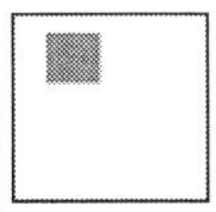	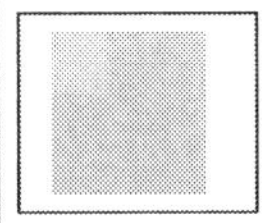	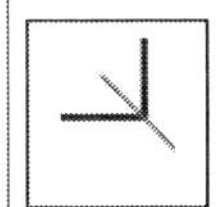
img094.jpg	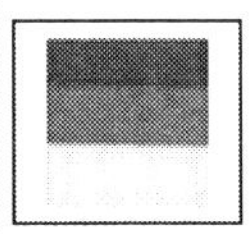	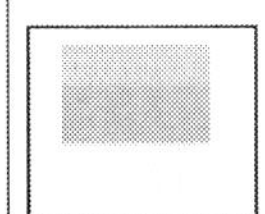	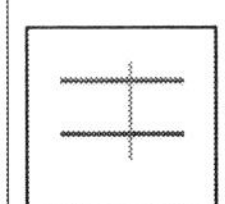
img101.jpg	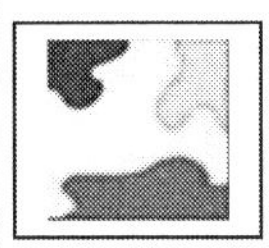		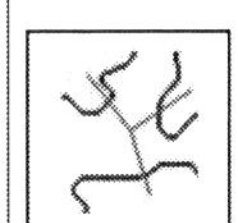
img156.jpg	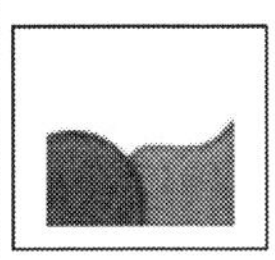	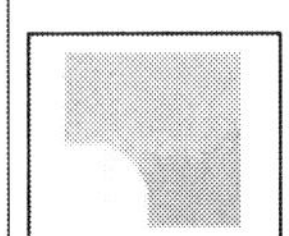	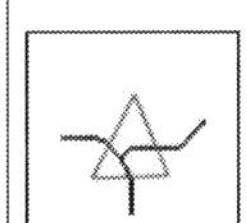
batik020.jpg		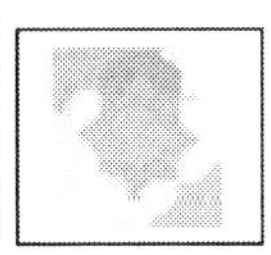	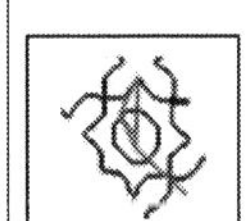
batik114.jpg		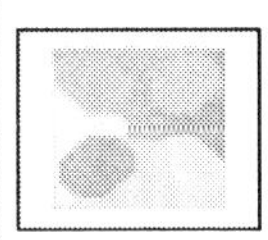	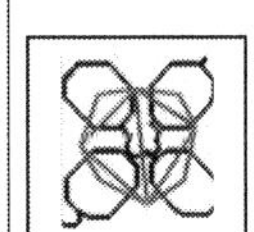
batik173.jpg	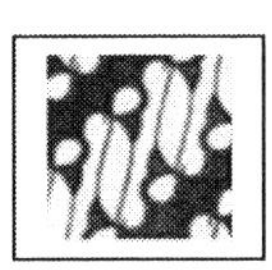	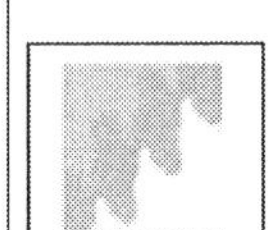	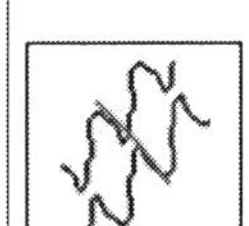

4.3. Experimental Results

The following is the result of a trial of the graph extraction process from an image. From the original image, the segmentation process is carried out and then extracted to get the region adjacency graph from the image. Graph feature extraction is obtained based on the label value given to each region resulting from segmentation. The label value provided is an integer value of 0 to (n_region-1). Each image will have one Region Adjacency Graph. Through these stages, there are several features that are generated from each graph, including:

1. Node: nodes will be formed representing each region of the image segmentation results
2. Centroid: Each region will have one centroid which is the midpoint value
3. Mean Color: Each region will have an average pixel value in RGB
4. Area: Each region will have a total of RGB pixels which is the area of the region
5. Edge: each pair of vertices from neighboring regions will form an edge
6. Weight: Each edge of the graph has a weight which is the value of the Euclidean distance between two neighboring regions

The graph of the RAG of each image will be stored on the storage media in a "*.gpickle" file format using the "networkx" library

Furthermore, from the resulting RAG, image retrieval testing was carried out using the VF2 algorithm. An overview of the process of retrieval is as follows and can be seen in Figure 6:

1. The user first enters a query image
2. The query is extracted from its graph feature (RAG). Graph extraction has also been carried out on image data stored in the database.
3. Comparisons will be made between the query image graph and the graph database image to look for similarities using VF2
4. After the comparison process, several isomorphism images will be selected
5. Then the data retrieval that has been selected is conducted

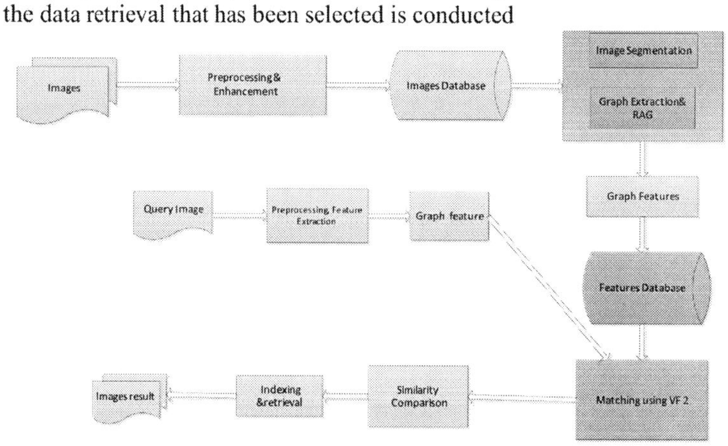

Figure 6. A framework of image retrieval

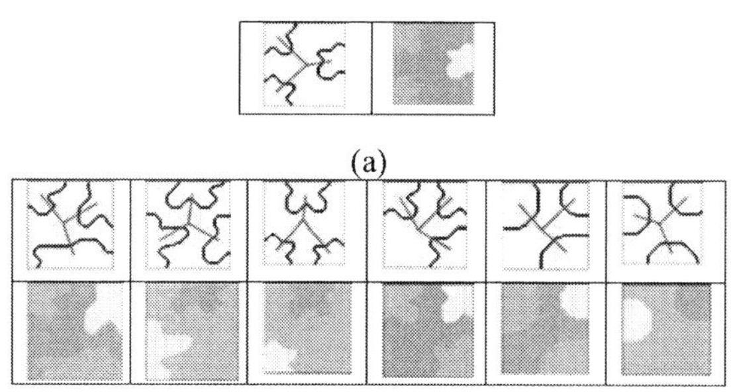

(b)

Figure. 7. Artificial image testing using the VF2 algorithm: (a). Artificial Image Query, (b). Test result using VF2

The following is an example of an image retrieval that meets the isomorphism requirements created using artificial images (figure 7) and batik images (figure 8). The number of retrievals can be determined by the user (1..n).

The next stage is performance evaluation using precision and recall values of image retrieval. Precision is the level of accuracy of the answers given by the system to the information requested by the user (Precision = TP / (TP + FP)). Meanwhile, recall is the success rate of the system in retrieving information (Recall = TP / (TP + FN)). The evaluation uses 200 synthetic images consisting of 4 classes and 180 real images consisting of 3 classes. Testing is carried out on all images. Table 3 are the evaluation results.

Based on table 2, the matching and retrieval evaluation performance, which includes precision and recall value on artificial image data, has good results. The values are 96.2% and 84.5%. However, a good precision result which is 87.8% is obtained for batik core motif data, while the recall result is 19.4%. This is due to the resulting RAG graph variation and noise so that the graph comparison does not provide the desired value. An isomorphic RAG graph due to rotation gives the correct result.

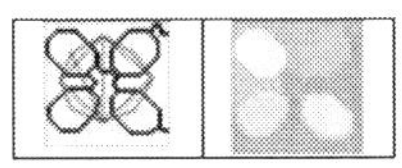

(a)

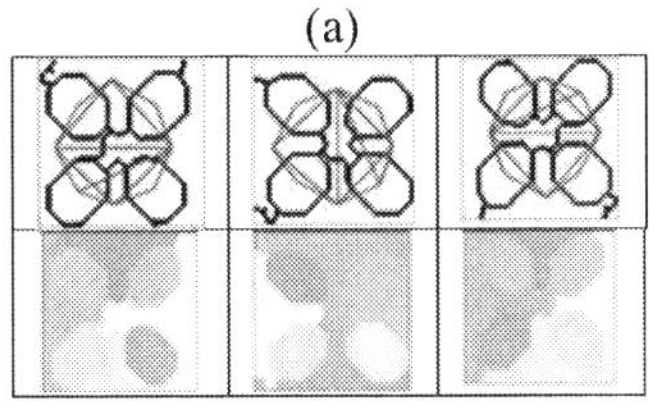

(b)

Figure. 8. Real image testing using the VF2 algorithm: (a). Batik image query, (b). Batik image query

Table 2. Performance results of matching and retrieval evaluation Tests

Dataset	Precision (%)	Recall (%)
Synthetic	96.2	84.5
Batik	87.8	19.4

5. Conclusions and Suggestions

It has shown the steps of graph extraction from batik images. The extraction of graphs from other types of images can follow the model that has been made. A simple image capture prototype has been created to test RAG batik. The Minimum Spanning Tree method is a technique used in segmentation. Many segmentation techniques have been developed, both graph-based and non-graph-based. One technique can work well with certain sets of images but can fail to segment other sets of images. Selection of the right technique to obtain Region Adjacency Graph (RAG) is an important and decisive step for extracting the graph from the image

An image Graph with RAG representation requires further exploration stage. There is a lot of information and features stored in this graph. The utilization of existing information is an important problem in the scope of graph-based classification and becomes the next challenge in image analysis with graphs. Various image analyses with graph methods have been developed for classification tasks including graph comparison techniques such as graph edit distance, kernel matching, or greedy matching. Image retrieval can also be done using this graph information by applying the graph matching technique

Although the graph is considered as a powerful representation, there is one major drawback that is the lack of a suitable method that can be used for classification and retrieval. Computational complexity is also a big problem in graph matching techniques. The graph mining method can be said as a way to overcome this deficiency.

Acknowledgments
This study is supported by BUDI-DN (No: PRJ-5777 /LPDP.3/2016) and P3MI STEI ITB 2020

References
[1] Lezoray O and Grady L 2012 Graph theory concepts and definitions used in image processing *Image Processing and Analysis with Graphs* pp 1–24
[2] Brun L, Foggia P and Vento M 2018 Trends in graph-based representations for pattern recognition *Pattern Recognition Letters* (Amsterdam: Elsevier) **129** pp 19–25
[3] Malmberg F 2015 *Image Processing Using Graph* (Uppsala: Centre for image analysis Swedish University)
[4] Camilus K S and Govindan V K, 2012 A Review on graph based segmentation *Int. J. Image, Graph. Signal Process.* **4** pp 1–13
[5] Holzinger A, Malle B and Giuliani N 2014 On Graph Extraction from Image Data in *Int. Conf. on Brain Informatics and Health* **2003** pp 552–63
[6] Strogatz S 2001 Exploring complex networks *Nature* **410** pp 268–76
[7] Dehmer M, Emmert-Streib F and Mehler A 2011 *Towards an Information Theory of Complex Networks: Statistical Methods and Applications* (Basel: Birkhaeuser)
[8] Holzinger A and Jurisica I 2014 Knowledge discovery and data mining in biomedical informatics: The future is in integrative, interactive machine learning solutions *Knowledge Discovery and Data Mining LNCS* **8401** (Heidelberg: Springer) pp 1–18
[9] Sharma H *et al* 2012 Determining similarity in histological images using graph-theoretic description and matching methods for content-based image retrieval in medical diagnostic *Diagnostic Pathology* **7**
[10] Boykov Y and Kolmogorov V 2004 An Experimental Comparison of Min-Cut/Max-Flow Algorithms for Energy Minimization in Vision *IEEE Trans. PAMI* **26(9)** pp 1124–37
[11] Trémeau A and Colantoni P 2000 Regions adjacency graph applied to color image segmentation *IEEE Trans. Image Process.* **9(4)** pp 735–44
[12] Saglam A and Baykan N A 2015 An Efficient Object Extraction with Graph-Based Image Segmentation in *Proc. - 2015 4th Int. Conf. on Advanced Computer Science Applications and Technologies ACSAT IEEE 2015* pp 86–91
[13] Peng B Zhang L and Zhang D 2013 A survey of graph theoretical approaches to image segmentation *Pattern Recognition* **46(3)** pp 1020–38
[14] Felzenszwalb P F and Huttenlocher D P 2004 Efficient graph-based image segmentation *Int. J. Comput. Vis.* **59(2)** pp 167–81
[15] Ullman J R 1976 An algorithm for subgraph isomorphism *J. ACM* **23(1)** pp 31–42
[16] Cordella L, Foggia P and Vento M 2001 An improved algorithm for matching large graphs *3rd IAPR Workshop Graph-Based Representations in Patern Recognition* pp 149–59

Middleware Development to Connect Telegram Messenger and Instant Messenger for the Elderly

Charles Eka Swandi[1], Kristian Adi Nugraha[2], Danny Sebastian[3], Restyandito[3]

Faculty of Information Technology, Informatics Department, Universitas Kristen Duta Wacana

E-mail: [1]charles.eka@ti.ukdw.ac.id, [2]adinugraha@staff.ukdw.ac.id, [3]danny.sebastian@staff.ukdw.ac.id, [4]dito@ti.ukdw.ac.id

Abstract. For the elderly, instant messengers can be a media for them to interact with other people so that they don't feel lonely. However, the elderly often find it difficult to interact with big instant messengers such as LINE, Whatsapp and Telegram, which is why an instant messenger designed specifically for the elderly is needed. Unfortunately, there are several limitations in which the messages can only be exchanged within the same instant messenger. To tackle this problem, a middleware that will be used to connect this instant messaging for the elderly with Telegram messenger. The middleware is developed with API and Webhook method. Other than that, Telegram bot is used so that messages can be sent out from Telegram. Through the testing done, the middleware developed has successfully sent out messages from Telegram to the instant messenger for elderly and vice versa.

1. Introduction

There are many things that change as people grow older [1]. For the elderly, one of the changes that happened is how they aren't able to interact with other people as much as they used to, which may result in loneliness. Studies have concluded that social isolation is often associated with an increased risk of death [2]. This shows the importance of social interaction for the elderly. The availability of instant messaging applications may help the isolated elderlies interact with other people by exchanging messages, either with their family or other people. However, due to their physical and cognitive limitations, the elders often find it hard to interact with common instant messengers such as Whatsapp, LINE, and Telegram, which is why an instant messenger designed specifically for the elders is needed. Unfortunately, users can only exchange messages within the same instant messenger [3]. This limitation is a huge factor on why bigger instant messengers are more appealing to the users [4]. The younger generation would prefer to use popular messengers such as LINE, Whatsapp or Telegram. This will limit the users' interaction only to those who are using the same instant messenger.

Based on the problems explained above, a solution is needed to create an instant messenger that is not only easy to use, especially for the elders, but can also be used to connect with users who are using different instant messengers. The writers proposed a solution by creating a middleware which is able to connect the messaging application the writers specifically develop for the elderly with various other instant messengers. This study is a part of a bigger study where the best interface for the elders and middleware to connect the instant messenger made for the elders to other instant messengers will be developed too. However, this paper will focus only on the development of middleware to connect the developed instant messenger for the elders and Telegram and dummy interface is used. The middleware is developed with API and Webhook method. Other than that, Telegram bot is used so that messages can be sent out from Telegram. Through the testing done, the middleware developed has successfully sent out messages from Telegram to the instant messenger for elderly and vice versa.

2. Related Work

Prior to the system development, the authors conducted a review to the research with focus on API or webhook. On one research [5] a mobile instant messenger with Microsoft Translator, an automatic translator is made. API is used on the instant messenger to integrate messages on Microsoft Translator. On another research [6] Dropbox API is used as a tool to exchange messages between two databases. The author stated that by using Dropbox API, Public IP is no longer needed in order to exchange data between databases. The next research [7] explains the use of LINE bot to exchange messages between Telegram and XMPP. Webhook and API are used to send messages between chat bots. However the test result shows several system problems. The test result shows that the system has weaknesses that it is very possible that message delivery failures occur because of large amounts of requests, and large database growth can quickly result in full storage space The same method is also used in this study [8], which uses Telegram bot to share campus information. Researchers stated that webhook is able to give zero latency in handling many requests at the same time. Another study [9] compares long polling and webhook methods on Telegram bot. This study found that the webhook method performs three times better than the long polling method. Another study [10] also compares long polling and webhook method performance, focusing on their speed and character length of the messages. The test result shows that webhook requires shorter time than long polling, and shorter message results in shorter response time. Based on these previous studies, this study will focus on using API and webhook to connect Telegram and the instant messenger for elderly. API is used on the Telegram side, while webhook is used on Telegram bot as an intermediary. Furthermore, in order to handle the problem found [7], data removal on the database will be done periodically.

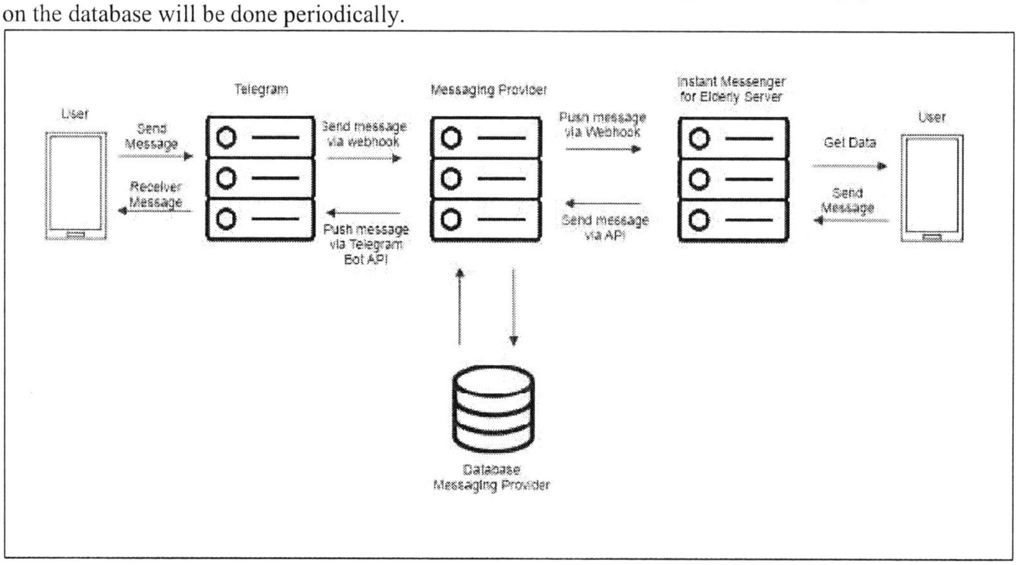

Figure 1. Design System

3. Design System

3.1. Design System

IOP Conf. Series: Materials Science and Engineering 1077 (2021) 012007 doi:10.1088/1757-899X/1077/1/012007

Figure 1 illustrates the system's design. Middleware acts as an intermediary between the instant messenger which sends messages from Telegram to the instant messenger for elderly and vice versa. Webhook method is used on the Telegram Bot to forward messages to the middleware and from the middleware to the instant messenger for elderly while API is used to forward messages from the instant messenger for elderly to the Telegram bot.

3.2. API Design

The API created for this study is designed for three different types of client, Telegram Bot Admin, Telegram Bot User, and instant messenger for elderly. Telegram Bot Admin's API Client is used to set connection between the corresponding Telegram account and the middleware. Telegram Bot User's API Client is used to send out messages to the instant messenger for elderly. instant messenger for elderly's API Client is used to send out messages to Telegram Bot User. The API created on the middleware functions as an intermediary between the Telegram with the instant messenger for elderly is listed below.

Table 1. List of Designed API on the Middleware

Metode	URI	Keterangan
POST	/TelegramBotAdmin	*Webhook* for *Telegram Bot Admin's* client
POST	/TelegramBotUser/<token>	*Webhook* for *Telegram Bot Admin's* client
GET	/<token>/sendMessage	API Client for instant messenger for elderly, used to forward messages to *Telegram*
GET	/<token>/getTelegramUserInfo	API Client for instant messenger for elderly, used to retrieve information from corresponding *Telegram* account

3.3. Telegram Bot Commands Design

Commands that are used in this study are specifically designed for two types of client, Telegram Bot Admin, and Telegram Bot User. The commands designed are listed below.

Table 2. List of Telegram Bot's Commands

Bot	Command	Fungsi
Telegram Bot Admin	/help	To provide information about the bot's functions and how to use them.
Telegram Bot Admin	/addBot	To register a Telegram account so that the middleware can be used, by adding the bot's token created before. Telegram Bot will then be added as Telegram Bot User Client
Telegram Bot User	/help	To provide information about bot's functions and how to use them
Telegram Bot User	/getOTP	To generate OTP that will be used to set connection between corresponding Telegram account and instant messenger for elderly
Telegram Bot User	/getKontak	To show the corresponding account from the connected instant messenger for elderly.
Telegram Bot User	/setGroupAs	To create a bot to forward messages from corresponding instant messenger for elderly's account.

3.4. *Database Design*
The database used in the middleware to connect Telegram and the instant messenger for elderly is made using MySQL.

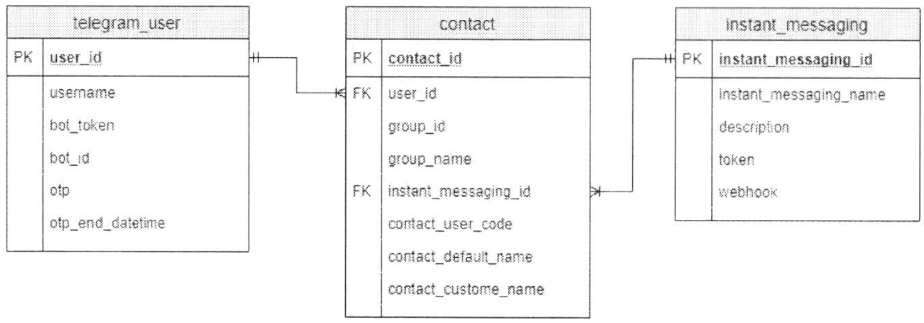

Figure 2. Entity Relationship Diagram (ERD)

4. Result and Discussion
Before the middleware can be used to exchange messages between Telegram, and the instant messenger for elderly, both accounts should be added as contacts first. On the Telegram side, the group chatting feature will be used to replace contact, as illustrated in Figure 4. There are three groups on Telegram that represent three accounts on the instant messenger for elderly. The usage of group as contact is done with the intention that the saving of more than one contact through the instant messenger for elderly on Telegram is possible. Every dedicated group will consist of only two members, the Telegram account and the Telegram bot. The Telegram bot will show messages from the corresponding account, and receives messages to then be forwarded by the middleware. After the contact has been made successfully, message exchange can be done between Telegram and the instant messenger for elderly. Figure 5 shows Telegram group modeling steps to become contacts on the instant messenger for elderly.

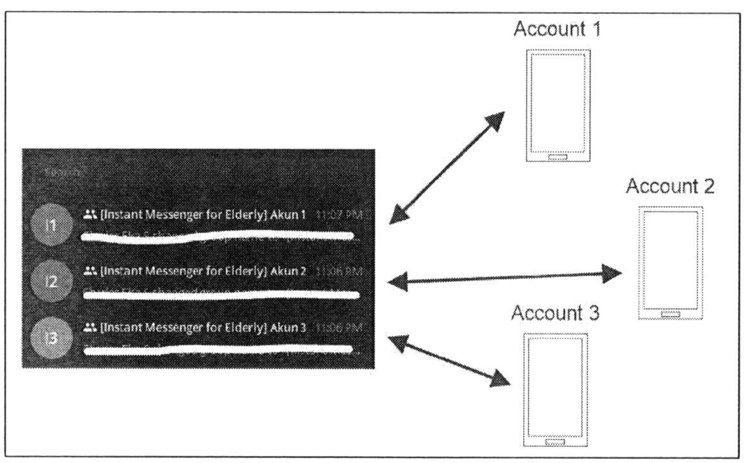

Figure 3. Illustration of group usage as contact replacement

Simple testing is done to measure the text messages exchanged between the two instant messengers. This test is done to make sure that the developed system runs well, by making sure that the messages sent have the same length as the messages received. The test result is presented on table 3. Information obtained from Telegram bot API documentation that the maximum character length that Telegram can store is 4096 characters. If the length exceeds 4096 characters, the rest of the characters will be sent on the next message. However, messages sent from the instant messenger for elderly cannot be forwarded to Telegram if the length of the character is more than 4096 characters.

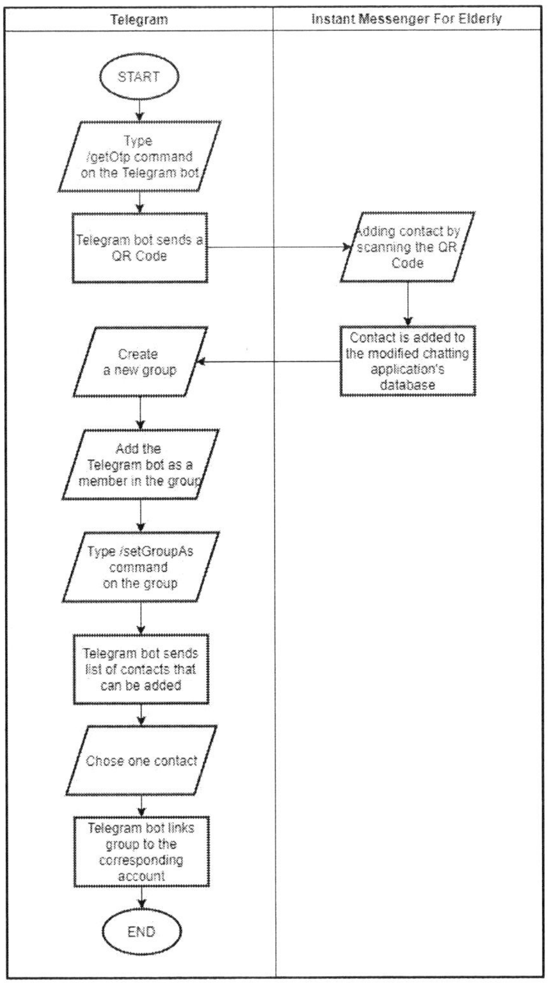

Figure 4. Add Contact Flowchart

5. Summary

The middleware has successfully sent out simple messages on different instant messenger with a maximum number of characters that can be sent in one message is 4096 characters. However, currently the middleware can only send out text messages. By using the webhook method and flexible API, sending messages other than text messages is very possible, and will be done in future works. Other than that, message storing features will be developed, so that unsent messages can be saved on the middleware's database, and to avoid the problems found on this study [7], messages that have been delivered successfully will be deleted.

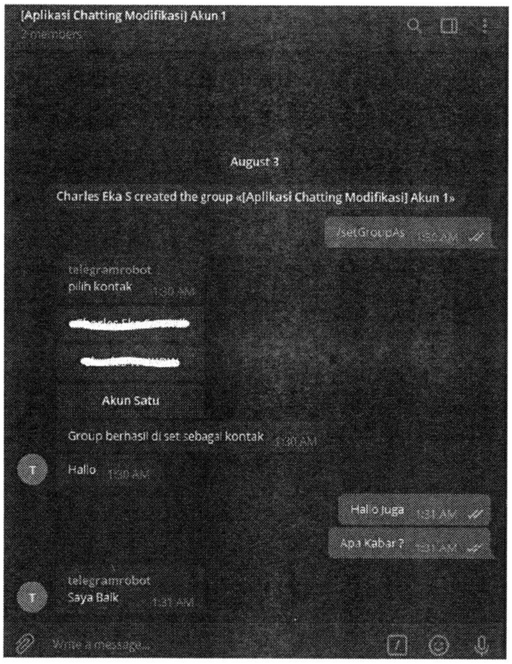

Figure 5. Message exchange from Telegram

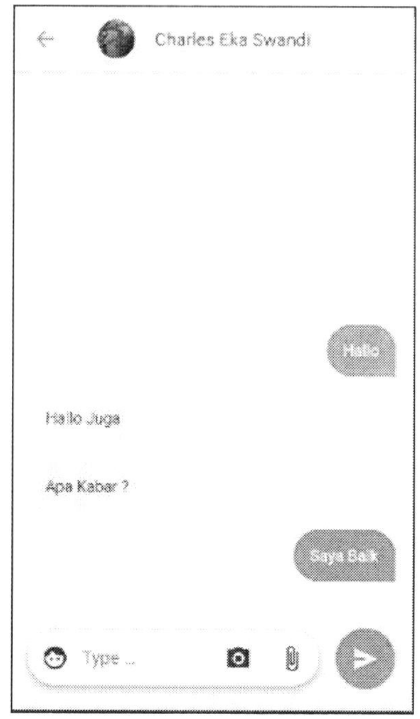

Figure 6. Message exchange from instant messenger for elderly

Table 3. Test Result

Character Length	From Telegram Bot	From instant messenger for elderly
100	v	v
500	v	v
1000	v	v
2000	v	v
3000	v	v
4096	v	v
4097	x	x

References

[1] Restyandito, Zebua J A and Nugraha K A 2019 Perancangan ikon pada aplikasi kesehatan untuk lansia berbasis mobile *Jurnal Teknologi Informasi dan Ilmu Komputer (JTIIK)* vol **6** no 6 pp 637–44

[2] Perissinotto C M, Cenzer I S and Covinsky K E 2012 Loneliness in older persons a predictor of functional decline and death *Arch. Intern. Med.* vol **172** no 14 pp 1078–83

[3] Restyandito, Kurniawan E and Widagdo T M 2019 Mobile application menu design for elderly in indonesia with cognitive consideration *J. of Phys.: Conf. Ser.* **1196** 012058

[4] Essays 2018 *How Did Whatsapp Become Popular?* Retrieved on November 2018 from https://www.ukessays.com/essays/technology/whatsapp-user-growth-8352.php?vref=1

[5] Musyafi F and Afrianto I 2015 Membangun aplikasi chatting dengan penerjemah otomatis berbasis mobile *Jurnal Ilmiah Komputer Dan Informatika* vol **4** no 2 pp 79–84

[6] Hanafi A, Sukarsa M and Wiranatha A K A C 2017 Pertukaran data antar database dengan menggunakan teknologi API *LONTAR KOMPUTE* vol **8** no 1 pp 22–30

[7] Sukarsa I M and Buana P W 2019 LINE Messenger as a transport layer to distribute messages to partner instant messaging *Int. J. of Modern Education and Computer Sci.* vol **11** no 3 pp 1–9

[8] Setiaji H and Paputungan I V 2018 Design of telegram bots for campus information sharing *Conf. Ser.: Mater. Sci. Eng.* **325** 012005

[9] Kavats E A and Kostenko A A 2019 Analysis of connection methods of telegram robots with server part *System Technologies* vol **3** no 122 pp 19–24

[10] Soeroso H, Arfianto A Z, Mayangsari N E and Taali M 2017 Penggunaan bot telegram sebagai announcement systempada intansi pendidikan *Seminar MASTER* vol **1** p 2

Comparison of K-Medoids and Self Organizing Maps Algorithm in Grouping Hydrometeorological Natural Disasters in Java Island

Novendri Isra Asriny*, **Desy Rizki Ramadhanty, Muthia Dishanur Izzati, Matiin Laugiwa Prawira Putra, Annisa Shofura, Chairani Uli Albi, Edy Widodo**

Universitas Islam Indonesia

E-mail: *17611040@students.uii.ac.id

Abstract. Indonesia is a country that is prone to disasters. Based on data from the National Disaster Management Authority, there are 6868 disasters in the last 3 years. Changes in weather, temperature, and wind direction that are quite extreme in Indonesia lead to one of the disasters namely hydrometeorological disasters such as floods, landslides, tornadoes, forest fires, and drought. According to the National Coordinating Agency for Disaster Management, this type of disaster is quite common in Indonesia. Over the past 3 years, Java Island received the highest number of disasters. The National Disaster Management Authority states that tornadoes contribute 21% of all disasters in Indonesia. Therefore, the researcher wants to grouping the cases of hydrometeorological natural disaster in Indonesia, especially in Java. This research is to determine the areas that are the focus of the government in disaster mitigation efforts using K-Medoids Clustering and Self Organizing Maps. Based on the results of this study, it was found that the suitable method used was the Self Organizing Maps Method because it was seen based on a smaller standard deviation value. Cluster one there are 108 members who are groups with relatively low disaster vulnerability, while cluster two there are 10 members are city and district groups that have relatively high disaster vulnerability.

1. Introduction

Disasters are occasions that undermine and disturb people's lives caused by normal, non-natural, and human variables that result in fatalities, natural harm, property misfortunes, and mental impacts [11]. Indonesia is a disaster-prone country caused by several factors. Based on data from the National Disaster Management Authority, the number of natural disasters in Indonesia is 6868 events [5].

Indonesia is a tropical locale with two seasons, to be specific warm and rain. Changes in climate, temperature, and wind heading are extraordinary sufficient to cause disaster, one of which is hydrometeorological calamities such as landslide, tornadoes, floods woodland fires, and dry spell [4]. These types of disasters occur quite frequently in Indonesia according to disaster data from the National Coordinating Agency for Handling Disaster (BAKORNAS PB).

Over the past 3 years, Java Island received the highest number of disasters. A tornado is the most frequent disaster, from BNPB data, it shows that a tornado contributed 21% of all disasters in Indonesia [12]. Therefore, the researcher wants to group of hydrometeorological natural disaster cases in Indonesia, especially in Java to determine areas that are the focus of the government in disaster mitigation efforts using K-Medoids Clustering and Self Organizing Maps which then can be identified groups of disaster-prone areas. In this research, a K-Medoids Clustering and Self Organizing Maps

(SOM) method will be compared to determine the best method in this study. K-Medoids is a clustering method that can overcome objects with large values that allow deviations from the data distribution [6]. SOM is a gathering strategy that gives organizing classes based on topology. SOM is prepared iteratively through a several of input design forms so that each input design will be prepared by as numerous age [2].

There are a few considerations that related to this inquiry about, to be specific Clustering the effect of Seismic tremors in Indonesia Utilizing Kohonen Self Organizing Maps, which comes about within the creation of 4 clusters. Cluster 1 is the highest number of earthquake events and damaged buildings. Cluster 2 is the number of damaged roads and land. Cluster 3 is the number of victims who died. Cluster 4 is the number of victims who suffer [8]. Implementation of K-Medoids Grouping Algorithm for Group Crime Patterns in Yogyakarta, which results in the creation of 3 clusters. The crime rate is high there are 4 members, the crime rate is medium there are 6 members and the crime rate is low there are 8 members [19]. Comparison of Weighted Fuzzy C-Means Method with Fuzzy C-Shell (Case Study: Disaster Events in Indonesia by Regency), which produces the best strategy utilized in catastrophe clustering is Fuzzy C-Means strategy since it has the littlest Xie Beni list and the comes about of clustering got is cluster 1 comprises of 39 locales, cluster 2 comprises of 314 locales, cluster 3 comprises of 77 areas, cluster 4 comprises of 14 locale [16].

2. Literature Study

2.1. K-Medoids Clustering
The K-Medoids algorithm or what is known as the PAM (Partitioning Around Medoid) algorithm was developed by Leonard Kaufman and Peter J. Rousseeuw, which is similar to K-means because of the two algorithms it is partitional which breaks the dataset into groups [22]. This method is based on the use of medoids rather than observations of the mean held by each cluster, to reduce the sensitivity of the partition for the extreme values that exist in the dataset [21]. K-Medoids is present to overcome the weaknesses of K-Means that are sensitive to outliers due to an object with a large value that may deviate substantially from the distribution of data [13]. K-Medoids uses the partition grouping method to group n sets of objects into some k clusters. This algorithm uses objects in a collection of objects that represent a cluster. The objects that represent a cluster are called medoids. Medoid is an object that is located centrally in a cluster so that it is robust to outliers.

2.2. Self-Organizing Maps (SOM)
Clusters validation that researchers use is internal validation, because this validation is based on evaluating the results of clustering in a quantitative concept in which the agricultural sector production data is also in the form of quantitative data. There are several methods or methods for internal validation, including connectivity, silhouette values, and Dunn index. Dunn index is the ratio of the smallest distance between observations in different clusters with the largest distance in each data cluster. Dunn index is obtained from the division between the smallest distance between observations in different clusters with the largest distance in each data cluster. Silhouette index is calculated as the degree of confidence in the process of clustering on observation with a cluster that is said to form well when the index value approaches 1 and the conditions conversely if the index value approaches the number -1. The value of silhouette in the range of -1 to 1. Connectivity index is seen based on characteristics that have the closest distance (neighbor). This is a parameter in determining the number of neighbors who contribute to the measurement of connectivity.

2.3. Silhouette Coefficient
Silhouette coefficient is used to see quality and cluster strength, how well an object is positioned in a cluster. This method is a combination of cohesion and separation method. Calculation stages of the Silhouette coefficient are as follows [1]:

1. Calculate the average distance from observation, for example i with all the other observations that are in one cluster:

$$\alpha(i) = \frac{1}{|A|-1}\sum j \in_{A,j\neq i} d(i,j) \tag{1}$$

2. Calculate the average distance from observation i to all observations in other clusters, and take the smallest value:

$$d(i,C) = \frac{1}{|A|}\sum j \in C\ d(i,j) \tag{2}$$

Where *d(i,C)* is the average distance of observation i with all objects in other cluster C where A $\neq$ C.

3. The Silhouette Coefficient values are:

$$s(i) = \frac{b(i) - a(i)}{\max\big(a(i), b(i)\big)} \tag{3}$$

2.4. Within Sum Square (WSS) Method

The WSS method is a method used to measure the average square distance from all points in the cluster to the cluster center point. Clusters with low WSS values have better closeness between objects than clusters with higher WSS values. The formula used in WSS is,

$$WSS = \Sigma_{i=1}^{k}\Sigma_{j=1}^{ni}\left(X_{ji} - \bar{X}_i\right)^2 \tag{4}$$

where WSS is the number of squares in, k is the number of clusters, n_i is the sample size of cluster i, X_{ij} is the j measurement of the i cluster, and X_{ij} is the average value of the overall distance of the data [9].

2.5. Standard Deviation (SD)

Standard deviation is a statistical value that has the benefit of determining the distribution of data in the sample, as well as how close the individual data is to the average sample value [10]. If the deviation value shows a large value, it can be seen that the individual data points are far from the average sample value [16]. Mathematically, the SD is [7]:

$$s = \sqrt{\frac{\Sigma(X-\bar{X})^2}{n-1}} \tag{5}$$

where

s = Sample Standard Deviation
X = Individual Value
$\bar{X}$ = Sample Mean
n = Sample Size

3. Methods

3.1. Types of Data Sources

The population in this study is data on the incidence of hydrometeorological natural disasters in Java, which are available on the website of the National Disaster Management Authority (BNPB). Meanwhile, 12 variables were sampled, consisting of provincial, district/city, flood, landslide, tornado, drought, and forest and land fires (Table 1).

3.2. Research Stages

The methods used in this study are the K-Medoids Method and Self Organizing Maps. Figure 1 shows a flowchart that will be used in the study.

Table 1. Research Variable

No	Variable	Dimension	Source	Measurement
1	Province	Province on the island of Java	BNPB	Nominal
2	District/city	District/city of each province in Java Island	BNPB	Nominal
3	Flood	Number of flood disasters	BNPB	Interval
4	Landslide	Number of landslide disasters	BNPB	Interval
5	Tornado	Number of tornado disasters	BNPB	Interval
6	Drought	Number of drought disasters	BNPB	Interval
7	Forest and Land Fires	Number of forest and land fires disasters	BNPB	Interval

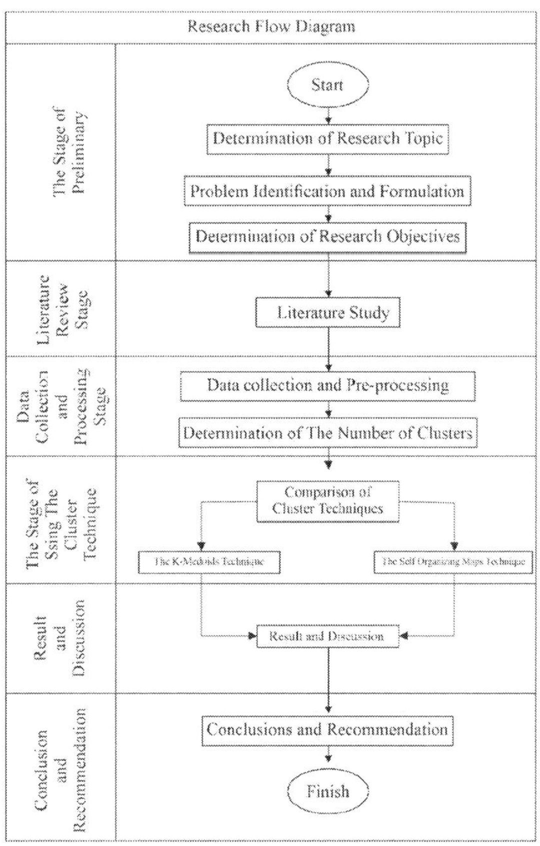

Figure 1. Research Flowchart

Based on Figure 1, the following are the stages of data collection and processing carried out:

1. Perform data preprocessing.
2. Determining the number of clusters that will be used using the Silhouette Score method. The silhouette score coefficient is a method used to evaluate clusters and see the quality of data placement in a cluster based on the Sum Square of Error (SSE) value. Silhouette Score is important to see if the resulting cluster is of good quality. The stages of calculating the silhouette coefficient are as follows [14].
   a. Calculate the average distance of object i to all objects in the group. We call this the mean distance i.
   b. Calculate the average distance of object i to all objects in the other cluster we call b (i), and take the smallest value.
   c. Silhouette coefficient values are obtained from:
   $$S(i) = \frac{b(i) - a(i)}{\max\big(b(i), a(i)\big)} \quad (6)$$
   and can be written
   $$S(i) = \begin{cases} 1 - \frac{a(i)}{b(i)}, if\ a(i) < b(i) \\ 0, if\ a(i) = b(i) \\ \frac{b(i)}{a(i)} - 1, if\ a(i) > b(i) \end{cases} \quad (7)$$
   where:
   S (i) = Silhouette Score
   a (i) = Average distance between data i and all objects in the cluster
   b (i) = Average distance between data i to all objects in other clusters
   The range of values of the silhouette coefficient is -1 to 1. If the value of the silhouette coefficient is close to 1 then the object is in the right cluster. Meanwhile, if it is around 0 then the object can be between the clusters, and if the result is negative then the object may be in the wrong cluster [18]. The best number of clusters or the number of optimal clusters is the number of clusters with the highest average silhouette score where the average is taken from the silhouette value of each cluster.
3. Using K-Medoids cluster technique and Self-Organizing Maps.\
   a. Stages in the K-Medoids cluster technique [13]:
      i. Initialize the cluster center (medoid)
      ii. Calculates Euclidean distances
      $$D_{ik} = \sqrt{\sum_{i=1}^{m} (x_{ij} - c_{kj})^2} \quad (8)$$
      $D_{ik}(x_i, x_j) = EuclideanDistance$
      $x_i = Data\ (i)$
      $x_j = Data\ (j)$
      $x_{ij} = Data\ (i) atribute\ (j)$
      $c_{kj} = Data\ (k) atribute\ (j)$
      iii. Calculates the total distance
      iv. Repeat steps (i-iii) until the new total distance minus the old total distance> 0
   b. Stages in the Self-Organizing Maps cluster technique [20]:
      i. Initialize a number of weights with random values

ii. For each sample randomly selected in the dataset,
   1. Looking for the weighting closest to the Euclidean distance which is then called the BMU (Best Matching Unit)
   2. Updates weight with formula

$$w_{ij} = w_{ij} + \alpha(x - w_{ij}) \tag{9}$$

   3. Perform iterations to update the weights until they reach the predetermined iteration value and reduce the learning rate (α).
   4. Repeat from step 2

c. Comparing the results of clustering using the K-medoids and SOM methods based on the standard deviation ratio between clusters. The standard deviation ratio (R) was successfully used by [17] to determine the best method between the complete linkage and average linkage methods. The R value is a comparison between the standard deviation in the S_w group and the standard deviation between the S_b groups, the minimum R value indicates the best cluster method algorithm. The following equation is used to calculate the values of S_w and S_b:

$$S_k = \sqrt{\frac{\sum_{k=1}^{n}(x_k - \bar{x}_k)^2}{n-1}} \tag{10}$$

$$S_w = \frac{1}{n}\sum_{k=1}^{n} S_k \tag{11}$$

$$S_b = \sqrt{\frac{\sum_{k=1}^{n}(\bar{x}_k - \bar{x})^2}{n-1}} \tag{12}$$

$$R = \frac{S_w}{S_b} \tag{13}$$

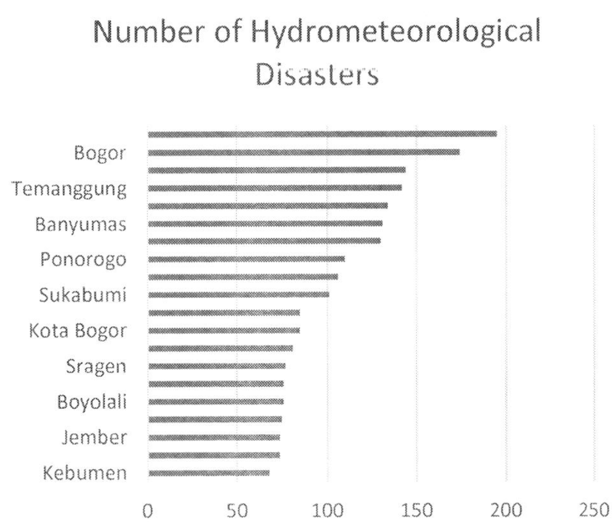

Figure 2. Descriptive analysis results

4. Results and Discussion

4.1. Descriptive Analysis

Researchers will conduct a descriptive analysis of the data on the number of hydrometeorological disasters. Descriptive analysis was carried out to find a general description of the development of hydrometeorological disasters during the 2017-2019 period.

Based on Figure 2, there are 20 districts with the highest number of hydrometeorological disasters among other districts in Java. Cilacap Regency is the district with the highest number of hydrometeorological disasters on Java Island of 195 incidents during the 2017-2019 period.

4.2. Determination of Cluster Optimal Number

The formation of the number of clusters is obtained using the silhouette and elbow method approach to see the comparison of Sum Square Error (SSE) values to get the optimal cluster formed.

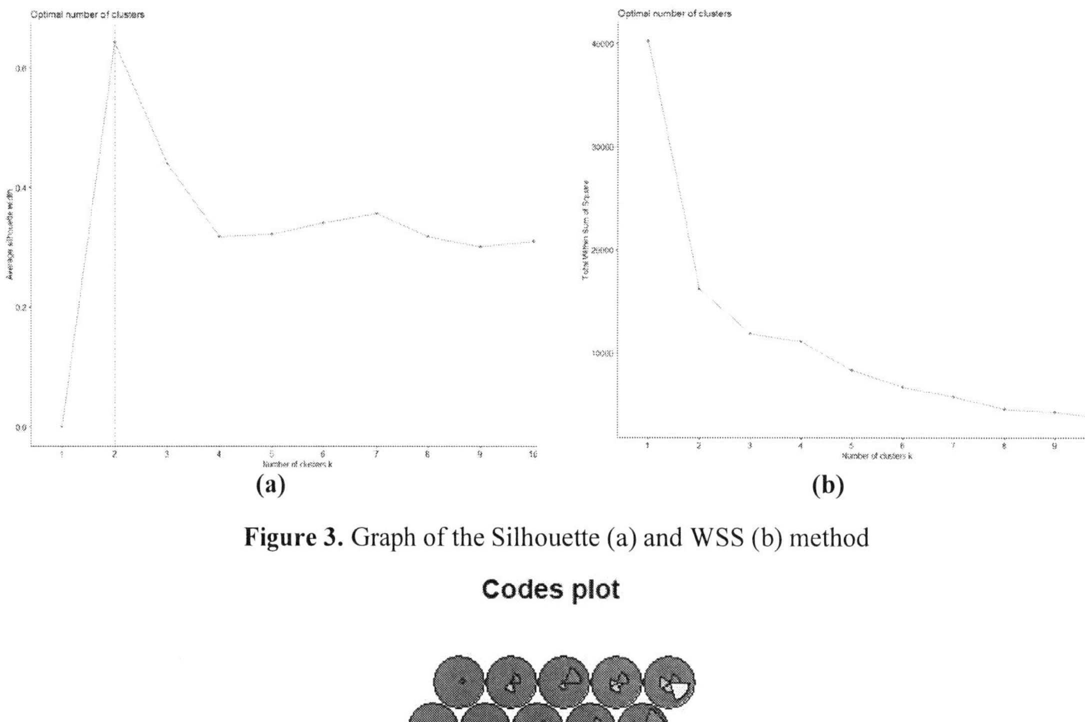

(a) (b)

Figure 3. Graph of the Silhouette (a) and WSS (b) method

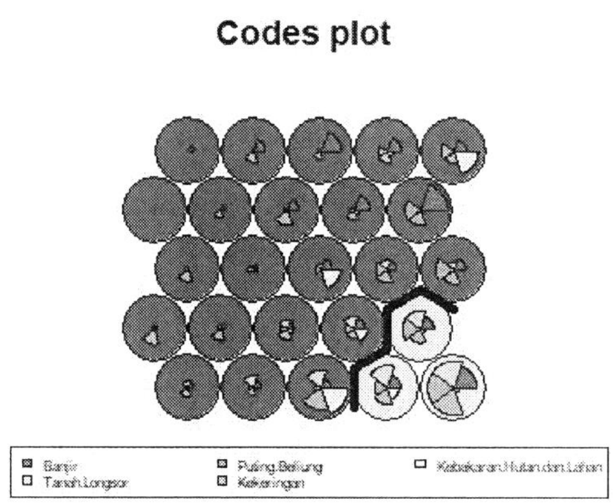

Figure 4. *Output* fan diagram

Based on the SSE silhouette and WSS graph output in Figure 3, an optimal cluster is obtained, namely, the two clusters seen from SSE in the silhouette method in the second cluster are the highest average silhouette values and if viewed based on the value within the sum square in the plot of the WSS method then seen in cluster two has a significant change to cluster three. So that the optimal cluster in this study uses the optimal cluster of two clusters on the number of Hydrometeorological disasters in Java.

4.3. Clustering Analysis with Self Organizing Maps

A model with an R program is obtained in the form of a fan diagram from the analysis of the SOM algorithm. The researcher uses a fan diagram with a hexagonal display with a 5 x 5 grid. The fan diagram is an illustration of the distribution of mapping variables. In the fan diagram below there are 2 different colors, which show the results of the clusters of each district in Java. Each color in the fan diagram has different characteristics. The SOM algorithm can be seen in Figure 4.

The researcher made a table containing the results of the clustering of regencies in Java, based on the number of hydrometeorological disasters.

Table 2. Clustering result using SOM

Cluster	Total	Cluster Member
1	108	Bandung, Bandung City, Bangkalan, Banjar City, Bantul, Banyuwangi, Batang, Batu City, Bekasi, Bekasi City, Blitar, Blitar City, Blora, Bojonegoro, Bondowoso, Boyolali, Brebes, Ciamis, Cianjur, Cilegon City, Cimahi City, Cirebon, Cirebon City, Demak, Depok City, East Jakarta, Garut, Gresik, Grobogan, Gunung Kidul, Indramayu, Jember, Jepara, Jombang, Karanganyar, Karawang, Kebumen, Kediri, Kediri City, Kendal, Kepulauan Seribu, Klaten, Kudus, Kulon Progo, Kuningan, Lamongan, Lebak, Lumajang, Madiun, Madiun City, Magelang City, Magetan, Main Jakarta, Majalengka, Malang, Malang City, Mojokerto, Nganjuk, Ngawi, North Jakarta, Pacitan, Pamekasan, Pandeglang, Pangandaran, Pasuruan, Pasuruan City, Pati, Pekalongan, Pekalongan City, Pemalang, Ponorogo, Probolinggo, Probolinggo City, Purbalingga, Purwakarta, Purworejo, Rembang, Salatiga City, Sampang, Semarang, Serang, Serang City, Sidoarjo, Situbondo, Sleman, South Jakarta, South Tangerang City, Sragen, Subang, Sukabumi City, Sukoharjo, Sumedang, Sumenep, Surabaya City, Surakarta City, Tangerang, Tangerang City, Tasikmalaya, Tasikmalaya City, Tegal, Tegal City, Trenggalek, Tuban, Tulungagung, West Bandung, West Jakarta, Wonosobo, Yogyakarta City
2	10	Banjarnegara, Banyumas, Bogor, Bogor City, Cilacap, Magelang, Semarang City, Sukabumi, Temanggung, Wonogiri

From the cluster using the Self Organizing Maps algorithm, the result is that cluster one consists of 108 cities and regencies that have a high similarity value and there are 10 cities and regencies in cluster two in Java that have high similarity.

The researcher segmented the calculation of the average occurrence value of each cluster group and the results were summarized as in Table 3.

Table 3. Cluster profilitation

Variable	*Cluster* 1	*Cluster* 2
Flood	6.592593	16.9
Landslide	7.490741	61.8
Tornado	10.26852	49.3
Drought	1	2.4
Forest Fires	1.898148	3.8

Table 3 shows the results of the average value of disaster events, cluster 1 is a cluster with areas that are not too vulnerable to hydrometeorological disasters. Meanwhile, cluster 2 is a cluster with areas that are prone to hydrometeorological disasters. Based on the calculation of the average of each group obtained in the first cluster tornado disaster to be the disaster that has the highest average means that the disaster occurs more frequently, while in the second cluster it is found that landslide disasters are more frequent because they have an average value higher.

4.4. Clustering Analysis with K-Medoids

Using cluster analysis with the K-Medoids algorithm the results are in the form of a cluster plot consisting of two clusters as in Figure 5.

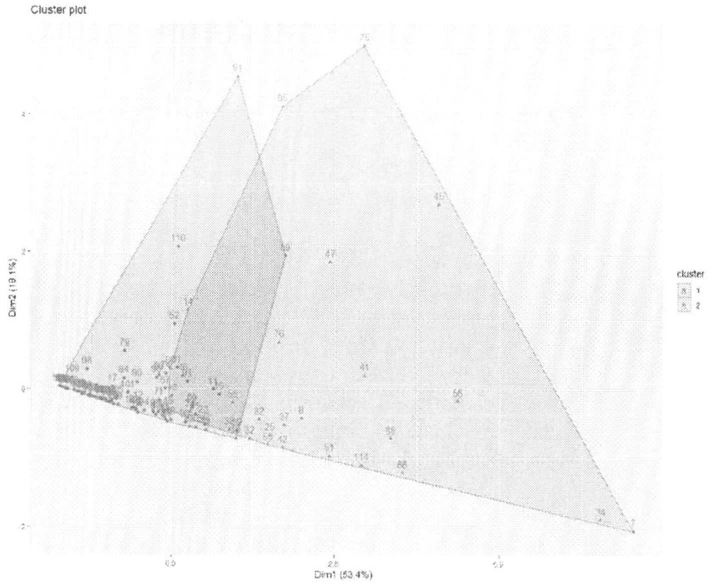

Figure 5. K-Medoids clustering output

Table 4. K-Medoids clustering results

Cluster	Total	*Cluster Member*
1	86	Bandung, Bandung City, Bangkalan, Banjar City, Bantul, Banyuwangi, Batang, Batu City, Bekasi, Bekasi City, Blitar, Blitar City, Blora, Bojonegoro, Bondowoso, Ciamis, Cianjur, Cilegon City, Cimahi City, Cirebon, Cirebon City, Demak, Depok City, East Jakarta, Garut, Gresik, Grobogan, Gunung Kidul, Indramayu, Jombang, Karawang, Kediri, Kediri City, Kendal, Kepulauan Seribu, Klaten, Kudus, Kulon Progo, Lamongan, Lebak, Lumajang, Madiun, Madiun City, Magetan, Main Jakarta, Malang, Malang City, Mojokerto, Nganjuk, Ngawi, North Jakarta, Pacitan, Pamekasan, Pandeglang, Pasuruan, Pasuruan City, Pekalongan, Pekalongan City, Probolinggo, Probolinggo City, Purbalingga, Purwakarta, Rembang, Salatiga City, Sampang, Serang City, Sleman, South Jakarta, Subang, Sukabumi City, Sukoharjo, Sumedang, Sumenep, Surabaya City, Surakarta City, Tangerang, Tangerang City, Tangerang South City, Tasik Malaya, Tegal, Tegal City, Tuban, Tulungagung, Yogyakarta City
2	32	Banjarnegara, Bayumas, Bogor, Bogor City, Boyolali, Brebes, Cilacap, Jember, Jepara, Karanganyar, Kebumen, Kuningan, Magelang, Majalengka, Pangandaran, Pati, Pemalang, Ponorogo, Purwerjo, Semarang, Semarang City, Serang, Sidoarjo, Situbondo, Sragen, Sukabumi, Tasikmalaya City, Temanggung, Trenggalek, West Bandung, Wonogiri, Wonosobo

K-Medoids output shows that there are two clusters obtained based on the calculation of similarity between observations of the number of hydrometeorological disasters that occurred in Java. Table 4 shows the results of the profiling are shown by clustering.

From the cluster using the K-Medoids algorithm, the result is that cluster one consists of 86 cities and regencies that have a high similarity value and there are 32 cities and regencies in cluster two in Java that have high similarity.

The researcher segmented the calculation of the average occurrence value of each cluster group and the results were summarized as in Table 5.

Table 5 shows the results of the average value of disaster events, cluster 1 is a cluster with areas that are not too vulnerable to hydrometeorological disasters. Meanwhile, cluster 2 is a cluster with areas that are prone to hydrometeorological disasters. Based on the calculation of the average of each group obtained in the first cluster tornado disaster to be the disaster that has the highest average means that the disaster occurs more frequently, while in the second cluster it is found that landslide disasters are more frequent because they have an average value higher.

Table 5. Clustering profilitation

Variable	*Cluster* 1	*Cluster* 2
Flood	5.58	12.5
Landslide	3.92	34.1
Tornado	6.27	33.2
Drought	0.872	1.78
Forest Fires	1.19	4.41

4.5. Calculation the Standard Deviation of the Self Organizing Maps Algorithm

From the grouping process using the SOM method described previously, 2 clusters have been obtained. Then the calculation of standard deviation in groups (S_b) and standard inter-group deviation (S_b) will then be made in the SOM method as follows:

1. Standard deviation in groups (S_w)
   Before calculating the S_w value the first k-group standard deviation is calculated. The following is an example of calculating the k-group standard deviation (S_k), by looking at the appendix, following the calculation of the standard deviation of cluster 1, where $\bar{x}_2 = 5.45$, and $\bar{x}_2 = 26.84$.

$$S_1 = \sqrt{\frac{(0.2-5.45)^2+(1.6-5.45)^2+(2.2-5.45)^2+\cdots+(0.2-5.45)^2}{108}} = 4.811416$$

$$S_2 = \sqrt{\frac{(34.8-26.84)^2+(20.2-26.84)^2+(17-26.84)^2+\cdots+(28.8-26.84)^2}{10}} = 6.255749$$

   For overall results the S_k value is as follows:

Table 6. Standard deviation of the SOM method

Cluster	Standard Deviation
1	4.811416
2	6.255749

   After getting the k-standard deviation value, then the standard deviation in the group can be calculated as follows:

$$S_w = \frac{4.811416+6.255749}{2} = 5.533583$$

2. Standard deviation between groups (S_b)

The standard deviation between groups (S_b) can be calculated by first calculating the average of the whole group $\bar{x}$ as follows:

$$\bar{x} = \frac{5.45+26.84}{2} = 16.145$$

After obtaining the average of the whole group $\bar{x}$, the standard deviation between groups (S_b) can be calculated as follows:

$$S_b = \sqrt{\frac{(5.45-16.145)^2+(26.84-1\ .145)^2}{2}} = 10.695$$

3. Ratio S_w and S_b
   According to Barakbah and Arai (2007) in Laraswati (2014), from the minimum ratio of S_w to S_b, in this case, a good method is seen from the minimum ratio value, as follows [3]:

$$S = \frac{5.533583}{10.695} x100\% = 0.517399$$

4.6. Calculation the Standard Deviation of the K-Medoids Method

From the grouping process using the K-Medoids method previously described, 2 clusters were obtained. Then the calculation of standard deviation in groups (S_w) and inter-group deviation (S_b) in the K-Medoids method will then be performed as follows:

1. Standard deviation in groups (S_w)
   Before calculating the S_w value the first k-group standard deviation is calculated. The following is an example of calculating the k-group standard deviation (S_k), by looking at the appendix, following the calculation of the standard deviation of cluster 1, where $\bar{x}_1$ = 3,565, and $\bar{x}_2$ = 17.2.

$$S_1 = \sqrt{\frac{(0.2-3.565)^2+(1.6-3.565)^2+(2.2-3.565)^2+\cdots+(0.2-3.565)^2}{86}} = 2.947477$$

$$S_2 = \sqrt{\frac{(34.8-1\ .2)^2+(20.2-17.2)^2+(17-17.2)^2+\cdots+(28.8-17.2)^2}{32}} = 7.898101$$

For overall results the S_k value is as follows:

Table 7. Standard deviation of the K-Medoids method

Cluster	**Standard Deviation**
1	2.947477
2	7.898101

After getting the k-standard deviation value, then the standard deviation in the group can be calculated as follows:

$$S_w = \frac{2.947477 + 7.898101}{2} = 5.422789$$

2. Standard deviation between groups (S_b)
   The standard deviation between groups (S_b) can be calculated by first calculating the average of the whole group $\bar{x}$ as follows:

$$\bar{x} = \frac{3.565 + 17.2}{2} = 10.38256$$

After obtaining the average of the whole group $\bar{x}$, the standard deviation between groups (S_b) can be calculated as follows:

$$S_b = \sqrt{\frac{(3.565 \quad .38256)^2+(17.2-10.38256)^2}{2}} = 6.817442$$

3. Ratio S_w and S_b
   According to Barakbah and Arai (2007) in Laraswati (2014), from the minimum ratio of S_w to $S_b$in this case a good method is seen from the minimum ratio value, as follows [3]:

$$S = \frac{5.422789}{6.817442} x100\% = 0.795429$$

4.7. Comparison of Standard Deviation of Self Organizing Maps and K-Medoids Method
According to Barakbah and Arai (2007) in Laraswati (2014), from the minimum ratio of S_w to S_b, in this case, a good method is seen from the minimum ratio value because the smaller the odds ratio S_w and S_b mean that the more similarities between observations or observations the closer to the cluster center point. [3]

Table 8. Comparison of Standard Deviation Ratio

No	Method	S_w	S_b	Ratio S_w and S_b
1	*Self-Organizing Maps* (SOM)	5.533583	10.695	**0.517399**
2	*K-Medoids*	5.422789	6.817442	0.795429

Based on Table 8, the standard deviation ratio value shows that the Self Organizing Maps (SOM) method has better performance among the K-Medoids method. This can be seen from the value of the ratio of σ_w to $\sigma_b$in the Self Organizing Maps (SOM) method which is smaller than the K-Medoids method.

5. Conclusion
Based on the Self Organizing Maps method, it is found that the first cluster of 108 members is a group with relatively low disaster vulnerability, while the second cluster of 10 members is a group of cities and districts that have relatively high vulnerability to hydrometeorological disasters. Whereas in the K-Medoids method, the results obtained from cluster one as many as 86 members are groups with relatively low disaster vulnerability, while cluster two as many as 32 members are groups of cities and districts that have relatively high vulnerability to hydrometeorological disasters. The best clustering method in this study is the Self Organizing Maps method with a standard deviation ratio value of 0.517399. Where this value is smaller than the K-Medoids method.

6. Recommendation
This study obtained results where there are several areas on the island of Java that have a high level of risk of hydrometeorological disasters, this can be overcome by mitigating hydrometeorological disasters in areas that are the focus of hydrometeorological disasters, so that damage or loss of life caused by hydrometeorological disasters can be minimized.

References
[1] Anggara M, Sujiani H and Nasution H 2016 Pemilihan Distance Measure Pada K-Means Clustering Untuk Pengelompokkan Member Di Alvaro Fitness *Jurnal Sistem dan Teknologi Informasi (JUSTIN)* vol **1** no 1
[2] Bação F L 2010 *Introduction to Kohonen's Self-Organizing Maps* (Lisboa: Instituto Superior de Estatistica E Gestao de Informacao Universidade Nova de Lisboa Portugal)
[3] Barakbah A R and Kohei A 2004 *Determining Constraints of Moving Variance to Find Global Optimum and Make Automatic Clustering* Retrieved from http://lecturer.eepis-its.edu/~ridho/papers/Barakbah_IES_2004.pdf
[4] BNPB 2019 *Potensi Ancaman Bencana* Retrieved from https://bnpb.go.id/potensi-ancaman-

bencana
[5] BNPB 2020 *Tabel Data Informasi Bencana Indonesia* Retrieved from https://bnpb.cloud/dibi/tabel1b
[6] Chrisnanto Y H 2015 Penerapan algoritma Partitioning Around Medoids PAM clustering untuk melihat gambaran umum kemampuan akademik mahasiswa *SENTIKA 2015*
[7] Dawson-Sanders B and Trapp R G 1990 *Basic and Clinical Biostatistics* (Norwalk, Connecticut: Appleton & Lange)
[8] Halim N N and Widodo E 2017 Clustering Dampak Gempa Bumi di Indonesia Menggunakan Kohonen Self Organizing Maps SOM *Seminar Nasional Integrasi Matematika dan Nilai-Nilai Islami* (Malang: Mathematics Department) pp 188–94
[9] Hardi M P W, Kusuma D A and Basuki P 2019 pengelompokan topik dokumen berbasis text mining dengan algoritme k-means: studi kasus pada dokumen kedutaan besar Australia Jakarta *Visi Pustaka* **21(1)** pp 67–76
[10] Hidayat R N, Sabri L M and Awaluddin M 2019 Analisis desain jaring GNSS berdasarkan fungsi presisi (studi kasus : titik geoid geometri Kota Semarang) *Jurnal Geodesi Undip* **8(1)** pp 48–55
[11] Hilmi E, Hendarto E, Riyanti and Sahri A 2012 Analisis potensi bencana abrasi dan tsunami di pesisir Cilacap *Jurnal Penanggulangan Bencana* **35**
[12] Iryanthony S B 2015 Pengembangan modul kesiapsiagaan bencana angin puting beliung untuk mahasiswa pendidikan geografi Unnes *Jurnal Geografi*
[13] Jiawei H and Kamber M 2006 *Data Mining: Concept and Techniques* Second Edition (San Fransisco: Morgan Kaufmann)
[14] Kaufman L and Rousseeuw P J 1987 Clustering by means of Medoids *Statistical Data Analysis Based on the Norm and Related Methods* ed Y Dodge (Amsterdam: North-Holland) pp 405–16
[15] Mikhail and Gracie 1981 *Analysis and Adjustment of Survey Measurement* (New York: Van Nostrand Reinhold Company Inc)
[16] Mulyadi S 2018 *Perbandingan Metode Fuzzy C-Means Terbobot dengan Fuzzy C-Shell Studi Kasus: Kejadian Bencana di Indonesia per-Kabupaten* (Yogyakarta: Universitas Islam Indonesia)
[17] Ningsih S, Sri W and Nasution Y N 2016 Perbandingan kinerja metode complete linkage dan average linkage dalam menentukan hasil analisis cluster: studi kasus produksi palawija Provinsi Kalimantan Timur 2014/2015 *Prosiding Seminar Sains dan Teknologi FMIPA UNMUL* vol **1** no 1 pp 46–50
[18] Prasetyo E 2014 *Data Mining Mengolah Data Menjadi Informasi Menggunakan Matlab* (Yogyakarta: ANDI)
[19] Sandy A E H 2019 Implementasi K-Medoids algoritma pengelompokan untuk pola kelompok kejahatan di Yogyakarta *Int. J. of Applied Sciences and Smart Technologies* vol **1** no 1
[20] Septianusa and Supriyaningsih 2014 Karakteristik dan segmentasi pertanian padi menggunakan algoritma Self Organizing Maps SOMs Kohonen *Prosiding Seminar Nasional Menuju Masyarakat Madani dan Lestari*
[21] Vercellis 2009 *Sistem Informasi* (Yogyakarta: Lokomedia)
[22] Wira B, Budianto A E and Wiguna A S 2019 Implementasi metode K-Medoids clustering untuk mengetahui pola pemilihan program studi mahasiswa baru tahun 2018 di Universitas Kanjuruhan Malang *Jurnal Terapan Sains & Teknologi* **1(3)** pp 54–69

IOP Conf. Series: Materials Science and Engineering 1077 (2021) 012009 doi:10.1088/1757-899X/1077/1/012009

Case-based Mobile Tourism Attractions Recommender System

Septia Rani[1], Dimastyo Muhaimin Arifin[2], Sheila Nurul Huda[3], Dhomas Hatta Fudholi[4]

[134]Department of Informatics, Universitas Islam Indonesia
[2]Undergraduate Program in Informatics, Universitas Islam Indonesia

E-mail: [1]septia.rani@uii.ac.id, [2]dimastyo.arifin@students.uii.ac.id

Abstract. The variations and the increasing number of tourism attractions give tourists more choices to enrich their travel experience. However, it also presents drawbacks, including confusing tourists when planning a trip, because they have to choose from many tourist attractions options that match their preferences. In this study, we propose an algorithm that can generate recommendations of tourist attractions to the user using a case-based reasoning approach. Case-based recommendation is chosen because it gives the user a more personalized recommendation that they can customize later on within the application. Besides, the implementation of this approach in the tourism domain is still hard to find. As many as 217 tourist attraction objects in Yogyakarta city are gathered as the case study. We label each object with three attributes, namely category, visitor type, and activities. A similarity function called the simple matching coefficient is used to calculate the item's score based on the user specification. For assessing the proposed algorithm, we implement the model as an Android application. To evaluate the recommendation result, we use mean absolute error calculation (MAE) and mean absolute percentage error (MAPE), resulting in 4.1 MAE and 5% MAPE, which is considered highly accurate. Thus, the recommendation from the system is well suited to the user's preference.

Keywords: case based reasoning, recommender system, tourist attraction.

1. Introduction

In the past few years, the tourism sector has received the attention of many people. According to the World Tourism Organization (UNWTO), tourism comprises the activities of persons traveling to and staying in places outside their usual environment for not more than one consecutive year for leisure, business, and other purposes not related to the exercise of an activity remunerated from within the place visited [1]. One of the factors that influence the tourism sector's progress is the innovation that causes the emergence of many variations in tourism destinations, tourism products, business models, and services.

The variations and the increasing number of tourism destinations bring advantages for both tourists and tourism providers/operators. However, sometimes the vast offer of options presents its drawback for future travelers. To gain the best experience during their travel, travelers must possess a good judgment or understanding of their potential destination. Travelers who do not have this trait may find it hard to decide their destination, given their lack of information. It becomes clear that this issue must be addressed.

This study aims to build a mobile application that recommends tourist attractions for future travelers. We propose case-based reasoning as the algorithm to generate a recommendation. In a case-based recommender system, specific cases are specified by the user as target points, and the similarity metrics are used to retrieve similar items to these cases [2]. During the registration step in our proposed system, users are asked to fill their profile and choose several samples of tourist attractions that they like. After that, each time users want to plan their trip and get a recommendation, they must specify their specifications. The user specification (known as the case) is then used as the baseline and combined with domain knowledge to recommend tourist attractions. One of the advantages of using a case-based recommendation is that these selected cases can be altered anytime by the user through their profile so that they may receive different recommendations if they are not satisfied by the current recommendation.

The rest of this paper is organized as follows. Section 2 describes related work. Section 3 describes the methodology which is being used in this study. Section 4 describes the result and discussion. Finally, section 5 concludes the overall remarks of this study.

2. Related Work

Recommender system is an area with many possibilities of application, rise to fame in the last decade along with many flourishing topics in the Artificial Intelligence field. Recommender system for tourism is one of many interesting applications, especially since the tourism sector is considered one of the biggest sectors globally [3]. A tourism recommender system can help the user get a personalized tourist attraction recommendation for a city that they have never visited before. Over the last decade, research efforts on the recommender system for tourism have been conducted by many researchers with a diverse and interesting idea. The most common categorization of the recommender system is content-based filtering, which recommends similar items and collaborative filtering that recommend items from similar people. There is also knowledge-based filtering [4]. Some add another touch to make the recommendation more interesting, such as serendipitous recommendation [5] or context-aware recommendation [6] to address specific issues.

Recommendation (or decision) making can sometimes be a challenging task, as the basis of recommendation is data which can be too scarce. For example, in a cold start application with no previous user preferences data or too much data (for example, an application that collects thousands of rating data from thousands of users). Among the researches which use rating data from the user to generate recommendation are [7, 8], known as collaborative filtering. The problem in collaborative filtering lies in the availability of sufficient and adequate number of data. Another approach is needed for a cold start application with no (or limited number) of user data.

Another approach using the case-based method on tourist recommendation falls under the knowledge-based category, for example, as described in [9, 10, 11]. In their research, they used a case-based recommender to generate recommendations. This method proved its capability by generating a bundled region as its recommendation for the user. Implementation of a case-based recommender on a mobile device was also conducted by [12]. They used a third-party travel planning system to retrieve a predetermined case and combine it with the user's constraint to generate recommendations. A different approach was taken by [13] to generate recommendations. They utilized a location-based social media's user activity, built clusters of activity, and used it to generate recommendations. Research [14] categorized tour destinations as Tour Building Block (TBB). Before recommending a destination, the similarity between the user's constraint and TBBs is calculated using a semantic matching algorithm. Alteration of basic case-based recommender was conducted by [15]. They extended the basic case-based recommender into a case-based conversational recommender, enabling users to generate similar cases or ask users to create their constraint. This alteration is actually a variation in the form of collecting user cases, or in a broader term known as user profile or preferences. There are other methods to collect such information, divided as implicit and explicit methods [16]. The implicit method works seamlessly when a user scrolls and views a tourist destination description that attracts him. This method keeps track of user behavior while exploring the application. Sometimes the user is just curious about some places. This curiosity will still be taken into account regardless of whether the user is really interested or just randomly clicks something on the screen. On the other hand, the explicit method asks the user what they prefer and explicitly determines the user profile. A fixed profile such as gender, date of birth (to collect information about age), and user basic preferences need to be input just one time only at the start of the application to reduce the repetitive actions. Some dynamic preferences should be entered dynamically.

Research about increasing case-based reasoning's efficiency was conducted by [17]. Their research aims to increase efficiency in retrieving recommendation by mimicking the bee's instinct when gathering food. By using this method, they can generate a recommendation after 500 iterations. Another efficiency effort was also conducted by [18]. They proposed a new approach to calculate similarity value between attributes and manage to outscore Support Vector Machine, Naïve Bayes, Artificial Neural Network, and Radial Basis Network by 96.69%.

This research explores and implements other approaches in case-based recommendation, where the recommendations are based on the specific user's preferences and constraints. The user preferences and constraints are driven by the derivation of local knowledge in tourism, which is then formed into attributes of tourism categories, activities, and types of visitors.

3. Methodology

In this study, we generate recommendations by using a case-based reasoning approach. It is a subset of a knowledge-based recommender system. This method was chosen due to its capability in solving cold start problems (a problem when the system cannot draw any inferences for users or items because there is no sufficient information). Users will need to specify their query by answering some questions in the application. A query or user's preference is then considered as a new case. Case-based reasoning solves a new case by retrieving items (in this case, the item is a tourist attraction) and adapting to fit the case [18].

3.1. Dataset

The tourist attractions which are used in this study were collected by using Google Maps API. Tourist attractions in the city of Yogyakarta, Indonesia, are used as a case study. The collected tourist attractions have three types of attributes: category, activity, and visitor type, affecting the recommendation. These attributes are derived from the specific knowledge base taken from local tourism authorities, which are then used as the base of user preferences and constraints. The values for each attribute are shown in Table 1. We have listed 217 tourist attractions and assigned values for each attribute. In our system, some attributes are likely to contain more than one value. For example, while in the park, the user can do the following activities: jogging, playing, walking around, enjoying the beauty, and relaxing. The labeling process is done by three persons who possess sufficient knowledge about the characteristics of tourist attractions in Yogyakarta.

Table 1. Values of recommendation attributes.

Attribute	Values
Category	reservoir, hill, lake, zoo, museum, garden, desert, bath, tomb, water boom, waterfall, valley, beach, park, mountain, cave, river, historical place, monument, building & structure, art gallery, temple, industrial heritage, creative art & workshop, tour park, amusement park, science, aquaria, library, culinary, mall, market, tourist village
Activity	climb, swim, go cave, jogging, camping, gardening, animal hunting, fishing, boat ride, playing, go around, outdoor sport, recreation, trekking, learn art, enjoy the beauty, study history, soak, take the picture, eat and drink, relax, learn to code
Visitor type	solo-travel, couple, friends, family-with-kids, family-with-elder-people

3.2. Recommender System

Figure 1 denotes the workflow of the recommender system used in this study. Starting from the "registration phase" and then followed by the "get recommendation phase".

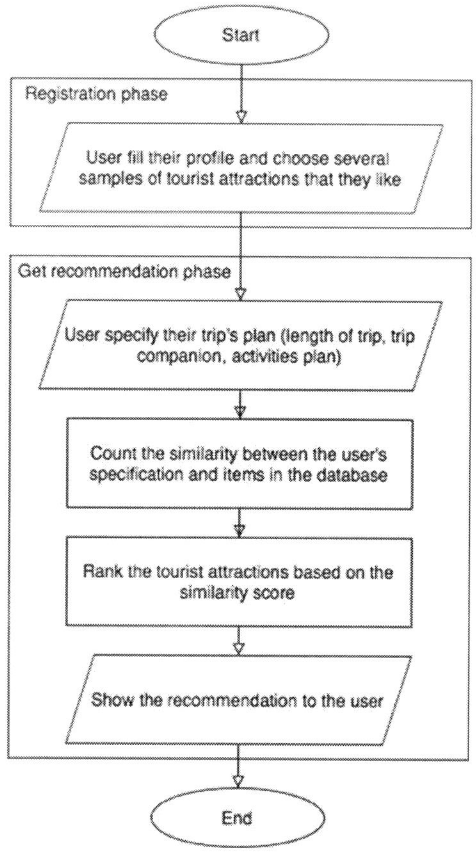

Figure 1. System's workflow.

In the registration phase, users are asked to fill their profile and pick several samples of tourist attractions that they like. This process is only done once at the beginning of registration, but users can change their choices through the edit profile menu provided in the application. The sample of tourist attractions that are presented to the users is from the category attribute. The chosen items will be stored in the database and attached to the user's profile.

Each time a user decides to create a new trip plan, they are asked to specify their trip's plan, including the length of the trip, trip companion, and activities plan. Trip duration information is important to determine the number of items to be recommended. The trip companion information will be matched with the visitor type attribute, while the activities plan information will be matched with the activity attribute attached to each tourist attraction item. To make it easier, users just need to pick the answer from the application's options. These steps are necessary to gain a more precise user's specifications. The user's specification is then used to define the new case, which will be solved using the case-based recommender system.

Upon completing the user's specification, the system will retrieve similar tourist attractions items that have been stored in the database. To generate a more precise recommendation, the system needs to calculate the similarity between tourist attraction items and user's specifications (user's query or new case). Equation (1) denotes the formula to calculate the similarity value. The attributes in the user's query and tourist attraction item use a categorical value type. Therefore a different approach is needed to calculate the similarity. We use the Simple Matching Coefficient to calculate these categorical attributes in our system, as shown in Equation (2).

$$f(t,x) = \frac{\sum_{i \in S} w_i \times Sim(t_i,x_i)}{\sum_{i \in S} w_i} \quad (1)$$

where

t = user's query
x = tourist attraction item
i = index of attribute
S = set of tourist attraction's attribute
w_i = weight of attribute i.
Sim(t_i, x_i) = similarity between user's query and tourist attraction item for attribute i

$$Sim(t_i,x_i) = \frac{M_{00}+M_{11}}{M_{00}+M_{10}+M_{01}+M_{11}} \quad (2)$$

where

M_{11} = the total number of attributes where t_i and x_i both have a value of 1
M_{01} = the total number of attributes where the attribute of t_i is 0 and the attribute of x_i is 1
M_{10} = the total number of attributes where the attribute of t_i is 1 and the attribute of x_i is 0
M_{00} = the total number of attributes where t_i and x_i both have a value of 0

To give a better perspective of our proposed similarity calculation, we will demonstrate a simple similarity calculation between a user query and a tourist attraction. Consider a tourist attraction in Table 2 and a user query in Table 3. Let us assume that there are 6 kinds of activity, 4 kinds of visitor type, and 7 kinds of category. The calculation of similarity for each attribute and similarity between the user's query and tourist attraction is shown in Table 4. We employ the Simple Matching Coefficient to count the similarity of activity attribute and category attribute. For the visitor type attribute, the similarity score is 1 if the value in the user's query can be found in the tourist attraction or 0 if it can be found. After all the similarity calculation of all attributes is done, we finally can calculate the similarity between the tourist attraction and the user's query.

Table 2. Tourist attraction sample.

Attraction	Activity (w = 3)	Visitor type (w = 5)	Category (w = 4)
Malioboro	Shopping, eat and drink, taking pictures	Family-with-kids, friends, couple	Monument

Table 3. User's query sample.

Query	Activity	Visitor type	Category
Q1	Shopping, adventure, taking pictures	Family-with-kids	Monument, science, zoo

Table 4. Similarity calculation.

Activity	Visitor type	Category	Similarity
$\frac{2+(6-4)}{(6-4)+1+1+2} = 0.67$	1	$\frac{1+(7-3)}{(7-3)+0+2+1} = 0.71$	$\frac{(3 \times 0.67)+(5 \times 1)+(4 \times 0.71)}{3+5+4} =$ 0.82

After calculating the similarity, the system will rank the possible candidate of tourist attractions based on their similarity value. Only tourist attractions with a similarity value above 75% will be shown in the application as the recommendation item. However, not all these recommended items will be presented to the user because the amount of presented items depends on the user's planned trip duration. This design is chosen to avoid a vast amount of recommendations that may overwhelm users.

3.3. Evaluation

To evaluate our recommendation model, 10 relevant users will be involved. These 10 users involved are the users who frequently travel so that they are considered enough as representation. Each user will be asked to use the application starting from the registration to getting the recommendation with the trip duration of two days. Users are asked to create their-own similarity value on each tourist attraction that is recommended by the application. This user-created similarity value will be subtracted with the generated similarity value and called the error. The evaluation method that we choose in this study is Mean Absolute Error (MAE) and Mean Absolute Percentage Error (MAPE).

Table 5. MAPE interpretation.

MAPE Value	Interpretation
Less than 10%	Highly Accurate Prediction
11% to 20%	Good Prediction
21% to 50%	Reasonable Prediction
More than 51%	Inaccurate Prediction

3.3.1. Mean Absolute Error (MAE). The MAE calculation involves the absolute value of the errors and then dividing the total error by n [19]. MAE indicates how much our model missed in predicting the similarity value. The calculation of MAE is shown in equation 3. To conclude the overall performance of our model, we calculate the average MAE of 10 users.

$$MAE = \frac{1}{n}\sum_{i=1}^{n} |y_i - y_i'| \quad (3)$$

where

y_i = user's similarity i-th value
y_i' = generated similarity i-th value
n = total amount of data

3.3.2. Mean Absolute Percentage Error (MAPE). Another useful evaluation is using the Mean Absolute Percentage Error (MAPE). MAPE is the mean or average of the absolute percentage errors of a prediction. Since MAPE is in percentage, it is more understandable than other evaluation metrics. Each recommended destination's absolute error is divided by the actual similarity value provided by the user to form it into an absolute percentage error. Absolute percentage errors are summed and averaged to compute MAPE. Equation 4 provides mathematical notation to calculate MAPE. This measure is easy to interpret because it provides the error in terms of percentages. Upon interpreting the MAPE, we use MAPE interpretation by [20], which is shown in Table 5.

$$MAPE = \frac{1}{n}\sum_{i=1}^{n} \left| \frac{y_i - y_i'}{y_i} \right| \quad (4)$$

where

y_i = user's similarity i-th value
y_i' = generated similarity i-th value
n = total amount of data

4. Results and Discussion

4.1. Application Implementation

In this study, we implemented the recommender system on an Android application. Java programming language is used to build a mobile application. A client-server architecture is used in the recommendation process. We built a separate web service that receives the input supplied by the Android application through API. The result of the recommendation from the web service is sent to the application through API and then the Android application presents the recommendation to the user.

Figure 2 and Figure 3 display the Android implementation based on the design made in section 3.2. Figure 2(a) displays the first activity that appears when a new user uses the application. It asks the user to use their Google account to log in to the application. The application will store the email and name from the Google account. When a new user logs in to the application, they will be asked to complete their personal information (shown in Figure 2(b) and Figure 2(c)). These data will be stored along with the user's email and name from the previous activity. Figure 2(d) represents the user preference activities, which is one of the most important keys in this study. It contains the list categories of tourist attractions (along with the sample of image) that the user needs to choose to satisfy the case-based reasoning requirement. The chosen categories will be stored in the database and will be used to help generate recommendations.

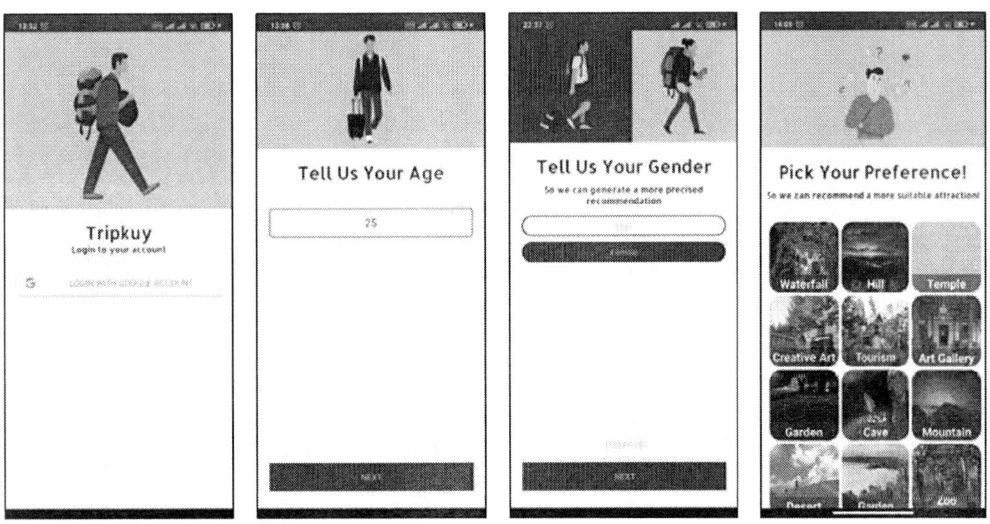

(a) (b) (c) (d)

Figure 2. Android implementation (registration phase)

After completing the registration steps, users will be redirected to the home activity. This activity contains bottom navigation to navigate between home, trip's history, and profile page. The trip history contains the trips that a user has done before, and the profile page contains the user's personal information and preferences. On the home page, users can choose to create a new trip plan. Figure 3(a) shows the activity that appears when a user chooses to create a new trip. In this activity, users are asked to complete some necessary information for their trip, including its length. After that, users are asked to choose the travel companion and activities plan (shown in Figure 3(b) and Figure 3(c)). This information (also called the user's query) is necessary to generate recommendation since it will be used to calculate similarities between visitor type attribute and activity attribute on the potential tourist attraction recommendation.

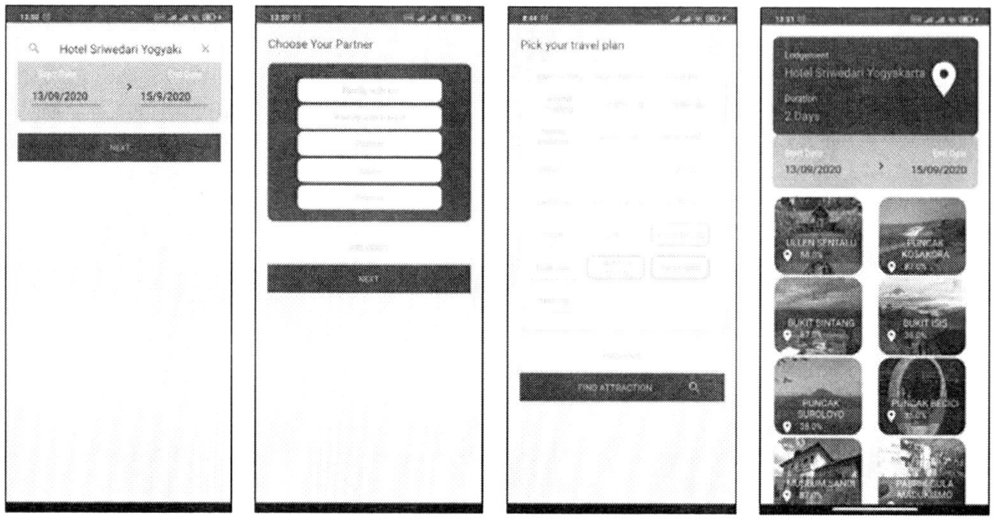

(a) (b) (c) (d)

Figure 3. Android implementation (get recommendation phase).

After completing all the required information, the application will send a POST request containing the trip specification to the web service. The web service will also retrieve the previously stored user profile that contains information about the tourist attractions categories that suit user preference. The web service then calculates similarities between potential tourist attraction items and the user's query. After completing the similarities calculation, the web service will send the result to the application through API. The application then presents recommendations to the user (shown in Figure 3(d)). The recommendation consists of the picture of the tourist attractions, its name, and the similarity score in percent. A higher similarity score indicates that the item suits more to the user specification.

Table 6. User assessment results.

User	MAE	MAPE
Person A	10.4	16%
Person B	1.79	2%
Person C	5.36	7%
Person D	2.31	3%
Person E	4.53	7%
Person F	4.8	6%
Person G	3.4	4%
Person H	3.27	4%
Person I	2	2%
Person J	3.2	4%
Average	4.1	5%

4.2. Recommendation Result

In our experiment, we managed to gather 10 travelers to assess our recommendation model. We asked them to compare our recommendation and their-own assessment regarding the recommended tourist attractions. We used Mean Absolute Error (MAE) and Mean Absolute Percentage Error (MAPE) to measure our recommendation model. The results of the test are shown in Table 6.

The MAE indicates how much error does our model generates in predicting the similarity between tourist attractions and the user's specifications. Table 2 shows various MAE and MAPE for 10 users, with person A has the highest MAE and MAPE, which indicates that our model does not suit well to her. Despite the large MAE and MAPE on person A, the average MAE for 10 persons is only 4.1, and the average MAPE for 10 persons is 5%. With MAPE less than 10%, according to [20], this recommendation is considered highly accurate.

5. Concluding Remarks

In this study, we implemented case-based reasoning as our recommendation model on the Android application. Each tourist attraction has 3 attributes (i.e., category, activity, and visitor type), which we use to calculate the similarity with the user's query. The application utilizes a separate web service to efficiently calculate the similarity between items (tourist attractions) and the user's preference. We evaluate our recommendation model using MAE and MAPE. It is shown that from 10 users, the average MAE is 4.1 and the average MAPE is 5%. With MAPE less than 10%, this recommendation is considered highly accurate.

References

[1] Tribe J 2009 *Philosophical Issues in Tourism* ed J Tribe (Bristol: Channel View)
[2] Aggarwal C C 2016 *Recommender systems* vol **1** (Switzerland: Springer International Publishing)
[3] Brandt T, Bendler J and Neumann D 2017 Social media analytics and value creation in urban smart tourism ecosystems *Inf. Manag.* **54(6)** pp 703–13 DOI: 10.1016/j.im.2017.01.004.
[4] Quijano-Sánchez L, Cantador I, Cortés-Cediel M E and Gil O 2020 Recommender systems for smart cities *Information Systems* **92** 101545 DOI: 10.1016/j.is.2020.101545
[5] Menk A, Sebastia L and Ferreira R 2017 CURUMIM: A Serendipitous Recommender System for Tourism Based on Human Curiosity *Proc. of the IEEE 29th Int. Conf. on Tools with Artificial Intelligence (ICTAI)* pp 788–95 DOI: 10.1109/ICTAI.2017.00124.
[6] Khallouki H, Abatal A and Bahaj M 2018 An ontology-based context awareness for smart tourism recommendation system *Proc. of the Int. Conf. on learning and optimization algorithms: theory and applications* pp 43:1–43:5 DOI: 10.1145/3230905.3230935
[7] Herzog D, Laß C and Wörndl W 2018 Tourrec: a tourist trip recommender system for individuals and groups *Proc. of the 12th ACM Conf. on Recommender Systems* pp 496–7
[8] Amoretti M, Belli L and Zanichelli F 2017 Utravel: smart mobility with a novel user profiling and recommendation approach *Pervasive and Mobile Computing* **38** Part 2 pp 474–89 DOI: 10.1016/j.pmcj.2016.08.008
[9] Herzog D and Wörndl W 2014 A travel recommender system for combining multiple travel regions to a composite trip *CBRecSys@ RecSys* **1245** pp 42–8
[10] Alptekin G I and Büyüközkan G 2011 An integrated case-based reasoning and MCDM system for web based tourism destination planning *Expert Systems with Applications* **38** pp 2125–32 DOI: doi:10.1016/j.eswa.2010.07.153
[11] Bouhana A, Fekih A, Abed M and Chabchoub H 2013 An integrated case-based reasoning approach for personalized itinerary search in multimodal transportation systems *Transportation Research Part C: Emerging Technologies* **31** pp 30–50 DOI: 10.1016/j.trc.2013.02.014
[12] Nguyen Q N, Cavada D and Ricci F 2003 Trip@dvice mobile extension of a casebased travel recommender system *Proc. of the 2nd Int. Conf. on Mobile Business*
[13] McKenzie G and Janowicz K 2014 Activities in a new city: itinerary recommendation based on user similarity *Spatial Knowledge and Information (SKI) Canada*
[14] Ten Hagen K, Kramer R, Hermkes M, Schumann B and Mueller P 2005 Semantic matching and heuristic search for a dynamic tour guide *ENTER* pp 149–59

[15] Mahmood T, Ricci F, Venturini A and Höpken W 2008 Adaptive recommender systems for travel planning *ENTER* **8** pp 1–11
[16] Hanani U, Shapira B and Shoval P 2001 Information filtering: Overview of issues, research and systems *User Modeling and User-Adapted Interaction* **11** pp 203–59
[17] Lorenzi12 F, dos Santos D S and Bazzan A L 2005 *Case-based recommender system inspired by social insects* Retrieved on November 2020 from https://www.inf.ufrgs.br/maslab/pergamus/pubs/eniabeeCR.pdf
[18] Rezvan M T, Hamadani A Z and Shalbafzadeh A 2013 Case-based reasoning for classification in the mixed data sets employing the compound distance methods *Eng. Applications of Artificial Intelligence* **26** pp 2001–9
[19] Willmott C J and Matsuura K 2005 Advantages of the mean absolute error (MAE) over the root mean square error (RMSE) in assessing average model performance *Climate research* **30** pp 79–82
[20] Lewis Colin D 1982 *Industrial and business forecasting methods* (Massachusetts: Butterwoth & Co Publishers Ltd)

Digital Village: Modelling of Public Wireless Internet Development using Supply Chain Structure Approach

Wijiyanto[1]**, Nurchim**[2]

[1,2]Faculty of Computer Science, Universitas Duta Bangsa Surakarta, Fivety five Bhayangkara Street, Solo, 57154, Indonesia

E-mail: [2]nurchim@udb.ac.id

Abstract. This study aims to describe the design of rural public wireless internet development based on a supply chain structure approach. Now, many new internet services have sprung up in the private, public, business, education, health, and even governmental sectors until considered as the leading global system. Especially in rural areas, the provision of cable internet networks and cellular internet networks have not been able to reach all rural areas, so a new breakthrough is needed, one of which is a public wireless internet network. This research was conducted with a Design-Based Research approach that combines scientific design methods with the problems that occur for internet development in rural areas of Indonesia. The finding that the development of wireless internet in rural areas can be carried out with the concept of a supply chain structure consisting of suppliers (ISP providers of internet sources), manufacturers (local government or village-owned enterprises managing and transmitting wireless internet to wholesaler), wholesaler (installed 5GHz wireless repeater in people residential areas), retailers (installed 5GHz wireless receiver devices from wholesaler and convert to 2.4GHz wireless transmitters installed in every resident's house) and customers (each resident's smartphones or notebooks device connected by wireless).

1. Introduction

A Digital Village is an idea that allows developers to reach undeveloped internet-connected village areas [1]. Now, many new internet services have emerged such as the private, public, business, academic, healthcare, and also governmental sector till considered as a leading global system [2]. The Digital Platform's general purpose is to present an accessible, user-friendly, modern and approach that will facilitate any village representative to improve their smartness [3]. Therefore, many of the government's intended to create digital villages by linking all schemes with internet [4].

In several projects about digital village, electrical infrastructure, sites vary by geographical location, and target population determine the kinds of devices that may be implemented in the field also amount of cost [5]. Now, the backbone of Access Point (AP) dominant still uses Unshielded Twisted Pair (UTP) cable media resulting in significant problem when to locations that are difficult to reach by cables as well as villages that have long distances from the City Center and quite difficult natural conditions [1]. Nonetheless, recent developments in wireless technologies are raising new hopes in rural areas of the developing world for sustainable Internet diffusion [6]. A wireless network is one of the best alternatives to build a practical, flexible, and high-mobility computer network [1].

Meanwhile, the availability of public Wi-Fi access was rarely found in poor residential areas and remote rural areas [7]. No need to be in doubt, the improvement of wireless communication technology is essential to use it for public safety, which could greatly increase the effectiveness of the exchange of information between emergency responders [8]. As soon as possible, it is necessary to develop a public wireless model that is easily accessible by rural communities in accordance with the characteristics of the village. This research aims to develop model of wireless public internet in rural areas using supply chain approach. Supply chain management used integrated one-way manufacturing process, in which the raw material from suppliers then delivered to the customer [9] like as send internet connection from provider to users.

2. Conceptual Basic

The increased uptake of the internet is seen as a vital part of strategies to avoid rural decline [10]. Effects of IT investment only achievable through a strong investment in internet infrastructure that provides IT adoption and application to support people to take advantage of the internet age in order to achieve increased well-being [11]. The geographical conditions of Indonesia's rural villages do not all have internet connections [12], even some of its inhabitants are in areas that are not yet affected by land network infrastructure via cable or fiber optics so that the need for the internet can reach remote villages is very important [13]. For this reason, it is necessary to have an integration between the internet supply process, internet distribution so that it can be consumed by the village community such as the supply chain process. These supply chain stages consist of suppliers, manufacturers, distributors, retailers, and customers [9] shown in Figure 1. It is inevitable for the sustainable supply chain to meet the aggressive change in customer requirements [14], similar with ensuring internet availability in rural areas have many challenges resulting in limited internet infrastructure wired or non-wired[15] due to disparities geographical and electricity access [16][17]. Thus required, research is carried out using a Design-Based Research (DBR) that combines scientific design methods with the problems about internet development in rural areas of Indonesia. DBR stages consist of (1) focusing on observing problems in building internet networks in digital villages, (2) understanding of problems in building internet network infrastructure in digital villages, (3) define solutions to problems, (4) conceive the needs for tools and materials, (5) making prototype solutions offered and (6) testing the prototipe [17].

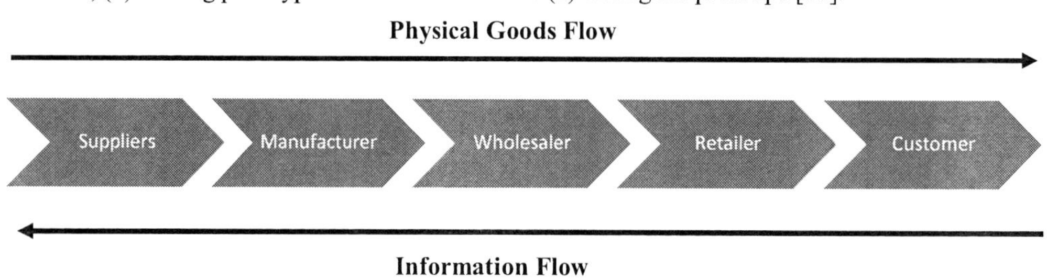

Figure 1. Supply chain structure approach

3. Results and Discussion

The next discussion will be discussed concerning the supply chain structure approach in providing internet infrastructure in order to realize digital village shown Figure 2.

3.1. Suppliers

The Supplier performance has the most effect [18] and plays important role on the successful supply chain management [19]. Therefore, process supplier selection is show a sophisticated, application oriented, decision-making (DM) problem and has received considerable attention [20]. Especially in the implementation of internet infrastructure, Internet Service Provider (ISP) and Mobile Network Operators (MNOs) is the main supplier to provide internet to the customers. However, challenges arise in the spread and expansion of the internet network in remote or rural areas caused like geographical, weather conditions, and infrastructure cost [21]. Wireless Internet Service Provider (WISP) networks as the solution connecting remote and rural areas to the internet [22] and potential building a low-cost sustainable Internet service [23].

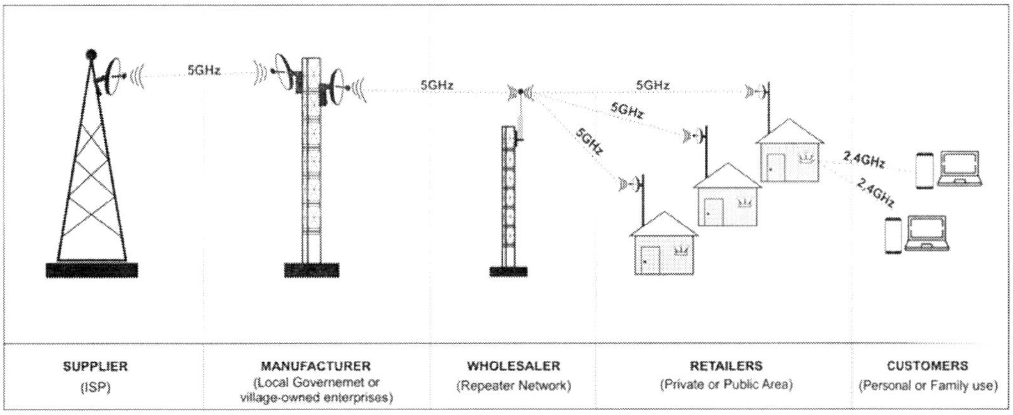

Figure 2. Design rural public wireless internet development use supply chain structure approach

3.2. Manufacturer

The Industrial Internet is often understood as the implementation of the general concept of Cyber Physical Systems (CPSs), through which data is closely collected, tracked from physical space and synchronized with cyber space [24]. Manufacturing services are more frequently shared among multiple stakeholders as various Industrial Internet platforms emerge, so that both the shared Manufacturing Service Composition (MSC) are optimally selected and promised [25]. In the rural areas, both the industrial internet and local government can working together [26] as manufacturer to increase internet service in there. In addition, rural local governments can also establish village-owned enterprises (BUM Desa) as manufacturer to provide their internet so that they can potentially play a vital role in the economic and community growth of villages to support community issues by mobilizing local resources that are difficult for profit-oriented businesses [27].

3.3. Wholesaler

Manufacturers usually distribute their products or services through a group of distributors who are approved to sell their products or services [28]. Under the leadership of the wholesaler who re-sells these goods to the customer through retailers, the manufacturer's profit perspective can be optimized [29]. The role of wholesalers is critically important in the supply chain, to choose the most suitable market [30]. Wholesalers play a role in distributing the system deployed in different geographical locations for internet networks to be closer to the end user, thereby reducing network congestion and accelerating distribution [31].

3.4. Retailer and customer

A retailer must produces value for its customers and acquires value from the markets in the business model [32]. In addition, this is important in the retail business model for directly customer interaction [33] well personal or family use and the last stage link among manufacturers and consumers [34]. This retailer concept can use to provide wireless internet network infrastructure for rural citizens so that easy to access it. Because of its fast data speed transmission and high coverage area, Wi-fi supports their to direct connection without having a central access point and more applicable to implement in disasters condition than other technologies [35]. Therefore, rural citizens as end consumers can connect their devices directly to internet access point in the home or rural public areas.

From this model shown in Figure 2, the stakeholders involved consist of ISP as internet connection supplier and rural local government or village-owned enterprises (BUM Desa) which is in charge of manage and control the internet network from the manufacturer stage to the customers. This model supports the implementation of digital villages in terms of infrastructure, while the issues of digital skills and literacy are that rural local government can build village-owned enterprises (BUM Desa) as digital telecentre. The digital telecentre must have aspect (1) as a social enterprise engaged in the application of acceptable ICT technologies, the establishment of telecentres must have this mission, (2) telecentres in the form of village-owned enterprises legal bodies, so that they have the ability to obtain external (CSR) and internal (village funds) financial capital and (3) in order to help the ensure the sustainability of ICT adoption, telecentre must include human resources the village community of origin [36].

4. Conclusion

A digital village is a great dream for a developing country to demonstrate justice in urban and rural life, especially in the development of technology. The internet has become a technology in recent years that plays an important role in the distribution of information and the economy's pace. However, there have been many barriers to the growth of internet connectivity in rural areas, such as geographical conditions, high costs of connecting an ISP fiber optic cable, mobile broadband signal has not yet penetrated all areas and fragmented residential areas. The findings of this study explain, the modelling of internet network infrastructure development in the rural areas using a supply chain approach. The process of connecting ISPs (suppliers) to rural citizens (end consumers) through a wireless network is based on this approach. Wireless networks have the advantage that, can connected without the need for a central access point so rural citizens can directly connect to the internet, greater coverage and more likely to be in disaster situations than other technology. Basically, this model involves the ISP as internet supplier and rural local government or village-owned enterprises (BUM Desa) which manage and control internet network from manufacturer until customers. This model can be solution to support digital village in terms of infrastructure, there is also for digital skills and literacy then a digital telecenter can be formed to ensure the sustainability of ICT adoption. For further study, it is necessary to study the performance of this wireless network design in accordance with the heterogeneous geographical conditions of the villages so that the disadvantages and challenges can be analyzed.

Acknowledgments

This study was supported by the Ministry of Science and Technology / National Research and Innovation Agency (Kemenristek/BRIN) of the Republic of Indonesia in 2020

References

[1] Rusdan M 2019 Design of wireless network system for digital village using wireless distribution system *J. Informatics, Inf. Syst. Softw. Eng. Appl.* **1** pp 51–9
[2] Paul P and Bhuimali A 2017 Emerging Internet Services Vis-à-Vis Development: A Theoretical Overview *Int. J. Recent Res. Sci. Eng. Technol.* **5** pp 19–25
[3] Beranič T, Zamuda A, Brezočnik L, Turkanovi´c T, Lentini G, Polettini F, Lué A, Vitale A C, Martinez-Gil J and Pichler M 2019 Facilitating the digital transformation of villages *Proc. of the Central European Conf. on Information and Intelligent Systems* pp 281–8
[4] Vijayan A 2019 Digital India – A roadmap to sustainability *Int. J. Innov. Technol. Explor. Eng.* **8** pp 571–6
[5] Verma S and Ryan J P 2016 A framework for information access in rural and remote communities *Int. Symp. Technol. Soc. Proc.* **2016-Decem** pp 96–9
[6] Pereira J P R 2016 Broadband access and digital divide *Adv. Intell. Syst. Comput.* **445** pp 363–8
[7] Wang M, Liao F H, Lin J, Huang L, Gu C and Wei Y D 2016 The making of a sustainable wireless city? Mapping public Wi-Fi access in Shanghai *Sustainability* **8**

[8] Xiang W 2017 Applicability of wireless communication techniques in public safety *Int. Conf. on Innovation & Management* pp 653–7
[9] Deshmukh A J and Vasudevan H 2014 Emerging supplier selection criteria in the context of traditional vs green supply chain management *Int. J. Manag. Value Supply Chain.* **5** pp 19–33
[10] Komorowski Ł and Stanny M 2020 Smart villages: Where can they happen? *Land* **9**
[11] Evans O 2019 Repositioning for increased digital dividends: internet usage and economic well-being in Sub-Saharan Africa *J. Glob. Inf. Technol. Manag.* **22** pp 47–70
[12] Setiawan T and Padmaningrum D 2020 Towards the design of village information systems as a villager communivation medium *Int. J. Multi Sci.* **1** pp 1–6
[13] Hidayat S, Ramdani T, Alam I F, Sfenrianto S and Kaburuan E R 2019 The Role of high throughput satellite as sky highway infrastructure to support the acceleration of internet entry into vilages in Indonesia *Int. J. Mech. Eng. Technol.* **10** pp 552–60
[14] Manavalan E and Jayakrishna K 2019 A review of Internet of Things (IoT) embedded sustainable supply chain for industry 4.0 requirements *Comput. Ind. Eng.* **127** pp 925–53
[15] Onitsuka K, Hidayat A R R T and Huang W 2018 Challenges for the next level of digital divide in rural Indonesian communities *Electron. J. Inf. Syst. Dev. Ctries.* **84** pp 1–25
[16] Sujarwoto S and Tampubolon G 2016 Spatial inequality and the Internet divide in Indonesia 2010–2012 *Telecomm. Policy* **40** pp 602–16
[17] Nurchim N and Pradana A I 2019 Prototipe green internet technology guna mendukung infrastruktur desa pintar *J. Appl. Informatics Comput.* **3** pp 102–6
[18] Valipour P S, Safaei G A and Fallah L H 2019 Resilient supplier selection and segmentation in grey environment *J. Clean. Prod.* **207** pp 1123–37
[19] Liu T, Deng Y and Chan F 2018 Evidential supplier selection based on DEMATEL and game theory *Int. J. Fuzzy Syst.* **20** pp 1321–33
[20] Chai J and Ngai E W T 2020 Decision-making techniques in supplier selection: Recent accomplishments and what lies ahead *Expert Syst. Appl.* **140** 112903
[21] Bhuiyan M 2020 *Solutions For Wireless Internet Connectivity in Remote and Rural Areas* Thesis (Oulun Yliopisto: University of Oulu)
[22] Cameron D, Valera A and Seah W K G 2020 ElasticWISP: Energy-proportional WISP networks *Proc. IEEE/IFIP Netw. Oper. Manag. Symp. 2020 Manag. Age Softwarization Artif. Intell. NOMS 2020*
[23] Hameed A, Mian A N and Qadir J 2016 Low-cost sustainable wireless internet service for rural areas *Wirel. Networks* **24** pp 1439–50
[24] Li J Q, Yu F R, Deng G, Luo C, Ming Z and Yan Q 2017 Industrial internet: a survey on the enabling technologies, applications, and challenges *IEEE Commun. Surv. Tutorials* **19** pp 1504–26
[25] Zhang Y, Tao F, Liu Y, Zhang P, Cheng Y and Zuo Y 2019 Long/short-term utility aware optimal selection of manufacturing service composition toward industrial internet platforms *IEEE Trans. Ind. Informatics* **15** pp 3712–22
[26] Oh S H and Mardis M A 2019 Rural broadband and advanced manufacturing: research implications for information studies *Lect. Notes Comput. Sci. (including Subser. Lect. Notes Artif. Intell. Lect. Notes Bioinformatics)* **11420 LNCS** pp 265–73
[27] Arifin B *et al* 2020 Village fund, village-owned-enterprises, and employment: evidence from Indonesia *J. Rural Stud.* **79** pp 382–94
[28] Yang X, Cai G, Ingene C A and Zhang J 2020 Manufacturer strategy on service provision in competitive channels *Prod. Oper. Manag.* **29** pp 72–89
[29] Nielsen I E, Majumder S and Saha S 2019 Exploring the intervention of intermediary in a green supply chain *J. Clean. Prod.* **233** pp 1525–44
[30] Tavana M, Amoozad M H, Beheshti M and Abbasi K A A 2020 Optimal strategic alliance in multi-echelon supply chains with open innovation *Manag. Decis. Econ.* **41** pp 1365–84
[31] Antonopoulos A, Kartsakli E, Perillo C and Verikoukis C 2017 Shedding light on the internet:

stakeholders and network neutrality *IEEE Commun. Mag.* **55** pp 216–23
[32] Sorescu A, Frambach R T, Singh J, Rangaswamy A and Bridges C 2011 Innovations in retail business models *J. Retail.* **87** pp S3–16
[33] Jocevski M, Arvidsson N, Miragliotta G, Ghezzi A and Mangiaracina R 2019 Transitions towards omni-channel retailing strategies: a business model perspective *Int. J. Retail Distrib. Manag.* **47** pp 78–93
[34] Bilgin S E and K Yücel E 2020 Distribution network design for e-retailing application: a model suggestion for local retailer in Izmir *J. Bus. Res. - Turk* **12** pp 2205–14
[35] Alnashwan R and Mokhtar H 2019 Disaster management system over Wifi direct *2nd Int. Conf. Comput. Appl. Inf. Secur. ICCAIS 2019* pp 1–6
[36] Nurchim N and Nofikasari I 2018 Analisis model pengembangan telecenter guna mewujudkan desa pintar di Indonesia *Fountain Informatics J.* **3** pp 36–40

Smart Bumdes: Innovation of E-Collaboration and E-Commerce for Village Economic Development Solution in the New Normal Era (Case Study in Batu City)

A Prabowo[1], F Arisha[2], D Triyana[3], A F Baharudin[4]

[1]Public Administration, Universitas Brawijaya, Indonesia
[2]Informatics Engineering, Universitas Brawijaya, Indonesia
[3]French Language and Literature, Universitas Brawijaya, Indonesia
[4]Information Systems, Universitas Brawijaya, Indonesia

E-mail: ardianprabowo11@gmail.com

Abstract. The COVID-19 outbreak in Indonesia has burdened society with enormous costs, from the number of deaths, limited activities, and loss of work. The negative effect of this causes global economic crisis. Businesses feel the negative impact of the Corona COVID-19 virus pandemic which threatens the status of the continuity of the business, especially businesses that do not have sufficiently good adaptations, such as businesses owned by BUMDes *(Village- owned enterprises)*. The purpose of this research is to find the latest innovations through Smart BUMDes based on technology (*application*). The research method used is the Empathy Map method with direct interview techniques with BUMDes and the efforts under them, and refined through the Experimental method to make the application. The results of this study present an online platform called Smart BUMDes which functions as e-collaboration and e-commerce. E-collaboration functions as a forum that facilitates training and business consultation, Controlling, Monitoring and Reporting by related Ministries, Local Governments, Village Governments (BUMDes and businesses under them). E-commerce has a function as a platform for selling local village products and services. From these two functions, it is hoped that they can build the village economy in the new normal era.

Keywords: BUMDes (Village-owned enterprises), Innovation, e-collaboration, e-commerce, management information systems

1. Introduction

Inequality has been a long major troubling issue for the human society, and is a core of academic inquiry in social sciences [1,2]. Differences in the level of economic development between regions will result in imbalances [3,4,5,1]. The differences population concentration between rural and urban area affected rural area become more lag [6,7]. For example, urbanization can increase income inequality because there are a lot of kind jobs with higher wage in urban than in rural [8]. Disparities in development between regions can be gaps in: (a) the capita income, (b) the quality of human resources, (c) the availability of infrastructure such as transportation, energy and telecommunications, (d) social services such as health service, education, etc, and (e) access to banking [9]. These factors that widen the urban-rural disparities resulted the imbalance between urban-rural inequality that stands out in regional economic development [10,11,12]. This is the case in many developing countries [13,14], Including Indonesia [15,16,17].

Economic inequality will increase the poverty rate in the village [18]. This figure is exacerbated again by the COVID-19 pandemic in 2020 [19,20,21]. Previously, the disease COVID-19 appeared in Wuhan-China in December 2019 [22,23,24]. COVID-19 is a disease caused by a highly contagious viral infection and caused the acute respiratory syndrome by coronavirus [25], coronavirus is one of the main pathogens that target the human respiratory system [26]. On March 11, 2020, COVID-19 was announced as a global pandemic by the General World Health Organization (WHO), which means that the infectious of coronavirus disease outbreak has spread throughout the world. It is driven by the speed with which we cross country and continent borders, so that it caused an economic crisis everywhere. Hence, the government needs to provide a policy especially in village rebuilding economy. The government must ensure the actions would take is sensitive to the needs of the poor and the most vulnerable people so that handling the impact of the crisis by he coronavirus outbreak is easier to overcome. The analysis of the past outbreaks such as SARS and Ebola shows that poverty has become an important factor in disease transmission [27,28].

According to Ministry of Villages, Disadvantaged Areas, and Transmigration, there is a need for strategic steps in village economic development through increasing the revitalization of Village-Owned Enterprises (BUMDes) [29]. However, the steps taken by the government are still not significant, namely providing numbers in revitalizing BUMDes. Revitalizing businesses that adaptable to the current pandemic conditions, so that the businesses can survive is a good strategy that needed. Modifying (updating) the village economic development program in the new normal era is one of the strategies to accelerate new development during COVID-19 pandemic [30]. Since this pandemic can be transmitted through direct contact [31], there is a need for work facilities that can reduce direct contact between stakeholders wich easy to apply. The use of technology is very helpful in the new normal era [32,33,34,35]. In addition, the government and village communities must undertake reforms, especially in utilizing information technology and digitalization in public services [36]. So that researchers use an approach through technology, in order to maximize the role of BUMDes in village economic development. The innovation is called Smart BUMDes with services in the form of *e-commerce* and *e-collaboration*.

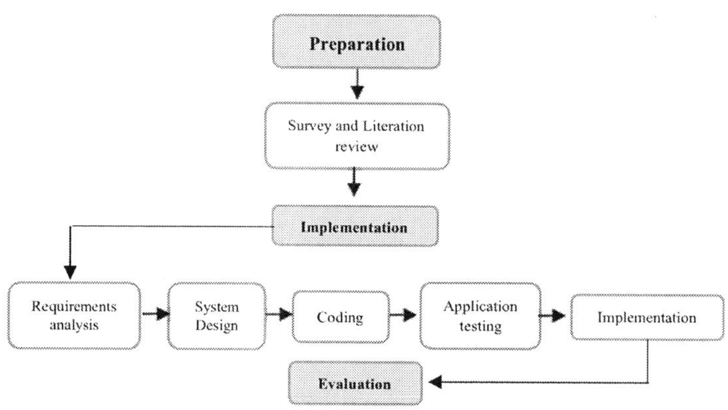

Fig 1. Flow diagram of experimental method

2. Experimental Method

Experiment method is follows (Fig. 1):

1. Preparation. This stage is the first step to make the possible survey. The location of the survey itself is located at Pesanggrahan, Batu City Government, in BUMDes's Small and Medium

Micro Units (UMKM). Types of data used are primary data (interview) and secondary data (citation). Interviews were done by adjusting the research focal using Empathy Map Theory.

2. Implementation. Problem analysis using Empathy Map Theory and also costuming requirement of BUMDes and UMKM is a first step in this stage. The requirements are divided into two parts; fuctional requirement and non-functional requirement. Functional requirements contain processes that can be operated by the system and nonfunctional requirements describe supporting features and tools that will be used for building and developing the system such as XAMPP, Notepad++, codeIgniter platform, bootstrap, and also detail of testing device. Then, the system design will be build using incremental model process. The advantages of incremental process are making development process faster, easier to know if has an addition user's requirement, and having minimum resource to make changes in software [37]. The implementation of the system is coding in the form of programme codes according from system design. If it was implemented, there will be a Smart BUMDes websites. This application need to be test to know bugs and error in the system and to check all requirements are implemented. Then, the system can be accessed by all BUMDes, UMKM, also Indonesia's society.
3. Evaluation. Evaluation is conducted to determine the effectiveness of Smart BUMDes application and to ensure that the application is running according to previous planning.

3. Result and Discussion

3.1. BUMDes (Village-Owned Enterprise) Innovation

BUMDes or Village-Owned Enterprises based on Permendes No. 4 of 2015 concerning BUMDes and Law No. 6 of 2014 concerning Villages. Article 87 states that BUMDes are formed based on the spirit of kinship and mutual cooperation to utilize all economic potential, economic institutions, as well as natural potential and human resources in order to improve the welfare of rural communities. BUMDes describes tourism goods and services. In addition, BUMDes collects several types of MSME below such as social business, leasing, intermediaries, trade, financial business, and holding by the kelompok sadar wisata (pokdarwis), goup of people who aware to tourism issue in village. Management of BUMDes is adjusted to local potential, for example at BUMDes in Pesanggrahan Village. Products produced are in the form of milk and hydroponic farming, financial businesses in the form of savings and loans for the community, and natural tourism like coban (waterfall trip), and climbing up the Mt.Panderman.

In 2014 there were 1,022 BUMDes units. This number increased to 18,446 units in 2015 spread across several regions in Indonesia. However, the increase in the number of BUMDes is not accompanied by good management and supervision by the government which has an impact to the product price or services from BUMDes which tend to lack consistency in quality and quantity so that these products or services are less desirable in the market. In addition, there are still many villages that do not have qualified human resources in planning, managing and marketing BUMDes product. Based on these, there are several unproductive areas, for example BUMDes in Batu City. According to Subhan, Head BUMDes Pesanggrahan Village stated that Batu city, because of the many inhibiting factors during the management proses of BUMDes, from 19 villages have BUMDes in Batu City, only few villages considered productive as BUMDes, namely Sidomulyo , Bumiaji, and few Guest Houses in there [38]. This is currently exacerbated by the occurrence of COVID-19 [36].

Smart BUMDes is an e-commerce and e-collaboration platform in building the village economy in the midst of the COVID-19 pandemic (New Normal Era). Electronic commerce is the process of buying and selling goods electronically by consumers and from company to company through computerized business transactions [39]. Businesses have decreased levels based on visits or the presence of consumers in places severely affected by the Covid-19 pandemic, such as public transportation, tourism, hotels, offline retail, shopping centre, transportation of people and goods. Businesses that can adapt to market dynamics is one that can continue to develop, one of them is online business. Interaction adjustments using online application platforms performed with telecommunications business and online shopping (basic needs and health products) [40], so it is necessary to develop e-commerce services in Smart BUMDes which are needed during the COVID-19 pandemic. Meanwhile, electronic collaboration (e-collaboration) is operationally defined here as collaboration using electronic technologies among different individuals to accomplish a common task [42]. This is a broad definition that encompasses not only computer-mediated collaborative work, but also collaborative work supported by other types of technologies that do not fit most people's definition of a "computer" [41]. Collaboration electronically facilitate stakeholders in accessing BUMDes like BUMDes party provide a report to the Regional Government of Batu City through its Community Empowerment, Women and Family Planning , the regional Government can provide BUMDes data in Batu City, and the Central Government through the Ministry of Villages, Disadvantaged Areas and Transmigration monitoring the implementation of Smart BUMDes. Each stakeholder can monitor and control any development data from BUMDes every day without meeting in person. In addition, Smart BUMDes can serve as a training facility and provide information from the government to BUMDes. So, through this application, Information and Communication Technology (ICT) can be used to support user groups [43].

From e-commerce innovation and e-collaboration, there is interaction between stakeholders in it. Various innovations can be applied to various forms of business interaction. Interactions in the business can be in the form of a B2B (Business to Business) model where business transactions occur between business and other business people. B2C (Business to Consumer) business carried out by producers to consumers directly, C2C (Consumer to Consumer) business interactions carried out by individuals (consumers) to other individuals (consumers), C2B (Consumer to Business) is a business model where consumers (individuals) create and shape value for business processes, B2G (Business to Government) this process occurs between business people and government agencies, G2C (Government to Consumer) is an interaction between the government and the public as consumers [44]. In addition, there are G2G (Government to Government) or government with Government, the interactions are tailored to the needs [45].

3.2. Empathy Map

Anyone who studies the business model should be able to parse the profile of the intended customer segment. A good way to start is to use the empathy map, a thought tool that helps us walk beyond customer demographic characteristics and develop a better understanding of the environment, behaviour, concerns and aspirations. With this tool we can find a stronger business model because customer profiles guide the design of better value propositions, a more convenient way to reach customers, and better customer relationships. Map empathy is a visual aid developed by visual thinking company called XPLANE. This one-page visual tool consists of six boxes of questions that enable the company to better understand what customers really want [46]. Anyone who studies the business model needs to be able to parse the profile of the intended customer segment. Empathy Map could help as a thought tool that helps us walk beyond customer demographic characteristics and develop a better understanding of the environment, behaviour, concerns and aspirations. With this application we can find a stronger business model because of customer profiles guide the design of better value propositions, a more convenient way to reach customers, and better customer relationships. Map empathy is a visual aid developed by visual thinking company called XPLANE. This one-page visual tool consists of six boxes of questions that enable the company to better understand what customers really want [46].

From the results of field study using Empathy Map in table 1, it can be concluded that some of the needs of BUMDes Management are as follows:

1. Needed innovation related to e-collaboration development in the form of facilities connecting Government with BUMDes in every village in Indonesia, so that government can monitor the productivity. It is important to do collaboration within the scope of government [47,48,49,50], because BUMDes is a business sector program within the government [51,52]. Based on research conducted [53], [54], [55], it is revealed that the importance of

collaboration between the central government, local government, village government and BUMDes. To maximize this e- collaboration, it is needed to make it easier between stakeholders in one application to anticipating the possibility if the collaboration is more likely competition in the future, because it is seen as less likely, where financial constraints tend to increase rather than reduce competitive pressure [56].

2. Currently, Generations X and Y have access to buy and to sell through the marketplace (e-commerce) on smartphones because it is more reliable, effective and efficient in the use of time, effort, and price. From previous research, the majority of generation Z teenager chose the online marketplace site as their shopping choice. The online marketplace presents information on service quality, benefits, offered, and a positive reputation on the site [57]. Even consumers can easily get product quality and price feasibility according to taste and costs so as to encourage repurchase [58,59].
3. The need for business consulting facilities by experts on product training such as Packaging of product, business finance, strategic market etc. This facility is indispensable for units under BUMDes such as Micro-Small and Medium Units. In previous research [60, 61], BUMDes has certain limitations and weaknesses in its management process so that monitoring and assistance is needed for improvement. These can be obtained from academics, government, and business [62, 63], and can be pursued by the government as a business management facilitator in BUMDes.

3.3. Operation of Smart BUMDes

BUMDes located in every sub-region or village in Indonesia. The purpose of BUMDes is to facilitate the society to improve or build a business through BUMDes unit, supervise performance of BUMDes units, and provide a consultation center and training about BUMDes. The community must register to BUMDes in order to join BUMDes unit, then BUMDes will verificate of its eligibility. If it is accepted then BUMDes unit can be operated. BUMDes can monitor BUMDes units progress through result of transaction history in every single BUMDes units. Requirements Analysis and System Design from BUMDes operational data can be arranged functional system requirement as follow:

1. The system provides view function of BUMDes units infomation for visitor
2. The system provides educative BUMDes videos function for visitor
3. The system provides register costumer account function for visitor
4. The system provides register BUMDes unit account function for visitor
5. The system provides view function of transaction status for costumer and BUMDes unit
6. The system provides update function of transaction status for costumer and BUMDes unit
7. The system provides update function of service or product information for BUMDes unit
8. The system provides update function to confirm registered visitor for BUMDes
9. The system provides buy function of product or service in BUMDes unit for costumer
10. The system provides update information function of BUMDes unit for BUMDes
11. The system provides delete function of BUMDes unit for BUMDes
12. The system provides add function of BUMDes unit for BUMDes
13. The system provides add function of educative BUMDes video collection for BUMDes
14. The system provides delete function of educative BUMDes video collection for BUMDes
15. The system provides detail profit result function of BUMDes unit for BUMDes
16. The system provides update function of profile for costumer
17. The system provides list of view BUMDes unit BUMDes based on categories (product or service, city, sub-region or village)
18. The system provides detail order function for costumer
19. The system provides view function of transaction history for BUMDes, BUMDes unit and costumer
20. The system provides login function for visitor
21. The system provides logout function for costumer, BUMDes, dan BUMDes unit

Table 1. Empathy map

Questions	The needs of BUMDes
What did he see? See Describe what customers see in the environment. - What does it look like? - Who surrounds it? - Who are his friends? - What's the problem?	1. Competition of products from domestic and abroad in large scale. 2. Products from small and medium micro units of BUMDes are still constrained on the consistency of quality and quantity. 3. The community is less actively participating 4. The problems faced by BUMDes are generally the same in the management of human resources and marketing and the lack of government and media participation.
What did he hear? Hear Describe how the environment affects customers - What did his friends say? The couple? - Which media channels have an effect? - Who influenced him and how?	1. Products produced less able to compete with medium and large industrial products. 2. Lack of local and national media participation in "Branding" products produced by BUMDes units. 3. Platforms such as Tokobagus.com, OLX.com, etc. as a product marketplace cannot be entered by the micro small unit product of BUMDes 4. Influential media in Indonesia is a medium that is managed privately and is currently being intensively-incessant marketing through social media such as Instagram, Line, Youtube etc. 5. Generations X and Y already use smartphones and supported internet channels to the village by the government. 6. Marketplace has been widely used by the public society.
What does it say and do? Say and Do - What's his biggest frustration? - What are the risks he fears?	1. Reporting the results of BUMDes to the village authorities completed reporting on accountability for a month. 2. Central Government (Ministry of Village) get data BUMDes results for a year due to the length of process of bottom-up procedural. 3. Limited human resources professionals in managing units under BUMDes 4. Inadequate planning in taking into account organizational strategy. 5. Productivity every year is reduced. 6. People are less interested in Tourism Services provided.
What kind of hurt does the customer feel? Pain - What does he really want to achieve? - How does he measure success?	1. Lack of government expenses because the government does not have media / facilities to accommodate or collect BUMDes so that the Government does not have data related to the development of BUMDes. 2. Potential Villages like in Pesanggrahan as an example; milk, has been taken over by large companies that result in the inclusion of a few breeders. This potential will be taken over and managed by BUMDes to increase revenue 3. Products and services can be recognized locally or nationally.
What are customer acquisitions? Gain	1. A percentage of increased activities such as milk management activities 2. External investors who want to invest in micro small unit from BUMDes 3. The large number of communities involved in the development and management of small micro units in BUMDes. 4. Increased productivity in the management and marketing of Products and services produced by Units of BUMDes.

Fig 2. Use case diagram of Smart BUMDes System workflow

Here is the steps of management information system on Smart BUMDes in general (Fig. 2):

1. Requirements:
   a. Customer, BUMDes Unit Administrator, BUMDes Administrator and visitor must have a mobile device or computer and have good internet connection
   b. Customer, BUMDes Unit Administrator, BUMDes Administrator and visitor must have a good internet connection.
2. Case 1:
   a. Visitor chooses a city and/or sub-region of BUMDes Unit place
   b. Visitor picks up a product or service
   c. Visitor becomes a costumer by register as costumer and login to system
   d. Costumer fills details of order and payment option and click submit
   e. BUMDes Unit gets notification about new order from costumer
   f. BUMDes Unit confirm an acceptance of costumer order
   g. Costumer gets notification and pay of billing order and update payment status in system
   h. BUMDes Unit gets notification and check the costumer payment, if it has paid then BUMDes Unit update status of order as in progress
   i. BUMDes Unit sends costumer order (in case it is product) or give an invoice (in case it is service)

   j. Costumer gets a product or an invoice
   k. BUMDes Unit changes order status as complete and it becomes a transaction history
3. Case 2:
   a. Costumer come directly to BUMDes Unit
   b. BUMDes Unit inputs costumer information and order data into database on the computer

3.4. Advantages and long-term benefits

Table 2 shows the advantages and long term benefits of Smart BUMDes and figure 3 shows the basic view of Smart BUMDes homepage. In the figure 4, we notice several things:

(a). Initial logo display of Smart BUMDes
(b). Display selects login or list menu
(c). Display of Log in for digital access
(d). Display of Sign up with email or phone number (complete your registration)
(e). The first home page of Smart BUMDes
(f). Product selection list view
(g). Selection of product and product quantity
(h). The total price of the product purchased
(i). The total price of the product purchased + Shipping Charges

Table 2. Advantages and long-term benefits of Smart BUMDes

Objects	Advantages of Smart BUMDes	Long Term Benefits
BUMDes	Can review progress of each BUMDes unit based on transaction history through application of management system	Each BUMDes unit can be controlled, so it can continue to evaluate
	Can provide information related to product and service updates	Time efficiency related to information delivery and administration
BUMDes Units	Can register with paperless document process	Time efficiency related to information delivery and administration
	Can receive a customer order information faster and provide customer order information progress directly through the system	Time efficiency related to information delivery and flexibility over time
Customer	Choose a product or service and can order directly through the system	Save cost, time and effort in ordering product or service
	Can receive reporting information of order through the website system	Time efficiency related to information delivery and flexibility over time
Visitor	Can know any information accessed through the website system	Time efficiency related to information delivery
Government	Be able to know the BUMDes development (good / bad) at any time by only through the application and provide immediate action when BUMDes is seen that is not growing properly.	Easier for the Government to monitor and control the BUMDes situation at any time.
		Efektive and efficient in the use of budget, time and energy.
	Can provide training online.	More knowledge to develop BUMDes.

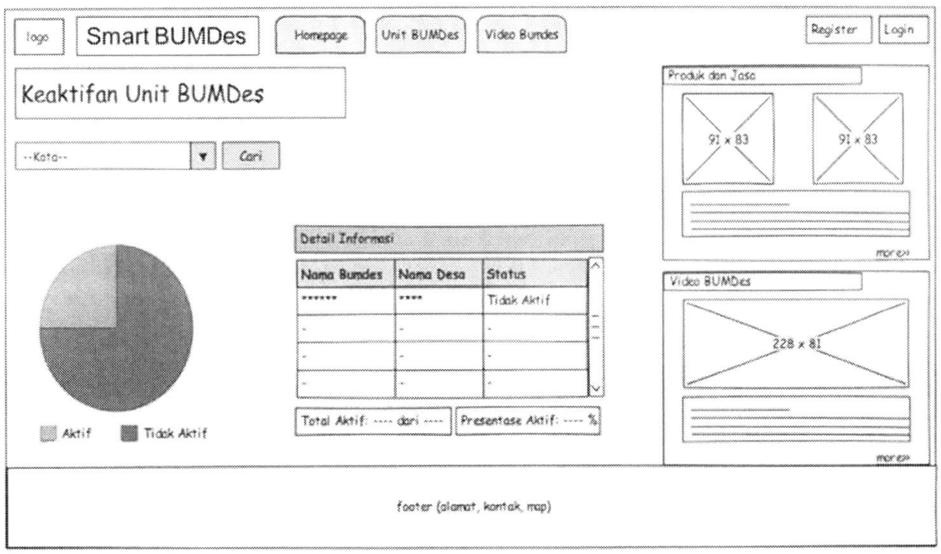

Fig 3. Basic View of Smart BUMDes Homepage

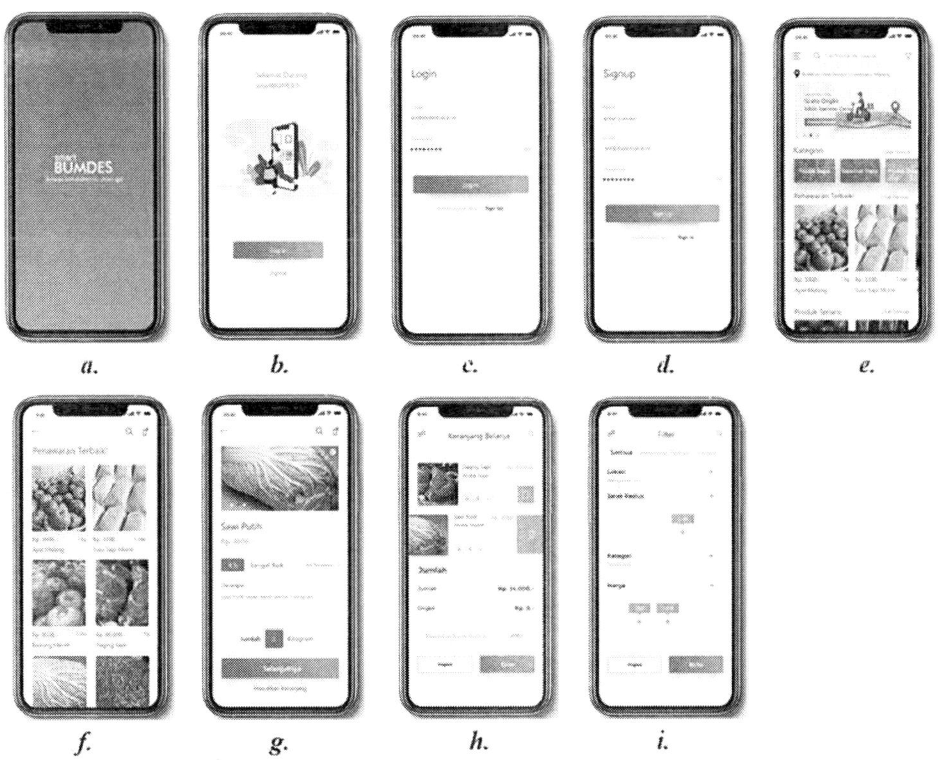

Fig 4. Display of Smart BUMDes in Smartphone

3.5. Village Economic Development in New Normal Era

The main problems of rural communities are poverty and underdevelopment [64,65,66,67,68]. On this basis, village economic development is needed which is hoped to be able to overcome various problems that occur in rural areas. The objective of the implementation of village economic development is to create an area where people can enjoy a creative, healthy life, and have a high life expectancy[69]. In order to achieve this, there will be several principles for village economic development that must be implemented [69], namely: 1) Transparanency. Opennes is needed in Smart BUMDes. The Government and BUMDes must have information disclosure so that both parties can know each other wether or not the business processes that are taking in BUMDes are running; 2) Partisipation. The active participation of BUMDes (include BUMDes unit) and the people community (visitor dan customer) to run the Smart BUMDes program, so that the desired business goals can be achieved inside and outside the village [70] so that e-commerce can really be useful in maximazing sales of local products [71] ; 3) Accountkuntabel. The planning, implementation, and evaluation processes that are carried out must be minimized so that business processes can run optimally; 4) Can be enjoyed by all people. Smart BUMDes facillitates the sale of BUMDes Unit products in the village that can be accessed by every community so as to create a broad trade market.; and 5) Sustainable.

The contribution of the BUMDes exixtence in strengthening the village economy includes sources of funds for increasing village income, meeting community needs, and developing village independently [72, 73]. Smart BUMDes is designed to run continuously and sustainably with the inclusion of Informastion Communication and Tehnology (ICT) principles in the form of applications which is a way for BUMDes to adapt well to the times so that, the improvement of the welfare of rural communities can be permanent. In addition, the application of information technology can also make it easier for BUMDes user to manage their business [74]. One of the ways to develop rural economy in the *new normal* era can be done through Smart BUMDes wich allows everyone to make transactions without face to face. Through e-collaboration and e-commerce, the Smart BUMDes application can support the strengthening of village economy during and after the *new normal* era.

4. Conclusion

Smart BUMDes (Village-owned enterprises) is one of the opportunities to village economic development solution in New Normal Era, based on Permendes no 4 of 2015 on BUMDes and Law No. 6 of 2014 on villages. Broadly speaking, BUMDes produces tourism goods and services. Starting in 2017, Village Owned Enterprises (BUM Desa / BUMDes) become one of the priority programs of the Ministry of Village, Disadvantaged Area Development and Transmigration (Kemendes). Through BUMDes, villagers are encouraged to manage the economy autonomously. However, BUMDes products tend to lack the consistency of quality and quantity so that the products or services are less desirable in the market than there are still many villages that do not have the qualified human resources in planning, managing and marketing the products in managing BUMDes. Smart BUMDes which is a marketplace (e-commerce) platform in BUMDes products and e-collaboration services is also aimed at facilitating product design and consultation training to determine BUMDes business by expert teams and other potential village promotion. In other words, Smart BUMDes can create BUMDes Holding to integrate data to the Ministry of Village so that the government can control the development of BUMDes. In developing this system, this idea requires advanced refinement so that the results are as expected.

Acknowledgments
Thanks to Universitas Brawijaya and Government of Batu City for facilitating this research.

References

[1] Liao F H and Wei Y D 2016 Sixty years of regional inequality in China: trends, scales and mechanisms working *Paper Series N 202* (Santiago, Chile: Rimisp)
[2] Stiglitz J 2009 The global crisis, social protection and jobs *Int. Labor Review* **148** pp 1–2

[3] UNESCO 1980 *Regional Disparities in Educational Development: A Controversial Issue* (Paris: United Nation Press)
[4] Kutscherauer A 2010 *Regional Disparities in Regional Development of The Czech Republic: Their Occurrance, Identification and Elimination* (Ostrava: VŠB-Technical University of Ostrava)
[5] Wang Z 2015 The imbalance in regional economic development in China and its reasons *Private Sector Development and Urbanization in China* (New York: Palgrave Macmillan) DOI: 10.1007/978-1-137-47327-1_4
[6] Avtar R, Tripathi S, Aggarwal A K and Kumar P 2019 Population – urbanization – energy *Nexus: A Review J. of Resources* **8(3)** 136
[7] Zhou Y, Kong Y, Wang H and Luo F 2020 The impact of population urbanization lag on eco-efficiency: A panel quantile approach *J. of Cleaner Production* vol **244** DOI: 10.1016/j.jclepro.2019.118664
[8] Ha N M, Le N D, and Trung-Kien P 2019 The Impact of urbanization on income inequality: a study in Vietnam *J. of Risk and Financial Management* vol **12** 146 DOI: Doi: 10.3390 / jrfm12030146
[9] Daryanto A 2004 Competitiveness advantages and superior commodity identification techniques in developing regional economic potential *Agrimedia* vol **9** no 2
[10] Wilonoyudho S, Rijanta R, Keban Y T and Setiawan B 2017 Urbanization and regional imbalances in Indonesia *Indonesian J. of Geography* vol **49** no 2
[11] Imai K S and Malaeb B 2018 Asia's rural-urban disparity in the context of growing inequality *27 IFAD Research series*
[12] Cao T C V and AkitaT 2008 Urban and rural dimensions of income inequality in Vietnam *GSIR Working Papers: Economic Development and Policy Series*
[13] Ran M, Chen L and Li W 2020 Financial deepening, spatial spillover, and urban – rural income disparity: evidence from China *Sustainability J.* vol **12** 1450 DOI: 10.3390 / su12041450
[14] Anríquez G and Stamoulis K 2007 Rural development and poverty reduction: is agriculture still the key ? electronic *J. of Agricultural and Development Economics* vol **4** no 1 pp 5–46
[15] Siswanto V K, Routray J K and Dewi R 2014 Rural urban disparity in and around Surabaya Region, Indonesia *Sci. and Technol., The J. for Technol. and Sci.* vol **25** no 2
[16] Laksono A D, Wulandari R D and Soedirham O 2019 Urban and rural disparities in hospital utilization among Indonesian adults *Iran J. Public Health* vol **48** no 2 pp 247–25
[17] Akita T and Pirmansah A 2011 Urban Inequality in Indonesia *IUJ Research Institute: Economics and Management Series*
[18] Syawie M 2011 Poverty and Social Inequality *J. of Information* vol **16** no 03
[19] ODI 2020 *From Pandemics to Poverty: Hotspots of Vulnerability in Times of Crisis* (London: ODI Press)
[20] FAO 2020 *COVID-19 and Rural Poverty: Supporting and Protecting the Rural Poor in Times of Pandemic* (Rome: FAO) DOI: 10.4060/ca8824en
[21] Han J H, Meyer B D and Sullivan J X 2020 Income and poverty in the COVID-19 pandemic *BPEA Conf. Drafts*
[22] Epidemiology Working Group 2020 The epidemiological characteristics of an outbreak of 2019 novel coroavirus diseases (COVID-19) in China *Chin. J. Epidemiol.* vol **2** pp 145–51
[23] Huang W H *et al* 2020 2019 novel coronavirus disease (COVID-19) in Taiwan: reports of two cases from Wuhan, China *J. of Microbiology, Immunology and Infection*
[24] Wu Y C, Chen C S and Chan Y J 2020 The Outbreak of COVID-19: An Overview *J. of the Chinese Medical Association* vol **83** issue 3 pp 217–20
[25] Adnan S M, Khan S, Kazmi A, Bashir N and Siddique R 2020 COVID-19 infection: origin, transmission, and characteristics of human coronaviruses *J. of Advanced Research*
[26] Rothan H A and Byrareddy S N 2020 The epidemiology and pathogenesis of coronavirus disease (COVID-19) outbreak *J. of Autoimmune* **109** Doi: 10.1016/j.jaut.2020.102433

[27] Fallah M P *et al* 2015 Quantifying poverty as a driver of ebola transmission *PLoS Neglected Tropical Diseases* **9(12)** e0004260
[28] Bucchianeri G W 2010 Is SARS a poor man's disease? Socioeconomic status and risk factors for SARS transmission *Forum for Health Economics & Policy* **13(2)** DOI: http://dx.doi.org/10.2202/1558- 9544.1209
[29] Ministry of Villages, Disadvantaged Regions, and Transmigration 2020 Revitalization of BUMDes *Ministry of Health Provides Registration Number* Retrieved on 28 March2020 from https://kemendesa.go.id/berita/view/detil/3339/revitalization-bumdes-kemendes
[30] Muhyiddin 2020 Covid-19, new normal and development planning in Indonesia *The Indonesian J. of Development Planning* vol **IV** no 2
[31] WHO 2020 *Clinical Management of Severe Acute Respiratory Infection When Novel Coronavirus (2019-nCoV) Infection is Suspected: Interim Guidance 2020* Retrieved on 28 March 2020 from https://apps.who.int/iris/bitstream/-handle/10665/330854/WHO-nCoV-Clinical-2020
[32] Azzahra N F 2020 Assessing barriers to distance learning in Indonesia during the Covid-19 pandemic *Policy Brief no 2* (Jakarta: CIPS)
[33] Herliandry L D, Nurhasanah, Suban M E and Kuswanto H 2020 Lessons during the Covid-19 pandemic *J. of Educational Technol.* vol **22** no 1
[34] Pakpahan R and Fitriani Y 2020 Analysis of the use of information technology in distance learning amid the corona COVID-19 virus pandemic *J. of Information Systems, Applied, Management, Accounting and Research* vol **4** no 2 pp 30–6
[35] Sampurno M B, Kusumandyoko T C and Islam M A 2020 Social media culture, public education and the COVID-19 pandemic *J. of Syar-i Social & Culture* (Jakarta: FSH UIN Syarif Hidayatullah) vol **7** no 6 pp 529–42 DOI: 10.15408/sjsbs.v7i5.15210
[36] FES Indonesia 2020 *Village Resilience during the Covid-19 Pandemic: Overcoming Problems and Seeking Opportunities* Retrieved on from March 2020 from https://www.fes-indonesia.org/id/news-events-publications-detail/ketahanan-desa-dalam-masa-pandemi-covid-19- menglasi-masalah-dan-mencari-peluang/
[37] Sommerville I 2011 *Rekayasa Perangkat Lunak* (Jakarta: Erlangga)
[38] Kompas.com 2017 *Generasi Milenial Didorong Kembangkan Pariwisata Nglanggeran* Retrieved on 11 May 2020 from https://travel.kompas.com/read/2017/11/16/091200427/generasi-milenial-didorong-kembangkan
[39] Laudon K C and Laudon J P 1998 *Management Information Systems: New Approaches to Organization & Technology* 5th edition (New Jersey: Prentice Hall)
[40] Taufik and Ayuningtyas E A 2020 The impact of Covid-19 pandemic on business and online platform existence *Jurnal Pengembangan Wiraswasta* vol **22** no 01
[41] Kock N 2005 *What is e-collaboration?* (USA: Idea Publishing Group)
[42] Schauer B, Zeiller M and Riedl D 2010 Reviewing The E-Collaboration Marketplace: A Survey of Electronic Collaboration Systems *Proc. of the Int. Conf. on e-Business* pp 69–75 DOI: 10.5220/0002991100690075
[43] Bouras C, Giannaka E and Tsiatsos T 2008 *e-Collaboration Concepts, Systems and Applications* DOI: 10.4018 / 978-1-59140-993-9.ch024
[44] Sandhusen E L 2008 *Marketing* ISBN:9780764139321 (New York: Barron's Educational Series)
[45] Klischewski R and Askar E 2010 Success factors of developing G2G services: The case of Egypt. ICEGOV 2010 *Proc. of the 4th Int. Conf. on Theory and Practice of Electronic Governance Beijing, China* October 25 - 28, 2010
[46] Osterwalder A dan Yves P 2012 *Business Model* (Jakarta : Elex Media Komputindo)
[47] Wanna J 2008 *Collaborative Government: Meanings, Dimensions, Drives and Outcomes* (Australia: ANU Press)
[48] Zadek S 2006 The Logic of Collaborative Governance: Corporate Responsibility, Accountability, and the Social Contract *Working Paper No 14 Harvard University*

(Massachusetts: The Joan Shorenstein Center on the Press)
[49] Miles E and Trott W 2011 *Collaborative Working* (London: Institute for Government Press)
[50] Sørensen E and Torfing J 2012 Introduction Collaborative Innovation in the Public Sector *The Innovation J.: The Public Sector Innovation J.* vol **17** no 1 pp 1–14
[51] Prasetyo R A 2016 Peranan Bumdes dalam pembangunan dan pemberdayaan masyarakat di Desa Pejambon Kecamatan Sumberrejo Kabupaten Bojonegoro *Jurnal Dialektika* vol **11** no 1
[52] Chintary V Q and Lestari A W 2016 The Role of village government in managing village-owned enterprises *Jurnal Ilmu Sosial dan Ilmu Politik* vol **5** no 2
[53] Efendi A 2019 *The Role of Village Governments in Community Empowerment through the Village-Owned Enterprise Program (BUMDes) in Bandar Sono Village Nibung Hangus, Batu Bara Regency* Thesis (Medan: Fakultas Dakwah dan Komunikasi, Universitas Islam Negeri of North Sumatera)
[54] Suman A P, Ronny E N, Amalia S K, Hardanto H, Rusmalina K, Candra A and Amir F 2019 *Local Economic: Empowerment and Collaboration* (Malang: UB Press)
[55] Suryati A and Nazaki 2019 *Collaborative Governance in Management of Bumdes in South Toapaya Village, Toapaya, Bintan Regency in 2018* Thesis (Tanjung Pinang: Universitas Maritim Raja Ali Haji) Retrieved from http://repository.umrah.ac.id/3715/
[56] Lacomba J A, Lagos F and Neugebauer T 2011 *Who makes the pie bigger? An experimental study on co-opetition New Zealand Economic Papers* vol **45** no 1–2 pp 59–68
[57] Prayekti A R 2018 *Perceptions of Generation Z Youth in Surabaya on Brand Preference on Bukalapak's Online Marketplace Site* Thesis (Surabaya: Universitas Airlangga)
[58] Palma M A and Andjarwati A L 2016 The Influence of Product Quality, Convenience, and Price on Repurchase Intention with Satisfaction as an Intervening Variable (Study on Customers of Fashion Products through Online Stores in Surabaya) *J. of Research in Economics And Management* vol **16** no 1 pp 84–104
[59] Tanwijaya P M 2017 *Comparison of the Effect of Perceptions of the Attributes of the Online Shop Tokopedia and Bukalapak on Repurchase Intention Among Millennials* Thesis (Bandung: Universitas Katolik Parahyangan)
[60] Zulkarnaen R M 2016 Development of village economic potential through village-owned enterprises (Bumdes) Pondok Salam, Purwakarta Regency *Dharmakarya: Jurnal Aplikasi Ipteks untuk Masyarakat* vol **5** no 1 pp 1 4
[61] Junaidi 2020 Assistance in the management and compilation of financial statements at Ladongi Village-Owned enterprises (Bumdes), Malangke, North Luwu Regency *RESONA: Junral Ilmiah Pengabdian Masyarakat* vol **4** no 1 DOI: 10.35906/resona.v4i1.286
[62] Hidayat A T, Pujiati L, Hidyati N, Hendrawan S A, Suprapto S and Nurali 2018 Assistance in the preparation of financial Statements at Village-Owned enterprises (Bumdes) Lestari in Bandung Village, Diwek, Jombang *Comvice: J. of Community Service* vol **2** no 1 pp 16–20
[63] Triaryanti I M 2019 *Collaborative Governance in the Management of Bumdes in the Tourism Village of Pujon Kidul, Pujon, Malang Regency* Thesis (Surabaya: Universitas Airlangga)
[64] Suryahadi A, Isdijoso W, Nurbani R I, Yumna A, Izzati R A and Warda N 2018 *Dynamics of Inequality in Rural Indonesia* Retrieved on 11 September 2020 from https://smeru.or.id/id/content/dinamika-ketimpangan-di-perdesaan-indonesia
[65] Abidin E E and Amruddin A 2020 Inequality of development in West Sulawesi *Jurnal Arajang* **3(1)** pp 1–17 DOI: 10.31605/arajang.v3i1.582
[66] Agusta I 2014 *Regional Inequality and Prevention Policies in Indonesia: A Study of Strategic, Historical, and Paradigmatic Issues since Pre-Colonial* (Jakarta: Yayasan Obor Indonesia)
[67] Djese T S 2016 Looking at East Nusa Tenggara measuring problems, offering solutions *Masyarakat Indonesia* **4(2)** pp 277–88
[68] Fauzi M R, Rustiadi E and Mulatsih S 2019 Inequality, spatial patterns, and regional development performance in East Java Province *Jurnal Perencanaan Pembangunan Wilayah*

Dan Perdesaan-JP2WD **3(3)** pp 157–71 DOI: 10.29244/jp2wd.2019.3.3.157-171
[69] Suleman A R *et al* 2020 *BUMDES Towards Optimization of Village Economy* (Medan: Yayasam Kita Menulis)
[70] Sanjaya P K A, Hartati N P S and Premayani N W W 2020 Empowerment of village owned enterprise managers (BUMDes) berdikari through digital marketing system implementation *CARADDE: Jurnal Pengabdian Kepada Masyarakat* vol **3** no 1
[71] Setioningtyas W P, Dwiarta I M B and Waryanto R B D 2020 Utilization of multimedia as a media for promoting tourism destinations and souvenir products for tourism villages in Kebontunggul Village, Gondang, Mojokerto Regency *Ekobis Abdimas* vol **1** no 1
[72] Agunggunanto E Y, Arianti F, Kushartono E W and Darwanto D 2016 Development of independent village through management of Village-Owned Enterprises (BUMDES) *Jurnal Dinamika Ekonomi dan Bisnis* vol **13** no 1
[73] Ramadana C B 2013 The existence of Village-Owned Enterprises (Bumdes) to strengthen the village economy *Jurnal Administrasi Publik* vol **1** no 6
[74] Hasanah U 2018 *Applications for Borrowing Bumdes, Poto Village, Moyo Hilir Sub- district based on WEB* Thesis (Sumbawa: Universitas Teknologi Sumbawa)

Indonesia Infrastructure Development Topic Discovery on Online News with Latent Dirichlet Allocation

Ahmad Fathan Hidayatullah[1], Muhammad Rifqi Ma'arif[2], Muhammad Habibie[3], Siti Khomsah[4]

[1]Department of Informatics, Universitas Islam Indonesia, Yogyakarta, Indonesia
[2,3]Center of Data Analytics Studies and Services, Universitas Jenderal Achmad Yani, Yogyakarta, Indonesia
[4]Department of Data Science, Institut Teknologi Telkom, Purwokerto, Indonesia

E-mail: [1]fathan@uii.ac.id, [2]muhammad.rifqi@gmail.com, [3]muhammadhabibi17@gmail.com, [4]siti@ittelkom-pwt.ac.id

Abstract. In the period of 2014-2019, the Indonesian government has defined infrastructure development as a priority. Related to this, various media, in both online and offline, routinely report the infrastructure development news to the public. However, it is quite difficult for people to obtain a summary of information from the internet about the infrastructure development that have been carried out by the government. This study aims to provide a brief summary about infrastructure development in Indonesia by performing topic modeling approach using Latent Dirichlet Allocation. We found that the use of bigram language model could help identify phrase from the corpus. Therefore, the keywords contained in the topics are more interpretable and acceptable. Moreover, a coherence score measurement was applied to find the best number of topics from our dataset. Based on the experiments, we obtained 40 topic models from our dataset. From those 40 topics, we inferred several topic labels, such as, oil and gas infrastructure; power plant infrastructure; information technology infrastructure and internet networks; road infrastructure; reservoir infrastructure, irrigation networks and water resources; railway infrastructure; and airport infrastructure.

1. Introduction

Infrastructure development plays an important role in a country. Good infrastructure development is one of an indicator which shows the growth, prosperity and economic stability. As a developing country, Indonesia is still increasing its infrastructure development in some fields. In addition, there are many regions in Indonesia, which need a lot of infrastructure development, such as road, bridges, airports, and ports. In the period of 2014-2019, the Indonesian government has defined infrastructure development as a priority. The purpose of this program is to push the economic growth and to realize the justice and equity for all people in all regions in Indonesia. Therefore, the Indonesian government has proposed a program which is called national strategic projects (in Indonesian: *Proyek Strategis Nasional,* or PSN)[1]. The national strategic projects are managed by the Committee for Acceleration of Priority Infrastructure Delivery (in Indonesian: *Komisi Penyediaan Percepatan Infrastruktur Prioritas*,

[1] https://kppip.go.id/proyek-strategis-nasional/

or KPPIP). KPPIP works across ministries and institutions as bridges, assisting project owners in preparing and implementing project development[2].

Related to this, various media, in both online and offline, routinely report the infrastructure development news to the public. However, it is quite difficult for people to obtain a summary of information from the internet about the infrastructure development that have been carried out by the government. This is caused by the number of news articles on the internet which provide lots of information about infrastructure development. Therefore, it is necessary to provide a brief summary about infrastructure development in Indonesia. The summary will be obtained by analyzing the topics from the online news articles. The analysis aims to obtain an overview and an illustration of the progress and problems. The analysis would be performed by extracting related topics through online news media, which provide news about infrastructure projects and development in Indonesia.

Topic modeling research in domains related to governance has attracted many researchers to discover hidden topic from texts. DiMaggio, et al [1] analyzed U.S. government arts funding for newspaper by applying Latent Dirichlet Allocation (LDA). Cogburn [2] applied topic modeling to illustrate the trends and topics in the internet governance and cybersecurity debates. Deng, et al [3] proposed a multi-level topic model with LDA to analyze the concern of citizens on social media during man-made disasters. Topic modeling on social media platform was also investigated by Driss, et al [4]. They applied both LDA and Latent Semantic Analysis (LSA) to bring messages and information from citizens to policy-makers. Sha, et al [5] presented a Hawkes binomial topic model to analyze COVID-19 conversations on Twitter among U.S. governors and presidential cabinet members.

This study aims to provide a brief summary regarding the infrastructure development in Indonesia by extracting topic from online news. In this work, the topic extraction will be carried out using topic modeling approach method which is called LDA. LDA has a whole coverage of important processes in the development of topic models, so that it can be used as a reference framework for other researchers in developing specific domains on other topics [6]. In addition, LDA has shown remarkable performance in various natural language processing tasks to find hidden topics in text documents [7].

2. Latent Dirichlet Allocation (LDA) Topic Modeling

Topic modeling provides a good way to analyze large amounts of text data that have not been classified [8]. Topic modeling is a text mining approach that is reliable in finding hidden text data and finding relationships between words from a corpus [9]. Topic modeling assumes that a document is a generative result of a set of hidden topics, where a topic is a probability distribution over words [8].

LDA was first introduced by [10] as a generative probabilistic model to automatically discover the hidden topics from a particular corpus data D. The probability calculation of LDA is represented as:

$$p(D|\alpha,\beta) = \prod_{d=1}^{M} \int p(\theta_d|\alpha) \left(\prod_{n=1}^{N_d} \sum_{z_{dn}} p(z_{dn}|\theta_d)\, p(w_{dn}|z_{dn},\beta) \right) d\theta_d \quad (1)$$

In equation (1), both α and β represent the hyperparameter to obtain the likelihood of modeling. The α represents the Dirichlet prior parameter for the topic distribution in the document level and β describes the word probability distribution for the particular topic. The topic distribution of document d is represented in a form of vector θ. The z notation refers to the hidden topics from document d. The notation M and N represent the length of documents and the number of terms in the document respectively. Figure 1 illustrates the LDA model representation [10].

[2] https://kppip.go.id/tentang-kppip/

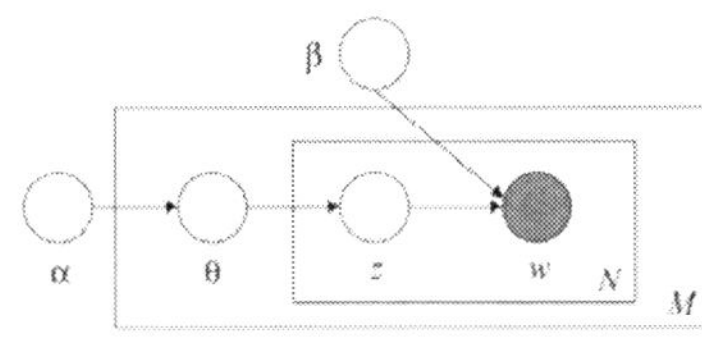

Figure 1. LDA Model

3. Methodology

3.1. Data Collection and Preprocessing

We retrieved the dataset from some articles of online news media in Indonesia, which published during the period of 2014-2019. Before downloading the articles, we first determined some keywords which related to infrastructure development and national strategic projects, such as development, infrastructure, road, bridge, oil and gas. Furthermore, those keywords will be used to find the relevant articles through the internet. To obtain the content of the articles, we copied the URL of the articles, as the input parameter to the scraper. Table 1 shows the sample of the article URLs from several online news media. We utilized a Python scraper library called newspaper[3] to extract articles from online news media. The newspaper library works for more than ten languages including Bahasa Indonesia. Therefore, there is no special needed when we find the articles in Bahasa Indonesia. We collected 301 articles in Bahasa Indonesia with the total of 5,243 sentences, 92,880 words and 9,566 unique words.

Table 1. Sample of the article URLs

No	URL
1	http://www.beritasatu.com/satu/506680-oktober-tol-pemalangbatang-jalani-uji-kelayakan.html
2	http://jambi.tribunnews.com/2018/02/20/12-proyek-infrastruktur-yang-ambruk-di-indonesia
3	http://kupang.tribunnews.com/tag/pembangunan-infrastruktur-jembatan
4	https://news.okezone.com/read/2017/06/13/340/1714620/perbaikan-jalan-trans-kalimantan-dikebut-jelang-mudik-lebaran
5	https://tirto.id/luhut-sebut-target-infrastruktur-minimal-tercapai-90-persen-di-2019-cyBg

The data preprocessing tasks in this work were slightly different from preprocessing data text in social media. In social media text, the text data are not clean and containing a lot of noisiness such as unstructured and non-standard words. On the contrary, the data of this study were more structured and did not contain non-standard words. In this preprocessing step, we performed some preprocessing tasks such as Unicode normalization, removing URLs, removing punctuations, removing numbers, case folding, removing additional white spaces and removing stop words.

3.2. Topic Modeling and Evaluation

The topic modeling process was conducted by utilizing a free Python library for topic modeling using Latent Dirichlet Allocation called Gensim[4]. The topic modeling process was started by reading the preprocessed dataset. From this preprocessed data, we created a list of words and split the document based on tokens. We also build a bigram model to help identify phrases from our dataset. The next stage is to build a dictionary, then followed by creating an LDA model with TF-IDF (Term Frequency-Indexed Document Frequency). Finally, we compute coherence values from the each LDA model to evaluate the model.

[3] https://newspaper.readthedocs.io/en/latest/
[4] https://radimrehurek.com/gensim/index.html

3.3. Topic Analysis and Visualization

We analyzed the topic model based on the best number of the topics. From this analysis, we expected to obtain comprehensive and brief illustration of the infrastructure development which obtained from online news. Topic analysis was conducted by analyzing the relationships between words in each cluster group in the resulting LDA model. In addition, it will be seen also the relevance of topics between clusters to one another. Finally, we employed PyLDAvis[5] as a visualization tools to analyze the extracted topics. PyLDAvis is built based on LDAvis [11], which is an interactive visualization tool for Latent Dirichlet Allocation. The LDAvis result is presented in html format and it provides a flexibility in term of exploring topic-term relationships utilizing relevance [11].

4. Result and Discussion

To build the topic model, we defined the number of topics as a parameter in LDA. We defined the number of topics from 5 until 50. Moreover, we also evaluated the topic coherence values in every 5 topics to obtain the best topic coherence value. Figure 2 shows the graph of the coherence score obtained from 5 to 50 topics.

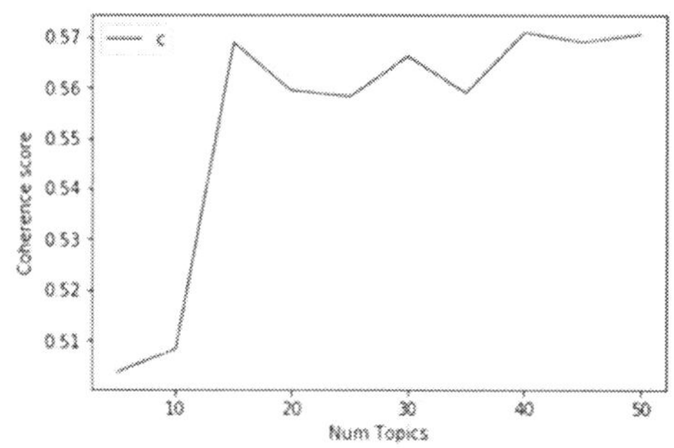

Figure 2. Coherence Score Graph

It can be seen from the graph in Figure 2 that the coherence value increased significantly starting from 5 topics to 15 topics. After reaching 15 topics, the coherence value has slightly decreased for 20 topics. However, the coherence value obtained continues to fluctuate up to 50 topics and reach the highest score on the topic 40. This result is then used as a reference for further analysis, so that in-depth discussion will focus on 40 topics. The exact coherence values are shown in Table 2.

Table 2. Coherence values based on the number of topics

Number of Topics	Coherence Value
5	0.5036
10	0.5083
15	0.5689
20	0.5594
25	0.5582
30	0.5662
35	0.5589
40	0.5709
45	0.5690
50	0.5705

[5] https://pyldavis.readthedocs.io/en/latest/

Table 3. The most interpretable topics

Topic	Keywords	Label
Topic#1	*gas, migas* (oil and gas), *gas_bumi* (natural gas), *pipa* (pipe), *pgn* (Perusahaan Gas Negara/national gas company), *bph_migas* (Badan Pengatur Hilir Minyak dan Gas/the Agency's Governing Body downstream Oil and Gas), *holding_bumn, badan_usaha* (business entity), *bumi* (earth).	Oil and gas infrastructure
Topic#3	*pembangkit_listrik* (power plants) , *kantor_kementrian* (ministry office), *pembangkit* (the ministry of power's office), *tenaga* (power), *listrik* (electricity), *produksi* (production), *berjalan_lancar* (running smoothly), *kapasitas* (capacity), *provinsi_jambi* (Jambi Province).	Power plant infrastructure
Topic#5	*internet_berkecepatan* (internet speed), *internet* (internet), *laporan_jateng* (Central Java report), *berkecepatan* (speed), *pemerintah_pusat* (central government), *tinggi* (high), *badan_pengatur* (regulatory body), *bbm* (fuel oil), *kuala_tanjung* (Kuala Tanjung village), *mnc* (internet network service provider company).	Information technology and internet networks infrastructure
Topic#6	*palapa_ring* (internet infrastructure project), *bendungan* (dam), *palapa* (internet infrastructure project), *ring* (internet infrastructure project), *jepara* (Jepara district), *pelanggan* (customer), *penyelesaian* (settlement), *rudiantara* (the name of Indonesian minister of communication and information), *efisien* (efficient), *kartini* (place name in Jepara).	Information technology and internet networks infrastructure
Topic#8	*perbaikan_jalan* (roadwork), *lanjut* (continue), *november* (November), *reguler* (regular), *mencapai_persen* (percentage achievement), *berfungsi* (functionate), *waktu* (time), *dipasang* (installed), *perbaikan* (repair), *pelemahan* (weakening).	Road infrastructure
Topic#9	*pertamina* (State Oil and Natural Gas Mining Company), *pembentukan_holding* (holding formation), *triwulan_pertama* (first quarter), *jangka_panjang* (long-term), *xl_axiata* (telecommunications operator company), *layanan_data* (data service), *miliar* (billion), *pgn_pertagas* (national gas company), *pmi* (purchasing manager index), *pipa_gas* (gas pipe).	Oil and gas infrastructure & Information technology and internet networks infrastructure
Topic#11	*pesawat* (aircraft), *pita_lebar* (broadband), *program_pita* (ribbon program), *dpr_ri* (Indonesian people's representative council), *serat_optik* (optical fiber), *program_padat* (solid program), *pita* (ribbon), *produk* (product), *optik* (optic), *broadband* (broadband).	Transportation & Information technology and internet networks infrastructure
Topic#12	*jaringan_irigasi* (irrigation network), *direktur_utama* (president director), *irigasi* (irrigation), *kementrian_pupr* (minister of public work and human settlements),*pembangunan* (development), *maret* (March), *operator* (operator), *daya_air* (water power), *jaringan* (networks), *tim* (team).	Reservoir infrastructure, irrigation networks and water resources
Topic#13	*kereta_api* (train), *percepatan_pembangunan* (accelerated development), *pemalang_batang* (Pemalang & Batang district), *api* (fire), *kereta* (train), *kota_semarang* (Semarang city), *solo* (Solo city), *jalan_tol* (toll road), *dpr_ri* (Indonesian people's representative council), *anggota_komisi* (commission members).	Railway infrastructure & Road infrastructure

Among those 40 topic groups, we tried to discover some topics that are most related to the national strategic projects undertaken by the Indonesian government. The topics were inferred by observing the correlation between the most 10 frequent words that appear in each topic. In this work, human judgement [12] was performed to help examine the keywords from the topic results and then give the topic label manually[13].

Table 3 and Table 4 show the most interpretable topics of this study. From the result, we found that the application of the bigram model has provided better results than without the bigram model. The use of the bigram model was able to capture some important phrase from our dataset. To determine phrase from the dataset, we set if two terms appear 10 times or more, then we will concatenate the two words and determine them as a phrase. As the result, we can see from the keywords for the resulting topic that the appearance of phrases could provide acceptable and more sensible topic label. For example, we can see from topic#1 that the appearance of the phrase *gas_bumi* (natural gas), *bph_migas* (Badan Pengatur Hilir Minyak dan Gas/the Agency's Governing Body Downstream Oil and Gas), *holding_bumn* and *badan_usaha* (business entity) is more acceptable than if both words of each phrase are separated.

Table 4. The most interpretable topics

Topic	Keywords	Label
Topic#25	*holding* (holding company), *holding_migas* (oil and gas holding company), *train* (train), *optimal* (optimal), *lrt_jakarta* (light rail transit Jakarta), *service* (service),*service_obligation* (service obligation), *asian_games* (Asian Games), *mandeh* (place name in West Sumatera), *presiden_jokowi* (president of Indonesia).	Oil and gas infrastructure & Railway infrastructure
Topic#35	*kmj* (The Karimunjawa islands), *penerbangan_perdana* (first flight), *menuju_karimunjawa* (towards Karimunjawa), *penerbangan* (flight), *arus_mudik* (homecoming flow), *media_sosial* (social media), *karimunjawa* (Karimunjawa), *semarang* (Semarang), *pertanian* (agriculture), *agus* (person's name).	Airport infrastructure
Topic#39	*pembangunan_waduk* (reservoir construction), *waduk* (reservoir), *mudik_lebaran* (homecoming), *selesai* (finished), *capaian* (achievements), *kepala_dinas* (head of department), *gus_irawan* (person's name/member of Indonesian people's representative council), *meningkatkan_kapasitas* (increase capacity), *tahun* (year), *keuntungan* (profit).	Reservoir infrastructure, irrigation networks and water resources

Based on Table 3 and Table 4, we concluded that there are two types of topic cluster. The first type of topic cluster is the topic which has one label only. This type of topic cluster only has one label because the majority of the most frequent words illustrate on a particular topic. On the other hand, the second one is the type of topic cluster, which has two topic labels. This type of topic clusters contains words that come from different topics. Among those topic clusters, based on the most 10 composed words, the topic number 9, 13, and 25 can be interpreted into two topics.

From the 40 topics obtained, we found 12 the most representative topic clusters that illustrate a brief summary about the national strategic project proclaimed by the Indonesian government in the period 2014-2019. Among those 12 topics, we inferred several topic labels, such as:

- Oil and gas infrastructure (Topic 1, 9 and 25).
- Power plant infrastructure (Topic 3).
- Information technology infrastructure and internet networks (Topic 5, 6, 9, and 11).
- Road infrastructure (Topic 13 and 8).
- Reservoir infrastructure, irrigation networks and water resources (Topic 12 and 39).
- Railway infrastructure (Topic 13 and 25).
- Airport infrastructure (Topic 35).

The topic model results produced were then visualized using the PyLDAvis library. PyLDAvis has two panels in its visualization. As shown in Figure 3, the left panel illustrates the inter-topic distance map via multidimensional scaling. We used the Jensen-Shannon divergence as the default computation metrics for the inter-topic differences [11]. The left panel depicts the overall view of the topic model result. In addition, we are also able to see how common each topic and how topic link with each other. The topic clusters generated in the left panel are indicated by circles with a specific number in each group/topic cluster.

The right panel is described by using bar charts. The bars show the most occurring terms of the topics. Here, we can see the 30 terms that are most relevant for a particular topic. From the right panel of our topic modeling result, we know from the bar chart that the terminology *'pembangunan infrasruktur'* (infrastructure development) and *'palapa ring'* are the most frequent appear in the corpora. This result indicates that the most discussed topics in our data are about infrastructure development and *'palapa ring'*. The term 'infrastructure development' is clearly explained that the Indonesian government places a high priority on infrastructure development during the period of 2014-2019. The second term, *'palapa ring'*, is a telecommunications infrastructure project in the form of fiber optic development throughout Indonesia. The appearance of the term *'palapa ring'* as the second occurring terms indicates about the development of infrastructure in the telecommunication domain.

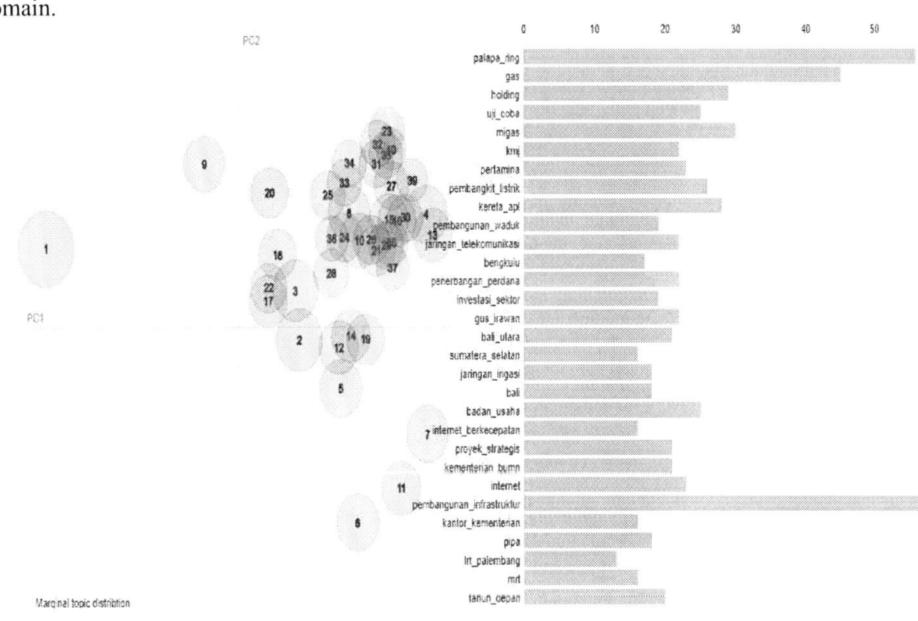

Figure 3. Topic Modeling Visualization using PyLDAvis

5. Conclusion

This research has succeeded in conducting topic modeling of the online news article data which contains information about infrastructure development in Indonesia. The application of topic modeling was performed utilizing the Latent Dirichlet Allocation method to discover topics related to infrastructure development. To build the topic model, this research has combined the LDA method with TF-IDF (Term Frequency-Indexed Document Frequency) and bigram language model. We found that the use of bigram language model could help identify phrase from the corpus. Therefore, the keywords contained in the topics are more interpretable and acceptable. From those 40 topics obtained, we inferred several topic labels, such as, oil and gas infrastructure; power plant infrastructure; information technology infrastructure and internet networks; road infrastructure;

reservoir infrastructure, irrigation networks and water resources; railway infrastructure; and airport infrastructure.

References

[1] DiMaggio P, Nag M and Blei D 2013 Exploiting affinities between topic modeling and the sociological perspective on culture: Application to newspaper coverage of U.S. government arts funding *Poetics* **41** pp 570–606
[2] Cogburn D L 2019 Analyzing trends and topics in internet governance and cybersecurity debates found in twelve years of IGF transcripts *Proc. of the 52nd Hawaii Int. Conf. on System Sciences* pp 902–11
[3] Deng Q, Gao Y, Wang C and Zhang H 2020 Detecting information requirements for crisis communication from social media data: An interactive topic modeling approach *Int. J. of Disaster Risk Reduction* **50** 101692
[4] Belkahla Driss O, Mellouli S and Trabelsi Z 2019 From citizens to government policy-makers: Social media data analysis *Government Information Quarterly* **36** pp 560–70
[5] Sha H and Hasan M A 2020 Dynamic topic modeling of the COVID-19 Twitter narrative among U.S. governors and cabinet executives *preprint arXiv:2004.11692*
[6] Verma A and Gahier A K 2015 Topic modeling of E-news in Punjabi *Indian J. of Sci. and Technol.* **8** pp 1–10
[7] Wang B, Liakata M, Zubiaga A and Procter R 2017 A hierarchical topic modelling approach for tweet clustering *Int. Conf. on Social Informatics* (New York: Springer) pp 378–90
[8] Alghamdi R and Alfalqi K 2015 A Survey of topic modeling in text mining *Int. J. of Advanced Computer Sci. and Application* **6** pp 147–53
[9] Jelodar H, Wang Y, Yuan C, Feng X, Jiang X, Li Y and Zhao L 2019 Latent dirichlet allocation (LDA) and topic modeling: models, applications, a survey *Multimedia Tools and Applications*
[10] Blei D M, Ng A Y and Jordan M I 2003 Latent dirichlet allocation *J. of Machine Learning Research* **3** pp 993–1022
[11] Sievert C and Shirley K 2014 LDAvis: A method for visualizing and interpreting topics *Proc. of the Workshop on Interactive Language Learning, Visualization, and Interfaces* (Baltimore, Maryland, USA: Association for Computational Linguistics) pp 63–70
[12] Chang J, Gerrish S, Wang C, Boyd-graber J L and Blei D M 2009 Reading tea leaves: how humans interpret topic models *Advances in Neural Information Processing Systems* pp 288–96
[13] Hu N, Zhang T, Gao B and Bose I 2019 What do hotel customers complain about? Text analysis using structural topic model *Tourism Management* **72** pp 417–26

Stacking Neural Network Models for Automatic Short Answer Scoring

Rian Adam Rajagede[1] **and Rochana Prih Hastuti**[2]

[1] Department of Informatics, Universitas Islam Indonesia, Jalan Kaliurang km. 14,5, Sleman, Yogyakarta, Indonesia, 55584

[2] Department of Computer Science and Electronics, Universitas Gadjah Mada, Bulaksumur, Yogyakarta, Indonesia, 55281

E-mail: `rian.adam@uii.ac.id`

Abstract. Automatic short answer scoring is one of the text classification problems to assess students' answers during exams automatically. Several challenges can arise in making an automatic short answer scoring system, one of which is the quantity and quality of the data. The data labeling process is not easy because it requires a human annotator who is an expert in their field. Further, the data imbalance process is also a challenge because the number of labels for correct answers is always much less than the wrong answers. In this paper, we propose the use of a stacking model based on neural network and XGBoost for classification process with sentence embedding feature. We also propose to use data upsampling method to handle imbalance classes and hyperparameters optimization algorithm to find a robust model automatically. We use Ukara 1.0 Challenge dataset and our best model obtained an F1-score of 0.821 exceeding the previous work at the same dataset.

1. Introduction

The short answer test is a type of exam in which participants are asked to answer questions with short answers which can consist of 1-2 sentences. Assessing student short answers on the exam is very challenging for the assessor. When a large number of students are assessed, for example on a national scale, assessors are required to remain consistent and objective in assessing hundreds or even thousands of student responses. The questions with short answer format also allow students to answer in their own style which can be varied for each student. Therefore, computer assistance in making automatic short answer scoring is deemed necessary to address these problems.

In 2019, NLP Research Group from Universitas Gadjah Mada, Indonesia, in collaboration with the Education Assessment Center, Ministry of Education and Culture of Indonesia, held the Ukara 1.0 Challenge[1]. In this challenge, participants from all over the nation were challenged to make an automatic short answer scoring system for Indonesian student exam. There are two short answer questions on this challenge with correct and incorrect labels. In this paper, we try to describe the improvement of our previous work on the Ukara 1.0 Challenge dataset.

There are several challenges that may arise in applying and making an automatic short answer scoring system. First, it is the number of models and hyperparameters that need to be searched.

[1] https://nlp.mipa.ugm.ac.id/ukara-1-0-challenge/

In conventional machine learning, a model is trained to solve a specific problem, or in our case, a model is only trained to assess one short answer question. When the number of questions to be assessed increases, the time required for searching the model and tuning its hyperparameters will also increase. The second challenge is the imbalance class. In an exam, the questions given are intended to test students' abilities, therefore the level of difficulty in each question will cause the number of correct responses to be much less than the number of wrong responses. Another challenge in automatic short answer scoring is the small amount of labeled data. The data labeling process is not easy because it requires an expert to validate the student responses and of course this is time consuming and costly.

In several previous studies, making short answer scoring or essay scoring was done using deep learning approaches, for example, using long-short term memory, convolutional neural network, or a combination of both [1, 2, 3]. In [4], the author uses random forest to create Automated Short Answer Scoring. By using random forest, the resulting model is also able to provide a simple explanation rather than just a score. The data used in [4] is the Automated Student Assessment Prize Short Answer Scoring (ASAP-SAS). In this data, the average student answers are slightly longer than the Ukara Challenge dataset.

In this paper, we take a different approach by using a simpler stacking model because of the small number of available data. We used the sentence-level feature as done by [3]. Without word sequence features, the automatic scoring process could be viewed as a text classification problem. The use of stacking models for text classification has been done in [5, 6] and shows better performance than a single model classifier. We also propose to use an upsampling method, Synthetic Minority Over-sampling Technique (SMOTE)[7], to handle imbalance classes and hyperparameters optimization algorithm, Tree-structured Parzen Estimator (TPE)[8] to find a robust model that performs well on each type of question. In this paper, we use *hyperparameters* term as model components which are preset and untrained (e.g. number of neural network layers). Meanwhile, the term *parameters* refers to the model components that can be trained (e.g. neural network weights) [9].

2. Dataset

We use short answer scoring dataset from Ukara 1.0 Challenge Dataset published in 2019 by NLP Research Group Universitas Gadjah Mada. In this dataset, there are two different question sets, question set A and question set B. Each question set consists of a stimulus text, a question, and a collection of student responses with its label. Question A relates to natural disasters and question B is related to social problems. Examples of responses given by students along with its label are shown in table 1.

Each question set is also divided into three parts, namely the train set, dev set, and test set. This division refers to the time the competition takes place. Participants are only allowed to make models using the train set. Dev set is used to test at the validation stage where participants can only use it 20 times. While the test set is given at the final stage where participants are only allowed to test 5 times. we use the same scheme for the experiment process.

2.1. Data Pre-processing

There are several pre-processing techniques that is applied to the data before extracting the feature set. The first is case folding, which consists of converting all words to lowercase and case-folding the punctuation and unnormal words such as "*yg*" to "*yang*". This is to get better vector representation in the feature extraction process.

The second step is remove some frequent words. We do word counting for all vocab in the training data, and then remove some words based-on its frequency. We do not immediately eliminate most frequent words because it is worried that these words are important keywords in the scoring.

Table 1. Examples of student response

Question	Response	Label
A	mereka akan kehilangan lahan pertanian mereka *(they will lose their agricultural land)*	Correct
A	sulit menyesuaikan diri dengan keadan alam di tempat tersebut *(difficult to adjust to the natural conditions in that place)*	Correct
A	akan mendapat suasana baru *(will get a new atmosphere)*	Incorrect
B	karena akan bermanfaat bagi yg membutuhkan *(because it will be useful for those in need)*	Correct
B	karena dengan menyumbang para konsumen dapat membantu para pekerja *(because by donating consumers can help workers)*	Correct
B	karena mengurangi sampah lingkungan dan polusi *(because it reduces environmental waste and pollution)*	Incorrect

2.2. Feature Extraction

We propose to use two features to represent a student response. The first feature is sentence embedding extracted from Fasttext [10] using Indonesian fasttext pretrained model [11]. Fasttext pretrained model originally trained to get the subword-level embedding using CBOW architecture. In sentence embedding, the process of embedding occurs at the sentence level by calculating the average of word embedding value of each word as seen in (1). Where w_i is the i-th word embedding vector, N is the number of words in that sentence, and $\|w_i\|_2$ is L2 norm of vector w_i. We assume that each student's response is one sentence. When there are several sentences in one response, we process them as one sentence.

$$s = \frac{1}{N}\sum_{i=1}^{N}\frac{w_i}{\|w_i\|_2} \tag{1}$$

The second feature used is response's word count. We assume that the length of the response matter to the scoring value. It is assumed that students with short responses convey less detailed information. For example. This feature was chosen following previous research by [12].The number of word in each response is normalized so that the value is between 0 and 1.

Using both features, each student's response is represented as a vector with a length of 301 values. The first 300 elements are result from sentence embedding and the last element is from the sentence length feature.

2.3. Class Balancing

It is natural in short answer questions that the number of correct responses is less than the number of incorrect responses. The question A consists of 191 correct responses and 77 incorrect responses, while question B consists of 168 correct responses and 137 incorrect responses. It can be seen that the two questions have an imbalance number of classes. To balance this condition, we use Synthetic Minority Over-sampling Technique (SMOTE) [7].

SMOTE performs data upsampling based on the vector data in the feature space. As shown in figure 1, SMOTE synthesizes new data that is on the line between the example from the minority class. The blue dots are examples of data from the minority class, while the smaller

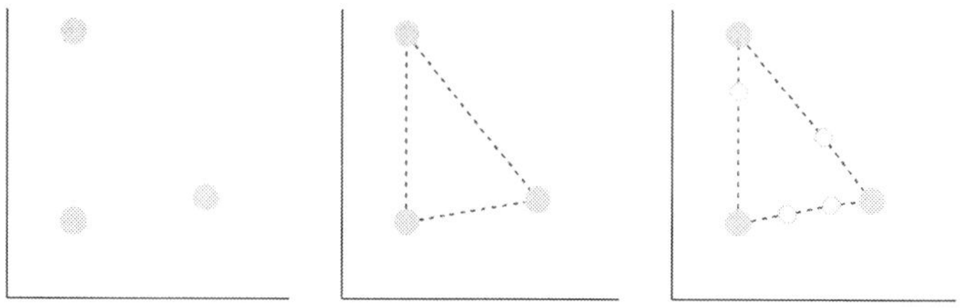

Figure 1. Illustration of SMOTE synthesizing new data.

dots are the new data. The synthetic data generated by SMOTE is data from correct responses with a total of 10% of the total data. This proportion is less significant in numbers compared to the former proportion, while still fit the real case condition.

3. Method

For the classification process, we combined two types of classifier models, namely Extreme Gradient Boosting (XGBoost) [13] and neural network into a stacking model. We chose these two methods in this study because both methods were proven to be superior in various machine learning cases. XGBoost is a tree boosting algorithm that has won several competitions in Kaggle[2] [14]. Although the stacking model can use various classifiers, we only use these two methods because the approaches of the two methods are very different. A small number of classifiers also reduce the training time and model size.

We propose to use the stacking model using both classifiers as base classifier (level-0 model) and other neural networks as the final classifier (level-1 model). We avoid using deep learning model in this experiment because the amount of labeled data is small, which is only 268 responses to question A and 305 responses to question B. Besides, it is worried that a high computation resource needed for deep learning training process will become an obstacle during the implementation in the real world.

3.1. Stacking model

Both XGBoost and neural networks have their own advantages, that is why we decide to combine these model in the stacking model [15]. We use the stacking method, which generally use several classifier model to produce output that later to be used together for the next stacking layer. The output from the latest classifier model is the one that decide each label. In this work, we use output probability of each class from base model as an input for final model [16]. Therefore, XGBoost and neural network as base classifiers produce probability of each class for each student response and then this probability values is used as input in the next neural network model that is a final classifier. The architecture can be seen in figure 2.

Stacking models usually use a more complex model for the base model and a simpler model for the final model. We use deeper neural network on the base model (two hidden layers) and the shallow neural network in the final model (zero or one hidden layer). Both neural networks models optimized using Adam optimizer [17].

3.2. Hyperparameters optimization

In this work, hyperparameters searching process was conducted automatically using one of the automatic hyperparameters optimization algorithms, Tree-structured Parzen Estimator (TPE) [8]. TPE algorithm uses a bayesian approach $P(score|hyperparameters)$ to find the best hyperparameter. TPE will create a probability model based on the history of the hyperparameters that have been used and then choose the best hyperparameters value for the next tuning iteration. Several frameworks can be used for hyperparameters optimizations, for

[2] https://www.kaggle.com/

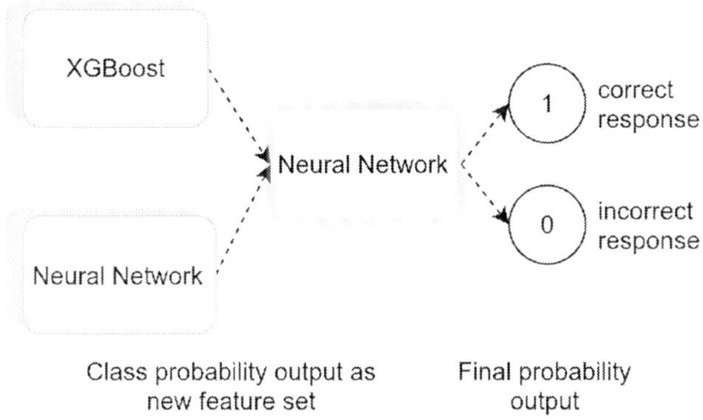

Figure 2. Stacking architecture using neural network and xgboost

Table 2. Search space of model hyperparameters

Hyperparameters	Search space
number of neuron in base neural net	$\{10k \mid k \in \{5, 75\}\}$
value of neural net base learning rate	$\{1 \times 10^{-5}k \mid k \in \{2, 10000\}\}$
base neural net iteration	$\{10k \mid k \in \{1, 100\}\}$
number of estimators in xgboost	$\{2k \mid k \in \{25, 350\}\}$
value of xgboost learning rate	$\{2 \times 10^{-5}k \mid k \in \{1, 5000\}\}$
value of xgboost subsampling	$\{2 \times 10^{-2}k \mid k \in \{40, 50\}\}$
number of layer in final neural net	$\{k \mid k \in \{0, 1\}\}$
number of neuron in final neural net	$\{2k \mid k \in \{5, 100\}\}$
value of neural net final learning rate	$\{1 \times 10^{-5}k \mid k \in \{1, 1000\}\}$
final neural net iteration	$\{10k \mid k \in \{1, 50\}\}$

example, Hyperopt [18] and Optuna [19]. In the experiment, we used Optuna as hyperparameters optimization framework with TPE algorithm.

Optuna will provide the set of hyperparameters $p \in P$ where P is the defined search space of hyperparameters. The search space for each hyperparameter can be seen in the table 2. The iteration process of training and tuning the model is shown in figure 3. Model $M_A(p, \theta)$ is a model with the initial hypeparameters p and parameters θ for question A, while $M_A(p, \theta)$ are the models that have been trained in the data for question A. The model is evaluated to calculate its F1-score, E, as objective function. Optuna will save the (E, p) pair to update its probability model and try to maximize the F1-Score in the next iteration. We did 200 iterations of hyperparameters searching for each model.

The experimental scheme that we conducted follows the Ukara 1.0 Challenge scheme. The model is trained only using the train set. To avoid overfitting, we used k-fold cross validation by dividing the train data into five folds. Evaluation is done by calculating the average F1-score between five folds. Twenty best models tested using the dev set. Then the five models with the highest F1-score on the dev set were tested using the test set. The results written here are the best F1-score results from the test set. F1-score is calculated for each question using (2).

$$F1 = 2 \times \frac{(Precision \times Recall)}{(Precision + Recall)} \tag{2}$$

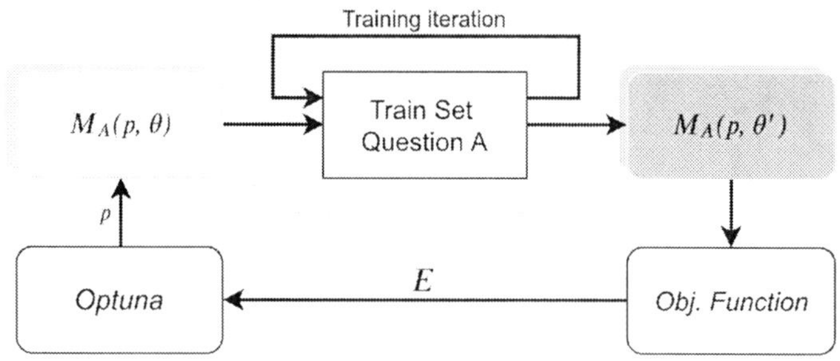

Figure 3. Training and hyperparameters tuning iteration using Optuna

Table 3. Performance comparison of stacking and single model

Models	F1-Score A	F1-Score B	Combined F1-Score
Single model XGBoost	0.88044	0.72695	0.80803
Single model Neural Network	**0.88461**	0.74343	0.81597
Stacking model	0.88192	**0.75951**	**0.81906**

Table 4. Performance comparison of SMOTE usage

Models	F1-Score A	F1-Score B	Combined F1-Score
Stacking model	0.88192	**0.75951**	0.81906
Stacking model + SMOTE	**0.88398**	0.75895	**0.82137**

4. Results and discussion

In the first experiment, we tried to compare the single model with the stacking model. We compare a single XGBoost model, a single neural network model, and a stacking model that combines both models. The hyperparameters of each model are from the same hyperparameters space as table 2. The results of this experiment are shown in table 3.

It can be seen in the table 3, that the stacking model is better than the single models at the combined F1-Score. We use combined F1-score to evaluate the model against a combination of all test types at the same time. This combined evaluation is done by combining the predicted results of questions A and B and then calculating using (2). This method was used in the Ukara 1.0 Challenge and the result may differ from averaging F1-score for all questions.

We also compare the effect of synthetic data on model performance. In table 4, we can see that SMOTE improves the overall model performance. Especially in question A where the class ratio is more unbalanced than question B. In question B there was a decrease in the F1-score when using SMOTE. This is probably due to the characteristics in question B which are more difficult than question A. It can also be seen from the large difference between the F1-Score of question A and question B. This decrease can also be due to the data ratio in question B which is sufficiently balanced and the use of SMOTE causes the model to be biased.

In table 5, we show the top-3 results from other participants published on the Ukara homepage. Our proposed model takes first place. The other two models for this challenge use different approaches. The participants who have secured second place use a deep learning method with bidirectional architecture LSTM with word embedding features. Meanwhile, the

Table 5. Performance comparison of our method with previously published methods

Models	Combined F1-Score
Stacking model + SMOTE (ours)	**0.821**
Bidirectional LSTM	0.81
Random Forest (A) and Logistic Regression (B)	0.81

other participants [20], who has secured third place, used different methods and features for each question, random forest method for question A and logistic regression for question B. The papers for both models can be accessed through the Ukara website. Unfortunately, the F1-Score results are displayed only up to two digits behind the comma on the website.

5. Conclusion

In this research, we have shown that the use of stacking model using neural network and xgboost makes it able to get a high score on the Ukara 1.0 challenge dataset. We also propose to use SMOTE and TPE to handle some problems in automatic short answer scoring and also improve the model performance. The model we produced received a combined F1 score of 0.821, better than previously published methods. From the results we obtained, it can be seen that there is still a big difference between F1 Score in question A and question B. This could be due to the different characteristics of the questions. In future research, it can be explored to use different features or methods for each question. Furthermore, it can also be explored the use of multi-task learning methods that train the model for all questions at the same time.

References

[1] Taghipour K and Ng H T 2016 A neural approach to automated essay scoring *Proc. of the 2016 Conf. on Empirical Methods in Natural Language Processing* pp 1882–91

[2] Riordan B, Horbach A, Cahill A, Zesch T and Lee C 2017 Investigating neural architectures for short answer scoring *Proc. of the 12th Workshop on Innovative Use of NLP for Building Educational Applications* pp 159–68

[3] Dong F, Zhang Y and Yang J 2017 Attention-based recurrent convolutional neural network for automatic essay scoring *Proc. of the 21st Conf. on Computational Natural Language Learning (CoNLL 2017)* pp 153–62

[4] Kumar Y, Aggarwal S, Mahata D, Shah R R, Kumaraguru P and Zimmermann R 2019 Get it scored using autosas—an automated system for scoring short answers *Proc. of the AAAI Conf. on Artificial Intelligence* vol 33 pp 9662–9

[5] Abuhaiba I S and Dawoud H M 2017 Combining different approaches to improve arabic text documents classification *Int. J. of Intelligent Systems and Applications* **9** pp 39

[6] ZhiWei M, Singh M M and Zaaba Z F 2017 Email spam detection: a method of metaclassifiers stacking *The 6th Int. Conf. on Computing and Informatics* pp 750–7

[7] Chawla N V, Bowyer K W, Hall L O and Kegelmeyer W P 2002 Smote: synthetic minority over-sampling technique *J. of Artificial Intelligence Research* **16** pp 321–57

[8] Bergstra J S, Bardenet R, Bengio Y and Kégl B 2011 Algorithms for hyper-parameter optimization *Advances in Neural Information Processing Systems* pp 2546–54

[9] Deisenroth M P, Faisal A A and Ong C S 2020 *Mathematics for Machine Learning* (Cambridge: Cambridge University Press)

[10] Bojanowski P, Grave E, Joulin A and Mikolov T 2017 Enriching word vectors with subword information *Transactions of the Association for Computational Linguistics* **5** pp 135–46

[11] Grave É, Bojanowski P, Gupta P, Joulin A and Mikolov T 2018 Learning word vectors for 157 languages *Proc. of the Eleventh Int. Conf. on Language Resources and Evaluation (LREC 2018)*

[12] Sultan M A, Salazar C and Sumner T 2016 Fast and easy short answer grading with high accuracy *Proc. of*

the 2016 Conf. of the North American Chapter of the Association for Computational Linguistics: Human Language Technologies pp 1070–5

[13] Chen T and Guestrin C 2016 Xgboost: A scalable tree boosting system *Proc. of The 22nd ACM SIGKDD Int. Conf. on Knowledge Discovery and Data Mining* pp 785–94

[14] Nielsen D 2016 *Tree Boosting with Xgboost-Why Does Xgboost Win "Every" Machine Learning Competition?* Master's thesis (Trondheim: NTNU)

[15] Wolpert D H 1992 Stacked generalization *Neural Networks* **5** pp 241–59

[16] Ting K M and Witten I H 1999 Issues in stacked generalization *J. of Artificial Intelligence Research* **10** pp 271–89

[17] Kingma D P and Ba J 2015 Adam: A method for stochastic optimization *3rd Int. Conf. on Learning Representations, ICLR 2015, San Diego, CA, USA, May 7-9, 2015, Conf. Track Proc.*

[18] Bergstra J, Yamins D and Cox D D 2013 Hyperopt: a python library for optimizing the hyperparameters of machine learning algorithms *Proc. of the 12th Python in Sci. Conf.* vol 13 (New Jersey: Citeseer) p 20

[19] Akiba T, Sano S, Yanase T, Ohta T and Koyama M 2019 Optuna: a next-generation hyperparameter optimization framework *Proc. of the 25th ACM SIGKDD Int. Conf. on Knowledge Discovery & Data Mining* pp 2623–31

[20] Septiandri A A and Winatmoko Y A 2020 Ukara 1.0 challenge track 1: automatic short-answer scoring in bahasa indonesia *Preprint arXiv:2002.12540*

IOP Conf. Series: Materials Science and Engineering 1077 (2021) 012014 doi:10.1088/1757-899X/1077/1/012014

Classical and Deep Learning Time Series Prediction Techniques in the Case of Indonesian Economic Growth

Hendra Bunyamin[1] and Meyliana[2]

[1]Informatics Engineering, Maranatha Christian University, Bandung, 40164
[2]Accounting Department, Maranatha Christian University, Bandung, 40164

E-mail: hendra.bunyamin@it.maranatha.edu, meyliana_oey@yahoo.com

Abstract. Gross Domestic Growth (GDP) as a proxy of an economic growth in one country is defined as the market value of services and goods produced by the country for a year. As the GDP of one country gets higher and higher through the years, this indicator points out that in general the economic growth of the country is growing; specifically, the values of services and goods accumulated in the market are increasing. To determine what indicators of a country that affects one's economic growth remains an open question. Therefore, this research attempts to study those indicators and particularly utilize them to predict the economic growth. To answer those questions, this research employs diverse time series techniques ranging from classic time series analysis to machine learning and deep learning. Subsequently, our dataset comprises World Development Indicators (WDI) of Indonesia from 1962 to 2016. By measuring Root-Mean-Square Error (RMSE), we show Seasonal Autoregressive Integrated Average (SARIMA) and Convolutional LSTM give the the best performance from classical and deep learning techniques respectively. Our analysis shows SARIMA's performance is boosted by its ability to capture trend and seasonality of the dataset; equally important, Convolutional LSTM equipped with convolutions as part of reading input into LSTM units significantly boost the performance over either Convolutional or LSTM networks. Furthermore, our analysis points out that the indicator which most strongly contribute to predict GDP is CO_2 emission. This result agrees to the fact that some countries with high CO_2 emission also has high GDP as well; also this finding should warn Indonesian government of the increasing CO_2 pollution.

1. Introduction

Research on sources of long-run economic growth and development has always been an interesting topic among economists. At the same time, there is still a huge gap between developed and developing countries based on observed patterns in economic growth and time across countries. This phenomenon drives economists to conduct more qualitative and quantitative research to find explanation of the phenomenon with the latest information and mathematical methods [1].

Why does economic growth provide such a valuable information for a country? The Organization for Economic Co-operation and Development (OECD) [2] states that economic growth represents a significant instrument to decrease poverty and improve quality of society lives in developing countries. Furthermore, economic growth studies of a single country and comparing multiple countries have shown that an increasing economic growth leads to an accelerated Millenium Development Goals (MDGs) achievement. The eight United Nations MDGs are related to one another and designed to meet the needs of the world's poorest. Concretely, eight MDGs declared by United Nations in 2015 are described in Figure 1.

Figure 1. Eight MDGs designed to achieve a better and sustainable future for all countries in 2015

Specifically, Rodrik [3] states an increasing economic growth shall create a circle of chances and prosperity; in other words, an accelerated increasing economic growth will open enough intensive jobs opportunities for parents to provide education funding for their children. This conducive circumstance encourages the emergence of entrepreneurs with new openings which, in the end, drives a higher economic growth than before.

Moreover, in line with statement from Rodrik, learning from the history of one country's economic growth has profound insights, should everyone study country development growth. Economic growth can enable ones to learn a decent living both for themselves and their family, including for those who are in the lowest level. Concretely, economic growth can eliminate poverty; moreover, previous studies have also come to the same conclusion. Therefore, economic growth information should be important to be known by a country.

Unfortunately most studies related to economic growth utilize simple linear regression of exogenous factors without involving endogenous factors. Whereas Kibritcioglu and Dibooglu [1] state economic growth is too complex to be analyzed only by simple linear regression, firstly this research is expected to be able to answer the information needs regarding the calculation of predictable economic growth for the future by using time series methods [4, 5, 6, 7, 8] and machine learning [9, 10, 11, 12] to analyze and predict time series. Secondly, this research also analyzes world development indicators (WDIs) that influence economic growth of a country, specifically of Indonesia. To the best of our knowledge, this article presents the first exploration on predicting Indonesian economic growth using classical and deep learning time series techniques. Moreover, this article provides several baselines for predicting Indonesian economic growth problem.

Our source code is published at `https://github.com/hbunyamin/hibah_internal-2019-world_development_indicators-codes`.

2. Literature Review

Related work of this research is divided into three parts as follows: functionality decomposition in temporal data mining, time series forecasting techniques, and deep learning techniques.

2.1. Functionality decomposition in temporal data mining

Temporal data is a type of data which varies through time and indexed by a time component. Pattern extraction processes from temporal data have been studied extensively in a work by Mahalakshmi et al. [13]. In short, functionality decomposition can be distinguished into temporal classification, temporal prediction, temporal association, temporal clustering, temporal regression, temporal summarization, and temporal outlier detection. Our research is categorized as temporal prediction; specifically, we look for a GDP model prediction with country indicators as attributes.

2.2. Time series forecasting techniques

In general, regression techniques can be applied into predicting temporal data. These techniques are capable of modeling various relationships between dependent variables and independent variables where, in our case, the dependent variable is GDP and the independent variables are world development indicators (WDI) which affect the GDP. Categorically, [8] describe that there are three types of regression techniques. Firstly, the regression technique which models the relationship between a dependent variable and independent variables without any time component. This model is commonly named an explanatory model [14] and depicted as

$$y = f(x_1, x_2, \ldots, x_n, \text{error}) \quad (1)$$

with y is an independent variable, x_1, x_2, ..., x_n are dependent variables. Specifically, error in Equation (1) denotes errors caused by random variation and other relevant variables which are not accommodated in the equation. In machine learning (ML), a dataset is divided into train and test set and a model which is similar to Equation (1) is trained on the train set and tested on the test set; therefore, the equation can also be a prediction model. Our research utilizes this regression type as machine learning models such as Ridge Regression [15], Lasso Regression [16, 17], Multilayer Perceptron [18], Random Forest [19, 20], and XGBoost [21]. Moreover, deep learning models which are more sophisticated than the ML ones, such as Convolution Neural Network (CNN) [22, 23, 24] are also employed for GDP prediction in this research.

The second type is a regression technique which takes into account the time component as described as follows:

$$y_{t+1} = f(y_t, y_{t-1}, y_{t-2}, y_{t-3}, \ldots, \text{error}), \quad (2)$$

with y_{t+1}, y_t, and y_{t-1} are the value of independent variable at time $t+1$, t, and $t-1$ respectively. The error in Equation (2) is the same as the one in Equation (1). Particularly, Equation (2) is commonly named a univariate time series model. Classical models employed in this research are simple mean, median, and persistence methods, and exponential smoothing for time series (ETS) [25, 26]. Additionally, deep learning techniques for this type which are utilized are Long Short-Term Memory (LSTM) [27] and CNN-LSTM [27, 22, 25, 28].

Finally, the third type regression technique is the one which combines features from the first and second regression types. Accordingly, the third regression type for our problem can be written as

$$y_{t+1} = f(y_t, x_1, x_2, \ldots, x_n, \text{error}). \quad (3)$$

where y_{t+1} and y_t are independent variables at time $t+1$ and t respectively. We also employ this regression type which represented by Seasonal Autoregressive Integrated Moving Average (SARIMA) [8, 26, 29] and a combination of CNN and LSTM (ConvLSTM) [22, 25, 27]. Specifically, SARIMA has the capability to model univariate time series with other features such as trend and seasonal components and ConvLSTM is a deep learning technique where the LSTM units of this model are modified into CNNs for reading input data.

3. Research Methodology

In general our approach consists of four steps [25] as follows: preprocessing the WDI dataset and designing experimental setup.

3.1. Preprocessing the WDI dataset

The Indonesian WDI dataset is divided into two parts that are the world development indicators (X) and the GDP (Y). The first part contains of 1,329 world development indicators (`https://data.worldbank.org/indicator`) where each indicator is a time series and the second part comprises 9 GDPs of Indonesia. Both parts have the same time span from year 1960 to 2016.

The preprocessing on the dataset has been done as follows: indicator time series whose information are incomplete through the time span are removed and time series that are white noise are also removed. Concretely, white noise time series are sequences of random number; therefore, they are unpredictable. We determine whether or not a sequence is white noise by checking the autocorrelation of the time series. Autocorrelation (a_k) computes the linear relationship between lagged values (x_t and x_{t-k}) in a time series with k is a parameter defining the lag between two consecutive values [8]. The value a_k is defined as

$$a_k = \frac{\sum_{t=k+1}^{T} (x_t - \overline{x})(x_{t-k} - \overline{x})}{\sum_{t=1}^{T} (x_t - \overline{x})^2}, \tag{4}$$

where T is the number of time components in the time series. Furthermore, the autocorrelations of a white noise time series for different k are approximately zero; however, not all values of autocorrelations are zeros because of random variation. Specifically, more than 95% of the autocorrelation values are within the range of $\pm\frac{2}{\sqrt{T}}$. Figure 2 shows an example of a white noise time series.

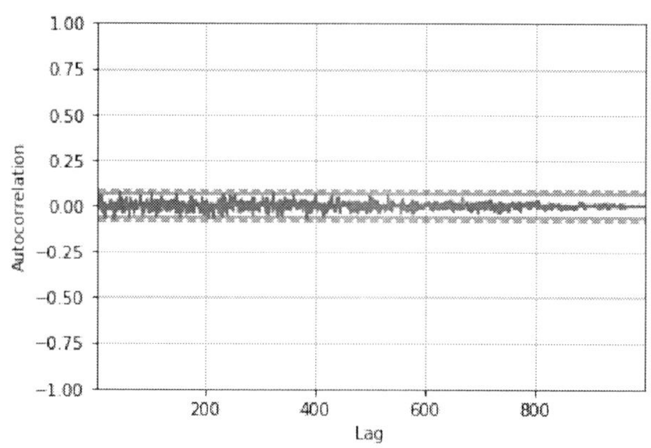

Figure 2. Autocorrelation of white noise time series (y). Both blue dotted lines represent $\pm\frac{2}{\sqrt{T}}$ and the autocorrelation values are within those values

On the other hand, Figure 3 shows the GDP time series of Indonesia is not white noise because less than 5% of autocorrelation values are within the range of $\pm\frac{2}{\sqrt{T}}$.

After white noise and incomplete world development indicators are removed, the number of indicators drops to 163 in a time span of 52 years. Additionally, in the second part of the dataset, there are nine different measurement metrics of GDP which are basically the same. We opt for GDP per capita with constant 2010 US$ inasmuch as it fairly reflects one country's economic

IOP Conf. Series: Materials Science and Engineering 1077 (2021) 012014 doi:10.1088/1757-899X/1077/1/012014

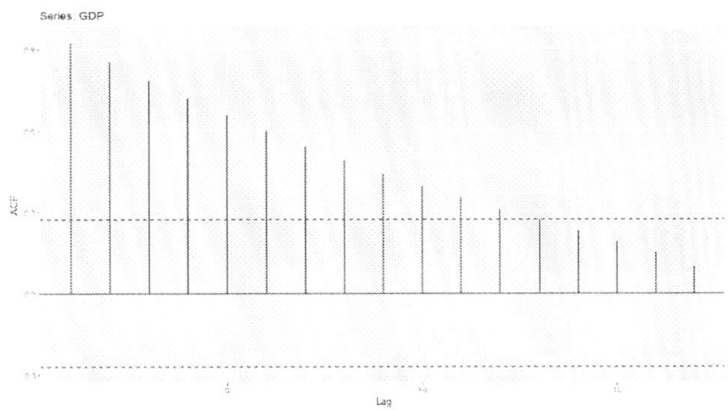

Figure 3. Autocorrelation of Indonesian GDP from year 1962 to 2013

output by dividing GDP by number of its people; moreover, the constant value of the currency makes GDP comparable during the time span of our experiment.

3.2. Designing experimental setup

Firstly all our prediction models are set up and summarized in Table 1. Specifically, we combine all classical, machine, and deep learning techniques.

Table 1. Classical, machine learning, and deep learning techniques from various different resources

Classical Techniques	Deep Learning Techiques
Simple mean, median [26, 29]	Ridge regression [15], Lasso regression [16, 17]
Simple persistence [8, 26, 29]	Random Forest [19], XGBoost [21]
Exponential smoothing (ETS) [8, 26, 29, 30]	Multilayer Perceptron [18]
SARIMA [8, 26, 29]	CNN [22], LSTM [27]
	CNN-LSTM, ConvLSTM [22, 25, 27]

After arranging the prediction models, we design an experimental setup to answer our first research question as follows:

- Dividing the dataset into train and test set. The train set consists of instances from year 1962 to 2003 (42 years) and the test set contains instances from year 2004 to 2013 (10 years).
- Train classical, machine learning, and deep learning models on the train set.
- Evaluate classical models with walk-forward validation. Walk-forward validation is a common practice in time series forecasting as it increases the number of train instances for models to learn at each time step [26].
- Evaluate machine and deep learning models on the test set.
- Both classical, machine learning, and deep learning models are measured by root mean squared-error (RMSE) evaluation metric.

Next, the second experimental setup aims to answer the question on the relationship between world development indicators and GDP. In this setup, we employ full dataset without any division and model both linear and non-linear relationships by multiple regression with Lasso regularization technique [31] and Random Forests [19].

4. Results

This section consists of three subsections which are results of GDP prediction with classical methods, deep learning methods, and finding relationships between world development indicators and GDP.

4.1. GDP prediction with classical methods

This article utilizes several classical time series forecasting methods as follows: Simple method, Exponential Smoothing (ETS), and Seasonal Autoregressive Integrated Moving Average (SARIMA). Table 2 describes hyperparameters and the RMSE results of these classical methods.

Table 2. Classical methods accompanied by their optimized hyperparameters

Methods	Hyperparameters
Simple method	config=`persist`, offset=1 & length=1
Simple method	config=`mean`, offset=1 & length=2
Simple method	`median`, offset=1 & length=2
ETS method	trend=`mul`, damped=True, seasonal=`None`, seasonal_periods=`None` use_boxcox=True, remove_bias=False
SARIMA method: Trend	order=`(1,1,0)`, seasonal_order=`(2,0,1,0)`, trend=`'t'`
SARIMA method: Trend & Seasonality	order=`(0,1,2)`, seasonal_order=`(0,0,1,12)`, trend=`'t'`

Table 3. GDP prediction results by classical methods measured by RMSE (the lower RMSE equals to the better performance of the model)

Methods	RMSE
Simple method	125.7454
Simple method	184.9804
Simple method	184.9804
ETS method	28.4467
SARIMA method: Trend	20.5524
SARIMA method: Trend & Seasonality	18.0193

In order to find hyperparameters that give the lowest RMSE of the models, we do grid search process on each time series model. Concretely, for simple method, the grid search process takes values of `mean`, `median`, and `persist` with `length` from 1 to 42. ETS method grid search process searches for optimized hyperparameters among trend=`['add', 'mul', None]`, damped=`[True, False]`, seasonal=`['add', 'mul', None]`, use_boxcox=`True, False`, remove_bias=`True, False`. Ultimately, SARIMA method has three hyperparameters which govern nature of the model, trend, and seasonality [25]. Specifically, the grid search process scans for the most optimized hyperparameters among

(i) `order`: a tuple of parameters which contains parameters (`p`, `d`, and `q`) to model the trend,
(ii) `seasonal_order`: a tuple which holds parameters (`P`, `D`, dan `Q`) to model seasonality,
(iii) `trend`: a parameter controlling the trend as one of these options: `'n'` (no trend), `'c'` (constant trend), `'t'` (linear trend), and `'ct'` (constant linear trend).

Grid search process done on SARIMA method covers hyperparameters as follows:
order=`{p=[0,1,2], d=[0,1], q=[0,1,2], t=['n','c','t','ct']}`, `P=[0,1,2]`, `D=[0,1]`, `Q=[0,1,2]`. After grid processes are done thoroughly, Table 2 finally shows that the lowest RMSE is achieved by SARIMA which considers Trend & Seasonality. Moreover, this combination of Trend & Seasonality of SARIMA models GDP Indonesia better than other classical methods as the more hyperparameters a model has, the better its performance is. This finding and M4 [32], a competition whose purpose is to advance predictions accuracy in theory and applications of time series analysis, are in accord.

4.2. GDP prediction with deep learning methods

Deep learning models specifically designed to predict Indonesian GDP in this research are described in Table 4.

Table 4. List of deep learning techniques to predict GDP. Train set and test set are the GDP and indicators from year 1962 to year 2003 and from year 2004 to year 2013 respectively.

Deep Learning Models	Hyperparameters
MP	# neurons in hidden layer=500, # epoch=100

CNN	# filter=64, # kernel size=2,
	activation=relu, optimizer=adam

LSTM	# stack in LSTM=50, activation=relu
	optimizer=adam

CNN-LSTM	# filter=64, # stack in LSTM=50 # kernel size=2, activation=relu, optimizer=adam
ConvLSTM	# filter=64, # stack in LSTM=50 # kernel size=2, activation=relu, optimizer=adam

Table 5 shows that the Convolutional LSTM model gains the minimum value of RMSE. Convolutional LSTM method reads subsequence blocks of time series observations through convolutional processes similar to CNN method. Apparently, this subsequence blocks reading give an advantage for the model to predict because the CNN operation on each step contributes additional information enriching the model to predict better. Based on the RMSE results, it appears that the Convolutional LSTM still have difficulty in predicting Indonesian economic growth due to its large RMSE values. Nevertheless, those models in Table 4 may be initialized as baselines for continuing research to pursue better performance. Visually, Figure 4 shows the comparison of our proposed deep learning techniques.

Table 5. GDP prediction results from deep learning techniques measured by RMSE

Deep Learning Models	RMSE ± Std
MP	190.298 ± 0.000
CNN	195.300 ± 56.693
LSTM	593.214 ± 211.292
CNN-LSTM	300.009 ± 224.893
ConvLSTM	179.130 ± 9.001

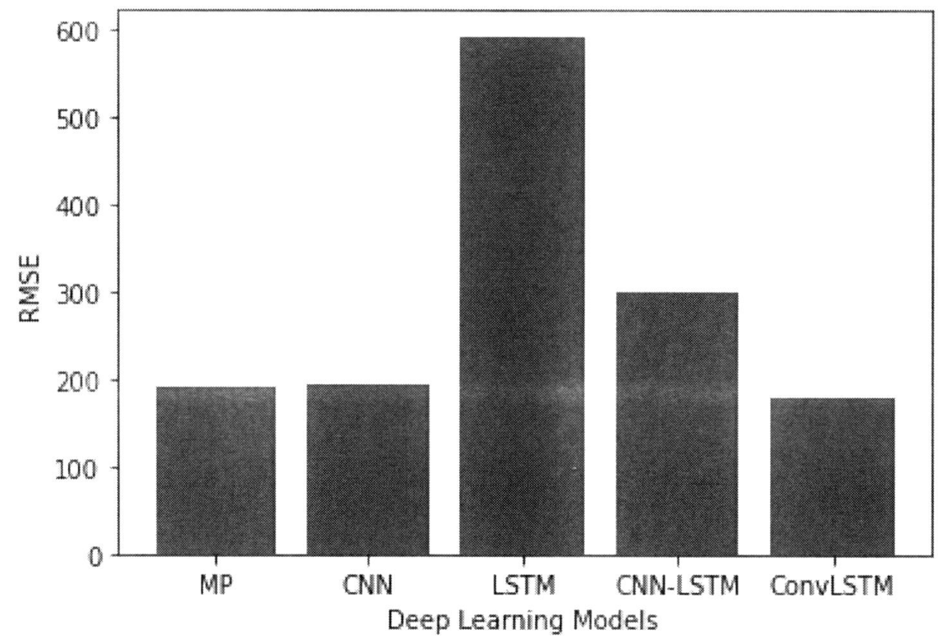

Figure 4. RMSEs from different deep learning techniques (the smaller the RMSE, the better the technique's performance)

4.3. Importance of world development indicators on predicting GDP

Our next analysis considers importance of the indicators on predicting GDP. In order to determine the importance of each indicator for predicting Indonesian GDP, we model the dataset by employing two models as follows:

- Random forest classifier [19] with number of tree estimator equals to 5. The importance of an indicator is calculated by the difference between impurity measures before reaching the indicator and at the indicator. Each calculation is weighted by its probability [33]. The more important an indicator is, the higher the difference value is.
- Lasso, a linear model designed to estimate sparse coefficients [16, 17] with $\alpha = 1$ and $\alpha = 0.01$. In this model, the importance of an indicator is measured by its coefficient [34]. The more important an indicator is the higher, the coefficient value is.

Ten most important indicators for predicting Indonesian GDP are shown in Table 6.

Interestingly, our research discovers CO_2 emissions are the most important indicator for predicting GDP. This finding suggests CO_2 emissions generated from heavy industries in Indonesia should be contributing the most to the GDP.

Table 6. Ten indicators which contribute the most for predicting Indonesian GDP

Ranking	World development indicators
1	CO_2 emissions
2	GNI (constant LCU)
3	GNI (current LCU)
4	GNI per capita growth (annual %)
5	GNI (constant 2010 US$)
6	General government final consumption expenditure (current LCU)
7	Net bilateral aid flows from DAC donors, United Kingdom (current US$)
8	Food production index (2004-2006 = 100)
9	Permanent cropland (% of land area)
10	GNI per capita (constant 2010 US$)

5. Conclusion

This research has modeled the growth of Indonesian economics represented by GDP from year 1962 to 2016. Moreover, we provide classical and deep learning time series models to predict the GDP. For the classical time series model, Seasonal Autoregressive Integrated Moving Average (SARIMA) gives the minimum RMSE; this finding is in accord with M4 Time Series competition [32]. Besides, Convolutional LSTM achieves the minimum RMSE for deep learning time series models as it takes an advantage of subsequence blocks reading to predict more accurate than other models. Lastly, CO_2 emissions are the most important world development indicator for predicting GDP. This finding can signify Indonesia relies on heavy industries to increase its GDP in coming years.

As to future work, we plan to study other potential signals which can be inferred from external sources for predicting economic growth.

References

[1] Kibritcioglu A and Dibooglu S 2001 Long-run economic growth: an interdisciplinary approach *Knowledge, Technology & Policy* **13** pp 59–70 ISSN 1874-6314 DOI: 10.1007/BF02693990
[2] OECD 2019 *Together, We Create Better Policies for Better Lives* Retrieved on 13 February 2020 from `https://www.oecd.org/about`
[3] Rodrik D 2008 *One Economics, Many Recipes: Globalization, Institutions, and Economic Growth* (New York: Princeton University Press)
[4] Gilbert P D 1993 *State Space and ARMA Models: An Overview of the Equivalence* Working Paper 93-4 (Ottawa, Canada: Bank of Canada) Retrieved from http://www.bankofcanada.ca/1993/03/publications/research/working-paper-199/
[5] Gilbert P D 1995 Combining var estimation and state space model reduction for simple good predictions *J. of Forecasting: Special Issue on VAR Modelling* **14** pp 229–50

[6] Gilbert P D 2000 A note on the computation of time series model roots *Applied Economics Letters* **7** pp 423–4 DOI: 10.1080/135048500351096
[7] Gilbert P D 2006 *Brief User's Guide: Dynamic Systems Estimation* Retrieved from http://cran.r-project.org/web/packages/dse/vignettes/Guide.pdf
[8] Hyndman R J and Athanasopoulos G 2018 *Forecasting: Principles and Practice* (Melbourne: OTexts)
[9] Ahmed N, Atiya A, Gayar N and El-Shishiny H 2010 An Empirical Comparison of Machine Learning Models for Time Series Forecasting *Econometric Reviews* **29** pp 594–621
[10] Fakhrazari A and Vakilzadian H 2017 A survey on time series data mining *2017 IEEE Int. Conf. on Electro Information Technol.(EIT)* pp 476–481 ISSN 2154-0373
[11] Aungiers J 2018 *Time Series Prediction using LSTM Deep Neural Networks* Retrieved on 7 November 2018 from https://www.altumintelligence.com/articles/a/Time-Series-Prediction-Using-LSTM-Deep-Neural-Networks
[12] Brownlee J 2018 *Deep Learning with Time Series Forecasting, Machine Learning Mastery* Retrieved on 18 December 2018 from https://machinelearningmastery.com/introduction-to-time-series-forecasting-with-python
[13] Mahalakshmi G, Sridevi S and Rajaram S 2016 A survey on forecasting of time series data *2016 Int. Conf. on Computing Technologies and Intelligent Data Eng. (ICCTIDE'16)* (IEEE) pp 1–8
[14] Shmueli G *et al.* 2010 To explain or to predict? *Statistical Sci.* **25** pp 289–310
[15] Rifkin R M and Lippert R A 2007 *Notes on Regularized Least Squares* Retrieved from https://www.researchgate.net/publication/37997883_Notes_on_Regularized_Least_Square
[16] Kim S j, Koh K, Lustig M, Boyd S and Gorinevsky D 2007 An interior-point method for large-scale l1-regularized logistic regression *J. of Machine Learning Research* **8** pp 1519–55
[17] Friedman J, Hastie T and Tibshirani R 2010 Regularization paths for generalized linear models via coordinate descent *J. of Statistical Software* **33** pp 1
[18] Rumelhart D E, Hinton G E and Williams R J 1986 Learning representations by back-propagating errors *Nature* **323** pp 533
[19] Breiman L 2001 Random forests *Machine Learning* **45** pp 5–32
[20] Jiang N, Fu F, Zuo H, Xiuping Z and Zheng Q 2020 A municipal pm2.5 forecasting method based on random forest and wrf model *Eng. Letters* **28** pp 312–21
[21] Chen T and Guestrin C 2016 Xgboost: A scalable tree boosting system *CoRR* **abs/1603.02754** pp 1 (*Preprint* 1603.02754)
[22] Lecun Y, Bottou L, Bengio Y and Haffner P 1998 Gradient-based learning applied to document recognition *Proc. of the IEEE* **86** pp 2278–324
[23] Bunyamin H, Heriyanto, Novianti S and Sulistiani L 2019 Topic clustering and classification on final project reports: a comparison of traditional and modern approaches *IAENG Int. J. of Computer Sci.* **46** pp 506–11
[24] Asriny D M, Rani S and Hidayatullah A F 2020 Orange fruit images classification using convolutional neural networks *IOP Conf. Series: Materials Sci. and Eng.* **803** pp 012020
[25] Brownlee J 2019 *Deep Learning for Time Series Forecasting: Predict the Future with MLPs, CNNs, and LSTMs in Python* Retrieved on 27 May 2019 for https://machinelearningmastery.com/deep-learning-for-time-series-forecasting
[26] Brownlee J 2019 *Introduction to Time Series Forecasting With Python: How to Prepare Data and Develop Models to Predict the Future* Retrieved on 27 May 2019 from https://machinelearningmastery.com/introduction-to-time-series-forecasting-with-python/
[27] Hochreiter S and Schmidhuber J 1997 Long short-term memory *Neural Computation* **9** pp 1735–80
[28] Aspiras L L F, Hanbal I F, Ingosan J S and Panlilio K B P 2020 Establishing the syntactic rules of the kankanaey dialect using RNN *IOP Conf. Series: Materials Sci.ience and Eng.* **803** pp 012023
[29] Brownlee J 2018 *11 Classical Time Series Forecasting* Retrieved on 8 January 2019 from https://machinelearningmastery.com/time-series-forecasting-methods-in-python-cheat-sheet
[30] NIST S *et al.* 2012 Engineering statistics handbook *E-Handbook of Statistical Methods* (NIST)
[31] James G, Witten D, Hastie T and Tibshirani R 2013 *An Introduction to Statistical Learning* vol 112 (New York: Springer)
[32] Makridakis S, Spiliotis E and Assimakopoulos V 2018 The m4 competition: results, findings, conclusion and way forward *Int. J. of Forecasting* **34** pp 802–8 ISSN 0169-2070
[33] Raschka S and Mirjalili V 2017 *Python Machine Learning Second Edition* (Mumbai: Packt Publishing Ltd)
[34] Hastie T, Tibshirani R and Friedman J 2008 *The Elements of Statistical Learning Second Edition* Springer Series in Statistics (New York: Springer New York Inc.)

Classifying Medical Document in Bahasa Indonesia using Semi-Supervised Learning

Dhomas Hatta Fudholi[1], Kiki Purnama Juwairi[2]

[1]Department of Informatics, Universitas Islam Indonesia, Yogyakarta, Indonesia,
[2]Master Program in Informatics, Universitas Islam Indonesia, Yogyakarta, Indonesia

E-mail: [1]hatta.fudholi@uii.ac.id, [2]20917020@students.uii.ac.id

Abstract. The medical domain has always been an all-time important domain since healthiness is everyone's purpose. People find medical document resources in the sea of data and information, such as the web. To support information retrieval and knowledge dissemination through the web, we analyze the use of semi-supervised learning to classify medical-related documents. The semi-supervised learning technique is chosen to show the possibilities of creating good classifiers with limited human supervision. In this research, we use the Naïve Bayes and Pseudo Labeling technique. We analyze different labeled:unlabeled data ratios of the training dataset in the experiment, starting from 4:3, 3:4, 2:5, and 1:6, to see the semi-supervised learning performance with different levels of human supervision. We get a relatively similar result in terms of classification average accuracy (81%-83%). Interestingly, in one experiment, the highest accuracy of the 1:6 ratio (85%) outperforms the 2:5 ratio (82%) and has the same accuracy as the 4:3 (85%). However, the standard deviation of the accuracy in the 1:6 ratio is the highest, amongst others (4.183). Finally, semi-supervised learning can be used to create a great classifier model of the medical domain in Bahasa Indonesia with less human supervision.

Keywords: Semi-Supervised Learning, Medical Document, Document Classification.

1. Introduction

Dissemination of information related to health is widely spread. Classification of medical documents will make it easier to find specific topics in health as needed. The purpose of classifying text documents is to get useful information from a set of documents. So that information can be organized and stored in a structured format [1]. In machine learning, a classification is a form of supervised learning. Modeling the classifier in classification is done by determining the number of categories or classes from a set of documents affixed as labels on each document. With the large data growth, manual labeling of documents was very time consuming and inefficient [2]. Therefore, the semi-supervised document classification can be one alternative in learning classifiers using labeled and unlabeled data in the training process, where most of the data available in unlabeled data [3].

Semi-supervised learning is a method to enable the inclusion of large numbers of unlabeled data as part of training data. Semi-supervised learning techniques can also utilize unlabeled data and labeled data to improve performance [4]. However, the development of many natural language applications considers this to be a challenge where the unlabeled data is relatively abundant, while the labeled data is somewhat limited [5]. The problem found in classifying documents using semi-supervised learning techniques is the use of data that is not labeled. One technique in semi-supervised learning to cope with the unlabeled data is to use the Pseudo Labeling method. The pseudo labeling method classifies unlabeled data by making pseudo labels using predictive results [6].

This study designed a classifier model to classify medical documents into health-related document categories or classes. The documents are written in Bahasa Indonesia and taken from trusted medical-related websites. Semi-supervised learning is used in this study to enable the possibilities to take a significant number of documents with minimum efforts of labeling tasks. The unlabeled data is pseudo-labeled, and the Multinomial *Naive Bayes* algorithm is used to model the classifier. The algorithm is chosen based on its ability to do multiclass classification through accurate yet straightforward and fast computation for a small dataset. We will also analyze different ratios of labeled:unlabeled data to understand the impact of different proportions in the accuracy of classification.

2. Literature Review

This research aims to create a classifier model to classify documents in the Indonesian language, especially in the medical field. Some of the techniques used for document classification in machine learning include Expectation-Maximization, Naive Bayes classifier, Support Vector Machine, Decision Trees, and Neural Network [7]. The Naive Bayes method is a simple probabilistic classification method. This method calculates a set of probabilities by adding up the frequency and value combinations from a given dataset. The Naive Bayes method assumes that all category attributes are independent of each other [8]. Research [9] used the Naive Bayes Multinomial, which is part of the Naive Bayes algorithm, for classifying news articles in Indonesian. Classification is done by combining TF-IDF features with the Naive Bayes Multinomial method. The highest precision and recall values in this research are 98.4%.

In classification, semi-supervised learning is an efficient method to add training data from unlabeled data [5] automatically. One of the advantages of semi-supervised learning techniques is unlabeled data to increase the accuracy value obtained from labeled data, often when labeling data manually [10]. Research [11] uses the Multinomial Naive Bayes method in semi-supervised learning to classify documents. One semi-supervised learning technique is to use Pseudo Labeling. Pseudo Labeling can create labels or pseudo labels for unlabeled data as learning processes [6]. Pseudo labels may be chosen based on the initial words which are representative of the category. It is considered that documents containing more core words from a category are more likely to be associated with specific categories [2]. However, other methods, such as clustering, are also used widely in semi-supervised learning. The clustering process is initially executed. Members in a cluster (unlabeled) follow labeled members within the cluster [12]. However, there are challenges in clustering techniques: (i) what if we have more than one class labeled member on one cluster, and (ii) what if we have no labeled member in one cluster [13].

Text or document classification may be highly valuable in the medical domain, such as medical coding [14]. Medical diagnosis needs to be assigned to a specific class. This information may be required to be available instantly. The probability is high that the information is presented in an unstructured form, resulting from the patient-physician communication through diagnosis and treatment [15]. Our research contributions are analyzing the use of semi-supervised learning techniques, combining Multinomial Naive Bayes and Pseudo Labeling to classify medical documents in Indonesian. Also, we divide the portions of labeled data and unlabeled data into several tests. Testing with different portions of labeled data and unlabeled data is intended to compare and analyze the performance of semi-supervised learning techniques based on classifier models that are built specifically in the Indonesian and medical domains.

3. Method

Classifying medical documents written in Bahasa Indonesia using semi-supervised learning and analyzing the model's performance trained with different labeled:unlabeled dataset ratios are the aims of this research. The document classification research method includes several phases: (i) *data collection*, (ii) *preprocessing and feature extraction*, (iii) *modeling*, and (iv) *analysis, evaluation, and implementation*.

3.1. Data Collection
The dataset used in this study consists of health-related articles taken from healthcare websites in Indonesia. We intentionally collect a small number of documents to see the performance of the model in this setup. We collect 800 documents in 10 categories. The ten categories are *kesehatan bayi (baby health), diabetes, diet, jantung (cardiac), kecantikan (beauty), kehamilan (pregnancy), kesehatan gigi dan mulut (dental and oral), kolesterol (cholesterol), kulit (skin), and mata (eyes).* The collected documents and categories are taken from health articles and their categories in alodokter.com, halodoc.com, sehatq.com, klikdokter.com, hellosehat.com, and doktersehat.com. The category of the article is based on the category given by the article sources. Every web page has different styles. Hence, the articles are saved without any automated scraping method to ensure the collected articles are within the correct category. 100 of 800 documents are taken out with an even amount of each document category, to be the test dataset. The rest of the 700 documents will then be divided into labeled and unlabeled data in different ratios during the experiments.

3.2. Preprocessing and Feature Extraction
Preprocessing is done to clean the documents. The documents often contain unnecessary characters that must be removed or commonly referred to as noise. All documents will go through the following preprocessing stages: (i) eliminating non-ASCII characters, (ii) eliminating punctuation, digits, URLs, white space; (iii) remove punctuation marks such as period punctuation, commas, exclamation points, and question marks; (iv) turn all letters into lowercase; (v) removing stopword which are non-descriptive words; and (vi) stemming.

Feature extraction uses information from a set of data to create features for the data before the training process. This study uses the TF-IDF weighting. TF-IDF gives high value if a word rarely appears in many documents, meaning that it has high discriminatory power in a document. On the other hand, it gives low value for the word that often appears in different documents, making the word unclear with the document category.

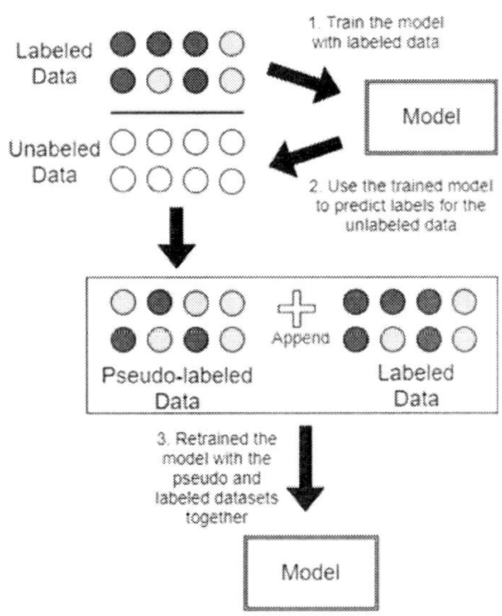

Figure 1. Modeling steps.

3.3. Modelling
The semi-supervised learning algorithm used in this study is Multinomial Naïve Bayes combined with the Pseudo Labeling technique. Naïve Bayes is a fully supervised learning method that requires a learning phase to build a probabilistic model. The probabilistic model will later be used to calculate

prior and conditional probability in determining the documents' categories. The Pseudo Labeling technique uses a model built from the labeled data to predict unlabeled data. Finally, by utilizing existing labeled data and pseudo-labeled data, a new classifier model is created. This classifier model will be used as a classifier model for the medical documents. Figure 1 shows the modeling steps that are elaborated before.

3.4. Analysis, Evaluation, and Implementation

The final phase of the study is to analyze, evaluate, and implement the model. The test dataset is used to measure the accuracy of the models. We design four different scenarios of semi-supervised training. The scenarios are divided based on the different ratio of training dataset between labeled and unlabeled data. In this study, the labeled:unlabeled ratios used are 4:3, 3:4, 2:5, and 1:6. The chosen ratios are based on the amount of the data and shifted slowly so that unlabeled data is getting higher than the labeled ones. The high number of unlabeled data becomes challenges in semi-supervised learning [5], and we tried to show the analytical finding using the shifted ratios. Table 1 shows the detailed four scenarios of training based on different training dataset combinations. The created final models from different scenarios are then implemented in a web-based application as a proof of concept in classifying medical documents written in Bahasa Indonesia.

Accuracy is the matrice used to evaluate the document classification model. The equation to measure the accuracy of the model is listed as Equation (1) as follows:

$$Accuracy = \frac{correctPrediction}{totalNumberOfPrediction} \times 100\% \quad (1)$$

correctPrediction refers to the correct classification, which consists of true positive, and true negative classifications. *totalNumberOfPrediction* refers to the number of all classification occurred during inferencing.

Table 1. Training Scenarios

Scenario	# labeled data	# unlabeled data	Ratio
SC1	400	300	4:3
SC2	300	400	3:4
SC3	200	500	2:5
SC4	100	600	1:6

4. Result and Discussion

One of the focuses of this study is to analyze the performance of semi-supervised learning in various scenarios. The analysis was carried out to determine whether the classifier model built by applying the Pseudo Labeling method using the Naïve Bayes Multinomial algorithm was able to produce a fairly good accuracy in classifying medical documents. The study has a total of 800 documents in 10 categories. For each scenario presented in Table 1, we take 100 documents to be the test dataset with an equal number of documents for each document category. Also, for each scenario, we run five different training processes. In each training process, we divide randomly labeled data and unlabeled in a certain ratio. Hence, we have a different accuracy score for each process. The highest and the lowest accuracy of each process are documented.

Table 2 shows the highest, the lowest, and the average accuracy of the final model created from labeled and pseudo-labeled data from five different training processes within each scenario. As seen in the table, the overall lowest and average accuracy values are pretty much anticipated, since the lower the number of labeled data (the higher the number of pseudo-labeled data) used in training decreases the accuracy value. In the average accuracy column, the decrement of value is quite small. The largest gap (between SC2 and SC4) is only 2.4%. We need to point out that the highest accuracy value of all scenarios is pretty similar, ranging from 82% to 87%. Interestingly, the model with the smallest number (SC4) of labeled data holds the same highest accuracy value (85%) with the model with the highest number of labeled data (SC1). However, SC4 has the highest standard deviation value, among others, which is 4.183. From the given analysis, we could say that semi-supervised learning using pseudo-labeled data has a big potential to be used as one alternative in designing document classification models, especially in Indonesian medical documents. We can reduce the number of labeled data to save resources during the data preparation without losing much accuracy and even have a chance to get higher accuracy.

Table 2. Accuracy of Classification Model

Scenario	Ratio	Highest accuracy	Lowest accuracy	Average accuracy	Standard Deviation
SC1	4:3	85.00%	81.00%	83.20%	1.643
SC2	3:4	87.00%	81.00%	83.40%	2.509
SC3	2:5	82.00%	79.00%	81.00%	1.581
SC4	1:6	85.00%	74.00%	81.00%	4.183

From the ten categories of the medical document used in this study, we found that the wrong predicted label is quite visible in some categories. Figure 2 shows the confusion matrix of one training process in SC4. We noticed that in many modeling/training processes, misclassification often happens between *kecantikan (beauty)* and *kulit (skin)*, and *jantung (cardiac)* and *kolesterol (cholesterol)*. In common sense, we can notice that both of them are related. Skincare is one factor in beauty and cholesterol affects the health of the heart (cardiac). Hence, the content of the document might be similar.

Figure 3 shows the prediction page in the classification app that is developed using the created model. The app shows the predicted results' output with its probability value made by the selected classifier model. Information on each category's probability value is sorted from the highest to the lowest value of the probability obtained.

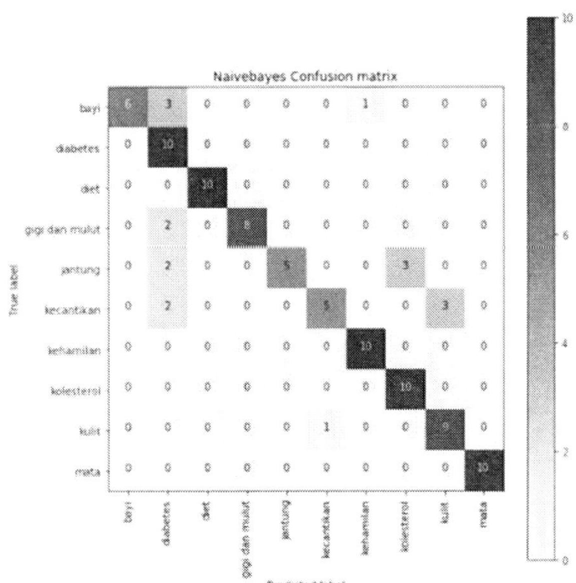

Figure 2. Confusion matrix in one training process in SC4.

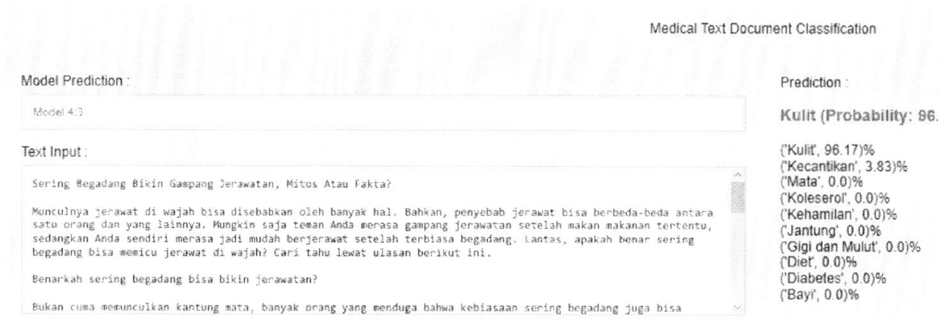

Figure 3. Medical document classification app.

5. Conclusion

The classification of medical documents using the semi-supervised learning technique utilizes two types of data: labeled and unlabeled data. In this study, the Naïve Bayes Multinomial method, combined with the Pseudo Labeling technique, is used. The domain taken as the case study in this study is medical document classification since the health domain is an all-time important domain. The classification of the medical documents supports information discovery, which is useful to various users. Creating the semi-supervised model involves training the labeled data to get a model to label the unlabeled data, which then becomes pseudo-labeled data. Both labeled data and pseudo-labeled data are then combined to get the final classifier model. We design different scenarios to show the method's performance in different ratios of labeled and unlabeled data. The accuracy value results show that even though the lower labeled data used in training might reduce the accuracy, the gap, or the difference is very small (less than 3% in our study). Moreover, the scenario with a lower number of labeled data still has a chance to bring higher accuracy rather than the model with a higher number of labeled data.

References

[1] Li X and Yang B 2018 A pseudo label based dataless naive bayes algorithm for text classification with seed words *Proc. Int. Conf. on Computational Linguistics* pp 1908–17

[2] Elhadad M K, Badran K M and Salama G I 2017 A novel approach for ontology-based dimensionality reduction for web text document classification *Int. Conf. on Computer and Information Science* pp 373–8

[3] Da Silva D F F, Coletta L F S, Hruschka E R and Jr Hruschka E R 2016 Using unsupervised information to improve semi-supervised tweet sentiment classification *J. Inf. Sci.* **355–356** pp 348–65

[4] Cascante-Bonilla P, Tan F, Qi Y and Ordonez V 2020 *Curriculum Labeling: Self-Paced Pseudo Labeling For Semi-Supervised Learning*

[5] Qiu Z, Cho E, Ma X and Campbell W 2019 Graph-based semi-supervised learning for natural language understanding *Proc. of the Thirteenth Workshop on Graph-Based Methods for Natural Language Processing* pp 151–8

[6] Arazo E, Ortego D, Albert P, O'Connor N E and McGuinness K 2019 Pseudo-labeling and confirmation bias in deep semi-supervised learning *Preprint* arXiv:1908.02983

[7] Basarkar A 2017 *Document Classification Using Machine Learning* Master Project (California: San José State University) DOI: https://doi.org/10.31979/etd.6jmu-9xdt

[8] Nafalski A and Wibawa A P 2016 Machine translation with javanese speech levels' classification *J. of Informatics, Control. Meas. Econ. Environ. Prot* vol **6** no 1 pp 21–5

[9] Wongso R *et al* 2017 News article text classification in Indonesian language *2nd. Int. Conf. on*

Comput. Sci vol **116** pp 137–43
[10] Leonandya R A, Distiawan B and Praptono N H 2016 A semi-supervised algorithm for indonesian named entity recognition *Proc. - 2015 3rd Int. Symp. Comput. Bus. Intell. ISCBI 2015* pp 45–50
[11] Zhao L, Huang M, Yao Z, Su R, Jiang Y and Zhu X 2016 Semi-supervised multinomial naive bayes for text classification by leveraging word-level statistical constraint *Proc. of the Thirtieth AAAI Conference on Artificial Intelligence* pp 2877–83
[12] Gowda H S, Suhil M, Guru D and Raju L N 2016 Semi-supervised text categorization using recursive k-means clustering *Int. Conf. on Recent Trends in Image Processing and Pattern Recognition* (Springer: Berlin/Heidelberg, Germany) pp 217–27
[13] Kowsari K, Yammahi M, Bari N, Vichr R, Alsaby F and Berkovich S Y 2015 Construction of fuzzy find dictionary using golay coding transformation for searching applications *Preprint arXiv*:1503.06483
[14] Kowsari K, Meimandi K J, Heidarysafa M, Mendu S, Barnes L and Brown D 2019 Text classification algorithms: A survey (MDPI) *Information* **10** 150
[15] Lauría E J and March A D 2011 Combining bayesian text classification and shrinkage to automate healthcare coding: a data quality analysis *J. Data Inf. Qual.* (JDIQ*)* **2** 13

IOP Conf. Series: Materials Science and Engineering 1077 (2021) 012016 doi:10.1088/1757-899X/1077/1/012016

A Heuristic Load Balancing Algorithm for Cloud Computing in Heterogeneous Resource Environment

Bagus Jati Santoso[1], Ary Mazharuddin Shiddiqi[2], Raldo Kusuma[3]

Department of Informatics, Faculty of Electrical Technology and Intelligent Informatics, Institut Teknologi Sepuluh Nopember, Surabaya, East Java, Indonesia

E-mail: [1]bagus@if.its.ac.id, [2]ary.shiddiqi@if.its.ac.id, [3]raldo16@mhs.if.its.ac.id

Abstract. Cloud computing is an alternative technology to develop web-based services. The continuous increase of internet users causes high web service traffic that leads to the requirement of good infrastructure in managing servers in cloud computing. Load balancing between servers is one of the main challenges in cloud computing across multiple nodes. The objective is to ensure that no single resource is overwhelmed nor underused. Load balancing is optimal when maximizing throughput, minimizing response times, and avoiding overloading on one of the connection lines. To solve problems in cloud computing infrastructure, we propose the utilization of two well-known heuristic algorithms to distribute loads on cloud computing. Experimental results indicated that the heuristic algorithm could distribute task loads proportionally to cloud computing resources. Our heuristic algorithm can also distribute tasks to resources with different computing resource specifications and different tasks weights.

1. Introduction

The Cloud computing application is multiplying each year that attracts many large organizations such as Amazon, Yahoo, and Google offer cloud services and have many users. The next generation of cloud computing will be measured on how effective the infrastructure and the availability of the offered services are. Load Balancing is one of the main challenges in cloud computing in distributing the workload across multiple nodes to ensure that no single resource is overwhelmed nor underused.

The characters of data transferred over cloud computing services are on-demand self-service, open network-access, resource pooling, rapid elasticity, and measured services. The on-demand self-service aims to provide services on user requests, the open network-access requires the cloud service to support heterogeneous clients, resource pooling provide a set of resources, resource allocation and practical physically supplied to the consumer, rapid elasticity enables the current resources in the cloud can be quickly provided to users, and measured services is used to monitor, control and report computing resources for planning.

A load balancing technique allocates workloads through multi-computers, clusters, links and nodes, central processing units (CPU), storage, and memory, to get optimal use of the resources, high throughput, less response time, and avoid overloading [8, 10]. This type of environment is not constrained by limited power (battery) as studied in [6, 2, 12]. Two well-known optimization algorithms can be used to meet these requirements: Genetic Algorithm (GA) and Particle Swarm Optimization (PSO). Both algorithms have similar characteristics, i.e. to randomly generate initial possible solutions and select solutions that meet a fitness function. GA's mechanism (combines the use of discrete and continuous parameters in one problem statement) is suitable for generating a large number of possible solutions. In contrast, PSO's mechanism is suited to handle discrete problem optimization [7]. GA finds the optimal solution by three steps: selection (chromosome selection of two parents from the population), crossover (crossover probability of parents to generate new offspring), and mutation (mutate parents to generate new offspring) [4, 15]. Each generation will be examined by a fitness function to select only genes with the most balanced loads between the computing resources to generate the next offspring. PSO optimizes computational resources by iteratively improving candidate solutions based on a given quality measure [3]. An improved PSO in cloud environment was proposed in [11] by considering complex networks when establishing a resource-task allocation model. PSO populates candidate solutions (particles) and moves these particles in the search space according to a simple mathematical. Each particle's movement is not only influenced by its best local position but also the best position in the search space. The best position is updated gradually which is found by other particles. Finding which algorithm is the best in performing load balancing is challenging. A challenging factor to determine a fair comparison between the two algorithms is designing the fitness function for each algorithm. The studies in [13, 7] compared GA and PSO algorithms for load balancing in cloud computing environment, however, it did not observe the performance and resource consumption of both algorithms.

We propose a heuristic algorithm to manage cloud computing loads to optimally distribute loads between cloud computing resources based on the GA and PSO algorithms. We design a fitness function for each algorithm to evaluate its performance. Then, we present the benefits and drawbacks of the two algorithms due to the fact that each algorithm does not dominate the other algorithm. The rest of this paper is as follows: we present our technique in Section II, experiments in Section III and conclusion in Section IV.

2. Method

Our load balancing technique consists of two main parts: the pre-processing and the main program. The pre-processing calculates Time to Compute (TTC), and the main program performs a heuristic algorithm (Figure 1).

The Required Computation Time (RCT) is the time to complete a task by a virtual machine or a server. Required Computation Time (RCT) calculation is performed by involving Size of job, Sj, and Computer Power of Resource i, CPi, as seen in Formula 2.

$$RCT_{[i][j]} = \frac{S_j}{C_{pi}} \quad (1)$$

2.1. Genetic Algorithm (GA)

We initiate the total number of jobs, resources, and forms of data randomly and then generate the next possible jobs, resources, and forms of data from the finest ones of the current generation [9]. We model the jobs as a series array (Figure 2). This array will later be used in the fitness function. When the RTC is obtained, the next part is to perform the heuristic algorithm. The j_i indicates an initiated task index, while the * indicates the job to be executed by the next resource.

The GA starts its process by populating an initial generation. Then, a fitness function is applied to select the best chromosomes in the current generation.

$$CT_{[i][j]} = w + c \quad (2)$$

$$\overline{CT} = \frac{\sum CT}{N} \quad (3)$$

$$F = \lambda \times maxCT + (1 - \lambda) \times \overline{CT} \quad (4)$$

where w is the waiting time, c is the time to compute a job, CT is the completion time, as depicted in Equation 2. totalCT is the total completion time, n is the total job, while averageCT is the average completion time., as written in Equation 3. maxCT is the maximum completion time, and λ is the minimizing factor (0-1). Chromosomes that pass the fitness function are selected to be used in populating the next generation, as written in Equation 4. The smaller the number of chromosomes that pass the fitness function, the better. Chromosomes with the highest fitness scores provide a more optimum load balancing solution. For this, a high threshold value is set.

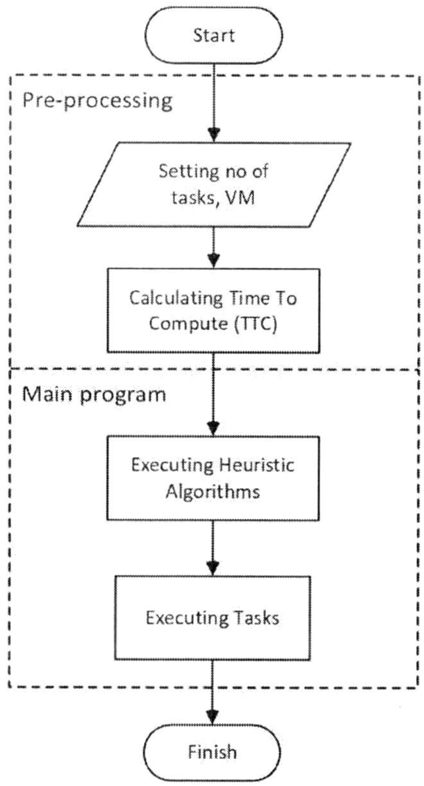

Figure 1. System architecture

j_1	j_3	*	j_4	*	j_2	j_5

Figure 2. An array representing the list of jobs in Genetic Algorithm

j_1	j_2	j_3	j_4	j_5	j_6
1	1	2	0	1	1

Figure 3. An array representing the list of jobs in Particle Swarm Optimization

2.2. Particle Swarm Optimization (PSO)

Particles are represented in an indexed array indicating the task and the server index where the job is allocated to [1, 5]. For example, j_1 allocated to r_1 (Figure 3). The PSO starts its process by populating initial particles containing a set of load balancing solutions. These particles are evaluated using a fitness function in every population generation. We apply the same fitness function with the one used in GA for the sake of comparison fairness. There are two references to evaluate the particles, i.e. PBest and GBest. PBest is the current best score of a particle, while GBest is the global best score of the current population. PBest of particles is continuously updated every iteration. If a fitness score of the particle in a generation is higher than before, then the PBest of the particle is updated. However, when a fitness score of the particle in a generation is not higher than before, then the PBest of the particle is not updated. Updating the Gbest considers the PBest of all particles. If there is a particle with higher PBest than the current Gbest, then the current GBest is updated with the Pbest of the particle. However, if there is no higher PBest than the current GBest, then the current Gbest remains.

Next, we calculate the velocity of particles that update the position of the particles. Velocity calculations are performed on each dimension inside a particle. Changing the particle position aims to find a better solution. This is performed by adding up the result of each particle (Formula 5) with the current particle's velocity. This procedure causes the velocity to keep changing every iteration if the best solution is not yet reached.

$$V_{i,d} = V_i - 1 + C_1 \times r_1 \times \left(p_{best} - p_{i,d}\right) + C_2 \times r_2 \times \left(g_{best} - p_{i,d}\right) \tag{5}$$

where vi,d is the new velocity, vi−1,d is the previous velocity, c1 is the learning rates for PBest, c2 is the learning rates for GBest, PBest is the best score of the particle, GBest is the global PBest score, pi,d is the particle i in dimension d, r1 is the random value 1, and r2 is the random value 2.

The procedures keep performing until the GBest value converges or has reached the maximum iteration. The latest GBest is regarded as the best solution to divide jobs into servers.

3. Experiments

We used dockers to simulate computing servers [14], an open-source application for consolidating files needed by software. Settings and file data supporters referred to as images. These images are collected in one location called the container. Docker is often used to develop applications due to its flexibility in various environments, e.g. different system configurations in a computer or a server. Details of the experimental environment are listed in Table 1.

We used artificial data containing a predetermined number of tasks, total computer power resources, and job sizes. We set the generation number of GA, chromosome number in a population, and parent numbers prior to running the algorithm. We also applied the same procedure in PSO by setting particles number and the maximum iteration.

Table 1. Experimental environment

Environment	Description
Operating system	Ubuntu 18.04.2 LTS
Programming	Python 3.6.9
Text Editor	Visual Studio Code 1.43.2
Virtualization	Docker 19.03.8

```
Computer Power          : [1, 1, 1]
Size Job                : [1 1 1 1 1 1 1 1 1]
Best solution           : ['7' '2' '5' '*' '8' '0' '6' '*' '4' '1' '3']
Best solution fitness   : 2.9545454545454546
ga done = 2020-05-21 16:27:11.700872
vm  0 total job  3
vm  1 total job  3
vm  2 total job  3
```

(a)

```
Computer Power          : [1, 3, 9]
Size Job                : [1 1 1 1 1 1 1 1 1]
Best solution           : ['7' '*' '4' '5' '2' '*' '0' '8' '6' '3' '1']
Best solution fitness   : 1.196969696969696968
ga done = 2020-05-21 16:32:44.632164
vm  0 total job  1
vm  1 total job  3
vm  2 total job  5
```

(b)

Figure 4. Genetic Algorithm's functionality experiment results

IOP Conf. Series: Materials Science and Engineering 1077 (2021) 012016 doi:10.1088/1757-899X/1077/1/012016

```
Computer Power          : [1, 1, 1]
Size Job                : [1 1 1 1 1 1 1 1 1]
Best Solution           : [1. 2. 0. 1. 2. 1. 0. 0. 2.]
pso done = 2020-05-21 16:40:58.236410
vm  0 total job  3
vm  1 total job  3
vm  2 total job  3
```

(a)

```
Computer Power          : [1, 3, 9]
Size Job                : [1 1 1 1 1 1 1 1 1]
Best Solution           : [1. 0. 1. 2. 2. 1. 2. 2. 2.]
pso done = 2020-05-21 16:42:10.986338
vm  0 total job  1
vm  1 total job  3
vm  2 total job  5
```

(b)

Figure 5. Particle Swarm Optimization's functionality experiment results

3.1. Experimental Scenario

We divided the experiments into two types: functionality and performance experiments. The functionality scenario aims to examine whether the system has met the requirements. The performance experiment aims to compare GA and PSO performances to provide the best solution to balance loads between servers.

3.2. Functionality testing

The GA shared the jobs with proportionally both in homogeneous and heterogeneous computational resources (Figure 4). Jobs are distributed equally in a homogeneous environment (Figure 4(a)) and proportionally in a heterogeneous environment (Figure 4(b)). For example, three identical computers with the same computational resources received the same number of jobs. On the other hand, jobs are distributed based on computational resources of each computer in a homogeneous environment. For example, computers with higher computational resources receive more jobs than other computers.

The PSO divided the jobs in a balanced manner with the same job size (Figure 5). Jobs are distributed equally in a homogeneous environment (Figure 5(a)) and proportionally in a heterogeneous environment (Figure 5(b)).For example, there are jobs with the same sizes that are all the same; then, all computers will receive an equal number of jobs. When various job sizes are given, they are shared proportionally. For example, a computer was only given one job with size nine that other computers with more number of smaller jobs.

3.3. Performance experiments

The addition of tasks requires more execution time and memory to find the best solution for both algorithms. We increased the number tasks to process and observed the performance of both algorithms (Figure 6). The execution time of GA is more stable and faster than PSO because due to its quick nature in finding the best solution (Figure 6(a)). However, GA requires more memory than PSO to perform the same number of jobs (Figure 6(b)).

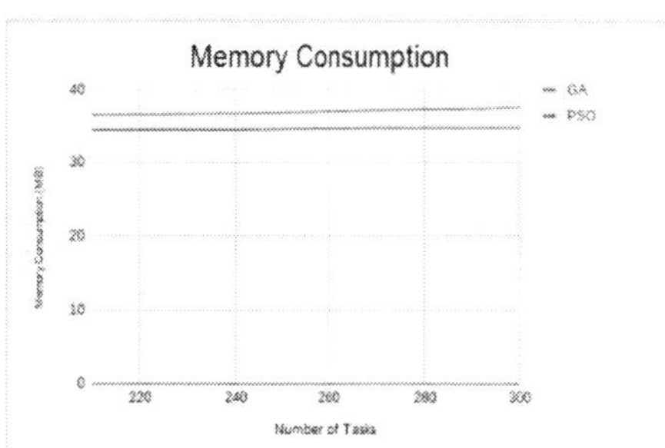

(a)

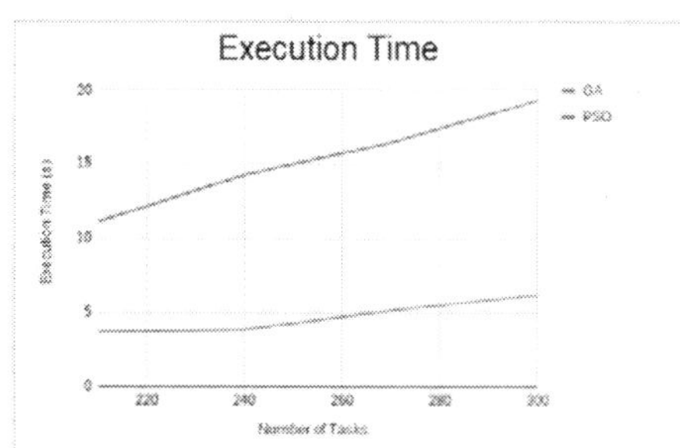

(b)

Figure 6. Genetic Algorithm and Particle Swarm Optimization experiments with increasing task numbers and steady chromosome or particle numbers (a) memory consumptions (b) execution times

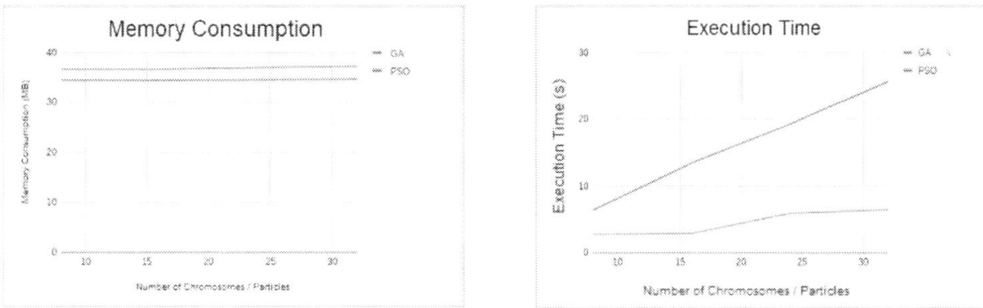

(a) (b)

Figure 7. Genetic Algorithm and Particle Swarm Optimization performances with increasing chromosome or particles and steady job number (a) memory consumptions (b) execution times

Next, we experimented with increasing chromosome or particle number and steady number of tasks to process (Figure 7). GA is still faster and more stable than Particle Swarm Optimization to converge (Figure 7(b)), but GA consumes more memory than PSO (Figure 7(a)). We conclude that crossover and mutation are the determining factors for the GA to outperform PSO. Lastly, we use the same number of jobs with an increasing generation or iteration numbers (Figure 8). Here, Genetic Algorithm still outperforms Particle Swarm Optimization in terms of execution time (Figure 8(a)), but with the same trade-off, i.e. GA requires more memory consumption than PSO (Figure 8(b)). We conclude that crossover and mutation are the determining factors for the GA to outperform PSO.

4. Conclusion

We proposed a load-balancing method based on GA and PSO. The GA based load balancing method starts by initiating populations containing solutions (chromosomes) of job distribution scenarios. Then, the chromosomes are evaluated using a fitness function. Chromosomes that pass the function are then crossovered and mutated to produce chromosomes in the next generation.

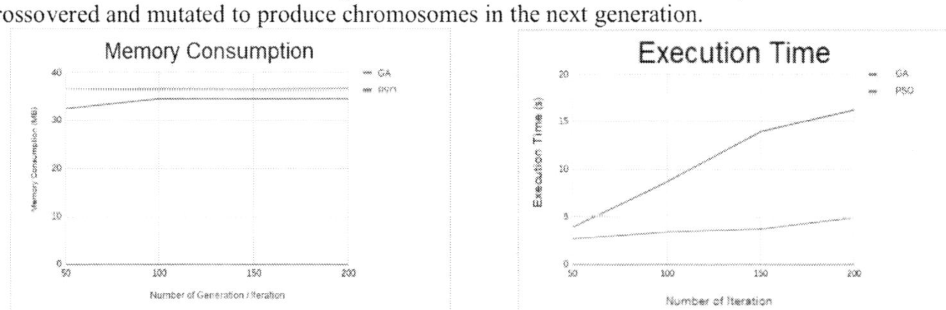

(a) (b)

Figure 8. Genetic Algorithm and Particle Swarm Optimization performances with increasing generation or iteration and steady job number (a) memory consumptions (b) execution times

This procedure is performed until we find convergent solutions or the maximum number of generations is reached. The PSO algorithm based load balancing method initializes particles with their movements to find the best solution for load sharing. Every particle will be updated in every iteration along with its velocity. The movements keep particles continue moving until the fitness values converge or iteration number is maximized. Experimental results indicate that the GA requires shorter execution time than PSO. This is due to the nature of GA that can produce convergent values faster than PSO.

However, the high performance of GA requires more memory than PSO to accommodate the generated chromosomes as well as the crossover and mutation processes. Based on the experiment results, we conclude that GA is more suitable for generating fast solutions, while the PSO is more suitable for lower resource environments.

References

[1] Alguliyev R M, Imamverdiyev Y N and Abdullayeva F J 2019 PSO-based load balancing method in cloud computing *Automatic Control Comp. Sci.* **53** pp 45–55 DOI: 10.3103/S0146411619010024

[2] Cuzzocrea A, Gaber M M and Shiddiqi A M 2014 Adaptive data stream mining for wireless sensor networks *Proc. of the 18th Int. Database Engineering & Applications Symp. (IDEAS '14)* (New York: Association for Computing Machinery) pp 284–7 DOI: 10.1145/2628194.2628213

[3] Dave A, Patel B, Bhatt G and Vora Y 2017 Load balancing in cloud computing using particle swarm optimization on Xen Server *Nirma University Int. Conf. on Engineering (NUiCONE) Ahmedabad* pp. 1-6 DOI: 10.1109/NUICONE.2017.8325618.

[4] Dasgupta K, Mandal B, Dutta P, Mandal J and Dam S 2013 A genetic algorithm (GA) based load balancing strategy for cloud computing *Procedia Technol.* vol **10** pp 340–7 DOI: 10.1016/j.protcy.2013.12.369

[5] de Almeida B S G and Leite V C 2019 Particle swarm optimization: a powerful technique for solving engineering problems *Swarm Intelligence - Recent Advances, New Perspectives and Applications* ed J D Ser *et al* (London: IntechOpen) DOI: 10.5772/intechopen.89633.

[6] Gaber M M and Shiddiqi A M 2010 Distributed data stream classification for wireless sensor networks *Proc. of the 2010 ACM Symp. on Applied Computing (SAC '10)* (New York: Association for Computing Machinery) pp 1629–30 DOI: 10.1145/1774088.1774439

[7] Hassan R, Cohanim B, de Weck O and Venter G 2012 A comparison of particle swarm optimization and the genetic algorithm *46th AIAA/ASME/ASCE/AHS/ASC Structures, Structural Dynamics and Materials Conf.* pp 804–7 DOI: 10.2514/6.2005-1897

[8] Lagwal M and Bhardwaj N 2017 Load balancing in cloud computing using genetic algorithm pp 560–5 DOI: 10.1109/ICCONS.2017.8250524

[9] McCall J 2005 Genetic algorithms for modelling and optimisation *J. of Computational and Applied Mathematics* Vol **184** no 1 pp 205–22 ISSN 0377-0427 DOI: 10.1016/j.cam.2004.07.034.

[10] Nuaimi K A, Mohamed N, Nuaimi M A and Al-Jaroodi J 2012 A Survey of load balancing in cloud computing: challenges and algorithms *2012 Second Symp. on Network Cloud Computing and Applications, London* pp 137–42 DOI: 10.1109/NCCA.2012.29

[11] Pan K and Chen J 2015 Load balancing in cloud computing environment based on an improved particle swarm optimization *2015 6th IEEE Int. Conf. on Software Engineering and Service Science (ICSESS), Beijing* pp 595–8 DOI: 10.1109/ICSESS.2015.7339128

[12] Parenreng J M, Syarif M I, Djanali S and Shiddiqi A M 2011 Performance analysis of resourceaware framework classification, clustering and frequent items in wireless sensor networks *Proc. of the Int. Conf. on e-Education, Entertainment and e-Management, Bali* pp 117–20 DOI: 10.1109/ICeEEM.2011.6137858

[13] Pathak K and Vahinde G 2015 Comparison of particle swarm optimization and genetic algorithm for load balancing in cloud computing environment *Int. J. of Research in Computer & Information* Vol **1** Issue 1 DOI: 10.5772/intechopen.89633

[14] Singh S and Singh N 2016 Containers & docker: emerging roles & future of cloud technology *2016 2nd Int. Conf. on Applied and Theoretical Computing and Communication Technol. (iCATccT), Bangalore* pp 804–7 DOI: 10.1109/ICATCCT.2016.7912109

[15] Thirumoorthy K, Selvakumar B and Umakanth N 2018 Genetic algorithm based task deployment approach for load balancing in cloud environment *Int. J. of Pure and Applied Mathematics,*

vol **118** no 20 pp 371–7

A Citation Data Collector Tool of Author's Profiles in Scopus Based on Web and Application Programming Interface (API)

Adian Fatchur Rochim[1], Tendi Nugeraha Wijaya[2], Dania Eridani[3]

Department of Computer Engineering, Diponegoro University, Tembalang, Semarang, Indonesia, 50275

E-mail: [1]adian@ce.undip.ac.id, [2]tendinw@students.undip.ac.id, [3]dania@ce.undip.ac.id

Abstract. Improving the quality of publications will enhance teaching, support community service, and improve the campus reputation. One of the measurements of research quality can be observed by looking at the number of citations. The more research cited, the greater the effect it will have on other research. In addition, the publication of research results is used by most researchers in relation to the promotion of their academic positions. The position of the author in the research affects the weight value of his/her publication. At this time, no application can be used to classify the author's position of each publication produced. This paper aims to create an application that acted as a search engine to retrieve and collect researcher data along with a list of their publications from the Scopus database. The application uses a REST API or Web Service of Scopus. Web Service allows the integration of various systems with different platforms or architectures for sharing data and services. This application was built using the PHP and JavaScript programming languages. The application's development uses the Agile Software Development method. Users can also save the search results for the researcher data and researcher documents list in the MySQL database. Furthermore, this application also allows users to download the search result in their computer devices. The System Usability Scale (SUS) method was used to test the application's usability and gained 80 as the final score.

1. Introduction

Measuring the quality of research activities is essential to determine an impact on the development of science, technology, social, and economic[1]. Good quality research can also improve the development of teaching, support community service, and enhance the campus's reputation. In today's digital era, an alternative to measure a scholarly output of research can be observed by looking at a number of citations and papers of authors' papers. The more a paper cited, the influence of the research on science and technology development is even more significant[2]. Similarly, the publication of research results is used by most researchers concerning the promotion of academic positions. The most significant weight of publication results is when, in publication, the researcher is the main author.

To obtain research publication data, researchers can open the citation database or scientific journals, such as Google Scholar, Scopus, Crossref, Web of Science, and many more. In the previous research, researchers also used the web scraping and Application Programming Interface (API) method to obtain data[2][3][4]. The use of web scraping or API is quite complicated to analyze the research. Another study used as a reference in this paper is the creation of a search engine to search researcher profiles and research publications[5]. It used a web scraping method on other literature databases to collect data.

Based on previous studies that have been mentioned above, there has not been found an application that can be used to classify the author number sequence of each published publication. And to make it easier for researchers to analyze their citation data statistics, an application that is integrated with the API is needed. The application can search for research statistics from database sources. Researchers can do a real-time live search for statistics of a study. This application's need is important for researchers to be used in finding and classifying the weight value of each publication produced. This application is important so that the academic promotion review team can easily classify each author's publication output.

In this paper, a search application is made to retrieve and collect research data from citation databases or scientific journals. This system uses REST API (Representational State Transfer Application Programming Interface) or Web Service. In this application, the database used is Scopus. The application connects to the Scopus database using the API provided by Elsevier. This application also allows users to save search results of researchers and researcher documents into a database. The database used in this application is MySQL. This application was built using the PHP and JavaScript programming languages.

This paper consists of six sections, i.e., introduction, literature review, methodology, result and discussion, acknowledgments, and conclusions. The introduction section has been presented above. The next section of this paper is the literature review regarding the concepts and theories used during this research. Then there will be sections of the methodology used in designing the application, results, and discussion about the application, acknowledgments, and the conclusions of this paper.

2. Literature Review

2.1. Application Programming Interface (API)

Representational State Transfer (REST) is a software architectural style for distribution or communication of web services focusing on system resources, first presented by Roy Fielding in 2000 in his dissertation[6]. Meanwhile, Application Programming Interface (API) is code for connecting one application with data and services from another application by giving all necessary permissions and allowing two software programs to communicate with each other[7]. REST API is a web service that uses Uniform Resource Identifier (URI), Hypertext Transfer Protocol (HTTP), and JavaScript Object Notation (JSON) for data format[7].

2.2. Scopus and Scopus API

Scopus is a database containing abstracts and citations of academic journals owned by Elsevier. Scopus contains approximately 24,600 titles from 5,000 publishers[8]. Scopus uses sophisticated tools and analytics to produce precise citation results, detailed researcher profiles, and insights that drive better decisions, actions, and outcomes. Scopus generates precise citation search results and automatically updates researchers and institutions' profiles, creating wider connections between people, published ideas, and organizations.

Scopus has an API that allows users to retrieve publication data via HTTP requests[9]. The Scopus API works in a machine-readable format that enables the software to find articles, authors, and institutions in the Scopus database[9]. This paper uses two Scopus API models. The first model is the author's search, which is used to search for researchers' profiles on Scopus. The second model is Scopus Search, which is used to search for researcher documents on Scopus.

2.3. Previous Research

In 2016, Adian conducted a research that aims to show a researcher's performance from their published articles. The application produced can crawl the data from Google Scholar, Crossref, and Mendeley[2]. In this application, the method used to retrieve data from Google Scholar is scrapping. In contrast, the method used to retrieve data from Crossref and Mendeley is by using API.

In 2017, Adian conducted a research that aims to analyze the number of readers of articles published by 100 Indonesian researchers in Mendeley[3]. This research used Mendeley's API to calculate the number of readers for each article. Then it combines with Google Scholar citations using the scraping method.

In 2017, Aditya conducted a research that aims to design a search feature with search engine for staff website of University of Indonesia. The application can perform a search for profile and publications of researchers[5]. This application collecting the data by web scraping on other literature databases.

3. Methodology

The Agile Software Development method is used to design the Data Collection Application. Its characteristics are to focus the developers on the developers' working conditions, the working software, the customers, and addressing changing requirements instead of focusing on detailed systems development processes, tools, all-inclusive documentation, legal contracts, and detailed plans[10]. Compared to other methods, the agile method was chosen because it is suitable for developing this application. In developing this application, the developer must be responsive in accepting changing requirements, the user is actively involved in the development, and focuses on working software instead of documentation.

After the application development is completed, the application was tested using two methods, namely the black box and the System Usability Scale (SUS) testing. Black box testing is used to test the functionality of the application, while SUS is used to test the usability of the application. SUS testing is done using ten respondents. Respondents were grouped based on gender, age, the latest education, and current occupation.

Based on the development method used, this section consists of four subsections, namely, planning, analysis, design, and implementation[10]. The testing result of the application will be discussed in the Result and Discussion section.

3.1. Planning

Software planning begins with determining the scope of the software, then continues by estimating the required resources. This app requires Scopus resources as a data provider. The application is built on a web basis using the PHP 7.2.10 and MySQL 5.0.12 to store and maintain the data. The local servers used to run the database are Apache and MySQL using XAMPP. Sublime Text 3 is used as a text editor. To access the application, it uses Google Chrome as a web browser. To run the search, it needs a university network, so to be used outside the university, it used OpenVPN application. This application is built by a writer who has the role of a full stack developer. It takes three months to build the Data Collection Application.

3.2. Analysis

After studying the references, there has not been found an application that can be used to classify the author number sequence of each published publication. Based on the analysis that has been done, we need an application that is integrated with a citation and scientific journal database that can search for researcher data and research documents. The application needed can also determine the author's position in the research. In this paper, a search application will be made to search for researchers' data and their publication from the Scopus database. This application uses the REST API to integrate the application with the Scopus database. Users can search for researcher data and a list of researcher documents on Scopus.

3.3. Design

The architecture of the Data Collection Tool can be seen in Figure 1. The architecture divides the system into four main components. The first part is the application that provides an interface for the user to send the search parameters, receive the information, and data processing. The second component is a web service that contains REST API which is used to retrieve data from the resource. The third component is the resource as a data provider. The last component is the database used to store the data.

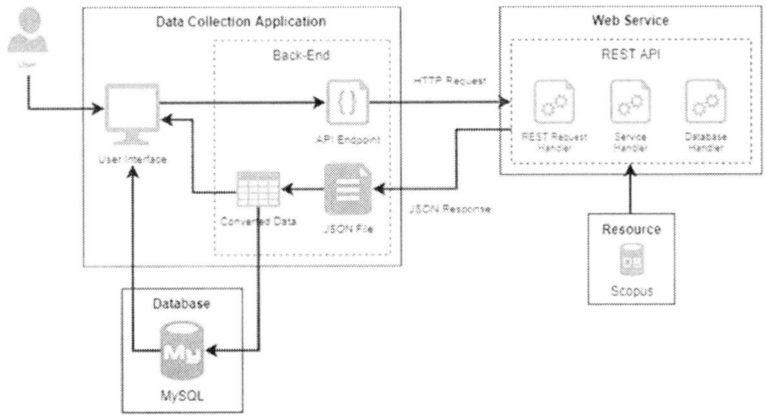

Figure 1. The Architecture of Data Collection Application

The main feature of the application is as a search engine. The search process starts when the user enters the keywords in a search form. The form will send parameters by adding data to the URL. API is used to retrieve data from a resource. This application uses Scopus API to access Scopus data in real-time.

To interact with the Scopus API, the system uses the cURL library. cURL is used to make an HTTP request. HTTP GET request method is used to get data information. The final result of the use of Scopus API and the cURL library is JSON data. The JSON data will be converted into an array. Furthermore, the array will be displayed in a table on the application interface so that it is easy for users to understand. Not only can be displayed, the search result that has been converted into an array can be saved into MySQL database. The process of inputting values from the array into the database uses SQL commands and PHP scripts.

Not all the data obtained from the Scopus API will be displayed in the application. This application will display data commonly used to analyze research, such as the number of citations, the number of authors, and the author's position. The data from the Scopus API that will displayed can be seen in Table 1 and Table 2.

Table 1. Author Search Used Field

No	Field	Name	Description
1.	preffered-name, surname	Name	First and last name of author
2.	affiliation-current	Affiliation	The author organization
3.	document-count	Documents	Number of author documents

Table 2. Scopus Search Used Field

No	Field	Name	Description
1	dc:title	Title	Article title
2	prism:doi	DOI	Digital Object Identifier
3	citedby-count	Cited by	Number of articles cited
4	author-count, @total	Number of Author	Number of article authors

The fields mentioned above can be obtained directly from the search result of the Author Search API and Scopus Search API. This application will display the researcher's position in research, but there is no author position field in the search result of Scopus Search API. Then a PHP script is used to determine the position of the author. The flowchart to determine the position of the author is shown in Figure 2. The author's name is searched in array using array_search function of PHP. It is searched on an array result from the JSON conversion from the Scopus Search API results The array_search result is the index number that contains the author's name. Indexes in arrays always start at 0. Then, to determine the author's position, the formula index+1 is used. For example, author x is the first author of the research. In the array, the author is at index 0. Thus, to display the position 1, the formula 0+1=1 is used.

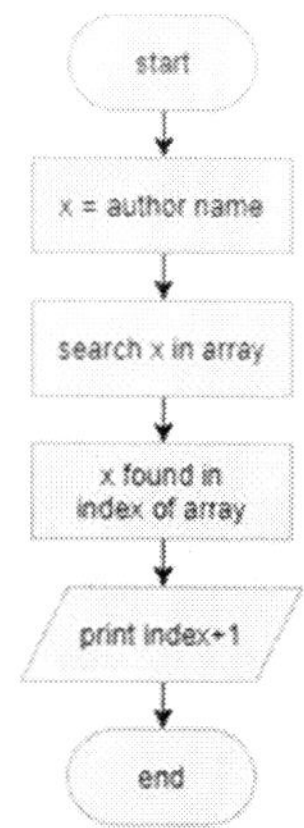

Figure 2. Flowchart to determine the position of the author

In the Scopus Search API, the maximum number of authors that can be displayed is 100. So that if the research has more than 100 authors, the author's position in the research can't be determined. In this application, when the author's position cannot be determined, it will display "N/A".

3.4. Implementation

The implementation is the last stage in the software development cycle. This stage consists of three parts, namely API Implementation, database implementation, and application implementation. The first two-part will be described in the next subsubsection, while the application implementation will be discussed entirely in the Result and Discussion subsection.

3.4.1. API Implementation. Implementation of the Scopus API is done using PHP. Scopus API is implemented along with cURL, a PHP library that can be used to make HTTP requests. There are two models of Scopus API used, namely Author Search API and Scopus Search API.

Author Search API is used to obtain the profile of the researcher. The parameters used in the Author Search are the first name, last name, and affiliation of the researcher. The flowchart for the implementation of Author Search API can be seen in Figure 3. The search parameters are read, then those will be added to the Author Search API URL. The system runs the HTTP GET request to read or get data from the source. It will get an HTTP status as a response to client requests. If the status code is 200, the server has successfully processed the request. A successful request will return JSON data. If the user performs a search more than once, the system will restart from reading the search parameters.

The second model that is implemented is the Scopus Search API. It uses the Author ID as the search parameter that obtained from the previous use of Author Search API. The flow diagram for the implementation of the Scopus Search API can be seen in Figure 4. The document search parameters will be read and added to the Scopus Search API URL. A HTTP GET request is executed and return a status code. The server successfully processed the request when the status code is 200. The application will check the number of documents. If those are less than 25, the search results will be displayed immediately. However, when the number of documents more than 25, a new HTTP GET request will be made for the next page. This is because the Scopus Search API will only display the most recent 25 documents in one execution. To display the 26th document onwards, a new HTTP GET request is needed for the 2nd page. This is an iterative process. A successful request will return JSON data.

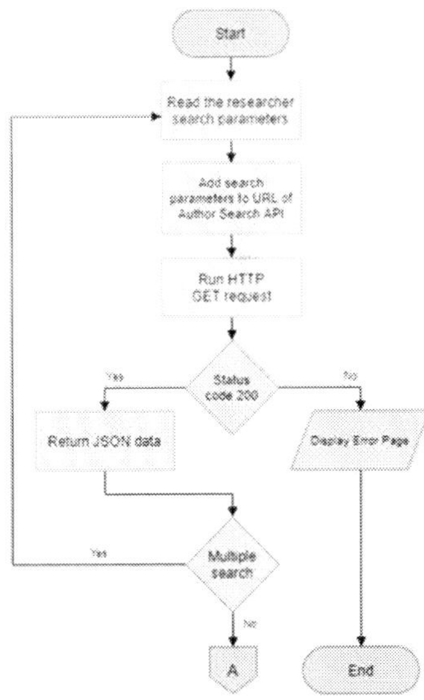

Figure 3. Flowchart for the implementation of Author Search API

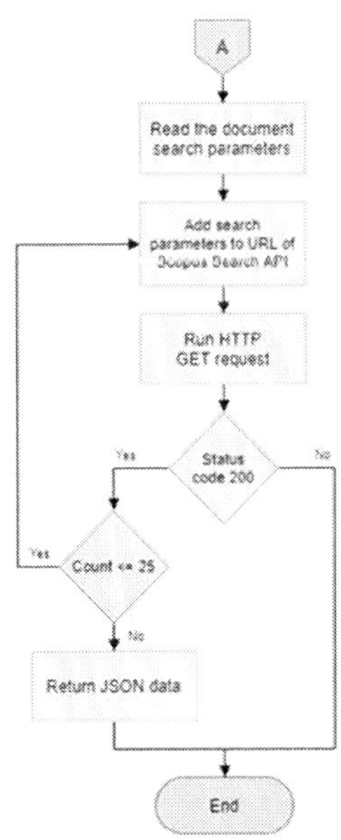

Figure 4. Flow diagram for the implementation of the Scopus Search API

3.4.2. Database Implementation. The database was created using SQL (Structured Query Language) using the MySQL DBMS (Database Management System). The local servers used to run the database are Apache and MySQL using XAMPP. The Data Collection Application database consists of three tables; there are users table, scopus_author table, and scopus_document table. The tables and their relation in the Data Collection Application database can be shown in Figure 5.

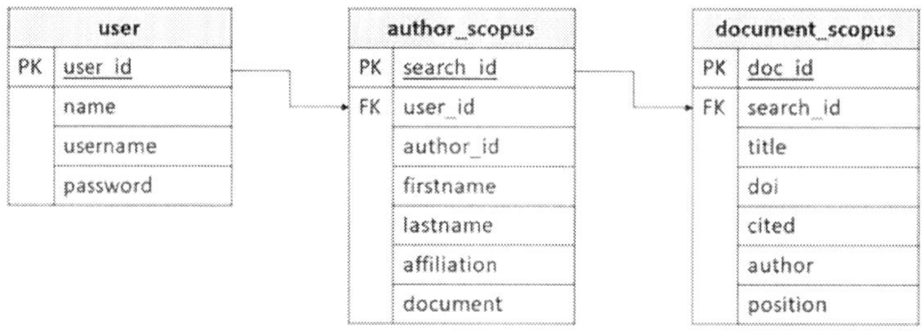

Figure 5. Tables and relation in the Data Collection Application database

4. Result and Discussion

This section will explain the application implementation. The application implementation will display the interface of the application. This section will also explain the test that aims to check whether the application made is following the design objectives that have been previously determined.

4.1. Application Implementation

To run the application, it requires an internet connection of an institution that subscribes to Scopus. The application is accessed via a web browser by entering the address of the application in the address bar. To be able to use the full features of the application, sign in is required for user. User searches for profiles and documents of researchers on the Search Page as shown in Figure 6.

Figure 6. Search Page of Data Collection Application. Users input name, last name, and affiliation of the researcher as search parameters. Author search can be done in multiple at one time.

The search results will be displayed on the Result Page. On that page, there are three options for each researcher, namely the Save Result button, the Download Docs button, and the View Docs button. The Save Result button is used to save the search result into the database. The Download Docs button is used to download the search result into the user's computer device in excel (.xls) format. The View Docs button is used to view the list of researcher's documents. A list of researchers' documents is shown in Figure 7.

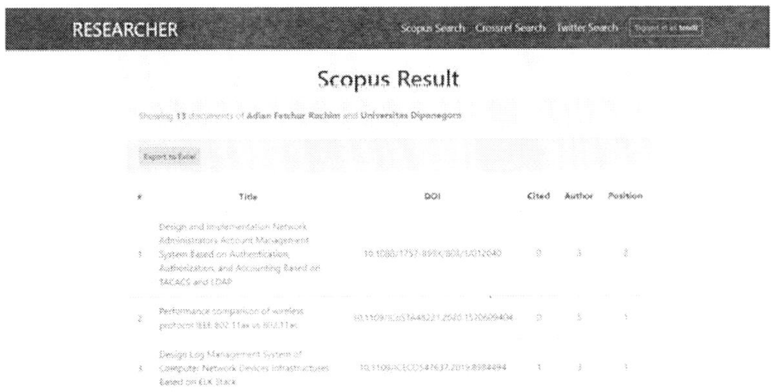

Figure 7. A list of researchers' documents. The information displayed is the research title, Digital Object Identifier (DOI), number of citations, number of authors, and author position.

4.2. Application Testing

Black box testing treats class as a black box[10]. In this test, the code's design and structure are unknown by the testers, technicians, and users who will perform this test. Based on the black box testing results that have been done, this application functionally can produce the expected output.

System Usability Scale (SUS) is also used to test the application. SUS is part of usability testing, which focuses on testing that is done directly by the end-user. SUS is a simple, ten-item scale giving a global view of subjective assessments of usability[11]. Respondents were asked to rate the usefulness of the product on a 5-point scale numbered from 1 (strongly disagree) to 5 (strongly agree)[12]. SUS testing is done in this application using ten respondents as the test sample. The processed SUS data can be seen in Table 3.

Table 3. The Processed SUS Data. The Data Collection Application has an average score of 80. Based on this assessment results, this application can be used easily by end-users as a software that can help users find and analyze the statistics of a researcher.

No	Respondent	Calculated Score										Total	Score
		Q1	Q2	Q3	Q4	Q5	Q6	Q7	Q8	Q9	Q10		
1	Respondent 1	3	4	4	4	4	4	4	4	3	1	35	88
2	Respondent 2	3	3	3	3	4	4	3	3	4	3	33	83
3	Respondent 3	3	4	4	3	4	4	4	4	3	0	33	83
4	Respondent 4	4	4	4	4	4	4	4	4	4	4	40	100
5	Respondent 5	3	3	4	4	4	2	2	4	4	3	33	83
6	Respondent 6	4	4	4	4	4	3	4	4	4	4	39	98
7	Respondent 7	3	3	3	3	3	3	3	3	2	3	29	73
8	Respondent 8	3	1	2	2	3	1	2	1	2	1	18	45
9	Respondent 9	3	3	3	3	3	3	3	3	3	1	28	70
10	Respondent 10	4	3	4	3	3	3	2	3	4	1	30	75
		Average Score (Final Result)											80

5. Conclusions

Based on the results of the analysis, design, implementation, and testing that have been done, it can be concluded that this study has succeeded in building the Data Collection Application. The application built has been able to fulfil its primary function to search for researcher data. The search for researcher data is a search for profiles and research documents on Scopus. This study succeeded in integrating an application with the Scopus API. The API is used as data sources in the researcher's data search. This application can be used to download the search results of researchers' documents into the user's computer device. In addition, this application is also successful in saving search results for research documents on Scopus in the MySQL database. Based on the SUS testing that has been done, this application has a final score of 80 means this application is easy to use by end-users in finding the statistics of a researcher.

Acknowledgments
This research was financially supported by the Ministry of Research and Technology, Republic of Indonesia through Fundamental Research Grant No. 225-98/UN7.6.1/PP/2020.

References

[1] Rochim A F, Muis A and Sari R F 2018 Improving fairness of H-index: RA-index *DESIDOC J. of Library & Information Technol.* **38** pp 378–86
[2] Rochim A F and Sari R F 2016 An exploration of Mendeley reader and Google Scholar citation for analysing indexed article *Asea Uninet Sci. Plenary Meet* pp 78–85
[3] Rochim A F and Sari R F 2017 Evaluation of articles published in Mendeley and CrossRef in relation to the Google Scholar pages *ARPN J. Eng. Appl. Sci.* **12**
[4] Alberto M M, Enrique O M, Mike T and Emilio D L C 2018 Google Scholar, Web of Science, and Scopus: A systematic comparison of citations in 252 subject categories *J. Infometrics* **12** pp 1160–77
[5] Putra A 2017 *Pengembangan Prototipe Fitur Pengumpulan dan Pencarian Data Profil dan Publikasi Penelitian pada Website Staff Universitas Indonesia* Undergraduate thesis (Depok: University of Indonesia)
[6] Fielding R 2000 *Architectural Styles and the Design of Network-based Software Architecture* PhD dissertation (California: University of California)
[7] Benharosh J 2018 *What is REST API? in plain English* Retrieved on 20 July 2020 from https://phpenthusiast.com/blog/what-is-rest-api
[8] Elsevier 2020 *About Scopus - Abstract and Citation Database* Retrieved on 20 July 2020 from https://www.elsevier.com/solutions/scopus
[9] Wan K 2020 *What are Scopus APIs and how are these used?* Retrieved on 24 July 2020 from https://www.elsevier.com/__data/assets/pdf_file/0007/917179/Scopus-User-Community-Germany-API-final.pdf
[10] Dennis A, Wixom B H and Tegarden D 2015 *System Analysis & Design An Object-Oriented Approach with UML* (Danvers: John Wileys & Sons, Inc.)
[11] Brooke J 1996 SUS-a quick and dirty usability scale *Usability Evaluation in Industry* (London: Taylor & Francis) chapter 21 pp 189–94
[12] Sharfina Z and Santoso H B 2016 An Indonesian adaptation of the system usability scale (SUS) *Int. Conf. Adv. Comput. Sci. Inf. Syst.* pp 145–8

Analysis and Designing Enterprise Architecture of PT. Adigas Jaya Pratama on Sales and Service Function Using TOGAF Framework

Alfian Mulya Rachmandany, Rd. Lugina L. Utama, Muharman Lubis*, Nia Ambarsari

Information System, Faculty of Industrial Engineering, Telkom University, Indonesia

E-mail: *muharmanlubis@telkomuniversity.ac.id

Abstract. PT. Adigas Jaya Pratama is a company that distributes various products of Pertamina, which focus on delivering LPG (Liquefied Petroleum Gas) to the distributors. There are still various problems occurred in term of service level agreement (SLA), which in many case merely based on promise, while at the same time there are also the incapability of monitoring and checking for updating items in the inventory and the storages. Meanwhile, the process of handling customer complaints often causes the decreasing in the level of satisfaction and loyalty to the services provided. To overcome these problems, the solution made must take explicit concern to the harmony between business and information technology aspects. It is necessary to design an enterprise architecture (EA) with the TOGAF framework to provide a suitable blueprint for the development of sales and service function that can present respective ability for the company to have well performance leading to the growth and profit. This research provide the respective artefacts from every design stage including preliminary phase to the migration planning, which is expected to be applied in accordance with the business needs of the company to provide solutions for the existing problems.

1. Introduction

PT. Adigas Jaya Pratama is an authorized company engaged in the business of selling and distributing of various Pertamina products, which mostly serving as an official LPG agent, which provides shipping for 12 hours a day, 365 days / year excluding for Sundays and holidays. The speed and accuracy of delivery to consumers has always been a priority for the company. In addition, there are several problems faced by the sales department, one of which is in the business process of making contracts with the customers in term of SLA, where commonly it have not written in the form. The next problem is during the making of a travel permit where the sales department still has to check the goods stock manually every time customer makes an order. Meanwhile, the sales department also needs to remember the prices and various activities manually within the schedule. In the customer service function, one of the problems is the absence of complaint management system, which the administration only handles complaints without recording or tracking their status. Thus, it is hard for the company to determine what kind of activities should be conducted in respect to the priority level. It also does not know about the overall level of customer satisfaction because the company rarely communicate with the customers to find out how the customer's response to the services provided. Meanwhile, it was found there is such a difficulty to collect the relevant data needed if several customers made a complaint using the telephone because there was no proper standard for recording and documentation within the company. In the end, these kind of problems can lead to the reduction of customer's loyalty. The analysis and documentation of an enterprise should be presented its current and future states from an integrated perspective by considering the strategic planning, organizational culture and historical event [1]. The concept in designing EA at PT. Adigas Jaya Pratama through implementing The Open Group Architecture Framework (TOGAF) can provides a detailed method of how to build, manage and conduct business process that synergize the technology. This framework is the result of continuous contributions from large number of architecture practitioners from various sector, which bring easier for the company to adopt it because the attribute for the open source present flexibility and customizability in the every stages [2]. This type of standard is mostly used for its process completeness, interoperability or flexibility in using the elements, availability of architectural knowledge, vendor-neutrality and the alignment within the industry standards [3]. The implementation of application used should be abide to the government regulation to avoid worst case scenario that might happened in the future [4, 5, 6].

2. Theoretical Basis

EA is dedicated to improving the organization's performance by allowing the company to look itself from the perspective of the company's strategic direction, business practices, information flows and a comprehensive as well as the integrated view of technology resources. This kind of practices has objective to improve the sustainability, maintainability and survivability of company in achieving profit, growth and satisfaction to engage the customers in relation to their performance. In addition, the goal of EA to improve many of the legacy system, which often have the fragmented processes to be an integrated environment that can respond automatically and simultaneously to change and deliver business strategies more accurate and meet with the customer demand. Meanwhile, the benefits of applying an EA within the company related to create more efficient business processes and efficient IT operations, increased return on investment (ROI), reduced investment risk and faster or easier or even cheaper purchases within overall domain of service function. TOGAF ADM (figure 1) is an architectural framework providing methods and tools to help accept enterprise structures, production, use and maintenance. It builds on an iterative process model supported by best practices and a reusable set of existing architectural assets [2]. The implementation of ICT within the organization often faces failure due to the absence of a standardized model and rules related to the business process according to the context of the company and the strategic alignment of Information Technology (IT) [14, 15].

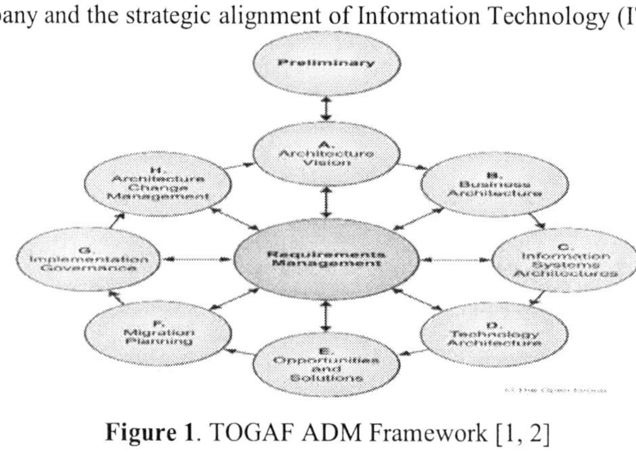

Figure 1. TOGAF ADM Framework [1, 2]

3. Methodologies

The study was conducted in 1.5 year from mid of 2018 until end of 2019, which the method of data collection is conducted in the qualitative approach, namely by collecting and analysing data through observation and interview to the head of sales, operation manager, driver, service employee and authorized expert at PT. Adigas Jaya Pratama. EA present holistic views which is top down approach of the system within the company to bring the alignment for business perspective and technology solution in handling the difficulties related data-driven. It is conducted in order to maximize organizational value and culture, to generate productivity and agility, supporting the cost reduction, deliver effective operational activities for profit growth and efficient service timeliness. The phases in EA demonstrates multiple layers of abstraction to develop a complete and shared view of information, extension and direction to understand the barriers and solutions that are considered best for the organization by adding testing method and constraints to those solutions.

4. Identification

From the company's main problems, an identification is done to find out the causes and consequences of these problems. The cause and the consequences are displayed in table 1.

Table 1. Problem Identification

Problems	Causes	Impact	Classification
Difficulties in reaching the monthly sales target set by Pertamina	No formal contract agreement between customer and company	A lot of customers don't follow the agreement made	Behavioral
	Not enough marketing done by the company	No increase in sales and customer's count	Behavioral
	Sales application cannot check the stock's good	Lack of efficiency in the sales process	Technical
	Sales application can't update the stock's good		Technical
	No price list data for customer who has contract with the company	There's a mistake in inputting the price which resulted in the customer doesn't pay the correct amount	Technical
The company does not know the level of customer satisfaction	No complaint recording and management	The company does not know which part of the company that has a lot of problems	Behavioral
	No application to support service process		Technical
	Company never does check up to the customer	Decrease in customer's loyalty	Behavioral

5. Analysis and Design

5.1. Preliminary Phase

It is necessary to describe the preparation to execute operational and required activities in order to fulfil the business orientation within the organization's structure, including defining organization structure and the fundamental principle for work culture [2]. The definition of architectural principles is a set of rules and guidelines that direct and support the approach for an organization to pursue its vision and mission concussively [2], which can be seen in table 2. Therefore, the negative implication of EA should be considering as well before the implementation take place such as the complexity of the used design after the intensive phase of familiarization and initial workshop or training, changing views and mind-set of the employee as the project progressed and the inapplicability of TOGAF's core prescriptions within business process lead to disruption of performance [8].

Table 2. Principle Catalogue

Architecture	Principle
Business Architecture	Business Continuity
	Legal Compliance
	Service Oriented
	Principle Priority
Data Architecture	Data is Asset
	Data Accuracy
	Data Security
	Data Sharing
Application Architecture	Application Availability
	Ease of use
	Integrated Application
Technology Architecture	Technology Security
	Change as needed
	Maintenance Technology

IOP Conf. Series: Materials Science and Engineering 1077 (2021) 012018 doi:10.1088/1757-899X/1077/1/012018

5.2. Architecture Vision

It is necessary to explain the initial phase of the ADM, which includes the information about defining the scope, identifying the stakeholders, creating the architecture vision and obtaining approvals [2]. The identification can lead to improve completeness of supporting data from the key performance indicators to target audience as well to increase the overall collection of data taken to produce and search time through reducing cost of conducting the desk evaluation [7].

5.2.1. Requirement Catalogue. Requirement catalogue explains the relationship between capability, problems, objectives and requirements. It captures every things that the enterprise needs to establish to meet its objectives [2], which can be seen in the table 3.

Table 3. Requirements Catalogue

Capability	Problems	Objective	Requirement
Sales	No formal contract agreement between customer and company	Improving the quality of cooperation between the company and customers	A clear contract making
	Not enough marketing done by the company	Increasing the customers count	Increasing the marketing
			Maintained communication with customers
	Sales application cannot check the stock's good	Improving the sales process	Optimal sales system
	Sales application can't update the stock's good		
	No price list data for customer who has contract with the company		
Service	No complaint recording and management	Increasing the customer service process	Communication between functions is working correctly
	No application to support service process		The flow of complaint management is clear
			Have a good service system.
	Company never does check up to the customer	Increasing the customer satisfaction level	Knowing the level of customer satisfaction.
			Have a good service system.
			Maintained communication with customers

5.2.2. Solution Concept Diagram. A solution concept diagram provides high-level guidance on possible solutions to achieve the objective of participating in the architecture [2]. In actual, it describes how the enterprise structure lifecycle is developed and managed and forms the core of the process standard for company to establish [3]. A solution concept diagram for PT. Adigas Jaya Pratama can be seen in figure 2.

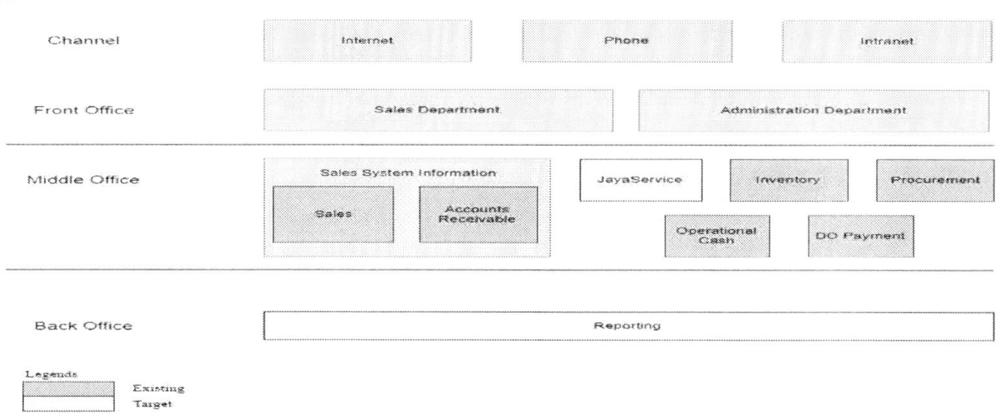

Figure 2. Solution Concept Diagram

5.3. Business Architecture

The Business Service/Function catalogue can be used to identify capabilities of an organization and to understand the level that governance is applied to the functions of an organization [2]. Table 4 shows the Business Service/Function.

Table 4. Business Service/Function Catalogue

Function	Business Service
Sales	Acceptance of company cooperation
	LPG Sales
	Gas Installation
Customer Service	Customer satisfaction survey
	after-sales service

5.4. Information System Architecture

It is also necessary to explain the data collaboration and application architecture regardless its order [2]. This phase objective is to develop the Target Data Architecture that enables the Business Architecture and the architecture vision to be identified [2]. Interestingly, the modelling and analysis approach can provides a means to effectively evaluate the infrastructure vulnerability and presents set of alternative for decision making based on the strategy and budget profile [9]. To show the correlations between each data, a logical data diagram (figure 3) can be used, which the key purpose to show logical views of the relationship between critical data entities within the enterprise but the set of viewpoints are required to describe architecture [10].

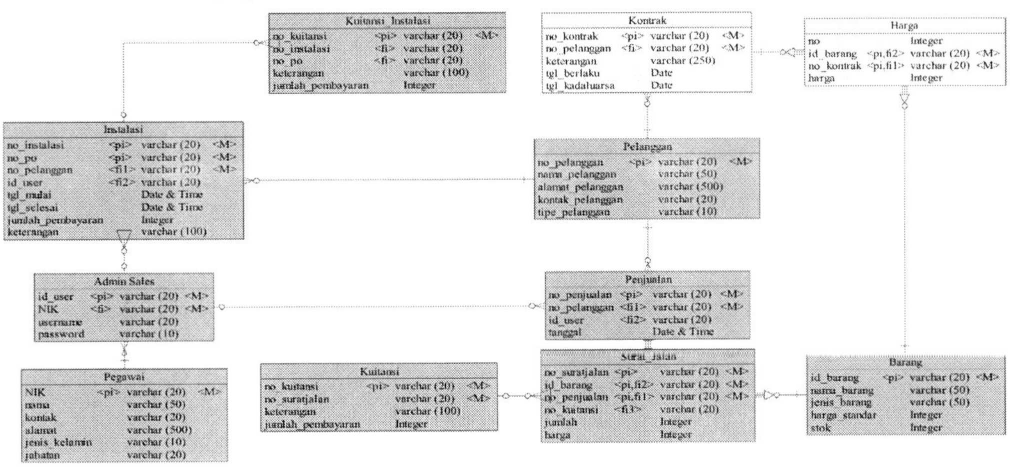

Figure 3. Logical Data Diagram Sales

5.5. Application Architecture

This phase objective is to develop the targeted application architecture that enables the business architecture and the architecture vision [2]. To show how applications interact with each other, an application communication diagram will be used (figure 4). Its objective related to discuss every visualization in order to map the relevant interaction between program or software [2, 3]. Project Architects work at the enterprise level and create the ultimate architectural team, which each of them being responsible for various area in the business, applications, integration, data and infrastructure throughout the organization to develop principles, standards, global roadmaps and other architectural documents relevant to the field. In contrast, solution engineers at the project level work with project teams and spend most of their time developing architectural documents for individual IT projects. Meanwhile, solution engineers work closely with project engineers to ensure that the project structure they create meets the values, principles, and standards set in the relevant field. [8].

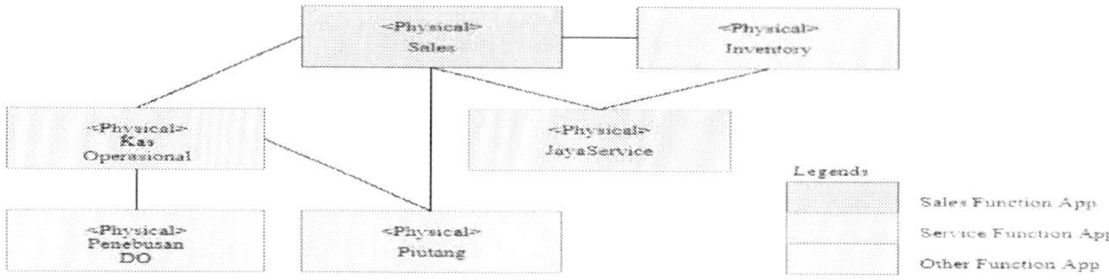

Figure 4. Application Communication Diagram

5.6. Technology Architecture

The purpose of this stage is to develop a targeted technology architecture that enables the provision of architectural insight, targeted business, data and application building components through technology services components [2]. The UI platform diagram shows the technology platform that supports the operation of the information system architecture, which this figure covers all aspects of the infrastructure system and provides an overview of the enterprise technology platform. The platform decomposition diagram for sales and service function can be seen in Figure 5.

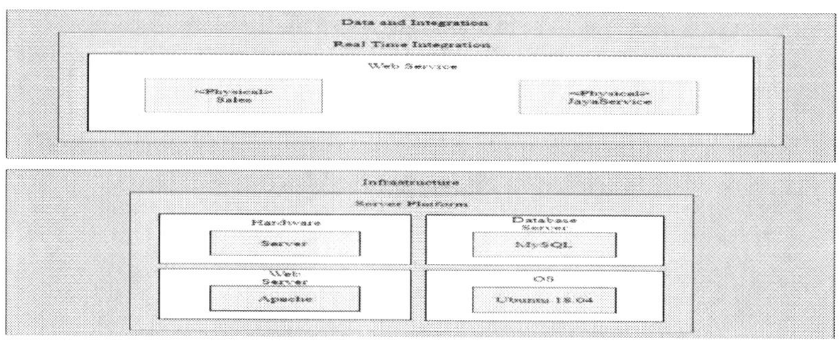

Figure 5. Platform Decomposition Diagram

Table 5. Target Architecture Project Catalogue

Project	Sub Project
Sales app modification	Business process improvement
	Application requirements identification
	Application improvement
	Application testing
	Application implementation
	Applications integration
	Making manual books
JayaService app creation	Business process improvement
	Application requirements identification
	Application improvement
	Application testing
	Application implementation
	Applications integration
	Making manual books
	Employee training
Infrastructure fix	Process business improvement
	Requiremens identifications
	Infrastructure improvement
	Infrastructure implementation
	Integration of new structures with applications
	Imfrastructure maintenance

5.7. Opportunities and Solutions
This phase describes the process of identifying delivery vehicles (projects, programs, or portfolios) that effectively deliver the Target Architecture identified in previous phases [2]. Table 5 shows the suggested project for the target architecture.

5.8. Migration Planning
The purpose of this phase is to choose an implementation project that varies into priority based on the interpretation of the risk and benefits of the project to be used as a basis for planning implementation and migration plans, which is shown in the Table 6. It meant for the suggested project based on company's business objective to improve their performance, which aligned to company vision and mission accordingly.

Table 6. IT Roadmap

Proyek	**2020**				**2021**			
	1	**2**	**3**	**4**	**1**	**2**	**3**	**4**
Infrastructure fix								
Sales app modification								
JayaService app creation								

6. Conclusions
According to the analysis and design of EA that has been conducted in the sales and service function, it can be concluded that the company can implement better service in delivering logistic and distribute the product to the client after the realization and identification of the architecture have been finished. Actually, TOGAF ADM provides sequenced inputs, stages and outputs for the respected company to follow, which the roadmap produced will evaluate the values and risks in the migration planning phase to help provide the direction and guidance in implementing the target architecture. It also consists of several projects to support the target architecture, namely the infrastructure improvement project that will be carried out at the end of 2020, the Sales application modification project in early 2021 and the JayaService application creation project from the beginning of 2021 to the middle of the year. Therefore, the verification and validation in the qualitative cannot be established due to different paradigm built, though this research focus on creating credibility and dependability to the source to generate the artefact to solve the research problem.

References
[1] Bernard S A 2012 *An Introduction to Enterprise Architecture* 3rd Edition (Bloomington: AuthorHouse)
[2] The Open Group 2018 *TOGAF® VERSION 9.2* (US: The Open Group)
[3] Cameron B H and McMillan E 2013 Analyzing the current trends in enterprise architecture frameworks *J. of Enterprise Architecture*
[4] Lubis M and Maulana F A 2010 Information and electronic transaction law effectiveness (UU-ITE) in Indonesia *IEEE ICT4M* p C-13-19
[5] Lubis M, Lubis A R, Lubis B and Lubis A 2018 Incremental innovation towards business performance: data management challenges in healthcare industry in Indonesia *MATEC Web of Conf. ICIEE* 218
[6] Lubis M, Fauzi R, Lubis A R and Fauzi R 2018 Analysis of project integration on smart parking system in Telkom University *Proc. of CITSM 2018 IEEE*
[7] Hodijah A, Sundari S and Nugraha AC 2018 Applying TOGAF for e-government implementation based on service oriented architecture methodology towards government governance *J. of Physics: Conf. Series* 1013
[8] Kotusev S 2018 TOGAF-based enterprise architecture practice: exploratory case study *Communications of the Association for Information Systems* **43(20)**
[9] Huff J, Medal H and Griendling K 2018 A Model-based systems engineering approach to critical infrastructure vulnerability assessment and decision analysis *Systems Eng.* **22(2)**

[10] Sohn M, Kang S and Lee H J 2014 Integrated data model development framework for the architecture descriptions *J. of Internet Services and Information Security* **4(4)**
[11] Aldea A, Febriani T R, Daneva M and Iacob M E 2017 The strategy blueprint: a strategy process computer-aided design tool *Int. Symp. on Business Modeling and Software Design*
[12] Jonkers H, Lankhorst MM, ter Doest H, Arbab F, Bosma H and Wieringa R 2006 Enterprise architecture: management tool and blueprint for the organization *Information Systems Frontiers* **8(2)** pp 63–6
[13] Ateetanan P, Usanavasin S, Shirahada K and Supnithi T 2017 From service design to enterprise architecture: the alignment of service blueprint and business architecture with business process model and notation *Serviceology for Services* pp 202–14
[14] Pratama N A and Sumitra I D 2019 Designing enterprise architecture for marketing adverstising media system based on TOGAF architecture development method *IOP Conf. Series: Materials Sciencce and Engineering* 662
[15] Yuliana R and Rahardjo B 2016 Designing an agile enterprise architecture for mining company by using TOGAF framework *Proc. of IEEE CITSM*

Clustering with Principal Component Analysis and Fuzzy Subtractive Clustering Using Membership Function Exponential and Hamming Distance

Annisa Eka Haryati[1] and Sugiyarto[2]

[1,2]Departement of Mathematics, Ahmad Dahlan University, Yogyakarta, Indonesia

Email: [2]sugiyarto@math.uad.ac.id

Abstract. The problem of dimension reduction in multivariate data is how to obtain a smaller number of variables but still be able to maintain most of the information contained in the data. One method that can be used is principal component analysis (PCA). Principal component analysis (PCA) is a technique used to reduce the dimensions of data consisting of several dependent variables while maintaining the variance in the data. PCA can be used to stabilize measurements in statistical analysis, one of which is cluster analysis. Fuzzy clustering is a method of grouping based on membership values that include fuzzy sets as a basis for weighting for grouping. One method of fuzzy clustering is Fuzzy Subtractive Clustering (FSC). The method used in this study is PCA and FSC. The purpose of this study is to compare the most optimal cluster results using PCAFSC and FSC methods. The results obtained indicate that the clustering using the PCAFSC method is better than the FSC method.

1. Introduction

The problem of dimensional reduction in multivariate data is how to get a smaller number of variables but still be able to retain most of the information contained in the data. One method that can be used is principal component analysis (PCA) [1]. This method is designed to reduce the spatial dimension of the data to find and interpret dependencies among variables or to help stabilize measurements in statistical analyzes such as regression analysis or cluster analysis [2].

Other According to [3], principal component analysis (PCA) is a technique used to reduce the dimensions of data consisting of several dependent variables while maintaining the variance in the data. This can be done by transforming the data into new variables (main components) that are not correlated to form a new coordinate system with maximum variance. Therefore, PCA can be used to reduce the dimensions of data without significantly reducing the characteristics of the data [4].

Cluster analysis is a technique used to look for patterns in a data set by grouping observations into groups. The aim is to find the optimal grouping in which the observations of objects in each cluster are similar, but the clusters are different from each other [5]. The basic principle in cluster analysis is to classify objects (observations) in a cluster that have very large similarities with other objects in the same cluster (similarity) but are not very similar to other objects in different clusters (dissimilarity). This means that a good cluster will have high homogeneity (similarity) between members in one cluster (within-cluster) and high heterogeneity (difference) between clusters (between-cluster) [6].

There are several methods that can be used in the grouping, one of which is the most frequently used is Fuzzy C-Means (FCM). In this method, the number of groups is known and uses a group membership matrix that was determined previously. However, the initial group membership matrix was initialized randomly so that the FCM method was complicated. Another grouping method is Fuzzy Subtractive Clustering. This method is used when the number of groups is unknown. Fuzzy Subtractive Clustering has more consistent results and better speed compared to FCM [7]. The concept of the FSC method is to determine the area in a variable that has the highest density of the surrounding points. The point with the most number of neighbors will be selected as the center of the cluster and its density will be reduced. Then the selection of another point with the most neighbors will be made to become the center of another cluster. This step will be repeated until all points are tested [8].

There are several studies on clustering, including research conducted by [9] used the FSC to classify polymers according to their similarity to various chemicals in the vapor phase. The polymer that determines the center of the cluster is selected to create the sensor array. The virtual sensor array is determined based on the polymer selected by FSC using a polymer-coated SAW oscillator as a chemical sensor. Research conducted by [10] on fuzzy clustering combined with PCA to reduce seventeen variables by producing three main components. Furthermore, the three main components will be grouped using the FCM as the basis for evaluating driver service levels. The PCA and FCM algorithms not only extend the function of grouping based on principal component analysis, but also increase the algorithm speed.

In addition, research conducted by [11] used the PCA method to reduce data dimensions, where the results of the main components will be used as input to the FCM method. The distance method used is the euclidean distance. Another study was also conducted by [12] using Fuzzy C-Means and Fuzzy Subtractive Clustering to compare the planning for the placement of the UMTS900 Node B network on the existing BTS of one of the operators in the Malang area using the Fuzzy C-Means and Fuzzy Subtractive Clustering methods. The results obtained are Fuzzy Subtractive Clustering which results in a more even placement of Node B in each of the 5 districts in Malang City compared to Fuzzy C-Means. Based on some of the studies above, that the PCA method can optimize clustering performance and the FSC method is better than FCM, so this study will use the PCA and FSC methods for clustering using exponential membership functions and hamming distance.

2. Literature Reviews

2.1. Principal Component Analysis (PCA)

Suppose that the random vector m is known, namely $X_1, X_2, \dots, X_m$. The requirements for the main components of the variables $X_1, X_2, \dots, X_m$ are the vectors of $PC_1, PC_2, \dots, PC_l$ which are obtained with the following condition $l < m$, $Var(PC_1) > Var(PC_2) > \cdots > Var(PC_l)$, and $Cov(PC_i, PC_j) = 0$, where $i \neq j$. The principal component depends only on the covariance matrix $\boldsymbol{S}$ of $X_1, X_2, \dots, X_m$. In general, the main component is obtained from the form of a linear combination, where the definition of a linear combination is as follows:

Definition [13] A vector $\boldsymbol{w}$ is said to be a linear combination of vectors $\mathbf{v_1}, \mathbf{v_2}, \dots, \mathbf{v_r}$ if there are scalars $\mathrm{c}_1, \mathrm{c}_2, \dots, \mathrm{c}_\mathrm{r}$ such that

$$\mathbf{w} = \mathrm{c}_1\mathbf{v_1} + \mathrm{c}_2\mathbf{v_2} + \cdots + \mathrm{c}_\mathrm{r}\mathbf{v_r} \tag{1}$$

Therefore, the form of the main components is as follows:

$$\begin{aligned} PC_1 &= \sum_{j=1}^{m} a_{j1}X_j = a_{11}X_1 + \cdots + a_{m1}X_m \\ &\vdots \\ PC_m &= \sum_{j=1}^{p} a_{jm}X_j = a_{1m}X_1 + \cdots + a_{mm}X_m \end{aligned} \tag{2}$$

where

$$\boldsymbol{a} = \begin{bmatrix} a_{11} & a_{21} & \cdots & a_{m1} \\ a_{12} & a_{22} & \cdots & a_{m2} \\ \vdots & \vdots & \ddots & \vdots \\ a_{1m} & a_{2m} & \cdots & a_{mm} \end{bmatrix}$$

and

$$\boldsymbol{X} = \begin{bmatrix} x_{11} & x_{12} & \cdots & x_{1m} \\ x_{21} & x_{22} & \cdots & x_{2m} \\ \vdots & \vdots & \ddots & \vdots \\ x_{n1} & x_{n2} & \cdots & x_{nm} \end{bmatrix}$$

To determine the number of main components, it can be done in 2 ways, namely:
First, through proportion of total variance (PTV)

$$PTV = \frac{\lambda_k}{\lambda_1 + \lambda_2 + \cdots + \lambda_p} \times 100\%,\ k = 1,2,\dots,p \tag{3}$$

If a large proportion of the total population of variants for m, say 80 to 90%, can be said to be with one, two or three components, then that component can replace the original m variable without losing much information [14].

Second, through the Cumlative Proportion of Variance (PKV)

$$PKV = \frac{\sum_{j=1}^{k} \lambda_j}{\lambda_1 + \lambda_2 + \cdots + \lambda_p} \times 100\% \tag{4}$$

If the majority of the total variance (70 - 80%), for large m can be connected by the first, second, or third principal components, then this principal component can interpret m the original variable without losing much information [15].

2.2. *Fuzzy Set*

Definition [16] If X is a collection of objects denoted by x, then the fuzzy set U in X is expressed as the set of consecutive pairs

$$U = \{(x, \mu_U(x)) |\ x \in X\}$$

where $\mu_U(x)$ is the degree of membership x in the fuzzy set U at interval $[0,1]$.

2.3. *Fuzzy Subtractive Clustering*

Fuzzy clustering is a method of grouping based on membership values that includes fuzzy sets as a weighting basis for grouping. Each data is given a value for the probability of being able to join each existing group, which means that the data is not absolute or firm to be a member of one group only, but also has a value for the probability of being a member of another group with the largest degree of membership showing a high tendency for data to be be a member of a particular group [17].

One of the fuzzy clustering methods is Fuzzy Subtractive Clustering (FSC), where the number of clusters to be formed is unknown. The basic concept of this method is to determine each point of the data which has a high density of the surrounding points. The point with the most number of neighbors will be used as the center of the cluster. Then the density of the point that will be used as the center of the cluster will be reduced and the algorithm will select another point that has the most neighbors to become the center of another cluster. This is done until all points are tested.

3. Methods

The research method used is clustering using Principal Component Analysis (PCA) and Fuzzy Subtractive Clustering (FSC) by using hamming distance and exponential membership function. Output of the PCA will be used as input for FSC, then, the quality of the clusters formed will be seen using the Partition Coefficient. Below will be one by one regarding the methods used in this research.

3.1. Principal Component Analysis Steps

The steps of the Principal Component Analysis are as follow:

- Data Normalization
- The data normalization used in this study is Softmax normalization, defined as follows [18]:

$$x'_{ij} = \frac{1}{1 + e^{-\left(\frac{x_{ij}-\mu_j}{\sigma_j}\right)}} \quad (5)$$

- Covariance Matrix
- The sample covariance for the m and n variables is

$$s_{mn} = \frac{1}{k-1}\sum_{j=1}^{k}(x_{jm} - \bar{x}_m)(x_{jn} - \bar{x}_n) \quad (6)$$

The covariance matrix (S) is as follows:

$$\boldsymbol{S} = \begin{bmatrix} s_{11} & s_{12} & \dots & s_{1p} \\ s_{21} & s_{22} & \cdots & s_{2p} \\ \vdots & \vdots & \ddots & \vdots \\ s_{p1} & s_{p2} & \cdots & s_{pp} \end{bmatrix} \quad (7)$$

- Eigen Value and Eigen Vector

  The eigen values (λ) of the matrix S measuring $n \times n$ can be found using equation (8) as follows:

$$det(S - \lambda I)x = 0 \quad (8)$$

The eigen vector (a) corresponding to λ can be found using equation (9):

$$det(S - \lambda I)x = 0 \quad (9)$$

- Determine the Number of Main Components

  The number of major components can be determined using equations (3) and (4).
- Compiling the Main Components

  In general, the main components are obtained from the form of a linear combination using equation (2).

3.2. Fuzzy Subtractive Clustering Steps

The steps of the Fuzzy Subtractive Clustering algorithm are as follow:

- Determine the value of the parameter, i.e. r (radius), q (squash factor), accept ratio, rejecect ratio.
- Converts data to fuzzy numbers using the exponential membership function [19]:

$$\mu(x) = \begin{cases} 1 & x \leq a \\ \dfrac{e^{-\left(\frac{x-a}{b-a}\right)} - e^{-s}}{1 - e^{-s}} & a \leq x \leq b \\ 0 & x \geq b \end{cases} \quad (10)$$

- Determine the potential of each data point $D_i; i = 1,2,3, \dots, n$ with the following steps:

  First, Calculate the distance of each data against μ_{A_j} using the hamming distance [20]:

$$Dist_{ij} = \left(\frac{|\mu_A(u_i) - \mu_B(u_i)|}{r}\right) \quad (11)$$

where $\mu_{A_j} = \mu_{B_{ij}}; j = 1,2, \dots, m; i = 1,2, \dots, n$

Second, determine the initial potential of each data point using :

$$D_i = \sum_{k=1}^{n} e^{-4\left(\sum_{j=1}^{m} Dist_{ij}^2(x_i)\right)} \quad (12)$$

- Search for data points with the greatest potential value:
  $M = max[D_i | i = 1,2, \dots,]$; for the first iteration.
  $Z = max[D_i | i = 1,2, \dots,]$; for the second, third iteration an so on.
- Calculate the ratio (R) of a candidate center cluster using
$$R = \frac{Z}{M} \tag{13}$$
  In the first iteration, the value Z = M.
- Check the eligibility of prospective cluster centers to cluster centers by using 3 conditions as follows:
  a) Condition 1: if the ratio > accept ratio, then the prospective cluster center can be accepted as a new cluster center.
  b) Condition 2: if the reject ratio < ratio ≤ accept ratio is checked the feasibility of the prospective cluster center can be accepted as a new cluster center, if not then the iteration is stopped because there are no more data considered to be prospective cluster centers. The procedure in condition 2 is as follows:
$$Md = -1$$
  For $k = 1,2, \dots, p$, where p = number of clusters
$$Sd_k = \sum_{j=1}^{m} \left(\frac{V_j - C_{kj}}{r} \right)^2 \tag{14}$$
  If $(Md < 0)$ or $(Sd_k < Md)$, then $Md = Sd_k$,
$$Mds = \sqrt{Md};$$
  where Mds is the closest distance from the prospective cluster data center to the cluster center. If (ratio + Mds) ≥ 1; prospective cluster centers are accepted as new cluster centers. Meanwhile, if (ratio + Mds) < 1 then prospective cluster centers are not accepted and will not be reconsidered as new cluster centers (the potential data is set zero) centers are not accepted and will not be reconsidered as new cluster centers (the potential data is set to zero).
  c) Condition 3: if the ratio is reject ratio, then there are no more data points that will be considered to be prospective cluster centers and the iteration is stopped.
- If a new cluster center is obtained, a reduction in the potential data around the previous cluster center is performed using
$$D_i^t = D_i^{t-1} - D_{c_{ki}} \tag{15}$$
  where
$$D_{c_{ki}} = Z * e^{-4\left[\sum_{j=1}^{m} \left(\frac{C_{kj} - x_{ij}}{r * q} \right)^2 \right]} \tag{16}$$
- Returns the cluster center to the original data using
$$x = (a - b) \ln(\mu - \mu e^{-s} + e^{-s}) + a \tag{17}$$
- Calculate the degree value of membership using
$$\mu_{k_i} = e^{-\sum_{j=1}^{m} \left(\frac{x_{ij} - C_{kj}}{\sqrt{2}\sigma_j} \right)} \tag{18}$$
  where
$$\sigma_j = \frac{r * \left(X_{max_j} - X_{min_j} \right)}{\sqrt{8}} \tag{19}$$
- Calculating the validity index using the Partition Coefficient [21]

$$PC = \frac{1}{N}\left(\sum_{i=1}^{N}\sum_{j=1}^{K}\mu_{ij}^{2}\right) \tag{20}$$

where N is the number of research objects, K is the number of clusters, and μ_{ij} is the value of membership of the i-th object with the center of the j-th group.

4. Result and Discussion

The PCA calculation process will produce output in the form of the main component, where the main component will be used as input in the FSC calculation process. The data used in this study is hypertension data. The amount of data is 500 data with 5 variables. The variables used are age (X_1), sex (X_2), systolic pressure (X_3), diastolic pressure (X_4), and body weight (X_5). Data processing was performed using Jupyter Notebook Software with the Python programming language. Then, the cluster results from the FSC with the PCA input will be compared with the cluster results from the FSC only.

Table 1. Eigen value and total variance.

Component	Eigen Value (λ)	Variance (%)	Kumulatif (%)
1	0.0689	30.45	30.45
2	0.0569	25.11	55.56
3	0.0459	20.30	75.86
4	0.0374	16.51	92.37
5	0.0173	7.62	100

4.1. Fuzzy Subtractive Clustering with Input Principal Component Analysis

The clustering process starts from PCA calculations by normalizing the data using (1) so that the data is in a certain range. Before determining the main component, each variable must be related to each other so that the covariance matrix (***S***) will be searched first. From the covariance matrix, a PCA analysis will be carried out by looking at the eigen values of each variable. The new variable (principal component) that is formed is based on the proportion of the total variance and the proportion of cumulates. The results of the calculation of eigen values and variance can be seen in Table 1.

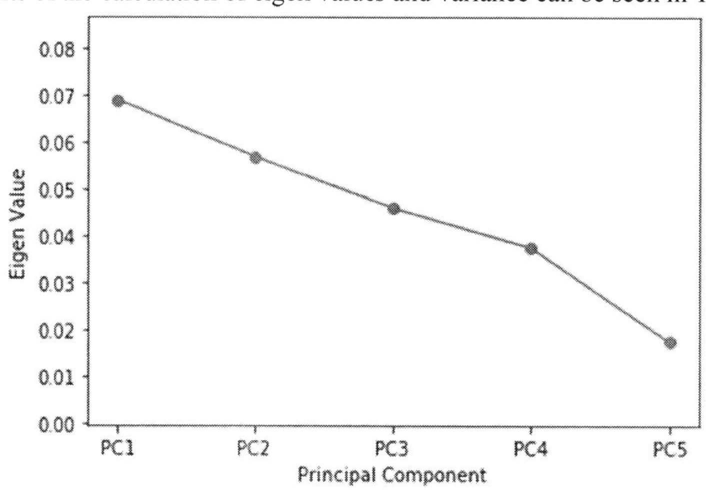

Figure 1. Eigen value.

In Table I, there are 3 new variables or main components that represent the analyzed variables. The ability of each component to represent the variables analyzed is indicated by the amount of variance described, which is called the eigenvalue. The value of the eigen values shows the contribution of the main components to the variance of all the original variables analyzed. In addition, Table 1 shows that the PKV value ≥ 70% is 75.86%. This means that the number of main components formed is 3 components. The amount of this variance also determines the principal component number to be selected based on the total proportion of the variance or eigen values. The eigen values graph can be seen in Figure 1.

Then, based on the previously obtained eigen vectors, it will show which variables are included in the 3 main components produced. The eigen vectors obtained from the 3 largest eigenvalues are the loading/coefficient values of the three main components formed. The loading value / coefficient of the 3 main components, namely.

Table 2. Loading/Principal Component Coefficient.

Variable	Component		
	1	2	3
1	-0.4380	**0.5087**	0.3117
2	-0.1385	-0.1556	**0.8373**
3	**0.7084**	0.6652	0.1942
4	**-0.5217**	0.5187	-0.2555
5	-0.0663	-0,0801	**0.3142**

Table 2 shows the three main components that are formed, the largest loading / coefficient value (in bold) will be chosen where this value is able to explain the variables that affect hypertension. The three new variables (main components) that were formed can be seen in Table 3.

Table 3. Summary of PCA.

Component	Variable	Loading/ coefficient	The variants describe
PC_1	X_3	0.7048	30.45 %
	X_4	-0.5317	
PC_2	X_1	0.5078	25.11 %
PC_3	X_2	0.8373	20.30 %
	X_5	0.3142	

Based on Table III, PC_1 consists of variables X_3 dan X_4 with a variance of 30.45 %, PC_2 consists of variable X_1 with a variance of 25.11 % dan PC_3 consists of variables X_2 dan X_5 with a variance of 20.30%.

Then the output from the PCA will be used as input to the FSC. The results of the comparison of the Partition Coefficient value and the number of iterations between the PCAFSC and FSC methods are shown in Table 4.

Table 4. Comparison of the results PCAFSC and FSC.

Number of clusters	PCAFSC			FSC		
	r	Partition coefficient	Number of iterations	r	Partition coefficient	Number of iterations
2	0.5	0.5874	3	0.72	0.5369	3
3	0.55	0.5301	4	0.79	0.4801	9
4	0.69	0.5974	5	0.97	0.5472	11

Based on Table IV, it is obtained that the largest Partition Coefficient value is found in the number of clusters 4, for each method of 0.5974 and 0.5472. The cluster center using PCAFSC is

$$C_{0.5} = \begin{bmatrix} 0.2250 & 0.3864 & 0.2464 \\ 0.3623 & 0.2964 & 0.6703 \\ 0.5189 & 0.5110 & 0.1633 \\ 0.2492 & 0.8169 & 0.2562 \end{bmatrix}$$

Meanwhile, the cluster center for the FSC method is as follows

$$C_{0.72} = \begin{bmatrix} 0.4340 & 0 & 0.4609 & 0.2862 & 0.4018 \\ 0.4634 & 1 & 0.4609 & 0.2862 & 0.6501 \\ 0.4196 & 1 & 0.4609 & 0.2862 & 0.1468 \\ 0.3919 & 0 & 0.5144 & 0.1597 & 1 \end{bmatrix}$$

5. Conclusion

Based on the PCA calculation results obtained 3 new variables (main components) of the 5 variables studied. The main components that are formed are PC_1, PC_2, and PC_3. The first main component (PC_1) consists of variables X_3 and X_4 with a total variance of 30.45%. The second main component (PC_2) consists of the variable X_1 with a total variance of 25.11%. The third main component (PC_3) consists of variables X_2 and X_5 with a total variance of 20.30%. The three main components formed resulted in a cumulative variance proportion of 75.86%, which means that the factors that influence hypertension can be explained only by the 3 new variables that are formed. For FSC calculations, the optimal number of clusters using the PCAFSC and FSC methods is 4 clusters, where the PC value for each method is 0.5974 with 4 iterations and 0.5473 with 11 iterations. So, it can be concluded that the clustering using the PCAFSC method is better than the FSC method.

References

[1] Mattjik A A and Sumertajaya I M 2011 *Sidik Peubah Ganda* (Bogor: IPB Press)
[2] Timm N H 2000 *Applied Multivariate Analysis* (New York: Springer)
[3] Jollife I T 2002 *Principal Component Analysis* (New York: Springer)
[4] Dafitri H, Asih M S and Astuti R I 2019 Media interaktif pengenalan angka dengan jari tangan menggunakan metode PCA *J. of Information System* **3** no 2
[5] Rencher A C 2002 *Methods Of Multivariate Analysis* (USA: Wiley Sons)
[6] Rahmawati, Wartono and Putri A Y 2019 Pengklasteran tingkat pendidikan pegawai menggunakan metode fuzzy subtractive clustering (Studi Kasus: Badan Kepegawaian dan Pelatihan Daerah Provinsi Riau) *J. of Mathematical Sci. And Statistics* **4** no 1
[7] Syahra Y, Syahril M and Yusnidah 2018 Implementasi data mining dengan menggunakan algoritma FSC dalam pengelompokkan nilai untuk menentukan minat belajar siswa SMP Primbana Medan *SAINTIKOM J.* **17** no 1
[8] Kusumadewi S and Purnomo H 2010 *Aplikasi Logika Fuzzy Untuk Pendukung Keputusan* (Yogyakarta: Graha Ilmu)
[9] Singh T S, Verma P and Yadava R D S 2017 Fuzzy subtractive clustering for polymer data mining for SAW sensor array based electronic nose *Proc. of Sixth Int. Conf. on Soft Computing for Problem Solving. Advances in Intelligent Systems and Computing* ed K Deep *et al* vol **546** (Singapore: Springer) pp 245–53
[10] Chen H and Li C 2019 Comprehensive service level analysis of online taxi drivers based on fuzzy clustering combined with principal component analysis *Proc. of the 2019 4th Int. Conf. on Humanities Science and Society Development (ICHSSD 2019)* pp 577–83
[11] Wang S, Wang J and Han Y 2019 Pattern matching strategy based on optimized fuzzy clustering algorithm for industrial internet *Preprint*
[12] Anggraini S, Kusumawardani M and Darmono H 2017 Perbandingan penempatan node B

UMTS900 pada BTS Existing menggunakan fuzzy c-means dan fuzzy subtractive clustering *J. of JARTEL* **4** no 1
[13] Anton H and Rorres C 2005 *Elementary Linear Algebra* (New Jersey: John Wiley Sons)
[14] Johnson R A and Wichern D W 2002 *Applied Multivariate Statitical Analysis* (USA: Prentice Hall)
[15] Gasperz V 1995 *Teknik Analisis Dalam Penelitian Percobaan* (Bandung: Tarsito)
[16] Zimmerman H 1996 *Fuzzy Set Theory And Its Applications* (London: Kluwer Academic)
[17] Jang J S R, Sun C T and Mizutani E 1997 *Neuro Fuzzy And Soft Computing: A Computational Approach To Learning And Machine Intelligence* (USA: Prentice Hall)
[18] Priddy K L and Keller P E 2005 *Artificial Neural Networks* (USA: SPIE Press)
[19] Mahajan S and Gupta S K 2019 *On Fully Intuitionistic Fuzzy Multiobjective Transportation Problem Using Different Membership Functions* (New York: Springer)
[20] Rezaei K and Rezaei H 2019 New Distance And Similarity Measures For Hesitant Fuzzy Soft Sets *Iranian J. of Fuzzy Systems* **16(6)** pp 159–76
[21] Bezdek J C 1981 *Pattern Recognition With Fuzzy Objective Algorthm* (New York: Plenum Press)

Developing a framework for self-isolation app: a case study of Jakarta's COVID-19 mitigation

R Gultom[1], A Hirzirahim[1], Y Nugraha[1,2*], Z Nursyufa[1], Hamdi[1], A L Suherman[1,3]

[1]Jakarta Smart City, Department of Communications, Informations, and Statistics, Jakarta, Indonesia
[2]School of Computing, Telkom University, Bandung, Indonesia
[3]Directorate of Research and Community Service, Telkom University, Bandung, Indonesia

E-mail: *yudhistira.nugraha@jakarta.go.id

Abstract. Self-isolation for COVID-19 suspects plays a crucial role due to a lack of medical personnel and hospital capacity in Jakarta. The ongoing COVID-19 pandemic has taken an unprecedented toll on health services because the number of infections continues to climb and is thought to be far higher because many people have not been tested. This phenomenon opens an opportunity to support innovative collaborations between government and city co-creators (e.g. start-up companies) to develop a digital application for monitoring the self-isolation activity of COVID-19 suspects. This paper proposes a conceptual framework for developing a self-isolation app by conducting in-depth interviews with 20 participants from the Jakarta Health Office and mobile health app start-up companies. The result of this paper figures out the key features of the app, such as suspects database, daily monitoring form, geotagging, chatroom, test scheduling, panic button, and daily notification. In this paper, Jakarta was selected as a case study since it is stated as a collaboration city which initiates the opportunity to business sectors for increasing access and public service quality. This paper uses design thinking methods as an evidence base to develop a self-isolation app. The conceptual framework provides the foundation for the input-process-output model and insights into what an appropriate approach for the requirements of the self-isolation app development. This paper provides a better understanding of common elements in building the app. Its conceptual framework fits in any government sector. Also, it can inspire other cities to consider the model in developing a similar application with structural and systematic phases in every development step. Further research should be carried out to build an extension model of the design thinking approach to get specific quantitative and qualitative results.

1. Introduction

Since Jakarta's first case of COVID-19 was reported in March 2020, the official data on corona.jakarta.go.id shows that the total number of COVID-19 cases has significantly increased to 18,230 infected patients on July 24, 2020 [1]. However, the Department of Health of Jakarta Provincial Government which hereinafter referred to Jakarta Health Office has limited services in terms of the capacity of medical personnel and the isolation rooms that only 657 places available [2]. As a result, the monitoring process of all the COVID-19 suspects (suspects) could not be accommodated. Such limitation causes the number of suspects referred for self-isolation to have significantly increased [1]. This issue led to the idea to execute self-isolation activities at the public facilities according to the health protocol. Self-isolation at home cannot provide a guarantee of the optimal prevention of the spreading from the coronavirus due to the lack of supervision of the medical workers [3].

On the other hand, smartphone contact tracing technology can be limiting the spread of infection through supervised location due to a significant percentage of the population can be quickly and precisely monitored [4]. Moreover, the COVID-19 symptoms are similar to the flu virus, and there are recorded cases of COVID-19, which are asymptomatic [5]. Thus, it is complicated problems to be addressed by the government, especially in Jakarta, to obtain and to monitor information regarding the spreading of COVID-19 virus at early as possible.

Developing an appropriate self-isolation app is a continuing concern within mobile health app start-up companies in handling COVID-19, especially suspect who require supervision from time to time [4]. Lack of business process understanding has existed as a critical component for developing a suitable application (the app). Data from several studies suggest that the digital intervention for monitoring suspects has a significant impact on the quarantine of individuals, and cooperation between government and private sectors is needed [4] [5] [6]. Such approaches and models, however, failed to address the development of the self-isolation app because they did not focus on the method, but only explained the features of the app. Interestingly, no previous study has investigated the development of such an app, especially in the context of Jakarta and Indonesia [6]. This condition indicates a need to understand the various perceptions of the self-tracking app that exist among government as a problem owner and start-ups as problem solvers.

This paper provides an insight into the conceptual framework underlying the foundation of the input-process-output model in building a self-isolation app. It seeks to explain the development of such a framework in the form of a system diagram. A design thinking approach was used to follow a case study design, with in-depth interviews of 20 participants from Jakarta Health Office, Jakarta Smart City, and private sectors (start-up) to investigate key features of such an app. This paper is the first study to undertake qualitative research with design thinking models. We use Jakarta as a case study because Jakarta is introduced as a collaboration city to initiate the opportunity for business sectors in increasing access and public service quality [7]. The importance and originality of this study are that it proposes a conceptual framework in building a self-isolation app in which such a framework can inspire other cities to consider the implementation of the self-isolation app. Also, the reader should bear in mind that the study is based on qualitative analysis of real cases in Jakarta.

The remainder of this paper is structured as follows. Section 2 describes the methodology for investigating insights into the development of the conceptual framework. Section 3 presents the results and analysis of the study. Section 4 discusses how the finding compares with related work and the implication for government, private, and academic sectors. Finally, the last section presents the conclusion.

2. Methodology

2.1. Aim

This paper investigates the requirements to develop a framework for developing self-isolation platform monitoring by employing 20 participants, consisting of 4 participants from the Jakarta Health Office, 7 participants from Jakarta Smart City, and 9 participants from private sectors.

The process of developing a self-isolation monitoring system starts from observing the requisites of the medical personnel from the Jakarta Health Office as a problem owner in the interest of mitigating the monitoring process of suspects. This study was conducted through interviews with participants from the Jakarta Health Office, Jakarta Smart City, and private sectors (mobile health app start-up companies). The recruitment process for participants was carried out by using existing connections via verbal, Zoom invitation, or official invitation letter from Jakarta's Department of Communications, Informatics, and Statistics. Before the research began, a group discussion was held between the Jakarta Health Office and the private sectors, which finally engaged 20 of 38 invited participants whereas 18 participants from private sectors lacked the technological capability to accommodate the requirement of

Jakarta Health Office. Group discussions were conducted via the Zoom Meeting due to the COVID-19 pandemic issues. The selected participants consisted of health and technology professionals as well as medical doctors, systems, and IT development experts. Table 1 presents a summary of the participants involved in this study.

Table 1. The participants' information and experience.

Participant			
Identifier	**Gender**	**Role**	**Category**
P1	Male	Doctor	Government sectors
P2	Male	Doctor	Government sectors
P3	Male	Doctor	Government sectors
P4	Female	Doctor	Government sectors
P5	Male	IT expert	Government sectors
P6	Male	IT expert	Government sectors
P7	Male	IT expert	Government sectors
P8	Male	IT expert	Government sectors
P9	Female	IT expert	Government sectors
P10	Female	IT expert	Government sectors
P11	Female	IT expert	Government sectors
P12	Male	Director/Doctor	Private sectors
P13	Male	IT expert	Private sectors
P14	Male	IT expert	Private sectors
P15	Female	IT expert	Private sectors
P16	Male	IT expert	Private sectors
P17	Male	IT expert	Private sectors
P18	Female	IT expert	Private sectors
P19	Male	IT expert	Private sectors
P20	Male	IT expert	Private sectors

2.2. Procedure

This study adopted a case study approach as an empirical inquiry method to understand a contemporary phenomenon and complicated situation faced by the Jakarta Health Office[8] [9], using focus group discussion (FGD) and in-depth interviews as the instrument for data collection of qualitative research [10]. It will give practical thoughts and opinions from participants. The activities of group discussion and interview took between 60 – 120 minutes per meeting.

There are six aspects of quality [11] and two elements of project management used to trigger the discussion and find the requirements of the Jakarta Health Office. For this purpose of this study, participants were asked to respond to the following aspects below:

- Performance refers to the functional product in operating characteristics to achieve its objectives.
- Reliability considers that the product will not fail within a specific period. It includes data security and integration.
- Serviceability appraises how convenient the product is to be repaired when it breaks down. It includes maintenance and cooperation agreement documents.
- Aesthetic indicates the impression of user response to the product. It includes ease of use.
- Perceived quality is based on the customer's experience of satisfaction with the product.
- Features are other functions that the product can perform besides the main functionality.

- A resource is the availability of human resources. It includes system maintenance, development, and users.
- Agreement cooperation: it includes cost, branding, period, test, and trial.

Finally, the participants had an opportunity to share any additional thoughts and opinions. Further, they were allowed to elaborate on their experiences beyond the aspects above, which were used for guidance to gain more information. This activity allowed for new elements to emerge, based on real experiences from participants in developing self-isolation app.

2.3. *Data Collection and Analysis*

This study used eight aspects to collect the requirements of system design and development. The design thinking approach of Stanford was used to create ideas and solve the problem [12]. It provides solutions for complex problems [13]. Moreover, it can help systems and strategies innovators to build a new concept because it will be the first step to develop others [14]. The design thinking process is an iterative process that has many detailed steps and intensive interaction with end-users. Each phase of the activities must achieve several results that can be delivered to the next stage [15]. The researchers must investigate and dig deeper to get the information needed from the problem owner to understand the context and problem [16]. Qualitative data was collected and analyzed through five-phases of design thinking. The result of focus group discussions and in-depth interviews will be mapped and analyzed in the design thinking method so that it can reduce the bias and obtain innovative works for the next study [16]. Figure 1 shows the five phases of the design thinking method were applied in this study.

Empathize	• To understand the situation and collect the information from the problem owner. • This phase uses an in-depth interview. • Applying the empathy map canvas as an instrument.
Define	• To identify the project objectives and understand the requirements of the problem owner. • This phase uses brainstorming with the stakeholders. • Applying the customer journey map and describing the requirements of the problem owner.
Ideate	• To explore ideas and solutions of the problem. • This phase uses brainstorming with the stakeholders. • Applying the sketches concept as an instrument.
Prototype	• To build a prototype based on ideas, and to present the result to the problem owner. • This phase uses brainstorming with the stakeholders. • Applying the storyboards as an instrument.
Test	• To test and fulfill the prototype, and to ensure the requirements of the problem owner.

Figure 1. Five-phases of the design thinking method

3. Result

This paper introduces a conceptual framework in the form of a system diagram, as shown in Figure 2. It begins with the problem owner, Jakarta Health Office, which has a goal to conduct a collaboration with private sectors that have competency in digital technology to develop a self-isolation monitoring app for Covid-19 suspects. Moreover, the stakeholders related to the problem are listed as Jakarta Health Office, Jakarta Smart City, and private sectors (start-up companies). Chosen stakeholders are the prominent policymakers and industrial actors that directly correlate with the problem owner goal. The strategies are derived from the crisis management of the Jakarta Health Office, in which such requirements will affect self-isolation monitoring app development. The government stakeholders such as Jakarta Health Office and Jakarta Smart City are then collaborating in framing the supportive policies to be considered in this model, such as:

- Jakarta Governor Regulation Number 306 of 2016. This policy governs Jakarta Smart City should develop coordination, partnership, and integration in information and technology dissemination with various parties [17].
- Jakarta Governor Regulation Number 159 of 2019. This policy states the obligation of the Jakarta Health Office in building a collaboration with private sectors to resolve health problems [18].

Further, the proposed framework consists of three main elements: Input-Process-Output model. First, the input variable is derived by the assumption that the number of COVID-19 suspects, the number of medical personnel, and the capacity of isolation rooms will be following the current trend that is independent of the process. The process, as the second aspect, is divided into five phases (which will be discussed deeper later). This study learned about the app users more deeply after going through the definition phase to make sure the needs and the requirements of the problem owner, and after going through the test phase, it obtained new ideas and revealed insight to redefine the needs of the problem owner. The last aspect, the output, is about suspects that can be isolated independently according to health protocols and well monitored using the developed application.

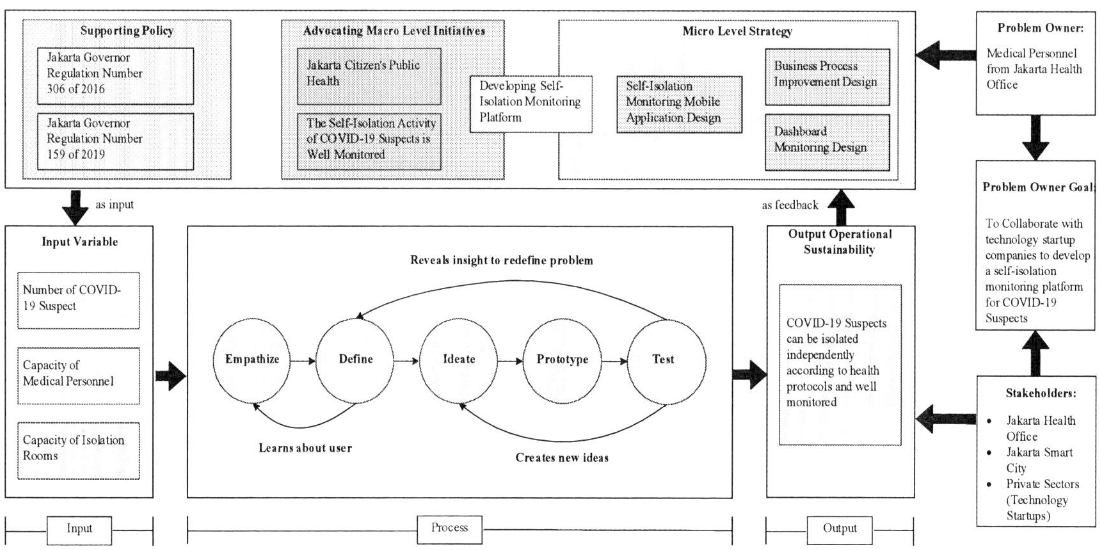

Figure 2. The conceptual framework of self-isolation app development.

3.1. *Empathize*

In the empathize phase, Jakarta Smart City tries to collect and understand the condition of Jakarta Health Office in handling COVID-19 cases in DKI Jakarta. The daily work of medical personnel includes giving medical services, doing research, monitoring and controlling suspects, giving treatment, testing in the laboratory, etc. However, the impact of COVID-19 makes medical personnel need to provide extra services to the suspects; even they have a high risk of COVID-19 exposure. From that perspective, the medical personnel is overwhelmed by the increasing number of suspects treated every day. Table 2 describes an empathy map of Jakarta Health Office.

Table 2. Jakarta health office empathy map.

A. Say:	B. Do:	C. Think:	D. Feel:
"The total number of COVID-19 cases has significantly increased during the pandemic that makes the medical staff has to provide extra service to the suspects."	"Medical officers supervise not only one COVID-19 suspect but more than that."	"It's expected the workload of medical staff can be stabilized. So it can accommodate services during the increasing number of COVID-19."	"The medical staff was overwhelmed by the increasing number of suspects treated every day."

3.2. Define

This phase begins with identifying the actual problem and trying to find the needs and requirements of the problem owner. Data were gathered from brainstorming activity, which is held among stakeholders to find the goals, and it was mapped into the customer journey map. Figure 3 provides the customer journey map based on group discussions with stakeholders. The customer journey map is used to illustrate the experience of the medical personnel from the Jakarta Health Office and the context of the problem.

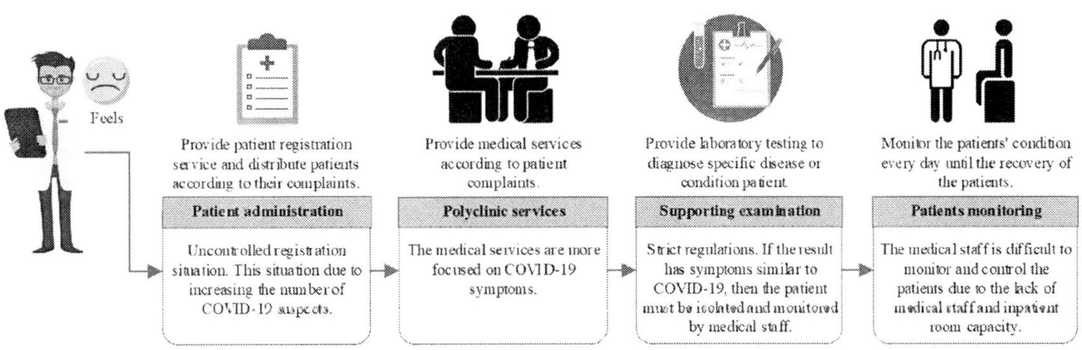

Figure 3. Customer journey map.

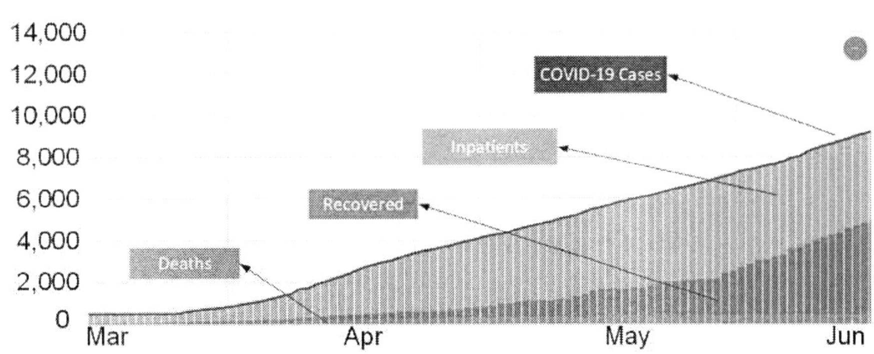

Figure 4. Graph of the increment COVID-19 inpatients.

In summary, the customer journey map confirms that the medical personnel feels overwhelmed due to the lack of medical staff and inpatient rooms capacity. This condition is supported by data of the level of COVID-19 inpatients increased day by day (March to June 2020), as can be seen in Figure 4 [19]. The Indonesian Ministry of Health also has given direction to conduct a self-isolation protocol by the Circular Letter of the Ministry of Health number 202 of 2020 [20].

After analyzing and understanding the problem, this phase formulates that the problem owner needs a self-isolation system based on technology to stabilize the workload of medical personnel, reduce the spread of COVID-19, and improve the medical services to the suspects. Table 3 explains the Jakarta Health Office requirements of the self-isolation monitoring of COVID-19 suspects.

Table 3. The Jakarta Health Office requirements of self-isolation monitoring of COVID-19 suspects.

Identifier	Needs	Description
N1	Collecting self-isolation data	Collecting and identifying data on the number of suspects, based on data COVID-19 patients from Jakarta Health Office and the public who require self-isolation for 14 days.
N2	Monitoring suspects' condition	Monitoring the activities and conditions of suspects for 14 days.
N3	Tracking the suspects	Tracking the movement of the self-isolation suspects for 14 days.
N4	Communicating with suspects	Two-way communication. If the suspects have a COVID-19 symptom for 14 days, the medical staff can immediately provide treatment.
N5	Scheduling suspects to come to the nearest public health centre	If the suspect doesn't get better during 14 days, the public health centre officer will provide a schedule to get medical services in the nearest public health centre.

3.3. Idea

From the various problems that have been outlined, then the Jakarta Health Office with Jakarta Smart City begin a focus group discussion activity between the private sectors and the Jakarta Health Office to brainstorm the new self-isolation process based on digital technology.

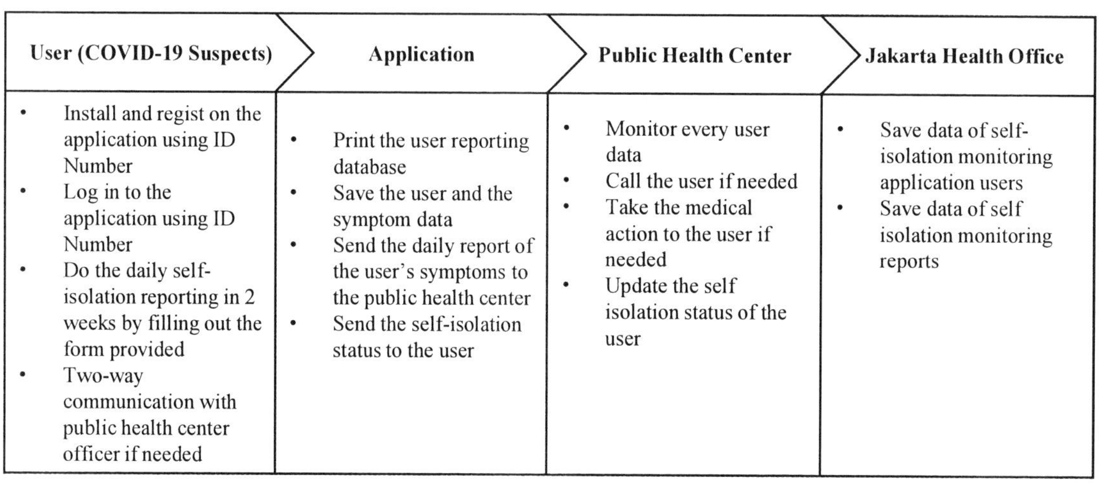

Figure 5. The business process of self-isolation monitoring application.

The result is a new integrated system solution in monitoring and reporting the self-isolation COVID-19 suspects. This system is expected to accelerate the work process of the related parties. The system is expected to accelerate the work process of related parties to monitor all suspects. The conditions, the locations, and the treatments to every suspect in DKI Jakarta still can be maximally facilitated as well. Figure 5 shows the collaborative business process system.

This self-isolation monitoring and reporting system are integrated with the Jakarta Health Office, public health centres, application providers (private sectors), and users (self-isolation COVID-19 suspects). Users need to create an account with the ID number (resident number in identity card) to register on the application. The user data will be verified and directed to the Jakarta Health Office for the daily monitoring requirements of the suspect's condition by the public health centre in two weeks. Users have to fill out the form provided as a monitoring process. The daily data will be automatically recorded in the Jakarta Health Office system. The application also allows for two-way communication between suspects and public health centre officers for further treatment if needed. The system contains confidential information of COVID-19 suspects which is be guaranteed by segregation of duties between government and private sectors to reduce opportunities for unauthorized or unintentional modification or misuse of the organization's assets. Jakarta Health Office as an authorized party has the responsibility to permit the data utilization based on Non-Disclosure Agreement (NDA).

Based on that business process, a solution requisite is formulated as the features that will be included in the application. Table 4 describes the result of features expected to resolve the problem.

Table 4. Self-isolation monitoring application features.

Needs	Feature	Description
N1	Suspect Database	The table which contains self-isolation suspects' data included their profile and status
N2	*Daily monitoring form*	It includes several questions based on the Jakarta Health Office standard for daily suspect condition checking
N3	*Geotagging*	the feature which can be tracked the location and the movement of the suspects during the self-isolation period
N4	*Chat room*	The feature which allows the suspects and the public health centre officer to have a two-way communication
N4	*Panic button*	The feature which enables the suspects to send an emergency to the public health centre officer to be rescued immediately
N4	*Daily Notification*	The daily notification feature which will be popped up on the suspects' device as a reminder of the daily reporting
N5	*Test scheduling*	It contains the calendar and the digital clock which allow the suspects to choose the Swab test schedule

3.4. Prototype

After the solutions are agreed upon by every participant concerned, the next pace is designing a prototype which includes a self-isolation monitoring and reporting system. In this phase, private sectors, Jakarta Smart City and Jakarta Health Office are collaborating to develop a prototype into two types of the app designed which are mutually integrated, the first one is a mobile application which will be used for suspects and public health centre officers, and the second one is a suspect monitoring dashboard used by public health centre officers. Figure 6 is the interaction system diagram of the self-isolation app.

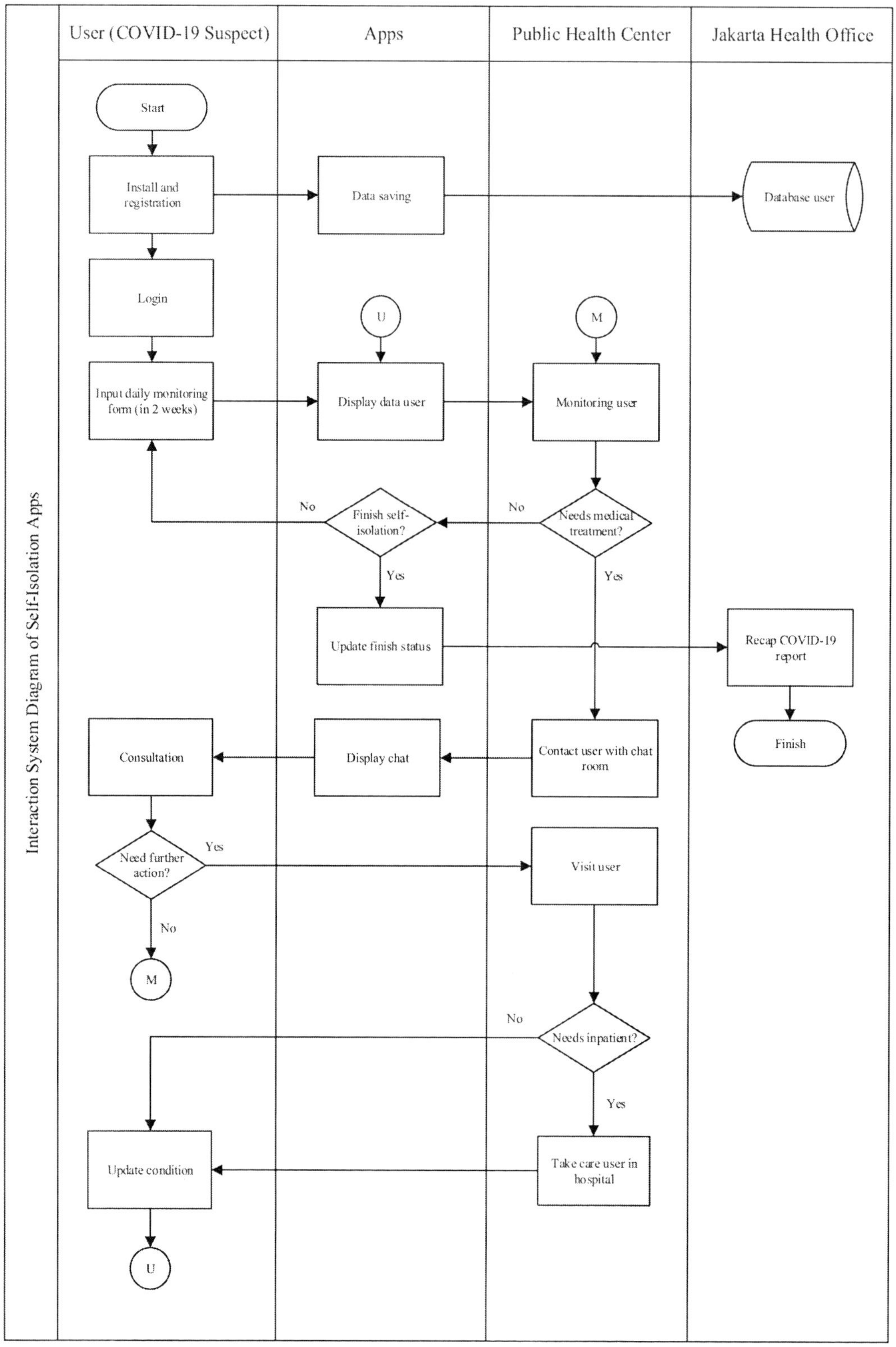

Figure 6. Interaction System Diagram of Self-Isolation Apps.

3.5. Test

The next pace after completing the prototype is an implementation phase which will be started by pilot testing activity for the selected medical personnel at the public health centre and continued with the cooperation agreement signing among the participants concerned. The application then will be introduced to the medical personnel widely in technical guidance activities and disseminated to the public. This app is easy to use for medical personnel and the public. Surprisingly, testing phase revealed that people visited public health centre only checked up their health condition. On the other hand, this activity can be conducted by a self-assessment monitoring form.

In summary, these results show that supporting policy is needed to support a collaboration between the government and the private sector. This case study that focused on building an end-user app can be accommodated by the design thinking method, which has a human-centred design and iteration process. The iterating process should deepen the result of the defined phase from the empathize phase to learn the medical personnel needs accurately. Furthermore, the testing phase result revealed some insights and created new ideas for features app, which can be deepened by the iterating process from define and ideate phases.

4. Discussion

This paper has several contributions to the literature and potential guidelines for developing the COVID-19 self-isolation monitoring app, in which different sectors, who have diverse roles and responsibilities, as well as different perspectives, are collaborating.

4.1. Reflection on Related Works

Prior studies have noted the importance of the conceptual framework for developing a system or an app [21] [22]. Still, they only focused on technical issues and less on raising the holistic view of a management and governance perspective. It differs from the results presented here confirms the use of design thinking method that facilitates the holistic view of management perspective to make an understanding not only how to create an app but also support the policies, advocate the macro-level initiatives and micro-level strategies that commonly asked from the managerial level in the organization. The use of the design thinking method enables collaboration with different stakeholders for developing an app and can make it more convenient in analyzing processes, integrating technical and non-technical factors, and applying user-centred design [23]. It allows participants to empathize with different stakeholders with open up discussion, intuitive, and rational minds, leading to innovative solutions [24].

Several reports have shown that there are few studies on developing an app for COVID-19, such as the study of portable health screening devices of respiratory infections development [25], quarantine tracker, and COVID-19 info-self reporting application in Bangladesh [6]. However, such studies did not represent the app development method used, but it only spelt out the features with their development goals. Moreover, another research only discussed the model for analyzing the data obtained from the implementation of smartphone contact tracing applications [4]. Thus, this paper is present as a pioneer of the academic research of developing the COVID-19 self-isolation app and hopefully can help other cities to build a similar app to reduce the spreading of COVID-19.

4.2. Implications

Implications include:

- Implications for Private Sectors: Collaboration with the government sector gives more value-added to strengthen company branding due to gaining a collaboration license from the government to their portfolio. Contrary to expectations, there are some participants from private sectors who are not sufficient to address government's standard operating procedures where an agreement shall be governed by the laws of the Republic of Indonesia to ensure that all sensitive government data are limited to authorized users. Some participants are not sufficient to address a standard daily monitoring form and chat room feature. This combination of findings provides some support for the premise that the private sectors should have a willingness to adjust their business process first while wanting to collaborate with government sectors.

- Implications for Government Sectors: The present research raises the possibility that smart collaboration improves the way of serving the public using technology, harness innovation to improve the outcomes, and engage young entrepreneurs who run start-ups in the technology sector to meet evolving needs. Thus, government sectors also can use this framework into many collaborations for developing an app in any public service sector such as education, public safety, public welfare, and other vital industries public sectors that can be implemented smart city initiatives. Unfortunately, this study needed extra time to complete government administration tasks such as a memorandum of understanding, meeting invitation letters, collaboration proposals from the private sector, and collaboration agreement.
- Implications for Researchers: this paper can provide the needs, features, and guidelines to develop the COVID-19 self-isolation monitoring app through the design thinking approach. However, the utility of design thinking research is not limited to develop technological aspects. It can, therefore, be assumed the design thinking approach can also be used to create user-centred business models or other systems. On the other hand, the most significant limitation lies in the fact that suspect diagnosis based on self-assessment potentially biases from the doctor's so that the assessment result becomes less accurate.

5. Conclusion

This paper has developed a conceptual framework that illustrates a holistic picture of smart collaboration in developing the COVID-19 self-isolation monitoring app that can be used as the pioneer reference framework for the next researches in the future. The current conceptual framework highlights the importance of the design thinking approach. This paper has also shown that the smart collaboration of government sectors and private sectors is playing a significant role to use technologies as a solution to help the problem owner finding the key features of the COVID-19 self-isolation apps, such as suspect databases, daily monitoring form, geotagging, chat room, test scheduling, panic button, and daily notification. Overall, this paper strengthens the idea that the government sectors should increase the collaboration with private sectors using a design thinking approach that needs willingness from private sectors to be flexible aligning their business process to government sector requirements.

A useful conceptual framework can support the next step of modelling in any government service sector. Further research should be carried out to build an extension model of the design thinking approach to get specific quantitative and qualitative results. Further studies need to be carried out more validated and verified quantitative methods.

Acknowledgments

The authors thank the Jakarta Health Office and Private Sectors involved in this paper for their supports and valuable discussion. This paper has no competing interests.

References

[1] Jakarta Smart City 2020 *Data Pemantauan COVID-19 DKI Jakarta* Retrieved from https://corona.jakarta.go.id/id

[2] Yunianto T K 2020 *Anies Mengaku RS di Jakarta Tak Dirancang Terima Ledakan Pasien Corona* Retrieved from https://katadata.co.id/berita/2020/04/16/anies-mengaku-rs-di-jakarta-tak-dirancang-terima-ledakan-pasien-corona

[3] Feng Z, Cheng Y, Ye L, Zhou M Y, Wang M W and Chen J 2020 Is home isolation appropriate for preventing the spread of COVID-19? *Public Health* **183** pp 4–5

[4] Hernández-Orallo E, Manzoni P, Calafate C T, and Cano J 2020 Evaluating how smartphone contact tracing technology can reduce the spread of infectious diseases: the case of COVID-19 *IEEE Access* **8** pp 99083–97

[5] Cai J, Sun W, Huang J, Gamber M, Wu J and He G 2020 Indirect virus transmission in cluster of

COVID-19 cases, Wenzhou, China 2020 *Emerging Infectious Diseases* **26** pp 1343–5
[6] Islam M N and Islam A K M N 2020 A systematic review of the digital interventions for fighting COVID-19: The Bangladesh perspective *IEEE Access* **8** pp 114078–87
[7] Documentation & Information Legal Network DKI Jakarta 2020 *Jakarta Development Collaboration Network* Retrieved from https://jdih.jakarta.go.id/uploads/default /produkhukum/PERGUB_NO._24_TAHUN_20201.pdf)
[8] Yin R K 2003 *Case Study Research, Design, and Methods* (Newbury Park: Sage Publications)
[9] Denscombe M 2010 *The Good Research Guide: for small-scale social research projects* (Berkshire: Open University Press)
[10] Denzin N and Lincoln Y 1994 *Handbook of Qualitative Research* (Thousand Oaks: Sage Publications)
[11] Montgomery D C 2009 *Introduction to Statistical Quality Control* (Arizona: John Wiley & SonsInc)
[12] Stanford 2012 *An Introduction to Design Thinking Process Guide* Retrieved from https://dschool-old.stanford.edu/sandbox/groups/designresources/wiki/36873/attachments/74b3d/ModeGuideBOOTCAMP2010L.pdf
[13] Kolko J 2015 Design thinking comes of age *Harvard Business Review* Retrieved from https://hbr.org/2015/09/design-thinking-comes-of-age
[14] Brown T 2008 Design thinking *Harvard Business Review* Retrieved from https://hbr.org/2008/06/design-thinking
[15] Chen P and Huang R 2017 Design thinking in app inventor game design and development *IEEE17th Int. Conf. on Advanced Learning Technologies* pp 139–41
[16] Haglund L 2020 Evidence basing the study environment needs at a small specialist university by using design thinking methods *J. of EAHIL* 16 pp 4–11
[17] Documentation & Information Legal Network DKI Jakarta 2016 *Formation, Organization, and Procedures of Jakarta Smart City Governance unit* Retrieved from https://jdih.jakarta.go.id/uploads/default/produkhukum/PERGUB_NO.306_TAHUN_.2016_.pdf
[18] Documentation & Information Legal Network DKI Jakarta, Formation 2019 *Organization, and Procedures of Jakarta Health Office* Retrieved from https://jdih.jakarta.go.id/uploads/default/produkhukum/PERGUB NO. 159 TAHUN 2019.pdf
[19] Gugus Tugas Percepatan Penanganan COVID-19 2020 *Data COVID-19* Retrieved from https://covid19.go.id/peta-sebaran
[20] Ministry of Health Republic Indonesia 2020 *Official Documents and COVID-19 Handling Protocol* Retrieved from https://www.kemkes.go.id/article/view/20031700001/Dokumen-Resmi-dan-Protokol-Penanganan-COVID-19.html
[21] Sluga A and Butala J P 2015 A Conceptual framework for collaborative design and operations of manufacturing work systems *CIRP Annal* **54** pp 437–40
[22] Wehde M B, Buren T J V and Benscoter M A 2018 A conceptual framework for developing an implantable research platform: a multimodality approach to stimulating and recording *IEEE Int. Symp. on Medical Measurements and Applications*
[23] Alahmari F and Anandhavalli M 2018 Using design thinking in information system development: a survey *21st Saudi Computer Society National Computer Conf. (NCC)* pp 1–6
[24] Levy M and Huli C 2019 Design thinking in a nutshell for eliciting requirements of a business process: a case study of a design thinking workshop *IEEE 27th Int. Requirements Engineering Conf.* pp 351–6
[25] Jiang Z, Hu M and Zhai G 2020 Portable health screening device of respiratory infections *IEEE Int. Conf. on Multimedia & Expo Workshops (ICMEW)* pp 1–2

Implementation of Socket Priority Module for Unmanned Aerial Vehicle Network using FlyNetSimulator

Jauzak Hussaini Windiatmaja, Johannes Calvin Tjahaja, Kgs. Al Amin, Riri Fitri Sari*

Department of Electrical Engineering, Faculty of Engineering, University of Indonesia

E-mail: *riri@ui.ac.id

Abstract. Unmanned aerial vehicle (UAV) is an aircraft without a human pilot onboard and a sort of unmanned vehicle. UAV Network research activities in the real world can pose several challenges, especially in urban environments or in densely populated environments. This is due to the nature of the UAV which requires a large area of operation. Therefore, digital simulation is vital in conducting research related to UAV and its networks. One of the UAV simulators recently developed is FlyNetSim. FlyNetSim is an open-source UAV Network Simulator based on NS-3 and Ardupilot. In the application, UAV can have some network-related problems. One of them is when a UAV is in an environment that has a lot of interfering networks because it is contending a network channel. This can cause network delay. These delays will raise with the increase in the number of nodes that interfere with the UAV. To overcome the problem of interfering nodes, we propose to add a priority set module for UAV and Ground Control Station (GCS). The proposed module will set priority for the UAV and GCS socket. This module is intended to give priority to packets sent by UAVs or GCS. Later the packet that has priority will be sent first to reduce the delay caused by the appearance of interfering nodes. Setting priority in the socket may reduce the average delay caused by contending nodes. Test result shows that there is a 16.2% reduction of average delay. Test result also shows that there is a 4% raise in packet delivery ratio.

Keywords: Unmanned Aerial Vehicles, NS-3, FlyNetSimulator, Socket Priority

1. Introduction

Unmanned Aerial Vehicle (UAV) research is emerging due to the rise of UAV usage in daily life. Research conducted by Skylogic Research shows that almost 68% of UAVs are purchased for professional purposes [1]. UAV usage for professional purposes is diverse, starting from agricultural surveillance activities, disaster management, to environmental monitoring [2][3]. To the extension of wireless coverage, UAV can be integrated into communication infrastructure [4]. Research related to UAV and Network is presented in [5][6]. UAV Network research activities in real life can pose some challenges, especially in the urban area or in a very densely populated area. This is due to the nature of the UAV which requires a large area of operation. Therefore, digital simulation is vital in conducting research related to UAVs and its networks.

One of the existing UAV network simulators is FlyNetSim. FlyNetSim is built on NS-3 and Ardupilot. NS-3 is an open-source network simulator with wide community support [7]. Ardupilot is a simulator that is used to model UAV with its characteristics such as control, navigation, and route plan.

ArduPilot also supports a broad type of vehicles such as helicopters, multi-copters, fixed-wing plane, ships, submarines, and rovers [8]. FlyNetSim can capture interdependencies between the operation of UAV and the environment of the network, such as sensing, navigating, and condition of battery [9].

In the application, UAV can have some network-related problems. One of them is when a UAV is in an environment that has a lot of interfering networks because it is contending a network channel. This can cause network delays. These delays will increase with the increase in the number of nodes that interfere with the UAV. Therefore, we propose a socket priority module. The proposed module set priority for the UAV and GCS socket. This priority is intended to give priority to packets sent by UAVs and GCS. Later the packet that has priority will be sent first to reduce the delay caused by the appearance of interfering nodes.

The rest of this paper is structured as follows. Section II reviews the simulation tools related to UAV, network, and a combination of the two. Section III discusses the details of the module we proposed and their implementation on FlyNetSim. Section IV discusses the results and analysis of the implementation carried out. Finally, Section V concludes the results of the research conducted.

2. UAV and Network Simulator

This section lay out the overview of network simulators and UAV simulators. We present the epitome of FlyNetSimulator, a joint UAV-Network simulator that is used to implement the proposed socket priority algorithm.

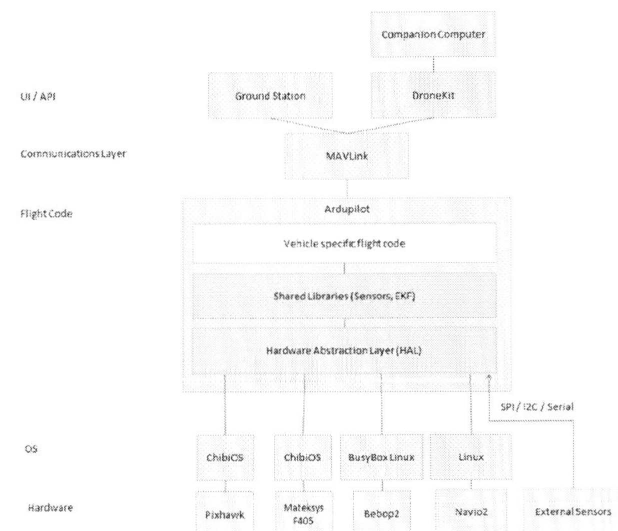

Figure 1. ArduPilot Basic Structure [14]

2.1. *UAV Simulator*

The main function of UAV simulators is to archetype functional aspects and aerodynamics aspects of UAV systems. Some simulators like H-SIM Flight Simulator only simulate human-controlled vehicles and autonomous vehicles [10]. PX4 and Ardupilot are the most used autopilot simulators. Ardupilot is a pioneer in open-source system of autopilot simulator that supports helicopters, multi-copters, fixed-wing aircraft, rovers, antenna trackers, and even submarines. Ardupilot has a wider range of supported platforms than PX4, such as Pixhawk [11], NAVIO [12], and Erle-Brain [13]. The fundamental structure of Ardupilot consists of five main components, i.e. vehicle code, hardware abstraction layer (AP_HAL), open-source libraries, tools directories, external support code (i.e. dronkekit and mavlink) [14]. The fundamental structure of Ardupilot is described in Figure 1.

Some UAV simulators are built based on Ardupilot Simulator to broaden its functionalities. For example, Gazebo [15] broaden the scope of simulated vehicles to include planes, small robots, and land rovers using Robot Operating System [16]. Gazebo is a very heavy simulator, so that it might be hard to run on one environment when doing simulation on many vehicles. Other simulators based on Ardupilot Simulator are XPlane-10 and RealFlight. However, most of them only focus on navigation and motion. To create FlyNetSim, Software In The Loop (SITL) is used [14]. SITL is a simple simulator build on Ardupilot that supports autopilot navigation. SITL allows ArduPilot to run on end-user computer instantaneously, without requiring any special hardware. It exploits the way that ArduPilot is a compact autopilot that can run on many platforms.

When running in SITL the sensor data originated from a flight dynamics model in a flight simulator. Ardupilot has a wide scope of vehicle simulators built-in. It can be interfaced with several external simulators. This allows ArduPilot to be used on a broad variety of vehicle types, i.e. ground vehicles, fixed-wing aircraft, multi-rotor aircraft, underwater vehicles, antenna trackers, camera gimbals, and some kind of sensors such as optical flow and lidar.

2.2. Network Simulator

There exist many simulators that offer full or a part of network simulation. Most simulators are discrete event-driven. Discrete event-driven means that it forms a timeline of events that are sequentially processed. One pioneer of network simulator is TOSSIM [17]. TOSSIM is created to model the transmissions of wireless sensors.

OPNET [18] is a commercial network simulator. OPNET provides wireless network simulation and supports a broad variety of technologies and protocols. NS-3 [7] and OMNET++ [19] are the most used network simulators by the researcher, among other open-source network simulators. NS-3 and OMNET++ simulators are also a discrete-event driven. NS-3 can simulate many technologies. NS-3 is capable to embed diverse statistical models for traffic generation, mobility, and channel gain. It provides comprehensive modeling of physical layer function. NS-3 supports 4G, 5G, and WiFi communication protocols. NS-3 can also support extended protocols and options at different layers of the network. Furthermore, NS-3 can be effectively used to interface with libraries, external systems, and external applications.

2.3. FlyNetSimulator

FlyNetSim is a non-proprietary UAV and Network Simulator built on the base of NS-3 and Ardupilot. FlyNetsim is capable to complement the experiment with a simulated edge server, UAV, and wireless networks. Unlike other integrated UAV-network simulators, FlyNetSim is capable to support various scenarios in terms of UAV controls, e.g., data stream transport, telemetry controls, and multi-UAV usage. FlyNetSim is also capable of emulating a wide range of network schemes, i.e. ad-hoc communications of device-to-device, heterogeneous networks system, multi-hop and multi-path communications, and also different mobility models. FlyNetSim simulator is more efficient compared to other simulators in terms of computational resource usage and scalability. Hence, FlyNetSim simulator can support the complex tasks of information autonomy framework. Using this simulator, we may simulate the real-world network conditions with varying load and mobility [5].

FlyNetSim architecture is described in Figure 2. The FlyNetSim architecture inheres four main components: Fly simulator to commence UAV, GCS, and its visualization tools; Network simulator to simulate real-life network infrastructure and its environment (i.e. contending nodes); Middleware to supports two-way communications based on ZMQ through the network; Autogen to produce a map for the UAV components (i.e. sensors) and the corresponding network nodes.

2.4. Socket Priority

Socket is an endpoint of a bidirectional transmission link connecting two applications run on a network. A socket is bound to a port number so that the TCP layer can identify the application that data is destined to be sent to. Commonly, a server runs on a dedicated computer and has a socket with specific port number. Server task is listening to the socket for a client to make a connection request. To establish a connection, client should know the hostname of the machine on which the server is running and the port number on which the server is listening. The client is also obligated to identify itself to the server so it binds to a local port number that will be used during a connection session. In the case of everything works out positively, the server acknowledges the connection [20].

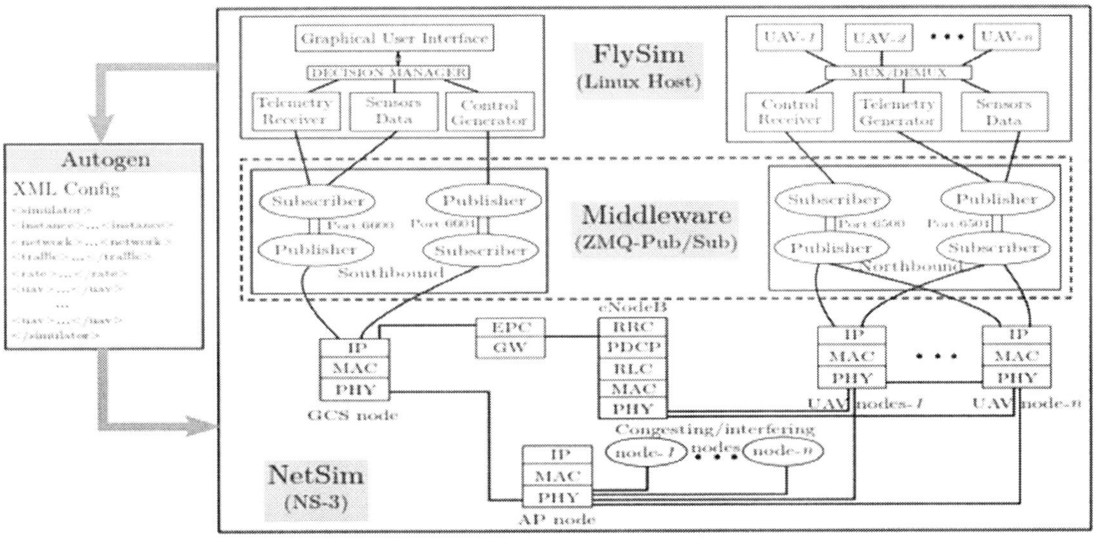

Figure 2. FlyNetSim Architecture [5]

Socket allows the network to set some options. One of them is Socket Priority. Socket priority allows the network to set the input message priority for this socket. Valid values for this parameter are usually in hierarchical integer (i.e 0-6). If we set the priority to 0, the application's priority defined in the network services database is used. For sockets with a priority value of 1, traffic is read in and processed even during input list shutdown and if traffic limiting is enabled for this server, traffic limiting checks are bypassed. When there are resource constraints, packets with the largest priority values are discarded first. The priority for an individual socket overrides the application's priority defined in the network services database. This option is valid only for TCP and UDP sockets [21].

3. Method

The packets are prioritized from trusted nodes by giving the socket from trusted node a priority. The priority for an individual socket overrides the application's default priority defined in the network. By giving higher priority to trusted node, we aim to out-weight the trusted node from the interfering node, so that the access point will process the trusted packet first. The proposed algorithms are implemented in FlyNetSimulator that is based on Ardupilot and NS3. The implementation is based on C++.

3.1. Proposed Algorithm for UAV Socket Priority Module

Figure 3 shows the flowchart of the proposed algorithm. The flowchart started with checking if the socket used to send the packets are trusted. The trusted sockets consist of UAV nodes and Ground Control Station node. If the socket came from trusted node the simulator then set a priority tag in the socket. The Socket then sent the desired packet, either in the form of a UAV command, GCS command, or just contending packets.

At the receiving side, when receiving packets, it first checks the incoming packets, whether it has a priority socket or not. If the packet came from the priority socket, then the receiver will prioritize the packet to be received. If the packet does not come from the priority socket, then the receiver will ignore the packet.

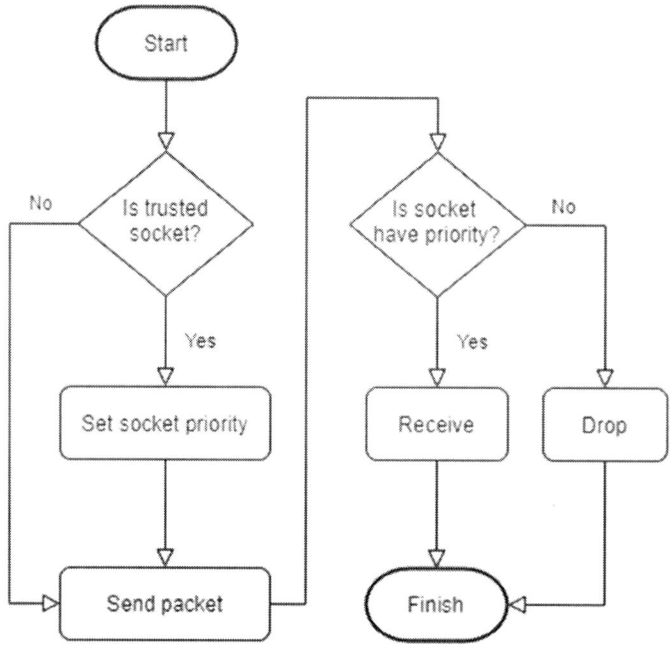

Figure 3. Flowchart of the algorithm for UAV Socket Priority

3.2. Implementation in FlyNetSim

The implementation of our proposed method is described in two algorithms: Algorithm 1 for packet send and Algorithm 2 for packet receive.

Algorithm 1. Send Packet

```
procedure SendMsg(socket, message)
while send request
   socket <- socket priority
   packet <- create packet(message)
   socket -> send packet
end while
```

Algorithm 1 shows the implementation of sending trusted packets. *SendMsg* is a procedure to send packet from UAV or Ground Control Station node. *SendMsg* expects an input of socket source and the message to send. The algorithm starts with checking if there is a packet send request. If send request exists, then *SendMsg* creates a socket priority for the socket source input. Subsequently, the message is encapsulated in the packet and the packets are sent via the socket created.

Algorithm 2. Receive Packet

```
procedure RcvPacket(packet, address)
while packet receive request
   if priority exist in socket
      receive the packet
      message <- parse packet message
   end if
end while
```

Algorithm 2 shows the implementation in receiving packets. *RcvPacket* is a procedure to receive packet, whether from a trusted node or a contending node. *RcvPacket* expects an input of packet received and sender address. The algorithm starts with checking if there is a packet receive request. If packet receive exist, then *RcvPacket* checks whether the packet came from the priority socket or not. If the packet came from a priority socket, *RcvPacket* then receives the packet and parse the message inside the packet.

3.3. Test Scenario

We measure the delay and packet delivery ratio (PDR) for evaluating the proposed method. Delay and PDR are measured in two scenarios which have various parameters of interfering nodes, traffic rates of interfering nodes (in Mbps), and packet size each of each interfering node (in bytes). Each scenario uses LTE as the network type. Each scenario runs with and without the implementation of socket priority module to find out if the module created successfully reduces delay and affects the packet delivery ratio.

Detailed scenarios are described in Table 1. The first scenario (S01) uses 1 interfering node, 4 Mbps traffic rates, and 100 bytes packet size on each contending node. The second scenario (S02) uses 2 interfering nodes, 6 Mbps traffic rates, and 120 bytes packet size on each contending node. The number of interfering nodes on two scenarios are defined to represent single interference (1 node) and multi interference (more than 1 node) situation. Traffic rates, packet size, and the number of interfering nodes are defined increasing (from 4 to 5, from 100 to 120, and from 1 to 2) to discover whether these three variables affect the performance of the UAV network.

Table 1. Scenario List for Testing.

Scenario ID	Interfering Nodes	Traffic Rates	Packet Size Each Node
S01	1	4	100
S02	2	6	120

4. Results and Analysis

The results and analysis of the implemented method of UAV Socket Priority Module in FlyNetSim are described as follows.

4.1. Results

Delay test scenario results are described in Figure 4. The result shows a reduction in the average delay during packet transmission from Unmanned Aerial Vehicle to Ground Control Station. This reduction in delay is caused by using priority socket on the trusted node, thus the GCS only accepts incoming packets from the UAV and not by other interfering nodes. In scenario 1, the network average delay reaches a peak at 20 ms when not using the set priority socket on the trusted nodes, whereas the average delay reached a peak at 17 ms when using the set priority socket on the trusted nodes which means there is a reduction by 3ms. There is an 18.6% reduction in the network average delay in scenario 1. In scenario 2, the network average delay reaches a peak at 50 ms when not using the set priority socket on the trusted nodes, whereas the average delay reached a peak at 39 ms when using the set priority socket on the trusted nodes which means there is a reduction by 11ms. There is a 13.9% reduction in the network average delay in scenario 2.

Packet Delivery Ratio test scenario results are described in Figure 5. The result shows a slight rise in packet delay during packet transmission from Unmanned Aerial Vehicle to Ground Control Station. This slight raise in packet delivery ratio is caused by using priority socket on the trusted node, thus the GCS only accepts incoming packets from the UAV and not by other interfering nodes. In scenario 1, the packet delivery ratio is 91% when not using the set priority socket on the trusted nodes, whereas the packet delivery ratio is 94% when using the set priority socket on the trusted nodes.

There is a 3% increase in the ratio. In scenario 2, the packet delivery ratio is 89% when not using the set priority socket on the trusted nodes, whereas the packet delivery ratio is 94% when using the set priority socket on the trusted nodes. There is a 5% increase in the ratio. The packet delivery ratio is similar to scenario 1 and scenario 2 when using a set priority socket on the trusted nodes.

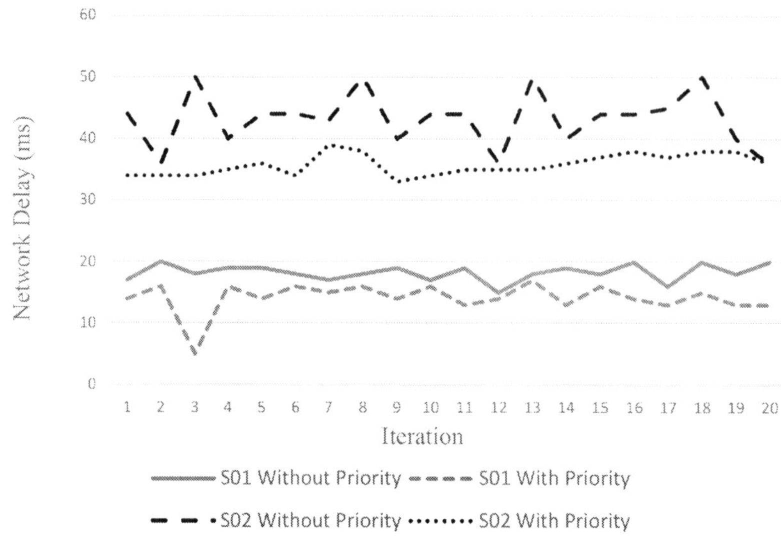

Figure 4. Result of Delay Test

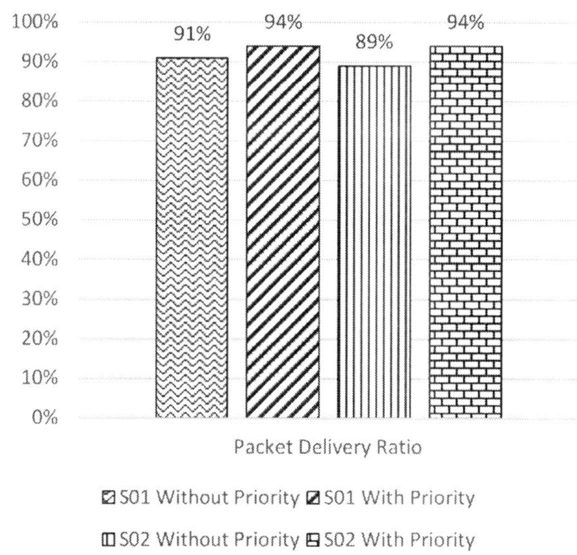

Figure 5. Result of PDR Test

4.2. Analysis

The average network delay can be reduced by blocking the incoming packet from interfering nodes, which means the Ground Control Station only receives packets that come from the socket that is prioritized. Hence the time needed to receive all incoming packets is reduced because the total accepted packets are also reduced. The packet delivery ratio is also increased when the priority method is applied. This might happen because the optimum rate is reached when using the priority method. When a socket is given a priority means that it will be prioritized so other incoming packets from the socket other than the priority socket will be blocked. There are 6 socket priority types in NS-3, i.e. best effort, filler, bulk, interactive bulk, interactive, and control. In these experiments, we use control to solve the problem of the network delay.

5. Conclusion

We have presented a Socket Priority Module for Unmanned Aerial Vehicle Network using FlyNetSimulator. The goal is to reduce packet delay caused by contending nodes that interfere with the UAV and GCS nodes. Scenario 1 averaging 3.3 ms delay reduction and 3% packet delivery ratio raise; and scenario 2 averaging 5.9ms delay reduction and 5% packet delivery ratio raise. The test shows that there is a 16.2% average delay reduction and a 4% average packet delivery ratio raise of both scenarios. From the experiment, we can conclude that using socket priority module in UAV and GCS node reduced the delay and raise the packet delivery ratio that is caused by contending nodes.

Acknowledgment

This work is supported by the PTUPT Grant of Ministry of Research, Technology, and Higher Education (Kemenristekdikti) of Republic of Indonesia number: NKB-356/UN2.RST/HKP.05.00/2020.

References

[1] Bezmenov V, Safin K and Stepanov S 2020 Application of unmanned aerial vehicles for solving engineering tasks *IOP Conf. Series: Materials Science and Engineering* **890(1)** DOI: 10.1088/1757-899X/890/1/012135

[2] Erdelj M, Natalizio E, Chowdhury K R and Akyildiz I F 2017 Help from the sky: leveraging UAVs for disaster management *IEEE Pervasive Computing* **16** (1) pp 24-32 DOI: 10.1109/MPRV.2017.11

[3] Sona G, Passoni D, Pinto L, Pagliari D, Masseroni D, Ortuani B and Facchi A 2016 UAV multispectral survey to map soil and crop for precision farming applications *Int. Archives of the Photogrammetry, Remote Sensing and Spatial Information Sciences* **XLI-B1** pp 1023–9

[4] Shaikh Z, Baidya S and Levorato M 2018 Robust multi-path communications for UAVs in the urban IoT *2018 IEEE Int. Conf. on Sensing, Communication and Networking (SECON Workshops)* pp 1–5

[5] Callegaro D, Baidya S, Ramachandran G S, Krishnamachari B and Levorato M 2019 Information autonomy: self-adaptive information management for edge-assisted autonomous UAV systems *MILCOM 2019 - 2019 IEEE Military Communications Conf.* pp 40–5

[6] Callegaro D, Baidya S and Levorato M 2020 Dynamic distributed computing for infrastructure-assisted autonomous UAVs *IEEE Int. Conf. on Communications (IEEE ICC 2020)*

[7] Henderson T R, Lacage M, Riley G F, Dowell C and Kopena J 2008 Network simulations with the ns-3 simulator *SIGCOMM demonstration* **14** 14 p 527

[8] Ardupilot 2020 *Ardupilot Autopilot Suite Documentation* Retrieved on 10 June 2020 from http://ardupilot.com/

[9] Baidya S, Shaikh Z and Levorato M 2018 FlyNetSim: an open source synchronized UAV network simulator based on ns-3 and ardupilot *ACM Int. Conf. on Modeling, analysis simulation of wireless and mobile systems (ACM MSWiM 2018)*

[10] H-SIM 2020 *H-SIM Flight Simulator* Retrieved on 10 June 2020 from http://www.h-sim.com

[11] Meier L, Tanskanen P, Heng L, Lee G H, Fraundorfer F and Pollefeys M 2012 PIXHAWK: a micro aerial vehicle design for autonomous flight using onboard computer vision *Autonomous Robots* **33**(1-2) pp 21–39

[12] Navio 2020 *NAVIO2 Autopilot* Retrieved on 10 June 2020 from https://emlid.com/navio/

[13] ErleRobotic 2020 *Erle Robotics* Retrieved on 10 June 2020 from https://www.erlerobotics.com

[14] Ardupilot 2020 *Ardupilot SITL* Retrieved on 10 June 2020 from http://ardupilot.org/dev/docs/sitl-simulator-software-in-the-loop.html

[15] Koenig N and Howard A 2004 Design and use paradigms for Gazebo, an open-source multi-robot simulator *Proc. of IEEE/RSJ Int. Conf. on Intelligent Robots and Systems (IROS 2004)* **3** pp

2149–54
[16] Quigley M, Conley K, Gerkey B, Faust J, Foote T, Leibs J, Wheeler R and Ng A Y 2009 ROS: an open-source robot operating system *ICRA workshop on open source software*
[17] Levis P, Lee N, Welsh M and Culler D 2003 TOSSIM: accurate and scalable simulation of entire TinyOS applications *Proc. of the 1st Int. Conf. on Embedded networked sensor systems* pp 126–37
[18] Chang X 1999 Network simulations with OPNET *Proc. of the 31st Conf. on Winter simulation: Simulation—a bridge to the future* **1** pp 307–14
[19] Varga A and Hornig R 2008 An overview of the OMNeT++ simulation environment *Proc. of the 1st Int. Conf. on Simulation tools and techniques for communications, networks and systems & workshops (ICST 2008)*
[20] Parekh C and Patel R M 2017 Encryption and decryption implementation method using network socket programming *Int. J. of Advanced Research in Computer Sci.* **8(5)** pp 2597–604
[21] IBM 2020 *IBM Knowledge Center* Retrieved on 10 June 2020 from http://ibm.com/support/knowledgecenter

The Implementation of User Centered Design Methods on Public Service Mapping Websites

Ginanjar Wiro Sasmito[1], M. Fikri Hidayattullah[2]

[1,2]Department of Informatics Engineering, Politeknik Harapan Bersama, Tegal, Indonesia

Email : [1]anjar.dosen@gmail.com, [2]fikri@poltektegal.ac.id

Abstract. Public services are basically held to increase the level of community satisfaction. To achieve the satisfaction, a city / country will be demanded the quality of professional public services. One of them is the city of Tegal. Tegal is one of the cities in Indonesia. In order to implement a smart city, the city government of Tegal strives to provide services to the public quickly, effectively and efficiently. The public services of Tegal City Government have been running quite well, but information on services has not been integrated between Regional-Owned Enterprises (BUMD) and the private sector. Therefore a website-based mapping media was created in the form of clustering of Tegal city public services, both organized by the government, BUMD and the private sector are integrated each other and can be accessed and used easily by the people of Tegal city. This website can provide the closest route and is equipped with a comparison chart of public service categories in the city of Tegal. This application was built with the User Centered Design method so that it can be adjusted to the needs of the user and speed up the completion time, which only takes 105 days or 15 weeks. The results of this mapping website test were carried out by using Black Box Testing with the equivalence partitions method, which is the effectiveness value for all ID Test results is 83%, and the user test results are carried out by using the Usability Testing method that applied the SUS scale, with a value of 89% or excellent.

1. Introduction

Cities are large human settlements[1], cities generally have housing, sanitation, utilities, land use and communications systems. The urban population originally constituted only a small part of the total human population, however after two centuries of urbanization, nowadays the world's population lives in urban areas which have major consequences for globalization [2]. The growth of a city is always directly proportional to population growth, technological developments, and the demands of a modern lifestyle. It makes the city is being unfriendly to the people, there are always various inequalities, both economic and social in every part of the city, congestion and various pollution also cannot be avoided.

Public service is defined as any form of services, either in the form of goods or public services [3] which should be implemented by the Central Government, Regions government, as well as State-Owned Enterprises (BUMN) and Regional-Owned Enterprises (BUMD) in order to fulfill the needs of the community and in terms of implementation of statutory provisions. Public services that are transparent, accountable and have high responsibility will generate high public trust as well. It takes public service ethics as a pillar and public trust to be substantial and fundamental in order to create good government [4]. The purpose of public service is basically for community satisfaction and to achieve the satisfaction, professional public service quality is necessary. These principles are reflected in: transparency, accountability, participation, security of rights, balance of rights and obligations. This is what every community wants, one of which is the city of Tegal community.

Tegal is a city in Central Java - Indonesia with an area of - + 39.68km^2 [5], In order to implement smart city which is an integration concept between information and communication technology with various physical devices connected to the Internet of Things (IoT) network to optimize the efficiency of city operations and services that are directly connected to its citizens [6], so the city government of Tegal makes every effort to provide services to the community quickly, effectively and efficiently.

Public services in the city of Tegal have been running quite well, but service information has not been integrated between those organized by BUMD and private. Besides that, not a few people in the city of Tegal do not know the right place to get public services in accordance with the problems they are experiencing and also not a few people who do not know the location of several public services in the city of Tegal, so people often experience difficulties and confusion in getting service for the problems they are experiencing.

These things can certainly hinder the progress of the city of Tegal to implement a smart city. Therefore a publication media with attractive visualization is needed, such as a website-based mapping application which is integrated between the government, BUMD and the private sector in the form of cluster and can be accessed or used easily by society of Tegal city.

This application is developed using the User Centered Design method, which is one of the methods of developing application software in which the user is the center of the system development process, objectives / characteristics, context, and system environment that are all based on user experience [7]. This method is one of the applicable methods used in Human-Computer Interaction, its relationship with user needs and satisfaction [8].

2. Related Research

Zheng, et all developed a local climate zone classification map and established an urban climate database as an information platform and planning decision-making in Hong Kong. The Geographical Information System is applied to synergize various types of data and classify the land surface in Hong Kong based on the local climatic zone classification [9].

Sasmito and Nishom developed a land use mapping system in the city of Tegal. The mapping system is a website platform application with mapping content for housing, government facility land, educational facility land, health facility land, trade facility land, industrial facility land, mixed service facility land, tourism facility land, and agricultural facility land which can be easily accessed [10].

The User Centered Design (UCD) method is applied for modeling the user interface in producing a "Panic Button" application that can be used by Deaf people. This application is designed to make the user interface can match the characteristics and limitations of the user. The USE questionnaire was used to test the prototype product and resulted a usability value of 87% with a very good category after the third iteration. These results indicate that the redesign product met the needs of the deaf, by using appropriate[11].

Ulya et all applies the User Centered Design (UCD) method to design a smartphone application to introduce culture that can be used by deaf people. The usefulness of the prototype was measured by the USE Questionnaire method. the result of usability test showed a value of 82.75%. This score indicates that the user interface design has a usability value with a very good category, and it can be said that the user interface has met user needs [12].

User Centered Design (UCD) is used as a method to develop e learning applications: Android-based reproductive learning which allows blind adolescents to easily operate it. The designed applications can involve several human senses in their utilization, which offers flexibility, efficiency and user friendly. The application allows users to interact through several inputs such as touch or gestures and get output in the form of sound. In this research, the design is focused on the user. The user interface model has been tested with the USE questionnaire parameter which reaches an average usability percentage of 87.4%, this shows that usability is obtained in the very good category[13].

User Centered Design (UCD) was utilized in the research of Luna, Daniel R et all, the study was used to redesign a Warning System at the Italiano de Buenos Aires Hospital in Argentina which focuses on electronic health records with a warning system for drug interactions which is given to patients. His research found that the UCD warning system was more efficient (faster warning resolution), more effective (tasks completed with fewer errors), and more satisfying. These results suggest that UCD techniques adhering to ISO 9241-210 can produce more useful warnings than traditional designs [14].

Korpershoek, et all in their research have applied the User Centered Design (UCD) method to develop Mobile Health (mHealth) application for the management of chronic obstructive pulmonary disease (COPD) which involves patients in managing their own health. The UCD method used consists of four iterative phases: (1) background analysis and design conceptualization, (2) reusability alpha testing, (3) iterative software development, and (4) field usability testing. His research shown that by following the UCD method, the mHealth intervention developed meets the needs and preferences of COPD patients, and has high potential to be effective in reducing the impact of exacerbations [15].

3. Methodology

3.1. Thinking Framework

The thinking framework for this study is shown in the figure 1.

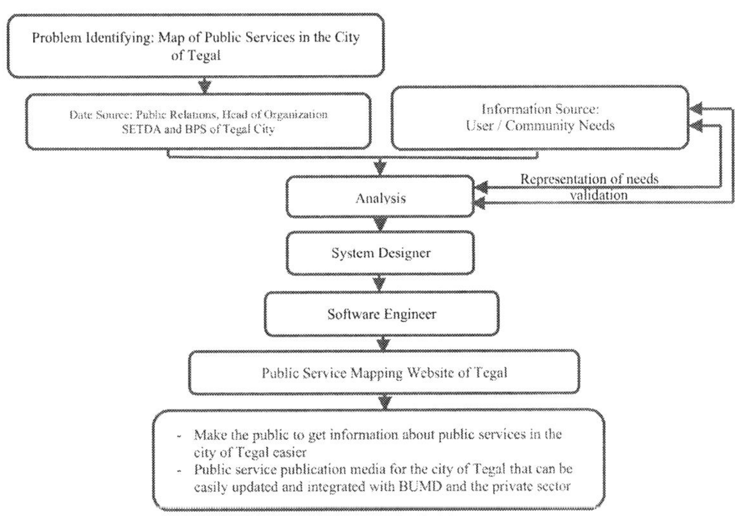

Figure 1. Thinking Framework

Based on Figure 1, the frame of thinking in this study begins with identifying problems regarding information on mapping public services in the city of Tegal. After the identification of the problem has been completed, it is followed by data collection, where the data sources come from Public Relations, the Head of the Organizational Section of the regional Secretariat (SETDA) and the Central Bureau of Statistics (BPS) of Tegal city. In addition, information on the needs of the community / users is also collected regarding the ease of access to public service information.

After the data and information sources have been collected and validated, the analysis is conducted to determine the right solution of existing problems, and determine the needs to produce a solution to these problems. The results of the analysis are submitted to the system designer as the main / reference material in producing the design, interface, application flow and prototype of an application. At the design stage, the user is used as the main source of knowledge, since the solution of the problems must be based on a perspective and oriented towards user needs. The design that has been completed produced and approved by the user, then submit to the software engineer to create a product as a solution of the existing problems; a public service mapping website, so the website is expected can make the society easily to get information about public services in the city of Tegal and can be easily updated and integrated with Regional owned enterprises (BUMD) and the private sector.

3.2. *Software Development Method*

The method applied in designing this mapping application is the User Centered Design (UCD) method, with the scheme [16] in figure 2.

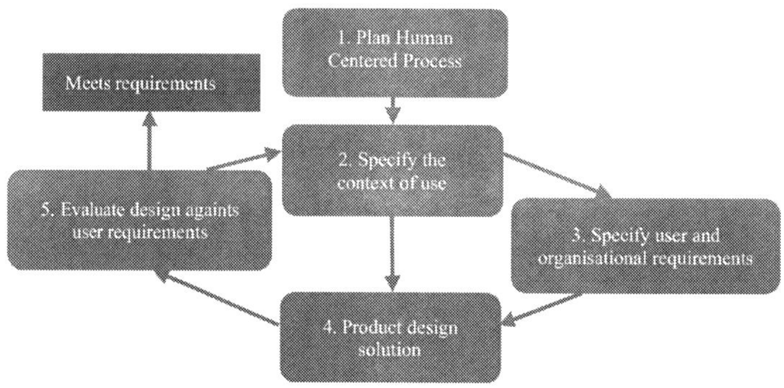

Figure 2. User Centered Design Model

Figure 2 can be explained in full as follows:

1. Plan the human centered process
   At this stage the researcher determines that the process of making this mapping website is user-centered, therefore to find out the substance and detail of the problem, and produce the user need applications, the user is involved the production of the application from the beginning to the end of the production process by providing data, information as well as arguments.
2. Specify the context of use
   This stage determines the user of the mapping website is generated. Public service mapping website of Tegal consists of 3 users:
   a. The community is the user who can access information on the mapping of public services in the city of Tegal in general.
   b. Admin is a user who manages data on every public service in the city of Tegal, whether organized by the government, BUMD, or the private sector.
   c. Super Admin, is a user who manages admin data
3. Specify User and Organisational requirement
   At this stage, identification of every user need is carried out through a questionnaire technique. Based on the results of the distributed questionnaires, the following results are:
   a. Information Needs
      - Public service agency data
      - Public service agency profile data
      - Public service information data that is owned by each agency
      - Location data and address of public services
      - Public service route data
   b. Functional Requirements
      - Connect with Google Maps servers
      - Administrators have public service data management facilities for each agency
      - Able to display every existing public service profile
      - People easily choose public services based on predefined keywords
      - Able to route to the location the intended public service

- Super admin has admin data management facility
- Able to display static data of the number public services according to the classification

c. Non-Functional Requirements

- The application has an authentication system to sign into the Admin and Super admin pages
- Able to display information in the form of area maps and statistical charts
- Applications can be accessed anywhere and anytime, while connected to the internet
- Applications can be accessed using a variety of browsers
- Applications can be accessed using various platforms, such as: Tablets, Laptops, Personal Computers, and Smart phones.

4. Product design solution
This stage is the solution design stage, which consists of:
a. Website design
The website design used in making this application is the Unified Modeling Language (UML) which is a design for object-oriented systems [17]. At this stage the user is really involved to provide advice and information about the needs of making user interface designs, user experience designs, system flows, and prototypes according to user needs. so that it is expected that the resulting website is really in accordance with the user's perspective and needs.
b. Website development
This stage is the coding stage / system design which refers to the system design, built by using the PHP programming language, My SQL database. PHP is a programming language which is widely used to create and develop a website and can be used together with HTML [18]. Meanwhile, MySQL is a database management system that uses SQL (Structure Query Language) basic commands [19].

5. Evaluate design againt user requirement
In this study, evaluation was carried out in two ways, they are:
a. Website evaluation
Website evaluation is carried out by using black box testing that focuses on existing functions in the application [20]. Black box testing tends to find several things such as: incorrect or absent functionality, database access errors, interface errors, performance errors, and initialization and termination errors [21].
b. User evaluation
In this study, user evaluation is carried out by using usability testing, this is conducted to determine the ease of using the application, the efficiency and effectiveness of the application, and user satisfaction to the result of website application [22].

4. Result and Discussion

4.1. User Interface

Design and build a website for mapping public services in the city of Tegal has been successfully created. The display of the design results page is in accordance with the design. It is shown in figures 3 and 4.

The application of the UCD method to develop this Public Service Mapping Website has been completed successfully in 15 weeks or 105 days according to the explanation in table 1. This is due to the vital role of users who are directly involved in providing the information needed from the initial process to the completion of this mapping website, both users in public service agencies as admin and community users. So this accelerates the completion of software development and ultimately has an impact to more optimal cost efficiency and resource productivity.

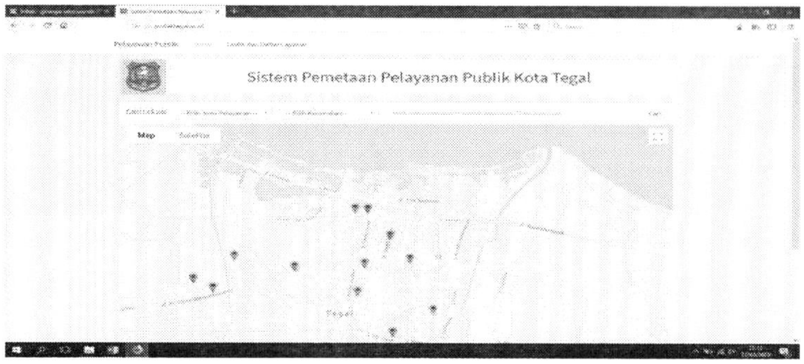

Figure 3. Main page of the Public Service Mapping Website

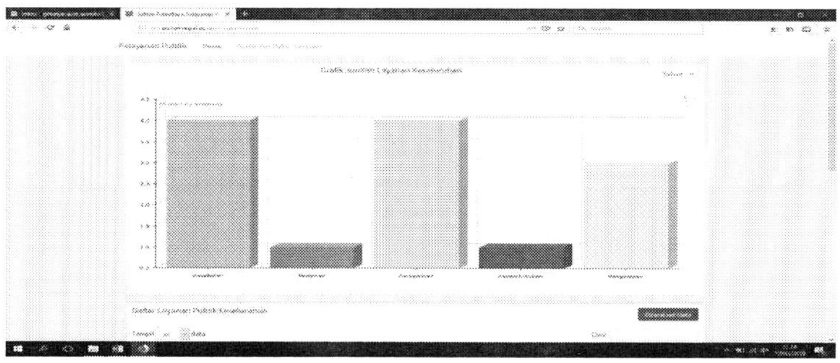

Figure 4. Visualization of Comparative Graph of Public Service Mapping

Table 1. Duration of Mapping Website Development

Stages	Duration	Number of Team
Plan the human centered process	1 weeks	1
Specify the context of use	2 weeks	1
Specify User and Organisational Requirement	2 weeks	2
Product design solution	8 weeks	2
Evaluate design againt user requirement	2 weeks	1

This result is a fact that the UCD method can accelerate the duration of system development with better software quality, since all the processes conducted have been adjusted to user needs [23]. This method is very appropriate for small to medium scale software development, However, smooth, effective and efficient communication between the team and users is one of the determinants of successful software development using the UCD method.

4.2. Application Test

The website application that has been produced is tested using Black Box Testing, by determining the input range and output range using the equivalent partitions method. Several fields on the application interface display are also in the database table structure. The application of Black Box Testing is used to test the input data sample. The guideline used is the input condition specifies the range, so a value of 1 is defined as a valid state, and a value of 2 is defined as an invalid state for its equivalence.

Based on the results of the tests that have been carried out in accordance with table 2, it can be seen whether the test results were successful or failed, to all fields on the menus of the website application. Test ID 1,2,4,5,6 all are True, so, the effectiveness value is 100%, while in Test ID 3, the field testing conducted is 1 false, and resulted in a 2:3 effectiveness value, namely 66%. So that the effectiveness value for all Test ID results obtained: 83%.

Table 2. Black Box Testing

Test ID	Description	Test data field	Test Result	Conclusion
1	Sign In Menu	Username, Password	True, True;	Success
2	Service Type Category Menu	Category Type Id, Category Name	True, True;	Success
3	Service Management	Service Id, Service Name, Number of Services	True, True, False;	Failed
4	Service Profile	Profile Id, Address, Point of Coordinates	True, True, True;	Success
5	Report Management	Report Id, Report Type, Report Name	True, True, True;	Success
6	User Management	User type, user account, user profile	True, True, True;	Success

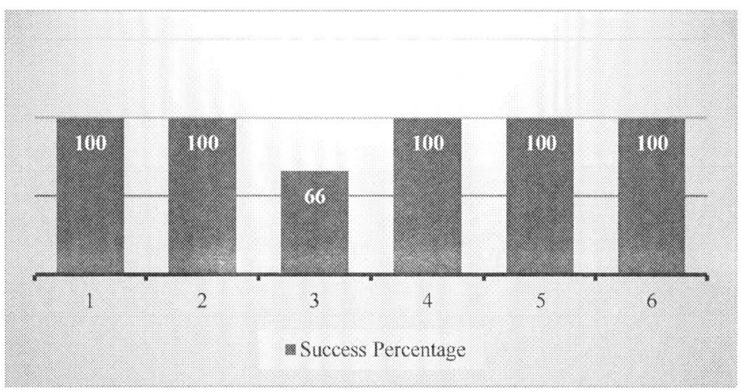

Figure 5. Graph of Black Box Testing Result

4.3. User Testing

After the website application has been successfully tested by using the Black Box Testing method, then testing is carried out to the user by using the Usability Testing method. This test was carried out by distributing questionnaire sheets randomly to 10 Mapping Website Admin, Society, Academics, and Software Developers with 10 questions, regarding: the ease of application, efficiency and effectiveness, and user satisfaction. With a list of questions according to table 3.

Table 3. List of Questions

No	List of Questions
1	I think this application is easy to use
2	I found this system is not very complicated
3	I think I will use this application
4	I think I don't need assistant to use this application
5	I find the integrated various functions in this application
6	I think the contents of this application are quite consistent
7	I have no difficulty using this application
8	I feel comfortable using this application
9	I think the quality of the information in the application is very good
10	I think users will quickly understand using this application

IOP Conf. Series: Materials Science and Engineering 1077 (2021) 012022 doi:10.1088/1757-899X/1077/1/012022

Before filling out the questionnaire, respondents were asked to operate the Mapping Website in accordance with their respective levels, except for academics and software developers who were asked to use it at all user levels. Then each respondent provides an assessment of the questions that have been given through a questionnaire sheet, with a description of the score according to table 4.

Table 4. Description of score value

No	Score	Description
1	5	Excellent
2	4	Good
3	3	Okay
4	2	Poor
5	1	Awful

The results of the respondent's score are in accordance with those in Figure 6.

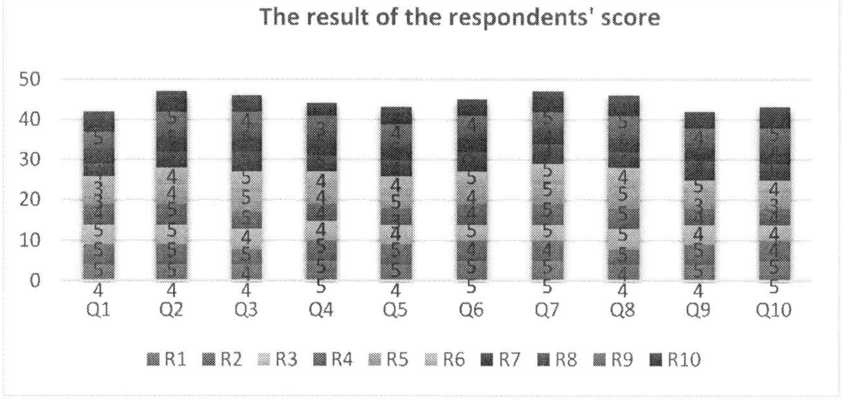

Figure 6. Results of Respondents' Score

The results of the total value in Figure 6 are calculated using the SUS (System Usability Scale). The calculation used the formula [22]:

$$Y = \frac{Q}{S} \times 100\% \tag{1}$$

where

Q = Number of respondents' scores for each question
S = Maximum total respondent score
Y = percentage value

So the results of Usability Testing using SUS are:

$$Y = \frac{445}{500} \times 100\% = 89\%$$

The Adjective Rating based on the SUS scale is according to table 5.

4.4. Security Information

Information security is an the protection of information from various threats in order to ensure the continuity of business processes, reduce risks, open opportunities and optimize return on investment (ROI) [24]. In designing the security system on this mapping website, there are aspects that have been conducted, they are:

1. Confidentiality: A concept to protect the confidentiality and privacy of information
2. Integrity: The concept of protecting the accuracy of processing information and data from improper modification

3. Availability: The concept of ensuring systems and data can be accessed
4. Control Access: The concept to ensure a person has full authorization in accessing information
5. Non Repudiation: The concept to ensure someone cannot deny for having done a transaction
6. Privacy: The concept to ensure that the privacy of a person's account or data on a is protected
7. Authentication: The concept to ensure that the authenticity of the message comes from the requested information
8. Authority: The concept to guarantee that existing information cannot be manipulated by unauthorized persons.

The series of these aspects have user id and password authorization when accessing the administrator menu.

Table 5. Adjective Rating

Score	Grade	Adjective Rating
>80.3	A	Excellent
68-80.2	B	Good
67	C	Okay
51-66	D	Poor
<51	E	Awful

5. Conclusion

Based on the results and discussion, it can be concluded that the Mapping Website can provide the information of location, route, profile, facilities, graphics and categories of public services of Tegal city, whether organized by the government, Regional-Owned Enterprises (BUMD), or the private sector.The mapping website is quickly completed, which is only 105 days or 15 weeks, since it is developed by using the User Centered Design (UCD) method where from the beginning to the end it always involves users in completing data for this mapping website to suit user needs. The test results of this website application are carried out by Black Box Testing with the equivalent partitions method in which the effectiveness value for all Test ID results is: 83%, and the user test results are carried out by using the Usability Testing method that applied the SUS scale, with a value of 89% or excellent.

References

[1] A and Kuper K 1996 *The Social Science Encyclopedia* 2nd Edition (London: Routledge)
[2] James P 2015 *Urban Sustainability In Theory And Practice : Circles Of Sustainbility* (London: Routledge)
[3] Anderfuhren-Biget S, Varone F and Giauque D 2014 Policy environment and public service motivation *Public Adm.* vol **92** no 4 pp 807–25 DOI: 10.1111/padm.12026
[4] Lewis C W and Gilman S C 2005 *The Ethics Challenge in Public Service: A problem-solving Guide* (New Jersey: John Wiley & Sons)
[5] BPS 2018 *Badan Pusat Statistik Kota Tegal Dalam Angka Tegal* (Tegal: BPS)
[6] Graham C and Peleg A 2017 *Will a Smart City Have a High Street?* (New York: Springer International Publishing) pp 143–57
[7] McLoone H E, Jacobson M, Hegg C and Johnson P W 2010 User-centered design and evaluation of a next generation fixed-split ergonomic keyboard *Work* vol **37** no 4 pp 445–56 DOI: 10.3233/WOR-2010-1109
[8] Ahamed M M and Bakar Z B A 2017 Triangle model theory for enhance the usability by user centered design process in human computer interaction *Int. J. Contemp. Comput. Res.* vol **1** no 2 pp 1–7
[9] Zheng Y *et al* 2018 GIS-based mapping of Local Climate Zone in the high-density city of Hong Kong *Urban Clim.* vol **24** pp 419–48 DOI: 10.1016/j.uclim.2017.05.008
[10] Sasmito G W and Nishom M 2018 Mapping of land use system in Tegal City *Int. J. Res. Eng.*

Innov. **2(5)** pp 498–502
[11] Gumay M R A, Effendy V and Junaedi D 2020 Modeling user interface design for panic button application for deaf people using user-centered design method *IOP Conf. Ser. Mater. Sci. Eng.* vol **830** no 2 DOI: 10.1088/1757-899X/830/2/022097
[12] Ulya A H, Effendy V and Junaedi D 2019 User interface modelling of Sundanese Culture introduction for deaf youth using user-centered design method *J. Phys. Conf. Ser.* vol **1192** no 1 DOI: 10.1088/1742-6596/1192/1/012037.
[13] Priowibowo D, Effendy B and Junaedi V 2020 Designing user interface using user-centered design method on reproductive health learning for visual impairment teenagers *IOP Conf. Ser. Mater. Sci. Eng.* vol **830** no 2 DOI: 10.1088/1757-899X/830/2/022092
[14] Luna D R, Rizzato-Lede D A, Otero C M, Risk M R and de Quirós F G B 2017 User-centered design improves the usability of drug-drug interaction alerts: experimental comparison of interfaces *J. Biomed. Inform.* vol **66** pp 204–13 DOI: 10.1016/j.jbi.2017.01.009
[15] Korpershoek Y J G, Hermsen S, Schoonhoven L, Schuurmans M J and Trappenburg J C A 2020 User-centered design of a mobile health intervention to enhance exacerbation-related self-management in patients with chronic obstructive pulmonary disease (copilot): mixed methods study *J. Med. Internet Res.* vol **22** no 6 pp 1–19 DOI: 10.2196/15449
[16] Albani L and Lombardi G 2010 (FIMI) *User Centred Design for EASYREACH* Retrieved from http://www.aal-europe.eu/wp-content/uploads/2019/12/D1.1-_User-Centred-Design-for-EasyReach_-v1-1.pdf
[17] Use B, Diagrams C, Diagrams S and Modeling V 2012 UML with rational rose 2002 *J. Surg. Orthop. Adv.* vol **21** no 4 pp v–vi doi: 10.1093/jhered/ess125
[18] Carr M and Gray D 2018 *Beginning PHP: Master the latest features of PHP 7 and fully embrace modern PHP development* (Birmingham: Packt Publishing)
[19] Mehta C *et al* 2018 *MySQL 8 Administrator's Guide: Effective guide to administering high-performance MySQL 8 solutions* (Birmingham: Packt Publishing)
[20] Dashti M T and Basin D A 2020 Theory of Black-Box Tests *Preprint arXiv:2006.10387*
[21] Jan S R, Shah S T U, Johar Z U, Shah Y and Khan F 2016 An innovative approach to investigate various software testing techniques and strategies *Int. J. Sci. Res. Sci. Eng. Technol.* vol **2** no 2 pp 682–9
[22] Sasmito G W, Zulfiqar L O M and Nishom M 2019 Usability testing based on system usability scale and net promoter score *2019 2nd Int. Semin. Res. Inf. Technol. Intell. Syst. ISRITI* pp 540–5 DOI: 10.1109/ISRITI48646.2019.9034666
[23] Couture B *et al* 2018 Applying user-centered design methods to the development of an mhealth application for use in the hospital setting by patients and care partners *Appl. Clin. Inform.* vol **9** no 2 pp 302–12 DOI: 10.1055/s-0038-1645888
[24] Shojae S, Manaf A B A and Zamani M 2012 Comparative analysis of master-key and interpretative key management (IKM) Frameworks *Cryptogr. Secur. Comput.* ed J Sen (Rijeka: IntechOpen) chapter 10 DOI: 10.5772/35329

Acquiring Commonsense Knowledge during Collaborative Storytelling

Jilyan Dy, Celina Brito, Vincent Tan, Julius Lola, Ethel Ong*

De La Salle University, Manila, Philippines

E-mail: ethel.ong@dlsu.edu.ph*

Abstract. Conversational agents that engage children in collaborative storytelling need a collection of domain knowledge to formulate its responses. However, the current manual processes of collecting, annotating and extracting data from a corpus of unstructured text is time-consuming that often impedes the population of a knowledge base. An alternative is to design the agent as a teachable peer that can continuously learn from its human users. In this paper, we describe a storytelling agent that can expand its domain knowledge base by learning new assertions from its story-based conversation with children. We also use conversations as a means of validating the acquired knowledge. Results showed that the agent can achieve a recall value from 83% to 94% in extracting capabilities, property and instance relations while encountering difficulty in identifying assertions that describe the location of story events. Analysis of the conversation logs showed that the performance of the agent is affected by how children describe the characters and events in their stories.

1. Introduction

Knowledge acquisition (KA) refers to the extraction of knowledge from a data source to feed the information requirements of an NLP task, such as question-answering systems [1], dialogue or conversational systems [2, 3], and text understanding and generation. Decades of research in KA have seen the utilization of varying approaches, including crowdsourcing [4, 5], information and relation extraction [6, 7, 8, 9], and more recently, knowledge embeddings [10]. These are aimed at addressing the manual process of collecting, annotating and extracting data from a corpus of unstructured text which often impedes the population of a knowledge base or an ontology.

Ontology learning is a type of KA that applies (semi-) automatic construction approaches to provide software agents with semantic knowledge in a particular domain [11]. It includes extending an existing ontology by adding new binary semantic relations or assertions, resolving duplicates and revising incorrect assertions. Thus, acquiring knowledge to populate ontologies has been repeatedly described as a time-consuming effort and often left to knowledge engineers.

Everyday human-to-human communication is a means by which we can share knowledge with one another. Children can learn new concepts and their relations from storytelling. Much like how a child learns new concepts from stories, we can enable a software agent to continuously learn new knowledge from its conversations with children. The emergence of conversational interfaces pose an opportunity for software agents to acquire commonsense assertions from children's input. Such unstructured text may contain useful knowledge about children's perception of the world they live in and which software agents can use as leverage when engaging in collaborative storytelling with children.

In this paper, we present our work in expanding the ontology of a conversational storytelling agent (CSA) by learning new assertions from its dialogue with children. Section 2 provides an overview of commonsense ontology and CSA. While children's input may be expressed in simple sentences, these still require correct sentence parsing to identify concepts and the type of semantic relation present in the text. Furthermore, in the process of extraction, the presence of duplicate and contradictory information cannot be avoided but should be properly handled [6]. Section 3 describes the identification, extraction and validation of assertions. Section 4 presents our results and findings, followed by lessons learned in acquiring commonsense knowledge through story-based conversations.

2. Related Works

A commonsense ontology contains concepts and relations that are relevant and make sense in a particular domain such as children's stories. Following ConceptNet [12], knowledge is represented as a semantic network containing binary assertions of the form *relation(concept1, concept2)*. Concepts are abstract or general ideas that can be classified according to semantic categories, namely entities, objects, locations, actions and attributes. In the context of storytelling, the semantic categories are used to model different real-world concepts that children are familiar with. These include story characters, places, objects, events and abstract ideas (emotions).

Ontology population is concerned with adding new assertions into the ontology for use in natural language processing tasks such as story generation. Of primary importance to ontology population is the set of semantic relations that can exist between two concepts. Among the 20+ relations defined in ConceptNet [13], a set of relations relevant to children's stories include those that describe entities (*isA*, *partOf*, *hasProperty*, *usedFor* and *locationOf*) and event causality (*capableOf*, *receiveAction* and *effectOf*).

Extracting relations from documents is crucial to ontology population in order to provide the knowledge requirements for automated story generation and conversational interfaces [3, 14]. The ConceptNet ontology used in MakeBelieve [15] sourced its assertions through the Open Mind Common Sense (OMCS) project [13]. SUMO Stories [16] represented story elements such as characters, objects, location and events in the Suggested Upper Merged Ontology (SUMO) [17]. Story Sense [18] generates stories with blanks and uses children's answers to gain new assertions that can be used to generate more stories. The work of [9] utilized a predefined set of relation extraction templates to extract conceptual relations from children's stories.

In these works, they found that children's narratives have various characteristics that affect the extraction quality and content of the ontology. OMCS [13] is crowdsourced from adults and may extract concepts that are not suitable for children. SUMO Stories [16] need to balance between what is truth and modeling fictional elements typically found in children's stories. Story Sense [18] found that children may have their own sense of what can be considered as true in their imaginative story world. Samson [9] had to work with dialogues as these contain relevant information about the story plot, character relationships and event causality.

Conversational storytelling agents (CSA) can bring life to virtual environments by formulating interesting and engaging responses to human utterances [14, 19]. They can also provide assistance to children during the story construction process [20, 21]. During the collaboration, CSAs can play varying roles including facilitator, tutor, co-author and teachable peer. A *teachable peer* is a kind of a pedagogical agent where learning can be achieved by teaching others [22]. In this role, the CSA has the knowledge level of a younger peer, giving the child the responsibility to teach new concepts and clarify misconceptions of the agent regarding elements of the story. CSAs also employ dialogue moves to elicit details about story characters and events, and then provide *feedback* to acknowledge the child's input, *pumps* to urge the child to expound on his/her input, and *hints* to help the child to formulate the next story text. These were adapted by [20] from the intelligent software tutor of [23].

IOP Conf. Series: Materials Science and Engineering 1077 (2021) 012023 doi:10.1088/1757-899X/1077/1/012023

3. Knowledge Acquisition

Knowledge acquisition aims to identify and extract new assertions of the form `relation(concept1, concept2)` from a child's input and to store these into the ontology that serves as the CSA's knowledge base. We present the design of the ontology and the knowledge acquisition process in this section.

3.1. Ontology Design

The ontology has two components: global and local ontology. The global ontology contains assertions from the work of [24] and crowdsourced assertions that have confidence scores above the set threshold. These assertions can be used by the CSA to contribute to storytelling by giving hints on possible story text.

The local ontology, shown in Table 1, acts as a temporary repository of new assertions. Each assertion has a corresponding *confidence score (CS)* which is initially set to 0. The score may increase or decrease during validation with children. The confidence score is the basis for deciding when an assertion can be moved to the global ontology. The *ID* indicates the source of the assertion and is tracked so that the child cannot participate in the validation of his/her own assertion.

Table 1. Structure of the local ontology.

ID	Concept 1	Relation	Concept 2	CS
1	laptop	hasProperty	black	0.3
1	grandmother	capableOf	love	0.5
2	cake	hasProperty	big	0.1

3.2. Acquisition Process

A conversation is comprised of a set of dialogue turns. The agent and the child take turns formulating an input or a response. The child can give four types of input, namely *command* to instruct the CSA to start or end the storytelling session, *answer* to respond to a question posed by the CSA, *silence* when the child does not know the answer or is stuck during storytelling, and *story text*. The agent then formulates corresponding dialogue moves (feedback, pump, hint) as a response to the child's input.

As a teachable agent, the CSA leverages on the story text input to dynamically learn new assertions. Knowledge acquisition proceeds as follows (see Figure 1). When a child's input has been identified as a story text, the CSA applies a set of patterns containing parts-of-speech tags and word dependencies to detect the presence of candidate semantic relations. Examples of these patterns are shown in Table 2.

Assertions are then extracted and compared against the local ontology. If the assertion already exist in the ontology, the corresponding confidence score is incremented by 1. Otherwise, the assertion is considered new and is added to the local ontology with a confidence score of 0.

3.3. Validation Process

Knowledge validation, shown in Figure 2, is performed using the *suggestion* and *follow-up* dialogue moves. The phrasing of these dialogues should give the impression that the agent is collaborating in the task by proposing story text embedded with one or more assertions that need to be validated. Given an input sentence `S`, the agent selects a word `w` (where `w` $\in$ `S` and `POS(w) = noun`) and uses this to perform the following query on the local ontology:

`(w, ?relation, ?concept2)` or `(?concept1, ?relation, w)`

The query result `R` is a set of assertions $\{A_1, A_2, ..., A_n\}$ that contains `w` as one of the concepts. The CSA selects A_i (where $A_i \in$ `R`) and formulates a suggested story text. For example, if `w = "baby"` and A_i = *capableOf(cry, baby)*, the response is *"What if 'baby cries'?"*

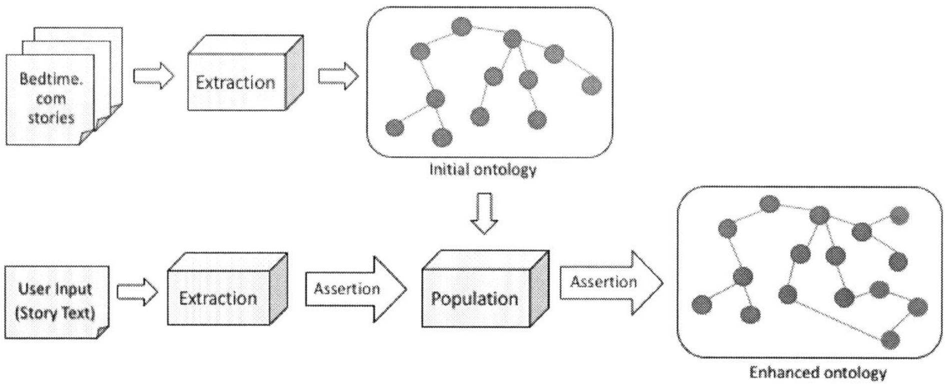

Figure 1. KA process to populate the local ontology. Orange nodes are new assertions from an input story text.

Table 2. Extraction patterns using both parts-of-speech tags and word dependencies.

Relation	Concept1	Keyword	Concept2
isA	nsubj	be	attr
hasA	nsubj	have	dobj
capableOf	nsubj	aux	ROOT
capableOf	nsubj		ROOT
capableOF	ROOT	agent	pobj
receivesAction	nsubjpass	auxpass	ROOT
receivesAction	ROOT	det	dobj
atLocation	nsubj	go	advmod
atLocation	nsubj	in | at	pobj

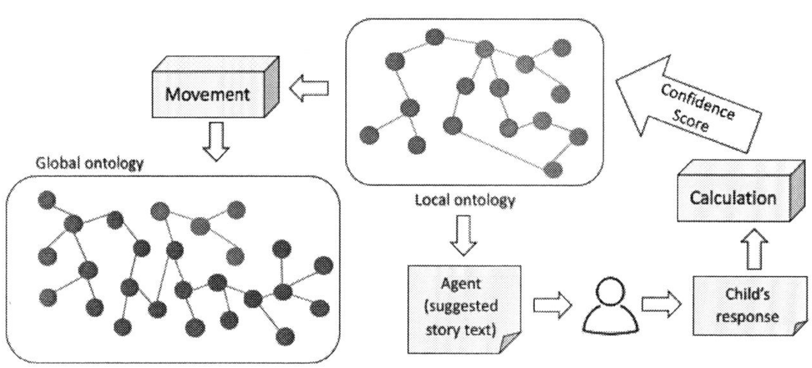

Figure 2. KV process. Orange nodes are new assertions; gray nodes are moved from the local to the global ontology.

Table 3. Annotated template for generating a multi-assertion suggestion.

Template	(1) (2) a (3) to make a (4).
Word Relations	capableOf(1, 2) receivesAction(3, 2), usedFor(3, 4)
Example Assertions	capableOf(Peter, carve) receivesAction(plane, carve), usedFor(plane, gift)
Text	(Peter) (carved) a (plane) to make a (gift).

A positive input from the child increases the confidence score for A_i. A negative input, however, triggers a *follow-up* dialogue move, specifically, *"Why not? Don't you like it or do you think it is wrong?"* This is necessary to differentiate between an invalid assertion (from the child's perspective) versus the child's non-acceptance of the suggested story text to be included as part of his/her story. An invalid assertion merits a decrease in the corresponding confidence score, while a non-acceptance of the suggested story text does not effect its value.

A suggested story text can contain more than one assertion sourced from the ontology using an annotated template as shown in Table 3. An annotated template includes a set of word relations indicating the semantic relations between the numbered variables to be used in filling up the slots. This approach is adapted from [25]. The agent iterates through all the word relations and queries the ontology to find matching assertions until it can fill-up the template slots to generate a response. When the child gives a negative response to a multi-assertion suggestion, the CSA generates a follow-up dialogue that also contains multiple assertions as illustrated in Table 4. The child then selects which of the assertions, i.e., `capableof(dog, study)` or `usedFor(book, study)`, is incorrect.

Table 4. An example log showing multiple assertions used in a follow-up dialogue.

Turn	Agent	Child
1	What if the dog used book for study?	No
2	Why not? Don't you like it or is it wrong?	Wrong
3	Which one is wrong? A. Dog can study B. Book used for study C. All of them	A

4. Test Results

12 children who are between the age of 7 to 11 years old were invited to conduct a one-on-one storytelling session with the CSA. In this section, we present the test results that assess the CSA's ability as a teachable agent by learning new assertions from children.

4.1. Knowledge Acquisition

To assess the performance of the CSA as a teachable peer, we measured its precision and recall values in extracting assertions from children's inputs. The results are shown in Table 5. *True positive* (TP) refers to correct extraction of assertions, *false positive* (FP) refers to incorrect assertions and *false negative* (FN) are assertions that were not extracted. Given these, *precision* is computed as the number of correctly extracted assertions `(TP)` over the total number of extracted assertions `(TP+FP)`. *Recall* is computed as the number of correctly extracted

Table 5. Performance results in extracting assertions from children's input.

Relation	TP	FP	FN	Precision	Recall	F-Score
isA	17	15	1	53.13%	94.44%	68.00%
hasA	10	4	3	71.43%	76.92%	74.07%
hasProperty	43	13	8	76.79%	84.31%	80.37%
capableOf	46	26	9	63.89%	83.64%	72.44%
atLocation	1	9	19	10.00%	5.00%	6.67%
receivesAction	44	14	14	75.86%	75.86%	75.86%
desires	0	0	2	0.00%	0.00%	-
instanceOf	12	1	42	92.31%	22.22%	35.72%
usedFor	1	3	2	25.00%	33.33%	28.57%
Total	**174**	**85**	**100**			

assertions over the total number of word relations (`TP+FN`). The *F-Score* is the harmonic mean of the precision and recall values and is used to measure the accuracy of the test.

Stories that children share mostly revolve around entity (characters and objects) descriptions and actions. The agent's corresponding responses using pumps such as *"What happens next"* and *"Tell me more then"* are specifically designed to elicit such story details. Story text expressed in `subject-verb-object` and `adjective-noun` phrases are then extracted as *hasProperty*, *capableOf* and *receivesAction* relations, thus their high recall and F-score values. In particular, the CSA performs well in extracting *hasProperty* (see Table 6).

Table 6. Sample story text with *hasProperty* relations.

Line	Story Text	Assertions
1	harry is a dark little boy while cookie is a tall handsome man	[isA(harry, boy)] [hasProperty(boy, dark)] [hasProperty(boy, little)] [isA(cookie, man)] [hasProperty(man, tall)] [hasProperty(man, handsome)]
2	pepper and cookie are the smartest	students in class [isA(pepper, student)] [isA(cookie, student)] [hasProperty(student, smart)]
3	students are mean to people who fail	[hasProperty(student, mean)]

The low precision and recall values can be attributed to the different ways that children express their ideas and the presence of grammar errors in their story text. Misspellings, lack of punctuation marks and the use of common nouns as proper nouns can lead to incorrect POS tagging and dependency parsing by spaCy. The agent missed a number of named entities in the story of P12 shown in Table 6, such as *pepper* and *cookie* in line 2, which could be extracted as `instanceOf(pepper, dog)` and `instanceOf(cookie, dog)`. This is because spaCy did not consider these common nouns as named entities. The agent also encountered difficulty in extracting *atLocation* relations using the extraction patterns in Table 2 with the lowest precision and recall values of 10% and 5%, respectively.

Another way to evaluate the CSA's performance is by looking at the precision and recall values for each participant's story, as shown in Table 7. P7 has the highest precision but the

Table 7. Precision and recall values in extracting assertions from participant's story.

Child	TP	FP	FN	Precision	Recall	F-Score
P1	2	3	1	40.00%	66.67%	50.55%
P2	16	10	6	61.54%	72.73%	66.67%
P3	5	2	2	71.43%	71.43%	71.43%
P4	31	13	10	70.45%	75.61%	72.94%
P5	10	2	5	83.33%	66.67%	74.07%
P6	31	7	23	81.58%	57.41%	67.39%
P7	2	0	11	100.00%	15.38%	26.66%
P8	15	5	6	75.00%	71.43%	73.17%
P9	11	10	9	52.38%	55.00%	53.66%
P10	16	11	7	59.26%	69.57%	64.00%
P11	10	7	18	58.82%	35.71%	44.44%
P12	25	15	2	62.50%	92.59%	74.63%
Total	**174**	**85**	**100**			

lowest recall. While the CSA did not commit any errors in extraction, it missed 11 sentences with candidate assertions which affected the recall value. A trace of the conversation log showed that P7 typed the whole story in a single dialogue turn:

> Once upon a time, there was a mermaid named Janesa. she wished to become a human one day ... she was told to drink a potion at dawn. when she drank it she became a human! she met a prince and they got married after a week.

However, the agent extracted assertions only from the last sentence, *receivesAction(prince, meet)* and *instanceOf(week, date)*, and missed a number of candidate assertions such as *capableOf(wish, mermaid)*, *capableOf(drink, mermaid)* and *receivesAction(potion, drink)*.

4.2. Knowledge Validation

Validating crowdsourced knowledge is a crucial step in the knowledge acquisition task. To validate the crowdsourced assertions in its local ontology, the agent uses the *suggestion* and *follow-up* dialogue moves. To vary the types of responses, the CSA uses a weighted randomizer to calculate the ratio between the frequency of use of a dialogue move and the total number of dialogue turns made by the agent. As seen in Table 8, feedback and pumps are used more often, comprising nearly 90% of the agent's dialogue turns. When the CSA cannot find an assertion from its local and global ontology for a given input word to fill-up an annotated template, it resorts to using general pumps (*"Tell me more then"*) and specific pumps (*"I don't know much about _. Please help me learn more."*).

Furthermore, only 29 (54.72%) of the total attempts by the agent to formulate a suggestion were realized as actual responses. Analysis of the logs showed that hints and suggestions often fail when the selected template requires more than one assertion. In the conversation logs of P5, P6 and P8, hints and suggestions were not used because the CSA encountered multiple failures during the session. When a failure occurs, the event chain and dialogue history are refreshed and the conversation has to start anew.

The last experiment aims to determine if the CSA is using assertions acquired from one child and validating this with another child. Two evidences were found in the logs shown in Table 9. In the first instance, the agent learned from P8 that a *dog* is *capableOf* performing the action *find*, which it validated by asking P11. In the second instance, the CSA learned from P4 that a *boy* can be located in the *countryside*, thus making the suggestion to P12.

Table 8. Frequency of use vis-a-vis CSA's attempts to formulate the dialogue moves.

Dialogue Move	**Actual Use**		**Attempts**	
	Count	Frequency	Total	Actual ÷ Attempt
Feedback	81	27.36%	119	68.06%
General Pump	72	24.32%	72	100.00%
Specific Pump	100	33.78%	100	100.00%
Hint	14	4.73%	43	32.56%
Suggestion	29	9.80%	53	54.72%
Total	296	100.00%		

Table 9. Excerpts of logs showing knowledge acquisition and validation across users.

Child	**Text**	**Assertion**
P8	Child: ... and the dog found a human ...	[capableOf(dog, find)]
P11	Child: (*talks about dog*) Agent: What if dog can find?	
P4	Child: a boy living in a countryside loved ...	[atLocation(boy, countryside)]
P12	Child: (*talks about boy*) Agent: What if boy was near countryside?	

5. Conclusion

We presented the use of story-based conversations as a means for a conversational agent to acquire and to validate assertions from children. A two-level ontology is used to separate validated knowledge (global ontology) from newly acquired knowledge (local ontology) that are yet to undergo validation. The agent suggests story text as a means of soliciting confirmation from a child on the correctness of an assertion. The confidence score associated with an assertion in the local ontology is updated through this exchange and is also used to determine when an assertion can be moved to the global ontology. The resulting ontology is mostly comprised of assertions that describe story entities and their actions. These are represented by the relations *hasProperty*, *capableOf* and *receivesAction*. Variances in the linguistic and storytelling abilities of children led to F-score values ranging from 75% - 95% for the most common semantic relations. Low recall values for *atLocation* and *instanceOf* relations can be attributed to the difficulty in detecting locations and named entities from the input text.

Current research in neural conversational models are investigating the use of knowledge graphs to provide conversational agents with domain knowledge [2, 3, 14]. These can be explored as strategies to extract commonsense knowledge from children's input text. Allowing children to provide explanations beyond the current *"no"* and *"don't like"* responses to follow-up questions can also be explored as another avenue for enriching the local ontology.

While suggestions are the primary means by which the agent can validate the correctness of an assertion in its local ontology, our experiments show that only 55% of the attempted suggestion dialogue moves were realized as responses. This is due to the limited number of assertions in the ontology that can be used to fill-up the annotated response templates and the variances in the topics and themes of the stories that children share with the agent. Further experiments should therefore be conducted to expand the knowledge base with new assertions and to validate existing assertions.

References

[1] Fan J, Kalyanpur A, Gondek D and Ferrucci D 2012 Automatic knowledge extraction from documents *IBM J. of Research and Development* **56** pp 1–10
[2] Liu Z, Niu Z Y, Wu H and Wang H 2019 Knowledge aware conversation generation with explainable reasoning over augmented graphs *Proc. of the 2019 Conf. on Empirical Methods in Natural Language Processing and the 9th Int. Joint Conf. on Natural Language Processing* (Association for Computational Linguistics) pp 1782–92
[3] Zhou H, Young T, Huang M, Zhao H, Xu J and Zhu X 2018 Commonsense knowledge aware conversation generation with graph attention *Proc. of the 27th Int. Joint Conf. on Artificial Intelligence* pp 4623–9
[4] Herdagdelen A and Baroni M 2009 Bagpack - a general framework to represent semantic relations *Proc. of the EACL 2009 Geometrical Models for Natural Language Semantics Workshop* pp 33–40
[5] Singh P, Barry B and Liu H 2004 Teaching machines about everyday life *BT Technol. J.* **22** pp 227–40
[6] Alani H, Kim S, Millard D, Weal M J, Lewis P H, Hall W and Shadbolt N 2003 Automatic extraction of knowledge from web documents *The 2nd Int. Semantic Web Conf. Workshop on Human Language Technol. for the Semantic Web and Web Services*
[7] Banko M, Cafarella M J, Soderland S, Broadhead M and Etzioni O 2007 Open information extraction from the web *Proc. of the 20th Int. Joint Conf. on Artificial Intelligence* pp 2670–6
[8] Etzioni O, Cafarella M, Downey D, Popescu A M, Shaked T, Soderland S, Weld D S and Yates A 2005 Unsupervised named-entity extraction from the web: An experimental study *Artificial Intelligence* **165** pp 91–134
[9] Samson B P and Ong E 2014 *PKAW 2014* ed Kang B (Cham: Springer International Publishing Switzerland) pp 195–208
[10] Goodwin T and Harabagiu S 2017 Embedding open-domain common-sense knowledge from text *Proc. of LREC Int. Conf. on Language Resources Evaluation* pp 4621–8
[11] Petasis G, Karkaletsis V, Paliouras G, Krithara A and Zavitsanos E 2011 Ontology population and enrichment: State of the art *Knowledge-Driven Multimedia Information Extraction and Ontology Evolution, LNCS* vol 6050 ed Paliouras G *et al.* (Heidelberg: Springer) pp 134–66
[12] Speer R, Chin J and Havasi C 2017 Conceptnet 5.5: An open multilingual graph of general knowledge *Proc. of the 31st AAAI Conf. on Artificial Intelligence* pp 4444–51
[13] Liu H and Singh P 2004 Commonsense reasoning in and over natural language *Proc. of the Int. Conf. on Knowledge-Based and Intelligent Information and Engineering Systems, LNCS* vol 3215 ed Negoita M *et al.* (Heidelberg: Springer-Verlag) pp 293–306
[14] Young T, Cambria E, Chaturvedi I, Zhou H, Biswas S and Huang M 2018 Augmenting end-to-end dialogue systems with commonsense knowledge *Proc. of the 32nd AAAI Conf. on AI* pp 4970–7
[15] Liu H and Singh P 2002 Makebelieve: Using commonsense knowledge to generate stories *18th National Conf. on Artificial Intelligence* (AAAI) pp 957–8
[16] Cua J, Ong E, Manurung R and Pease A 2010 Representing story plans in sumo *Proc. of the NAACL HLT 2010 Second Workshop on Computational Approaches to Linguistic Creativity* (Association for Computational Linguistics) pp 40–80
[17] Niles I and Pease A 2001 Towards a standard upper ontology *Proc. of Formal Ontology in Information Systems* pp 2–9
[18] Chua Jr R C and Ong E 2012 Commonsense knowledge acquisition through children's stories *Knowledge Management and Acquisition for Intelligent Systems, LNCS* vol 7457 ed Richards D and Kang B (Springer, Berlin Heidelbert) pp 244–50
[19] Tarau P and Figa E 2004 Knowledge-based conversational agents and virtual storytelling *Proc. of the 2004 ACM Symposium on Applied Computing* pp 39–44
[20] Ong D T, De Jesus C R, Gilig L K, Alburo J B and Ong E 2018 A dialogue model for collaborative storytelling with children *Proc. of the 26th Int. Conf. on Computers in Education* (Asia-Pacific Society for Computers in Education) pp 205–10
[21] Sun M, Leite I, Lehman J F and Li B 2017 Collaborative storytelling between robot and child: A feasibility study *Proc. of the 2017 Conf. on Interaction Design and Children* pp 205–14
[22] Zhao G, Ailiya and Shen Z 2012 Learning-by-Teaching: Designing teachable agents with intrinsic motivation *J. of Educational Technol. & Society* **15** pp 62–74
[23] Graesser A, Chipman P, Haynes B and Olney A 2005 Autotutor: An intelligent tutoring system with mixed-initiative dialogue *IEEE Transactions on Education* **48** pp 612–8
[24] Ong D T, De Jesus C R, Gilig L K, Alburo J B and Ong E 2018 Building a commonsense knowledge base for a collaborative storytelling agent *Knowledge Management and Acquisition for Intelligent Systems, LNCS* vol 11016 ed Yoshida K and Lee M (Springer, Cham) pp 1–15
[25] Hong B A and Ong E 2009 Automatically extracting word relationships as templates for pun generation *Proc.*

of the NAACL HLT Workshop on Computational Approaches to Linguistic Creativity (ACL) pp 24–31

IOP Conf. Series: Materials Science and Engineering 1077 (2021) 012024 doi:10.1088/1757-899X/1077/1/012024

Comparison between Butterworth Bandpass and Stationary Wavelet Transform Filter for Electroencephalography Signal

Xiaoxi Kang[1], Dini Oktarina Dwi Handayani[2], Hamwira Yaacob[3]

[1,2]School of Computer Science & Engineering, Faculty of Innovation & Technology, Taylor's University, 47500 Subang Jaya, Selangor, Malaysia
[3]Computer Science Department, Kulliyyah of Information and Communication Technology, International Islamic University Malaysia, 53100 Gombak, Selangor

E-mail: [1]kangxiaoxi@sd.taylors.edu.my,
[2]DiniOktarina.DwiHandayani@taylors.edu.my, [3]Hyaacob@iium.edu.my

Abstract. The Electroencephalography (EEG) signals able to obtain the information from the brain signals. Reduce the noise of the raw EEG data can improve the accuracy of the result. The pre-processing step of the raw EEG data can generate a clean signal and improve the accuracy. The purpose of this paper is to compare the Butterworth bandpass (BB) and stationary wavelet transform (SWT) method for the pornography addiction EEG data. The data was collected from Yayasan Kita dan Buah Hati (YKBH), Jakarta, Indonesia, using the Brain Maker EEG machine with 19 channels. We used mean square error (MSE) and peak-to-noise ratio (PSNR) to compare the quantitative value for the filtered EEG signals. The result shows that the BB filter is more effective in removing the noise and keep the original information.

1. Introduction

Brain activities have been used by the scientist to understand human behaviours. Potenza [1] proved that brain activities reflected with addictive behaviours. A specific region of our brain correlates with different behaviours. However, the raw EEG signals are consist of noise and artefacts. To process the EEG signals, people usually transfer the raw data into different frequencies before analyzing the data. Mostly there are four essential bands are used for the researcher measure the brain activity which are δ (1 to 4Hz), θ (4 to 7Hz), α(8 to 15Hz), and β (12.5 to 28Hz) [2]. Before training any machine learning algorithm using EEG information, researchers usually generate a specific code (in the form of a script) to retrieve peak values, wave frequencies, or translate raw EEG data into a Fourier transform continuum diagram, or turns to a wavelet. Then the data can be used to develop a classification model.

According to Caglar [3], there are four steps to process the EEG signals after collecting the raw EEG data.

1. The first step is the pre-processing, which reduce the noise of the data. The less noise of the data will provide higher accuracy in the following steps.
2. The second step is the feature extraction to minimize the loss of valuable information.
3. The third step is the feature reduction to remove the unimportant information.
4. The last step is the classification which determines the group of the input data.

The pre-processing step will improve the accuracy of the result. Daud, S.S, et al., 2015, used BB and SWT as a pre-processing method [4]. To check the quality of the filter, MSE and PSNR used for the testing [5].

The MSE and PSNR aim to measure the quality of the signal. The score shows the degree of the similarity/fidelity and level of error/distortion between the original signal and filtered signal.

The MSE between the signals can be defined as

$$MSE(x, y = \frac{1}{N}\sum_{i=1}^{N}(xi - yi)^\wedge 2 \quad (1)$$

where x and y are two finite lengths (discrete signals), N is the number of signals samples, and x and y are the values of the samples in x and y [5]. The MSE value usually converted into the PSNR value for determining the quality of an image in decibel value. PSNR can be defined as

$$PSNR = 10 \log 10 \frac{L^\wedge 2}{MSE} \quad (2)$$

where L is the dynamic range of allowable image pixel intensities [5].

Moshfeghi, M et al., 2020, suggested that the bandpass is more suitable for the EEG signals filtering process [6]. The bandpass filters able to filter specific frequency for analysis, and another reason is able to retain the original signal with removing the artifacts. We expected that our result would also show a better outcome after applying bandpass.

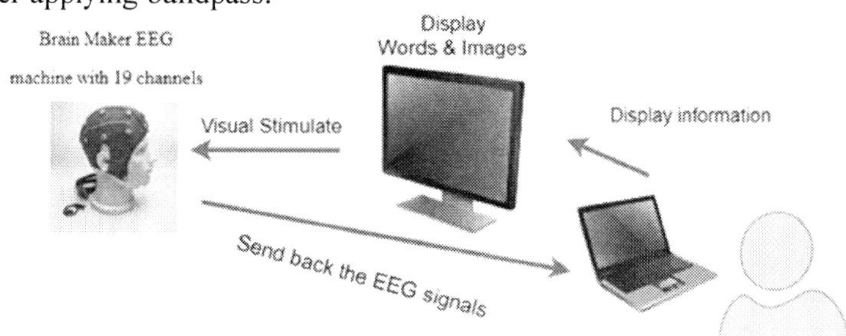

Figure 1. Experiment set up.

2. Material & Methodology

2.1. Material

The EEG signal data was collected from YKBH, Jakarta, Indonesia. The data was collected using the Brain Maker EEG machine with 19 channels. Figure 1 shows the experiment set up for this study. The participants wear the EEG device while watching a stimulus that displayed on the monitor. The signals will be collected when the participants were doing the tasks.

Table 1. Participants Details.

No	Subject ID	Sex	Result
1	S1	M	Addicted
2	S2	F	Not Addicted
3	S3	F	Not Addicted
4	S4	M	Not Addicted
5	S5	M	Addicted
6	S6	M	Addicted
7	S7	M	Not Addicted
8	S8	M	Not Addicted
9	S9	F	Addicted
10	S10	F	Addicted
11	S11	F	Addicted
12	S12	M	Not Addicted
13	S13	M	Not Addicted
14	S14	M	Addicted

There are 14 participants (girl =5, boy = 9) are joining in this experiment. The ages varied from 13 to 15 years old. The psychologist has verified among the 14 subjects that 7 of them have a porn addiction, and 7 of them do not have a porn addiction. Table 1 and figure 2 displayed the details of the participants. In our research, we choose S12 and S11 for our pre-processing steps since one is addicted subjects, and another one is non-addicted subjects.

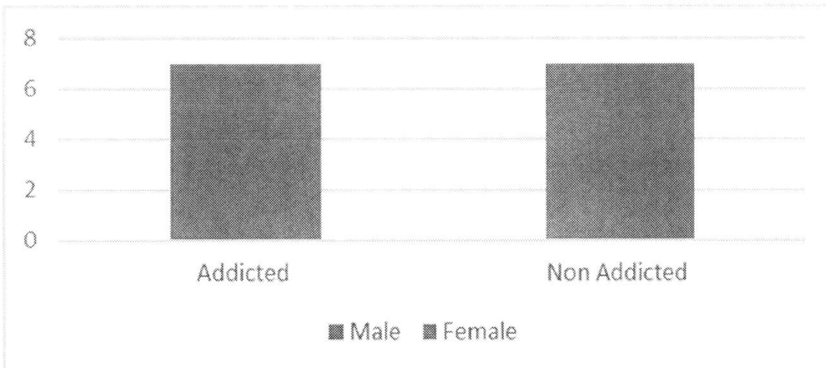

Figure 2. Participants Details.

There are six tasks in the data collection process. The total times for one participant are 12 minutes. The six tasks are baseline, emotional state, task 1, task2, task 3, and baseline extension. There are two baseline data collections, one at the beginning and one at the end with eyes open and eyes closed. The emotional state data were collected for happy, calm, sad, and fear, each emotion for 1 minute. The emotional state was based on the International Affective Picture System (IAPS) [7]. It provides a set of colour photographs that set of normative emotional stimuli. The stimuli will show a picture of happy, calm, sad, and fear photos for the participants. Task 1 is memorizing 15 words in 1 minute, and task 2 is an executive task in 2 minutes where the participants watch a video given by psychologist. Task 3 is the recall 15 words form task 1 in 1 minute. Table 2 displays an overview of the protocols.

Table 2 Overview of the Protocols.

Time	Item	Item details
2 minutes	Base Line	Eyes Closed (1 minute) Eyes Open (1 minute)
4 minutes	Emotional State (IAPS)	Happy (1 minute) Calm (1 minute) Sad (1 minute) Fear (1 minute)
1 minute	Task 1	Memorize 15 words (1 minute)
2 minutes	Task 2	Executive Tasks (2 minutes)
1 minute	Task 3	Recall 15 words (1 minute)
2 minutes	Base Line Extension	Eyes Closed (1 minute) Eyes Open (1 minute)
Total	six tasks	12 minutes

2.2. Methodology

There are 19 channels from the original signal (see figure 3), however, we will only use three channels for our pre-processing experiment as the preliminary study for the pre-processing methods. The channels are Fp1, Fp2 and Fz. We selected data from two different participants. One is an addiction, and one is non-addiction. For this experiment, we choose the first baseline, one minute for eye closed.

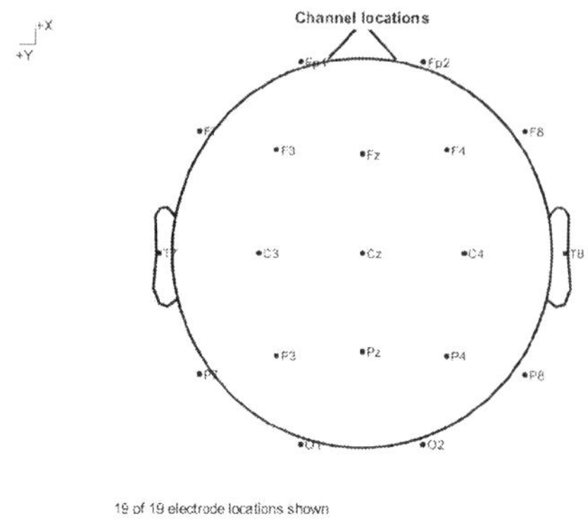

Figure 3. The 19-channel electrode.

Table 3. Channel Information.

Number	labels	theta	radius	X	Y	Z	sph_theta	sph_phi	sph_radius
1	"P4"	140.0000	0.3440	-0.0677	-0.0566	0.0469	-140.0000	28.0000	0.0999
2	"O2"	162.0000	0.5150	-0.0949	-0.0307	-0.0047	-162.0000	-2.7000	0.0999
3	"P8"	126.0000	0.5280	-0.0587	-0.0804	-0.0088	-126.0000	-5.0500	0.0999
4	"T8"	90.0000	0.5330	0.0000	-0.0994	-0.0104	-90.0000	-5.9700	0.0999
5	"C4"	90.0000	0.2670	0.0000	-0.0743	0.0669	-90.0000	42.0000	0.1000
6	"Cz"	0	0	0.0000	0	0.1000	0	90.0000	0.1000
7	"Fz"	0	0.2530	0.0714	0	0.0699	0	44.4000	0.0999
8	"F4"	39.9000	0.3440	0.0677	-0.0566	0.0469	-39.9000	28.0000	0.0999
9	"Fp2"	17.9000	0.5150	0.0949	-0.0307	-0.0047	-17.9000	-2.7000	0.0999
10	"F8"	53.9000	0.5280	0.0587	-0.0804	-0.0088	-53.9000	-5.0500	0.0999
11	"Fp1"	-17.9000	0.5150	0.0949	0.0307	-0.0047	17.9000	-2.7000	0.0999
12	"F7"	-53.9000	0.5280	0.0586	0.0804	-0.0088	53.9000	-5.0500	0.0999
13	"F3"	-39.9000	0.3450	0.0677	0.0567	0.0469	39.9000	28.0000	0.1000
14	"C3"	-90.0000	0.2670	0.0000	0.0742	0.0668	90.0000	42.0000	0.0998

We selected 15 000 data points with 60 seconds duration. The raw data saved in the tdms file and we have converted into a csv file. The sampling rate of the data frequency is 250 HZ. We loaded the data into MATLAB software for data processing process. Firstly, we used QtiPlot to convert the raw EEG data into a csv file. After we converted the raw EEG data, we loaded csv file in MATLAB and save the data into vectors. We selected three channels for our processing which are: Fp1, Fp2 and Fz. The channel information is generated from the EEGLAB. Table 3 displayed the details of each channel. We choose the Fp1 and Fp2 and Fz because it is the prefrontal area which is most related to the addiction behaviour[8].

Secondly, after we imported the data and got the channel information, we have selected the Fp1, Fp2 and Fz for both addicted and non-addicted subjects. The following figure displays the steps for the pre-processing step.

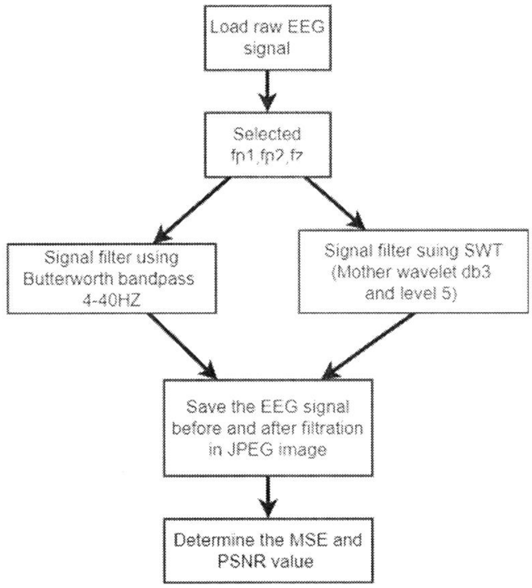

Figure 4. Pre-processing step.

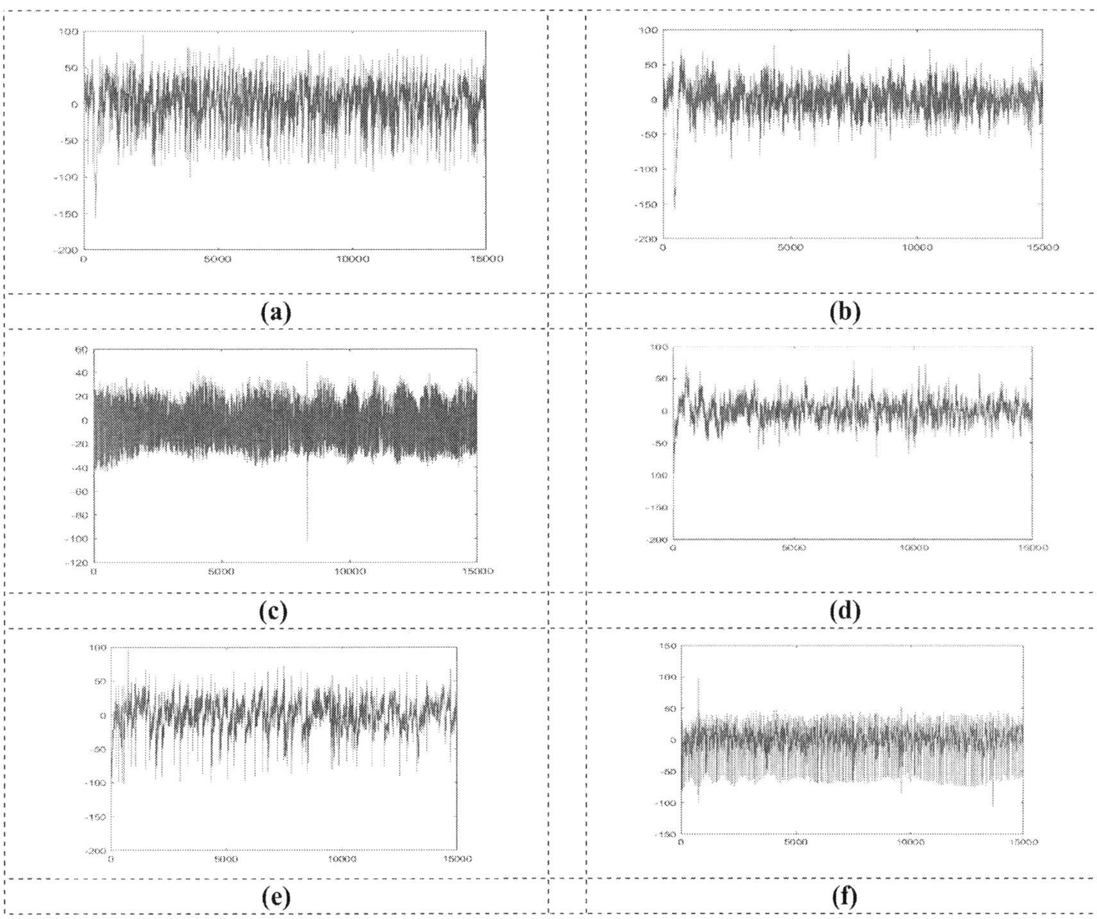

Figure 5. (a) Raw addiction in Fp1 (b) Raw addiction in Fp2 (c) Raw addiction in Fz (d) Raw non-addiction in Fp1 (e) Raw non-addiction in Fp2 (f) Raw non- addiction in Fz.

The first filter we used is a 4th order of BB. This type of filter is chosen because it has a linear response compare to others. The cut-off frequency was 4 – 40HZ. The delta frequency is below 4 HZ, and gamma frequency is more than 40HZ. These two frequencies were eliminated because it considers as noise. Hence, we choose the frequency band between 4-40 HZ.

The second filter is the SWT filter. The dB3 wavelet mother with five decomposition levels of decomposition is chosen for this filter [9]. This wavelet was selected because it can detect and localize the spike in the EEG signal [10].

After obtaining a clean signal from these two filters, we will save the clean signals in jpeg image. After we get the original signal image and filtered signal image, we will use the MSE and PSNR value to determine the quality.

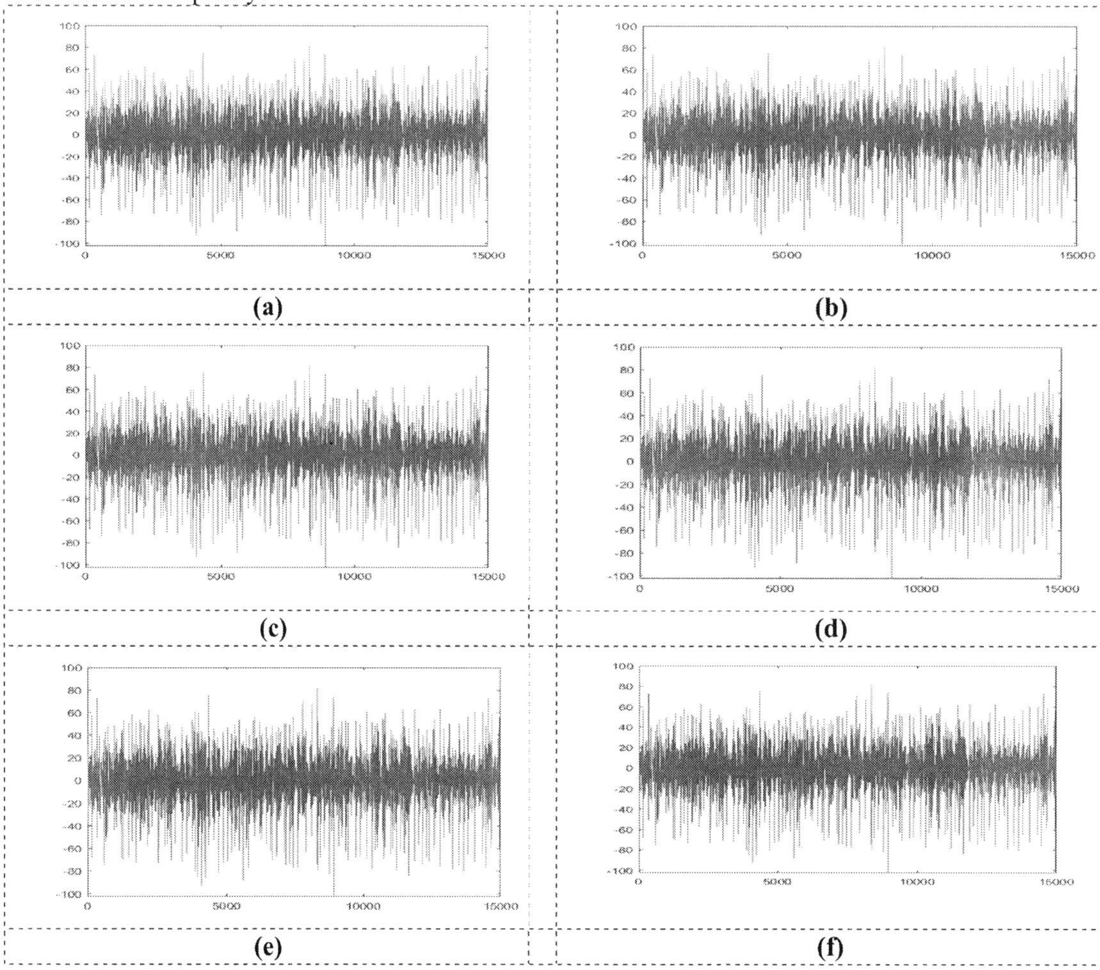

Figure 6. (a) Clean Fp1 signal after BB filter for addiction subjects (b) Clean Fp2 signal after BB filter for addiction subjects (c) Clean Fz signal after BB filter for addiction subjects (d) Clean Fp1 signal after BB filter for non-addiction subjects (e) Clean Fp2 signal after BB filter for non-addiction subjects (f) Clean Fz signal after BB filter for non-addiction subjects.

3. Result and discussion

3.1. Raw EEG Signal

Figure 5 displays the original signal image for Fp1, Fp2 and Fz for addicted and non-addicted subjects. This image is generated from the MATLAB. The image (a), (b), (c) is the addiction subject for signal Fp1, Fp2, and Fz. The image (d), (e), (f) is the raw signal for the non-addiction subject.

3.2. Signal after BB Filter

The cut off frequency for the filter is 4 – 40HZ. The function is used the BB filter function in the MATLAB. The low pass is 4hz, and the high pass is 40. The order of the bandpass filter is 4. This number is a reference from [4]. The code is used for the bandpass filter in MATLAB to get the filtered data. The frequency of the data is 250hz. Figure 6 displays the result for the BB filter for Fp1, Fp2 and Fz:

```
[b,a]=butter(order,[fcutlow,fcuthigh]/(fs/2),'bandpass');
```

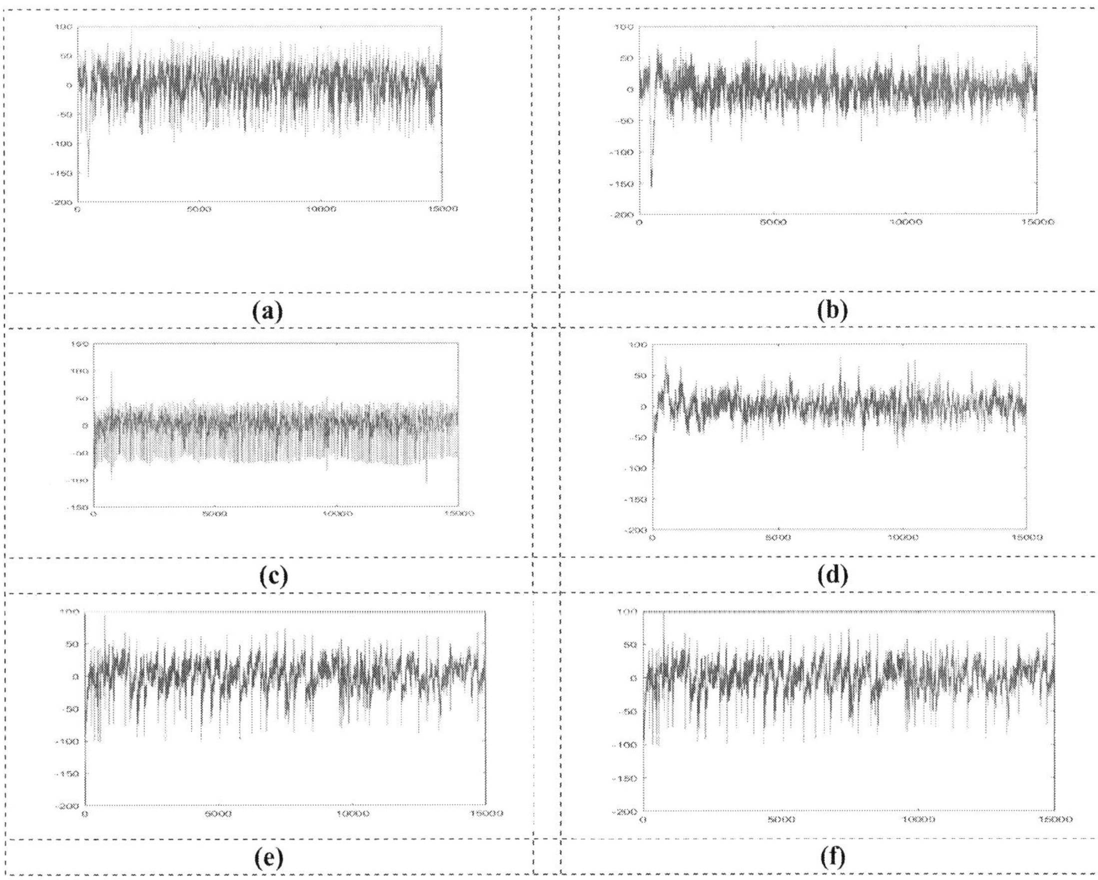

Figure 7. (a) Clean Fp1 signal after SWT filter for addiction (b) Clean Fp2 signal after SWT filter for addiction (c) Clean Fz signal after SWT filter for addiction. (d) Clean Fp1 signal after SWT filter for non-addiction (e) Clean Fp2 signal after SWT filter for non-addiction (f) Clean Fz signal after SWT filter for non-addiction.

3.3. Signal after SWT filter

Figure 7 displays the EEG signals after using the SWT. We used the wavelet in db3 and level of 5. The following script is the code of the SWT filter from MATLAB.

```
result = wdenoise(inputSignal,5, ...
    'Wavelet', 'db3', ...
```

```
'DenoisingMethod', 'Bayes', ...
'ThresholdRule', 'Median', ...
'NoiseEstimate', 'LevelIndependent');
```

3.4. MSE and PNSR

Figure 8 and Table 4 display the MSE for BB filter and SWT at channel Fp1, Fz and Pz for addiction and non-addiction subjects.

Table 4. MSE at channel Fp1, Fp2 and Fz

	Filter/Channel	Fp1	Fp2	Fz
Addiction	BB	4.0005	16.0010	36.0015
	SWT	144.0059	144.0088	144.0176
Non-addiction	BB	64.0020	100.0024	144.0029
	SWT	144.0017	144.0146	144.0205

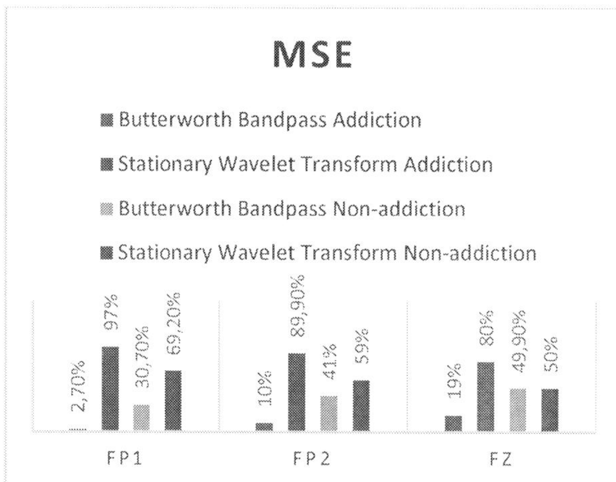

Figure 8. MSE of the BB and SWT filter

The BB filter has a lower MSE value compared to the SWT filter from the figure and table above. The most difference is shown from the addiction subject on channel Fp1. The BB is only 4.005, and SWT is 144.0059. The most percentage difference we can see from the addiction subjects for the Fp1, Fp2 and Fz channels.

Table 5. PSNR at channel Fp1, Fp2 and Fz

	Filter/Channel	Fp1	Fp2	Fz
Addiction	BB	42.1437	36.1233	32.6016
	SWT	26.5810	26.5809	26.5806
Non-addiction	BB	30.1029	28.1647	26.5811
	SWT	26.5808	26.5807	26.5806

Based on Table 5 and Figure 9, it shows that the BB filter has more significant PSNR value at the three channels for both addiction and non-addiction subjects. It means that it has more noise filtered out.

Compare to the Daud [4]; our result shows the different result of MSE and PSNR value, we can see that the BB filter at cut off frequency 4 to 40 HZ can filter out more noise compared to the SWT with dB3 and 5 decomposition level. In addition, the original information does not loss for the BB Techniques.

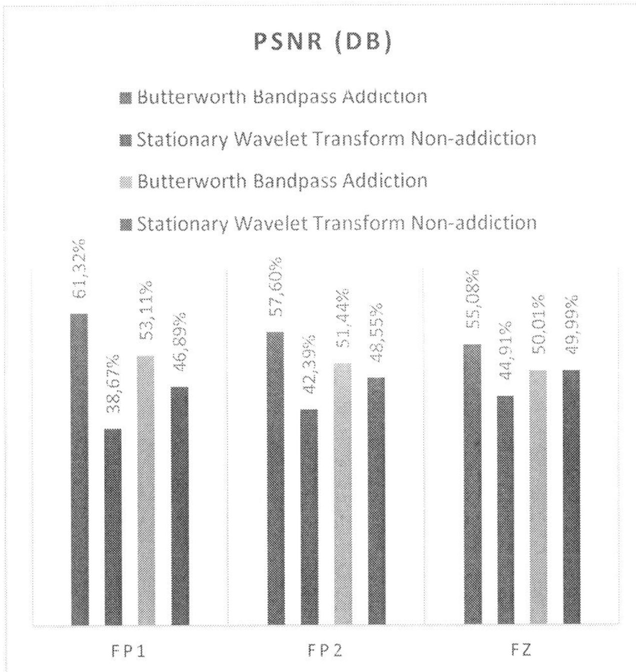

Figure 9. PSNR of the BB and SWT filter

4. Conclusion

There are many noise data that affect the accuracy of the EEG signal. In this paper, we have compared two different filters to reduce the noise from the original data. The MSE and PSNR is a popular parameter to determine the quality of signal after filtering. BB is able to remove all artifact and keep the original information from our research results.

Acknowledgement

This work is supported by Fundamental Research Grant Scheme (FRGS) funded by the Ministry of Higher Education (Grant code: FRGS/1/2019/ICT02/TAYLOR/03/1).

References

[1] Potenza M N 2014 Non-substance addictive behaviors in the context of DSM-5 *Addict. Behav.* **39** pp 1–2 DOI: 10.1016/j.addbeh.2013.09.004

[2] Li G, Lee C H, Jung J J, Youn Y C and Camacho D 2019 Deep learning for EEG data analytics: A survey *Concurr. Comput. Pract. Exper.* (New Jersey: John Wiley and Sons Ltd) DOI: 10.1002/cpe.5199

[3] Uyulan C and Erguzel T T 2017 Analysis of time - frequency EEG Feature Extraction Methods for Mental Task Classification *Int. J. Comput. Intell. Syst.* **10** 1280 DOI: 10.2991/ijcis.10.1.87

[4] Daud S S and Sudirman R 2015 Butterworth Bandpass and Stationary Wavelet Transform Filter Comparison for Electroencephalography Signal *Proc. - Int. Conf. Intell. Syst. Model. Simulation ISMS* pp 123–6 DOI: 10.1109/ISMS.2015.29

[5] Wang Z and Bovik A C 2009 Mean squared error: Lot it or leave it? A new look at signal fidelity measures *IEEE Signal Process. Mag.* **26** pp 98–117 DOI: 10.1109/MSP.2008.930649

[6] Moshfeghi M, Tuke A, Jyoti B and Bartaula P 2013 *Emotion Recognition from EEG Signals using Machine Learning* Retrieved on 7 October 2020 from www.bth.se/ing

[7] Bradley M M and Lang P J 2007 The International Affective Picture System (IAPS) in the study of emotion and attention *Handbook of Emotion Elicitation Assessement* ed J A Coan and J J B Allen (Oxford: Oxford University Press) pp 29–46 Retrieved on 10 March 2020 from https://psycnet.apa.org/record/2007-08864-002)

[8] Sun Y, Wang H and Bo S 2019 Altered topological connectivity of internet addiction in resting-state EEG through network analysis *Addict. Behav.* **95** pp 49–57 DOI: 10.1016/j.addbeh.2019.02.015

[9] Palendeng M E 2011 *Removing Noise From Electroencephalogram Signals For BIS Based Depth of Anaesthesia Monitors* Thesis (Queensland: University of Southern Queensland)

[10] Jain P and Govindaiah P K 2014 Removal of Artifacts and Analysis of EEG Signal Retrieved on 12 March 2020 from https://www.researchgate.net/publication/260600286_REMOVAL_OF_ARTIFACTS_AND_ANALYSIS_OF_EEG_SIGNAL_USING_DATA_DRIVEN_PARAMETERS

Word Embeddings Evaluation on Indonesian Translation of Al-Quran and Hadiths

Muhammad Zidny Naf'an[1,2], Yunita Sari[1], and Yohanes Suyanto[1]

[1]Department of Computer Science and Electronics, Universitas Gadjah Mada, Yogyakarta, Indonesia
[2]Department of Informatics, Institut Teknologi Telkom Purwokerto, Central Java, Indonesia

E-mail: [2]muhammadzidny@mail.ugm.ac.id

Abstract. Word vectors are an important part of machine learning. Word vectors are a numerical representation of text data. One of the methods that can be used to convert text into numerics is word embeddings. The word embeddings algorithm that researchers often use is Continuous Bag of Word, Skip-Gram, and FastText. This paper will discuss the transformation of textual data from Islamic knowledge domain documents into numerical forms using these three algorithms, then evaluate the word vector results using intrinsic and extrinsic evaluation techniques. We conduct intrinsic evaluations by determining the words to be evaluated, then checking for the existence of synonyms, antonyms, related words, and derived words from the nearest set of words based on vector values. We also tried to use vector words to solve word analogy problems. The best word vector in extrinsic evaluation is the result of the CBOW algorithm which is integrated with Binary Relevance and Multilayer Perceptron, with an accuracy value of 77.56% and a hamming loss value of 8.14%.

1. Introduction

Word embeddings are the vectors that represent words as a point in a multidimensional semantic space [1]. Some words that have semantic relations (synonym, antonym, word similarity, word relatedness, etc.) will have close vectors. Word embeddings will avoid the sparse matrix form in word representation, as happens when TF-IDF is used to represent a word [2]. That is, word embeddings will produce dense vectors with the predefined size of dimensions. The dense vector is also more effective while used as input features in the machine learning algorithm and prevent overfitting conditions. Dense vectors are also better in capturing synonyms than sparse vectors [1].

Word embeddings have been widely used in natural language processing and text mining, such as text similarity [3], text classification [4,5], multi-label text classification [6,7], sentiment analysis [8], machine translations [9,10], etc. Moreover, previous research works have applied word embedding for domain-specific text analysis (i.e. social science [11], religion [12], health and medical [13,14], mining [15], and cybersecurity [16]).

Based on searches, there are currently no pre-trained word embeddings for Islamic domain-based documents. In addition, there is also no research on evaluating word embeddings from documents based on the Islamic domain. Evaluation on word embeddings is quite important. Because word embeddings can affect the performance of machine learning algorithms. We hope that this word embeddings evaluation research can be used as a reference in choosing the right word embeddings algorithm for Islamic domain text.

In this paper, we extract word embeddings from Indonesian text (i.e. Al-Quran translation, Al-Quran interpretation, and hadith translation) using the Continuous Bag of Word (CBOW), Skip-Gram, and FastText algorithms. Word embeddings will be evaluated using intrinsic and extrinsic evaluation methods. Intrinsic evaluation is the word embeddings quality measures based on experiments in which word embeddings are compared to human judgments on word relations (synonym, antonym, etc.). Extrinsic evaluation is an evaluation technique that measures word embeddings quality based on the performance of the machine learning algorithm or other NLP tasks with word embeddings as the feature vectors. The main objective of this paper is to find the best word embeddings model from textual data of Islamic domain documents in the Indonesian language.

The main contributions of this paper are:

- developing word embedding Indonesian Quran and Hadiths Translation,
- provide a thorough evaluation on the developed word embeddings.

2. Literature Review

Nooralahzadeh, et. al. [15] evaluated domain-specific word embeddings from oil and gas textual data. About 8 million texts were trained on CBOW and Skip-Gram algorithm with a hyperparameter tuning experiment. They conducted two evaluations approach, which are intrinsic and extrinsic evaluations. On intrinsic evaluation, they randomly chose 100 unique words on the oil and gas domain then retrieved the 10-most-similar words for each unique word provided in the word embeddings. They evaluated word embeddings based on synonym, antonymy, and alternative form from the 10-most-similar words to evaluate the tuning experiment. After getting the best parameter, they analyzed the error of the 10-most-similar words using several categories: spelling variant, alternative form, references-synonym, human-judge synonym, antonym, hypernym, hyponym, co-hyponym, holonym, meronym, related, unrelated/unknown. The frequent errors occur in related categories, hyponym, and co-hyponym. On extrinsic evaluation, word embeddings were used as the features of the multi-label classification task. They ran basic CNN by Kim (2014) and several modified CNN models for the experiment. The best result obtained was from CNN models with two embedding layers (randomly embedding vector and domain-specific vector) integrated with the retrofitting method and Out of Vocabulary (OOV) handling.

Similar to Nooralahzadeh, et. al. work's, Sarma, et. al. [14] trained "Substances User Disorders" (SUDs) dataset with a modified word embeddings algorithm for resulting high-quality word embeddings vector. They combined large scale corpora embeddings and domain-specific embeddings using linear Canonical Correlation Analysis (CCA) or a nonlinear kernel CCA (KCCA). They found that the CCA/KCCA with combined word embeddings improves substantially over the generic embeddings. They evaluated combined word embeddings in extrinsic evaluation tasks on binary sentiment classification with the Logistic Regressor method. These architectures tested on four public datasets: Yelp, Amazon, IMDB, and A-Chess.

Another work by Roy, et. al. on [16] evaluated their proposed method on two cybersecurity text corpora: a malware description corpus and a "Common Vulnerability and Exposure (CVE)" corpus. The authors developed a novel Annotation and Word Embedding (AWE) algorithm. They stated that there was a diversification of the types of domain knowledge. To overcome this, the authors built the protection of text annotations in the form of predicate structural arguments. The basis of AWE is the Word2Vec system with input to the AWE algorithm is the text and annotations that will produce output in the form of vector representations of words and annotations.

3. Methodology

We define the steps of this research in this section. Then we describe the technical implementation steps of the algorithms CBOW, Skip-gram, and FastText. We use the Word2Vec and FastText libraries in the Python programming language.

3.1. Methodology
The research methodology of this study can be seen in Fig. 1. We began by collecting text data from the Indonesian translation of Al-Quran, the Indonesian interpretation *(tafseer)* of the Al-Quran, and the Indonesian translation of the hadith in the book of Sahih Al-Bukhary.

Figure 1. The research methodology

After collecting text data, we performed pre-processing the text, namely: case folding, sentence parsing, removing numbers and punctuations, repeated words checking (example: "buku-buku" *(books)*), and words tokenisation. From sentence parsing part, we obtained 157,870 sentences with 45,327 unique words. In Bahasa Indonesia, plural words are formed by repeating nouns and adding dashes between the two words [17], for example, the word "buku-buku" which means "many books". Therefore, it is necessary to check for repeated words so that the meaning of the words does not change and they are not split into two tokens.

Specifically, for the hadith text, we applied the Named Entity Recognition (NER) process to recognize the names of narrators. We use rule-based NER based on word writing patterns. Words that indicate the name of the narrator are written in square brackets ([...]). This process is intended so that the series of narrators names are not separated when doing the tokenization process. Example of original text and NER result on Table 1.

Table 1. Example of original text and text after NER

Original text	NER result
Telah menceritakan kepada kami **[Abdullah bin Yusuf]** berkata, telah mengabarkan kepada kami **[Malik]** dari **[Hisyam bin 'Urwah]** dari **[bapaknya]** dari **[Aisyah]**…………."	Telah menceritakan kepada kami **abdullah_bin_yusuf** berkata, telah mengabarkan kepada kami **malik** dari **hisyam_bin_urwah** dari bapaknya dari **aisyah** ……………."

3.2. Implementation
There are two global architectures in Word2Vec [18,19], those are CBOW and Skip-Gram. We implemented both of them and FastText algorithm [20], then choose the best one.

Mikolov, et. al. [19] create Word2Vec to reduce computational complexity on the Neural Network Language Model (NNLM), developed by Bengio, et. al (2013). On NNLM, There is a probability distribution calculation on the hidden layer for all words in the vocabulary (V as the vocabulary set and |V| as the number of words in the dictionary) and produces an output with a V dimension length. So Mikolov et. al. used hierarchical softmax, where the vocabulary is represented as a Huffman binary tree. With this model Mikolov et. al. can minimize computing costs [19]. Mikolov tries two architectures to build word vectors, that are Continous Bag-of-Word (CBOW) and Continous Skip-Gram. The difference between these two architectures is seen in the purpose of architecture. The CBOW predicts the target word (w_i) based on the context (surrounding word from w_i), and the Skip-gram aims to predict the surrounding words given the word w_i.

Joulin, et. al. built FastText architecture for extracting word embeddings from texts [5]. FastText is similar to the CBOW model, where the middle word is replaced by a label, and they use a bag of n-grams as features to get some partial information from the local word.

First, we build the architecture of word embeddings, with default values for hyper-parameters, i.e. *dim* = 100, *win* = 5, *min_count* = 5, and *neg* = 5, while *dim* is the dimensionality of vector or vector size, *win* is the size of windows for grabbing context words, *min_count* is the minimum frequency of the word, and *neg* is negative sampling size. Gensim library [21] was used for implementing CBOW, Skip-Gram (SG), and FastText architectures. Based on Table 2, it can be seen that the CBOW algorithm has the fastest training time than the other algorithms.

Table 2. Time consuming when building bag of word and model (without phrase). The value of window = 5, size = 100, and min_count = 5

Architecture Name	Time for Building BOW (seconds)	Time for Building Model (seconds)
CBOW	4.00	846.91
Skip-Gram (SG)	3.82	2414.06
FastText	5.52	3542.96

4. Evaluation

Based on our best knowledge, there is still no gold standard resource available to evaluate the semantic distribution of words in Bahasa as they already exist for English, such as Simlex-999 [22], SimVerb-3500 [23], etc. Therefore, we built our domain-specific gold standard in the field of Islam based on the glossary available in Islamic religious education books [24]. We chose 75 unique words from [24].

In this evaluation, we used an evaluation model defined by Bakarov [25] in which he divided the word embeddings evaluations into two major parts, namely extrinsic and intrinsic evaluation. Extrinsic evaluation methods are based on the ability to use word embeddings as the feature vectors of supervised machine learning algorithms or as used in one of NLP tasks. The performance of the supervised model as a measure of the quality of word embeddings.

Table 3. Examples of words and its related words from CBOW algorithm. Syn = synonym, Ant = antonym, Rel = Related Word, Der = Derivative, and Conc = Conclusion

Word	Most Related Word	Syn	Ant	Rel	Der	Conc
zhalim	aniaya	1	0	0	0	1
zhalim	adillah	0	1	0	0	1
zhalim	sewenang-wenang	1	0	0	0	1
zhalim	semenamena	1	0	0	0	1
zhalim	zalim	1	0	0	0	1
zhalim	kejam	1	0	0	0	1
zhalim	berwenang	0	0	0	0	0
zhalim	lalim	1	0	0	0	1
zhalim	curang	1	0	0	0	1
zhalim	adil	0	1	0	0	1

4.1. Intrinsic Evaluation

The intrinsic evaluation method will evaluate word embeddings based on human judgment (expert judgment). In this evaluation, we used human evaluators who would evaluate interrelated words produced by the word embedding architecture.

On [15], three kinds of word relation for evaluating word embedding were used, i.e. synonym, antonym, and alternative form. We replaced alternative forms to related and derivative forms. Furthermore, we used Indonesian Thesaurus by Kateglo.com for synonym (ex. *adil → jujur*), antonym (*lemah → kuat*), a related word (*nabi → rasul*), and derivative form (*iman → keimanan*).

For each word in the glossary (supposed word *g*), we extract 10 related words (words w_n) and group them into one of these following categories: synonym, antonym, related words, or derived words. If there is one category that matches, between *g* and w_n then this pair is given the conclusion value of 1, otherwise 0. Table 3 shows the example of 10 related words from "zhalim" (harsh) and the category of each related word.

We use Equation (1) to calculate true presentation for each word embeddings algorithm.

$$\frac{1}{N}\sum_{i=1}^{N}\max(Syn_i, Ant_i, Rel_i, Der_i) \quad (1)$$

Where Syn_i, Ant_i, Rel_i, $Der_i \in \{0,1\}$, and N is the number of rows.

Table 4 describes the results of the evaluation of each algorithm. The algorithm with the best results is Skip-Gram and FastText that managed to get 52.53% of the most related words that can be categorized.

Table 4. Result of human judgment intrinsic evaluation

Model	True Presentation For Each Row	Error
CBOW	51.73%	48.27%
Skip-Gram	52.53%	47.47%
FastText	52.53%	47.47%

4.2. Word Analogy

The second most popular intrinsic evaluation is the word analogy method (some researchers also calls it linguistic regularities, analogical reasoning, or word semantic coherence) [25]. The idea of word analogy is solving mathematical operation with the operands are word vectors. For example, given a pair of words ***a*** and ***a****, and the third word is ***b***. Then the relation based on the analogy between ***a**** with ***a*** can be used to predict ***b**** from ***b*** which has the same relationship ***a**** with ***a***. For example, *car:wheel* :: *bird:b**, meaning that the *wheel* is "a part of" the *car*, so *b** is "part of" the *bird* (wings, beaks, claws, etc.) [26].

To get the prediction of the word ***b**** it is necessary to know the vector value of ***b****. Mathematically, to get the vector ***b****, it can be done with a simple vector operation as described in Equation (2). Furthermore, based on vector ***b****, we can search for the word with the closest vector to vector ***b**** as shown in Equation (2) [27].

$$w_{b^*} \approx w_{a^*} - w_a + w_b \quad (2)$$

In the study of English linguistics, it is easier to find a public dataset as the ground truth of analogy issues. Gao, et. al. [28] proposed WordRep dataset which is divided into 26 semantic classes, WordRep was developed from the "Google Analogy" data by Mikolov, et. al. [19] which combines morphological and semantic elements in pairing words. However, it is different for Bahasa Indonesia. The resources supporting Indonesian language studies are still very rare. Likewise, for the analogy dataset, based on our research no one has yet made an analogy dataset for Indonesian. Therefore, in testing with this analogy approach, we created our own dataset. The dataset we made are consisted of 38 pairs which contained 6 categories, namely antonyms, synonyms, hyponyms, hypernym, derivative words, and related words.

Based on the words analogy example, given a pair of words ***a***, ***a****, and ***b*** we determine the word ***b**** by applying Equation (2) and looking for words in vocabulary that have a similar vector to vector ***b****. The determination of similar words uses the method "3CosMul" [29] from Gensim library as described in Equation (3).

$$\underset{b^* \in V}{\text{argmax}} \frac{\cos(b^*, b)\ \cos(b^*, a^*)}{\cos(b^*, a) + \varepsilon} \quad (3)$$

with V is vocabularies and $\varepsilon = 0.001$ use for preventing division by zero.

The results of the analogy test are shown in Table 5. Overall, the three embedding algorithms can solve analogy problems, especially in the class of synonyms, antonyms, derivative forms, and related words. Based on Table 5, it can be seen that the best algorithm that is able to solve word analogy problems is Skip-Gram. However, it failed when completing the hyponym and hypernym classes.

Because hypernym and hyponym or vice versa are rarely appeared together as contexts and target words. In the derivative class, the FastText algorithm has 100% accuracy, because when FastText conducts training on words, this algorithm takes subwords based on *n-grams* word. This makes FastText succeed in retrieving similar words from morphologically similar words. For example on pair of words:

budaya:kebudayaan :: *bangsa:b**
(cultural:culture :: nation:b*).

The ***a*** is a basic word, and ***a**** is a word ***a*** that is affixed with "ke-an". So ***b**** should be "kebangsaan". FastText obtained top-5 prospective answers as follows: "bangsa-bangsa" (nations), "berbangsa-bangsa" (nations), "kebangsaan" (nationality), "sebangsa" (countrymen), "mangsa" (prey).

Table 5. Evaluation result using analogy problem on CBOW architecture

Class	CBOW	Skip-Gram	FastText
antonym	60.00%	60.00%	60.00%
derivative	60.00%	80.00%	100.00%
hypernym	40.00%	40.00%	0.00%
hyponym	40.00%	80.00%	40.00%
related	46.20%	61.50%	30.80%
synonym	100.00%	100.00%	100.00%
Average	57.70%	**70.25%**	55.13%

4.3. Extrinsic Evaluation

The next evaluation uses an extrinsic evaluation approach, we used word vectors from the CBOW algorithm, Skip-Gram, and FastText as the feature of multilabel text classification. For this experiment, we use the dataset from [30]. The contents of this dataset are the Indonesian translation text of Hadiths Shahih Bukhary. This dataset consists of 1064 rows where each row has 3 labels: [suggestion, prohibition, information] and each label represented in binary value {0, 1}. We split dataset into three parts: 45% for training, 22% for validation, and 33% for testing.

We implemented pre-processing steps to the text including case folding, word normalization, remove punctuation, and tokenizing.

Table 6. The best parameter of learning methods

Word Embedding	Learning Method	Validation Accuracy	Best Parameters
CBOW	BR+MLP	78,09%	alpha = 1e-05 hidden layer = (100, 10)
CBOW	BR+SVM	75,00%	kernel = RBF
Skip-Gram	BR+MLP	78,37%	alpha = 1e-06 hidden layer = (100,)
Skip-Gram	BR+SVM	75,00%	kernel = RBF
FastText	BR+MLP	78,23%	alpha = 1e-06 hidden layer = (100, 10)
FastText	BR+SVM	75,00%	kernel = RBF

In this evaluation, we solve the Multilabel Classification (MLC) task using the Binary Relevance (BR) problem transformation approach. In BR, we combined two different learning algorithms, Multi-Layer Perceptron (MLP) and Support Vector Machine (SVM). BR will treat each label as a separate single-label classification problem. We implemented the classification algorithms using Scikit-Multilearn library [31].

We performed hyperparameter tuning for MLP with this following parameter values: *alpha* = {0.00001, 0.000001} and *hidden_layer_sizes* = {(100,), (100.10,), (100,10,3,)}. While for the SVM, we tried three different kernels which are Radial Base Function (RBF), sigmoid, and linear. The learning algorithm input is an average of word vectors for each document.

Table 6 shows the optimal parameter for each scenario. After getting the optimal parameters, then we re-train the learning algorithm using the optimal parameters. We used accuracy score, hamming loss, f1-micro average score, and f1-macro average score as the evaluation metrics.

Hamming loss calculates the symmetric difference between the predicted label and the ground truth then calculate the number of mismatches label set. The F1-macro-averaging evaluates each label by summing the number of true positives, false positives, true negatives and false negatives, and independently computes the F1-measure for each label. While the F1-micro-averaging counts once after the F1-measures for all labels have been collected [32].

Based on Table 7, the average testing accuracy of Multilabel Classification (MLC) is 76% and the average F1-macro average can reach 89%. This result proves that word vectors produced in this study are promising. Based on testing accuracy and hamming loss values, the best word vectors on extrinsic evaluation is the result of the CBOW algorithm integrated with Binary Relevance and Multilayer Perceptron.

Table 7. Result of multi-label classification using the best parameter

Word Embedding	Learning Method	Testing Accuracy	Hamming Loss	F1-micro average	F1-macro average
CBOW	BR+MLP	77.56%	8.14%	91.00%	77.00%
CBOW	BR+SVM	71.88%	10.42%	88.00%	73.00%
Skip-Gram	BR+MLP	77.56%	8.33%	91.00%	77.00%
Skip-Gram	BR+SVM	75.28%	9.75%	88.00%	56.00%
FastText	BR+MLP	77.56%	8.52%	90.00%	78.00%
FastText	BR+SVM	75.28%	9.75%	88.00%	56.00%

5. Conclusion

This paper has discussed intrinsic and extrinsic evaluations of the corpus with the specific domain of religious texts. Based on our research results, CBOW is the fastest algorithm when conducting training on the data we use. Meanwhile, based on the results of the intrinsic test, the best algorithms in categorizing related words are Skip-Gram and CBOW with an accuracy rate of 52.53%. We also tested word vectors extracted from the text to solve word analogy problems, where the best performance was achieved by the Skip-Gram algorithm with an average accuracy rate of 70.25%.

Apart from intrinsic testing, we also perform extrinsic testing. In this type of test, word vectors become feature learning on the Multilabel Classification (MLC) task for hadith text. The best algorithm for this MLC task is CBOW which is combined with Binary Relevance and Multilayer Perceptron (MLP). The level of accuracy obtained by CBOW+BR+MLP is 77.56% with a hamming loss value of 8.14%.

From this study, we conclude that special pre-processing (such as NER) are needed for certain data. For example, in the hadith text, it is necessary to separate the narrators and contents of hadith *(matn)* parts. For further research, we plan to perform fine-tuning in the word embeddings architecture so that the expected results of the word vector become more optimal.

References

[1] Jurafsky D and Martin J H 2008 *Speech and Language Processing* (New Jersey: Prentice Hall)
[2] Chen G, Ye D, Xing Z, Chen J and Cambria E 2017 Ensemble application of convolutional and recurrent neural networks for multi-label text categorization *2017 Int. Jt. Conf. Neural Networks* pp 2377–83
[3] vor der Brück T and Pouly M 2019 Text similarity estimation based on word embeddings and

matrix norms for targeted marketing *Proc. of the 2019 Conf. of the North* (Stroudsburg, PA, USA, Minnesota: Association for Computational Linguistics) pp 1827–36
[4] Kim Y 2014 Convolutional neural networks for sentence classification *Proc. of the 2014 Conf. on Empirical Methods in Natural Language Processing (EMNLP)* (Doha, Qatar) pp 1746–51
[5] Joulin A, Grave E, Bojanowski P and Mikolov T 2017 Bag of tricks for efficient text classification *Proc. of the 15th Conf. of the European Chapter of the Association for Computational Linguistics* vol 2 (Melbourne:Association for Computational Linguistics) pp 427–31
[6] Georgakopoulos S V., Vrahatis A G, Tasoulis S K and Plagianakos V P 2018 Convolutional neural networks for toxic comment classification *Proc. of the 10th Hellenic Conf. on Artificial Intelligence* (Patras, Greece: Association for Computing Machinery)
[7] Okky Ibrohim M, Sazany E and Budi I 2019 Identify abusive and offensive language in Indonesian Twitter using deep learning approach *J. of Physics: Conf. Series* vol **1196** (Bristol: IOP Publishing)
[8] Petrolito R and Dell'Orletta F 2016 Word embeddings in sentiment analysis *Proc. of the Fifth Italian Conf. on Computational Linguistics CLiC-it 2018* (Torino: Accademia University Press)
[9] Cholakov K and Kordoni V 2016 Using word embeddings for improving statistical machine translation of phrasal verbs *Proc. of the 12th Workshop on Multiword Expressions* (Berlin, Germany: Association for Computational Linguistics) pp 56–60
[10] Zheng Z and Chen J 2018 Modeling past and future for neural machine translation *Trans. Assoc. Comput. Linguist.* **6** 145–57
[11] Kozlowski A C and Taddy M 2019 The Geometry of culture : analyzing the meanings of class through word embeddings *Am. Sociol. Rev.* **84** pp 905–49
[12] Alturayeif N S 2017 *Text Mining and Similarity Measures of the Quran and the Bible* (Leeds: University of Leeds)
[13] Jiang Z, Li L, Huang D and Jin L 2015 Training word embeddings for deep learning in biomedical text mining tasks *2015 IEEE Int. Conf. Bioinforma. Biomed.*
[14] Sarma P K, Liang Y and Sethares W A 2018 Domain adapted word embeddings for improved sentiment classification *56th Annual Meeting of the Association for Computational Linguistics (Short Papers)* (Melbourne: Association for Computational Linguistics) pp 37–42
[15] Nooralahzadeh F, Øvrelid L and Lønning J T 2018 Evaluation of domain-specific word embeddings using knowledge resources *Proc. of the Eleventh Int. Conf. on Language Resources and Evaluation (LREC-2018)* pp 1438–45
[16] Roy A, Park Y and Pan Sh 2017 Learning domain-specific word embeddings from sparse cybersecurity texts *Comput. Res. Repos.* **abs/1709.0**
[17] Tim Pengembang Pedoman Bahasa Indonesia Badan 2016 *Pedoman Umum Ejaan Bahasa Indonesia* (Jakarta, Indonesia: Badan Pengembangan dan Pembinaan Bahasa Kementerian Pendidikan dan Kebudayaan)
[18] Mikolov T, Sutskever I, Chen K, Corrado G and Dean J 2013 Distributed representations of words and phrases and their compositionality *NIPS*
[19] Mikolov T, Chen K, Corrado G and Dean J 2013 Efficient estimation of word representations in vector space *CoRR* **abs/1301.3**
[20] Bojanowski P, Grave E, Joulin A and Mikolov T 2017 Enriching word vectors with subword information *Trans. Assoc. Comput. Linguist.* **5** pp 135–46
[21] Řehůřek R and Sojka P 2010 Software Framework for topic modelling with large corpora *Proc. of LREC 2010 workshop New Challenges for NLP Frameworks* (Valletta, Malta: University of Malta) pp 46–50
[22] Hill F, Reichart R and Korhonen A 2015 SimLex-999: Evaluating semantic models with (genuine) similarity estimation *Comput. Linguist.* **41** 665–95

[23] Gerz D, Vulic I, Hill F, Reichart R and Korhonen A 2016 Simverb-3500: a large-scale evaluation set of verb similarity *EMNLP 2016 - Conf. on Empirical Methods in Natural Language Processing, Proc.* (Austin, Texas: Association for Computational Linguistics) pp 2173–82

[24] Marzuki 2012 *Pembinaan Karakter Mahasiswa Melalui Pendidikan Agama Islam di Perguruan Tinggi Umum* (Yogyakarta: Pusat Mata Kuliah Universiter, Universitas Negeri Yogyakarta)

[25] Bakarov A 2018 A survey of word embeddings evaluation methods *Preprint arXiv:1801.09536*

[26] Wang B, Wang A, Chen F, Wang Y and Kuo C-C J 2019 Evaluating word embedding models: methods and experimental results *APSIPA Trans. Signal Inf. Process.* **8** pp 1–13

[27] Allen C and Hospedales T 2019 Analogies Explained: towards understanding word embeddings *Preprint arXiv:1901.09813*

[28] Gao B, Bian J and Liu T-Y 2014 WordRep: a benchmark for research on learning word representations *Preprint arXiv:1407.1640*

[29] Levy O and Goldberg Y 2014 Linguistic regularities in sparse and explicit word representations *Proc. of the Eighteenth Conf. on Computational Language Learning* (Baltimore, Maryland, USA) pp 171–80

[30] Bakar M Y A, Adiwijaya A and Al Faraby S 2018 Multi-label topic classification of hadith of Bukhari (Indonesian Language Translation) using information gain and backpropagation neural network *Proc. of the 2018 Int. Conf. on Asian Language Processing, IALP 2018* (IEEE) pp 344–50

[31] Szymański P and Kajdanowicz T 2019 Scikit-multilearn : a scikit-based Python environment for performing multi-label classification *J. Mach. Learn. Res.* **20** pp 1–22

[32] Charte F, Rivera A J, Charte D, del Jesus M J and Herrera F 2018 Tips, guidelines and tools for managing multi-label datasets: The mldr.datasets R package and the Cometa data repository *Neurocomputing* **289** pp 68–85

IOP Conf. Series: Materials Science and Engineering 1077 (2021) 012026 doi:10.1088/1757-899X/1077/1/012026

Restaurant Recommendation System Based on User Ratings with Collaborative Filtering

Achmad Arif Munaji[1], **Andi Wahju Rahardjo Emanuel**[2]

[1,2]Magister Informatika, Universitas Atma Jaya Yogyakarta, 44 Babarsari Street, Yogyakarta, Indonesia

E-mail: [1]arief.munaji65@gmail.com, [2]andi.emanuel@uajy.ac.id

Abstract. Recommendation systems are widely used as a reference in marketing products or businesses. The number of choices given sometimes makes a person confused in making choices. The recommendation system provides a solution in overcoming this. Collaborative filtering is a fairly popular algorithm used in recommendation systems that are based on references and information obtained from users. User rating is a variable that is widely used as a reference in establishing a recommendation system. User rating can influence other users in making their choice of the same product. The purpose of this study is to conduct an experiment in the application of a collaborative filtering algorithm and use the Pearson correlation function as a method used to see the level of user similarity in making a restaurant recommendation system. The methods used in this research are: (1) datasets preparation, (2) datasets filtering, (3) collaborative filtering, (4) pearson correlation, (5) recommendations, and (6) evaluation. Collaborative filtering combined with the Pearson correlation produces recommendations that are suitable for users based on characteristics or the same level of interest between users, making it easier for users to make choices. This is also indicated by a small error rate based on the RMSE results.

1. Introduction

In living everyday life, humans are always faced with a condition where they must make a choice. The many choices offered often make us confused in making choices. To overcome this, we often ask for suggestions or recommendations from other people. In the modern era, many problems like this have been overcome by implementing a recommendation system. The use of recommendation systems will often be encountered in marketing systems, such as online applications and online trading. This application is carried out with the aim of seeing potential customers and providing recommendations for a product related to the common interests of users [1]. Providing recommendations for certain products or goods will have a certain impact, such as increasing the level of popularity because the product or goods will often be seen by users and increasing the chances of the product or item being sold [2].

Collaborative filtering is an algorithm used to create a recommendation system. This algorithm is based on the following opinion: someone has the same views and interests, their preferences and interests are strong, and can make choices based on reference to their previous preferences [3]. Collaborative filtering uses certain aspects as a basis for making recommendations, such as user behavior in the past, items purchased by users, and ratings of items that have been purchased [4].

The purpose of this study is to conduct experiments in the application of collaborative filtering algorithms and use the Pearson correlation function as a method used to see the level of user similarity in making restaurant recommendation systems based on ratings given by visitors or users in the past. The stages of research carried out include: (1) datasets preparation, (2) datasets filtering, (3) collaborative filtering, (4) pearson correlation, (5) recommendations, and (6) evaluation.

2. Literature review

A recommendation system is a system designed for users with the aim of providing recommendations for a particular item. There are many areas where recommendation systems can be applied, such as social media, e-learning, web news articles, e-commerce, and many more [5]. Collaborative filtering is an algorithm used in making recommendations based on user personaliation [6]. The use of collaborative filtering is very helpful in making recommendation systems on large datasets [7] [8]. In its implementation, the calculation of user similarity is carried out and by using this concept as a reference that users who have the same characteristics will tend to assess the same items as well. In collaborative filtering, the use of grouping techniques is done by grouping the same users into several groups [9]. After the similarity between items is obtained, collaborative filtering will calculate the predicted value of the missing data [10].

In measuring the value of similarity, there are many methods that can be used, such as Pearson Correlation, Spearman Rank, Discounted Similarity, and others. Previous research has tried to see how popular and good these methods are in measuring the similarity value. Based on the results of research that has been done, the Weighted Pearson Correlation produces a good accuracy value in prediction. In addition, the results of these studies also suggest that data differences also affect the choice of methods to be used [11]. The completeness of the data used is an important issue that must be considered for many researchers. Data sparsity is a problem related to data that is often found. A new collaborative filtering method is proposed to overcome this problem by adopting a clustering algorithm [12]. Improvements to the similarity equation used in calculating user portraits, item characteristics, and user activity data were proposed in previous studies with the aim of increasing the accuracy of the recommendation results [13]. Although Pearson Correlation is very well used to calculate similarity, previous studies have suggested that, in certain cases, the Pearson correlation shows inaccurate results [14].

3. Method

In this section, the author will explain the stages used in research. The stages of the methodology can be seen in Figure 1.

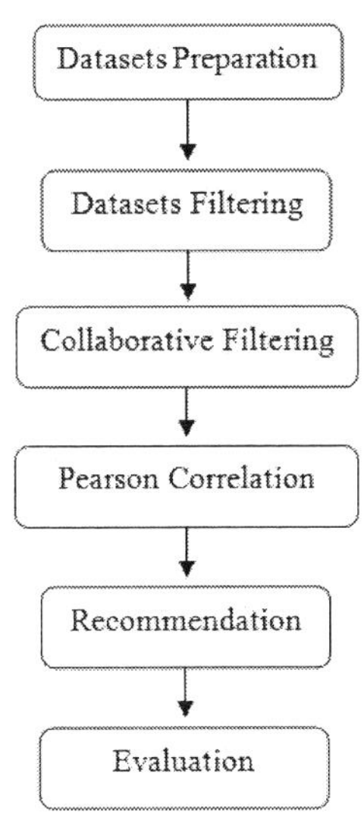

Figure 1. Research methods.

As shown in Figure 1, the research methodology used in this study consists of six stages. These stages include: datasets collection, datasets filtering, collaborative filtering, pearson correlation, recommendation, and evaluation. The following is an explanation regarding the stages of the research methodology:

3.1. Datasets preparation

At this stage the authors collect a dataset that will be used as research training data. The dataset used is a dataset obtained from one of the dataset provider websites kaggle.com [15]. There are two datasets used in this study, namely the restaurant dataset and the rating dataset.

3.2. Datasets filtering

At this stage, the author will filter the content contained in each dataset. This is done because not all of the content on the dataset will be used during the recommendation making process and lighten when the process takes place. The following is a brief overview of the dataset content used in table form:

Table 1. Restaurant dataset after selection.

placeID	*name*	*address*
134999	Kiku cuernavaca	Revolucion
132825	Puesto de tacos	Esquina santos degollado y leon guzman
135106	El rincn de san Francisco	Universidad 169
132667	Little pizza emilio portes gil	calle emilio portes gil
132613	Carnitas mata	lic. Emilio portes gil

Table 1 displays the contents of the restaurant dataset that will be used and is the result of the selection that has been done. The five data rows represent the first five rows contained in the dataset. In total, the restaurant dataset contains 130 lines and is a restaurant in Mexico. The following is a brief explanation of the contents of the dataset: the placeID column contains the restaurant location id, the name column contains the name of the restaurant, and the address column contains the address the restaurant is located.

Table 2. Rating dataset after selection.

userID	*placeID*	*rating*
U1077	135085	2
U1077	135038	2
U1077	132825	2
U1077	135060	1
U1068	135104	1

Table 2 displays the ranking dataset that will be used in this study. Similar to the dataset in Table 1, the Table 2 dataset is also a selected table. Table 2 represents the first five rows in the ranking dataset. The ranking dataset contains 1161 rows and three columns. The userID column contains the user id, the placeID column contains the restaurant location id, and the rating column contains the rating given by the user.

3.3. Collaborative filtering

At this stage, the authors apply the algorithm in making a recommendation system. The recommendation system will be created using collaborative filtering based on filtering users. This technique uses other users as a guide in recommending an item to user input. Search for users who have the same reference or opinion as input data, then provide recommendations for the appropriate item to the input. Following are the steps for collaborative filtering in making a restaurant recommendation system based on user rating references:

- Select users with restaurants the user has visited.
- Based on the ratings given for each restaurant, find the top X neighbors.
- Get a recording of the restaurant users have visited for each neighbor.
- Calculate restaurant similarity scores using the Pearson Correlation Function.
- Provides restaurant recommendations based on the highest score.

Table 3. Experimental data.

No	x	y	$(x_i - \bar{x})(y_i - \bar{y})$	$(x_i - \bar{x})^2$	$(y_i - \bar{y})^2$
1	3	5	-0.224489796	0.020408163	2.469387755
2	4	2	-1.224489796	0.734693878	2.040816327
3	5	3	-0.795918367	3.448979592	0.183673469
4	1	4	-1.224489796	4.591836735	0.326530612
5	3	5	-0.224489796	0.020408163	2.469387755
6	4	4	0.489795918	0.734693878	0.326530612
7	2	1	2.775510204	1.306122449	5.897959184
	sum		-0.428571429	10.85714286	13.71428571
			$\sqrt{}$	3.295017884	3.703280399

3.4. Pearson correlation

At this stage, the author also uses the Pearson Correlation Function as a method to compare all similar users and find which users are most similar. In its application, the Pearson Correlation will later calculate the linear relationship between the two variables:

$$r = \frac{\sum_{i=1}^{n}(x_i - \bar{x})(y_i - \bar{y})}{\sqrt{\sum_{i=1}^{n}(x_i - \bar{x})^2}\sqrt{\sum_{i=1}^{n}(y_i - \bar{y})^2}} \quad (1)$$

For example, there are two variables x and y, where each of the two variables represents a rating given by two users and has a value of n. $\bar{x}$ as the mean of the x and $\bar{y}$ as the mean of the y. The following is a simple example of applying Pearson (x, y) using experimental data in table 3:

$$r = \frac{-0.428571429}{3.295017884 \times 3.703280399} = -0.03512$$

3.5. Recommendation

At this stage the author will provide the best restaurant recommendations with the highest score based on the results of the recommendation system.

3.6. Evaluation

At this stage an evaluation will be carried out using the Root-Mean-Square Error (RMSE). RMSE is an evaluation technique that is often used in collaborative filtering [16]. RMSE is formulated as follows:

$$RMSE = \sqrt{\frac{1}{N}\sum\left(P_{u,i} - r_{u,i}\right)^2} \quad (2)$$

In the above formula, $P_{u,i}$ represents the rating prediction, $r_{u,i}$ is the observation of the actual rating or rating, and N is the rating that users have given the restaurant. The rating prediction is a rating from the results of the recommendations produced, while the observation rating is a restaurant's current rating before the recommendation is made. The smaller the resulting RMSE value, the more accurate the predictions will be [17].

4. Result and discussion

4.1. Implementation process and recommendations

Implementation process and recommendations is as follows:

- Create a new user input containing 10 lines of restaurant name, restaurant address, and rating. This data can be seen in Table 4. This new data is data that will be used as testing data in the recommendation system creation experiment

Table 4. User input.

Name	*address*	*rating*
Little pizza Emilio Portes Gil	Calle Emilio Portes Gil	5
Rockabilly	Agustin De Iturbide	1
Restaurante 75	Villa de Pozos 4497 Villa de Pozos	2
Abondance Restaurante Bar	Industrias 908 Valle Dorado	3
Sirlone	Carr. Mexico	4
Unicols Pizza	Plaza Del Carmen	4
Restaurant El Muladar de Calzada	De Guadalupe 460 San Miguelito	3
La Posada del Virrey	Av. V. Carranza	3
Restaurant and Bar and Clothesline Carlos N Charlies	Domingo 10 711 El Empleado	2
Giovannis	40 Norte Civac 1RA. Seccion	5

- Adds the placeID from the restaurant dataset to the newly created user input data. Extract the restaurant place id from the restaurant dataset and then add it to the user enter as shown in Table 5.

Table 5. Adding placeID to user input.

placeID	*name*	*address*	*rating*
132667	Little pizza Emilio Portes Gil	Calle Emilio Portes Gil	5
132937	Rockabilly	Agustin De Iturbide	1
135070	Restaurante 75	Villa de Pozos 4497 Villa de Pozos	2
135069	Abondance Restaurante Bar	Industrias 908 Valle Dorado	3
132854	Sirlone	Carr. Mexico	4
132856	Unicols Pizza	Plaza Del Carmen	4
135033	Restaurant El Muladar de Calzada	De Guadalupe 460 San Miguelito	3
132862	La Posada del Virrey	Av. V. Carranza	3
134983	Restaurant and Bar and Clothesline Carlos N Charlies	Domingo 10 711 El Empleado	2
135013	Giovannis	40 Norte Civac 1RA. Seccion	5

- With the id location of the restaurant on user input, filtering of users who have visited the same restaurant can be done. After that, the grouping is done based on the user id and ends with the creation of groups so that the user who has the most input has a greater priority. Figure 2 shows an example of a group display that has been created based on the userID.

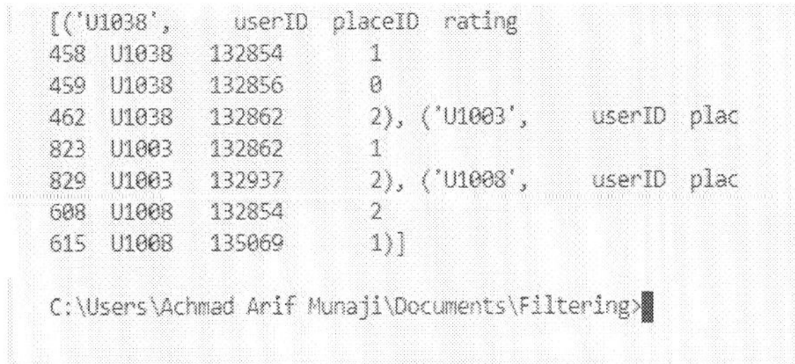

```
[('U1038',      userID  placeID  rating
458  U1038  132854       1
459  U1038  132856       0
462  U1038  132862       2), ('U1003',      userID  plac
823  U1003  132862       1
829  U1003  132937       2), ('U1008',      userID  plac
608  U1008  132854       2
615  U1008  135069       1)]

C:\Users\Achmad Arif Munaji\Documents\Filtering>
```

Figure 2. Grouping by userID.

- See the similarity of users in the dataset and enter the user. The next step will be to compare all users with the users who have been determined and choose the most similar using the Pearson Correlation Coefficient. The Pearson Correlation will be calculated between the user and group sections that were previously created.

Table 6. Example of similarity results.

similarityIndex	*userID*
-0.866025	U1038
-1.000000	U1003
1.000000	U1008
-1.000000	U1014
0.000000	U1016

Table 6 shows an example of the calculation results to see the similarity of users. In this table the similarity value of each user is shown in the SimilarityIndex column.

- Users with top x that have similarities to user input. In this step, the top 50 rows of user data will be retrieved that have the same level of user input.

Table 7. Example of the top user row.

similarityIndex	*userID*
1.0	U1008
1.0	U1037
0.0	U1045
0.0	U1101
0.0	U1059

Table 7 shows an example of the top 5 rows that have similarities with entries. After this process is complete, we can start making restaurant recommendations to enter users.

- Selected user ratings for each restaurant. In this step, we will calculate the weighted average of the restaurant ratings that have been given using the Pearson Correlation as a weight. Weights obtained from multiplication of restaurant ratings with the results of previous Pearson similarities that have a similarity index.

Table 8. Example of rank and weight multiplication results

similarityIndex	*userID*	*placeID*	*rating*	*weightedrating*
1.0	U1008	132866	1	1
1.0	U1008	132858	1	1
1.0	U1008	132854	2	2
1.0	U1008	135108	1	1
1.0	U1008	132851	1	1

After ranking the ranking with weights is complete as shown in Table 8, the next step is to proceed to collect new rankings and divide the rank by the number of weights to produce a recommendation score

- Recommendation results. The ranking is done based on the recommendation score obtained to see which reports have the best recommendation score and are recommended by the algorithm. Table 9 shows the 10 best restaurants based on algorithm results.

4.2. Evaluation

After the recommendation making process is complete, the next step is to conduct an evaluation. RMSE is used at this stage. The following is the implementation of evaluation using RMSE:

$$RMSE = \sqrt{\frac{1}{N}\Sigma\left(P_{u,i} - r_{u,i}\right)^2} = 1.359589956010579$$

The RMSE results above are the result of cumulative errors that are rooted. Then we can find the average error of each data by dividing the RMSE results by the amount of data that has been normalized or data that has the same interest with user input, namely 462, then the average error per data is 0.003. These results indicate a small error rate in the data or a good level of accuracy.

Table 9. Results of restaurant recommendations

placeID	*name*
132854	Sirlone
132851	KFC
132847	Don Burguers
135054	Restaurante y Pescaderia Tampico
132869	Dominos Pizza
135108	Potzocalli
132858	Hamburguesas Valle Dorado
132846	El Lechon Potosino
132830	Rincon Huasteco
132866	Chaires

5. Conclusion

The use of collaborative filtering based on user references has been widely applied in making recommendation systems [18]. In this study, collaborative filtering is used based on filtering users. User ratings for certain restaurants are used as a reference for other users to create a recommendation system. The calculation of the similarity between one user and another user is done using the Pearson Correlation. After obtaining the similarity value for each user, the user's rank is multiplied by the Pearson similarity value to obtain the weight. Equality and weight values are added based on their respective indices. Then to get the best restaurant recommendations, the division is carried out between the number of similarity scores and the weight of each index and ends with a sorting process based on the highest score of each index. The last step that needs to be done is to check the error rate against the resulting prediction rating. Based on the evaluation using RMSE, it was found that this experiment has good accuracy.

References

[1] Wei J, He J, Chen K, Zhou Y and Tang Z 2017 Collaborative filtering and deep learning based recommendation system for cold start items *Expert Syst. Appl.* vol **69** pp 1339–51

[2] Kawasaki M and Hasuike T 2018 A recommendation system by collaborative filtering including information and characteristics on users and items *2017 IEEE Symp. Ser. Comput. Intell. SSCI 2017 - Proc.* vol **2018-January** pp 1–8

[3] Jiang L, Cheng Y, Yang L, Li J, Yan H and Wang X 2019 A trust-based collaborative filtering algorithm for E-commerce recommendation system *J. Ambient Intell. Humaniz. Comput.* vol **10** no 8 pp 3023–34
[4] Mathew P, Kuriakose B and Hegde V 2016 Book recommendation system through content based and collaborative filtering method *Proc. 2016 Int. Conf. Data Min. Adv. Comput. SAPIENCE 2016* pp 47–52
[5] Elssaedi M M, Salih O M and Ahmeed A 2019 Optimizing selection using Pearson correlation coefficient: a case study in movie industry *AIP Conf. Proc.* vol **2146** issue August
[6] Liu X 2019 Improved collaborative filtering algorithm based on multi-dimensional fusion similarity *Proc. - 2019 IEEE Int. Conf. Smart Internet Things, SmartIoT 2019* pp 440–3
[7] Samundeeswary K and Krishnamurthy V 2018 Comparative study of recommender systems built using various methods of collaborative filtering algorithm *ICCIDS 2017 - Int. Conf. Comput. Intell. Data Sci. Proc.* vol **2018-January** pp 1–6
[8] Mu Y, Xiao N, Tang R, Luo L and Yin X 2019 An Efficient Similarity Measure for Collaborative Filtering *Procedia Comput. Sci.* vol **147** pp 416–21
[9] Koohi H and Kiani K 2016 User based collaborative filtering using fuzzy C-means *Meas. J. Int. Meas. Confed.* vol **91** issue May pp 134–9
[10] Moe W W, Aye N and Htwe A 2017 Performance comparison of collaborative filtering prediction methods on recommendation system *Am. Sci. Res. J. Eng. Technol. Sci.* vol **28** no 1 pp 2313–4410
[11] Al Hassanieh L, Jaoudeh C A, Abdo J B and Demerjian J 2018 Similarity measures for collaborative filtering recommender systems *2018 IEEE Middle East North Africa Commun. Conf. MENACOMM 2018* pp 1–5
[12] Ma L, Wang X, Chen A, Gao R, Tang Y and Xiao L 2017 A new collaborative filtering algorithm based on data smoothing *Proc. - 2017 2nd Int. Conf. Multimed. Image Process. ICMIP 2017* pp 305–8
[13] Ju Y, Bailin L and Zhixiang Z 2017 A collaborative filtering recommendation algorithm with improved similarity calculation *2nd Int. Conf. Sens. Netw. Comput. Eng.* vol **79** pp 158–61
[14] Liu Y, Feng J and Lu J 2017 Collaborative filtering algorithm based on rating distance *Proc. 11th Int. Conf. Ubiquitous Inf. Manag. Commun. IMCOM 2017*
[15] Kaggle 2017 *Restaurant Data with Consumer Ratings | Kaggle* Retrieved on 15 July 2020 from https://www.kaggle.com/uciml/restaurant-data-with-consumer-ratings?select=geoplaces2.csv
[16] Zhang J, Wang Y, Yuan Z and Jin Q 2020 Personalized real-time movie recommendation system: practical prototype and evaluation *Tsinghua Sci. Technol.* vol **25** no 2 pp 180–91
[17] Jiang L, Cheng Y, Yang L, Li J, Yan H and Wang X 2019 A trust-based collaborative filtering algorithm for E-commerce recommendation system *J. Ambient Intell. Humaniz. Comput.* vol 10 pp 3023–34
[18] Chen VX and Tang T Y 2019 Incorporating singular value decomposition in user-based collaborative filtering technique for a movie recommendation system: A comparative study *ACM Int. Conf. Proc. Ser.* pp 12–5

IOP Conf. Series: Materials Science and Engineering 1077 (2021) 012027 doi:10.1088/1757-899X/1077/1/012027

Comparative analysis of fuzzy database Tahani model and Fuzzy Multi-Attribute Decision Making TOPSIS method in cotton product for determination recommendations in textile industry (Case study: PT. Pandatex)

Erizta Alifa Prabandani[1], Shofwatul 'Uyun[2]

[1]Informatics Department, Universitas Islam Negeri Sunan Kalijaga, Yogyakarta 55281, Indonesia
[2]Magister of Informatics Department, Universitas Islam Negeri Sunan Kalijaga, Yogyakarta 55281, Indonesia

E-mail: [2]shofwatul.uyun@uin-suka.ac.id

Abstract. PT. Pandatex is a company engaged in the textile industry with raw materials derived from fibers. Product selection based on supplier offers makes it difficult for companies to make choices that match the quality classification and production needs. This research was made by comparing the fuzzy Tahani method and the FMADM TOPSIS method with the same parameters like fiber strength, fiber length, micronaire, price, quantity, and shipment. The fuzzy Tahani method is used to find the value of the degree of membership through the fuzzification process and the value of fire strength. Meanwhile, the FMADM TOPSIS method is used to find the best alternative from alternatives that have the shortest distance from the positive ideal solution, and have the longest distance from the negative ideal solution. The results of this study were conducted with case trials. The fuzzy Tahani method produces 4 recommendations and the FMADM TOPSIS method results in a ranking of all cotton products. From the final results of the two methods used, the calculation of the fuzzy Tahani method is more accurate with accuracy in calculating the value of fire strength.

1. Introduction

The textile industry is a labor-intensive industry where in terms of labor, the development or addition of industrial capacity can easily be accommodated by an abundance of workers with more competitive wages. The ability of the textile industry to be increasingly competitive, both in the domestic and global markets, can be seen in the growth rate of the textile industry throughout 2018 which was recorded at 8.73 percent or able to exceed the national economic growth of 5.17 percent [1].

PT. Pandatex is a company engaged in the textile industry with raw materials derived from fibers. Fabric is the product of the company. One of the processes of processing materials in fabrics that must be considered in the selection of the right cotton product as the main raw material, so as not to hamper the production process at the company. The increase in fabric production was offset by the large supply of raw materials with various kinds of product submissions offered by suppliers. Selection of products based on offers supplier makes it a little difficult for the company to make choices that match the quality classification and production needs because the product selection process is initially determined only based on price and the quantity that are not taken into account so that the results obtained are less than optimal.

The MCDM (Multi-Criteria Decision Making) method is a method for selecting the best alternative from several mutually exclusive alternatives based on general performance in various criteria determined by decision-makers [2]. However, this method has a weakness, it is inefficient for resolving incorrect, uncertain, and unclear data. Usually, this method assumes that the final decision on an exclusive alternative is expressed in real numbers so that the ranking stage is less representative of certain problems and problem-solving is only centered on the aggregation stage [3]. Fuzzy AHP (Analytic Hierarchy Process) is combined with an approach AHP concept. Fuzzy AHP closes the weaknesses that exist in AHP, namely problems with criteria that have more subjective characteristics. The uncertainty of numbers is represented by an order of scale [4]. However, the disadvantage of fuzzy AHP is the need for additional information, namely optimistic values and pessimistic values [5].

Along with technological developments, there is an idea that can define the concept of relations fuzzy in a Database Management System (DBMS) by using membership degrees. This idea was put forward by Tahani in 1977. On the fuzzy Tahani uses set fuzzy theory to obtain information in queries [6]. The Method Fuzzy Multi-Attribute Decision Making (FMADM) is a development of the Multi-Attribute Decision Making (MADM). In decision making, if a problem cannot be presented accurately into a crisp value, the application of logic Fuzzy can be a solution. The application of logic fuzzy in the MADM method is better known as Fuzzy Multi-Attribute Decision Making [7].

In this study, a comparative analysis was carried out by fuzzy Tahani and the FMADM TOPSIS method. The parameter to be used are fuzzy criteria including fiber strength, fiber length, fiber fineness or micronaire, price, quantity, and shipment. The purpose of this study was to determine the most appropriate method to use in determining cotton product recommendations.

2. Literature Review

2.1. Related work

These are the following related research to comparative analysis and determination of recommendations using the fuzzy database Tahani model or the fuzzy MADM TOPSIS. First related research is about comparison analysis of poverty assistance recipients with WP and TOPSIS method [8]. The results of this study include the main factors that influence the calculation results. In addition, the use these method is part of MADM which aims to rank the alternatives that are compared in the form of the resulting total preference value. Another research related to a comparative analysis of methods is the comparison analysis of TOPSIS and SAW methods in determining recipients of assistant community development [9]. The results of this study are being able to solve cases with the TOPSIS and SAW methods with several conditions that have been determined by sensitivity testing that the most optimal method is SAW, with a change in SAW value of 14.65% while TOPSIS is 4.02%.

Comparison of AHP and fuzzy Mamdani for decision support system for house selection in Cepu area was introduced by Widyassari and Yuwono. The comparison between AHP and fuzzy Mamdani with the case of house selection is 0.90298. The accuracy level of AHP is better than fuzzy Mamdani in the case of house selection with MSE of 0.048538 [10]. Research conducted by Astari and Komarudin regarding the selection of the best employees uses the criteria of knowledge, expertise, personality, work discipline, and loyalty [11]. The results of this study are using Tahani fuzzy logic, the assessment is more accurate and reduces subjectivity in selecting the best employees in the decision support system research at PT. Culture Royale. Further research was introduced by Rochman in his research which examines recommendations for purchasing auction items using the criteria of price, quality, quantity, bidder interest, and claims for data analysis using the Tahani model fuzzy database method. The results from this study are the application can assist companies in remodeling the choice by showing that the decision support system can work properly with a 70% [12].

2.2. Fuzzy database Tahani model

The Fuzzy database Tahani model still uses standard relations, but the model uses a set fuzzy theory to get information on its queries. Therefore, the initial data that is processed is data that has a value crisp (certain or clear existence), and when searching for cryptic data, the process is called a process fuzzy query through a fuzzy database the Tahani model. The idea of the fuzzy database Tahani model is to use membership degrees to define the concept of relations fuzzy in a database system [13].

2.3. *Fuzzy Multi-Attribute Decision Making TOPSIS method*

The MADM method refers to making decisions based on the selection of several options. In decision making, if a problem cannot be presented accurately into a value crisp, the application of fuzzy logic can be a solution, which is better known as Fuzzy MADM.

The TOPSIS method uses the principle that the chosen alternative must have the closest distance from the positive ideal solution and the farthest distance from the negative ideal solution. This concept is widely used in several MADM models to solve decision problems because the concept is simple and easy to understand, the computation is efficient, can measure the performance of decision alternatives in a simple mathematical form [7].

2.4. *Textile*

The industry is a business to process semi-finished goods or raw materials into finished goods that have added value for profit. Meanwhile, the textile is material from fibers that are processed in the form of cloth or yarn as a material for making clothing or other products [14]. Textiles are defined as goods or objects whose raw materials come from fibers (cotton, polyester, rayon) by spinning them into threads and then weaving them, or knitting them into fabrics after finishing (refinement) using raw materials, textile products [15]. The textile industry which uses cotton fiber as raw material, the quality of the raw material needs to be known in order to predict the quality of production (yarn or textile). The quality of cotton fiber is stated based on the staple length, grade, and character determined by a cotton classer well-trained. Characteristics of cotton fiber are influenced by strength, smoothness, maturity, and stretchability [16][17].

3. Research Method

The stages for designing the fuzzy database Tahani model and fuzzy MADM TOPSIS method to solving the problem by choosing the best solution are as follows:

3.1. *Fuzzy database Tahani model*

In the fuzzy Tahani method, there are parameters and fuzzy sets, each fuzzy set has a domain. There are 6 parameters used in this study, namely: price, fiber length, fiber strength, micronaire, quantity, and shipment. The stages for using the fuzzy Tahani method can be seen in Figure 1.

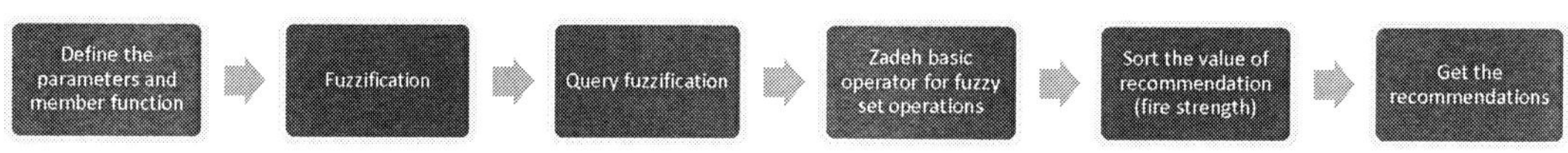

Figure 1. Stages for using the fuzzy Tahani method

3.1.1. Define the parameters and member function. These following are the parameters and membership functions used by the fuzzy database Tahani model:

1. Price parameters
   The price parameters consist of 3 fuzzy sets, namely low, normal, and high. Low and high sets use a linear function approach. While in the normal set uses a triangular membership function. The membership function in the price parameter can be formulated as in the equation (1).

$$\mu_{\text{LowPrice}}[x_1] = \begin{cases} 1; & x_1 \le 60 \\ \frac{(75 - x_1)}{(75 - 60)}; & 60 \le x_1 \le 75 \\ 0; & x_1 \ge 75 \end{cases}$$

$$\mu_{\text{NormalPrice}}[x_1] = \begin{cases} 0; & x_1 \le 70 \text{ or } x_1 \ge 90 \\ \frac{(x_1 - 70)}{(80 - 70)}; & 70 \le x_1 \le 80 \\ \frac{(90 - x_1)}{(90 - 80)}; & 80 \le x_1 \le 90 \end{cases} \quad (1)$$

$$\mu_{\text{HighPrice}}[x_1] = \begin{cases} 0; & x_1 \le 65 \\ \frac{(x_1 - 85)}{(100 - 85)}; & 85 \le x_1 \le 100 \\ 1; & x_1 \ge 100 \end{cases}$$

2. Fiber length parameters
The fiber length parameters consists of 3 fuzzy sets, namely short, medium, and long. Short and long sets use a linear function approach. While in the medium set uses a triangular membership function. The membership function in the fiber length parameter can be formulated as in the equation (2).

$$\mu_{\text{FiberLengthSHORT}}[x_2] = \begin{cases} 1; & x_2 \le 0{,}8 \\ \frac{(1{,}1 - x_2)}{(1{,}1 - 0{,}8)}; & 0{,}8 \le x_2 \le 1{,}1 \\ 0; & x_2 \ge 1{,}1 \end{cases}$$

$$\mu_{\text{FiberLengthMEDIUM}}[x_2] = \begin{cases} 0; & x_2 \le 1 \text{ or } x_2 \ge 1{,}1 \\ \frac{(x_2 - 1)}{(1{,}12 - 1)}; & 1 \le x_2 \le 1{,}12 \\ \frac{(1{,}23 - x_2)}{(1{,}23 - 1{,}12)}; & 1{,}12 \le x_2 \le 1{,}23 \end{cases} \quad (2)$$

$$\mu_{\text{FiberLengthLONG}}[x_2] = \begin{cases} 0; & x_2 \le 1{,}13 \\ \frac{(x_2 - 1{,}13)}{(1{,}36 - 1{,}13)}; & 1{,}13 \le x_2 \le 1{,}36 \\ 1; & x_2 \ge 1{,}36 \end{cases}$$

3. Fiber strength parameters
The fiber strength parameters consists of 3 fuzzy sets, namely intermediate, average, and strong. Intermediate and strong sets use a linear membership function. While in the average set uses a triangular membership function. The membership function in the fiber strength parameter can be formulated as in the equation (3).

$$\mu_{\text{FiberStrengthINTERMEDIATE}}[x_3] = \begin{cases} 1; & x_3 \le 24 \\ \frac{(26 - x_3)}{(26 - 24)}; & 24 \le x_3 \le 26 \\ 0; & x_3 \ge 26 \end{cases}$$

$$\mu_{\text{FiberStrengthAVERAGE}}[x_3] = \begin{cases} 0; & x_3 \le 25 \text{ or } x_3 \ge 29 \\ \frac{(x_3 - 25)}{(27 - 25)}; & 25 \le x_3 \le 27 \\ \frac{(29 - x_3)}{(29 - 27)}; & 27 \le x_3 \le 29 \end{cases} \quad (3)$$

$$\mu_{\text{FiberStrengthSTRONG}}[x_3] = \begin{cases} 0; & x_3 \le 28 \\ \frac{(x_3 - 28)}{(30 - 28}; & 28 \le x_3 \le 30 \\ 1; & x_3 \ge 30 \end{cases}$$

4. Micronaire parameters
The micronaire parameters consists of 3 fuzzy sets, namely fine, medium, and coarse. Fine and coarse sets use a linear membership function. While in the medium set uses a triangular membership function. The membership function in the micronaire parameter can be formulated as in the equation (4).

$$\mu_{\text{MicFINE}}[x_4] = \begin{cases} 1; & x_4 \le 3 \\ \frac{(4 - x_4)}{(4 - 3)}; & 3 \le x_4 \le 4 \\ 0; & x_4 \ge 4 \end{cases}$$

$$\mu_{\text{MicMEDIUM}}[x_4] = \begin{cases} 0; & x_4 \le 3{,}9 \text{ or } x_4 \ge 5 \\ \frac{(x_4 - 3{,}9)}{(4{,}45 - 3{,}9)}; & 3{,}9 \le x_4 \le 4{,}45 \\ \frac{(5 - x_4)}{(5 - 4{,}45)}; & 4{,}45 \le x_4 \le 5 \end{cases} \quad (4)$$

$$\mu_{\text{MicCOARSE}}[x_4] = \begin{cases} 0; & x_4 \le 4{,}9 \\ \frac{(x_4 - 4{,}9)}{(6 - 4{,}9)}; & 4{,}9 \le x_4 \le 6 \\ 1; & x_4 \ge 6 \end{cases}$$

5. Quantity parameters
The quantity parameters consists of 3 fuzzy sets, namely a few, normal, and a lot of. A few and a lot of sets use a linear membership function. While in the normal set uses a triangular membership function. The membership function in the quantity parameter can be formulated as in the equation (5).

$$\mu_{\text{QuantityAFew}}[x_5] = \begin{cases} 1; & x_5 \le 50 \\ \frac{(100 - x_5)}{(100 - 50)}; & 50 \le x_5 \le 100 \\ 0; & x_5 \ge 100 \end{cases}$$

$$\mu_{\text{QuantityNormal}}[x_5] = \begin{cases} 0; & x_5 \le 100 \text{ or } x_5 \ge 300 \\ \frac{(x_5 - 100)}{(200 - 100)}; & 100 \le x_5 \le 200 \\ \frac{(300 - x_5)}{(300 - 200)}; & 200 \le x_5 \le 300 \end{cases} \quad (5)$$

$$\mu_{\text{QuantityALotOf}}[x_5] = \begin{cases} 0; & x_5 \le 250 \\ \frac{(x_5 - 250)}{(500 - 250}; & 250 \le x_5 \le 500 \\ 1; & x_5 \ge 500 \end{cases}$$

6. Shipment parameters
The shipment parameters consists of 3 fuzzy sets, namely fast, normal, and slow. A few and a lot of sets use a linear membership function. While in the normal set uses a triangular membership function. The membership function in the shipment parameter can be formulated as in the equation (6).

$$\mu_{\text{ShipmentFast}}[x_6] = \begin{cases} 1; & x_6 \le 6 \\ \frac{(40 - x_6)}{(40 - 6)}; & 6 \le x_6 \le 40 \\ 0; & x_6 \ge 40 \end{cases}$$

$$\mu_{\text{ShipmentNORMAL}}[x_6] = \begin{cases} 0; & x_6 \le 30 \text{ or } x_6 \ge 100 \\ \frac{(x_6 - 30)}{(65 - 30)}; & 30 \le x_6 \le 65 \\ \frac{(100 - x_6)}{(100 - 65)}; & 65 \le x_6 \le 100 \end{cases} \quad (6)$$

$$\mu_{\text{ShipmenSlow}}[x_6] = \begin{cases} 0; & x_6 \leq 90 \\ \frac{(x_6 - 90)}{(150 - 90)}; & 90 \leq x_6 \leq 150 \\ 1; & x_6 \geq 150 \end{cases}$$

3.1.2. Fuzzification. Fuzzification is the first phase of fuzzy calculation, which is the process of converting firm values to values fuzzy. The strict value referred to in this case is the parameter value of cotton products. The value of the cotton product parameter is used as a value fuzzy which will be calculated as the value of its membership degree to the set fuzzy. The calculation is performed using the membership function expression fuzzy contained in each parameter. The membership function used the linear membership function and the triangular membership function. These are the following of cotton product data used for fuzzification calculations, as shown in Table 1.

Table 1. Cotton product data

ID	Price	Fiber length	Microaire	Fiber Strength	Quantity	Shipment
46	72	1,06	4,9	26	325	165
57	75,69	1,1	4,9	27	500	134
77	75,37	1,17	4,9	28	100	43
129	74,57	1,17	44,6	30,5	500	74
...	...	...	...	...	...	...
402	70,75	1,1	4,9	28	110	45

3.1.3. Query fuzzification. Query fuzzification by taking parameter data fuzzy including price, fiber length, fiber strength, micronaire, quantity, and shipment to perform criteria selection on cotton products according to input user. The concept of relation fuzzy in a DBMS uses the degree of membership which is calculated based on the respective membership functions [18].

3.1.4. Zadeh basic operator for fuzzy set operations. In this study, the operator used is the AND operator to combine variables. The AND operator is obtained by taking the smallest membership value between elements in the relevant set using the min function. The recommended cotton products are those that have a recommendation of $0 <$ fire strength ≤ 1. Cotton products that have a fire strength equal to 0 are not recommended. The higher the value, fire strength the greater the chance for the cotton product to be recommended.

3.2. *Fuzzy Multi-Attribute Decision Making TOPSIS method*

The fuzzy MADM TOPSIS method is used to rank by the following stages can be seen in Figure 2.

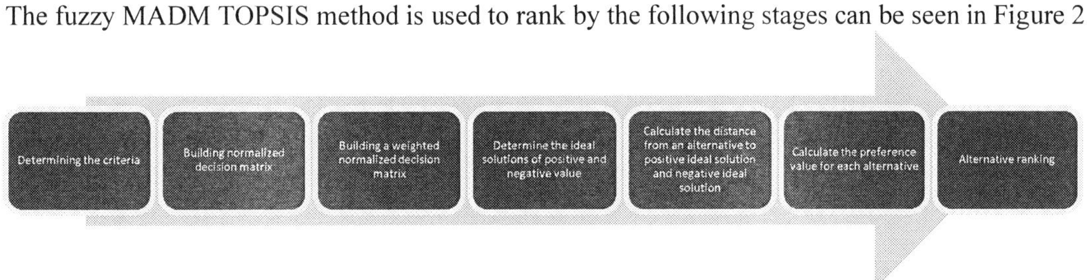

Figure 2. Stages of the fuzzy MADM TOPSIS method

3.2.1. Determining the criteria. The criteria used are as follows: price (C1), fiber length (C2), fiber strength (C3), micronaire (C4), quantity (C5), shipment (C6). These criteria are then determined by the weights. This weight is in the form of a fuzzy number which is converted to a crisp number. The crisp number that is closer to number 1, then the higher of the dependency level, while the crisp number that is closer to the number 0, then the lower of the dependency level.

Price, quantity, and shipment criteria each consist of three fuzzy numbers, namely: low (R), normal (N), and high (T) for the price criteria. Furthermore, for the quantity criteria, namely: little (S), normal (N), and a lot (B). Then for the shipment criteria, namely: fast (C), normal (N), and long (L).

The criteria for fiber length, fiber strength, and micronaire consist of five fuzzy numbers, namely: short (SR), intermediate (I), medium (M), long (L), and very long (VL). Furthermore, for fiber strength criteria, namely: weak (W), intermediate (I), average (AVG), strong (ST), and very strong (VST). Then for the micronaire criteria, namely: very fine (VF), fine (F), average (AVG), coarse (C), and very coarse (VC).

From each of these weights, a variable is created which will be converted into a fuzzy number with a formula as in equation (7).

$$\frac{var_n}{(m-1)} \tag{7}$$

with n = 0, 1, 2,…, n; and m = 1,2,…, m.

As in equation (7), var_n is the n-th variable and m is the number of variables.

Table 2. Variables and weight values on the criteria C1, C5, and C6

Fuzzy number	Crisp Value
Low (R), A few (S), Fast (C)	V0 / (3-1) = 0/2 = 0
Normal (N)	V1 / (3-1) = 1/2 = 0,5
High (T), A lot of (B), Slow (L)	V2 / (3-1) = 2/2 = 1

Table 3. Variables and weight values on the criteria C2, C3, and C4

Fuzzy number	Crisp Value
Short (SR), Weak (W), Very Fine (VF)	V0 / (5-1) = 0/4 = 0
Intermediate (I), Fine (F)	V1 / (5-1) = 1/4 = 0,25
Medium (M), Average (AVG)	V2 / (5-1) = 2/4 = 0,5
Long (L), Strong (ST), Coarse ©	V3 / (5-1) = 3/4 = 0,75
Very Long (VL), Very Strong (VS), Very Coarse (VC)	V4 / (5-1) = 4/4 = 1

3.2.2. Building normalized decision matrix. The TOPSIS normalized decision matrix requires a performance rating of each alternative A_i on each normalized weight of the C_j criterion. These are the following of the process of normalizing the attribute values to form a normalized matrix (R):

$$r_{ij}=\frac{x_{ij}}{\left(\sum_{i=1}^{m} x_{ij}^2\right)^{1/2}} \tag{8}$$

with i = 1,2,…,m; and j = 1,2,…, n.

As in equation (8), r_{ij} is the element of the normalized decision matrix and x_{ij} is the ith alternative performance rating for the j-th attribute.

3.2.3. Building a weighted normalized decision matrix. These are the following process of multiplying the weight with the value of each attribute to form a matrix (y) which is then used to determine the ideal positive solution A^+ and the ideal negative solution A^-:

$$y = \begin{bmatrix} y_{11} & y_{12} & \dots y_{1j} \\ y_{21} & y_{22} & \dots y_{2j} \\ \vdots & \vdots & \\ y_{i1} & y_{i2} & \dots y_{ij} \end{bmatrix} \tag{9}$$

for $y_{ij} = w_i r_{ij}$; with i = 1, 2, ..., m; and j = 1, 2, ..., n.

As in equation (9), yij is the element of the weighted normalized decision matrix, while w_i is the weight of the j-th criterion and r_{ij} is the element of the normalized decision matrix.

3.2.4. Determine the ideal solutions of positive and negative value. The positive ideal solution is denoted by the A^+ symbol and the negative ideal solution is denoted by the symbol A^-. The positive ideal solution is defined as the sum of all the best scores that can be achieved for each attribute, while the negative ideal solution consists of all the worst scores achieved for each attribute.

$$A^+ = (y_1^+, y_2^+, \dots, y_n^+), \tag{10}$$

$$A^- = (y_1^-, y_2^-, \dots, y_n^-), \tag{11}$$

$$y_j^+ = \begin{cases} \max_{ij} y_{ij} & ;\text{if j is an attribute of benefit} \\ \min_{ij} y_{ij} & ;\text{if j is an attribute of cost} \end{cases} \tag{12}$$

$$y_j^- = \begin{cases} \max_{ij} y_{ij} & ;\text{if j is an attribute of benefit} \\ \min_{ij} y_{ij} & ;\text{if j is an attribute of cost} \end{cases} \tag{13}$$

with j = 1, 2, ..., n.

3.2.5. Calculate the distance from an alternative to positive ideal solution and negative ideal solution. The distance between the value of each alternative with the positive ideal solution matrix and the negative ideal solution matrix. The distance between the alternative A_i and the positive ideal solution is defined as:

$$D_i^+ = \left(\Sigma_{j=1}^{n} (y_i^+ - y_{ij})^2\right)^{1/2} \tag{14}$$

with i = 1, 2, ... , m.

The distance between the alternative A_i and the negative ideal solution is defined as:

$$D_i^- = \left(\Sigma_{j=1}^{n} (y_{ij} - y_i^-)^2\right)^{1/2} \tag{15}$$

with i = 1, 2, ... , m.

3.2.6. Calculate the preference value for each alternative. The preference value for each alternative (V_i) is given as:

$$V_i = \frac{D_i^-}{D_i^- + D_i^+} \tag{16}$$

with i = 1, 2, ... , m. A larger V_i value indicates that the alternative A_i is prioritized.

3.2.7. Alternative ranking. Alternatives can be ranked according to the order of V_i from the largest to the smallest. Therefore, the best alternative is nearest to the ideal positive solution and inmost away from the negative ideal solution. The alternative with the enormous V_i is the best solution.

4. Results and Analysis

4.1. Trial case with fuzzy database Tahani model

The cotton product desired by the user has the grade/type LM (Low Middling), SM (Strict Low Middling), and MID (Middling). The selection of cotton product data on the input grade/type according to all cotton product data (421 data) has been selected as many as 270 cotton product data with grade/type LM, SM, and MID. This trial case is using the AND operator.

After the grade/type of the selected cotton product has been selected, then choose the criteria for cotton products with low prices, large quantities, strong fiber strength, medium fiber length, and coarse miconaire (coarse) as well as normal delivery times. Then calculate the value of the degree of membership in each selected cotton product data. These are the following calculation of the membership value from cotton product data with id 46.

Calculation of price parameters using equation (1). The membership value for the price parameter is = 0.2. Then for the calculation of the fiber strength parameters using the equation (3). The membership value for the fiber strength parameter is = 0. Then for the calculation of the quantity parameters using the equation (5). The membership value for the quantity parameter is = 0,3. Then for the calculation of the shipment parameters using the equation (6). The membership value for the shipment parameter is = 0,2. Then for the calculation of the micronaire parameters using the equation (4). The membership value for the micronaire parameter is = 0,2. Then for the calculation of the fiber length parameters using the equation (2). The membership value for the fiber length parameter is = 0,5217.

The calculation of the recommendation value by taking the smallest membership value between elements in the relevant set uses the min function. The data with id 46 has value 0 and data with id 129 has 0,029 according to the smallest membership value.

If the final result of the recommendation value is 0 then it is not recommended. Then the value of fire strength is ranked from highest to lowest. These are the following final result of product recommendations, as shown in Table 4.

Table 4. The end result of product recommendations

ID	Name	Grade	Value of fire strength
114	USA MOT RECAPS	SLM	0,177
119	USA SOUTH TEXAS RECAPS	MID	0,1753
279	USA M/E RECAPS	MID	0,1
129	USA M/E RECAPS	LM	0,029

4.2. Trial case with fuzzy Multi-Attribute Decision Making TOPSIS method

These are the following stages for the trial case carried out with the fuzzy Multi-Attribute Decision Making TOPSIS method:

The decision matrix is calculated based on the alternative value for each criteria. These are the following calculation result of the decision matrix on cotton products, as shown in Table 5.

Table 5. Decision matrix

ID	Price	Fiber length	Fiber strength	Micronaire	Quantity	Shipment
46	0,5	0,5	0,5	0,75	1	1
57	0,5	0,5	0,5	0,75	1	1
77	0,5	0,75	0,5	0,75	0,5	0,5
129	0,5	0,75	1	1	1	0,5
...	...	...	...	...	...	...
402	0,5	0,5	0,5	0,75	0,5	0,5

Determine a preference weights by assigning properties to each criterion. Attribute consists of benefits and costs. C1=1 (benefit), C2=0,75 (benefit), C3=0,5 (benefit), C4=0,5 (benefit), C5=1 (benefit), and C6=0,25 (cost).

Build a normalized matrix and produce a normalized matrix R. Here is a calculation to determine the normalized matrix in alternative 1 using equation (8):

$$r_{11}=\frac{0{,}5}{(x_{1\,1}{}^2+x_{2\,1}{}^2+x_{3\,1}{}^2+x_{4\,1}{}^2+\ldots+x_{270\,1}{}^2)^{1/2}}=\frac{0{,}5}{8{,}33667}=0{,}05998$$

Build a weighted normalized matrix. These are the following calculation to determine a weighted normalized matrix in alternative 1 using equation (8):

$$y_{11}=w_1 r_{11}=1(0{,}05998)=0{,}05998$$

Determine the positive ideal solution and the negative ideal solution. These are the following results of calculating the positive ideal solution using equations (10), (11), (12), and (13). Positive ideal solution for C1=0,11995, C2=0,05478, C3=0,03987, C4=0,03754, C5=0,07625, and C6=0,01821. While negative ideal solution for C1=0, C2=0,01826, C3=0,01994, C4=0, C5=0, and C6=0,0091.

Table 6. Ranking

No	ID	Name	Preference value
1	35	19/20 SJV ACALA SAW GIN	0,92019
2	37	19/20 SJV ROLLER GIN	0,92019
3	38	19/20 SJV ROLLER GIN	0,92019
4	210	USA M/E 2019/20	0,87475
5	293	USA SOUTH TEXAS RECAPS	0,79394
...	...	...	...
270	92	USA RECAPS M/E 2018/19 CROP	0,38888

Calculate the positive ideal distance and the negative ideal distance, which is the distance between the value of each alternative with the positive and negative ideal solution matrix. These are the following of calculation to determine the positive and negative ideal distances in alternative 1 using equations (14) and (15):

$$D_1^+=\left(0{,}11995\text{-}0{,}05998)^2+(0{,}05478\text{-}0{,}03652)^2+\ldots+(0{,}01821\text{-}0{,}01821)^2\right)^{1/2}=0{,}06645$$

$$D_1^-=\left((0{,}05998\text{-}0)^2+(0{,}03652\text{-}0{,}01826)^2+\ldots+(0{,}01821\text{-}0{,}00910)^2\right)^{1/2}=0{,}10305$$

Calculate the preference value for each alternative using equation (16).

$$V_1=\frac{0{,}10305}{0{,}10305+0{,}06645}=0{,}60796$$

Display the alternative ranking of the product, by the highest to lowest preference values. These are the following final result of the cotton product ranking, as shown in Table 6.

Table 7. Comparison results

ID	Fuzzy Tahani		FMADM TOPSIS	
	Fire Strength	Rank	Preference value	Rank
46	0	-	0,60796	94
57	0	-	0,60796	95
77	0	-	0,53055	193
129	0,029	4	0,64873	9
114	0,177	1	0,62915	64
119	0,1753	2	0,62322	80
279	0,1	3	0,62322	82
...	...	...	...	...
402	0	-	0,50462	257

4.3. Comparison results from the fuzzy Tahani method and the fuzzy MADM TOPSIS method

In the Tahani fuzzy method, alternatives with id 46, 57, and 77 have a fire strength value of 0, so the alternative is not recommended. Meanwhile, the alternative fuzzy MADM TOPSIS method with id 46, 57, and 77 has a preference value of 0.60796; 0.60796; and 0.53055 and the three of them are ranked in the 94th, 95th, and 193th places (Table 7). The final results obtained from the calculations by these two methods can be concluded that there are differences. The two methods used in this study can play a role in determining the recommendation for cotton products.

5. Conclusion

In this study, it can be concluded that the fuzzy database method Tahani model and the fuzzy MADM TOPSIS method can be used in completing the selection of several alternatives based on several predetermined criteria. Comparison method of fuzzy Tahani and fuzzy MADM Topsis is obtained from the results of fire strength values and preference values.

The results of the calculation on fuzzy Tahani carried out with a trial case that resulted in 4 recommendations for cotton products, namely USA MOT Recaps with a value of 0.177, USA South Texas Recaps with a value of 0.1753, USA M/E Recaps with a value of 0.0853 and USA M/E Recaps with a value of 0.0286. Where as for the results of the calculation on fuzzy MADM TOPSIS method produces a ranking for all recommended cotton products starting from 19/20 SJV Acala Saw Gin, 19/20 SJV Roller Gin, 19/20 SJV Roller Gin with the same value, 0.92019, USA M/E 2019/20 with a value of 0.87475, USA South Texas Recaps with a value of 0.79394 and so on. It can be seen from the final results of the two methods used, the calculation of method is fuzzy database Tahani model more accurate with accuracy in calculating the value of fire strength and the parameters used in choosing this cotton product are relative and have multiple meanings.

References

[1] Ministry of Industry 2019 *Lampaui 18 Persen Industri Tekstil dan Pakaian Tumbuh Paling Tinggi* Retrieved from http://kemenperin.go.id/artikel/20666/

[2] Febistian H, Andreswari D and Erlansari A 2015 Implementasi metode MCDM dalam pemilihan Kantor Urusan Agama (KUA) teladan dengan menggunakan PROMETHEE (studi kasus: Kementerian Agama Kepahiang) *Jurnal Rekursif.* **3.2** pp 196–207

[3] Maulana M R 2012 Penilaian kinerja karyawan di Ifun Jaya Textile dengan metode Fuzzy SAW *Jurnal Ilmiah ICTech* **10.1** pp 1–12

[4] Hermawan T 2013 *Penerapan Fuzzy Analytic Hierarchy Process untuk menentukan siswa teladan tingkat Sekolah Menengah Atas* Thesis (Semarang: Universitas Dian Nuswantoro)

[5] Rahardjo J and Sutapa I N 2002 Aplikasi fuzzy analytical hierarchy process dalam seleksi karyawan *Jurnal Teknik Industri* **4.2** pp 82–92

[6] Kusumadewi S and Purnomo H 2013 *Aplikasi Logika Fuzzy untuk Pendukung Keputusan* (Yogyakarta: Graha Ilmu) chapter 4 pp 177–241

[7] Kusumadewi S, Hartati S, Harjoko A and Wardoyo R 2006 *Fuzzy Multi-Attribute Decision Making (Fuzzy MADM)* (Yogyakarta: Graha Ilmu)

[8] Sukerti N K 2016 Analisis perbandingan penerima bantuan kemiskinan dengan metode Weighted Product (WP) dan TOPSIS *Jurnal Ilmiah DASI* **17.3** pp 1–7

[9] Fatkhurrochman and Astuti D 2018 Analisis perbandingan metode TOPSIS dan SAW dalam penentuan penerima bantuan pembangunan rumah masyarakat kurang mampu *Semnas TI & Multimedia* **2.8** pp 67–72

[10] Widyassari A P and Yuwono T 2018 Perbandingan Analytical Hierarchy Process dan Fuzzy Mamdani untuk sistem pendukung keputusan pemilihan rumah di daerah Cepu *J. Comp. Information Syst. & Technol. Management* **1.2** pp 50–4

[11] Astari A P and Komaruddin R 2018 Sistem pendukung keputusan pemilihan karyawan terbaik dengan metode Fuzzy Tahani *PIKSEL* **6.2** pp 169–78

[12] Rochman T 2019 *Decision Support System Rekomendasi Pembelian Barang Lelang Menggunakan Metode Fuzzy Database Model Tahani Studi Kasus CV Prima Engineering*

(Gresik: Universitas Muhammadiyah Gresik)
[13] Efendi R, Ernawati E and Hidayati R 2014 Aplikasi Fuzzy Database model Tahani dalam memberikan rekomendasi pembelian rumah berbasis web *Pseudocode* **1.1** pp 32–43
[14] Asmara A, Purnamadewi Y L, Mulatsih S and Novianti T 2013 Faktor-faktor yang memengaruhi perkembangan investasi pada industri tekstil dan produk tekstil (TPT) Indonesia *Jurnal Manajemen Teknologi* **12.2** pp 140–60
[15] Ferdyansyah D and Santoso E B 2013 Pola spasial kegiatan industri unggulan di Propinsi Jawa Timur (studi kasus: subsektor industri tekstil, barang kulit, dan alas kaki) *Jurnal Teknik ITS* **2.1** pp C37–C42
[16] Soeprijono P, Poerwanti, Widayat and Jumaeri 1973 *Serat-serat tekstil* (Bandung: Institut Teknologi Tekstil Bandung)
[17] Saroso B and Darmono 2002 *Pascapanen, Gining, dan Klasifikasi Mutu Serat Kapas* (Malang: Monograf Balittas Buku 2) pp 2013–210
[18] 'Uyun S and Madikhatun Y 2011 Rekomendasi berbasis Fuzzy untuk pemilihan Sekolah Lanjutan Tingkat Atas *Jurnal Informatika Ahmad Dahlan* **5.2** 101756

Identifying User Characteristics of the Top Three E-Wallet Services in Indonesia

Anggraeni Dias Saputri[1], Ahmad R Pratama[2]

[1]Master Program in Informatics, Universitas Islam Indonesia, Sleman, DIY, Indonesia
[2]Department of Informatics, Universitas Islam Indonesia, Sleman, DIY, Indonesia

E-mail: [1]anggraeni.saputri@students.uii.ac.id, [2]ahmad.rafie@uii.ac.id

Abstract. Electronic wallet (e-wallet) services have been the most commonly used electronic payment (e-payment) method in Indonesia within the past few years. By using survey data from 409 e-wallet users collected in the first half of 2020 and analyzed with descriptive statistics and logistic regression models, this study explores the top three e-wallet services in this country (i.e., GoPay, OVO, and DANA) to identify the user characteristics of the top three e-wallet which currently being used in Indonesia and to investigate which user groups, if any, tend to go with one e-wallet service over another. The results show that GoPay has the largest user base but OVO is the one with the most loyal user base while DANA has the most different user base of them all. The findings in this study can complement and enrich the literature on the use of ICT in the financial sector in Indonesia and shed light on the use of e-wallet and e-payment industry in this country. Apart from laying out some ground work for further research, the insight from this study can also be used by each e-wallet service to help make decisions as part of their marketing strategy.

Keywords: Indonesia; E-Wallet; Users; GoPay; OVO; DANA; Marketing

1. Background

Technological developments have reached various aspects of life, including in financial aspects. One good example is the development of payment instruments. Like in any other part of the world, payment methods in Indonesia began with a barter system, followed by the discovery of coins and then paper-based money. The first paper money owned by Indonesia was devised in 1946 after Indonesia's independence day [1]. Afterwards, payment systems continued to develop with more sophisticated innovations. Nowadays, the technological development at play is in the form of the electronic payment (e-payment) system, also known as the cashless payment system [2].

Many institutions were involved in the development and administration of payment systems in Indonesia [3]. E-payment requires an Internet connection to work, similar to other electronic environments such as electronic banking (e-banking), electronic shopping (e-shopping), or electronic learning (e-learning) [4]. E-payment itself is categorized into several types, including credit cards, electronic wallets (e-wallets), electronic cash, digital checking systems and wireless payment systems [5]. As per Bank Indonesia, the use of paper checks as a traditional payment has dropped by 30% over the last four years while the use of e-payment instruments has increased over the last ten years. Additionally, the use of cash with denomination of smaller than IDR 50,000 has decreased. In 2018, transaction values increased by 22% with 99% of them were related to the e-payment system while the rest is about personal finance and business finance [6].

In China, the world's most populous country with the highest economic growth, e-wallets have become the primary financial instrument. By 2018, 56.1% financial transactions made by Chinese people were done through e-wallet services [7]. Meanwhile in Indonesia, e-wallets have also become the most commonly adopted e-payment systems [8]. As shown in a previous study, university students preferred e-wallets over e-money when paying for food, beverages, clothes and entertainment [9].

A physical wallet in general contains coins, paper money and some cards such as identity cards that can help the transaction process. E-wallets were developed with the same function in mind in the form of software modules, portable devices, or smart cards [10]. Another definition of e-wallet is the online container that stores users 'payment information, along with the users' account and password. It also allows users to make e-commerce transactions integrated into their mobile devices [11]. The e-wallet service is a prepaid account so there is no denial of services and goods unlike in the case of credit cards [12]. E-wallet has several advantages, are as follows the transaction of e-wallet is a secured payment system, users need to be registered so any suspicious transaction can be tracked, all transactions will be recorded therefore can be monitored to increase safety for purchasing activities, e-wallet payment system is reliable, robust and user-friendly [13].

There are many e-wallet service providers in Indonesia ranging from large companies, bank institutions, to digital start-ups. According to Bank Indonesia that issued on 27 May 2020, there are 48 e-wallet services in Indonesia with the official license [14]. In 2018, e-wallet transactions in Indonesia have reached USD 1.5 billion and are predicted to continue rising to USD 25 billion by 2023. The number of e-wallet users keeps increasing in line with the continued development of financial technology in Indonesia. A previous study shows that most active users of e-wallet services in Indonesia are in the age group of 20-30 years old (52.3%), followed by teenagers (33.3%), and the elderly at the bottom of the list (13%) [15]. Based on their popularity, the top e-wallet services in Indonesia in Q4 of 2017 are GoPay, LinkAja and OVO. In Q1-Q3 of 2018, OVO and LinkAja were racing for the second place while the first position was held strongly by GoPay. Between Q4 of 2018 and Q2 of 2019, the order seemed to begin stabilizing with GoPay as the market leader, followed by OVO and LinkAja in the second and third place [16]. However, based on the most recent data, the top three highest ranked e-wallets in Indonesia based on active users are GoPay, OVO, and the newcomer, DANA that successfully leapfrogged LinkAja in the process [17].

There are many factors behind every choice made by a person, including when deciding to use certain e-wallet services. Each e-wallet service has certain features that might appeal to different groups of users. In Malaysia, where approximately 81.9% respondents have some experience using e-wallet services by 2020, the age of the users was found to significantly affect their e-wallet service preferences [18-19]. In Thailand, another Southeast Asian country, both age and sex affect e-wallet preferences of the users [20]. Furthermore, another study in Nigeria found out that the characteristics of e-wallet users were also affected by their level of education in addition to their age [21].

Meanwhile, there is little to no research trying to identify the characteristics of e-wallet users in Indonesia. Understanding users' characteristics is important because this information is useful in helping with the product development and marketing strategy. Each country has its own unique features and by their extension, different challenges from other countries, even the neighbouring ones. In the case of e-wallet services, there are different players in almost every country. Therefore, the objective of this study is to identify users' characteristics of e-wallet services specific to the Indonesian market. Specifically, this study aims to answer two research questions, i.e. 1) what are the user characteristics of the top three e-wallet currently being used in Indonesia? and 2) what user groups, if any, tend to go with one e-wallet service over another? While there are more than five e-wallet services available in Indonesia, this study will only focus on the top three based on their popularity.

2. Methods

An online survey was conducted in Indonesia within the first half of 2020. A total of 409 users (188 males, 45.97% and 221 females, 54.03%) participated in the study. Their age ranged between 17 and 59 years old (mean = 25.7, SD = 5.01). They came from different places of origin, simplified into Java (n=284, 69.44%) and outside Java (n=125, 30.56%). The same simplified category was used for their place of residency with 334 participants living in Java (81.66%) and 75 participants living outside Java (18.33%). The participants also reported their education, profession and income. For their educational attainment, 86 participants reported having secondary education (21.02%) while the remaining 323 had some college degree (78.97%). The profession was categorized into Information, Communication and Technology field (ICT) (n=113, 27.63%), Finance field (n=13, 3.18%), Students (n=42, 10.27%) and the rest was put together as others (n=241, 58.92%). Based on their income, participants were categorized into three groups, 232 low-income respondents (56.72%) earning equal or less than IDR 2,999,999 (approximately USD 200) per month, 73 middle-income respondents (17.85%) earning between IDR 3,000,000 and IDR 4,999,999 (approximately USD 333) per month, and 104 high-income participants (25.43%) earning IDR 5,000,000 or more per month.

Data analysis includes descriptive statistics, logistic regression, and multinomial logistic regression. For the dependent variables in this study, participants were asked what e-wallet services they were using and if there were more than one, which one they most frequently use was. Meanwhile, participants' demographic information, including age, sex, origin, residency, profession, income, and education was used as the independent variables. In terms of profession, three dummy variables were generated to represent the three most relevant professions that were selected to be included in the model (i.e., ICT, finance, and students) and to be compared with everything else. For the income variable, two dummy variables were generated to represent the middle-income and the high-income individuals to be compared with the low-income group used as a reference category.

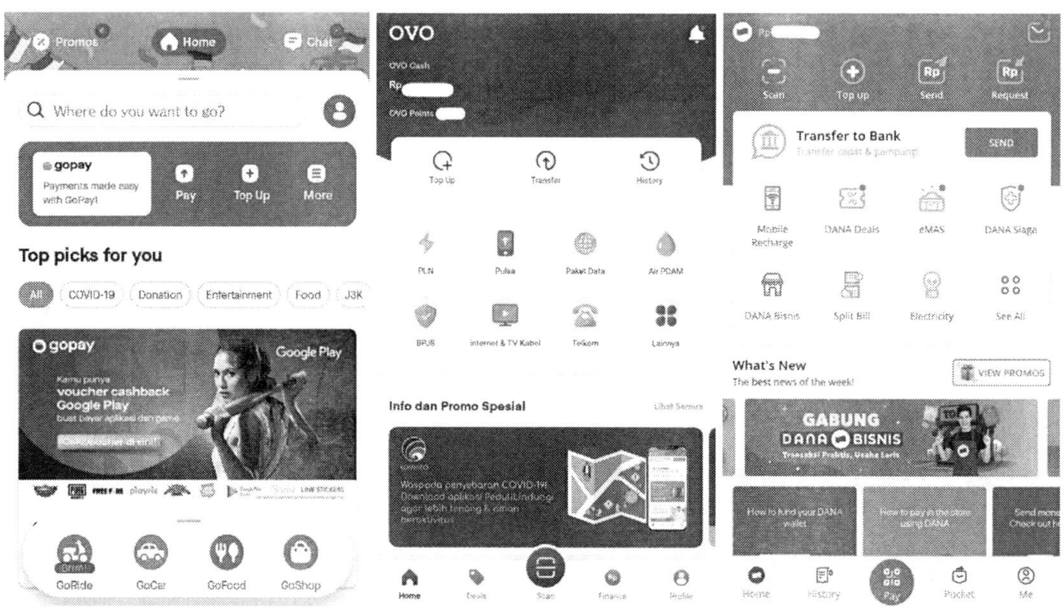

Figure 1. User interface of GoPay, OVO and DANA mobile apps.

3. Results

As it turned out, the top three e-wallet services used by participants in this study are GoPay, OVO, and DANA. Figure 1 shows the interface of the three e-wallet mobile applications on the Android platform.

As mentioned earlier, a total of 409 e-wallet users participated in this study. Table 1 provides the demographic information (i.e., age, sex, place of origin, place of residency, profession, income, and educational attainment) of these e-wallet users. In addition, this demographic information is also presented separately for each of the top three e-wallet services to see whether there are some differences in the characteristics of their users.

Table 1. Descriptive Statistics of e-wallet users participating in this study

Variables	**Category**	**E-Wallet Services**							
		Any (100%)		GoPay (71.64%)		OVO (65.28%)		DANA (43.28%)	
		freq	**%**	**freq**	**%**	**freq**	**%**	**freq**	**%**
Age	17-25	288	70.42%	208	70.99%	192	71.91%	135	76.27%
	26-35	99	24.21%	68	23.21%	62	23.22%	36	20.34%
	>35	22	5.38%	17	5.80%	13	4.87%	6	3.39%
Sex	Male	188	45.97%	131	44.71%	124	46.44%	96	54.24%
	Female	221	54.03%	162	55.29%	143	53.56%	81	45.76%
Origin	Java	284	69.44%	200	68.26%	185	69.29%	119	67.23%
	Non-Java	125	30.56%	93	31.74%	82	30.71%	58	32.77%
Residency	Java	334	81.66%	242	82.59%	221	82.77%	139	78.53%
	Non-Java	75	18.34%	51	17.41%	46	17.23%	38	21.47%
Profession	ICT	113	27.63%	87	29.69%	84	31.46%	61	34.46%
	Finance	13	3.18%	4	1.37%	6	2.25%	4	2.26%
	Student	42	10.27%	31	10.58%	26	9.74%	19	10.73%
	Others	241	58.92%	171	58.36%	151	56.55%	93	52.54%
Income	Low	232	56.72%	145	49.49%	136	50.94%	98	55.37%
	Medium	73	17.85%	61	20.82%	49	18.35%	28	15.82%
	High	104	25.43%	87	29.69%	82	30.71%	51	28.81%
Education	Secondary ed.	86	21.03%	57	19.45%	56	20.97%	44	24.86%
	College degree	323	78.97%	236	80.55%	211	79.03%	133	75.14%
All Samples		**409**	**100%**	**293**	**100%**	**267**	**100%**	**177**	**100%**

The descriptive statistics presented in Table 1 may have suggested some differences in characteristics of each e-wallet service across different groups of users. To see if those differences are statistically significant, a multivariate approach, which in this case is a logistic regression analysis, is needed. The results of logistic regression analysis for each of the top three e-wallet services are summarized in Table 2. Each model shows the likelihood of participants to use the respective e-wallet service based on their demographic information used as the predictors.

Despite having the largest user base among all other e-wallet services, it turned out that GoPay was not the most frequently used e-wallet among all participants. That title went to OVO, instead. As shown in Table 3, more than 40% participants in this study picked OVO as their most frequently used e-wallet, beating GoPay at slightly more than 30%. Just like in the previous case, DANA ranked the third as it was the most frequently used e-wallet by just shy of 11% participants.

Table 2. Logistic Regression results of E-wallet Services

Variables	**E-wallet Services**		
	GoPay	**OVO**	**DANA**
Age	1.022	0.976	**0.918** **
	(0.027)	0.021	**(0.025)**
Sex			
(Female)	1.326	1.078	**0.564** **
	(0.321)	0.241	**(0.123)**
Origin & Domicile			
(Not from Java, currently in Java)	**2.550** *	1.844	**0.418** *
	(1.161)	(0.736)	**(0.159)**
(From Java, not currently in Java)	1.976	3.453	0.448
	(1.696)	(2.883)	(0.313)
(From Java, currently in Java)	1.246	1.326	0.632
	(0.399)	(0.396)	(0.189)
Profession			
(ICT)	1.176	1.506	1.503
	(0.337)	(0.404)	(0.372)
(Finance)	**0.139** **	0.492	0.636
	(0.092)	(0.293)	(0.405)
(Student)	1.500	0.850	0.983
	(0.651)	(0.336)	(0.384)
Income			
(Medium)	**3.230** ***	1.488	0.991
	(1.184)	(0.443)	(0.295)
(High)	**3.098** ***	**2.603** ***	1.582
	(1.024)	**(0.779)**	(0.431)
Education			
(College)	1.043	0.758	0.699
	(0.328)	(0.230)	(0.212)
Constant	0.585	2.191	15.015 ***
	(0.447)	(1.485)	(11.582)
χ^2	**42.62 *****	**25.44 ****	**35.89 *****
df	**9**	**9**	**9**
Observations	**409**	**409**	**409**

Note: The numbers reported were odds ratios with standard errors between parentheses. Signif. cosdes: '***' 0.001 '**' 0.01 '*' 0.05

Again, the descriptive statistics presented in Table 3 may have suggested some differences in characteristics of users who choose one e-wallet service over another. However, a multivariate approach in the form of multinomial logistic regression analysis can help uncover some significant factors behind that decision. The results of multinomial logistic regression analysis are summarized in Table 4. The model shows the likelihood of participants to choose one of the three most frequently used e-wallet services compared to any other e-wallet services used by the participants in this study combined, including but not limited to ShopeePay and LinkAja. In this multinomial logistic regression model, only ICT and students are included as dummy variables for professions since in DANA's case, nobody from the financial sector picked it as their most frequently used e-wallet service.

Table 3. Descriptive Statistics the most frequently used e-wallet in this study

Variables	Category	Any (100%)		GoPay (31.30%)		OVO (40.10%)		DANA (10.76%)	
		freq	%	freq	%	freq	%	freq	%
Age	17-25	288	70.42%	87	67.97%	123	75.00%	34	77.27%
	26-35	99	24.21%	31	24.22%	36	21.95%	9	20.45%
	>35	22	5.38%	10	7.81%	5	3.05%	1	2.27%
Sex	Male	188	45.97%	56	43.75%	71	43.29%	29	65.91%
	Female	221	54.03%	72	56.25%	93	56.71%	15	34.09%
Origin	Java	284	69.44%	91	71.09%	112	68.29%	26	59.09%
	Non-Java	125	30.56%	37	28.91%	52	31.71%	18	40.91%
Residency	Java	334	81.66%	105	82.03%	141	85.98%	31	70.45%
	Non-Java	75	18.34%	23	17.97%	23	14.02%	13	29.55%
Profession	ICT	113	27.63%	33	25.78%	54	32.93%	13	29.55%
	Finance	13	3.18%	2	1.56%	5	3.05%	0	0.00%
	Student	42	10.27%	17	13.28%	14	8.54%	5	11.36%
	Others	241	58.92%	76	59.38%	91	55.49%	26	59.09%
Income	Low	232	56.72%	68	53.13%	84	51.22%	34	77.27%
	Medium	73	17.85%	28	21.88%	28	17.07%	4	9.09%
	High	104	25.43%	32	25.00%	52	31.71%	6	13.64%
Education	Secondary ed.	86	21.03%	25	19.53%	36	21.95%	12	27.27%
	College degree	323	78.97%	103	80.47%	125	78.05%	32	72.73%
All Samples		**409**	**100%**	**128**	**100%**	**164**	**100%**	**44**	**100%**

4. Discussion

Based on the descriptive statistics in Table 1, the number of participants who have some experience using e-wallet services is consistent with the market share discussed earlier in the introduction. GoPay led the market with 293 users (71.64%), OVO came in second place with 267 users (65.28%), and DANA ranked third with 177 users (43.28%). However, it turns out that OVO has the most loyal user base where 164 of 267 (61.42%) users picked it as their most frequently used e-wallet service, far more than 128 of 293 (43.69%) users in GoPay's case and 44 of 177 (23.86%) users in DANA's case.

In terms of age, most e-wallet users came from the younger group between the age of 17 and 25 (70.42%). However, DANA is much more popular among the younger users compared to all other e-wallet services as confirmed in the logistic regression model in Table 2 and the multinomial logistic regression model in Table 4. It is also discovered that DANA is much more popular among male users than any other e-wallet services. In fact, it is the only e-wallet service that is used by more males (n = 96, 54.24%) than females (n = 81, 45.76%) in this study, almost the opposite of the general landscape of e-wallet users in this study with 54.03% females (n = 221) and 45,97% males (n = 188). The difference is even higher among those who picked DANA as their most frequently used e-wallet where the male users (n = 29, 65.91%) are almost twice as many as the female users (n = 15, 34.09%).

Table 4. Multinomial Logistic Regression results of Most Frequently Used E-wallet Services

Variables	E-wallet Services		
	GoPay	OVO	DANA
Age	.999	0.946	**0.871** *
	(0.026)	0.028	**(0.056)**
Sex			
(Female)	1.084	1.258	**0.423** *
	(0.335)	0.381	**(0.178)**
Origin & Domicile			
(Not from Java, currently in Java)	2.478	**3.509** *	1.444
	(1.559)	**(2.139)**	(0.849)
(From Java, not currently in Java)	1.147	0.621	0.434
	(0.962)	(0.574)	(0.554)
(From Java, currently in Java)	1.224	1.306	0.599
	(0.506)	(0.532)	(0.297)
Profession			
(ICT)	1.176	1.855	1.742
	(0.337)	(0.689)	(0.850)
(Student)	1.885	0.832	0.975
	(1.082)	(0.486)	(0.721)
Income			
(Medium)	1.487	1.268	0.445
	(0.599)	(0.511)	(0.287)
(High)	1.551	**2.217** *	0.581
	(0.630)	**(0.865)**	(0.337)
Education			
(College)	0.910	0.620	0.939
	(0.402)	(0.262)	(0.531)
Constant	1.038	5.864	43.453 *
	(0.944)	(5.523)	(69.767)
χ^2		**55.37** **	
df		**30**	
Observations		**409**	

Note: The numbers reported were relative-risk ratios with standard errors between parentheses.
Other e-wallet services (combined) were used as the reference category.
Signif. codes: '***' 0.001 '**' 0.01 '*' 0.05

Regarding the user's geographical location, more e-wallet users are originated from Java (69.44%) and currently reside in Java (81.66%). Nonetheless, it is interesting to see that compared to those who are both originated from and currently living outside Java, participants who are originated from outside Java but currently living in Java are much more likely to be GoPay's users while much less likely to be DANA's users (see Table 2), but they are most likely to pick OVO as their most frequently used e-wallet service (see Table 4). It is an indication that compared with any other e-wallet services, OVO is more appealing especially to users from this group.

The findings, especially from Table 2, also suggest that both GoPay and OVO are much more appealing to the middle to upper class economy. One most plausible reason is because of their cooperation with the most popular online transportation applications in Indonesia, GoJek and Grab [22]. Specifically, GoPay as an e-wallet service is part of GoJek, while OVO has collaborations with Grab. Both GoJek and Grab main service is transportation, but they also offer multiple other services such as food and package delivery currently operating in more than 50 cities in Indonesia [23]. Nonetheless, the results from Table 4 shows that the high-income class people are much more likely to pick OVO as their most frequently used e-wallet service. This phenomenon can be explained by two reasons. First, unlike GoPay that is part of GoJek mobile app, OVO has its own separate app, making it easier to use it for anything else outside the services provided by the online transportation app. Second, it also has a collaboration with Tokopedia, the most popular e-commerce platform in Indonesia with USD 16.5 billion worth of transactions in 2019 [24]. As for DANA that seems to be more appealing to the middle to lower class economy is perhaps due to the fact that DANA is used for disbursement of the Kartu Prakerja Indonesia, a kind of social security net program for those who are unemployed [25].

Based on these findings, it can be inferred that GoPay as the market leader is suffering from lacking uniqueness compared with the other top three e-wallet services. As suggested in the multinomial logistic regression model in Table 4, there is not even a single factor that differentiates GoPay users from the users of other e-wallet services. Meanwhile, OVO and DANA got strong loyalty from at least two different groups of users. The former can rely on the migrants currently living in Java and the high-income people, whereas the latter got strong support from the younger generation and male users. If anything, both OVO and DANA might have a better chance at focusing their marketing strategy towards these specific groups of people. As for GoPay, the market leader itself, it might be a good time to revisit their marketing strategy. Without some innovation, perhaps even a radical one, it might only be a matter of time that they will be taken over by OVO as the market leader in the e-wallet industry in Indonesia.

5. Conclusion

This study confirms the status of GoPay, OVO, and DANA as the top three e-wallet services in Indonesia by Q2 of 2020. The results show that GoPay has the largest user base (71.64%) but OVO, used by 65.28% of e-wallet users, is the one with the most loyal user base. More than 40% participants in this study chose OVO as their most frequently used e-wallet compared to the 31% of participants who stick with GoPay. Meanwhile, DANA as the third ranked e-wallet both in terms of user base and user loyalty has the most different user characteristics of them all. It is much more appealing to the younger users and male users than any other e-wallet services in Indonesia. The findings from this study can complement and enrich the literature in the financial technology sector and shed light on the use of the e-wallet and e-payment industry in Indonesia. Apart from laying out some ground work for further research, the insight from this study can also be used by each e-wallet service to help make decisions as part of their marketing strategy. Understanding the user's characteristics is important, especially for the industry as this information is useful in helping with the product development and marketing strategy.

One limitation in this study has something to do with the distribution of geographical location of participants that ended up with a simplification of Java and outside Java. A larger and more normally distributed sample from different regions of Indonesia would provide better and more comprehensive insight. Also, considering there are many more e-wallet services in Indonesia, it would be a good idea to investigate them all to see if there are some other e-wallet services that serve a niche market much more different than the other e-wallet services. Nonetheless, by identifying users' characteristics of the top three e-wallet services in Indonesia, this study shows that each e-wallet service has the potential to be more appealing to different groups of users especially with their unique features that differentiate them from each other. Considering the lack of such information in the literature, this study can lay out a good foundation for further understanding of the use of e-wallet services and the characteristics of e-wallet users in Indonesia.

References

[1] Banindro S B 2017 *Sejarah Uang Kertas "Oeang Republik Indonesia Masa Revolusi 1945-1949* (Yogyakarta: BP ISI)
[2] Pramono B 2006 Dampak pembayaran non tunai terhadap perekonomian dan kebijakan non moneter *Working Paper of Bank Indonesia* p 11
[3] Nabila M, Purwandari B, Nazief B A A, Chalid D A, Wibowo S S and Solichah I 2018 Financial technology acceptance factors of electronic wallet and digital cash in indonesia *ICITSI* pp 284–9
[4] Teoh W M Y, Chong S C, Lin B and Chua J W 2013 Factors affecting consumers' perception of electronic payment: an empirical analysis *Internet Research* **23(4)** pp 465–85
[5] Junadi and Sfenrianto 2015 A model of factors influencing consumer's intention to use *Procedia Computer Sci.* **5** pp 214–20
[6] Bank Indonesia 2020 *Digital Payment Transformation Bank Indonesia: Demographic Structure and Digital Vision Indonesia* Retrieved from https://www.bi.go.id/id/institute/kegiatan/Flagship/Contents/Digital%20Payment%20Transformation.pdf
[7] Korella J L and Wenwei L 2018 Retail payment behaviour and the adoption of innovative payments: a comparative study in china and germany *J. of Payments Strategy and Systems* pp 245–65
[8] Angelina C and Rahadi R A 2020 A conceptual study on the factors influencing usage intention of e-wallet in Java, Indonesia *IJAFB* **5(27)** pp 19–29
[9] Angelini K and Koesrindartoto D P 2019 E-money or e-wallet? a study of university students preference in choosing cashless payment systems *ICMEM IICIES* pp 64-8.
[10] Mainwaring S D, Anderson K and Chang M F 2005 What's in your wallet? implications for global e-wallet design *CHI Late Breaking Result: Posters*
[11] Singh S 2019 E-wallet: an overview *Int. J. of Technol. and Innovation Research* **1** pp 50–7
[12] Kalyani P 2016 An empirical study about the awareness of paperless e-currency transactions like e-wallet using ICT in the youth of India *JMEIT* **3(3)** pp 94–110
[13] Abu B N, Rosbi S and Uzaki K 2020 E-wallet transactional framework for digital economy: a perspective from Islamic financial engineering *Int. J. of Management Science and Business Administration* **6(3)** pp 50–7
[14] Bank Indonesia 2020 *Informasi Perizinan Penyelenggaraan dan Pendukung Jasa Sistem Pembayaran* Retrieved on 27 May 2020 from https://www.bi.go.id/id/sistem-pembayaran/informasi-perizinan/uang-elektronik/penyelenggara-berizin/Contents/Default.aspx
[15] Soegoto D S and Tampubolon M P 2020 E-wallet as payment instrument in the millennial era *IOP Conf. Series: Materials Science and Engineering* **879** pp 1–7
[16] Widiyanti W 2020 Pengaruh kemanfaatan, kemudahan pengguna dan promosi terhadap keputusan pengguna e-wallet ovo di depok *Moneter Jurnal Akuntansi dan Keuangan* **7**
[17] Devita V D 2020 *E-wallet lokal masih mendominasi q2 2019-2020* Retrieved on 12 August 2020 from https://iprice.co.id/trend/insights/top-e-wallet-di-indonesia-2020/.
[18] Teoh T T, Melissa, Yew H C and Heang L T 2020 E-wallet adoption: a case in malaysia *Int. J. of Research in Commerce and Management Studies* pp 216–23
[19] Subaramaniam K, Kalandaisami R and Jalil A B 2020 The impact of e-wallets for current generation *J. of Adv. Research in Dynamical & Control Systems* **12** pp 751–9
[20] Intarot P and Beokhaimook C 2018 Influencing factor in e-wallet acceptance and use *Int. J. of Business and Administrative Studies* **4(4)** pp 167–75
[21] Akinbile L A, Akwiwu U N and Alade O O 2014 Determinants of farmers' willingness to utilise e-wallet for accessing agricultural information on Osun State Nigeria *Nigerian J. of Rural* **15(1)** pp 105–13
[22] Mulyono H and Helmi S 2018 e-crm and loyalty: a mediation effect of custom experience and satisfaction in online transportation of Indonesia *J. of Economic Studies* **4(3)** pp 96–105

[23] Azzuhri A A, Syarafina A, Yoga F T and Amalia R 2018 A creative, innovative and solutive transportation for indonesia with its setback and how to tackle them: a case study of the phenomenal gojek *Integrative Business and Economics Research* **7(1)** pp 59–67
[24] Jayani D H 2019 *Tokopedia dengan Nilai Transaksi Terbesar* Retrieved on 15 Oktober 2019 from https://databoks.katadata.co.id/datapublish/2019/10/15/2014-2023-nilai-transaksi-tokopedia-terbesar-dibandingkan-e-commerce-lainnya.
[25] Evandio A 2018 *Dana Resmi Ditunjuk Jadi Mitra Prakerja* Retrieved on 20 August 2020 from https://teknologi.bisnis.com/read/20200820/266/1281291/dana-resmi-ditunjuk-jadi-mitra-kartu-prakerja

IOP Conf. Series: Materials Science and Engineering 1077 (2021) 012029 doi:10.1088/1757-899X/1077/1/012029

Indonesian Sign Language Recognition using YOLO Method

Steve Daniels, Nanik Suciati*, and Chastine Fathichah

Department of Informatics, Faculty of Intelligent Electical and Informatics Technology, Institut Teknologi Sepuluh Nopember, Surabaya, Indonesia

E-mail: *nanik@if.its.ac.id

Abstract. Sign language is a form of communication commonly used by people with hearing impairment or people with speech impediments. Not all ordinary people understand the language. The translation of sign language into the alphabet/text automatically will facilitate the communication of the deaf with ordinary people. This research aims to develop a sign language recognition system that can process input from video data using You Only Look Once (YOLO) in real-time. YOLO is an object detection method based on Convolutional Neural Network (CNN), which is accurate and fast. Retraining the Yolov3 pre-trained model is performed with adjustments to the number of channels and classes according to the sign language recognition requirement. In this research, we collect datasets independently based on the Indonesian Sign Language (BISINDO). In the experiment using image data, the system achieves 100% precision, recall, accuracy, and F1 score. While using video data, the system's performance gets precision 77.14%, recall 93.1%, accuracy 72.97%, and F1 score 84.38%, with a speed of 8 fps.

1. Introduction

Sign language is a form of non-verbal human communication expressed through hand movements to convey information to the recipient visually. Sign language is commonly used by people with hearing impairment or people with speech impediments. Most normal people do not learn sign language because it is not required to do. But if there's chance of a need to understand sign language, people cannot just be able to interpret sign language instantly without learning process. Therefore, a system that automatically recognizes and translates the sign from images or videos to alphabet/text will provide a way to understand sign language and ease communication between deaf/mute and ordinary people.

There are several previous researches that have done in sign language recognition. The type of approaches can be divided into two categories. The first approach is special equipment based, such as accelerometer [1], sensory glove [2], and Kinect [3]. In special equipment based approach usually there are two work steps. The step are capturing sensory signal from sensory equipment and recognizing sign from captured signal. This approach can provide accurate information, but there is a need for using particular equipment which can be either expensive or impractical for daily use.

The second approach is vision based. This approach is focused on vision features, which is lower cost in terms of equipment, requires only a camera for taking images or videos, and highly practical for daily usage because we can use webcam or smartphone camera as well. For example, there is a research using kNN classifier to recognize sign [4]. From computer vision and machine learning techniques combined, Convolutional Neural Network (CNN) was proposed and has become common method for vision related machine learning and been applied to various task such as object detection and recognition.

The CNN method was once used to translate American Sign Language (ASL) into the alphabet [5]. The CNN architecture used in that research is GoogLeNet, which has been trained on the 2012 ILSVRC dataset. The experiment obtained an accuracy of 72%. This lack of accuracy is due to the misclassification of alphabets with similar handshapes, for example, g / h and m / n / s / t.

One of the deep learning methods that have recently been widely used for object detection is "You Only Look Once" (YOLO). YOLO is developed based on Convolutional Neural Network (CNN) and can produce fast and effective object detection [6]. YOLO is used in various studies that require object detection, such as real-time localization of Bhutan number plates [7], pedestrian detection [8], and recognition of traffic signs [9].

In this research, we implement CNN with YOLO architecture to recognize Indonesian Sign Language (BISINDO) from video data. Transfer learning is used to retrain the YOLOv3 pre-trained model with adjustments to the number of channels and classes according to the Indonesian Sign Language requirement. We collect a dataset independently consists of 4,547 images with 24 classes of static sign language. We use 80% of the dataset for training 20% for testing. Performance targets measured include precision, recall, accuracy, F1 score, mAP, and fps.

This article is organized into five chapters. Chapter 1 contains the background and research objectives; Chapter 2 explains the theoretical basis; Chapter 3 provides an overview of the system design; Chapter 4 describes the results of the experiment and analysis, and Chapter 5 represents conclusions.

2. Theory

2.1. *Convolutional Neural Network*

Convolutional Neural Network (CNN) is a type of neural network that is commonly used in image data. CNN consists of neurons that have weight, bias, and activation functions. Generally, CNN consists of two parts, namely feature extraction layer and fully connected layer.

Feature extraction layer consists of 2 layers, convolutional layer and pooling layer. Convolutional layer consists of neurons that are arranged to form filters. This filter is shifted to all parts of the image, a dot operation is performed between the input and the value on the filter. The result of the process at this layer is a feature map. The feature map from the convolutional layer then passes through the pooling layer, a layer consisting of a filter with a certain size and stride which is shifted to all parts of the feature map. The purpose of the pooling layer is downsampling, which is to reduce the dimensions of the feature map. The feature map generated from the feature extraction layer is still in the form of a multidimensional array. Therefore, flattening is performed to change the shape of the feature map into a vector to be used as input for the fully connected layer. The output of this layer is the output of the network.

2.2. *You Only Look Once*

You Only Look Once (YOLO) is an object detection method known for its speed. Only a single convolutional network used to predict what objects will be in the image. This can happen because YOLO divides the image into cells/grids, each cell is responsible for predicting the number of bounding boxes, the confidence level of each cell, and the class probability.

2.3. *YOLOv3*

YOLOv3 is an improvement of the basic idea of YOLO, dividing an image into cells that are responsible for predicting objects. Changes from YOLO include feature extraction networks, use of bounding boxes, and detection at various scales [10]. YOLOv3 architecture consists of a feature extraction network and a detection layer with three different scales, which can be seen in Figure 1.

The feature extraction network used only consists of a Convolutional Layer, followed by batch normalization and the Leaky ReLU Activation function. Batch normalization is useful for helping the convergence model process during training. This network also has a residual block that passes results from one layer to the next layer and a deeper layer directly, so there is no degradation problem in a network that has many layers.

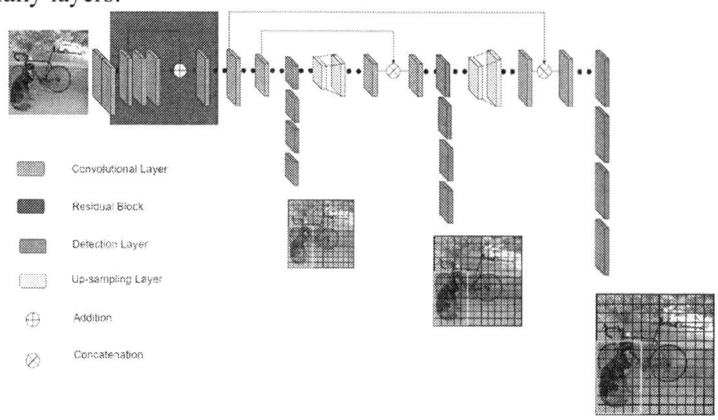

Figure 1. YOLOv3 Architecture [11]

The use of anchor boxes is done to replace predicting width and height of the bounding box directly because predicting the width and height of the bounding box causes an unstable gradient during training. Bounding box prediction is done by transforming the anchor box that has been assigned from the start. The object score and class score will enter the sigmoid function so that the results are between 0 to 1.

YOLOv3 performs predictions on three different scales. The detection layer is used for three feature maps of different sizes, each having a stride of 32, 16, and 8. With an input of 416×416, detection is carried out at a scale of 13×13, 26×26, and 52×52.

2.4. Non Maximum Suppression
Non-Maximum Suppression (NMS) is used by YOLO to eliminate bounding boxes that predict the same object. The way it works is to take the bounding box with the highest confidence then eliminate other bounding boxes that make predictions in the same class and have a similar location/overlap that exceeds the specified threshold.

2.5. Bahasa Isyarat Indonesia
Bahasa Isyarat Indonesia (BISINDO) is one of the sign languages used in Indonesia and is a language that grows naturally among the Deaf community in Indonesia. BISINDO alphabet signs are in the form of a static hand, except for the letters J and R which use gestures. BISINDO sign can be seen in Figure 2.

Figure 2. BISINDO sign [12]

3. Design
The design of the system to recognize Indonesian sign language is shown in Figure 3. The training process is performed to get a classifier model used later in the testing process to recognize sign language on image or video.

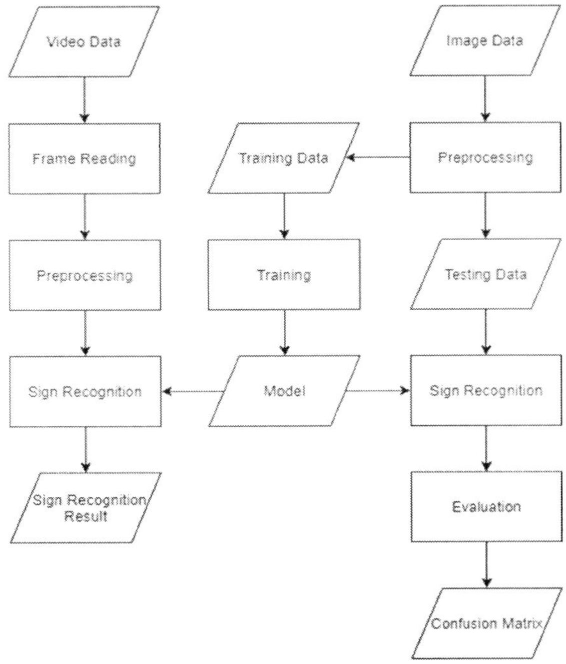

Figure 3. Diagram of the Indonesian sign language recognition

3.1. Dataset

The data used as input for the sign language recognition system is a dataset taken by the author in the form of photos and videos, consisting of 24 classes where each class there are about 160 to 220 images. This data is a set of images of characters, A to Z except J and R. Dataset specification is given in Table 1.

Table 1. Dataset Specification

Specification	**Value**
Resolution	3024 × 3024, 640 × 480
Extension	.jpg
Number of images	4,547
Number of class	24
Number of images per class	160-220
File size	50-500 kB
Channel	3 (RGB)

The dataset is split into training and testing data with a ratio of 8: 2; 8 for training data and 2 for trial data. The dataset is taken with different subjects and environments, a total of 5 conditions which are described in Table 2, and sample of each condition can be seen in Figure 4.

Table 2. Dataset Condition

Condition	**Description**
Condition 1	Slightly dim, taken from 20-40 cm away
Condition 2	Bright, taken from 20-40 cm away
Condition 3	Bright, taken from 20-80 cm away
Condition 4	Bright, taken from 20-40 cm away
Condition 5	Bright, taken from 150 cm away

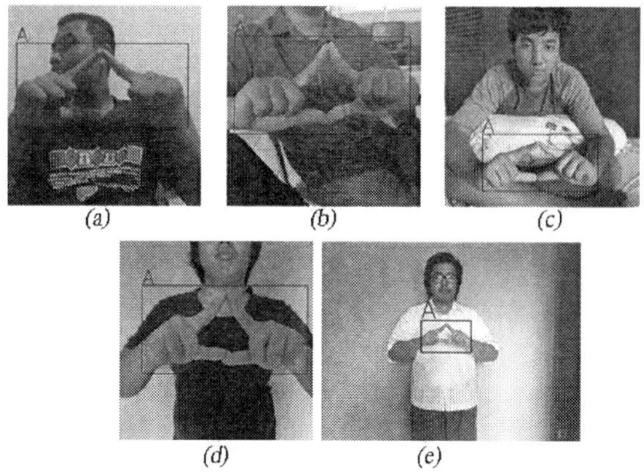

Figure 4. Sample of: (a) Condition 1; (b) Condition 2; (c) Condition 3; (d) Condition 4; (e) Condition 5

3.2. *Preprocessing*

At this stage, data preprocessing will be carried out before becoming input to the training process. The preprocessing data includes resizing to 416 × 416, conversion from RGB to grayscale, and adding data by rotating the available images by 1 and 2 degrees clockwise and counterclockwise. Preprocessing is done so that the data varies and equates the size of the training data. The result example can be seen in Figure 5.

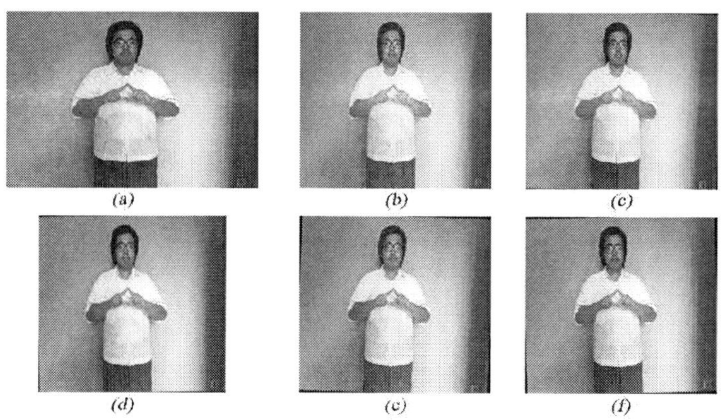

Figure 5. (a) Original image; (b) After resizing and converting to grayscale; (c) and (d) After rotating 1° counterclockwise (CCW) and clockwise (CW); (e) and (f) After rotating 2° CCW and CW.

3.3. *Training*

The training stage uses 80% of preprocessed dataset, while the rest is used for testing. The split is made so that train and test dataset has portions of all conditions. Adjustments are made from the original YOLO architecture, where changes occur in the number of channels and the number of filters in the convolutional layer before entering the Yolo layer. The training process utilizes training data to build a YOLO model. Training is done by using pretrained weight darknet53, which was trained on ImageNet, with various learning rates. Training will be stopped when the loss value does not decreasing anymore.

3.4. *Testing*

This phase has two types of testing, namely, testing of image data and video data. In image testing, several images are detected. At the end of the testing, a comparison of the detection results with the ground truth will be carried out, creating a Confusion Matrix to obtain accuracy, precision, recall, and

IOP Conf. Series: Materials Science and Engineering 1077 (2021) 012029 doi:10.1088/1757-899X/1077/1/012029

F1 score. Mean Average Precision (mAP) is also used to evaluate object detection task. In video testing, detection will be performed on each frame of the video data. If a class is detected sequentially more than the specified limit, it will be considered a recognized sign. The results of the recognized sign will be compared with the ground truth.

4. Experiments

In this chapter, there are several experiment scenarios: training results, image testing results, and video testing results.

4.1. Training Results

The training was carried out by forming three models with different learning rate parameters, namely 0.1, 0.01, and 0.001. The model is trained until the loss value does not decrease. Each model is compared based on the lowest loss value obtained. The training results can be seen in Table 3.

Table 3. Training Results

Learning Rate	Loss
0.1	0.942662
0.01	0.15379
0.001	0.143697

From Table 3, the model trained using a learning rate of 0.001 achieves the lowest loss value. This model will be used for subsequent test scenarios.

4.2. Image Testing Results

Testing on image data is done by running recognition on all images from the testing data. Some of result images can be seen in Figure 6.

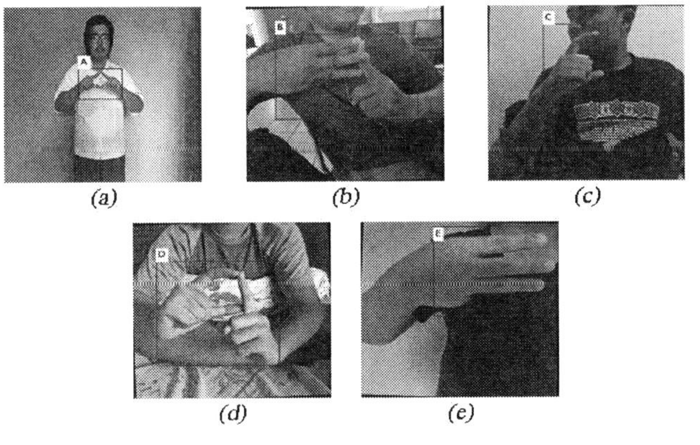

Figure 6. Prediction result examples: (a) A Sign, (b) B Sign, (c) C Sign, (d) D Sign, and (e) E Sign

Table 4. Object Detection evaluation

IoU Threshold	mAP (%)
0.75	71.16
0.5	99.91
0.25	99.96

Object detection task is tested with mAP as the evaluation metric. Three different Intersection over Union (IoU) threshold values is being used for this test. The test results can be seen in Table 4. The prediction results will be compared with the ground truth provided by the author. Detection is limited to

1 detection with the highest confidence for each image. The assessment will be carried out using precision, recall, accuracy, and F1 score. The test results can be seen in Table 5.

Table 5. Image Testing Results

Threshold (%)	Precision (%)	Recall (%)	Accuracy (%)	F1 Score (%)
100	100	41.67	42.20	58.83
90	100	76.53	76.97	86.71
80	100	83	83.33	90.71
70	100	87.72	87.90	93.46
60	100	91.56	91.62	95.59
50	100	94.53	94.57	97.19
40	100	96.70	96.72	98.32
30	100	98.27	98.28	99.13
20	100	99.51	99.52	99.75
10	100	99.98	99.98	99.99
9	100	100	100	100

As shown in Table 5, all the detections carried out resulted in the correct class. However, some images have low confidence so that they are not detected at a certain threshold.

Class O has the highest number of detections with confidence value below 80%. The detection result that has the smallest confidence is an image of O sign with a confidence of 9%.

Apart from being based on class, observations were also made based on image conditions. The image condition that has the most amount of low confidence is condition 5, where the cue image is the farthest away. From 247 cases of undetectable images at the threshold of 50%, 246 cases came from images with condition 5.

Table 6. Video Testing Results

Threshold (%)	Target	Result	Explanation
6	GLEAM	GLEAM	-
	IVY	OIOY	Prediction error of I becomes O, and V becomes O
	NICE	MICE	Prediction error of N becomes M
	SHIZA	SLZA	Alternating prediction of H between D and H without sequential detection of 5, the prediction error I becomes L
	WAX	WADA	Prediction error X becomes D, then becomes A
	BDFPUV	BPDBPDUVUV	The prediction error of D becomes P, F becomes B, P becomes D, U becomes V
	QUEST	QUEIST	Detection of I reached 6 sequentially when the transition of signals between E to S
7	WAX	WA	Prediction error of X becomes A
	BDFPUV	BPDBPDVUV	The prediction error of D becomes P, F becomes B, P becomes D, U becomes V
	QUEST	QUEST	Number of detected I sequentially is below 7

4.3. Video Testing Results

Testing on video data was carried out on several videos taken by the author, separated from the data taken by training in conditions similar to condition 5 (taken from 150 cm away). Testing videos have the same dimensions, namely 640 × 480. The video consists of two or more signs that are displayed consecutively, forming a series of letters according to the target letters of the video.

The number of objects that will come out as detection is limited to 1 with the highest confidence. The detection of signs in a sequence reaching the specified limit will be considered as a recognized sign and will be compared with the ground truth provided.

FPS calculation is done by calculating the time needed to perform signal recognition in one frame, then counting how many frames can be recognized in one second.

The results of testing on video data show a precision value of 80.56%, a recall of 93.1%, an accuracy of 72.97%, and an F1 score of 84.38%. Details of the experiment results is given in Table 6.

From the test results shown in Table 6, two errors were found. The first error is a class prediction error, for example, the letter 'N' in the 'NICE' target is known as' M ', also the' H 'in the' SHIZA 'target is not known because the prediction results obtained keep changing between' H 'and' D 'without successive predictions of reaching the specified streak limit. The second error is, sign recognition which occurs during the transition from one sign to another, for example, the prediction of the sign 'I' on the target 'QUEST' during the transition between signs 'E' to 'S' for 6 consecutive times.

The prediction error occurs is estimated for class N and M because they have similar terms, as well as P with D, U with V, I with L at a distance, and W with A at a distance.

For video detection, the speed ranges from 8-9 FPS for video input with dimensions of 640 × 480. 8 FPS speed cannot process all live input frames, which are generally 20-30 frames per second, in real-time. Therefore, for real-time usage, it is necessary to select a frame that will be recognized so that the frame recognized by the system is the newest frame that has just been obtained from the input. For example, if the system has a speed of 8 FPS and an input of 24 FPS, the frames to be detected are multiples of 3. With 8 frames detected every second and streak limit of 7, real-time sign language recognition can be done with sign language approaching a natural speed whereby the sign is carried out for about 1 second for each letter.

5. Conclusion

In this research, a recognition system for The Indonesian Sign Language using YOLO has been implemented. The experiment on image and video data got 100% and 72.97% accuracy, respectively. Recognition of the transition frames between one sign to another on video data contributes to the misrecognition error. An algorithm to distinguish a transition frame and a sign frame should be considered in the future to improve the accuracy of the system. With the processing speed of 8 fps, real-time recognition of video data can be performed by adjusting the speed of the recognition process and frames rate.

References

[1] Zafrulla Z, Brashear H, Yin P, Presti P, Starner T and Hamilton H 2010 American sign language phrase verification in an educational game for deaf children *2010 20th Int. Conf. on Pattern Recognition* pp 3846–9

[2] Oz C and Leu M C 2011 American Sign Language word recognition with a sensory glove using artificial neural networks *Eng. Applications of Artificial Intelligence* **24** pp 1204–13

[3] Lang S, Block M and Rojas R 2012 Sign language recognition using kinect *Int. Conf. on Artificial Intelligence and Soft Computing* pp 394–402

[4] Aryanie D and Heryadi Y 2015 American sign language-based finger-spelling recognition using k-Nearest Neighbors classifier *2015 3rd Int. Conf. on Information and Communication Technol. (ICoICT)* pp 533–6

[5] Garcia B and Viesca S A 2016 Real-time American sign language recognition with convolutional neural networks *Convolutional Neural Networks for Visual Recognition* **2** pp 225–32

[6] Redmon J, Divvala S, Girshick R and Farhadi A 2016 You only look once: Unified, real-time object detection *Proc. of the IEEE Conf. on Computer Vision and Pattern Recognition* pp 779–88

[7] Chen R C 2019 Automatic license plate recognition via sliding-window darknet-YOLO deep learning *Image and Vision Computing* **87** pp 47–56

[8] Qu H, Yuan T, Sheng Z and Zhang Y 2018 A Pedestrian detection method based on YOLOv3 model and image enhanced by Retinex *2018 11th Int. Congress on Image and Signal Processing, BioMedical Eng. and Informatics (CISP-BMEI)* pp 1–5

[9] Zaki P S, William M M, Soliman B K, Alexsan K G, Khalil K and El-Moursy M 2020 Traffic signs detection and recognition system using deep learning *Preprint arXiv:2003.03256*
[10] Redmon J and Farhadi A 2018 Yolov3: An incremental improvement *Preprint arXiv:1804.02767*
[11] Valdez P 2020 Apple defect detection using deep learning based object detection for better post harvest handling *Preprint arXiv:2005.06089*
[12] Peduli Kasih ABK (2018) *Mengenal Bahasa Isyarat* Retrieved on 18 December 2019 from https://www.ypedulikasihabk.org/2018/11/09/mengenal-bahasa-isyarat/

Handwritten Text Recognition using Fully Convolutional Network

Dewi Ayu Nirmalasari, Nanik Suciati*, Dini Adni Navastara

Department of Informatics, Faculty of Intelligent Electrical and Informatics Technology, Institut Teknologi Sepuluh Nopember, Surabaya, Indonesia

E-mail: *Corresponding author's e-mail address: nanik@if.its.ac.id

Abstract. Handwritten text recognition from images is challenging because there are many variations in handwriting as each person has a different writing style. This research implements multilevel recognition to solve this problem. In the first level, a Lexicon Convolutional Neural Network (CNN) model is used to recognize words (containing a set of word) that often appear in the text. If a word is not recognized by the Lexicon CNN model, which is designed for a limited number of words, then it goes to the next level. A character sequence recognition consisting of predicting the number of characters using CNN, cropping each character using a sliding window, and character recognition using Fully Convolutional Network (FCN), is applied at the second level. Experiments show that the system performance is promising. The experiment conducted using NIST Special Database 19 as a training dataset and a handwritten text on screen as a testing dataset. The best accuracy of word recognition, character number prediction, and character recognition is 99.98%, 98.56%, and 83.52%, respectively.

1. Introduction

Along with the development of technology, handwriting activities have not been abandoned. This is because handwriting activities are still easier, more efficient, and cheaper. Moreover, many important documents still need to be filled out manually and need to be digitized. Automatic recognition of handwritten text images is difficult because there are many variations in writing styles. Letters are often different even though the same person writes them.

Convolutional Neural Network (CNN) is often used in the image recognition process. CNN itself has many modifications, one of which is the Fully Convolutional Network (FCN). FCN still uses the convolution layer when predicting, however FCN does not depend on image size.

There are several deep learning approaches to recognize handwritten images. One of these approaches is to create a word image recognition system by training a system with word images [1]. This system itself will later be able to produce highly accurate recognition, but it is limited to word selection in the training data. In this approach, a large enough dataset is needed because of its dependence on word images.

Another approach is a character by character recognition of handwritten images [2]. This method itself is still a challenge because of the frequent errors in reading the characters from an image. Common mistakes are missing words or duplicated characters. Another challenge with word recognition is the difference in the width of the characters in one image. In this case, the use of FCN will be advantageous because it can accept input without any length limitation [3].

Based on those things, this research implements multilevel recognition system to solve the problem of handwritten text recognition. The first level recognizes word image using the Lexicon CNN model trained on a predefined word dictionary. When the system is unable to provide convincing recognition results, the word image will be recognized character by character using FCN.

2. Literature Review

Multiple works have been done to recognize handwriting text. Doetsch et al. [4] used RNN-HMM for recognizing offline handwriting. HMM is applied on training data to get frame-wise labeling, while the result of HMM will be feed into RNN. Wue et al. [5] used the character level convolutional method in feed-forward and RNN for recognizing handwritten Chinese character. In this experiment, it showed an improvement when combining neural network with back-off N-gram language models. Xie et al. [6] used CNN and multilayer LSTM for recognizing handwritten Chinese character. Other works of recognizing text from a natural image have shown using CNN to do feature extraction RNN give impressive results, such as Yin et al [7], and Su and Lu [8]. Wigington et. al [9] used bidirectional LSTM with normalization and augmentation. Pharm et. al [10] proposed multidimensional RNN with dropout to prevent over-fitting on the training dataset.

FCN method [11] takes an arbitrary size image and output region level classification. It benefits this research because the handwritten word has a different length. Bai et al. [12] using FCN and it shows it can outperform RNN on sequence modelling problem.

3. Our Approach

The system will perform handwritten text recognition using three neural networks. The three neural networks are Lexicon CNN, Character Count CNN, and Character Prediction FCN. Initially, after the image is pre-processed, the image will enter the word recognition process from image on a predefined dictionary. This process is called the Lexicon CNN. When the Lexicon CNN prediction results have confidence that exceeds the specified limits, the Lexicon CNN prediction results are recognized as a whole system output.

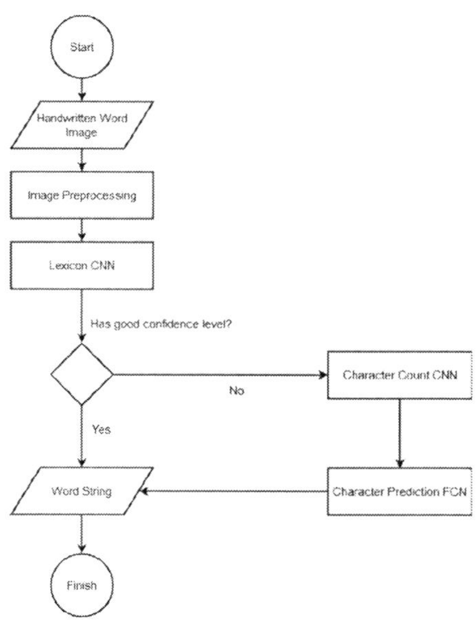

Figure 1. Flowchart of System

However, when the Lexicon CNN has low confidence, the image will be forwarded to the Character Count CNN. The Character Count CNN will provide numeric output that represents amount of characters in the word image. Later, before the image enters the Character Prediction FCN, the image will be resized to a size of 32x16 N, where N is the output of the Character Count CNN. In Character Prediction FCN, the image will be read using a sliding window and a series of prediction results that have high confidence will be said. This complete flow can be seen in Figure 1.

3.1. Lexicon CNN

Lexicon CNN is a word image recognition process based on predefined word dictionary that used in training process. These words are taken from the 50 words that appear most frequently in Brown corpus and 50 words that appear most frequently in Gutenberg corpus. Brown and Gutenberg are corpus provided by NLTK (Natural Language Toolkit), one of the famous libraries for natural language [13]. Brown corpus has 15 categories, 500 articles and 1.15 million words in English. Gutenberg Corpus contains approximately two million words from 18 e-books.

Fifty words from each of corpus are combined to create a new dictionary containing 59 different English words. In the training process, Lexicon CNN uses synthetic training data by combining the characters on the NIST Special Database 19 (SD19) into words. NIST (SD19) is a dataset of numeric and alphabetic characters, both upper and lowercase letters. The dataset itself is drawn from 3,669 handwritten forms and has a total of 814,225 characters [14].

Lexicon CNN architecture itself can be seen in Figure 2. Max pooling has a stride of 2x2. All layers use ReLU. Lexicon CNN training process uses batch size 64 with 30 epochs. In this training process, training data is divided into 9:1 with validation data. In this process, 1,000 images for 59 classes were created so that there were 59,000 training data on each word list.

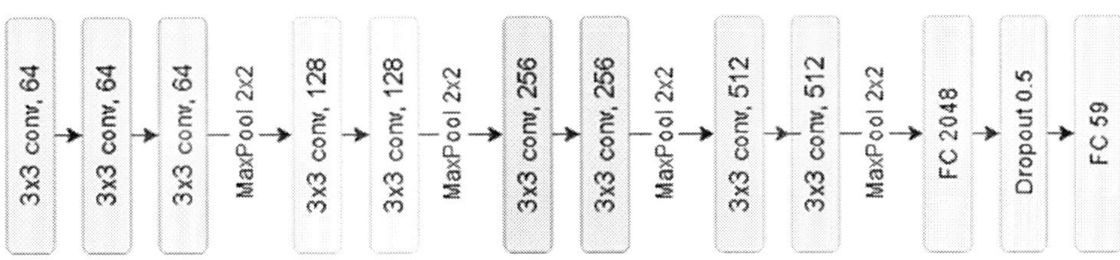

Figure 2. Architecture of Lexicon CNN

3.2. Character Count CNN

Character Count CNN works to recognize the number of characters in word image. In Character Count CNN, there are 16 classes, which represent the number of characters from 1 to 16. Character Count CNN also use synthetic word image as training data as in Lexicon CNN. In Character Count CNN, it uses words from a randomize characters so it may have no meaning. The characters used in making this word string are taken from 62 classes that will be recognized by the system, which can be seen in Figure 3.

Character Count CNN architecture itself can be seen in Figure 4. Max pooling has a stride of 2x2. Unlike Lexicon CNN, convolutional layer in Character Count CNN uses maxout. Character Count CNN training process uses batch size 32 with 50 epochs. In this training process, training data is divided into 9:1 with validation data. In this process, 1,000 images were also created for 16 classes so that there were 16,000 training data.

0	1	2	3	4	5	6	7	8
9	A	B	C	D	E	F	G	H
I	J	K	L	M	N	O	P	Q
R	S	T	U	V	W	X	Y	Z
a	b	c	d	e	f	g	h	i
j	k	l	m	n	o	p	q	r
s	t	u	v	w	x	y	z	

Figure 3. Character Classes to be recognized

3.3. Character Prediction FCN

Character Prediction FCN works to recognize word images by reading character by character in the image using a sliding window. Recognizable characters can be seen in Figure 3. The training data image used is the character image on NIST SD19.

Later in the testing process, the input word image will have a size of 32x16N where *N* is the output of the Character Count CNN. Character Prediction FCN architecture can be seen in Figure 5. Max pooling has a stride of 2x2. All layers use ReLU. Character Prediction FCN training process uses batch size 64 with 50 epochs. In this training process, training data is divided into 9:1 with validation data. In each class, 1,467 images were used so that there were 90,954 training data images.

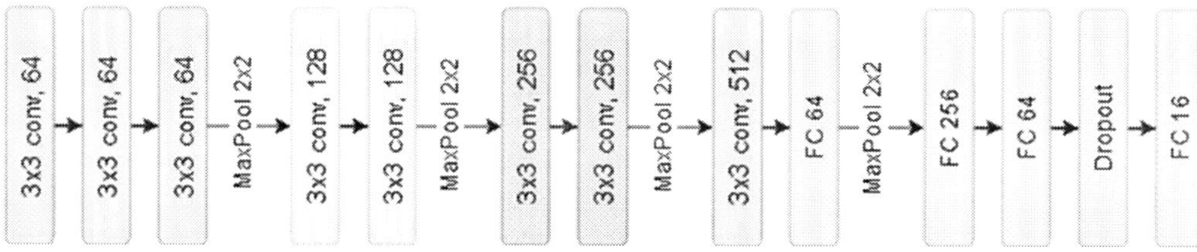

Figure 4. Architecture of Character Count CNN

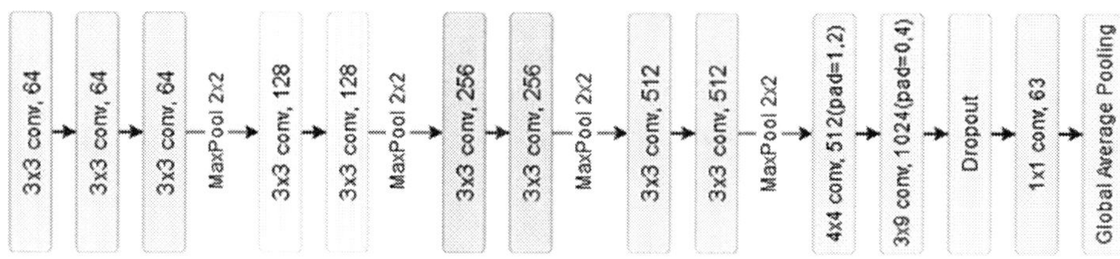

Figure 5. Architecture of Character Prediction FCN

3.4. Evaluation

The evaluation process is carried out on each neural network and as a whole system. The evaluation process for each neural network is carried out using validation data from each neural network. For the evaluation of a single system, the evaluation uses test data written by 10 different people who each write 60-62 English words. This word image is written on the touch screen using black ink with different thicknesses and levels of transparency and has a white background. These images then go through a selection process in which the image with overlapping characters will be deleted from the test data.

In these four evaluations, the evaluation process is carried out by measuring accuracy, precision, recall, and F1-score. In Lexicon CNN and the whole system, the evaluation process is also carried out by calculating the character error rate (CER) to measure how far the word image read error. The CER equation can be seen in Equation (1) [3].

$$CER = \frac{R + D + I}{R + D + I + C} \quad (1)$$

where *R* is the number of characters that have changed, *D* is the number of characters that are deleted, *I* is the number of characters that are increased and *C* is the number of correct characters.

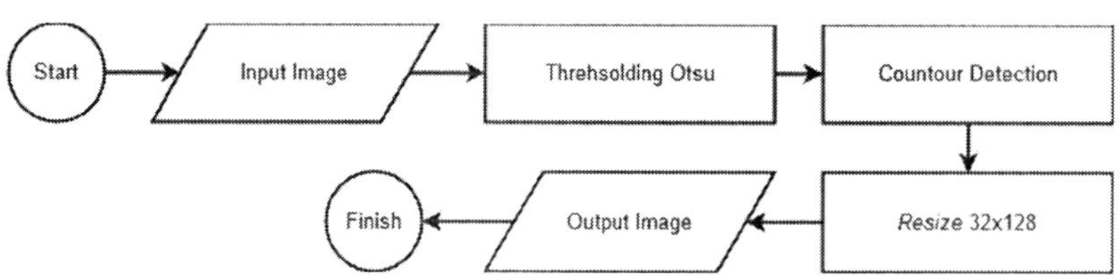

Figure 6. Flowchart of Pre-processing Testing Image

Before the evaluation process is carried out, the test data image will go through a pre-processing first so that the system can better recognize the word image. The pre-processing can be seen in Figure 6. This pre-processing is carried out to improve the ink used in writing, which is sometimes more transparent, by using Otsu thresholding. After that the word image will also be cut right into the word bounding box and then resized to 32x128, adjusted to the input size that has been installed on the neural network.

4. Results

4.1. Parameter Testing on Lexicon CNN

At Lexicon CNN, three trial scenarios will be carried out, namely testing on the regularizer type, testing on the optimizer type, and testing on the learning rate. The trials will be carried out on the Lexicon CNN on both Brown and Gutenberg word lists. The regularizers tested were L1 and L2. The optimizers tested were Adam, SGD, and RMSProp. The learning rates tested were 0.01, 0.001, and 0.0001. All these evaluation result will be shown in Table 1. In Lexicon CNN system, we use parameter that has best accuracy, which is using L2 regularizer, Adam optimizer, and a learning rate of 10e-4 in recognizing images based on a word dictionary. The best accuracy resulting from the use of parameters is 99.98% and the average CER is 0.01%.

Table 1. Evaluation Result on Lexicon CNN

	Accuracy (%)	Precision (%)	Recall (%)	F1-score (%)	CER Avg (%)
Regularizer					
L1	99.76	99.77	99.76	99.76	0.14
L2	**99.93**	**99.93**	**99.93**	**99.93**	**0.05**
Optimizer					
Adam	**99.93**	**99.93**	**99.93**	**99.93**	**0.05**
SGD	99.81	99.82	99.81	99.81	0.01
RMSProp	99.78	99.79	99.78	99.78	0.02
Learning Rate					
10e-3	99.93	99.93	99.93	99.93	0.05
10e-4	**99.98**	**99.98**	**99.98**	**99.97**	**0.01**
10e-5	99.97	99.97	99.97	99.97	0.03

4.2. Parameter Testing on Character Count CNN

In Character Count CNN, five trial scenarios will be carried out, namely testing the architecture, testing the regularizer type, testing the optimizer type, testing the learning rate, and testing the dropout. Architectural testing was carried out by comparing the performance of the character count on Lexicon CNN architecture and Character Count CNN architecture. The regularizers tested were L1 and L2. The optimizers tested were Adam, SGD, and RMSProp. The learning rates tested were 0.01, 0.001, and 0.0001. The dropouts tested were 0.1, 0.3, 0.5, 0.7, and 0.9. All these evaluation result will be shown in Table 2. In Character Count CNN, we use parameter that has best accuracy, which is using architecture same with Lexicon CNN, L2 Regularizer, Adam optimizer, learning rate 10e-4, and dropout 0.1.

The system is more able to provide good accuracy when using the same CNN architecture as the Lexicon CNN. This shows that the maxout activation function processes too many neurons, causing overfitting and causing the convolution process to take more time. The use of ReLU is considered sufficient because ReLU will give a value to all neurons that have a negative activation function, so that these neurons are considered inactive. In Character Count CNN, the best accuracy is obtained when using the L2 regularizer with Adam optimizer, learning rate 0.001 and dropout 0.1.

4.3. Parameter Testing on Character Prediction FCN

In the Character Prediction FCN, three trial scenarios will be carried out, namely testing the optimizer type, testing the learning rate, and testing the dropout. The optimizers tested were Adam, SGD, and RMSProp. The learning rates tested were 0.01, 0.001, and 0.0001. The dropouts tested were 0.3, 0.5, 0.7, and 0.9. All these evaluation result will be shown in Table 3.

In Character Prediction FCN, we will use parameter that has best accuracy, which is using Adam optimizer, a learning rate of 10e-4 with an L2 regularizer and a dropout of 0.9. The best accuracy is 83.52%. In contrast to the CNN Character Count, the smaller the dropout, the better the accuracy will be for the model. The character prediction accuracy is not high enough because FCN has not been able to distinguish uppercase, lowercase and numbers properly.

Table 2. Evaluation Result on Character Count CNN

	Accuracy (%)	**Precision (%)**	**Recall (%)**	**F1-score (%)**
		CNN Architecture		
Count Character CNN	6.25	0.39	6.25	0.74
Lexicon CNN	**96.25**	**96.33**	**96.25**	**96.25**
		Regularizer		
L1	96.06	96.20	96.06	96.06
L2	**96.25**	**96.33**	**96.25**	**96.25**
		Optimizer		
Adam	**96.25**	**96.33**	**96.25**	**96.25**
SGD	70.31	69.74	70.31	69.61
RMSProp	96.06	96.12	96.06	96.06
		Learning Rate		
10e-3	96.25	96.33	96.25	96.25
10e-4	**97.75**	**97.76**	**97.75**	**97.74**
10e-5	59.00	59.72	59.00	58.22
		Dropout		
0.1	**98.56**	**98.57**	**98.56**	**98.56**
0.3	98.19	98.21	98.19	98.18
0.5	97.75	97.76	97.75	97.74
0.7	95.69	95.97	95.69	95.97
0.9	94.19	94.26	94.19	94.18

4.4. Parameter Testing on the Whole System

In testing one whole system, three test scenarios will be carried out, namely testing the use of Lexicon CNN, testing on Lexicon CNN confidence levels, and testing on the sliding windows' size. All these evaluation results will be shown in Table 4.

In Character Prediction FCN, the best accuracy is generated using the Adam optimizer, a learning rate of 0.001 with an L2 regularizer and a dropout of 0.9. The best accuracy is 83.52%. In contrast to the Character Count CNN, the smaller the dropout, the better the accuracy will be for the model. The character prediction accuracy is not high enough because FCN has not been able to distinguish uppercase, lowercase and numbers properly.

Table 3. Evaluation Result on Character Prediction FCN

	Accuracy (%)	**Precision (%)**	**Recall (%)**	**F1-score (%)**
		Optimizer		
Adam	**82.55**	**83.81**	**82.55**	**82.29**
SGD	81.87	82.36	81.87	81.79
RMSProp	81.45	81.92	81.45	81.37
		Learning Rate		
10e-3	82.55	83.81	82.55	82.29
10e-4	**82.86**	**83.48**	**82.86**	**82.78**
10e-5	82.29	82.96	82.29	81.96
		Dropout		
0.3	81.85	84.98	82.85	82.54
0.5	82.86	83.48	82.86	82.78
0.7	82.95	83.51	82.95	82.90
0.9	**83.52**	**84.42**	**83.52**	**83.29**

Table 4. Evaluation Result on Overall System

	Accuracy (%)	**Precision (%)**	**Recall (%)**	**F1-score (%)**	**CER Avg (%)**
		Lexicon CNN Usage			
Without Lexicon CNN	4.43	12.36	4.43	5.49	63.08
Using Lexicon CNN	**57.31**	**83.17**	**57.31**	**64.98**	**29.86**
		Sliding Windows' Size			
32x16	**57.31**	**83.17**	**57.31**	**64.98**	**29.86**
32x32	56.91	82.59	56.98	64.67	35.19
		Confidence Level Limit			
65%	**68.80**	**83.45**	**68.80**	**73.97**	**22.25**
70%	57.31	83.17	57.31	64.98	29.86
75%	46.80	79.65	46.80	55.60	36.81

5. Conclusion

Based on the trials that have been carried out, we can obtain many information. First, based on the test parameters on Lexicon CNN, the model that was built produced the best accuracy of 99.98% with an average CER of 0.01%. The best model was obtained with L2 regularizer, learning rate 0.001, and using Adam optimizer. Second, based on the test parameters in the Character Count CNN, the model that was build has the best accuracy of 98.56%. This model was obtained using L2 regularizer, same architecture with Lexicon CNN, learning rate 0.001, Adam optimizer, and dropout 0.1. Third, in Character Prediction, the model that was built produced the best accuracy of 83.52% for the introduction of 62 classes. The FCN Character Prediction Model has not been able to predict characters well, because it is still unable to distinguish between uppercase and lowercase letters and numbers. The best model was built using Adam optimizer, learning rate 0.001 and dropout 0.9. Fourth, based on testing on a single system, the system is only able to produce an accuracy of 68.80% with an average CER of 22.25%. This was built using best model on Lexicon CNN, Character Count CNN, and Character Prediction FCN. This shows that the sliding windows algorithm has not been able to minimize the addition, replacement, and deletion of characters in reading handwritten images. In addition, here it can be seen that the recognition based on a word dictionary is indeed very helpful in handwritten text recognition. From all of these trials, we can see that Lexicon CNN helps a lot on handwritten recognition while we need to improve character prediction and the reading algorithm for every character on image. Moreover, there are some future works that need to be explored in the handwritten text recognition problem, such as how to handle text images with various colour ink, reading offline handwritten text, and automatic segmentation of word or letter in a sentence or paragraph image.

References

[1] Stuner B, Chatelain C and Paquet T 2017 LV-ROVER: Lexicon Verified Recognitzer Output Voting Error *Preprint ArXiv abs/1707.07432*

[2] Sueiras J, Ruiz V, Sanchez A and Velez J F 2018 Offline continous handwriting recognition using sequence to sequence neural networks *Neurocomputing* **289** pp 119–28

[3] Ptucha R, Such F P, Pilla S, Brockler F, Singh V and Hutkowski P 2018 Intelligent character recognition using fully convolutional neural network *Pattern Recognition* **88** pp 604–13

[4] Doetsch P, Kozielski M and Ney H 2014 Fast and robust training of recurrent neural networks for offline handwriting recognition *2014 14th Int. Conf. on Frontiers in Handwriting Recognition, IEEE* pp 279–84

[5] Wu Y-C, Yin F and Liu C-L 2017 Improving handwritten chinese text recognition using neural network language models and convolutional neural network shape model *Pattern Recognition* **65** pp 251–64

[6] Xie Z, Sun Z, Jin Z, Feng Z and Zhang S 2016 Fully convolutional recurrent network for handwritten Chinese text recognition *23rd Int. Conf. on Pattern Recognition (ICPR), IEEE* pp 4011–6

[7] Yin F, Wu Y-C, Zhang X-Y and Liu C-L 2017 Scene Text Recognition with Sliding Convolutional Character Models *Preprint arXiv:1709.01727v1*

[8] Su B and Lu S 2017 Accurate recognition of words in scenes without character segmentation using recurrent neural network *Pattern Recognition* **63** pp 397–405

[9] Wigington C, Stewart S, Davis B, Barrett B, Price B and Cohen S 2017 Data augmentation for recognition of handwritten words and lines using a cnn-lstm network *2017 14th IAPR Int. Conf. on Document Analysis and Recognition (ICDAR), IEEE* pp 639–45

[10] Pharm V, Bluche T, Kermovant C and Louradour J 2014 Dropout improves recurrent neural networks for handwriting recognition *2014 14th Int. Conf. on Frontiers in Handwriting Recognition* 285

[11] Shelhamer E, Long J and Darrel T 2015 Fully convolutional networks for semantic segmentation *2015 IEEE Conf. on Computer Vision and Pattern Recognition (CVPR)*

[12] Bai S, Kolter J Z and Koltun V 2018 An Empirical Evaluation of Generic Convolutional and Recurrent Networks for Sequence Modeling *Preprint arXiv: 1803.01271*

[13] Bird S, Loper E and Klei E 2009 *Natural Language Processing with Python* (California: O'Reilly Media Inc) chapter 1 p 3

[14] National Insitute of Standards and Technology, US Department of Commerce 2019 *NIST Special Database 19* DOI: 10.18434/T4H01C

IOP Conf. Series: Materials Science and Engineering 1077 (2021) 012031 doi:10.1088/1757-899X/1077/1/012031

Development of Unified Neuro-Affective Classification Tool (UNACT)

Md Farhad Hossain[1], Hamwira Yaacob[2], Azlin Nordin[3]

[1,2,3]Department Computer Science, International Islamic University Malaysia, Jalan Gombak, 53100, Selangor, Malaysia

E-mail: [1]emamulfarhat@gmail.com

Abstract. Brain signals have been analysed to understand the affective state of different cognitive and mental conditions. For example, through the analysis, we can visualize the changes of emotion while driving, identify an autistic kid, understand the conditions that stimulate attention while studying, and many more, because emotion has a strong impact on cognitive processes in humans' activities. This can be done through a machine learning technique, which includes data acquisition, pre-processing, feature extraction, and training. However, no existing tool integrates all supervised machine learning processes for affective state classification, which makes the process tedious and time-consuming for an analyst by doing programming. Therefore, this project aims to develop a brain analysis tool, namely Unified Neuro-Affective Classification Tool (UNACT). It consists of 3 main functions including training, classifying, and analysis. In the study of affective state electroencephalogram (EEG) signals have used, which measures brain signals. UNACT uses the Butterworth Bandpass filter for EEG signal filtering, the Power Spectral Density method for feature extraction, and the Multi-layer perceptron (MLP) for emotion classification. This tool can be used by a non-technical person to perform affective-emotional state analysis without having programming knowledge.

1. Introduction

Emotions play a crucial role in our life. It helps us to think and behave. Emotion regulation includes a variety of psychological processes including attitude, personality, mood, and motivation. The emotional effect on the learning process has been found to have a positive or negative influence on cognitive abilities for storing and retrieving knowledge, processing information and motivational factors, among many others [1]. That's' why recent days, there are growing efforts for technical and scientific advances in neuroimaging to understand neural processes associated with brain functions for clinical evaluation. For this purpose, numerous experimental models and techniques have been proposed to understand the affective-emotional state from brain signals. This will be helpful in biomedical research, mainly for the treatment of psychological disorders.

Brain signal can be measured through EEG signals. The study of brain-computer interfaces (BCI) has shown that emotions can be measured through EEG emotional recognition systems [2]. EEG signals generate significant temporal resolutions compared to other instruments. Brain waves can be inputs to calculate valence and arousal. All EEG channels can capture the emotional experience.

Emotions are complex and progressive, as shown by the valence degree and the arousal strength [5]. One of the recognized models for quantifying emotion is the circumplex model [4]. This representation is known as the Affective Space Model (ASM). ASM places an emotion as a point on a two-dimensional plane: (i) valence (x axis) and (ii) arousal (y axis). The representation of the ASM is shown in Figure 2. In the study of affective-emotional profiling through EEG signals by applying a supervised machine learning technique, which includes data collection, pre-processing, extraction of features and modelling. EEG signals are recorded during emotion stimulation activities for the data collection process. Preprocess includes filtering signals for removing noise. After that feature extractions for performing the trained model.

There are many techniques and existing tools to analyze EEG signals. Although different techniques applied for emotion profiling, but no existing tool analyze EEG signals for emotion profiling. Analysts need to program each time for analysis, which makes time-consuming and tedious. It is also very difficult for the non-programmer analyst to perform emotion affective-emotional state analysis. Thus, this project aims to develop a brain analysis tool, namely Unified Neuro-Affective Classification Tool (UNACT). UNACT is an integrated analysis tool consisting of EEG signal processing, feature extraction, supervised machine learning for training and affective state classification from EEG signals. The preprocessing involves the filtering of EEG signals using the Butterworth bandpass method. Then, power spectral density (PSD) used for feature extraction. PSD is calculated by using a short-time Fourier transform to take the average spectral power. Multilayer perceptron (MLP) used to conduct emotional state classification from a collection of input traits derived from EEG signals. Classification of emotions is measured by the valence and arousal. The performance is described in an affective space model (ASM) with four quadrants.

2. Review of previous works

There are several contributions to emotion profiling using supervised machine learning techniques. In [5], the authors illustrated two-dimensional affective state distribution of the brain during emotion stimulation. EEG data recording involved 6 children. Photos taken from the Radboud Faces Database (RafD) for stimulation of emotions [6] for the basic emotions. Preprocessing involves noise filtering using a bandpass filtering. The theta and alpha bands were extracted for emotion classification. Kernel Density Estimation (KDE) used for feature extraction and a multilayer perceptron (MLP) algorithm for emotion classification. Their findings prove that all basic emotional states are well distributed throughout the respective quadrants in ASM where the classifier signifies the arousal and valence. The findings also showed the dynamics of emotional responses described by the arousal and valence.

Then, the paper [7] compares the results of emotion profiling on original interrupted and adjusted EEG signals by casting using a supervised machine learning technique. The signals collected for 64 channels using A.N.T. EegoSport by following the international standard 10-20 EEG location of the electrode [8]. Power spectral density applied for feature extraction. A short-time Fourier transform used to take the average spectral power to calculate PSD. Emotion classification involved a multilayer perceptron (MLP) algorithm. Classification of emotions calculated on the basis of valence and arousal. The findings are shown in the Affective Space Model. This paper proved that the proposed technique is useful for emotion profiling using interrupted EEG signals because the accuracy increased in both situations with the total variant segments from 1 to 19 segments with 400 seconds of EEG collection.

Additionally, in [9] illustrated, multiple approaches for EEG signal collection during emotion stimulation, then various features extraction strategies and classification models for affective-emotional profiling. Furthermore, besides the discussed papers over the past few years a various new EEG signal processing methods proposal increasing for affective-emotional state analysis. In order to support the proposed approaches, new tools are needed to enable the comprehensive analysis of EEG data for affective state analysis using more sophisticated experimental designs. However, there are several EEG signal processing tools. The most widely used tool is EEGLAB [10] which was introduced in 2001. EEGLAB is an integrated graphical user panel and a script-based command-line tool [11]. It was implemented in MATLAB. Since EEGLAB is a collection of MATLAB functions, it needs MATLAB runtime and some understanding of MATLAB data structures. Reprogramming systems are not easy to use in EEGLAB. Its' features include (i) structure of events and functions to import, edit, and manipulate, (ii) degradation of electroencephalographic data by independent component analysis (ICA),

(iii) adaptability ready for anyone with multi-levels of sophisticated programming, (iv) enable any user to produce and modify plug-in functionality that appear automatically in the EEGLAB menu windows [11].

The NFT (Neuroelectromagnetic Forward Head Modeling Toolbox) [12] is accessible as an EEGLAB plug-in through the EEGLAB graphical interface [10]. The NFT software handles almost all of the functions required to produce a practical head model from magnetic resonance (MR) and/or measured EEG sensor coordinate, and provide advanced boundary element method (BEM) and finite element method (FEM) solvers to estimate the projected scalp fields for a given range of possible brain source areas, thereby estimating the EEG modeling solutions [12]. The following measures are performed by NFT: (i) segmentation of MR images, (ii) high-quality head models, (iii) deforming the template header model, (iv) electrode locations registration with the mesh of the head and (v) an effective and efficient forward-looking approach to the dilemma.

The SIFT (Source Information Flow Toolbox) is a toolbox for visualization and modeling data flow between EEG data points and is likely to be divided (immediately) into the full independent process using ICA. Currently, the toolbox consists of four main phases, (i) preprocessing, (ii) fitting of the model and estimating of connectivity, (iii) statistical analysis, and (iv) visualization. An important aspect of SIFT is that it focuses on predicting and visualizing an efficient multivariate connection within the source domain but also between scalp electrodes impulses. SIFT can help to locate temporary, network topology events that connect static element processes spatially [10].

The ERICA (Experimental Real-Time Interactive Control and Analysis) architecture is based on a particular data processing and real-time multiple synchronization infrastructures called DataRiver developed from the ADAPT data collection and stimulation control system. [13]. The key technology driving ERICA growth is the advancement of data acquisition through mobile brain imaging (MoBI) and methods of analysis [14] – simultaneous analysis of what the brain does, what the brain feels, and what the brain regulates while performing naturally inspired acts in ordinary 3-D activities. Another tool known as BCILAB [10] is a platform for the development of Brain-Computer Interfaces. Since data stream integration and pre-processing results have been achieved inside the ERICA framework, BCILAB is the open-source MATLAB toolkit and the EEGLAB plug-in can be used to support brain-computer interface analysis and, more generally, to develop, learn, use and test real-time prediction model that operates on the signal.

The existing tools are strong at their performance and fast analysis. However, this platform-dependent tool requires technical knowledge and expert level understanding to perform analysis. Therefore, the existing tools are not integrated with machine learning techniques for emotion profiling. For this purpose, the UNACT has developed to support analysts without having technical support for affective state analysis. By reviewing several papers and existing tools the current version of UNACT implemented using the Butterworth Bandpass filter, Power Spectral Density feature extraction technique, Multilayer Perception (MLP) algorithm as an emotion classifier and affective space model (ASM) to represent findings from the supervised machine learning technique.

3. Theoretical support for UNACT

3.1. Emotion profiling

Emotions describe human nature because humans' activities determine by emotions. A person how mentally strong can be identified by his emotional experience. An individual's activities towards society also can determine through emotional experience. We can feel the emotions through emotional experience because emotions link to our thought, actions, and belief. Different emotions are responsible for different actions. Emotions help us to survive. Take the example of cooking food. while people cook for, they become very conscious about cutting vegetables to avoiding cut hair and avoid burning from fire. Another example could be watching a horror movie. Although we are sitting in a safe place, we try to make ourselves strong-minded, sometimes people feel like hiding and very focused to control our body movement. Besides these many individuals are weak emotionally, by means, they are scared easily and may lead to death. While people are driving, they become more conscious to avoid accidents. Somehow the drivers may feel sleepy and cause accidents. Emotions are important in critical thinking because they can influence people's learning progress. There are different opinions on the number of basic emotions in research, as in paper [15-16] discussed different research about basic emotions theory. However, UNACT is working with four basic emotions happy, sad, calm, and fear. The affective-emotional profiling technique through brain waves by applying a supervised machine learning technique that includes data acquisition, preprocessing, feature extraction, and training illustrated in Figure 1. Brain signals can measure using the EEG signal. The recent version of UNACT does not integrate data acquisition techniques.

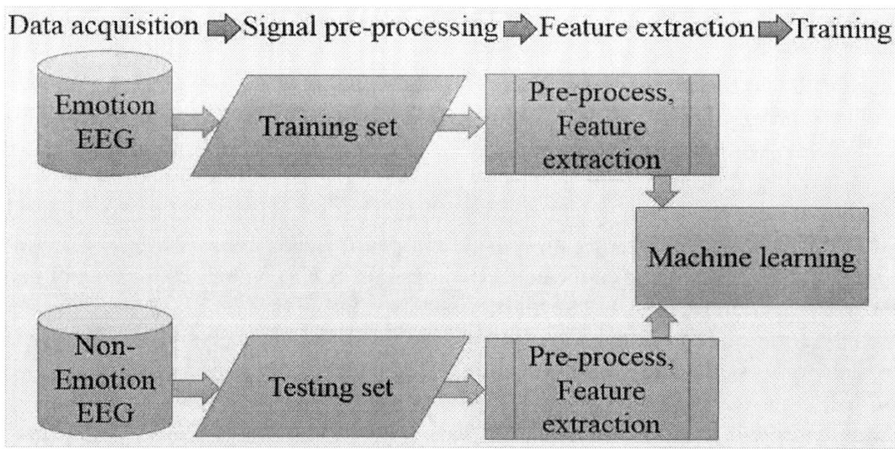

Figure 1. Emotion profiling

3.2. Affective Space Model (ASM)

Emotions are categorical, so every emotion is controlled by a distinct and autonomous neural [17]. Figure 2 illustrates Affective Space Model (ASM), it is a two-dimensional circumplex model of affect [4] to distinguish emotions by placing them in a dimension, focusing on individual assessment of specific emotional events.

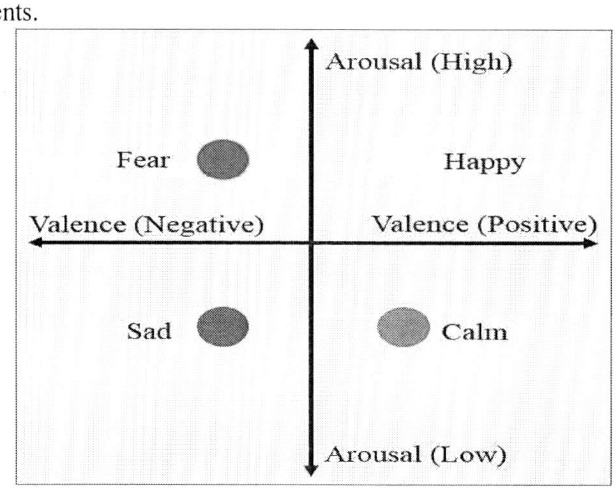

Figure 2. Affective Space Model (ASM)

ASM allows measuring emotions by mapping emotions on it, as illustrated in Figure 2. In [18] defined that the impact of arousal is the psychological-physiological state of being triggered and highly reactive to emotional responses. The physiological state involves high blood pressure and heart rate, among some other changes. The psychological state involves sensations of vigor, strength, and anxiety. Arousal is one dimension of emotion. To represent emotion in ASM needs another one dimension namely valence. Valence refers to positive or negative emotions. Positive emotions are joy, happiness, curiosity, interest, gratitude, excitement, and love. Whereas, negative emotions are like sadness, fear, anger, loneliness, jealousy, self-criticism. However, advanced neuroimaging technique uses brain waves to measure the values of arousal and valence. EEG tool allows measuring brain waves.

3.3. Electroencephalogram (EEG)

The electroencephalogram (EEG) is a neuroimaging tool to measure brain waves recorded by sensors on the scalp. It can capture electrical current when neurons are activated during brain activities. The EEG signals are high spatial resolution compared to other tools. EEG is known to be a secure and non-aggressive device [3][19]. The EEG sensor is an electrode placed on the top of the pericranium using a conductive gel. This helps to detect and direct the voltage to an amplifier. The electrodes are usually located in compliance with the international framework 10-20. The distance is specified between nasion, inion, and the latitude by the ten separations. The system faces an issue with ground circuit noise which is handled using differential amplifiers [21]. The brain waves measured in EEG are associated with the numerous oscillating sub-components and their subsequent breakdown. These are measured from peak value (between 0.5 v - 100 v amplitude). They are categorized as per their frequency, namely delta (0.5 Hz - 4 Hz), theta (4 Hz - 8 Hz), alpha (8 Hz - 13 Hz), beta (13 Hz - 30Hz) and gamma (> 30 Hz). However, theta and alpha bandwidths are commonly used to determine emotions [20-22].

3.4. Butterworth Bandpass Filter

The Butterworth filter is a signal processing filter to provide maximum flat frequency response in the pass-band. It is known as a maximally flat magnitude filter. The bandpass filter is used for wireless transmission and reception. It is used in a transmitter to limit the frequency of the output signal to the band assigned to the transmitter. This prohibits the transmitter to communicate with all other stations [23]. UNACT pre-processing involves filtering signals to remove noise and unnecessary artifacts. Bandpass filtering technique has been used to extract theta and alpha bands. Signal filtering involves a fifth-order Butterworth bandpass filter with a 4 Hz (lower frequency band) and a 13 Hz (upper-frequency band) cutoff frequency range.

3.5. Power Spectral Density

One of the most widely used feature extraction techniques is the power spectral density (PSD). The PSD is a measure of the power quality of the signal versus frequency. The power spectrum of the time series defines the power distribution to the signal frequency elements [24]. The Fourier analysis [25] defines that physical signals should be transformed into a variety of distinct frequencies or a frequency spectrum throughout a continuous range. As the signal energy is distributed across a limited interval of time, particularly when its total amount of energy is finite, the spectral energy density can be measured. The power spectral density applies to signals that have been existing for all time, or large enough over a time frame. Then, the PSD corresponds to the distribution of spectral energy that will be observed per unit time [24]. The power spectral density is used to describe the power of EEG signal components [26]. In UNACT, the PSD is used for the extraction of features. The PSD is calculated using the short-time Fourier transform (STFT) to take the average spectral power. The STFT is a Fourier-related transformation that used to evaluate the sinusoidal phase and frequency quality of the regional signal components as they travel over time [27].

3.6. Multi-layer perceptron (MLP)

Multilayer perceptron (MLP) is a feedforward artificial neural network (ANN). A single hidden layered MLP is known as the Vanilla neural network [28]. MLP uses a supervised machine learning approach known as back-propagation for instruction [29]. The MLP consists of a minimum three-node layer, namely an input layer (first layer), a hidden layer (second layer) and an output layer (third layer). The Input layer contains the same number of neurons and features. The second layer consists of a series of hidden layers, which is defined during the optimization process. MLP is being used for supervised machine learning, including emotion classification [5] [7] [30-33]. In UNACT, Multilayer perceptron (MLP) used to perform an emotional state classification based on the set of input features obtained from EEG signals. The classification of emotions is determined by the valence and the arousal.

4. Unified Neuro-Affective Classification Tool (UNACT)

4.1. Methods

UNACT is an affective-emotion profiling tool using the collected EEG signal. UNACT has developed under Agile software development methodology as illustrated in Figure 3. Every single part of UNACT design, coding and testing have done right after completion. If there were any changes required in terms of design and development, then done immediately.

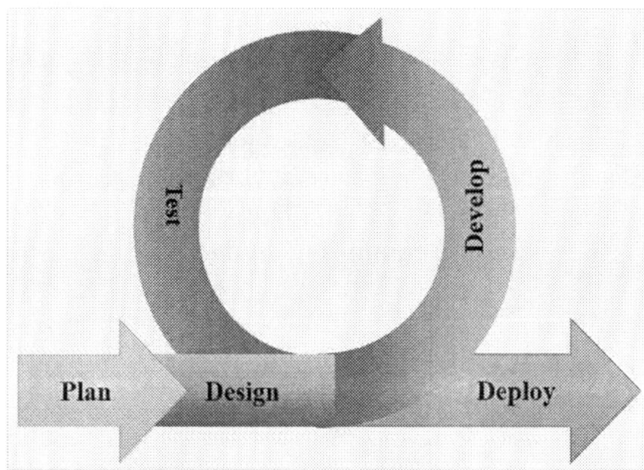

Figure 3. Agile methodology

The 4+1 [34] view model is designed to create a highly configurable UNACT architecture. This model includes a user view (Use case diagram), a logical view (Class diagram), a process view (Sequence diagram), a development view (Component diagram) and a physical view (Deployment diagram). Furthermore, the developed application possesses three modules includes (i) Modelling, (ii)Affective Classification and (iii) Analysis.

An important part of the system is the input and output design. UNACT only accepts the EEG data. Regarding training purposes, the collected EEG data should be in the sequence of Happy, Sad, Calm and Fear. Then, the data can be emotional or non-emotional EEG data to perform classification. Finally, the obtained data from the classification process should be used for analysis. UNACT produces the required output in every phase. In the modeling phase, the output is a trained valance, a trained arousal in the format of SAV. and emotion labeled data produced in CSV format. During the Classification phase, it produces only classified data in the format of CSV. Finally, the analysis part produces different required graphs (Arousal, Valence and Valence-Arousal) in the format of JPG.

UNACT integrated all supervised machine learning techniques for emotion profiling (described in section 3.1) except data acquisition (described in section 3.3) technique. Machine learning technique has been applied to the trained model, classify emotions and then visualize the classified emotions. Preprocessing purposes used the Butterworth bandpass filter technique (described in section 3.4) and used the Power Spectral Density (PSD) technique for the extraction of features. The Short-Time Fourier Transform used to take the average spectral power for PSD calculation (described in section 3.5). Then apply Multilayer perceptron (described in section 3.6) to conduct emotional state classification from a collection of input traits derived from EEG signals. Classification of emotions is measured by the valence and arousal (described in section 3.2). The performance is described in an affective space model (ASM) with four quadrants (described in section 3.2).

4.2. Tools

Python programming language has been used in the UNACT development. Python libraries used for system implementation. The libraries are NumPy, Pandas, SciPy, Sklearn, TkInter, and Matplotlib. However, NumPy, Pandas, SciPy, and Sklearn are used to applying machine learning technique. Matplotlib used for visualization, and TkInter used in the creation of user interfaces.

4.3. User Interface

UNACT user interfaces are illustrated in Figure 4 including popup windows of the respective phase.

Figure 4. UNACT user-interface

5. UNACT Evaluation

The UNACT produces several outputs in different phase as mentioned bellow.

1. Modelling:
   - ✓ Trained arousal and Trained valence in the format of SAV.
   - ✓ Emotion labeled data in the CSV format.
2. Affective classification:

- ✓ Classified data in CSV format
3. Analysis:
   - ✓ Graph for arousal and valence in JPG format- depicted in Figure 5.
   - ✓ Graph for valence (x-axis) vs arousal (y-axis) in JPG format- depicted in Figure 5.

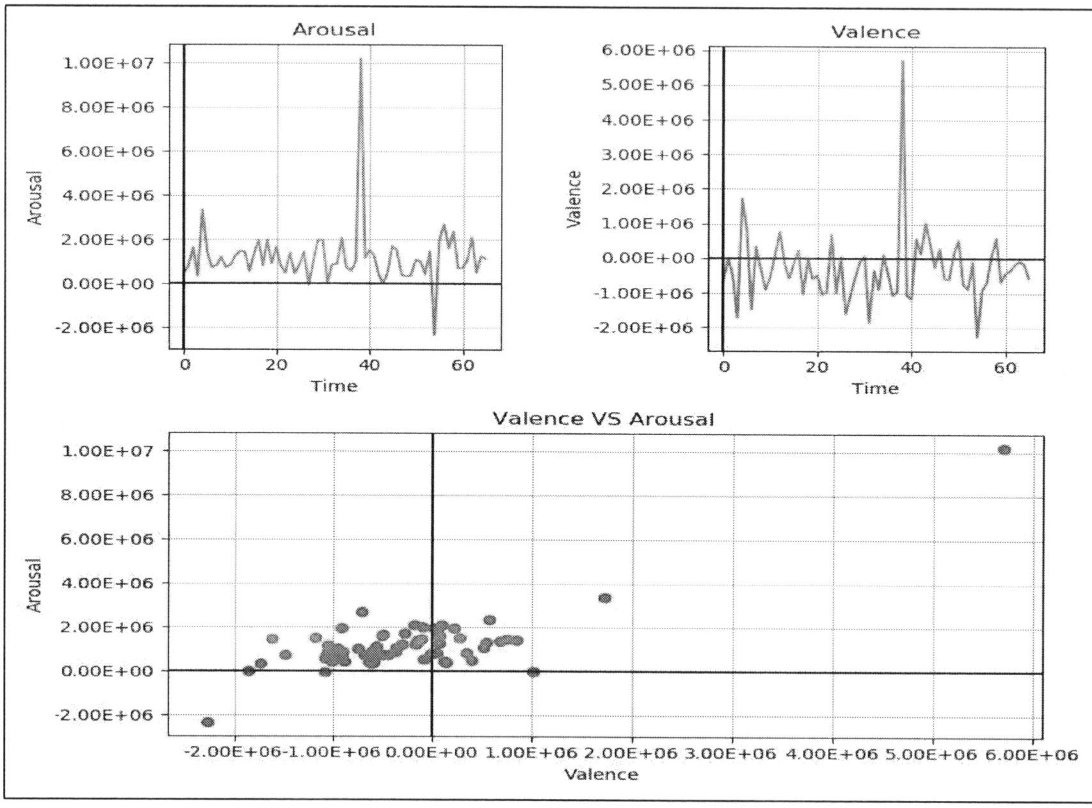

Figure 5. Graphical output from UNACT

The UNACT implementation has done under agile software development. The design, development (coding) and testing was repeated steps. While any block or part of the UNACT completed, it was tested. After the final product completion, the UNACT involved testing using EEG data (Collection process described in section 3.3). However, to compare the outcomes between the standard approach and the UNACT is used same Butterworth Bandpass filter, Power spectral density features, Multilayer perception algorithm. The output for valence classification accuracy is 88% and the arousal classification accuracy is 85%. The accuracy is slightly lower than a standard approach which was for valence is 89% and arousal is 88%. It may happen because of the PC power that the system utilizes for running the total process. The obtained results showed that UNACT is a real and reliable tool for the classification of emotions.

6. Conclusion

Emotion is a vital aspect of human behavior and attitude. Brain waves brain areas are known to be correlated with many cognitive and behavioral activities. Emotion can be used as a guide for those related activities. In order to analyze need tools which will save time and easy to use. UNACT is a tool that allows us to perform emotion profiling.

UNACT is user-friendly for non-technical persons. The user interface is simple, viable and easy-to-navigate for modeling, classification and analysis. Future work will focus on windowing and extension functionality for brain waves analysis using EEG signals including data acquisition technique, real-time EEG signal processing, multiple machine learning techniques to compare and get the best results. The UNACT aims to help in biomedical research because Neuro-Feedback therapy as an innovative procedure can help to recognize the basic emotional neuronal pathways by teaching the brain to regulate the actions of the brain and nervous system.

The result from UNACT Shows that it is a trustworthy tool to perform emotion profiling. However, future works will enhance UNACT as a self-analytic tool for the neurofeedback system.

References

[1] Pekrun R 1992 The impact of emotions on learning and achievement: towards a theory of cognitive/motivational media-tors *Appl. Psychol.* **41(4)** pp 359–76
[2] Murugappan M *et al* 2013 Time-Frequency Analysis of EEG Signals for Human Emotion Detection *Proc. Kuala Lumpur 4th Int. Conf. on Biomedical Engineering* ed R Magjarevic (Berlin, Heidelberg: Springer) **21** pp 262–5
[3] Sanei S and Chambers J A 2007 *EEG Signal Processing* 1st ed (New Jersey: Wiley-Interscience)
[4] Russell J A 1980 A circumplex model of affect *J. of Personality and Social Psychology* **39** p 1161–78
[5] Yaacob H, Karim I, Wahab A and Kamaruddin N 2012 Two-dimensional affective state distribution of the brain under emotion stimuli *34th Annual Int. Conf. of the IEEE EMBS San Diego, California, USA* pp 1–4
[6] Langner O, Dotsch R, Bijlstra G, Wigboldus D H J, Hawk S T and van Knippenberg A 2010 Presentation and validation of the Radboud Faces Database *Cognition and Emotion* vol **24** no 8 pp 1377–88
[7] Yaacob H, Omar H, Handayani D and Hassan R 2019 Emotional profiling through supervised machine learning of interrupted EEG interpolation *Int. J. of Advanced Computer Research* **9(43)** pp 242–51
[8] Niedermeyer E da Silva F L 2005 *Electroencephalography: basic principles, clinical applications, and related fields* (Amsterdam: Lippincott Williams & Wilkins)
[9] Alarcao S M and Fonseca M J 2017 Emotions recognition using EEG signals: a survey *IEEE Transactions on Affective Computing*
[10] Delorme A, Mullen T, Kothe C, Acar Z A, Shamlo N B, Vankov A and Makeig S 2011 EEGLAB, SIFT, NFT, BCILAB, and ERICA: New Tools for Advanced EEG Processing *Computational Intelligence and Neuroscience* volume **2011** pp 1–12
[11] Delorme A and Markeig S 2004 EEGLAB: an open source toolbox for analysis of single-trial EEG dynamics including independent component analysis *J. of Neuroscience Methods* vol **2** pp 9–21
[12] Acar Z A and Makeig S 2010 Neuroelectromagnetic forward head modeling toolbox *J. of Neuroscience Methods* vol **190** no 2 pp 258–70
[13] Vankov A 1987 *Adapt © 1987-2003 and Variete © 2000, 2001 are property of EEG Solutions LLC, and are used under free license for scientific non-profit research*
[14] Makeig S, Gramann K, Jung T P, Sejnowski T J and Poizner H 2009 Linking brain, mind and behavior *Int. J. of Psychophysiology* vol **73** no 2 pp 95–100
[15] Gu S, Wang F, Patel N P, Bourgeois J A and Huang J H 2019 A Model for Basic Emotions Using Observations of Behavior in Drosophila *Front. Psychol.* **10:781** DOI: 10.3389/fpsyg.2019.00781
[16] Picard R W 2000 *Affective Computing* (Cambridge: The MIT Press)
[17] Bestelmeyer P E G, Kotz S A and Belin P 2017 Effects of emotional valence and arousal on the voice perception network *Soc. Cogn Affect. Neurosci.* **12** pp 1351–8 DOI: 10.1093/scan/nsx059
[18] Niven K and Miles E 2013 Affect Arousal *Encyclopedia of Behavioral Medicine* ed M D Gellman and J R Turner (New York: Springer) DOI: https://doi.org/10.1007/978-1-4419-1005-9_1089

[19] Teplan M *et al* 2002 Fundamentals of EEG measurement *Meas Sci Rev* **2(2)** pp 1–11
[20] Luck S J 2014 *An introduction to the event-related potential technique* (New York: MIT press)
[21] Jaušovec N, Jaušovec K and Gerlič I 2001 Differences in event (related and induced EEG patterns in the theta and alpha frequency bands related to human emotional intelligence *Neuroscience Letters* vol **311** no 2 pp 93–6
[22] Il'iuchenok I R, Savost'ianov A N and Valeev R G 2001 Dynamics of spectral characteristics of theta (and alpha (range EEG during negative emotional reactions *Zh Vyssh Nerv Deiat Im I P Pavlova* vol **51** no 5 pp 563–71
[23] Butterworth S 1930 On the theory of filter amplifiers *Wire Eng* **7(6)** pp 536–41
[24] Stoica P and Moses R 2005 *Spectral Analysis of Signals* (New Jersey: Prentice Hall)
[25] Howell K B 2001 *Principles of Fourier Analysis* (Florida: CRC Press) ISBN 978-0-8493-8275-8
[26] Wang R, Wang J, Yu H, Wei X, Yang C and Deng B 2015 Power spectral density and coherence analysis of Alzheimer's EEG *Cogn. Neurodyn.* vol **9** no 3 pp 291–304
[27] Sejdić E, Djurović I and Jiang J 2009 Time-frequency feature representation using energy concentration: An overview of recent advances *Digital Signal Processing* **19(1)** pp 153–83 DOI: 10.1016/j.dsp.2007.12.004
[28] Hastie T, Tibshirani R and Friedman J 2009 *The Elements of Statistical Learning: Data Mining, Inference, and Prediction* (New York: Springer)
[29] Rosenblatt F 1961 *Principles of Neurodynamics: Perceptrons and the Theory of Brain Mechanisms* (Washington DC: Spartan Books)
[30] Othman M, Wahab A, Karim I, Dzulkifli M A and Alshaikli I F T 2013 EEG emotion recognition based on the dimensional models of emotions *Procedia-Social and Behavioral Sciences* **97** pp 30–7
[31] Bhatti A M, Majid M, Anwar S M and Khan B 2016 Human emotion recognition and analysis in response to audio music using brain signals *Computers in Human Behavior* **65** pp 267–75
[32] Nor N M, Salleh S H and Zubaidi A 2016 Understanding teacher stress when teaching the developed technology by using electroencephalogram (EEG) signals *J. of Applied and Physical Sciences* **2(3)** 65–76
[33] Murugappan M, Juhari M R, Nagarajan R and Yaacob S 2009 An investigation on visual and audiovisual stimulus-based emotion recognition using EEG *Int. J. of Medical Engineering and Informatics* **1(3)** pp 342–56
[34] Kruchten P 1995 The 4? 1 view model of architecture *IEEE Software* **12(6)** pp 42–50 DOI: 10.1109/52.469759

Document Searching of EPPS Test Result Using Indexing Method

Kahfiara Krisna[1], **Sri Mulyati**[2]

Department of Informatics Universitas Islam Indonesia, Yogyakarta

E-mail: [1]16523234@students.uii.ac.id, [2]mulya@uii.ac.id

Abstract. This study focuses on documents searching on the EPPS test result, mainly administered in the education sector for selecting prospective students in universities' admissions process. There has been a considerable increase of admission data along with annually administered admission test for prospective students in every beginning of an academic year. Some faculties may administer the EPPS test in the second stage test of admission process after the prospective students pass the first stage in addition to having the general admission test. These several tests lead to a huge data accumulation with different information, making it difficult for the organizers to locate the document containing 15 aspects of required information in a comprehensive information display. Thus, it is necessary to apply an indexing method by having a predetermined index to retrieve required information in a particular document or to find other related documents by way of indexing method. This research used 74 pdf documents as the dataset. In particular, 60 datasets were used as references to retrieve phrases on the specified data, and 14 datasets were used to test data. The results revealed the most significant data with an average precision of 0.89, recall of 0.93 and f-measure of 0.91.

Keywords: indexing method, documents, searching

1. Introduction

The psychological test administrator has several documents stored in a computer directory. One of the documents stored is the EPPS test result. the number of test result documents is in accordance with the number of participants who completed the test. The EPPS (Edward Personal Preference Schedule) test is a psychological test that is performed by an individual test taker or by an organization. The EPPS test is widely used in the education sector, mainly to select prospective students in universities admission. Normally, the result of the EPPS test file will be stored as a pdf document containing the psychogram matrix and the concluding result of test taker performances.

As the new academic year begins, multitudes of prospective students apply for university admission based on their desired majors. The admission of countless applicants' data along with the result of their EPPS test will result in a huge pile of data that require large data storage in every test period. Given the large number of new prospective applicants' data and new information, it is difficult for the organizers to provide prospective students with accurate result of the selection process. The ineffective and incomprehensive data display requires the organizers to re-filter a number of documents and locate the particular search result according to the predetermined assessment. Moreover, this condition is worsened by absence of recorded data that can display new information according to the existing data in a more efficient presentation to be easily understood by both experts and organizers, which remains an obstacle at this time. At certain times, it is also difficult to locate certain documents containing required words. The only way to find certain information from a set of documents is by looking for the similar words contained in a document. This process is normally done manually, which is definitely time consuming, especially when there is a need to locate certain documents in large numbers.

Searching a certain document by way of an index method helps in finding indexes that contain the required aspect of EPPS test results and locating the index of the words searched for in the document. It is possible to classify the document based on its similarity through the method of word indexing from a set of documents. The indexing process aims to change an unstructured document into a structured document to ease the process of finding the index of the necessitated word and thus to display the document [1].

Given the above explanation, this research aims to propose a method to ease the organizers in locating required data and filter the number of potential applicants by finding the important and required information of the EPPS psychological test conclusions using the predetermined aspects of needs and by locating the position of the stored documents. This research was targeted for high school students majoring in Science who would apply for the Medical Faculty of UII.

On this account, the researchers tried to propose the indexing method as an efficient way for searching for information and document similarities to find suitable information regarding the required aspects contained in the EPPS test conclusions and to look for the similarity of recorded data with new data information in order to get the highest results based on the aspects needed by the organizer. This system , a dictionary or related meaning to aspects of need is developed in the measuring instrument used so that searches can be spesifict to the content of the document in the referenced direktory. This research is formulated to answer the following research question, (1) how to find and display the information contained in the psychogram matrix results of the EPPS test to facilitate the search for the desired needs?

This paper is written based on the following structure of a research paper: the background, review of related works (literature), research methodology, the results and discussion, and the conclusions of this study.

2. Literature Review

Indexing method is the process of finding information from a large and unstructured document database in response to a required query [2]. The indexing process is carried out to display the desired information in the database by organizing the unorganized set of documents into a more structured document for an easy retrieval and easier documents finding based on the information contained in the required document [3].

This study addressed the EPPS (Edward Personal Preference Schedule) test as a form of personality test consisting of several answer choices to be selected by individuals according to the available questions. This test aims to find out personal needs that reflect the individual's personality [4]. This test was developed by Allen L. Edward by requiring us to select fifteen required aspects out of twenty compiled points that refer to Murray's personality concept. This personality test contains 225 questions, namely the compelling, objective and non-projective. The EPPS test is administered to observe the specific needs of an individual through 15 required aspects [5].

There have been several previous researchers on this topic as references of this study. The first research discussed the use of the indexing method in the search process for documents generated by manual search results. The advantage of index retrieval lies on structured data storage in order to store large amounts of various documents. The search process is done by creating a word index from the document [1].

The second study raised a topic on the search for of text documents of archived letters using indexing and query methods. This research discussed the process of restoring information with an index so that documents can be retrieved when needed and can be located easily. It also used the index method to retrieve an index that has been registered in the database to find the desired sentence. With more detailed processing in the query development process, word grouping can be summarized more simply [2].

The third research entitled "Single Document Keywords Extraction in Indonesian using Phrase Chunking" [6]. This research discussed the formation of keywords from documents automatically using phrase concatenation. Phrase chunking was done using the Part of Speech (POS) pattern to extract phrases and keywords. Keywords were selected from the candidates based on the number of words in the phrase and several types of phrase data forms. The shortest keywords with reduced strings were extracted from the abstract and sorted by their highest frequency values. This study conducted an experiment using types of phrases such as different sources of keywords, variations in phrase patterns and provided performance comparisons between types of phrases.

The fourth research discussed the use of a retrieval approach as a solution to existing problems by relying on concepts rather than keywords for indexing and retrieval. It aimed to retrieve documents that were relevant to a specific user request. Indexing and weighting of documents and queries were done using semantic entities, concepts, and keywords. In this approach, concepts were identified from the document and measured according to their frequency distribution and their similarity to other concepts in the document [7].

The technique used for text documents is widely used with the IR (Indexing and Retrieval) technique. The purpose of an IR system is to retrieve relevant material from a document database in response to a query [8]. The five research discussed the design of information retrieval system by employing Vector Space Model in the tracking and also similarity measures of Cosine Similarity as the method to rank the documents found that match with the keywords/query[9]

3. Research Metodology

We proposed an automatic keyword extraction technique using generate frase in Indonesian document. Figure 1 shows the methodology of this research.

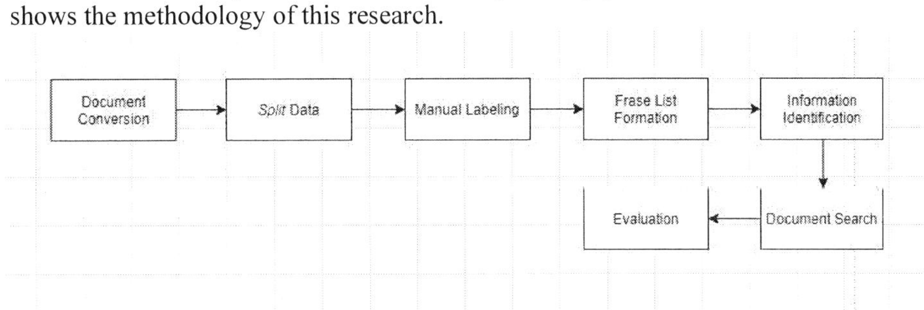

Figure 1. Research Metodology

3.1. Document Conversion

At this stage, the prospective applicants' psychological test results are collected into one EPPS folder in pdf form. The document will be an input data to be processed further. The EPPS test results contain two paragraphs of sentences relating to the applicant's most and least required aspects. The data on the EPPS test results will be taken by indexing the paragraph "description" that appears on the document and it will be processed further through the preprocessing process.

3.2. Preprocessing

The preprocessing is an early stage process that is prepared to carry out further processing of the text in the data before going through the indexing process. The process in this preprocessing stage only includes several steps, namely casefolding (changing the front letter to lowercase) and tokenizing (separating each word into one line only). The results of the preprocessing will later be saved and

IOP Conf. Series: Materials Science and Engineering 1077 (2021) 012032 doi:10.1088/1757-899X/1077/1/012032

converted to csv files. The CSV file is the result of the document conversion process and preprocessing of the EPPS test results.

3.3. Split Data
At this stage, the dataset is divided randomly into training data and test data with a ratio of 80%: 20%. The training data will be used as a source of formation of a list of phrases and labels, which will later be matched with the document. Meanwhile, test data is useful for looking for occurrences with a value similar to the recorded data.

3.4. Manual Labeling
At this point, the converted csv file will contain a table of "phrases" and "labels". The tables have been filled by phrases by the system from the conversion results, while the label table is still left empty that will require manual labeling by the researcher. The label will be entered manually based on the abbreviated name of the aspect of need. The phrase that is labeled contains the initials "other" or "O" if there are phrases not related to the label and initials from 15 aspects of the need if the phrase has something to do with the label.

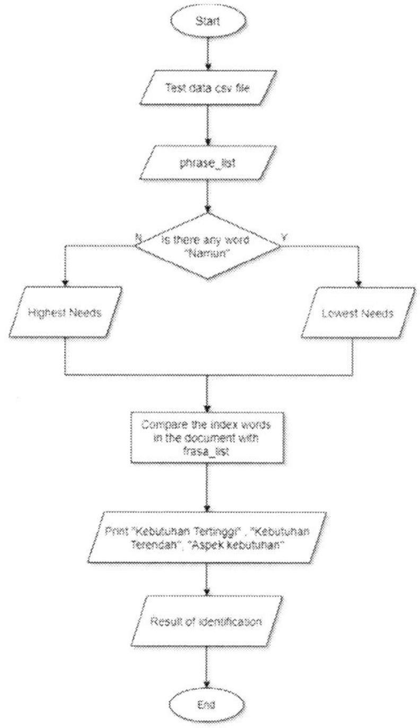

Figure 2. Information Identification process

3.5. Phrases List Formation
At this stage, the labeled training data will take the index of every word that contains a phrase with initials other than "other" to separate the phrases that will be stored in a new csv file named "phrase_list.csv". The phrase takes each pair of phrases and labels that have been manually inputted by the expert. The phrases in the phrase list will be sorted randomly based on the long and short-labeled words that are labeled by the researcher in the manual labeling process. In the manual labeling process, the author labels three or more words/tokens on the file, which will come out at the top of the list. In contrast, the words with fewer labels will appear in the bottom line of thephrase list.

3.6. Information Identification

In this study, the word indexing method was used to find information related to potential applicants. The information to find is the label on each file. At this stage, the system will identify the phrase by matching the label specified in the phrase_list. To carry out this process, test data is needed to determine the suitability of the created model to find the information that appears along with the names of aspects of needs. The process is identified based on the word "but" after the sentence, which will make the system classify the particular sentence as the highest need (most prioritized need) and the lowest need (least prioritized need). Afterwards, the system will check the index based on any phrases that appear in the document to identify sentences that refer to the label on phrase_list. The output will provide a description of the phrases and labels that appear on each file along with the label and a summary of the aspects of the needs that come out of the file. The txt results are stored in a different folder with the document folder. Figure 2 shows information identification process.

3.7. Document Search

Once the search results of the desired aspects of needs are known at this stage, these results will become a reference to see which prospective applicants of FK UII with the highest need in that aspect. Document searching can make it easier for the organizers to find documents of prospective applicants of FK UII based on the 15 desired needs as a component of the next assessment. For example, the system can recognize the need for "ACH" as the highest need, and thus the system will display data on which applicants have the highest needs for these aspects of needs. figure 3 shows document search process

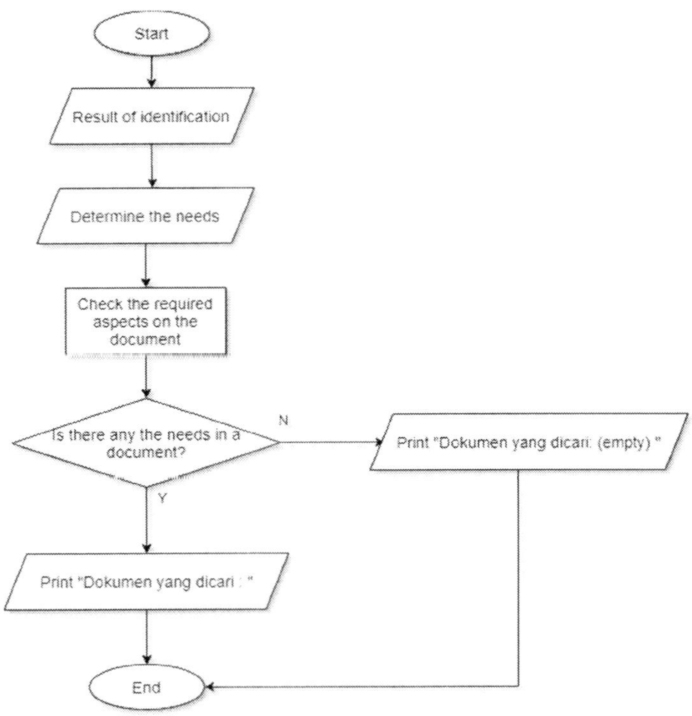

Figure 3. Document search process

3.8. Evaluation

Furthermore, to conduct a model testing, we need original data and new data to be tested will to obtain similar degree value. After obtaining a document containing relevant information, the test data will evaluate the original data against the new data by looking for the similar degree value. Then, it will evaluate each type of phrase form based on its precision, recall, and f-measure. The data to be used as test data amount to 20% of the entire data, which is equal to 0.20.

IOP Conf. Series: Materials Science and Engineering 1077 (2021) 012032 doi:10.1088/1757-899X/1077/1/012032

4. Result

The previous researches indicate the indexing method as a suitable method to solve the problems in document searching. The modeling stage of the obtained data aims to produce a system that is able to identify every word that appears related to 15 aspects of predetermined needs and provides information on which aspects of needs arise at the conclusion of the EPPS test. To help create this model, the researchers used the python3 programming language through the terminal and IDE Visual Code.

It is explained in formula 3.1 that the conversion process of the EPPS test results document, which is saved in the form of a pdf file into a csv file that takes the paragraph in the "description" index, will be carried out. Before converting the data into a CSV file, the text data is converted to a txt format to get all the text in the PDF file. The document conversion process requires a preprocessing process of casefolding using the text.lower () function and tokenization using the word_tokenize function. The output of one of the 74 documents converted to documents and preprocessing is presented in the figure 4.

kata	label
hasil	
pemeriksaan	
psikologis	
menunjukkan	
bahwa	
saudara/i	
maulfi	
natsir	
asyari	
memiliki	
kebutuhan	
yang	
tinggi	
untuk	
menyesuaikan	
diri	
dan	
mengikuti	
aturan/konvensi	
yang	
telah	
ditetapkan	

Figure 4. The output of document conversion

Once we obtained the converted documents, the 74 documents will be split into two parts, namely train data and test data with a default ratio of 20% of the train data. The train data and test data that have been converted will be labeled manually by the researcher by looking at several criteria from the aspect of requirements on the EPPS test. Labeling is done directly from the csv file of each document. The criteria for aspects of need along with keywords are presented in Table 1.

Furthermore, in the process described in formula 3.4, the text in the "description" is separated and stored in the csv "phrase_list" form with a table containing the pairs between the phrases and the labels. The sorting process uses the "sorted" function to sort the phrases that are labeled by authors. Phrases in phrase_list will be sorted based on phrases labeled with three or two words or only one word so that they are stored according to the longest and shortest word entered by the author. A list of the list phrases can be seen in Figure 5.

Table 1. Aspect of needs and the keywords.

Aspect of Needs	Keyword
Achievement (ACH)	Memperoleh sesuatu, mencapai prestasi, mencapai, prestasi, meraih sesuatu.
Deference (DEF)	Menyesuaikan diri, mengikuti aturan, patuh terhadap aturan.
Order (ORD)	Mengatur hidup, barang,ide, keteraturan, sistematis.
Exhibition (EXH)	Mengekspresikan diri, kepercayaan diri, yakin dengan kemampuan.
Autonomy (AUT)	Mandiri, tidak tergantung, pendirian/pendapat, tergantung pada orang.
Affiliation (AFF)	Memberikan perhatian, pergaulan harmonis, lingkungan sosial, menaruh perhatian, berinteraksi
Intraception (INT)	Berempati, bersimpati, kepekaan, memahami, merasakan, memahami orang.
Succorance (SUC)	Dimanjakan, diperhatikan, diurus
Dominance (DOM)	Mendominasi, memerintah, mempengaruhi, mengatur orang, menonjolkan diri, ketua kelompok.
Abasement (ABS)	Mengakui kesalahan, perasaan bersalah, menyalahkan diri
Nurturance (NUR)	Menolong, merawat orang/benda, menjaga barang, merawat orang.
Change (CHG)	Melakukan perubahan, menghindari rutin, sesuatu yang tetap.
Endurance (END)	Menyelesaikan pekerjaan, ketekunan, keuletan, kegigihan, ketahanan, teratur.
Heterosexual (HET)	Seksual, mengikuti pendapat
Aggression (AGG)	Mendobrak, berperilaku agresif

4.1. Information Identification

In the process of identifying information, the system will display the meaning contained in the EPPS test psychogram to make it easier to find out which aspects of the needs appear in the EPPS test conclusions of prospective applicants. Proses information identification we need to look for an index to separate paragraphs describing the highest and lowest needs. In this case, the word that distinguishes between the highest and lowest needs is the word "however" in the beginning of sentences.

Some characters that appear will also be replaced using the regular expression library in python. Figure 6 shows the output of the identification information based on phrase_list.csv. The phrase "Highest need" displays the phrases list containing the highest need of the EPPS test result, while the phrase "Lowest need" displays the phrase list containing the applicant's lowest need. The phrase list shows an information about the phrases and labels written in the sentence that concludes the EPPS test results. "Aspects of needs" contained in this output show briefly the aspects of needs based on the predetermined index phrases and labels.

frase	label
yakin dengan kemampuan	EXH
sesuatu yang tetap	CHG
tergantung pada orang	AUT
patuh terhadap aturan	DEF
pengambilan keputusan	AUT
mengikuti aturan/konvensi	DEF
mengekspresikan dirinya	EXH
memberikan perhatian	AFF
merawat orang/benda	NUR
melakukan perubahan	CHG
berperilaku agresif	AGG
merawat orang/benda	NUR
mengikuti pendapat	HET
memperoleh sesuatu	ACH
pendirian/pendapat	AUT
kehidupan seksual	HET
mencapai prestasi	ACH
perasaan bersalah	ABS
menyesuaikan diri	DEF
menaruh perhatian	AFF
kepercayaan diri	EXH
tidak tergantung	AUT

Figure 5. Phrase_list

```
1831150236.pdf
-------Kebutuhan Tertinggi-------
berempati -> INT
bersimpati -> INT
yakin dengan kemampuan -> EXH
menonjolkan diri -> DOM
ketua kelompok -> DOM
mengatur orang -> DOM
mempengaruhi orang -> DOM

Aspek Kebutuhan :  INT, EXH, DOM

=======Kebutuhan Terendah=======
patuh terhadap aturan -> DEF
patuh -> DEF
aturan -> DEF
mengikuti pendapat -> HET
berperilaku agresif -> AGG
agresif -> AGG
mencapai prestasi -> ACH
mencapai -> ACH
prestasi -> ACH
memperoleh sesuatu -> ACH
sesuatu yang tetap -> CHG
rutin -> CHG
ketahanan -> END
kegigihan -> END
ketekunan -> END

Aspek Kebutuhan :  DEF, HET, AGG, ACH, CHG, END
```

```
1831150091.pdf
-------Kebutuhan Tertinggi-------
mandiri -> AUT
tidak tergantung -> AUT
tergantung pada orang -> AUT
pendirian/pendapat -> AUT
menaruh perhatian -> AFF
bergaul -> AFF
harmonis -> AFF
menolong -> NUR
merawat orang -> NUR
benda -> NUR

Aspek Kebutuhan :  AUT, AFF, NUR

=======Kebutuhan Terendah=======
ketahanan -> END
kegigihan -> END
ketekunan -> END
mencapai prestasi -> ACH
mencapai -> ACH
prestasi -> ACH
memperoleh sesuatu -> ACH
patuh terhadap aturan -> DEF
patuh -> DEF
aturan -> DEF
mengikuti pendapat -> HET
keteraturan -> ORD
sistematis -> ORD
kehidupan seksual -> HET

Aspek Kebutuhan :  END, ACH, DEF, HET, ORD
```

Figure 6. An example of a document resulting from the information identification process

4.2. Document Search

Once the system finds and identifies the information, it will be easier for us to search for documents through the aspects of the needs listed in the results of each document that was identified in the previous process. This document search can be done everytime a user wants to find an information from multiple documents based on which of the 15 aspects of the desired need is required. If the user requires the search for "ACH" and "DEF", the system will randomly search for the document containing those words as the highest aspect of needs according to the desired result. Figure 7 shows the results for documents search process using the desired label of "ACH" and "DEF" which can be stored in the form of a txt file as what the experts required.

```
=======Dokumen yang dicari : =======
1831150091.pdf
1831150239.pdf
1832150031.pdf
1832150046.pdf
1831150104.pdf
1831150115.pdf
1832150084.pdf
1832150073.pdf
1832150027.pdf
1831150207.pdf
1831150214.pdf
1831150002.pdf
1831150202.pdf
1831150230.pdf
1831150177.pdf
1832150022.pdf
1832150069.pdf
1831150020.pdf
1832150091.pdf
1831150108.pdf
1831150090- Inkonsisten 9.pdf
```

Figure 7. The results of file implementation based on aspect of needs “ACH” and “DEF”

4.3. Evaluation

The next stage is to test and evaluate the test data with the original data on the system to look for the appearance of similar degree values. It showed 14 test data that had been split previously in the 3.2 process. In this evaluation, precision, recall, and f-measure are used as a form of evaluation of the system. The final results of the model testing are resulted from the calculation of the average value of each document shown in **Table 2**.

Table 2. The average per documents

	Precision	Recall	F-Measure
Mean	0,89	0,93	0,91

Table 2 displays the average results of the evaluation using different types of phrase forms. This search obtained the best results of all reducted phrase form types with an f-measure value of 0.97 for all data. The above test result does not contain the aspect of accuracy because accuracy is more appropriate for class distribution, while f-measure is a better metric for cases with an imbalance in evaluating the model.

5. Discussion

In this study, the system was designed using an indexing method based on a list of phrases the researcher created to obtain information in the text "description" of the EPPS test results. Based on the literature review, the indexing method in this study is done by adding a list of phrases as a dictionary to help identify information and to find out which documents are needed by the organizer. Figure 3 displays 60 lists of phrases that are formed. Phrase list formation is done by looking at examples of cases of phrase truncation and name entity recognition, where the type of truncation is based on the phrase tag used in the parse tree and the name of the phrase tag and the method of classifying text phrases with certain rules based on word classes. (POS) [10], [11].

The indexing method obtained a quite good result since it only used small amount of data. These results are obtained from a list of identified phrases, which are used as a reference. As a result, the system can match the information that appears in the EPPS test "description" text by producing phrases and labels along with the appearing aspects of needs to ease users in reading the information. However, such method is still constrained when it has to deal with a huge amount of data because it will need a throrough update to check the accuracy of manual labeling. Nevertheless, since this case used manual labelling, it is necessary to add an updated system to check system performance to avoid any errors in the inputted data. Therefore, it is pivotal that an expert in this field does manual labelling to avoid misinformation.

In this case, the system does not focus on using stopword removal, but on the removal of the word "other" contained in documents that have been manually labeled "O" to create a list of phrases or phrase_list. The development of stopword removal is a considerable topic for future research to make sure whether it should be used or not. In the evaluation, the results are not counted into the calculation of accuracy because in the search results the accuracy will be close to 100% or the "True Negative" result, which has an infinite value although in reality the results are just the reverse. Some existing examples, as shown by the TN, contain words that cannot be identified and should not be identified, but are still calculated using an accuracy formula that produces results close to 100%.

6. Conclusion

Based on the initial stages until the final stage of the research, we can conclude that the method of identifying information through the indexing process is an applicable method. This method can be used to find information related to the aspects of the needs required by the structured EPPS test for prospective applicants, since it can search for information by referring to the information identified in the phrase_list. The f-measure calculation obtained the result of 0.91, the precision of 0.89, and the recall of 0.93.

However, this method is not a benchmark to create a good system, since it still needs further updates and development. Therefore, it is expected that there will be further research to develop the current research with some improved elements such as by increasing the number of datasets and continuously updated list of phrases in order to get better results. It is also necessary to develop the current research by constantly checking and correcting keywords in the converted PDF file. Other possible developments can take the form of using other more suitable methods out of the indexing method.

References

[1] Hidayat W 2013 *Indexing and Retrieval Engine untuk Dokumen Berbahasa Indonesia dengan Menggunakan Inverted Index*

[2] Heriyanto 2015 Pencarian dokumen teks arsip surat *Seminar Nasional Informatika (SEMNASIF)* pp 173–7

[3] Broder A Z, Eiron N, Fontoura M, Herscovici M, Lempel R, McPherson J, Qi R and Shekita E 2006 Indexing Shared content in information retrieval systems *Advances in Database Technology - EDBT 2006. EDBT 2006 Lecture Notes in Computer Science* vol 3896 ed Y Ioannidis *et al* (Berlin, Heidelberg: Springer) pp 313–30

[4] Lubin B, Larsen R M, Matarazzo J D and Seever M 1985 Psychological test usage patterns in five professional setting *American Psychologist* **40(7)** pp 857–61

[5] Piedmont R L and Costa P 1992 An Assessment of the edwards personal preference schedule from the perspective of the five-factor model *J. of Personality Assessment* vol **58** no 1 pp

67–78
[6] Trisna I N P and Nurwidyantoro A 2020 Single document keywords extraction in Bahasa Indonesia using phrase chunking *Telecommunication, Computing, Electronics and Control (TELKOMNIKA)* **18** pp 1917–25
[7] Boubekeur F and Azzoug W 2013 Concept-based indexing in text information retrieval *Int. J. of Computer Science & Information Technol. (IJCSIT)* vol **5** no 1 pp 119–36
[8] Lu Guajun 1999 *Multimedia Database Manajemen Systems* (London: Artech House, Inc)
[9] Bakala N 2019 Information retrieval system by using vector space model *Int. J. of Scientific and Technol. Research* vol **8** issue 10 pp 1563–8
[10] Yang J 2004 Phrase chunking for efficient parsing in machine translation *MICAI 2004: Advances in Artificial Intelligence, MICAI 2004, Lecture Notes in Computer Science* vol 2972 R Monroy R *et al* (Berlin, Heidelberg: Springer) pp 478–87
[11] Arman A A , Putra A B N, Purwarianti A and Kuspriyanto 2013 Syntactic phrase chunking for Indonesian Language *Procedia Technol.* **11** pp 635–40

DCGAN-generated Synthetic Images Effect on White Blood Cell Classification

C A Hartanto[1], S Kurniawan[2], D Arianto[3], A M Arymurthy[4]

Faculty of Computer Science, Universitas Indonesia, Depok, Indonesia

Email: [1]cahyo.adhi91@ui.ac.id, [2]sandy.kurniawan@ui.ac.id, [3]dian.arianto@ui.ac.id, [4]aniati@cs.ui.ac.id

Abstract. White blood cell can give information about someone's health. Imbalanced amount of white blood cells indicates someone's disease. The disease detection using white blood cell can be done by using flow cytometry, however the use of the device has many drawbacks. The drawbacks can be solved by using computer aided classification. However, another problem arises, that the available white blood cell dataset has small dataset and the class distribution is imbalanced. This problem can be solved by using classic data augmentation and DCGAN to create synthetic image in order to balance the amount of white blood cell dataset. The balanced dataset then trained using ResNet50 model as the classification model. Accuracy, precision, recall and F1-score are used as the performance measures of the classification. The result shows that the model obtain accuracy of 82.5%. ResNet50 was chosen because it can produce good accuracy when used in medical image classification.

1. Introduction

White blood cell is part of the human immune system. As part of the human immune system, they protect the human body from the infection of various bacteria, viruses and fungi [1]. Based on the percentage of content in the blood, white blood cells are divided into five types namely basophil (0 - 1%), eosinophil (1 - 5%), lymphocyte (20 - 45%), monocyte (2 - 10%) and neutrophil (50 - 70%). Each white blood cell type has different nucleus shape and cytoplasm [2].

Lack or excess of white blood cells in the body may lead to many diseases. The disease can be diagnosed by doing a blood test. A blood test is done by counting the amount of each white blood cell type in the blood, usually called complete blood count test / hemogram test [1]. White blood cell classification and counting generally was done by hematology expert using a microscope. However this process was time consuming and prone to miscounting and misclassification of the white blood cells [3].

Given the high possibility of errors in the process, a computer aided classification system is needed to reduce the possibility of the error occurs. There are already devices such as flow cytometry that can be used to measure the physical attributes and properties of the white blood cells, but these laser-based devices are very expensive, require high maintenance and require direct blood samples [4]. Therefore, a white blood cell classification method was developed using image processing strategies to overcome these deficiencies.

Works related to the classification of white blood cell images include [3], [2], and [5]. Authors in [3] used the CNN models of AlexNet, GoogleNet, and ResNet-101 for the classification of white blood cells. In that study, it was found that the CNN AlexNet model obtained the best results on the task of classification and counting of white blood cells with an accuracy of 96.63%. Authors in [2] used the DeCA approach, Fine-tuned network approach, and WBCsNet. In that study, the authors used various image sizes to be tested to achieve the best performance. From the results of the experiment, the accuracy obtained by the method proposed by the author, WBCsNet, is 96.1%, better than the DeCA approach and Fine-tuned network approach. Authors in [5] used the method of deep learning called Inception Recurrent Residual Convolutional Neural Network (IRRCNN) method. By using WBC data from CellaVision, the experimental results show an accuracy of 99.94% for the WBC classification using the proposed method. There is an increase in accuracy of 1.4% and 2.35% compared to existing deep learning approaches. In that study, the authors conducted several data augmentation techniques such as flipping, rotation, and shearing to increase the amount of data to overcome imbalanced data. Using these techniques, the authors achieved an increase in the amount of training data from 352 to 9,957.

Authors in [6] used the DCGAN model for data augmentation in the classification of liver lesion images. From the results of experiments, the authors achieved higher accuracy when using DCGAN for data augmentation. Authors in [7] classified fish images data using ResNet50. From the experimental results, the authors obtained higher accuracy when compared to VGG and faster when compared to Inception. In this case, we use DCGAN to augment white blood cell image data to overcome imbalanced and insufficient data.

The image class on the image dataset can be balanced or imbalanced. However, the available white blood cell dataset has imbalanced dataset amount. Imbalanced dataset is a condition where the amount of dataset in each class is not balanced. This dataset condition can affect the training process and testing results. This problem can be solved by using classic data augmentation and Generative Adversarial Networks (GAN). In this study, we used classic data augmentation and Deep Convolutional Generative Adversarial Networks (DCGAN) to overcome the imbalanced data problem.

The contribution of the study is as follows:

1. Creating synthetic white blood cell image using DCGAN to overcome the imbalanced dataset problem.
2. Using a combination of synthetic images and real images on the training process of the ResNet50 model

The rest of this paper is organized as follows. The methodology used in the paper is described in section 2. In section 3, the result and discussion are presented. Finally, section 4 presents the conclusion of this paper.

2. Methodology

In this section, the dataset used in the study, classic data augmentation and the DCGAN architecture that will be used to overcome the imbalanced dataset problem will be described. The classification model will also be described in this section.

2.1. Dataset

The dataset used in this study was obtained from [8]. They provided two kinds of dataset, but we only used one of the two datasets that was obtained from the CellaVision blog [9], consists of 100 300x300 white blood cell images. The white blood cell color on the images is generally purple and may contain red blood cell around the white blood cell. Image classes from the dataset consists of neutrophil, lymphocyte, monocyte, eosinophil and basophil. The image class distribution from the dataset is shown in Table 1. The white blood cell image samples for this study are shown in Figure 1. The white blood cell image samples for this study are shown in Figure 1.

2.2. Classic Data Augmentation

Classic data augmentation is a technique used to overcome overfitting on the training process [10]. Overfitting can affect the accuracy obtained on a deep learning model. It indicates that the training process produce a model that does not learn the dataset well [11]. Data augmentation is the most simple technique to avoid overfitting by increasing the amount of the dataset [6]. In this study, three classic data augmentation is implemented, namely salt and pepper, random gaussian blur, and random rotation (between 5° and 355°). Flipping and scaling techniques are not implemented in this study. The use of flipping may produce the same image dataset. The scaling technique was not used because the training process already use the Region of Interest (ROI) of the image. The result of classic data augmentation is shown in Figure 6.

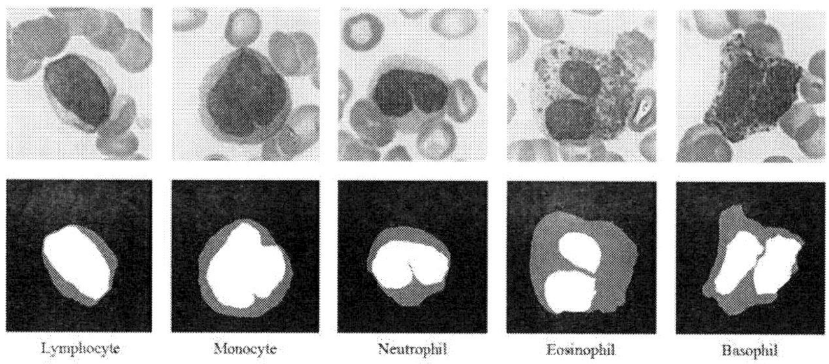

Figure 1. White Blood Cell Image Samples

Table 1. Image Class Distribution

Class	Dataset
Eosinophil	12
Lymphocyte	37
Monocyte	15
Neutrophil	33
Basophil	3

2.3. DCGAN to Produce A Balance Dataset

Generative Adversarial Networks (GAN) is a model that can be used to create synthetic images by learning from dataset [8], [6]. GAN consists of generator and discriminator. The generator's task is to create synthetic images from the real images, while the discriminator's task is to differentiate the synthetic image from the real image [12]. The generator is trained in stages, so that it can create synthetic images that are similar to the real images with greater amount and different from the dataset [13]. Whereas the discriminator is trained until it cannot distinguish the synthetic images from the real images. In this study, we used Deep Convolutional Generative Adversarial Networks (DCGAN), which was proposed by Radford et al. in 2016 [14]. They stated that DCGAN was known to be more stable in training phase than GAN and it generated better images. The DCGAN training was done for lymphocyte, monocyte, neutrophil and eosinophil class dataset. Basophil class dataset was not train since it has the least dataset images.

DCGAN was chosen since it has better accuracy results when used for image classification based on the study done in [6]. In addition, DCGAN has better convergence speed and is easier and more efficient in sampling [15]. Our proposed augmentation model using DCGAN for this study are shown in Figure 2 and Figure 3.

2.3.1. Generator. The generator uses a vector consisting of 100 random numbers that are trained to generate white blood cell images with a size of 128 x 128 x 3. Generator inputs are samples of z (1), ..., z (m). The images generated by the generator is used to fool the discriminator. The generator is trained in order to generate a synthetic image that resembles the original image until finally able to fool the discriminator.

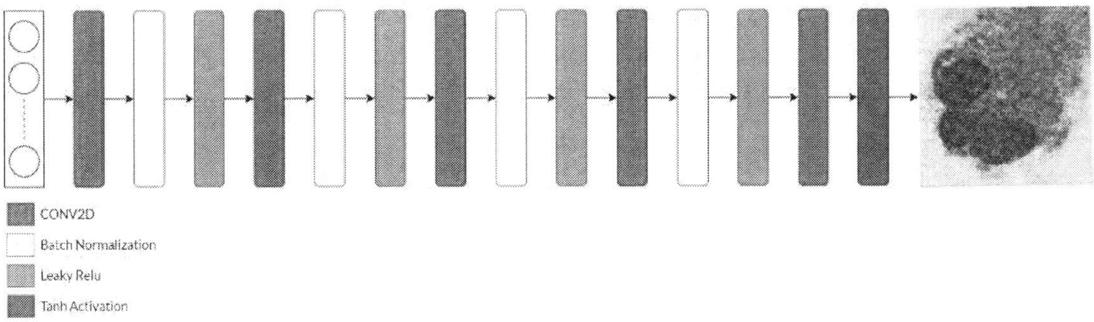

Figure 2. DCGAN Generator Architecture

2.3.2. Discriminator. The discriminator uses input in the form of two images, those are the original image from the dataset and synthetic image measuring 128 x 128 x 3 with the output in the form of decision of a real or fake image. That part of DCGAN is optimized so that it can recognize synthetic images as well as possible. Once the discriminator is unable to distinguish synthetic images and original images, the generator will not generate different synthetic images [16].

Once the discriminator and generator reach the Nash Equilibrium, so the DCGAN training is completed and the resulting synthetic image is used in the training process [16]. The selected optimizer was Adam since it prevents overfitting, achieved faster convergence, and more efficient computing [15].

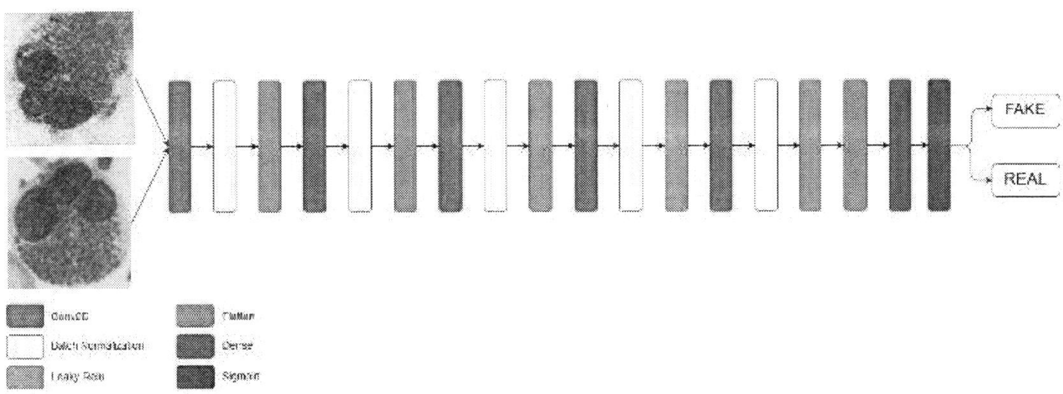

Figure 3. DCGAN Discriminator Architecture

2.4. *Transfer Learning*

Transfer learning is a knowledge transfer model that has studied the data to use in the target dataset [17]. The use of transfer learning can reduce training time because there is no need to build a model from the beginning [18], has a good feature extraction ability, and not easily overfit [7]. The pretrained model is a model that has been trained using a base dataset. The use of pre-trained models can achieve good accuracy, especially in terms of image classification. Several models can be used for transfer learning such as VGG, Inception Model and ResNet50 [19]. In this study, ResNet50 was used to train the white blood cell dataset.

2.5. *ResNet50*

ResNet is one of the deep learning models that is frequently used today for the problem of image recognition and image classification. One of the deep learning problems is vanishing gradient problem and degradation problem. Residual learning on ResNet can overcome these problems [20]. In addition, this part of ResNet can prevent the degeneration model problem during the training process [7]. In this study, ResNet was applied because it has good results in the classification of medical images [19].

ResNet50 is one variant of the ResNet models. Other variants of ResNet are ResNet101 and ResNet512. Residual learning is presented in Figure 4.

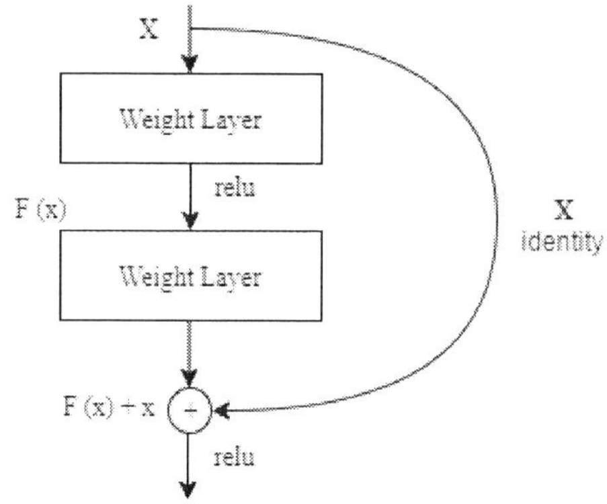

Figure 4. Residual Learning

ResNet50 consists of 50 convolutional layers. The fundamental difference between ResNet50 and other models is that ResNet50 have shortcut connection. Based on [7], ResNet50 model was chosen because it produces better accuracy dan VGG model and faster process than Inception. The identity shortcut in Figure 4 is used to passing information from input X to be combined with the output result from weight layer. It can be concluded that the function of shortcut connection is to retain important information of the input X.

The image size used is 128x128x3. The weight used is obtained from pre-trained ResNet50 model which has been trained using ImageNet dataset. All of the layers in ResNet50 have been frozen. Batch Normalization is used after the convolution layer and before the ReLU activation function. However, at the last convolution layer, the convolution block and the identity block are only followed Batch Normalization and the activation function is used after all information from the block and the shortcut connection are combined. There are 2 kinds of shortcut connection namely identity shortcut and projection shortcut. The identity shortcut is used when the dimension size of the input image and the output image are the same, while the projection shortcut is used when the dimensions size increase. The dimensions equalization are done by means of 1x1 convolution with stride 2 [21].

On the last layer of the ResNet50 base model, 2 Dense layers are used. The first Dense layer consists of 128 fully connected neurons. This layer is followed by ReLU activation function. The second Dense layer consists of 4 fully connected neurons followed by Softmax activation function. Softmax activation functions were chosen for its good performance on multi class classification problem [19]. Figure 5 describe the ResNet50 architecture in detail. We used Adam as the optimizer. Various parameters are used in the ResNet50 training process in order to obtain good accuracy in the white blood cell images classification.

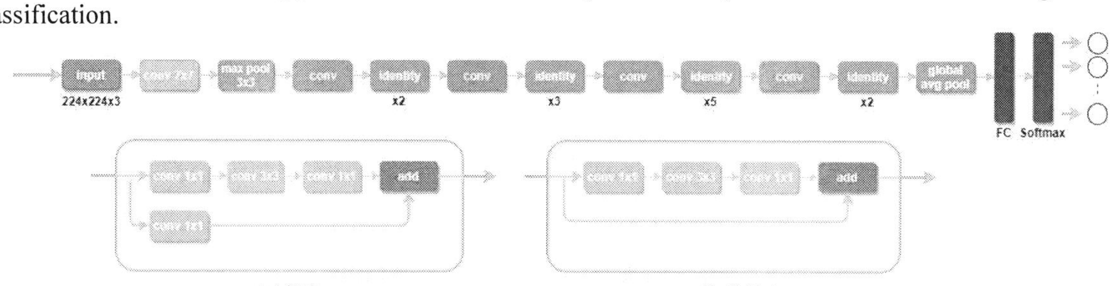

Figure 5. ResNet50 Architecture

2.6. *Evaluation*

In the evaluation phase, we used a confusion matrix to determine the accuracy, precision, recall, and F1-score of the classification model that we used. Confusion matrix is a tabular representation that can be used to visualize the performance of a technique [22]. Confusion matrix describes the number of True Positives (TP), False Positives (FP), True Negatives (TN) and False Negatives (FN) in two class problems. For example, there are two classes of A and B. Class A represents Positives and class B represents Negatives. So here, TP represents the amount of data classified correctly in class A, TN represents the amount of data correctly in class B, FP represents the amount of data classified incorrectly in class A and FN represents the amount of data classified incorrectly in class B. Accuracy is the ratio between the amount of data that is classified correctly and the total amount of data tested. Formulas for calculating accuracy, precision, recall, and F1-score are shown in equations (1), (2), (3), and (4) [23]:

$$\text{Accuracy} = \frac{TP+TN}{TP+TN+FP+FN} \tag{1}$$

$$\text{Precision}= \frac{TP}{TP+FP} \tag{2}$$

$$\text{Recall} = \frac{TP}{TP+FN} \tag{3}$$

$$\text{F1-Score} = \frac{2x(RecallxPrecision)}{(Recall+Precisio\ \ )} \tag{4}$$

3. Result and Discussion

This section describes the dataset scenario used in the study, the experimental setup, and the result and analysis.

3.1. *Dataset Scenario*

The synthetic images generated using GAN for each class is 2000 images. Every GAN generated images (The synthetic images) is rotated randomly between 0° and 360°. The synthetic image generated from GAN is shown on Figure 7. The synthetic images then combined with the classic augmented images on the training process. The real image from the dataset is used for ResNet50 testing process. The augmented image sample is shown on Figure 6.

The dataset is divided into 70% training dataset and 30% testing dataset. For the training dataset, it is split into 90% for training and 10% for validation. We used two dataset scenarios for this study. The difference between the scenarios is the ratio of synthetic images and real images used in the training process. The detail for the scenario is shown in Table 2.

Table 2. Dataset Scenario for Training

Scenario 1	Scenario 2
2,571 Train data (70% augmented data from GAN and 30% augmented data from dataset) each class	3,600 Train data (50% augmented data from GAN and 50% augmented data from dataset) each class
286 validation data each class (70% augmented data from GAN and 30% augmented data from dataset)	400 Validation data each class (50% augmented data from GAN and 50% augmented data from dataset)

3.2. *Experimental Setup*

Parameter used for DCGAN training is shown in Table 3. Various parameters are used in ResNet50 training process. The purpose of this various parameters is to find the best parameter for white blood cell classification. The parameter used in ResNet50 training process is shown in Table 4. Each parameter is used in the dataset scenario training process.

IOP Conf. Series: Materials Science and Engineering 1077 (2021) 012033 doi:10.1088/1757-899X/1077/1/012033

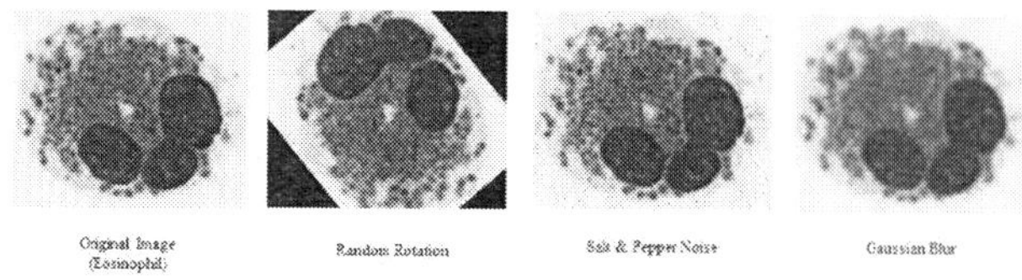

Figure 6. Classic Augmentation Results

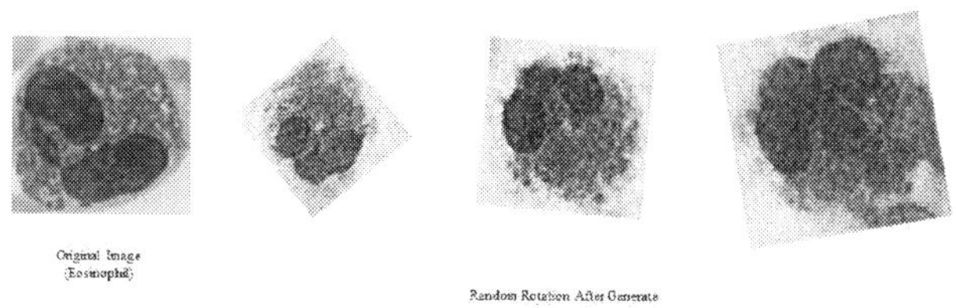

Figure 7. GAN Results

3.3. Experimental Results and Analysis

Before the white blood cell classification process using ResNet50, first we generate synthetic images using DCGAN. Using real images for each white blood cell class, we generate 2.000 synthetic images for each class. The synthetic image samples are shown in Figure 8.

The dataset is divided into training dataset and testing dataset. For the training dataset, it is split into 90% for training and 10% for validation. We used two dataset scenarios for this study. The difference between the scenarios is the ratio of synthetic images and real images used in the training process. The detail for the scenario is shown in Table 2.

Table 3. DCGAN Dataset Parameter

Parameter	Value
Learning Rate (Discriminator)	0.00005 (1e-5)
Learning Rate (Generator)	0.001 (1e-3)
Epoch	200

Table 4. ResNet50 Parameter

No.	Learning Rate	Epoch
1	0.0001	5
2	0.0001	10
3	0.00001	5
4	0.00001	10

Table 5. Best Testing Results

Scenario	Accuracy
Scenario 1	80.4%
Scenario 2	82.5%

Experiments were done using data obtained from [9] that had been augmented with classic data augmentation and DCGAN (synthetic images). There are two scenarios of model training in this study. The first scenario is by dividing the training data with 70% for synthetic images and 30% of original images using classic data augmentation. We used various parameter values such as learning rate and epoch. After conducting training and testing, we get the best results with a learning rate parameter value of 0.00001 and epoch 5, with an accuracy of 80.4%. Our second scenario uses training data of 50% for synthetic images and 50% for original images using classic data augmentation. After conducting training and testing, we get the best results with a learning rate parameter of 0.00001 and epoch 10, with an accuracy of 82.5%. Table 5 shows the best testing result for these experiments. Detailed confusion matrix for both scenarios are presented in Figure 9 and Figure 10. Classification errors occur because the GAN image has low quality even though it resembles blood cells in the dataset in terms of shape. This can be seen in Figure 8 lymphocyte class. Based on the shape, the lymphocyte class' GAN image matches the dataset, but the output is still blurry. Based on Table 5, the more data is used, the ResNet50 architecture can have better accuracy. The results of the evaluation of precision, recall and F1-score of each class for each scenario are presented in Table 6 and Table 7.

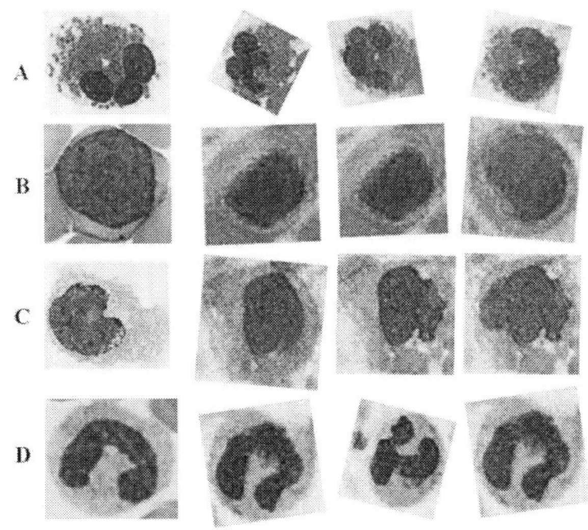

Figure 8. Synthetic Image Samples (A) Eosinophil, (B) Lymphocyte, (C) Monocyte and (D) Neutrophil.

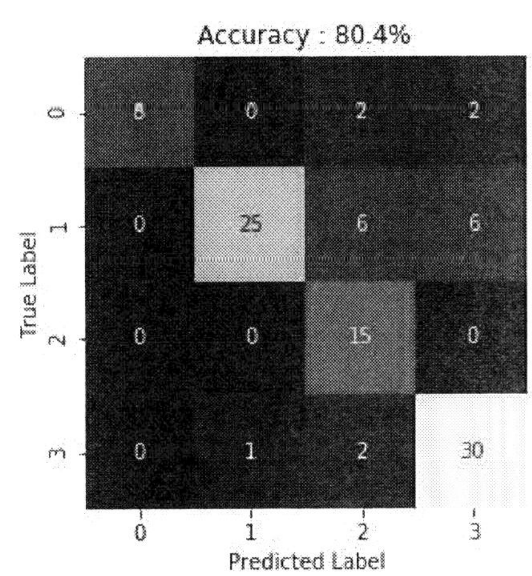

Figure 9. Scenario 1 Best Confusion Matrix

Table 6. Testing Result Dataset Scenario 1 for Best Parameter

Class	Precision	Recall	F1-Score
Eosinophil	100%	67%	80%
Lymphocyte	96%	68%	79%
Monocyte	60%	100%	75%
Neutrophil	79%	91%	85%

Table 7. Testing Result Dataset Scenario 2 for Best Parameter

Class	Precision	Recall	F1-Score
Eosinophil	71%	100%	83%
Lymphocyte	93%	68%	78%
Monocyte	82%	93%	87%
Neutrophil	81%	88%	84%

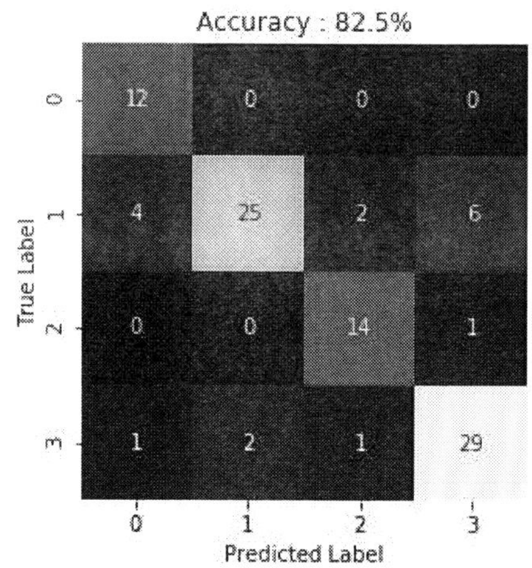

Figure 10. Scenario 2 Best Confusion Matrix

4. Conclusion

We have implemented DCGAN to generate synthetic white blood cell images in order to increase the training dataset amount on white blood cell classification. The classification model used for the white blood cell classification is ResNet50. There are two scenarios used in the training process. The first dataset scenario is training process using synthetic images and real images with 70:30 images ratio. The second dataset scenario is training process with 50:50 images ratio of real images and synthetic images. The real images are augmented using classic augmented method while the synthetic images are DCGAN generated and augmented images result. We only used four kinds of white blood cells as the dataset class, namely Eosinophil, Lymphocyte, Monocyte, and Neutrophil. The first dataset scenario obtain accuracy of 80.4% while the second dataset scenario obtain accuracy of 82.5%.

In this study, the used of DCGAN is not optimal so as to produce synthetic images that have poor quality. To improve the quality of the synthetic images produced by DCGAN, for future work, it is recommended to optimize the DCGAN to produce better quality synthetic images.

References

[1] Kutlu H, Avci E and Özyurt F 2019 White blood cells detection and classification based on regional convolutional neural networks *Med. Hypotheses* **135** 109472

[2] Shahin A I, Guo Y, Amin K M and Sharawi A A 2019 White blood cells identification system based on convolutional deep neural learning networks *Comput. Methods Programs Biomed.* **168** pp 69–80

[3] Macawile M J, Quiñones V V, Ballado A, Cruz J D and Caya M V 2018 White blood cell classification and counting using convolutional neural network *2018 3rd Int. Conf. on Control and Robotics Engineering (ICCRE), Nagoya* pp 259–63

[4] Rawat J, Singh A, Bhadauria H S, Virmani J and Devgun J S 2018 Application of ensemble artificial neural network for the classification of white blood cells using microscopic blood images Jitendra Virmani *Int. J. Comput. Syst. Eng.* **4** pp 202–16

[5] Alom Z, Yakopcic C, Taha T M and Asari V K 2018 Microscopic blood cell classification using

inception recurrent residual convolutional neural networks *NAECON 2018 - IEEE Natl. Aerosp. Electron. Conf.* pp 222–7
[6] Frid-adar M, Diamant I, Klang E, Amitai M, Goldberger J and Greenspan H 2018 GAN-based synthetic medical image augmentation for increased CNN performance in liver lesion classification *Neurocomputing* **321** pp 321–31
[7] Ma Y, Zhang P and Tang Y 2018 Research on fish image classification based on transfer learning and convolutional neural network model *14th Int. Conf. Nat. Comput. Fuzzy Syst. Knowl. Discov.* pp 850–5
[8] Zheng X, Wang Y, Wang G and Liu J 2018 Fast and robust segmentation of white blood cell images by self-supervised learning *Micron* **107** pp 55–71
[9] CellaVision 2020 *Anon CellaVision News Blast – CellaVision AB* Retrieved from http://blog.cellavision.com/
[10] Sirazitdinov I, Kholiavchenko M, Kuleev R and Ibragimov B 2019 Data augmentation for chest pathologies classification *2019 IEEE 16th Int. Symp. Biomed. Imaging* pp 1216–9
[11] Li Z, Kamnitsas K and Glocker B 2019 Overfitting of neural nets under class imbalance: analysis and improvements for segmentation *Medical Image Computing and Computer Assisted Intervention – MICCAI 2019* Lecture Notes in Computer Science vol **11766** ed D Shen *et al* (Cham: Springer)
[12] Madani A, Moradi M, Karargyris A and Syeda-mahmood T 2018 Semi-supervised learning with generative adversarial networks for chest x-ray classification with ability of data domain adaptation *2018 IEEE 15th Int. Symp. Biomedical Imaging* pp 1038–42
[13] Mahmoud M A B and Guo P 2019 A novel method for traffic sign recognition based on DCGAN and MLP with PILAE Algorithm *IEEE Access* **7** pp 74602–11
[14] Radford A, Metz L and Chintala S 2016 Unsupervised Representation Learning with Deep Convolutional *Preprint arXiv:1511.06434*
[15] Su P L, Sappa A D and Vintimilla B X 2017 Infrared image colorization based on a triplet DCGAN Architecture *2017 IEEE Conf. Comput. Vis. Pattern Recognit. Work.*
[16] Yang J, Li T, Liang G, He W and Zhao Y 2019 A simple recurrent unit model based intrusion detection system with DCGAN *IEEE Access* vol **7** pp 83286–96
[17] Tan C, Sun F, Kong T, Zhang W, Yang C and Liu C 2018 A Survey on Deep Transfer Learning *Preprint arXiv:1808.01974*
[18] Lagunas M and Garces E 2017 Transfer learning for illustration classification *CEIG - Spanish Comput. Graph. Conf.*
[19] Sai Bharadwaj Reddy A and Sujitha Juliet D 2019 Transfer learning with RESNET-50 for malaria cell-image classification *Proc. 2019 IEEE Int. Conf. Commun. Signal Process. ICCSP 2019* pp 945–9
[20] He K, Zhang X, Ren S and Sun J 2016 Deep residual learning for image recognition *Proc. IEEE Comput. Soc. Conf. Comput. Vis. Pattern Recognit.* pp 770–8
[21] Rezende E, Ruppert G, Carvalho T, Ramos F and De Geus P 2017 Malicious software classification using transfer learning of ResNet-50 deep neural network *Proc. - 16th IEEE Int. Conf. Mach. Learn. Appl. ICMLA 2017* pp 1011–4
[22] Lohita K, Sree A A, Poojitha D, Renuga Devi T and Umamakeswari A 2015 Performance analysis of various data mining techniques in the prediction of heart disease *Indian J. Sci. Technol.* **8** pp 1–7
[23] Ghoneim S 2019 *Accuracy, Recall, Precision, F-Score & Specificity, which to optimize on?* Retrieved from https://towardsdatascience.com/accuracy-recall-precision-f-score-specificity-which-to-optimize-on-867d3f11124

An evidence-based culture: implementing a large-scale social restriction policy on the COVID-19 prevention protocol in Jakarta

H Wiguna[1], Y Nugraha[1,2], J I Kanggrawan[1], A L Suherman[1,3], R S Akbar[4], P A Pradana[4]

[1]Jakarta Smart City, Jakarta, Indonesia
[2]School of Computing, Telkom University, Bandung, Indonesia
[3]Directorate of Research and Community Service, Telkom University, Bandung, Indonesia
[4]Electrical Engineering Faculty, Telkom University, Bandung, Indonesia

E-mail: [1]hansenwgn@gmail.com, [2]yudhistira.nugraha@jakarta.go.id

Abstract. COVID-19 has a severe and widespread impact, especially in Indonesia. One of the efforts of the DKI Jakarta Provincial Government in reducing the risk of spreading the virus is by implementing a Large-Scale Social Restriction (hereafter referred to as PSBB) policy. The hope of implementing this policy is to detain or even reduce active (positive) cases in Jakarta. In this paper, we conduct Exploratory Data Analysis (EDA) to analyze and evaluate the effect of the PSBB policy that has been made by the DKI Jakarta Provincial Government in minimizing the active cases and in increasing the number of tests in Jakarta. The results show that the PSBB policy was not directly able to limit the rate of active cases, but specific cases of people in intensive care were able to be suppressed. The results suggest that the other local governments can use EDA method in making further policies and interventions for handling COVID-19. The results also indicate that the public is expected to continue to take preventive actions, such as complying with health protocols established by the government.

Keywords. PSBB, COVID-19, Jakarta, exploratory data analysis, testing, data-driven policy.

1. Introduction

In December 2019, the COVID-19 outbreak began to spread in Wuhan, China. The World Health Organization (WHO) declared it a "Public Health Emergency of International Concern" on 30 January 2020 and upgraded the status to a pandemic on 11 March 2020 [1]. COVID-19 is a new disease that can cause respiratory problems and pneumonia. This disease is caused by infection with Severe Acute Respiratory Syndrome Coronavirus 2 (SARS-CoV-2). Clinical symptoms that appear vary, ranging from common cold symptoms (cough, runny nose, sore throat, muscle aches, headache) to severe complications (pneumonia or sepsis). From various studies, the primary method of transmission of this disease can be through respiratory droplets and close contact with sufferers [2]. People who are infected may have mild symptoms, such as fever, cough, and difficulty breathing. Indonesia as the fourth most populous country in the world is predicted to face a significant threat to this pandemic, especially Jakarta as the capital city and the centre of the national economy [3] [4]. The spread of COVID-19 cases in Jakarta continues to increase, for instance, people who have been confirmed positive and died due to the COVID-19 virus, from early March 2020. There was quite a stable period between 1 May – 7 July 2020 however, from 8 July to this day, active cases been in an upward trend.

Various efforts by the Indonesia and Jakarta governments have been made in dealing with the transmission of this virus. Governor Regulation 33/2020 concerning the implementation of Large-Scale Social Restrictions (PSBB) was issued on 9 April 2020. A virologist in the United States noted that stopping travel is the best way to stop the spread of the virus [5]. Apart from the social restriction policy, the DKI Jakarta Provincial Government has also increased the testing capacity. According to the World Health Organization (WHO), Jakarta is the only city in Indonesia that has achieved the minimum case detection benchmark, which is 1 per 1,000 populations per week [6]. The increasing capacity of the testing performed, the more active cases were found, both self-isolation patients and under-treatment patients. This means that the hospital bed capacity is reduced. The bed occupancy rate (BOR) has become our concern. BOR is a measure of utilization of the available bed capacity in a hospital. It indicates the percentage of beds occupied by patients in a defined period.

Currently, the development of research on this virus continues to progress and change rapidly until now, including efforts to find a vaccine. On 27 May 2020, WHO published updated interim guidance on the clinical management of COVID-19 [7] [8] and provided updated recommendations on the criteria for discharging patients from isolation. The updated criteria reflect recent findings that patients whose symptoms have resolved may still test positive for the COVID-19 virus (SARS-CoV-2) by RT-PCR for many weeks. Despite this positive test result, these patients are not likely to be infectious and therefore are unlikely to be able to transmit the virus to another person [9] [10] [11]. The reason for the change is that the initial recommendation of two negative RT-PCR tests at least 24 hours apart, has been complicated in light of limited laboratory supplies, health care facilities, and health care workers in areas with highly infectious or transmission. One of the challenges is long periods of isolation for individuals with prolonged viral RNA detection after the resolution of symptoms, affecting individual well-being, society, and access to healthcare [12]. Moreover, insufficient testing capacity to comply with initial discharge criteria happen in many parts of the world. In the Indonesia context, not many cities have adequate health care facilities that could provide all the requirements from WHO.

There was a slight change in terms of terminology base on the Decree of the Minister of Health Number HK.01.07/MENKES/413/2020 [13]. Started 30 July 2020, Jakarta official COVID-19 website also changed the design of the website by focusing more on providing active cases and testing capacity data. As a government, Jakarta Province plays a significant role in making policies related to the COVID-19 pandemic. The government of DKI Jakarta Province considers the balance of various aspects, such as the health side, the limitations of health facilities, the development of science and research, as well as other elements such as economic, social and psychological aspects. In this case, the role of data and analysis can help the government. As written in the journal *The Use of Big Data in The Public Policy Process: Paving the Way of Evidence-Based Governance* which shows that data can be used to increase efficiency, effectiveness, government transparency, and better service providers based on increasing insight into needs and citizen demands, as well as more informed policy-making [14].

This paper aims to analyze the impact of the large-scale restriction on the COVID-19 cases and testing capacity in Jakarta. Data-based evaluation means, in monitoring and evaluating the policies that have been implemented, the Jakarta Provincial Government does not rely solely on intuition or the subjective opinion of policymakers but makes use of the results of data analysis. The approach chosen in this article is exploratory data analysis (EDA). EDA was selected because its application adapted to the context and details of the study. Data is explored from various perspectives so that EDA does not stick to a standard technique. EDA has flexible characteristics and is needed to identify and investigate a phenomenon that occurs when conducting empirical research [15]. Due to balancing the health reason, considering the unnecessarily prolonged the isolation period that excluded from work or other responsibilities, limitation of the healthcare facilities and personnel, the analysis and pattern of the data are expected to give an invaluable contribution to the decision and the making of the better policy.

The structure of the writing of this paper is as follows: Section 2 provides information regarding the data set and methodology used in the study. Section 3 presents the result of the analysis related to the use of EDA method and other research related to the COVID-19 trends. Section 4 discusses how the findings could help with the implication for a concrete recommendation. Finally, Section 5 concludes this paper with conclusions and recommendations.

2. Methodology

Data processing for analysis of the spread of COVID-19 cases and testing in the DKI Jakarta Province consists of several stages. The first stage is getting the data from data sources. The primary data comes from the DKI Jakarta Health Office. The data consists of daily COVID-19 data which comes from the COVID-19 database used by the Health Office, Puskesmas (community health centres), and hospitals. Data collection was carried out by the DKI Jakarta Provincial Health Office Surveillance at each level of health facilities spread across Jakarta. The data period used is from 3 March to 24 August 2020. The data is in a table format and is updated every day. The second stage is data processing by the Jakarta Smart City and the Spatial Planning and Land Service Agency (*Dinas Cipta Karya, Tata Ruang, dan Pertanahan or Citata*). The third stage is the process of data preparation and cleaning, which is ensuring that there are no redundant data or inconsistent data conditions. Finally, the fourth stage is data visualization to the corona.jakarta.go.id microsite, and it is updated every day. Figure 1 below shows a flow diagram of the data processing process.

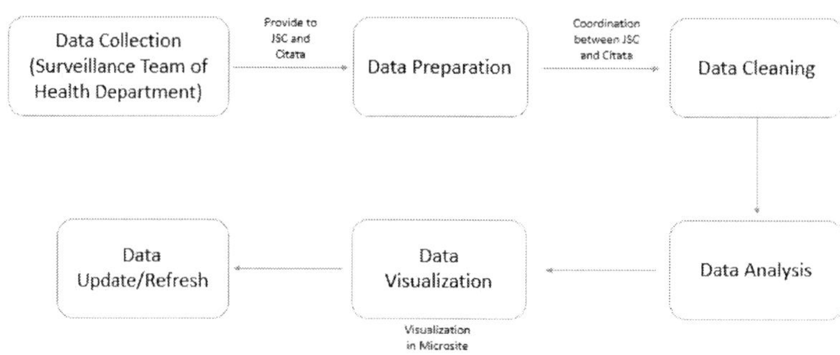

Figure 1. Data Processing Diagram

The data consists of COVID-19 patient's data along with information on gender, age group, and location. Also contains daily time series summary tables, including the number of active cases, new daily cases, self-isolation patients, under-treatment patients, and the number of testing in the past number of days from which the pandemic started in Jakarta, Indonesia. The data period used is from 3 March to 24 August 2020. These datasets are good quality data because each dataset has the following characteristic elements: credibility, availability, reliability, relevance, and readability [16].

After the quality of the dataset is guaranteed, the next step is to apply the Exploratory Data Analysis (EDA) method. EDA is a data exploration method using basic statistical techniques and data visualization techniques to summarize observational data [17].

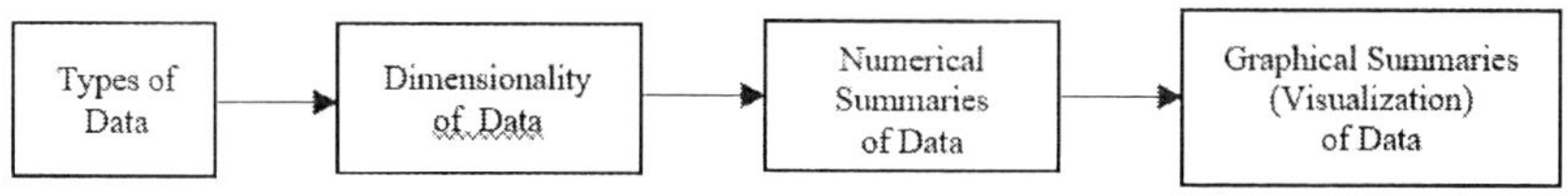

Figure 2. Exploratory Data Analysis Process

Figure 2 shows how the EDA approach process will be carried out. The first step, namely 'Types of Data' or knowing the data type of each field (column) to be used, whether the data is categorized as categorical or quantitative. The next step is 'Dimensionality of Data' or knowing the dimensions of the data, whether the data is one-dimensional (univariate), two-dimensional (bivariate) or multi-dimensional (multivariate). Then the 'Numerical Summaries of Data' is carried out or look at the numerical summary of the datasets such as the mean, median, standard deviation and others to find out the statistical conditions of the data. The final stage is 'Graphical Summaries (Visualization) of Data' or visualization to get interesting patterns and information related to the dataset being analyzed [18]. Selection of charts is adjusted to the purpose of the information to be obtained, the types and dimensions of data that have been known from the previous process.

With this approach, data can be explored from multiple perspectives because EDA is not fixated on a standard tool or technique [15]. The EDA approach is considered adequate in this context because this study does not require complex analytical methods. It is sufficient to observe patterns in general and see the relationship between several variables.

3. Results

This paper shows data analysis and exploration from Jakarta's COVID-19 data. The policy timeline can be divided into several periods: 1) PSBB Stage I: 10 April – 23 April 2020; 2) PSBB Stage II: 24 April – 22 May 2020; 3) PSBB Stage III: 23 May – 4 June 2020; 4) PSBB Transition Stage I: 5 June – present (on progress extended). Figure 3 shows the gender of COVID-19 patients.

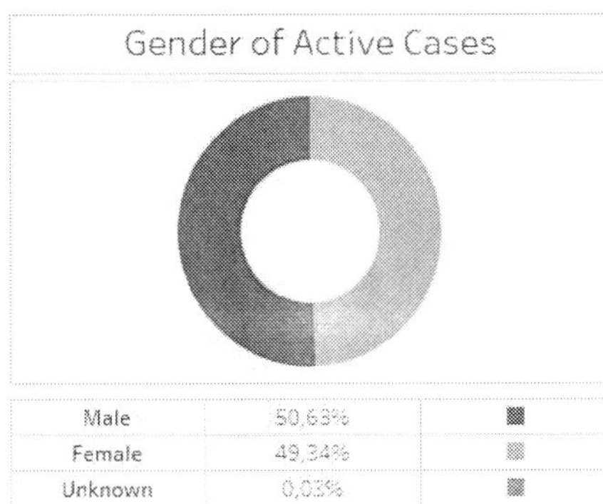

Figure 3. Active Case Gender (cut off 24 August 2020)

As shown in Figure 3, there is no significant difference between the gender affected by COVID-19. It can be assumed that the COVID-19 virus does not have more impact or transmit to a specific gender. Furthermore, Figure 4 shows the affected age groups per gender. The result indicates that up to the cut off on 24 August 2020, women who are involved in the age range of 20-39 years reach 49.17% while men who are involved in the age range of 20-39 years reach 44.18%. From the data, it can be seen that the productive age population dominates the affected age, this is possible because the community of that age is still actively engaged in mobilizing or working activities outside the home.

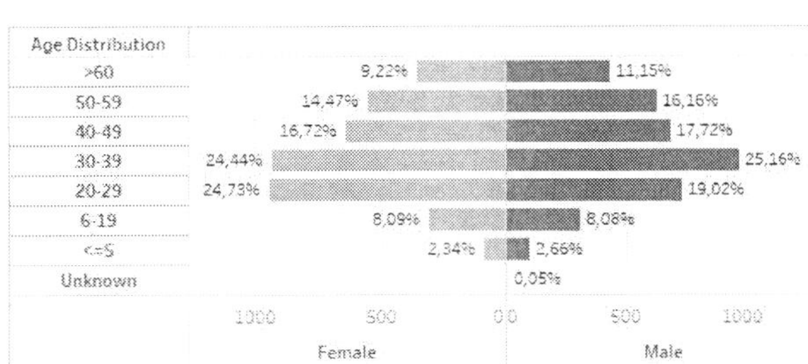

Figure 4. Active Cases Gender and Age Distribution (cut off 24 August 2020)

From the location data, it can be seen that the most active cases are in the Cilincing sub-district (Figure 5). The map in Figure 5 shows the information by way of color grading. Where the darker the color gradation is shown on the map (sub-district boundary), indicates the more positive case population. Figure 6 shows the distribution of active cases per districts in Jakarta. East Jakarta is the city area with the most active cases (23.87%) if it excludes unknown data.

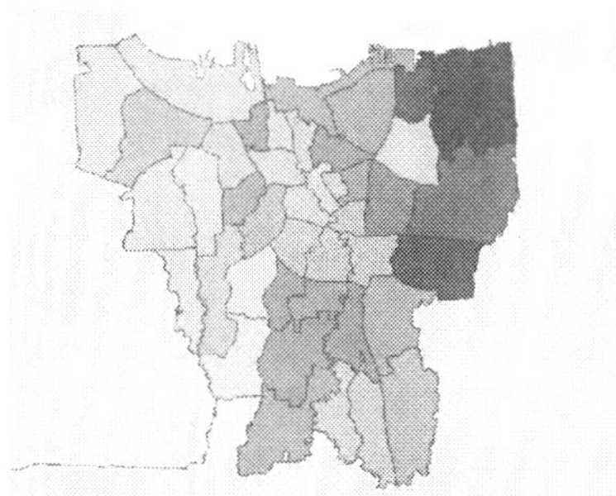

Figure 5. Mapping of Active Cases per District (cut off 24 August 2020)

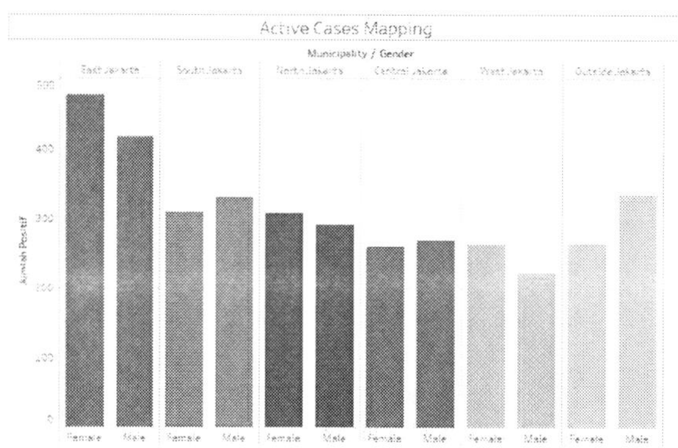

Figure 6. Active Cases per Municipality per Gender (cut off 24 August 2020)

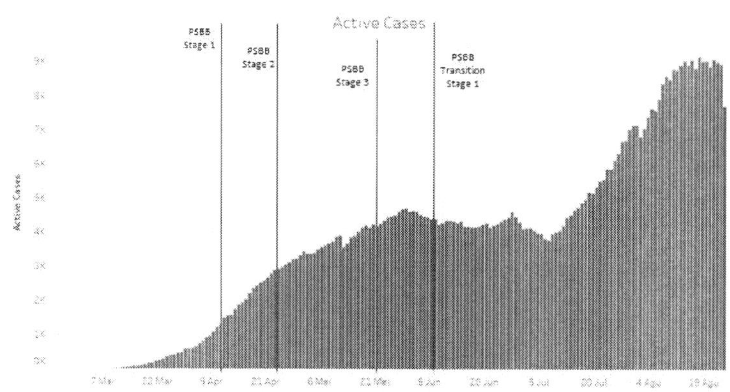

Figure 7. Active Cases Trend (3 March – 24 August 2020)

Started 30 July 2020, the design of corona.jakarta.go.id had been changed. The DKI Jakarta Province focuses on giving information about active cases and testing data. From Figure 7, we could see the active case number. In each period of Large-Scale Restriction, it does not appear to be very helpful in reducing

IOP Conf. Series: Materials Science and Engineering 1077 (2021) 012034 doi:10.1088/1757-899X/1077/1/012034

the development and increase of active cases. Active cases are patients with status under treatment (intensive care) or self-isolation. From 1 May – 7 July 2020, it can be said that active cases in Jakarta are relatively stable (there has not been a significant decrease or increase). However, from 8 July until now, active cases have an upward trend. Active cases continue to rise due to new cases continue to increase every day, which we can see in Figure 8.

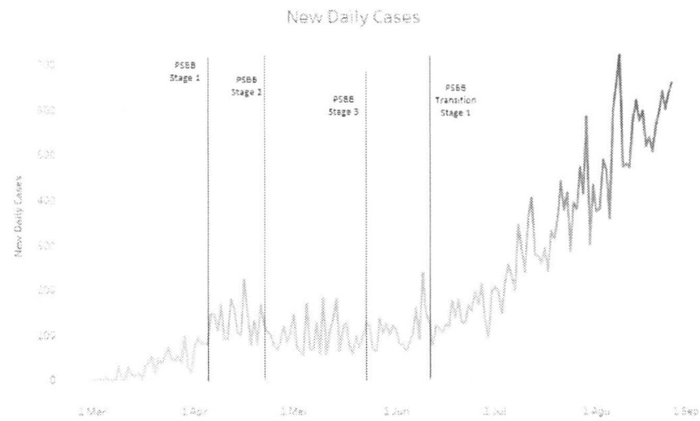

Figure 8. New Daily Cases (3 March – 24 August 2020)

As shown in Figure 8, the addition of new cases was relatively constant from the time the PSBB Stage I was implemented to the end of the PSBB Stage 3. However, since the PSBB Transition Stage I until now, daily new cases tend to experience a sharp increase. This trend does not merely mean the influence of social restriction policies that are not functioning correctly; of course, various factors influence this daily increase. The increment in new daily cases was mainly due to increased testing capacity. On the other hand, we might see from different perspectives such as people and transportation mobility during PSBB, whether we might conclude that people obey this policy. There needs to be further research as well as to how effective the people in Jakarta apply these social restrictions.

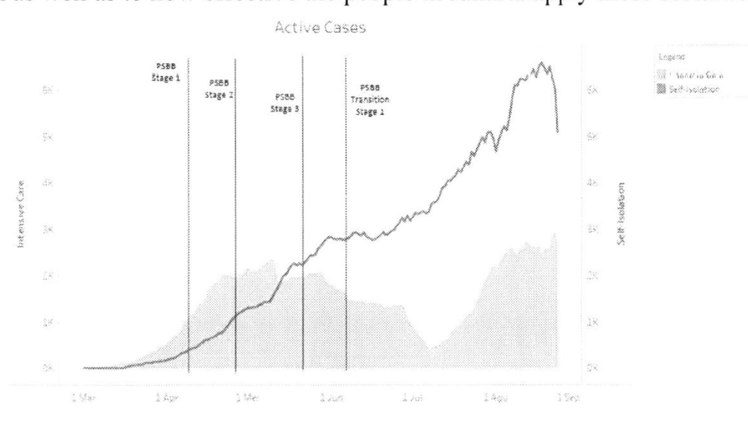

Figure 9. Self-Isolation & Intensive Care Cases (3 March – 24 August 2020)

Interestingly, when we look at a fraction of the active case. We can see active cases consisting of self-isolation and under treatment (intensive care) in Figure 9. The data shows that people who were in intensive care gradually decreased in the period from 10 May to 7 July 2020, which reached its lowest point of 405 people. But then it increased until it reached its peak, 2922 people, on 23 August 2020. This result suggests that from the PSBB stage II policy, it was able to reduce the number of people in intensive care, although the last 1.5 months had increased again. The initial hypothesis why patients in intensive care increased again at the end was that health workers were increasingly overwhelmed in

IOP Conf. Series: Materials Science and Engineering 1077 (2021) 012034 doi:10.1088/1757-899X/1077/1/012034

handling COVID-19 patients. We can also see through the chart availability of beds in the hospital below.

In contrast to people in intensive care, people with self-isolation status, in general, continued to experience an exponential increase from the start of cases to the period 21 August 2020, there was a significant decrease on 22–23 August 2020. However, the numbers were still far above those in intensive care. People with self-isolation status indicate the patient's condition is not experiencing severe enough symptoms so that they can carry out self-care and quarantine at home. Also, it is beneficial for overloaded health workers, so that health workers can take care of people who are more in need of the hospital.

The increasing number of active cases is due to the DKI Jakarta Provincial Government implementing active case finding. The number of PCR and RDT tests conducted in DKI Jakarta is the highest compared to other cities in Indonesia and reaches WHO standards. The ability of the number of tests that meet this standard cannot be separated from the cooperation of various parties, both government-owned and private laboratories.

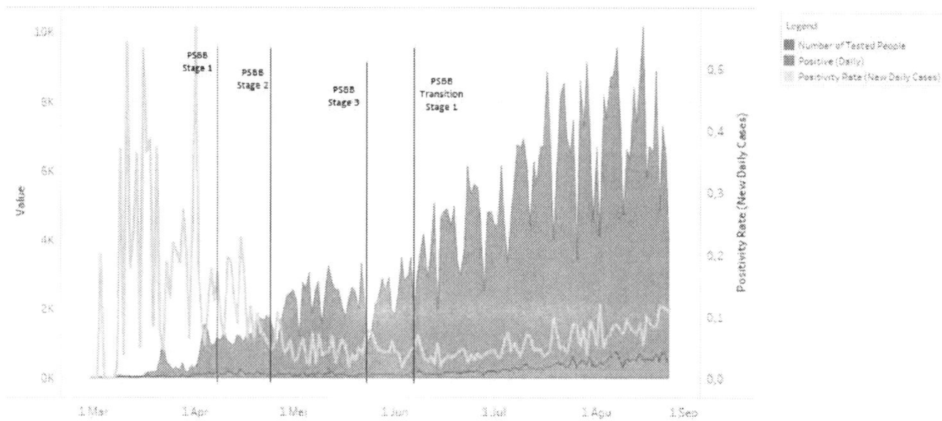

Figure 10. Positivity Rate and Number of Tested People (cut off 23 August 2020)

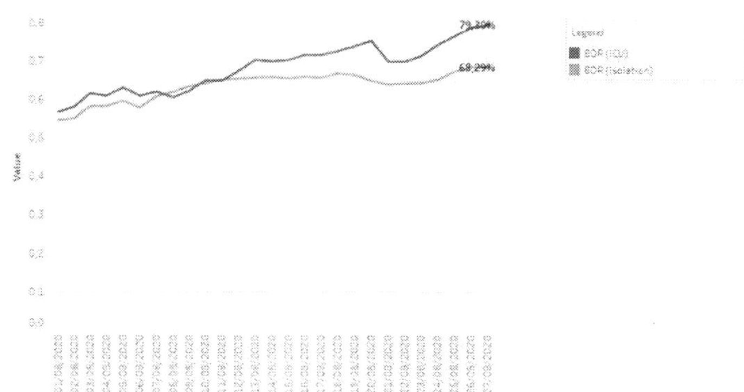

Figure 11. Bed Occupancy Rate ICU & Isolation (1 – 27 August 2020)

The positivity rate is the ratio between positive people and the number of people tested. As shown in Figure 10, the number of people tested in Jakarta continues to increase. Due to minimum testing, the positivity rate peaked at 56.71%, on 1 April 2020. We could see after the testing capacity increased, it was able to decline and stabilize, the average positivity rate until 23 August was 8.79%.

The data that is also related and needs attention is the bed occupancy rate. The Bed Occupancy Rate (BOR%) is a measure of utilization of the available bed capacity in a hospital. It indicates the percentage of beds occupied by patients in a defined period. From the data from the Jakarta Health Office, as shown in Figure 11, the use of beds in ICU is 79.3% and in isolation 68.29%. This data is data from a total of 67 COVID-19 Reference Hospitals in Jakarta. If it gets higher, of course, it indicates that the beds in the hospital for COVID-19 patients are running low, and this needs to be a common concern.

4. Discussion

This paper shows data analysis and exploration from Jakarta's COVID-19 data. In addition to exploring the Jakarta COVID-19 data, we also saw the impact of large-scale social restriction (PSBB) on the results of active cases, the number of tests, and the implicated bed capacity. The policy timeline can be divided into several periods: 1) PSBB Stage I: 10 April – 23 April 2020; 2) PSBB Stage II: 24 April – 22 May 2020; 3) PSBB Stage III: 23 May – 4 June 2020; 4) PSBB Transition Stage I: 5 June – present (on progress extended).

As research on COVID-19 continues to progress to date, the efforts and policies that are being made by the government of DKI Jakarta Province can be more effective. From the data analysis carried out, we can say that the PSBB policy did not influence efforts to hold back the rate of the addition of active cases in general. However, we can see that specific cases of people in intensive care have been successfully suppressed and even decreased. This information is a good thing considering the capacity or availability of beds in hospitals is also increasingly limited. So that we have to prioritize those who need to be hospitalized, it can be seen that people with an increasing status of self-isolation are a good sign.

Apart from active cases, along with the implementation of this PSBB policy, the Jakarta Provincial Government continues to increase the number of tests. As long as a vaccine has not been found, one indicator of a country or region that can be said to be successful in efforts to deal with COVID-19 is to pay attention to and increase testing capacity, especially PCR. Not only Jakarta, the official website of Tokyo, Japan also emphasized their test number data [19]. The more people who are confirmed positive, the treatment for those in need will be maximized. It is also not dismissed, of course, that the increasing number of tests can increase the number of positive cases (daily and accumulated). Another challenge is how to determine whether these self-isolating patients can completely escape without using multiplex PCR tests, given the limited cost and number of health facilities. In addition, self-isolation for workers who are affected and unable to do work from home needs to be considered for their mental health. We must balance the isolation time, which is not required so that people can also return to their work and their responsibilities; we must keep abreast of the latest research developments.

In this case, the government takes a significant role in making policy. Various aspects are, of course, taken into consideration. So far, the DKI Provincial Government has carried out the Phase I Transition PSBB Extension for the fifth time. How to make policies that emphasize health aspects, limited health facilities, mental and psychiatric health, economy, and various other essential elements is a difficult task and needs consideration and discussion from multiple parties.

5. Conclusion

This study aims to analyze Jakarta's COVID-19 data. Since the beginning of the outbreak of COVID-19 cases in Jakarta in March 2020 until now (cut off 24 August 2020), the age range of those infected with COVID-19 is consistent in the age range of 20-39 years. The PSBB policy carried out by the government was not directly able to restrain the rate of the addition of active cases in general. Still, specific cases of people in intensive care were able to be suppressed and even decreased. This information indicates a good sign considering that there are limited health facilities in the hospital. The number of tests that continues to increase is also a good indicator because the more people who are confirmed positive are found, the treatment for those in need will be maximized. Of course, the increasing number of tests also directly impacts the number of active cases. This paper presents an overview of the government in making additional policies. Since the research on COVID-19 still progressing, thus the challenge is how the government could take a balanced system and pay attention to various aspects such as health, limited health facilities, mental health, economy, and many other things.

References

[1] Askitas N, Tatsiramos K and Verheyden B 2020 Lockdown strategies, mobility patterns, and COVID-19 *Discussion Paper Series IZA Institute Labor Economics* issue 23
[2] Seto W H *et al* 2003 Effectiveness of precautions against droplets and contact in prevention of nosocomial transmission of severe acute respiratory syndrome (SARS) *Lancet* **361** no 9368 pp 1519–20
[3] Fadly F and Sari E 2020 An approach to measure the death impact of Covid-19 in Jakarta using autoregressive integrated moving average (ARIMA) *Unnes J. of Public Health* **9** no 2
[4] United States Census Bureau 2020 *U.S. Census Bureau Current Population* Retrieved on 8 July 2020 from https://www.census.gov/popclock/print.php?component=counter
[5] Upton J 2020 Pandemics are the dark side of global mobility *Nautilus* Issue no 84
[6] World Health Organization 2020 *Coronavirus disease 2019 (COVID-19) Situation Report 14 World Health Organization* Retrieved from https://www.who.int/indonesia
[7] World Health Organization 2020 *Clinical management of COVID-19 (Interim Guidance) World Health Organization* Retrieved from https://www.who.int/publications-detail/clinical-management-of-covid-19
[8] World Health Organization 2020 *Coronavirus disease (COVID-19) Situation Report – 129 World Health Organization* Retrieved from https://www.who.int/docs/default-source/coronaviruse/situation-reports/20200528-covid-19-sitrep-29.pdf?sfvrsn=5b154880_2
[9] Korea Disease Control and Prevention Agency 2020 *Findings from Investigation and Analysis of re-positive cases KDCA* Retrieved from https://www.cdc.go.kr/board/board.es?mid=a30402000000&bid=0030&act=view&list_no=3 6 7267&nPage=1
[10] Li N, Wang X and Lv T 2020 Prolonged SARS-CoV-2 RNA shedding: not a rare phenomenon *J. Medical Virology* **92** pp 2286–87 DOI: 10.1002/jmv.25952
[11] Xiao F *et al* 2020 Infectious SARS-CoV-2 in feces of patient with severe COVID-19 *Emerging Infectious Diseases* **26(8)** pp 1920–2 DOI:10.3201/eid2608.200681
[12] Cheng H Y *et al* 2020 Contact Tracing Assessment of COVID-19 Transmission Dynamics in Taiwan and Risk at Different Exposure Periods Before and After Symptom Onset *JAMA Intern Med.* **180(9)** pp 1156–63 DOI: 10.1001/jamainternmed.2020.2020
[13] Ministry of Health Decree of Republic Indonesia 2020 *HK.01.07/Menkes/413/2020 COVID-19 Prevention and Control Guidelines* Retrieved from https://covid19.go.id/storage/app/media/Regulasi/KMK%20No.%20HK.01.07-MENKES-413-2020%20ttg%20Pedoman%20Pencegahan%20dan%20Pengendalian%20COVID-19.pdf
[14] Studinka J and Guenduez A A 2018 The use of big data in the public policy process: paving the way for evidence-based governance *EGPA Conf. 2018*
[15] Behrens J T 1997 Principles and procedures of exploratory data analysis *Psychol. Methods* **2** pp 131–60
[16] Cai L and Zhu Y 2015 The challenges of data quality and data quality assessment in the big data era *Data Sci. J.* **14** no 2 pp 1–10
[17] Makridakis S, Wheelwright S C and McGee V E 1999 *Metode dan aplikasi peramalan* (Jakarta: Binarupa Aksara)
[18] UW Genome Sciences 2020 *Lecture 2: Descriptive Statistics and Exploratory Data Analysis* Retrieved from www.gs.washington.edu/academics/courses/akey/56008/lecture/lecture2.pdf
[19] Tokyo Metropolitan Government 2020 *Updates on COVID-19 in Tokyo* Retrieved from https://stopcovid19.metro.tokyo.lg.jp/en/

Forecasting demand factors of tourist arrivals in Indonesia's tourism industry using recurrent neural network

Melati Puspa Anisa*, Herry Irawan and Sri Widiyanesti

School of Economics and Business, Telkom University, Bandung, Indonesia

E-mail: *puspaanisa@student.telkomuniversity.ac.id

Abstract. Tourism is one of the sectors that have contributed to a wide variety, not just economically, but also socially political, culturally, regionally and environmentally. In Indonesia, the tourism sector has an important role in contributing to economic growth, especially in foreign exchange earnings. Therefore, it is very important to maintain and encourage the growth of tourism in Indonesia with the need for a model of forecasting the arrival of foreign tourists to Indonesia to assist the government in developing a tourism plan strategy. There are factors in demand forecasting which affect a demand in the tourism sector. In this study data in the form of fuel prices, exchange rates, per capita GDP, and the volume of bilateral trade from five countries over the span of January 2012 to December 2019 are used as variables that affect the arrival of foreign tourists. The method used to create a forecasting model is one of recurrent neural network architectures namely long short-term memory (LSTM). Three models are put in test and each model uses four types of parameters that are look-back value, hidden layer, number of epochs, and batch size. The first prediction model gives 97.21% of the highest accuracy. The second predictive model provides 99.17% of the highest accuracy. Lastly, the third prediction model provides 99.21% of the highest accuracy.
Keywords: *Forecasting, Tourism, Demand, Recurrent Neural Network, LSTM.*

1. Introduction

Indonesia is well known as a country in which nature, culture and historical heritage have a range of beauty. Such circumstances indicate there are high potentials for Indonesia in tourism [1]. Indonesia's tourism industry has become a significant part of national growth because it can hold a lot of human resources, boost the local economy, and state income [2]. In Indonesia's economic development, tourism has become a GDP main contributor. It is expected to be one of the critical drivers of accelerating economic growth in Indonesia by creating jobs and businesses, export revenues and improving infrastructure [22]. In developing international tourism, to increase the flow of foreign visitors, it is essential to undertake programs that are targeted and relevant. Tourism's contribution has a broad dimension, not only in economic terms but also in social-political, cultural and regional conditions. Economically, the tourism sector contributes significantly to the foreign exchange earnings of the state, local own-source revenue and also community revenue generated from tourism businesses that are built and open up several job opportunities and great employment opportunities [1]. The number of foreign tourists can increase the foreign exchange of the country. It is because of the high foreign currency exchange against the local currency [23]. Additionally, Noersasongko *et al.* (2016) point out increasing the number of foreign tourists is one of the ways to improve the currency value.

The government can prepare strategic steps to build the tourism industry with the knowledge of the number of tourists coming.

As a leading sector, tourism is identified as an essential sector to develop in synergy. Through a sustainable tourism approach, efforts to conserve nature and culture and their legacies need to be synergized to support the acceleration of national development [3]. The factor tourist arrivals are still the most widely held indicator of tourism demand over the past few years [4]. According to Filippini and Hunt (2011), the need for precise tourism forecasts is therefore particularly important because of the significant contribution of the industry to the economy. To reduce the uncertainty in the decision-making process, preparation relies heavily on reliable forecasts [5]. Any data about potential tourist traffic volumes and trends is vital for hoteliers, tour operators and other tourism and transportation-related industries to prepare their investment and marketing strategies. Government institutions also need precise demand forecasts for tourism to plan and provide the necessary tourism infrastructure, such as accommodation site planning and development of transportation, among other needs. The tourist arrivals and demand forecast help the company to plan better. Long-term and short-term projections serve multiple roles, ranging from personnel and other resources to public infrastructure or equipment investments [24].

Recent studies (e.g. [6], [2], and [7]) have shown such tourism forecasting was carried out by the researchers with different methods. Similarly, Alamsyah and Friscintia (2019) [6] reviewed tourism demand forecasting as well as its determinant factors from 5 major Indonesia tourist markets by artificial neural network model. Vahikan *et al.* (2017) [2] used three different factors of foreign tourist arrivals from ten countries to Bali using recurrent neural network backpropagation through time. The research conducted by comparing two forecasting methods of RNN-BPTT with single moving average (SMA). It is proven that RNN-BPTT outstanding the SMA accuracy. Furthermore, Balli *et al.* (2018) [7] adapted and applied SARIMAX/(E)GARCH volatility models to forecast visitor arrivals by air from its eight main tourist source markets to New Zealand, and to monitor macroeconomic factors along with global and regional structural changes.

The recurrent neural network approach is appropriate to resolve the problem of time-series prediction. RNN can overcome the non-linear problem prediction. The RNN model is a forecasting model that utilizes past values to predict future values of time-series data [8]. The application of the RNN method in forecasting is expected to produce figures that close to the actual data, so that, the results can be used as decision-making material to take action in minimize the decline occurs in foreign tourist arrivals in Indonesia [9].

2. Literature review

There are some literatures review in which became a highlight in this study.

2.1. Forecasting the demand factors in tourism industry

Tourism is made up of ideas and opinions held by people who form their decisions about going on a trip, where to go or where not to go, and what to do or not to do, how to relate to other visitors, locals and services personnel [10]. In principle, tourism can cover all kinds of trips, as long as it is both sightseeing and recreation. Tourism as a trip that takes place for a while, organized from one place to another, with the aim not to do business or to make a living in the places visited, but solely to enjoy a trip for sightseeing and recreation or to fulfil diverse desires [11]. Almost every organization, large and small, private and public, make explicit and implicit assumptions because an organization should plan to meet the potential conditions that have imperfect knowledge. Furthermore, modern tools for forecasting, together with computer capabilities, have become essential for organizations operating in the modern world [12]. Demand, in this case, more generally referred to as market demand, implies there is a demand on the market for certain goods at a certain price and at some time in any event [13]. Forecasting is very important for the tourism industry. Accurate forecasting provides direct assistance to government and industry stakeholders to help them make important decisions, prevent waste and inefficiency of tourism capital while reducing risk and uncertainty [14], [15]. The determinant of macroeconomic factors such as GDP, bilateral trade volumes, fuel prices, and exchange rate can be used as a guideline if it becomes a supporting variable for coming to a certain country [7].

2.2. Forecasting tourism demand with recurrent neural network architecture

A neural network is a pattern classifier that, in some ways, behaves like a human mind. A neural network consists of artificial neurons linked by the edges, organized in a collection of layers. The number of layers and neurons per layer determines the strength of the classifier and the number of fine-grain patterns that can be identified. Each edge that connects a node (neuron) has a weight. It is the value of these weights, modified for each pattern that stimulates the process of recognition within the neural network [16]. Deep learning is a method of learning which uses multi-layered neural artificial networks. Such artificial neural networks are made identical to the human brain, in which neurons are attached to form a very complicated neuron network [17]. Jeff Elman developed the first Recurrent Neural Network (RNN) in 1990. The use of a recurrent neural network is due to the improved adaptability of this type of network to processing sequences of time data compared to other types of neural networks [18],[19]. Allows the network to map to each output from all previous inputs [20]. The LSTM is a specific form of RNN capable of learning long term dependencies. It is specifically designed to prevent the gradient problem from vanishing / exploding. An LSTM is well suited for classifying and/or predicting data from the time series. There are several LSTM-unit architectures. A common architecture consists of a memory cell, an input gate, an output gate and a gate to forget gate [25], [26]. Moreover, each LSTM cell is calculating new hidden state and cell sate values. The figure 1 below [2], [18] illustrates the RNN being unrolled (or unfolded) in a full network when a previous time step is needed to determine the current time step which is shown as t-1 and t respectively.

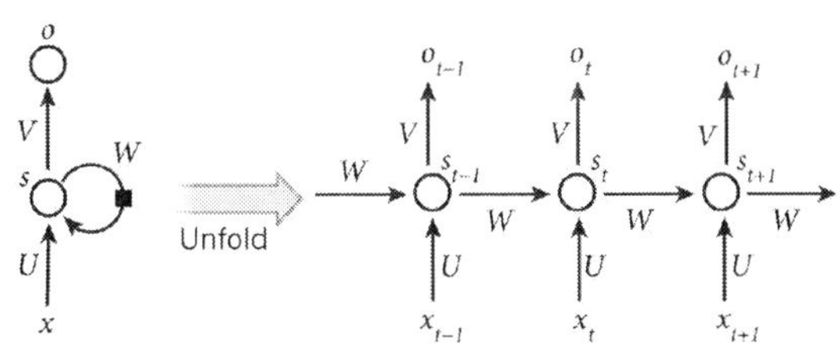

Figure 1. RNN structure.

3. Methodology

The data containing five main variables in which the explanatory variables or independent variable was determined from previous research about tourism industry as well. In this study, the data used is secondary data because the required data is available for public and can be retrieved from the government or institutional site. The data and its source for this research are listed as follows:

1) Tourist arrivals data was obtained from the site of statistic bureau of Indonesia
2) Per capita GDP data was obtained from Imf.org
3) Bilateral trade volume data was obtained from the site of statistic bureau of Indonesia
4) Fuel price data was obtained from Bloomberg.com
5) Exchange rate data was obtained from exchangerates.org.uk

The four determinant of macroeconomic factors such as per capita GDP, bilateral trade volumes, fuel prices, and exchange rate are believed to be useful as a guideline if it becomes a supporting variable for coming to a certain country as can be seen in table 1 [7]. Distribution of training data and testing data is divided into 80:20, where the split was done with the help of python script.

3.1. Data pre-processing

The pre-processing step is required before to make a forecast. Data Pre-processing is the stage in which the data is transformed or encoded to get to such a state that it can be easily parsed by the machine. In other words, the data characteristics can be readily interpreted by the algorithm [21]. These steps include:

1. Interpolation is intended for transforming the data from quarterly to monthly data.
2. Data normalization using MinMaxScaler in python programming language to normalize data distribution by rescale the range of data to (0, 1).
3. Success condition obtained after the pre-processing phases where the data can be processed easily by the machine.

Table 1. Data characteristics

No.	Data	The amount of data	Indicator
1.	Visitor Arrivals	480 data	The response variable or the variable that might affected by the explanatory variable.
2.	Per capita GDP	480 data	The 1st explanatory variable that affect the response variable.
3.	Exchange Rate	480 data	The 2nd explanatory variable that affect the response variable.
4.	Fuel Price	480 data	The 3rd explanatory variable that affect the response variable.
5.	Bilateral Trade Volume	960 data	The 4th explanatory variable that affect the response variable.
	TOTAL DATA	2,880 data	

3.2. Model building

The model of this research requires parameters such as shown in table 2. The look-back determine the number of past data taken for predicting the future value. Hidden layer for the layer needed for the model. The number of hidden layer is actually free to determine on how many layer needed but the average number of hidden layer that are often used starts from 1 at least and 4 layers at most. Epochs for number of iteration and batch size for number of batch in the model.

Table 2. Parameters for model building.

No.	Parameters	Values
1.	Look-back	7, 14, and 21
2.	Hidden layer	4
3.	Epochs	200, 300, and 400
4.	Batch size	1 and 3

The number of hidden layer consisting only one value was because the author previously made an attempt to run on 4 different hidden layers (1, 2, 3, and 4 hidden layer) and turned out using 4 hidden layers in several attempts yield a good result in accuracy. This condition applied to batch size as well from attempted using 3 different kinds of batch size (1, 3, and 7). This was done because if attempting to try for more numbers for testing then there will be many combinations of model and therefore, took longer time to process. Each parameters will be tested based on a predetermined value which then will be combined with the values of other parameters. The combination of each parameters is shown in table 3.

From the table above, several combinations of parameters have been obtained. For this study, the author decided to take 3 optimal model (those with low loss value) determined from all the parameters, especially based on number of iteration (epoch) in case one model is not suitable for particular dataset. The combinations that have been obtained will then be used as a model for forecasting the arrival of foreign tourists to Indonesia along with forecasting other variables that affect the arrivals. The 1st model has configuration of 21 look-back, 300 epochs, 4 hidden layers, and 3 batch size. Then 2nd model has combination of 21 look-back, 400 epochs, 4 hidden layers, and 1 batch size. Lastly the 3rd model with a configuration of 21 look-back, 400 epochs, 4 hidden layers, and 3 batch size.

Table 3. Testing results.

No.	*Look-back*	*Epochs*	*Hidden Layer*	*Batch Size*	*Loss*
1.	7	200	4	1	0.0019
2.	7	200	4	3	0.0010
3.	7	300	4	1	0.000753
4.	7	300	4	3	0.000689
5.	7	400	4	1	0.000586
6.	7	400	4	3	0.000527
7.	14	200	4	1	0.0011
8.	14	200	4	3	0.000920
9.	14	300	4	1	0.000623
10.	14	300	4	3	0.000851
11.	14	400	4	1	0.000897
12.	14	400	4	3	0.000629
13.	21	200	4	1	0.0015
14.	21	200	4	3	0.0011
15.	21	300	4	1	0.000613
16.	21	300	4	3	0.000349
17.	21	400	4	1	0.000502
18.	21	400	4	3	0.000292

Table 4. Prediction performance measurement.

	China	Malaysia	Singapore	Australia	Japan
Number of arrivals	RMSE:24,943.66 MAPE:11.24%	RMSE:22,604.23 MAPE: 7.86%	RMSE:23,119.53 MAPE: 14.70%	RMSE:9,867.22 MAPE. 7.15%	RMSE:6,398.90 MAPE:8.72%
Per capita GDP	RMSE:0.340 MAPE:2.33%	RMSE:0.245 MAPE:2.88%	RMSE:0.699 MAPE:0.83%	RMSE:0.408 MAPE:0.80%	RMSE:1.111 MAPE:1.70%
Fuel price	RMSE:0.178 MAPE:2.89%	RMSE:0.111 MAPE:3.52%	RMSE:0.429 MAPE:3.15%	RMSE:0.215 MAPE:3.70%	RMSE:0.185 MAPE:2.33%
Exchange rate	RMSE:0.004 MAPE:2.29%	RMSE:0.00525 MAPE:2.80%	RMSE:0.00788 MAPE:1.15%	RMSE:0.0432 MAPE:3.09%	RMSE: 0.000148 MAPE: 1.17%
Bilateral trade volume (import)	RMSE: 467,121,825.77 MAPE:10.01%	RMSE: 75,277,362.56 MAPE:10.01%	RMSE: 193,241,143.37 MAPE:8.03%	RMSE: 93,884,751.20 MAPE:9.69%	RMSE: 146,125,505.08 MAPE:7.27%
Bilateral trade volume (export)	RMSE: 293,566,690.52 MAPE:6.76%	RMSE: 75,277,362.56 MAPE:11.08%	RMSE: 171,394,218.64 MAPE:9.33%	RMSE: 38,390,112.46 MAPE:13.96%	RMSE: 86,513,049.89 MAPE 6.83%.

4. Result and discussion

The prediction of number of arrivals showing quite similar graphic trend for China, Singapore, and Japan as shown in figure 2. The prediction model between predicted data and actual data does not make a big difference for any of the countries. It was known that Australia have the lowest MAPE with 7.15% or equal to 92.85% accuracy of prediction from configuration model of 21 look-back, 4 hidden layer, 400 epochs, and one batch size.

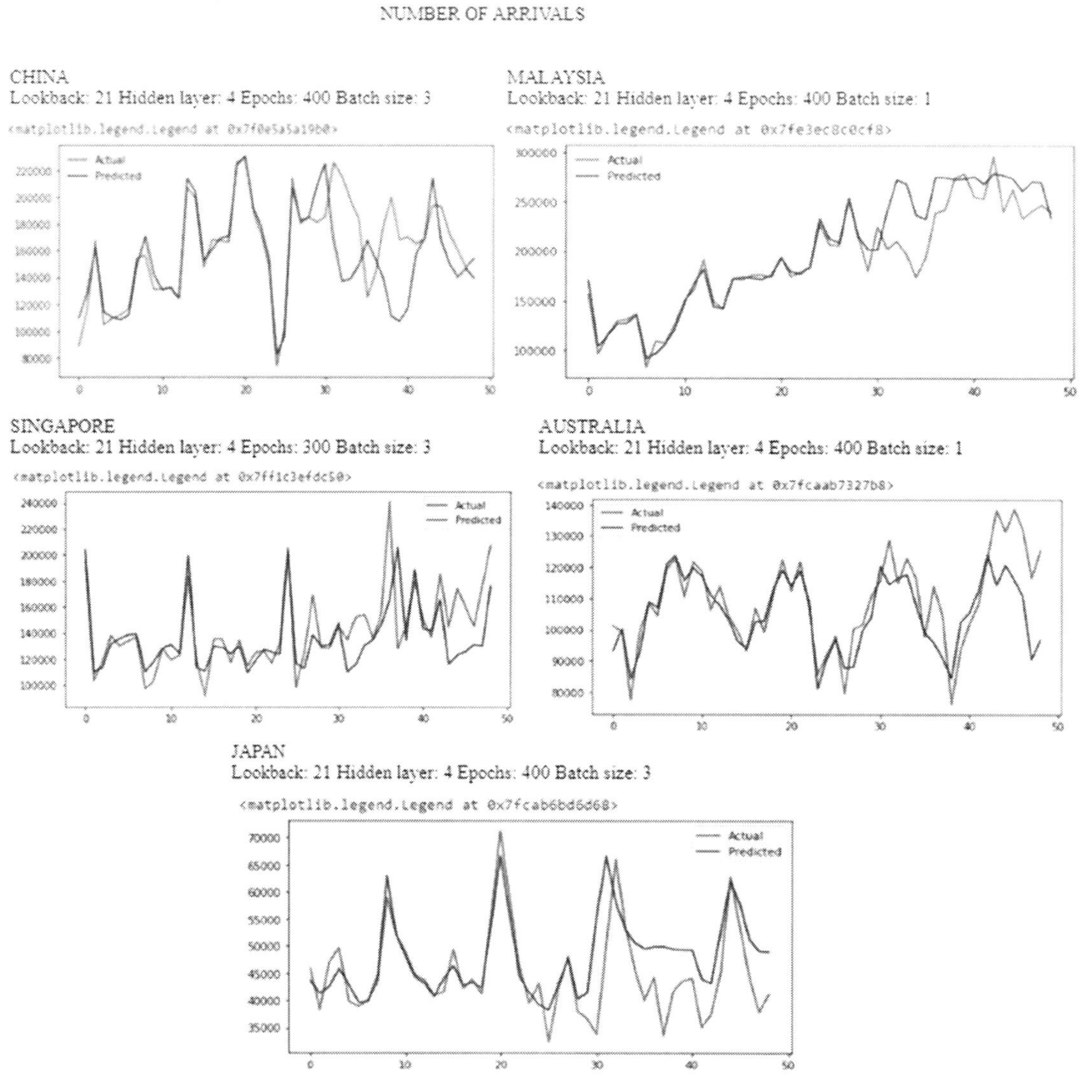

Figure 2. The results of the predictions.

When the prediction was done, it was expected to have an output of model performance in the form of RMSE value and MAPE value as can be seen in table 4. Root Mean Square Error (RMSE) is commonly used in many recommendation systems for calculating the difference between the expected scores (the real data) and the actual scores (the predicted data). While MAPE indicates how many errors in prediction (In percentage) relative to the real value [27], [28]. Each variable represents the performance of each country with their optimal combination model. After predicting all visitor countries and getting numeral from the results of the predictions then the next thing is to make regression and its analysis of the economic variables towards the number of tourist visits to Indonesia. Regression is conducted with the objective of seeing how variables such as per capita GDP, fuel price, exchange rate, and bilateral trade volume influence each country towards the number of international tourist arrivals to Indonesia.

Furthermore, can be seen in table 5. The regression result of per capita GDP variable shows a positive effect for China and Malaysia. This means per capita GDP has positive correlation on increasing those two countries visitor arrivals to Indonesia. This finding indicates that China's and Malaysia's per capita GDP growth increased the monthly arrivals of visitors to Indonesia. However, the insignificant positive growth coefficient of per capita GDP, shows in Australia, Singapore, and Japan. The potential interpretation of this negative per capita GDP correlation factor would be that these countries are classified as the top tourist source markets that are comparatively richer concerning their actual per capita GDP making the number of arrivals to Indonesia not affected by the variable.

Table 5. Regression coefficient results.

	China	Malaysia	Singapore	Australia	Japan
Per capita GDP	53,906.52	75,850.60	-7,919.24	-3,621.84	-1,158.67
Fuel price	-36,362.05	-19,763.21	45,780.49	-14,639.10	-4,298.26
Exchange rate	335,162.78	-568,330.02	-206,814.47	784,404.05	7,202,231.84
Bilateral trade volume (import)	-0.0000565	-0.0000307	0.0000137	-0.00000532	0.0000297
Bilateral trade volume (export)	0.0000430	0.0000776	-0.0000199	0.000000332	-0.0000273

The impact of fuel price showing positive coefficients for Singapore. This probably due to changes in fuel price affecting the transportation ticket price (for instance, like airfares) in which Singapore's visitor arrivals to Indonesia heavily affected by. Meanwhile, the rest of the countries such as China, Malaysia, Australia, and Japan were negatively affected by the changes in fuel prices. There were two countries presented negative parameters for the change in the exchange rate variable. These countries are Malaysia and Singapore. The reason may arises from factors beyond the macroeconomic determinants that cause the negative effect of the monthly visitor arrivals of Malaysia and from Singapore to Indonesia, such as closer proximity and flying distances for both countries to Indonesia. On the other hand, China, Australia, and Japan showing a positive coefficient. This indicating that as the exchange rate relative to these countries' currency against US dollar increase, the number of arrivals variable tends to increase too.

The bilateral trade volume for import was found to have negative effect, reducing the number of visitor arrivals from three countries that are China, Malaysia, and Australia at the moment when import was increase. Still with the same variable, Singapore and Japan had a positive coefficient in increasing visitor arrivals to Indonesia. Last variable to evaluate is bilateral trade volume for export. Three countries (China, Malaysia, and Australia) were positively affected by the variable. Indicating visitor arrival from these three countries are expected to increase when export was increase. Meanwhile, the negative impact was found for Singapore and Japan as well. It can be assumed when the export variable decreased, the number of arrivals from Singapore and Japan were increasing.

Besides regression coefficient, P-value is also evaluated in this study to determine whether a variable from each country is significant or not significant to the number of tourist arrivals. The results can be seen in table 6. The confident level was set to 95% with 5% of significance level. The P-value of per capita GDP variable was found significant on Malaysia meanwhile the rest of countries (China, Australia, Singapore, and Japan) was found to be insignificant. The variable of fuel price was not found to be significant for all the countries, indicating there is insufficient evidence to say that there exists a non-zero correlation. For exchange rate variable, it was reported to have significant on Australia only and insignificant for four countries proving there is insufficient evidence of a correlation exists. Japan was founded to be significant on bilateral trade volume for import variable. Showing there is sufficient evidence that a non-zero correlation exists. However, four other countries

(China, Malaysia, Singapore, and Australia) founded to be insignificant for the variable. Last variable, bilateral trade volume for export, were reported to have significant on Japan and China while the rest of the countries reported to have insignificant.

Table 6. P-value results.

	China	Malaysia	Singapore	Australia	Japan
Per capita GDP	0.0536201	0.0000013	0.240166	0.056733	0.635532
Fuel price	0.3410368	0.6105161	0.115543	0.15802	0.347144
Exchange rate	0.8103558	0.3175477	0.484417	0.01213	0.081532
Bilateral trade volume (import)	0.0695099	0.7583385	0.597859	0.852451	0.001794
Bilateral trade volume (export)	0.0379312	0.3760957	0.57907	0.952242	0.005488

5. Conclusion

The preceding chapter has been used to forecast the number of foreign tourist arrivals to Indonesia and their economic variables, followed by a regression analysis of the results of the forecast to see the relationship between dependent variables and independent variables. The conclusion of the research that has been carried out is the prediction performance measurement from model 1 with weight configuration of 21 look-back, 4 hidden layers, 300 epochs, and 3 batch size generates the lowest error value (MAPE) of 2.795%, providing 97.21% predictability. The model 2, comprises a weight combination of 21 look- back, 4 hidden layers, 400 epochs, and 1 batch size. The model yields the lowest error value (MAPE) of 0.832%, thus providing 99.17% predictability. And last with model 3 consisting a combination of 21 look-back, 4 hidden layers, 400 epochs, and 3 batch size. The model eventually success to generates the lowest error value (MAPE) of 0.795% making of 99.21% predictability.

The study's primary research findings have many important consequences for policy-makers strategic decision-making and policies, such as identifying the best strategies and solutions based on reliable tourism demand forecasting for Indonesia from its main tourism source markets. The forecasting of visitor arrivals finding would be useful in a way government as policy-maker establish policies in maintaining the tourism sector. Suggestions can be more directed towards strategic decision-making about an action that needs to be taken or prevented when an increase or decrease in tourist numbers happened. Utilization by the government can be done in the form of scheduling repair and maintenance of a city's infrastructure or any public facilities (like ATM and money changer) to avoid inconvenience for the tourist. The government can also pay attention to several economic variables are having a certain impact on several countries that visit Indonesia, especially from those 5 major tourist market.

The study’s suggestion for tourism related operators would be providing a useful insight on how tourism demand from perspective of economic variables affecting Indonesia’s visitor arrivals. Thus, to overcome the demand a forecasting model were build. The industry related to tourism then could have made an improvement in such sector as necessary. Examples on how to leverage the tourism sector from this visitor arrivals would be creating a strategic decision to utilize or marketing a potential tourist destination. Tourism operators could also enhance the industry throughout promotion on tourist destination or in the form of airline fares.

References

[1] Ministry of Tourism 2018 *Rencana Strategis 2018-2019: Kementrian Pariwisata* Retrieved

from http://www.kemenpar.go.id/
[2] Vihikan W O, Putra I G and Dharmaadi I A 2017 Foreign tourist arrivals forecasting using recurrent neural network backpropagation through time *TELKOMNIKA* pp 1257–64 DOI: 10.12928/telkomnika.v15i3.5993
[3] Astuti M T and Noor A A 2016 Daya tarik Morotai sebagai destinasi wisata sejarah dan bahari *Jurnal Kepariwisataan Indonesia: J. Penelitian Dan Pengembangan Kepariwisataan Indonesia* pp 25–46
[4] Kodituwakku W H, Hettiarachchi C, Wijesundara D M, Dias G K and Karunaratne D D 2014 Forecasting tourist arrivals in Sri Lanka *2nd Int. Conf. on Hospitality and Tourism Management*
[5] Filippini M and Hunt L C 2011 Energy demand and energy efficiency in the OECD countires: a stochastic demand frontier approach *Energy J.* **32** pp 59–80 DOI: 10.5547/issn0195-6574-ej-vol32-no2-3
[6] Alamsyah A and Friscintia P B 2019 Artificial neural network for Indonesian tourism demand forecasting *7th Int. Conf. on Information and Communication Technol. ICoICT 2019* DOI: 10.1109/icoict.2019.8835382
[7] Balli H O, Tsui W H and Balli F 2018 Modelling the volatility of international visitor arrivals to New Zealand *J. of Air Transport Management* **75** pp 204–14 DOI: 10.1016/j.jairtraman.2018.10.002
[8] Rizal A A and Hartati S 2016 Recurrent neural network with Extended Kalman Filter for prediction of the number of tourist arrival in Lombok *2016 Int. Conf. on Informatics and Computing (ICIC)* Mataram 2016 pp 180–5 DOI: 10.1109/iac.2016.7905712
[9] Hatzakis I and Wallace D 2006 Dynamic multi-objective optimization with evolutionary algorithms: a forward-looking approach *Proc. of the 8th Annual Conf. on Genetic and Evolutionary Computation* Seattle, Washington pp 1201–8 DOI: 10.1145/1143997.1144187
[10] Leiper N 1995 *Tourism Management* (Collingwood: TAFE Publications)
[11] Yoeti O A 1993 *Pengantar Ilmu Pariwisata* (Bandung: Angkasa)
[12] Sarjono H and Abbas B S 2017 *Forecasting: Aplikasi Penelitian Bisnis QM for Windows vs Minitab vs Manual* (Jakarta: Mitra Wacana Media)
[13] Yoeti O A 2003 *Tours and Travel Marketing* (Jakarta: Pradnya Paramita)
[14] Chen C F, Lai M C and Yeh C C 2012 Forecasting tourism demand based on empirical mode decomposition and neural network *Knowledge-Based Systems* **26** pp 281–7 DOI: 10.1016/j.knosys.2011.09.002
[15] Chen R, Liang C, Hong W C and Gu D X 2015 Forecasting holiday daily tourist flow based on seasonal support vector regression with adaptive genetic algorithm *App. Soft Computing* **26** pp 435–43 DOI: 10.1016/j.asoc.2014.10.022
[16] Cox E 2005 *Fuzzy Modeling and Genetic Algorithms for Data Mining and Exploration* (San Francisco: Elsevier Inc) DOI: 10.1016/b978-0-12-194275-5.x5000-2
[17] Primartha R 2018 *Belajar Machine Learning: Teori dan Praktik* (Bandung: Informatika)
[18] LeCun Y, Bengio Y and Hinton G 2015 Deep learning *Nature* **521** pp 436–44 DOI: 10.1038/nature14539
[19] Zagrebina S A, Mokhov V G and Tsimbol V I 2019 Electrical energy consumption prediction is based on the recurrent neural network *13th Int. Symposium "Intelligent Systems" (INTELS'18)* pp 340–6 DOI: 10.1016/j.procs.2019.02.061
[20] Graves A 2012 *Supervised Sequence Labelling with Recurrent Neural Networks* (New York: Springer) DOI: 10.1007/978-3-642-24797-2_2
[21] Pandey P 2019 *Data Preprocessing: Concepts* Retrieved on 25 November 2019 from https://towardsdatascience.com/data-preprocessing-concepts-fa946d11c825
[22] BPS-Statistics Indonesia 2017 *International Visitor Arrivals Statistics 2017* (Jakarta: BPS-Statistics Indonesia)
[23] Noersasongko E, Julfia F T, Syukur A, Purwanto, Pramunendar R A and Supriyanto C 2016 A

Tourism arrival forecasting using genetic algorithm based neural network *Indian J. of Sci. and Technol* **9** issue 4 pp 1–5 DOI:10.17485/ijst/2016/v9i4/78722
[24] Gunter U and Önder I 2014 Forecasting international city tourism demand for Paris: accuracy of uni and multivariate models employing monthly data *Tourism Management* **46** pp 123–35 DOI: 10.1016/j.tourman.2014.06.017
[25] Hochreiter S and Schmidhuber J 1997 Long short-term memory *Neural Computation* **9** issue 8 pp 1735–80 DOI: 10.1162/neco.1997.9.8.1735
[26] Saud A S and Shakya S 2020 Analysis of look back period for stock price prediction with RNN variants: A case study on banking sector of NEPSE *Procedia Computer Sci.* **167** pp 788–98 DOI: 10.1016/j.procs.2020.03.419
[27] Wang W and Lu Y 2018 Analysis of the Mean Absolute Error (MAE) and the Root Mean Square Error (RMSE) in assessing rounding model *IOP Conf. Ser.: Materials Science and Engineering* **324** DOI: 10.1088/1757-899x/324/1/012049
[28] Khair U, Fahmi H, Al-Hakim S and Rahim R 2017 Forecasting error calculation with mean absolute deviation and mean absolute percentage error *J. of Phys.: Conf. Ser.* **930** DOI: 10.1088/1742-6596/930/1/012002

Data Quality Assessment on Lecturer Primary Data: A Case Study on Higher Education Database at Ministry of Education and Culture Republic of Indonesia

Rizky T. Prasetyo[1], Yova Ruldeviyani[2], Editha D. Purnamasari[3], Apriliana F. Wibowo[4]

Magister of Information Technology, Faculty of Computer Science, Universitas Indonesia

Department of Information Technology, Faculty of Computer Science, Universitas Indonesia, Jl. Salemba Raya No.4, DKI Jakarta, Indonesia, 10430

Email: [1]rizky.tito91@ui.ac.id, [2]yova@cs.ui.ac.id, [3]editha.dewi@ui.ac.id, [4]apriliana.fajri@ui.ac.id

Abstract. The quality of an organization's data is an important component that can influence various policies taken in technical and strategic terms. This study aims to analyze the quality of basic data for lecturers on resource information systems whose data is integrated with the Higher Education Database (PDDikti). Measurement of data quality based on the rules that have been set in the main unit of higher education data management which is then determined in this study includes 4 dimensions of data quality, namely completeness, validity, accuracy, and currency. The application of quantitative methods by processing data and then analyzing the quality of the data using qualitative methods based on the dimensions that have been determined for the basic data quality assessment for lecturer in every tertiary institution in Indonesia. The results showed that the quality of the data had been resolved by 91.94% so that it can be used as a basis for decision making according to predetermined standards. The results of this data quality research can be used as supporting information in policy making in terms of developing and improving data quality in the scope of higher education, especially those related to the quality of basic data for lecturers.

1. Introduction

Data is one of the important components as a driving force for an organization's business that can have a significant impact on achieving the organization's strategic objectives [1]. Knowledge of the quality conditions of data can be an important factor in producing valuable data for the organization. This can be in line with the organization's strategy to determine data quality improvement policies and in terms of maintaining data quality so that it can provide benefits to the organization [2]. Some of the positive impacts of data quality management include increasing the value of organizational data, reducing the risk and cost of poor data quality, increasing organizational efficiency and productivity and increasing the reputation of the organization [1]. To have a positive impact it is necessary to identify the prevailing business processes in an organization that can have an impact on data quality and then be linked to find the root causes of the problem [3].

Various studies on data quality measurement have been carried out in various fields such as in the health sector, education sector to information technology sector. D. Effendi conducted research by measuring the quality of data and information at UNIKOM using the CALDEA and EVAMECAL methodologies [4]. Wijayanti, et al. Conducted a study on the assessment of data quality in higher education using a case study from a tertiary institution in Indonesia [2]. In addition, Herawani also conducted research on data quality improvement strategies covering student and learning activities in the Higher Education Database at the Ministry of Research, Technology and Higher Education [5]. From various studies on data quality, it can be seen that it is important to measure the quality of data from an organization.

Referring to the research on data quality in the Higher Education Database conducted by Herawani which is still in the scope of student and learning data, while in the Higher Education Database there are still several scopes of data that are important to know the quality of the data. In this case, related to the main data of teaching staff who are in the scope of the main unit of the ministry of education and culture organization is the main thing that will be measured the quality of the data. Measuring the quality of data in the context of teaching staff data is quite important to be able to find out how good the quality of the data you have so that it can support policy making by leaders in the main unit to the university level. In addition, there are still reports, especially from data owners, namely lecturers regarding the correctness of data that are not in accordance with the actual conditions, which also underlies the need to measure the basic data quality of teaching staff.

This study aims to analyze the quality of lecturer data, especially the primary data on lecturer registration contained in the Higher Education Database (PDDikti) which is reviewed from several strategic and technical rules such as regulations regarding the governance of the Higher Education Database or regulations regarding National Higher Education Standards as well as other regulations relating to the management of higher education data management in Indonesia. The methods used include quantitative and qualitative in measuring data quality. Quantitative method is used as one of the things in obtaining measurement results from data that has been obtained through SQL Query, while qualitative method is focused on analyzing the results of measuring data quality. The application of stages in the concept of Total Data Quality Management (TDQM) is based on the rules for the dimensions of data quality that are processed based on the provisions contained in PDDikti.

2. Literature Review

2.1. Data Quality

Data quality is the level of meeting expectations or expectations of data users on the data generated [6]. These expectations are defined into related terms with data characteristics that ensure data conformity for a specific purpose. Data quality requirements can be outlined in general and in more detail [6]. Requirements load dimensions like accuracy, completeness, validity and etc. Then, from these requirements will be compared with existing data for measured levels suitability, the higher the suitability level it will be the higher the data quality level.

2.2. Data Quality Dimension

The Data Quality dimension is a characteristic used as a reference for measuring the quality of the data. The data experts have provided a set of quality related dimensions data, according to DAMA DMBOK there are 11 (eleven) dimensions related to the data. Although nothing was agreed upon to set the dimensions of data quality, this formulation contains ideas in general [1]. To that end, every dimension is applied to an agency or dependent company at their respective policies.

The following are some data quality dimensions according to DAMA – DMBOK:

- Accuracy, refers to the level of data that is correctly represented the “real life” entity being modeled.

- Completeness, refers to an expectation of completeness indicates that a particular attribute always has a value defined in the data set. Complete is when all attribute values that should have been collected have collected [7].
- Validity, refers to whether data instances are stored, exchanged, or presented in a consistent format with the value domain, and consistent with the attribute value other similar.
- Currency, refers to the extent of current information with the world he is modeling.
- Timeliness, refers to the estimated time to accessibility and availability of information.

2.3. *Data Quality Management*

Data quality management is one of the activities from data governance to manage data quality. These activities include planning, implementation, and control for ensure the data is suitable for consumption and can meet the data needs of consumers. In general, it is applied using a methodology or technique [1]. Several methodologies have been developed to try describe procedures so that data quality can continue improved.

The most popular methodologies in data quality management are Total Information Quality Management (TIQM) and Total Data Quality Management (TDQM) [8]. Both TIQM and TDQM have a similar goal of providing a methodology for improving the data quality continuously. While TIQM is heavily influenced by practical experience, TDQM is the result of several years of research and can propose to identify and document information production processes and information product characteristics [9]. The methodology is applied according to the organization's data needs.

2.4. *Data Quality Measurement*

Measurement or assessment of data quality is one of the stages of data quality management. In the TDQM methodology, this measurement or assessment stage enters the second stage. The phases owned by the TDQM methodology are:

- Definition Phase, during this phase, the characteristics of the information product (IP) are captured, such as information requirements, core information objects and components, and their relationship [10]. In addition, the importance of the data quality dimension in the perceptions of IP suppliers, producers, consumers, and managers was identified through a survey that captures the first assessment of the quality of the underlying IP [10]. Furthermore, the information generation system is documented through the so-called "information generation analysis matrix" as the basis for further analysis and improvement [10].
- Measurement Phase, this phase is the beginning of developing data quality metrics. Metrics do not have to directly relate to data, but also with production or access processes, e.g. which updates how much data or how many unauthorized access occurred [10]. The developed metrics are implemented in a system and applied to the data to measure the quality of IP data periodically.
- Analysis phase, based on the measurement results, the root causes of identified data quality problems are analysed during this phase [10]. Metrics are also subject to further analysis, because sometimes there are metric results that need to be adjusted, extended, or improved [10].
- Repair or Upgrading Phase, in this phase the identified causes of quality problems need to be eliminated during this phase. Therefore, it is necessary to identify the necessary improvements, such as adjusting information and workflows to the infrastructure or modifying IP characteristics according to business needs [10]. The improvement phase activities are supported by the "information generation analysis matrix", which was originally created during the definition phase [10].

3. Methodology

The research was conducted with a case study of teaching staff data in higher education databases using quantitative data calculation methods and then analyzed using qualitative methods to obtain measurement results from the quality of teaching staff data in the higher education database. Assessment of the quality of the data uses direct SQL queries (Structured Query Language). The data used to measure the quality of the data is taken with limitations in the form of data for lecturers who are recorded as active, leave, study assignments, study permits and assignments in other agencies until May 2020 in the resource information system database that is integrated with the higher education database. The steps taken in conducting this research include the first three stages contained in the TDQM method. Furthermore, several stages used in this study were compiled and explained as follows:

Stage 1: Identify the required data based on the existing regulations.

Stage 2: Determine the dimensions of data quality that are used to measure data quality. Determination of the dimensions used is based on the Regulation of the Minister of Research, Technology and Higher Education Number 61 of 2016 concerning Higher Education Database [11].

Stage 3: Define criteria according to the rules applied to the assessed data.

Stage 4: Collect data based on the need for a data quality assessment from the database management system (DBMS). At this stage, PDDikti data are prepared related to lecturer data.

Stage 5: Measuring or assessing the quality of the data based on predetermined criteria. The result of measuring the quality of this data is in the form of a metric of the percentage of data conformity with the rules in the main data for lecturers.

Stage 6: Analyze the results of the data quality assessment based on predetermined dimensions. The analysis was carried out by means of observation and interviews with several officials related to the higher education database.

Stage 7: Provide conclusions and recommendations on the results of data quality assessments based on the results of data processing described in the previous stage.

4. Results and Discussion

Measurement of the quality of teaching staff data is carried out using the TDQM method. The data for lecturers is measured based on the Regulation of the Minister of Research, Technology and Higher Education Number 61 of 2016 Article 12 paragraph 2, this regulation is related to the dimensions applied to higher education data consisting of completeness, truth, accuracy, and up-to-date data [10]. The measurement results display the percentage of data conformity from each dimension with existing business rules, as illustrated in Table 1.

Table 1. Results of Data Quality Measurement

Dimension	The number of business rules		Business rules are met	Business rules that cannot be measured
	Total	Percentage	Percentage	Percentage
Completeness	86	100%	93.33%	6.67%
Validity	13	100%	88.91%	11.09%
Accuracy	18	100%	95.45%	4.55%
Currency	3	100 %	90.07%	9.93%

4.1. Measurement for Completeness Dimension

In the completeness dimension, anomalies are traced to the data, by ensuring that the expectation of the necessity of data values for certain attributes can be met. In this study, the attributes that are expected in the data of lecturers contained in the higher education database, as many as 86 attributes are required to be filled in or cannot be null. If the attribute contains zero, '-' or contains only spaces, the attribute is grouped into attributes that do not meet the criteria for a business rule. The rules for the completeness dimension are:

C1. A Resource entity that includes 23 attributes that cannot be null.

C2. A Certification Entity that includes 15 attributes that cannot be null.

C3. Power entity that includes 15 attributes that cannot be null.

C4. Functional Entity that includes 12 attributes that cannot be null.
C5. A formal education entity that includes 21 attributes that cannot be null.

Figure 1 displays the measurement results for this completeness dimension that has been met at 93.33%. The discrepancy between data and business rules is 6.67%.

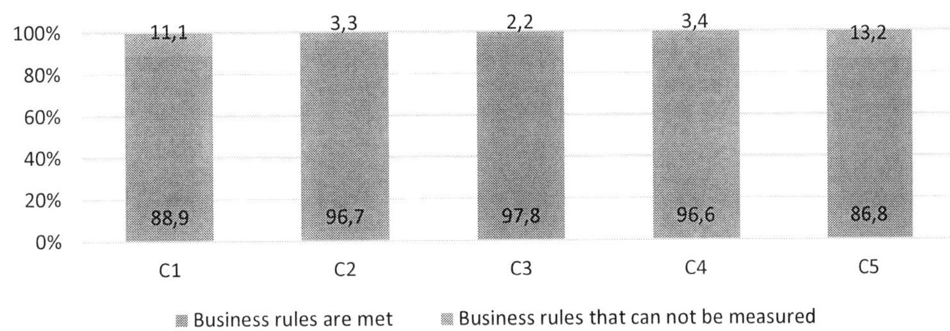

Figure 1. Graph of Measurement Results for Completeness Dimension

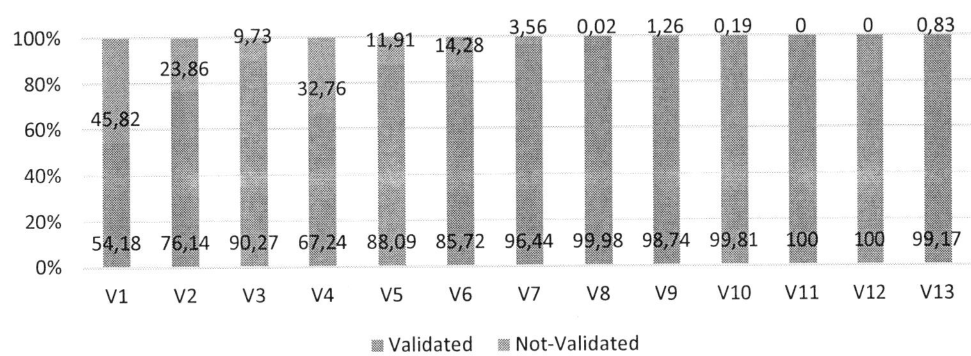

Figure 2. Graph of Measurement Results for the Validity Dimension

4.2. Measurement for the Validity Dimension

In the validity dimension, anomalies are traced to the data, by ensuring that there are data anomalies that are inconsistent with the domain values that have been made appropriate in the higher education database. In this study, measuring the validity dimension can be done by ensuring that the data already in the database management system (DBMS) is in accordance with applicable business rules. The business rules in the validity dimension are:

V1. Lecturers of diploma one and diploma two programs must have the minimum academic qualifications of a master's degree or applied master's degree.
V2. Lecturers of diploma three and diploma four programs must have academic qualifications at least as a graduate of a master's degree or an applied master's degree.
V3. Lecturers of undergraduate programs must have academic qualifications of at least a master's degree or applied master's degree.
V4. Lecturers of professional programs must have academic qualifications of at least a master's degree or applied master's degree relevant to the Study Program.
V5. Lecturers of master's programs and applied master's programs must have academic qualifications of a doctoral or applied doctoral graduate.

V6. Lecturers of specialist and subspecialty programs must qualify as subspecialty graduates, doctoral graduates, or applied doctoral graduates relevant to the Study Program.
V7. Lecturers of doctoral programs and applied doctoral programs must have academic qualifications of doctoral or applied doctoral graduates relevant to the study program, and can use professionally certified lecturers that are relevant to the study program and are qualified.
V8. The provision of a registration number adheres to the principle of 1 (one) lecturer only having 1 (one) registration number.
V9. Lecturers who have a NIDN are given a starting number ranging from 00 - 87.
V10. Lecturers based on a work agreement who have a NIDK are given an initial number 88.
V11. NUP for Tutors is given a starting number 97.
V12. The NUP for Instructors is assigned a starting number 98.
V13. NUP for Non-Permanent Lecturers is given a starting number 99.

Figure 2 displays the measurement results for the validity dimension, which has been fulfilled by 88.91% on the measured data for the 13 predetermined business rules, while there are 11.09% for data that are not validated. The following is a percentage graph for data invalidity in the measures that have been carried out.

4.3. Measurement for Accuracy Dimension
On the accuracy dimension anomalies are traced to the data, by ensuring that the data has a consistent, correct, and unambiguous value. In this research, the measurement of accuracy dimension can be done by ensuring that the data already in the database is in accordance with the current rules in the higher education database. The rules for accuracy dimension are:

A1. Year of Graduation is less than 2020
A2. Year of Entry is over 1900
A3. Year of Graduation is more than 1900
A4. Entry year is less than 2020
A5. Graduation Date is less than or equal to 26 May 2020
A6. TMT SK rank is less than or equal to 2020
A7. The date of SK rank is less than or equal to 2020
A8. The working period of the month group ranges from 0 to 12
A9. TMT SK Functional Position is less than or equal to May 26, 2020
A10. ID of functional position between the numbers 40 to 51
A11. The year of certification is less than or equal to 2020
A12. The date of the Assignment Letter is less than or equal to the date of May 26, 2020
A13. TMT Assignment Letter less than or equal to May 26, 2020
A14. The date of the CPNS SK is greater than the date of birth
A15. TMT SK Appointment is greater than the date of birth
A16. Year of Entry is less than the year of graduation
A17. TMT SK Rank is greater than the date of SK Rank
A18. TMT Assignment Letter is greater than the date of the Assignment Letter

Figure 3 displays the measurement results for this accuracy dimension, which is 95.45% of the data measured for the 18 specified business rules, while there are 4.55% for inaccurate data. The following is a percentage graph for data accuracy on the system.

4.4. Measurement for Currency Dimension
In the currency dimension, anomalies are traced to the data, by ensuring that the level of updating of the data matches the data object being modeled. In this study, dimensional measurements can be made by ensuring that the recording of time when saving data changes is adjusted to the applicable data format. The current data format contained in several attributes for teaching staff data [11] [12], as follows:

CR1. PNS NIP with a compiler code consisting of 18 (eighteen) digit numbers.

CR2. NIK with a compiler code consisting of 16 digits or numbers.
CR3. NPWP has a serial code with 15 digits or numbers.

Figure 4 displays the measurement results for this currency dimension as a whole that has been met by 90.07%. Meanwhile, the discrepancy was found to be 9.93%.

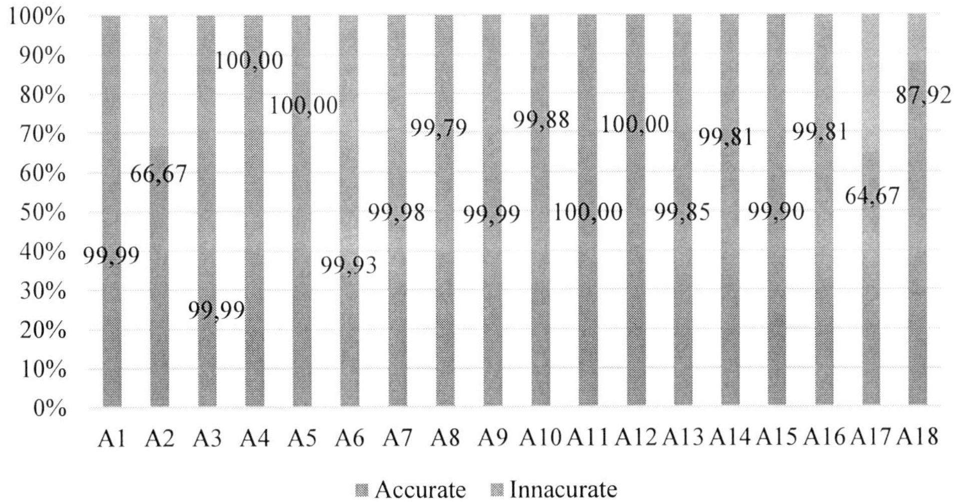

Figure 3. Graph of Measurement Results for Accuracy dimension

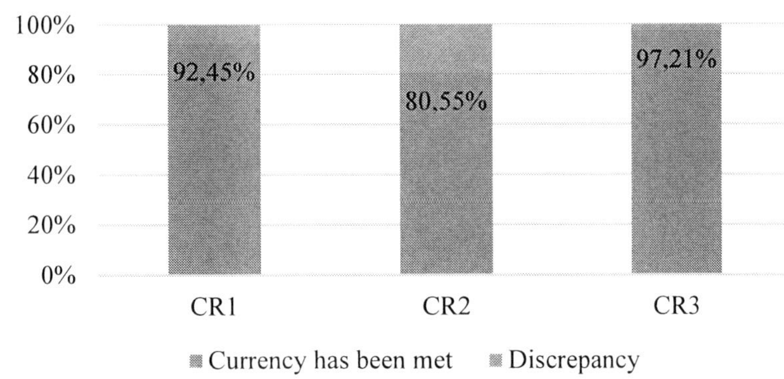

Figure 4. Graph of Measurement Results for Currency Dimension

4.5. Analysis of Causes of Data Quality Problems

Based on the problems identified in each data quality dimension, several causes of data problems were found. The cause of the problem is traced by conducting observations and interviews with information technology staff in charge of managing the integrated resource information system. Problems that occur in data quality are because:

- In the completeness dimension for attributes which cannot be null, there are still 18 attributes that are empty or '-' or spaces. This is because the data entered through the current system by the end-user has not been maximally implemented, besides validation that is still being carried out by users in the main unit can also cause errors in data verification. Another thing that causes incompleteness of data is still incompleteness of the data on the migration results.

- In the validity dimension, data is found that are validated by the current business rules. This is because the data has not used the latest validation rules. For example, validation for non-permanent lecturers' NUPs still has a bug in the system which causes the number given by the system to users not validated in accordance with the applicable rules.
- In the Accuracy dimension, data is found that are inaccurate and not in accordance with the rules previously stipulated in Regulation of the Minister of Research, Technology and Higher Education Number 61 of 2016 Article 12 paragraph 2 [13]. This is due to the lack of awareness of users to update data, besides the accuracy of checking by validators also affects the level of data accuracy so that it cannot be known that the data of lecturers has been updated massively.
- In the Currency dimension, data is found that have not been latest or updated on the attributes of NIDN, NIK, NIP and NPWP. This is due to several factors, such as errors when the system gave the lecturer registration number to lecturers. In addition, regulatory updates from the bureaucracy can also affect data changes that must be made, especially in database structures.

5. Conclusion

Based on the results and data analysis on the research that has been carried out on the main data of lecturer registration at PDDikti, the whole data has been fulfilled by 91.94% with compliance for the Completeness dimension of 93.33%, Validity 88.91%, Accuracy 95.45% and Currency 90.07%. The data anomalies found in this study are caused by several things such as awareness of updating data for each user, the ability of validator resources in the main unit, and some adjustments in the system related to bugs found and policy changes that affect data changes and database structures from higher education database.

Adjustments to the database structure need to be made in relation to adjustments to the bureaucratic structure of higher education which is back under the Ministry of Education and Culture. In addition, it is necessary to socialize the importance of maintaining data quality to each user from the university level to the validator user in the main unit according to the needs of PDDikti.

This study has limited data in terms of measuring data quality. The data used includes data on teaching staff which includes basic data on lecturer registration at PDDikti. The data taken includes data for teaching staff who are active, study assignments, study permits, leave and other agency assignments up to the deadline of May 2020 which are recorded in the resource information system integrated with PDDikti. This is caused by licensing issues from the organization regarding data security.

This research only reached the third stage of the TDQM methodology, namely the analysis of the causes of data quality problems in a case study of primary data on lecturer registration at PDDikti. For this reason, it is necessary to have research into the next stage of TDQM in order to be able to make strategies in improving the quality of basic data on teaching staff owned by PDDikti. In connection with the rules that have been made regarding each dimension of data quality, analysis of several documents contained in PDDikti and PP Number 61 of 2016 Article 5 paragraph 1. Furthermore, this research can be reviewed and re-analyzed by paying attention to several regulations issued due to bureaucratic restructuring in higher education. In addition, this research can be used as a reference for other people who want to evaluate the quality of higher education data from universities in Indonesia that follow the PDDikti requirements in accordance with the applicable rules in PDDikti and other related regulations.

References

[1] Dama International 2017 *DAMA - DMBOK Data Management Body of Knowledge, Second Edition* (New Jersey: Technics Publications)

[2] Wijayanti W, Hidayanto A N, Wilantika N, Adawati I R and Yudhoatmojo S B 2018 Data quality assessment on higher education: a case study of institute of statistics *Int. Seminar on Research of Information Technol. and Intelligent Systems (ISRITI)* pp 231–6

[3] Loshin D 2011 *The Practitioner's Guide to Data Quality Improvement* (USA: Elsevier)

[4] Effendi D 2012 Measurement and improvement of data and information quality in higher education using CALDEA and EVAMECAL *Seminar Nasional Rekayasa Sains Dan Teknologi (RESATEK)* **2** 2087–526

[5] Herawani P 2016 *Data Quality Improvement Strategy: A Case Study of Higher Education Database at The Ministry of Research, Technol., and Higher Education* (Jakarta: Universitas Indonesia)

[6] Coleman and Sebastian L S 2013 *Measuring Data Quality for Ongoing Improvement A Data Quality Assessment Framework* (USA: Elsevier)

[7] Liu L and Chi L 2002 Evolutionary data quality: a theory-specific view *Proc. of the 7th Int. Conf. on Information Quality* pp 292–304

[8] Huang K, Lee Y W and Wang R Y 1999 *Quality information and knowledge* (USA: Prentice Hall)

[9] Fürber C 2015 *Data Quality Management with Semantic Technologies* (Munchen: Springer)

[10] Wang R Y 1998 A product perspective on total data quality management *Communications of the ACM* **41** pp 58–65

[11] Kemenristekdikti 2016 *Appendix Number 2 of 2016 concerning Amendments to the Regulation of the Minister of Research, Technology and Higher Education Number 26 of 2015 concerning Registration of Lecturers in Higher Education* [in Indonesian] (Jakarta: Kemenristekdikti)

[12] BKN 2007 *Regulation of the Head of BKN Number 22 of 2007 concerning Civil Servant Identity Number* [in Indonesian] (Jakarta: BKN)

[13] Kemenristekdikti 2016 *Regulation of the Minister of Research, Technology and Higher Education Number 61 of 2016 on Higher Education Database* [in Indonesian] (Jakarta: Kemenristekdikti)

Traces Of The Brain's Learning Potential Present Within "Uneducational" Video Games.

Ayub bin Abdul Rahman[1], Mohd Syarqawy bin Hamzah[2], Hamwira Yaacob[3], Abdul Wahab bin Abdul Rahman[4]

Kulliyyah of Information and Communication Technology, IIUM, Malaysia

E-mail: [1]ayub@iium.edu.my, [2]syarqawy@iium.edu.my, [3]hyaacob@iium.edu.my, [4]abdulwahab@iium.edu.my

Abstract. Video games, especially in mobile devices, have the reputation of being negative to the mass consumer. They are thought to be addictive, mind-numbing time waster. But is that really true? A study has been conducted to analyze video games design styles based on neuro-affective computational model using electroencephalogram (EEG). Although the objective is to classify abstract and realistic design styles, a different off-topic materials have been discovered. During the data collection, subjects completed an aptitude test using an online instrument provided by the Cambridge University's Brain Lab. The EEG instruments recorded specific brain functions (cognitive, memory, verbal, creativity) separately. After classifying the data with emotional response, each brain functions give unique valence versus arousal scatter plot. What is interesting, however, is that when all of them plotted together in a single chart, they form a pattern similar to the data collected for playing video games. Could it possibly be that playing video games indeed stimulate specific brain functions? Or are these similarities in valence versus arousal plots just plain coincidences? A future research designed for brain function correlation using other model than emotions can confirm these findings. But for now, the consistency of these findings for all the subjects are hard to ignore.

1. Introduction

Educational value in video games is nothing new. Many production studios already made educational video games that are designed to educate the end users – who are the players – with a specific topic. Most educational video games are designed to combine educational philosophy and game design principles. Such games range from spelling activities to mathematical challenges. Games like these are known as educational games. Most of the time, they are categorized as edutainment games.

Video games outside the edutainment category, however, are perceived to have no educational values [2]. They are designed purely for entertainment and fun. Such games can be a good time-killer for some situations such as during long hours of commute in a public transportation. Due to the big impressions that these games serve nothing but pure enjoyment, some people think that they are time waster. Such negative perception gave these kind of video games a bad reputation of being destructive to the society[15].

Despite the mixed views of their place in the society, video games are something that no one can ignore. While non-educational games were never intended to be educational, there are published articles that use these kinds of games to stimulate children learning [2], [4], [5]. Such research findings provoke questions whether video games have any psychological effect to the human mind/brain. Are there any correlation between non-educational video games and learning activities? Can something that was designed purely for enjoyment can be valuable to the growth and potential of the brain?

2. Previous Works

Diving into details of the games used for experiments in the psychology of the human mind, it is startling that quite a lot of these researches use non-educational games. One of the most significant one is done by Franceschini [8] who applied action video game as a training medium to improve dyslexic children's literacy ability. The game being used is called *Space Invader* which is a 2D platform shooter with no literacy elements whatsoever. Training those children by playing *Space Invader* results in improved literacy performance. In fact, it is also claimed that action video games can actually cure dyslexia. The author did not specify why that particular game was chosen. The only criteria were that the game was chosen because it falls under the genre category.

Video games related researches are many and each has a different approach. Nevertheless, most that studied their relationship to the human psychology always found that video games have an effect on the player's mind regardless whether they are educational by design or not [2]. Their influence over the human psyche is too consistent to be assumed as coincidental. Although these video games were never meant to be educational, those researches that found it to be a 'wonder tonic' that cures psychological symptoms and boost learning potential [7], [9] are too significant to be ignored.

Those findings by other researches support the claims that video game is a medium that stimulates the mind. And it opens a huge question as to whether they have any correlation at all with specific brain's cognitive functions and their performance. A question that this paper is not to answer yet, but to reveal traces and evidences of specific learning aptitude might be present during a game play session.

3. Data Collection and Analysis

The research that this document discusses is a research to analyze two different design styles common in all video games production. The scope of the analysis is the classification of the design styles with the player's emotional state. To be able to measure emotions and apply computational models and algorithm for classification, the approach of using electroencephalogram (EEG) was proposed. By recording the brainwave signals of the player during a game play, quantitative analysis can be applied.

It must also be noted that the games chosen for this experiment has met the criteria to avoid data noise. Any attributes that can contribute to false signals or noise have been taken into account. For instance, game mechanics that are too complicated may disturb the experience factor between experienced and non-experienced players. The sounds are switched off so that the background music will not interfere with the data emotions stimulated by the video game designs. Most important of all, the video games used are not designed to be educational or put the players into any kinds of learning activities.

The concept of the data collection is simple. EEG data while playing video games was recorded so that it can be classified with a known model. In this case, the emotional model is used. Prior to the game play session, the player is exposed – and the EEG signals recorded – with emotional stimulant. The stimulated emotions of the player were recorded to serve as a working model so that the unknown signal during game play can be classified. After the emotional stimuli was administered, however, an added bonus of learning aptitude test was added just before the game play. Figure 1 is the overall flowchart of the data collection process.

Once the signals data have been acquired, they can then be analyzed quantitatively. Figure 2 is the work flow of the data analysis.

The EEG signals during game play session cannot yet identify the participants' emotional valence and arousal when they are playing the game. These unknown data are to be analyzed with the emotional model established earlier during the EEG working protocol session. By using machine learning algorithm, the emotional valence and arousal of the participants during play time can be determined.

The machine learning classification of unknown data determines their emotional valence and arousal by comparing the signals with the established model. The algorithm looks at the unknown signals for trends and characteristics that mimic the emotions model. Once it recognizes the emotions of the unknown data, it automatically labels them accordingly. Thus, there are two sets of data:

1. Training data: which is obtained during protocol, is used to build the emotions model.
2. Testing data: which is obtained during game play, is used to test the built model.

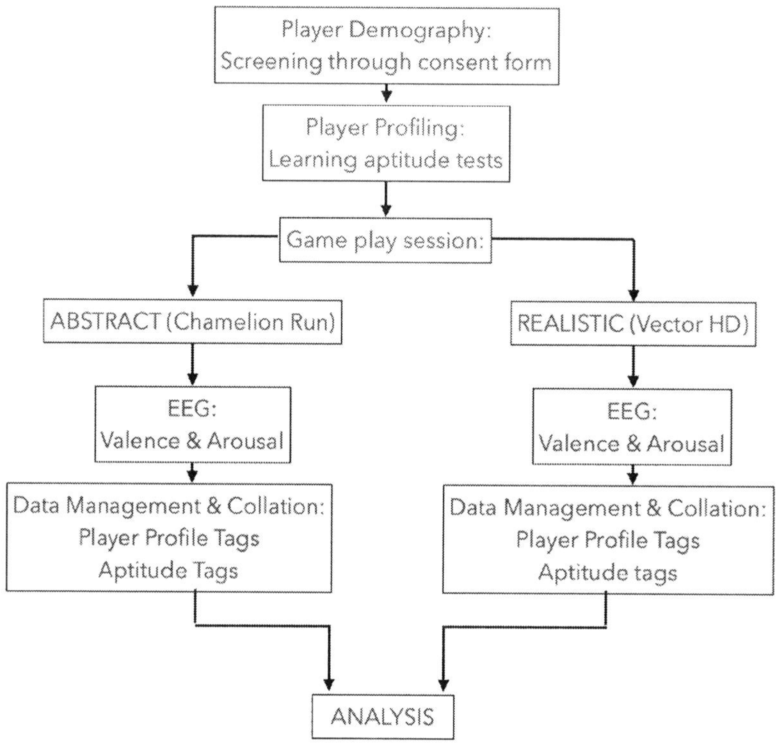

Figure 1. Data Collection Flowchart

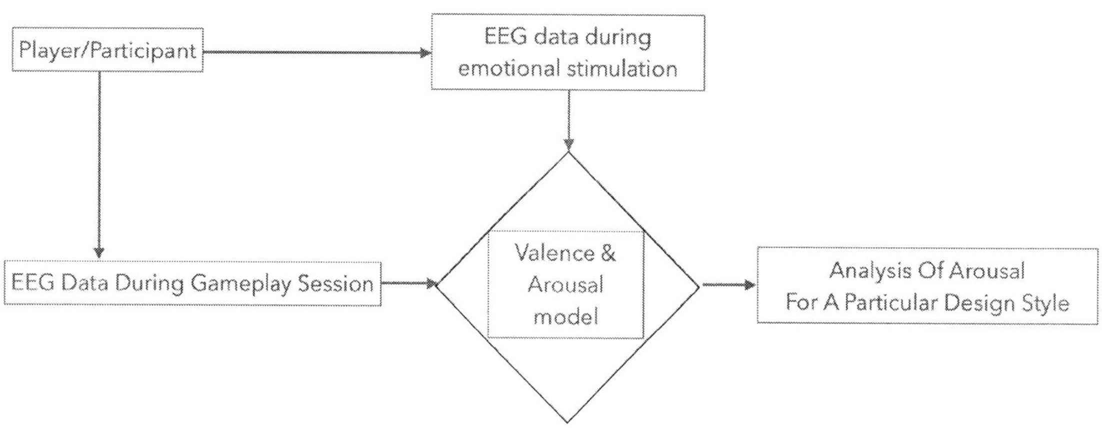

Figure 2. Data analysis work flow. Emotionally stimulated data were used as the known variable and trained to create a working model. Unknown variable from game play can then be classified.

The learning aptitude test was originally meant to be as a component for profiling the subjects that undertake this experiment. The test instrument measures four different brain functions and ideally the test result should be obtained just before the game play session. This is the reason why EEG data of the

IOP Conf. Series: Materials Science and Engineering 1077 (2021) 012037 doi:10.1088/1757-899X/1077/1/012037

aptitude test was recorded. The EEG device was already in place when the aptitude test was taking place, thus their signals were recorded anyway.

The test instrument for measuring learning aptitude was provided by the Cambridge University's Brain Lab. It is available online at https://brainlabs.me. The instrument measures four brain functions:

1. Cognitive
2. Verbal
3. Creativity
4. Logical thinking and reasoning

Each brain functions was measured with a score. These scores can be accumulated over time to monitor changes of the brain performance after every 24 hours. This instrument was a valid and credible instrument and is backed by numerous articles and publications.

For data analysis, raw data out of the EEG device was input and ran into Python codes that applies supervised machine learning to classify the unknown variable. Data from the learning aptitude was also ran with the code even though their result may not be useful for the research's objective. However, when the resulting analysis laid out in a scatter plot graph, a constant pattern of similarities is noticed for all subjects of the experiment.

There are two things that can be observed from these scatter plots. First, for each aptitude tests that measures a specific brain function, their scatter plot is almost unique to each other. The plots shows the emotional valence versus arousal (VA) of the player during each tests. The plots reveals that when a specific brain function is at work, their valence vs arousal is in a specific state. The scatter plots in figure 3 demonstrate one subject's valence versus arousal analysis.

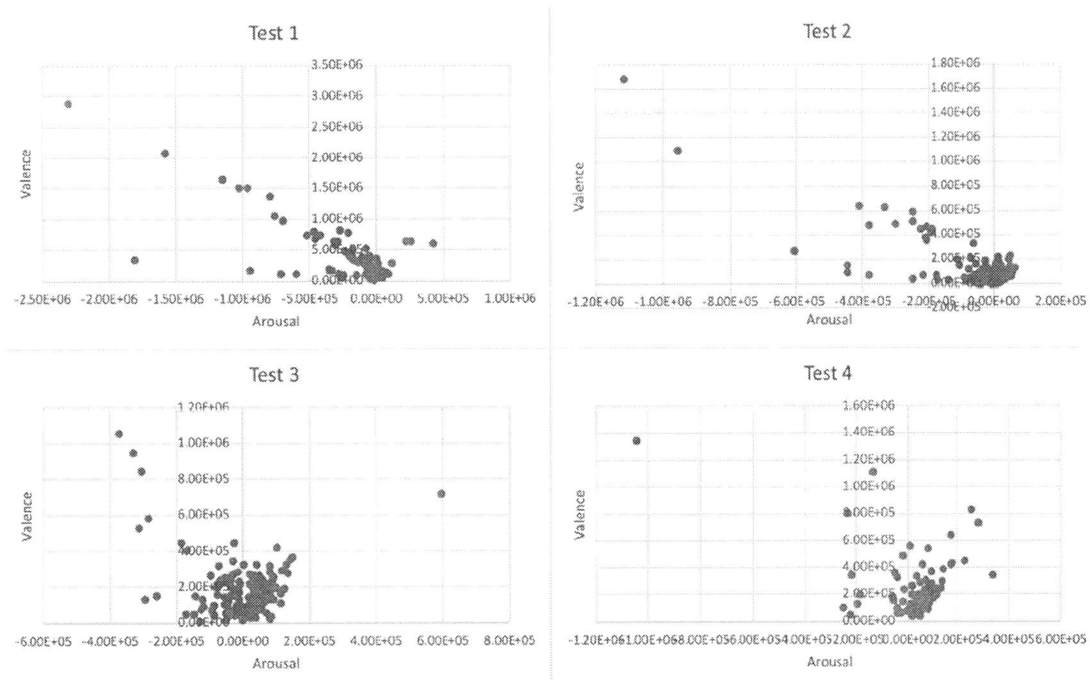

Figure 3. Four scatter plots exhibit unique patterns for each brain functions (cognitive, creativity, memory, and verbal).

Second, when these scatter plots are superimposed or combined into a single plot, it resembles the pattern of the game play emotional state. The plots in figure 3 are used again for demonstration and the result can be seen in figure 4.

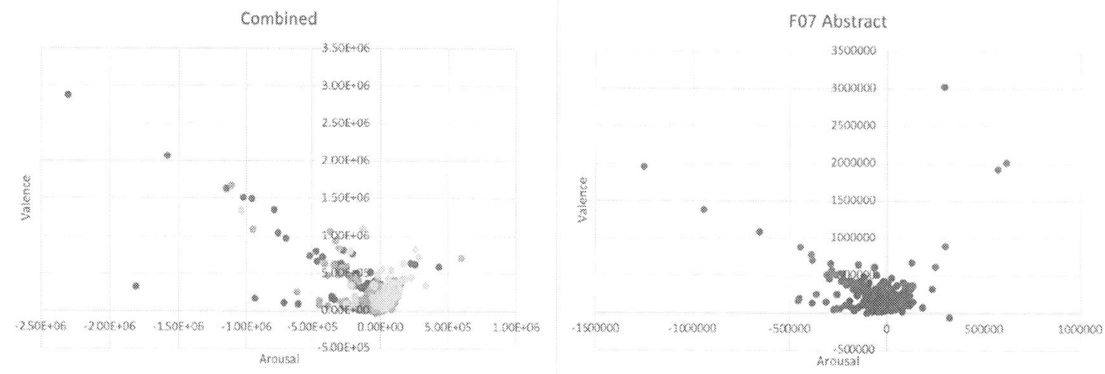

Figure 4. Combined aptitude test VA plots (left) versus gameplay VA plots (right).

The similarities are obvious when they are seen in visual observation. At this point, it is tempting to converge into a conclusion that four different components of the brain functions are present in a game that were not designed to stimulate any of those functions at all.

One might say that this finding is obvious since the brain have to work using all of its function all the time. But to what degree? Aptitude test results measure varying result of the brain performance. And in this experiment, the same subject may only shows not all of the brain's function when playing a different video game. This can be observed in figure 5.

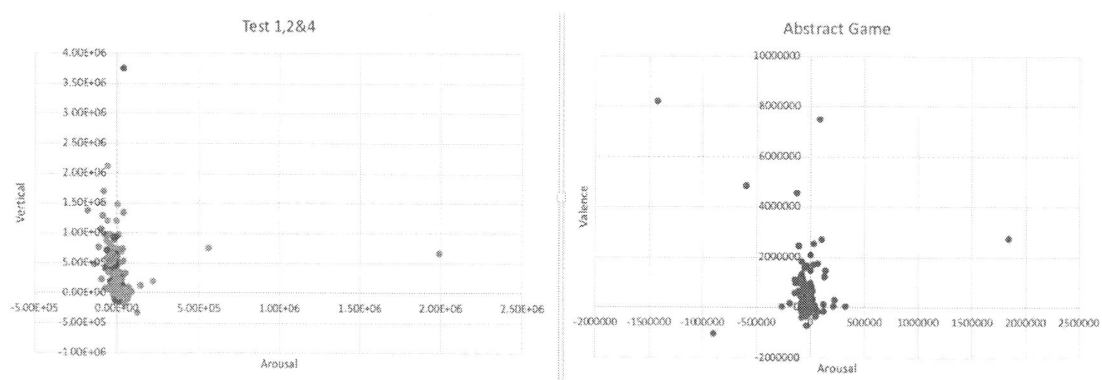

Figure 5. Aptitude tests 1, 2, and 4 (left) resembles scatter plot of gameplay of an abstract design scatter plot (left) when combined.

As in Figure 4, Figure 5 shows another subject's similarities of valence versus arousal scatterplot between aptitude tests and playing a video game. Only this time, only three brain functions are apparent. However, when the subject played another similar game but with a different design style, the scatter plot only reminiscent to only one of the brain functions, as can be seen in Figure 6.

These scatter plots are based on the emotional valence versus arousal data. Thus, it is not a concrete correlation between design styles in video games and learning aptitudes. However, Figure 5 and 6 both reveals as if it can be assumed that different games lit up different aspect of brain functions. The same experiment but using the aptitude tests as the machine learning model may provide a more solid evidence of this assumption.

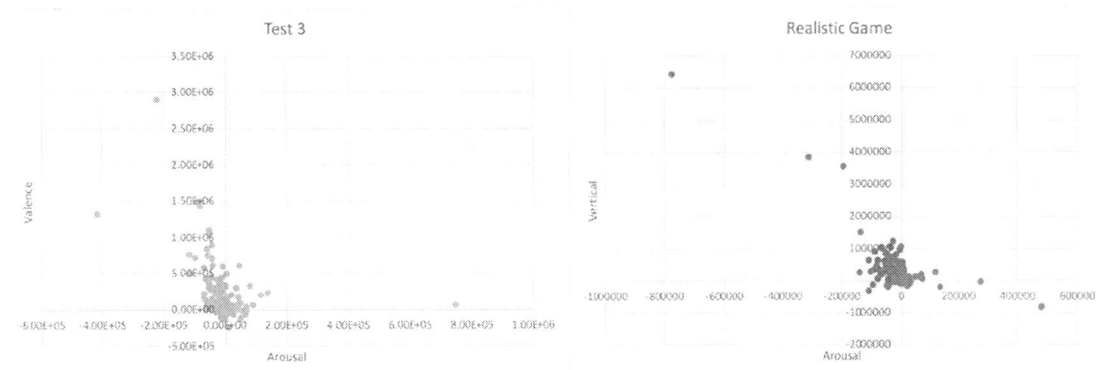

Figure 6. Only one of the aptitude test resembles the gameplay of a realistic game. The other patterns from other tests seem absent.

4. A Hypothesis For A Conclusion

The distribution of the valence and arousal signals from the EEG signals provided by the emotional scatter plots exhibit clear visual correlation between brain functions and video games designs. It is amazing to see that multiple brain functions indeed respond to video games that are not even designed to stimulate them. Unfortunately, the lack of quantitative analysis and statistical correlation means that it is not a concrete proof. This is due to the following:

a. The research was not designed to reach this conclusion and it is actually outside the research scope. It is merely a bonus discovery during data analysis.
b. The scatter plots similarities are only visual. No statistical correlation has been done to confirm this claim yet.
c. A more accurate research design is needed to verify this finding. Instead of emotional valence and arousal model, a specific computational model of the brain functions can be applied in the machine learning process.

Despite the above, the findings can be synthesized into a hypothesis for future research: Specific cognitive and learning aptitude functions of the brain are present and respond in varying degrees in a video game that are not even designed to stimulate them. Each brain functions emits unique EEG signals and they are possible to be correlated with video game designs. The brain's cognitive, creative, memory, and verbal functions may each be individually stimulated by the visual elements of the game. This includes design styles, colors, textures, shapes, and other design elements.

The idea of having fun and entertaining activities that simultaneously improve learning potential may not be a far-fetched concept after all. While this paper cannot give a concrete proof and concludes the presence of brain functions within a video game that was not designed to stimulate them, the trace of evidence can at least be observed visually in the research this paper discusses. And that trace is hard to ignore.

Acknowledgement

This study is funded by Malaysia Ministry of Education through the Research Acculturation Of Early Career Researcher Grant, RACER19-003-0003 (RACER/1/2019/ICT01/UIAM//1). Special thanks to all members of the Pervasive Computing Brain Development Group (PCBDG) from the Kuliyyah of ICT, IIUM for assistance in data collection.

References

[1] Alvarez J, Djaouti D, Ghassempouri R, Jessel J-P and Methel G 2006 Morphological study of the video games *3rd Australasian Conf. on Interactive Entertainment* December 2015 pp 36–43
[2] Ashinoff B 2014 The potential of video games as a pedagogical tool *Frontiers in Psychology* **5** September DOI: 10.3389/fpsyg.2014.01109

[3] Benlamine M S, Dombouya R, Dufresne A and Frasson C 2017 Game experience and brain based assessment of motivational goal orientations in video games *Lecture Notes in Computer Sci. (Including Subseries Lecture Notes in Artificial Intelligence and Lecture Notes in Bioinformatics) 10512 LNAI* pp 118–32 DOI: 10.1007/978-3-319-67615-9_11
[4] Blacker K. J, Curby K M, Klobusicky E and Chein J M 2014 Effects of action video game training on visual working memory *J. of Experimental Psychology: Human Perception and Performance* **40(5)** pp 1992–2004 DOI: 10.1037/a0037556
[5] Boot W R and Hommel B 2015 Video games as tools to achieve insight into cognitive processes *Frontiers in Psychology* **6(3)**
[6] Bunian S, Canossa A, Colvin R and El-Nasr M S 2018 Modeling Individual Differences in Game Behavior using HMM *Preprint Arxiv: abs/1804.00245*
[7] Chandra S, Sharma G, Salam A A, Jha D and Mittal A P 2016 Playing action video games a key to cognitive enhancement *Procedia Computer Sci.* **84** issue May pp 115–22 DOI: 10.1016/j.procs.2016.04.074
[8] Franceschini S *et al* 2017 Action video games improve reading abilities and visual-to-auditory attentional shifting in English-speaking children with dyslexia *Scientific Reports* **7(1)** pp 1–12 DOI: 10.1038/s41598-017-05826-8
[9] Gori S *et al* 2016 Dyslexia prevention by action video game training: behavioural and neurophysiological evidence *J. of Vision* **16(12)** 489 10.1167/16.12.489
[10] Järvinen A 2002 Gran stylissimo: the audiovisual elements and styles in computer and video games *Computer Games and Digital Cultures Conf.* pp 113–28 DOI: http://dx.doi.org/10.1177/1468796811398833
[11] Keo M 2017 *Graphical Style in Video Games*
[12] Khairuddin R N H R, Malik A S and Kamel N 2014 EEG topographical maps analysis for 2D and 3D video game play *2014 5th Int. Conf. on Intelligent and Advanced Systems: Technological Convergence for Sustainable Future ICIAS 2014 – Proc.* DOI: 10.1109/ICIAS.2014.6869517
[13] Kuk K, Jovanovic D, Jokanovic D, Spalevic P, Caric M and Panic S 2012 Using a game-based learning model as a new teaching strategy for computer engineering *Turkish J. of Electrical Engineering and Computer Sciences* **20** (SUPPL.2) pp 1312–31 DOI: 10.3906/elk-1101-962
[14] McMahan T, Parberry I and Parsons T 2015 Evaluating electroencephalography engagement indices during video game play *Proc. of the Foundations of Digital Games Conf. 2015*
[15] Palaus M, Marron E M, Viejo-Sobera R and Redolar-Ripoll D 2017 Neural basis of video gaming: a systematic review *Frontiers in Human Neuroscience* **11** issue May DOI: 10.3389/fnhum.2017.00248
[16] Poels K, Hoogen W van den, Ijsselsteijn W and de Kort Y 2012 Pleasure to play, arousal to stay: the effect of player emotions on digital game preferences and playing time *Cyberpsychology, Behavior, and Social Networking* **15(1)** pp 1–6 DOI: 10.1089/cyber.2010.0040
[17] Rapeepisarn K, Wong K W, Fung C C and Khine M S 2008 The relationship between game genres, learning techniques and learning styles in educational computer games *Edutainment 2008* pp 497–508 DOI: 10.1117/12.958586
[18] Schneider E, Wang Y and Yang S 2007 Exploring the uncanny valley with Japanese video game characters *Digital Games Research Association (DiGRA)* 546549
[19] Sharan L, Neo Z H, Mitchell K and Hodgins J K 2013 Simulated motion blur does not improve player experience in racing game *Proc. of Motion on Games* **149** pp 149–54 DOI: 10.1145/2522628.2522653

Image Captioning with Attention for Smart Local Tourism using EfficientNet

Dhomas Hatta Fudholi[1], Yurio Windiatmoko[2], Nurdi Afrianto[3], Prastyo Eko Susanto[4], Magfirah Suyuti[5], Ahmad Fathan Hidayatullah[6], Ridho Rahmadi[7]

[1,6,7]Department of Informatics, Universitas Islam Indonesia, Yogyakarta, Indonesia,
[2,3,4,5]Master Program in Informatics, Universitas Islam Indonesia, Yogyakarta, Indonesia

E-mail: [1]hatta.fudholi@uii.ac.id, [2]yurio.windiatmoko@students.uii.ac.id, [3]nurdi.afrianto@students.uii.ac.id, [4]prastyo.susanto@students.uii.ac.id, [5]maghfirah.suyuti@students.uii.ac.id, [6]fathan@uii.ac.id, [7]ridho.rahmadi@uii.ac.id

Abstract. Smart systems have been massively developed to help humans in various tasks. Deep Learning technologies push even further in creating accurate assistant systems due to the explosion of data lakes. One of the smart system tasks is disseminating 'users needed information', which is crucial in the tourism sector to promote local tourism destinations. In this research, we design a local tourism specific image captioning model, which will later support the development of AI-powered systems that assist various users. The model is developed using a visual Attention mechanism and uses the state-of-the-art feature extractor architecture EfficientNet. A local tourism dataset is collected and used in the research and two different captions: captions that describe the image literally and captions that represent human logical responses when seeing the image. The two kinds of captions make the captioning model more humane when implemented in the assistance system. We compared two different models using EfficientNet architectures (B0 and B4) with other well-known VGG16 and InceptionV3. The best BLEU scores we get are 73.39 and 24.51 for the training set and the validation set, respectively, using EfficientNetB0. The captioning result using the developed model shows that the model can produce logical caption for local tourism-related images.

Keywords. Image Captioning; Attention; Smart Tourism; EfficientNet

1. Introduction

Tourism has a significant influence on economic development. The increase in population creates vast potential in tourism and impacts tourism development [1], including the rapid growth in tourism villages. Tourists are becoming more interested in natural and cultural experiences [2]. Yogyakarta is one of Indonesia's leading tourist destinations, which improves the economy around its tourism destination [3]. Tourists are an asset [3][4] to develop regional economics. Yogyakarta's rural tourism destinations may face difficulties competing with other popular tours due to a lack of information [5]. Also, various forms of service facilities (lodging, equipment, etc.) and information should be maintained [2]. Therefore, one of the Indonesian government's programs in the tourism sector is to reduce social inequality between rural and urban areas (Gini Index up to 0.4) through community-based rural development programs [6].

Artificial Intelligence (AI) and the emergence of Deep Learning technologies have become the core of smart systems in various domains, which help people in many ways. One implementation of AI-powered systems is a chatbot. Chatbots are a valuable technology that can help collect and manage information related to the company's services or a corporation [7]. Chatbots provide a new perspective on doctor-patient (DP) and clinic-patient (KP) services in terms of cost and time [8]. Chatbots handle problems related to getting customer service requests quickly and accurately [9]. Chatbots can operate 24/7 (24 hours and 7 days) to help retailers improve customer service and increase efficiency [10]. A chatbot with skilled nurses supervision can be used for patients undergoing chemotherapy remotely, of course [11]. Chatbots are used in the university environment to answer student FAQs interactively using the AIML and LSA approaches [11]. The chatbot's ability that changes the interaction between customers and organizations in all sectors involved [12] may be one of the solutions in providing tourism information ubiquitously.

The interaction within chatbot systems may be pushed further to not only text-based communication. In the tourism industry, it is exciting to have fluid conversations based on photos as well. Image captioning technology enables this feature and uses human vision to collect information in a visual scene. Humans tend to focus on important and interesting objects in the visual scene. One of the works in image captioning using the SILICON dataset is [13]. Besides, describing an image may be challenging because it recognizes objects in an image and converts visual understanding into sentences [14]. However, giving information through photos can attract tourists to their favorite destination.

This research focuses on developing a model of image captioning, which is specialized in captioning local tourism-related photos. We use a local tourism-related dataset from tourism destinations in Yogyakarta, Indonesia. To develop the model, we use a Deep Learning technique that uses state-of-the-art feature extraction architecture EfficientNet in the encoder side and GRU in the decoder side. The visual Attention mechanism is also applied to model architecture. Furthermore, we compare different architectures and measure their performance as the evaluation.

2. Literature Review

Image captioning aims to set the image description automatically in the right term language [15]. An interesting thing comes when it not only reads the image based on its visual content but also from the image's linguistic context [16]. Some of the benefits of image captioning are to help create a description of an image for blind people in seeing the meaning of images [17]. The ability to translate visual content into natural language can be useful in various fields, such as human interaction with robots, real-time descriptions of car scenes for the visually impaired, and instantaneous display of traffic conditions to the public [18]. The image description creation model captures objects in the image and reveals the relationship between objects into a sentence [19]. Besides being useful for blind people, image captioning helps search images using text [20]. The description of an image in the industrial world can be used to share photos on social media and interact with robots [21]. Combined with smart wearable devices, image captioning can create personal notes based on pictures collected from supporting devices [22]. Image captioning is a multidisciplinary knowledge of computer vision and Natural Language Processing that is attractive to academics and the industrial world [23].

Deep Learning techniques become the core in building an image captioning model. A study in [14] focuses on captioning images using bidirectional LSTM. In the study, augmentation techniques such as multi-crop and multi-scale were used to prevent overfitting. In comparison, other research tends to use semantic model attention to generate image descriptions, such as using the top-down and bottom-up strategy to extract more image information, which has good results because it observes all the information contained in the image [17]. Another approach in image captioning is to cluster the area of attention with structural proximity. This approach models the dependencies between drawing image, word's description, and language model by RNN [19]. Another study of image captioning which takes other languages besides English is [24]. This study presented two Nepali-based models. The first model comprises an encoder (ResNet-50) and a decoder (a plain LSTM). The second model includes an encoder (InceptionV3) and a decoder (GRU network with attention). However, the result of this paper shows perplexity in comparing model A and model B. The results of the test samples depend on the priors seen during the training regime. The lack of datasets and the fact that Nepali translation is not very accurate to make this research could be improved for both the models with the further tuning of the architectures and training.

An attention mechanism can be applied to enhance the contextual aspect of the natural language sequence. The description of the image content with attention is compatible with human intuition. The accuracy of attention to an image using the evaluation matrix has a positive correlation. It needs to be improved to what extent the accuracy of attention is consistent with human perceptions [25]. The captioning model with attention area consists of three parts: combining image areas, word captions, and the NLP natural language model (RNN). Usually, the MS COCO dataset is used for testing the trained model [20]. Other research shows the AoA (Attention on Attention) method for encoder and decoder in image captioning with a score of 129.8 (CIDEr-D) on the COCO dataset [26].

Generating a caption that sensibly describes an image is key to understanding scenes in artificial intelligence [24]. Extracting coherent features of an image using an image-based model is one of the processes in image captioning. An image is described in the natural language from the extracted features of the image-based model. We propose to develop a model of image captioning in the domain of local tourism to support better engagement in smart tourism systems. We use state-of-the-art architecture EfficientNet for the feature extraction process and GRU architecture in caption generating decoder. The Attention layer is applied to aim to get a more sensible caption of the image.

3. Method

The local tourism becomes our domain in the research. Hence, we firstly collect a dataset of local tourism-related images and write captions for them. Secondly, we develop an image captioning model using EfficientNet as the base architecture. Finally, we test and evaluate the model using the BLEU score and compare the EfficientNet-based model with other models developed using well-known architecture. Each research stage is elaborated in detail below.

3.1. Data Collection and Caption Annotation

A total of 1696 local tourism-related images are collected through Google search engines using various keywords that are related to the local tourism destination in Yogyakarta, Indonesia. These keywords are 'monumen tugu' (Tugu monumen) , 'arung jeram' (rafting), 'malioboro' (Malioboro), 'pendaki wanita' (female hikers), 'wisata jeep kaliadem' (Kaliadem jeep tours), 'merapi park' (Park Merapi), 'air terjun sri gethuk' (Sri Gethuk waterfall), 'parangtritis' (Parangtritis), 'hutan pinus' (pine forest), 'bukit paralayang' (Paralayang hill), 'heha sky view' (Heha Sky View), 'ayunan langit watu jaran' (Watu Jaran sky swing), 'ice cream world', 'grojogan watu purbo' (Grojogan Watu Purbo), 'bukit isis' (Isis hill), 'bukit mojo gumelem' (Mojo Gumelem hill) and 'puncak segoro' (Segoro peak). All these keywords are closely related to tourism destinations in the local area of Yogyakarta, Indonesia.

The collected images have already been filtered by their high-resolution quality, which is considered as a sharp image. After collecting images, the next step is making a unique name for each image in a folder by running a randomized filename script. Finally, we made a list of annotations in .txt format and created a cross-checked list that maps images with their captions, as in Figure 1.

JSON
{} 0
caption : "many types of flowers in Merapi Park"
image_id : "0cv1krzup6x45sq3i8hb.jpg"
{} 1
caption : "the place is wide enough and the view makes you don't want to go home"
image_id : "0y6txkj1b7eqmw9gpunc.jpg"
{} 2
caption : "want to see miniature tours in various parts of the world, let's go to Merapi pa
image_id : "2ziuvk0d9aylptscgxj3.jpg"

Figure 1. Cross-check files of images and their captions.

Speaking of the captions in the datasets, the vocabulary size of the captions is 291. Figure 2 shows the top 50 and the least 50 most frequent words in the vocabulary. From the figure, we can see that there are conjunctions that we kept, such as *'yang'* (that) and *'dan'* (and). The image captioning task requires that the output caption forms complete logical sentences. It is different from other text analytics tasks, such as sentiment analysis, where those words may be omitted through stopword removal tasks since it is not necessary and becomes noises.

The dataset is divided into a training set and a validation set with an 80:20 ratio. We get 1356 data within the training set and 340 data for the validation set, transformed into tensor during the training through the Keras Tensorflow pipeline.

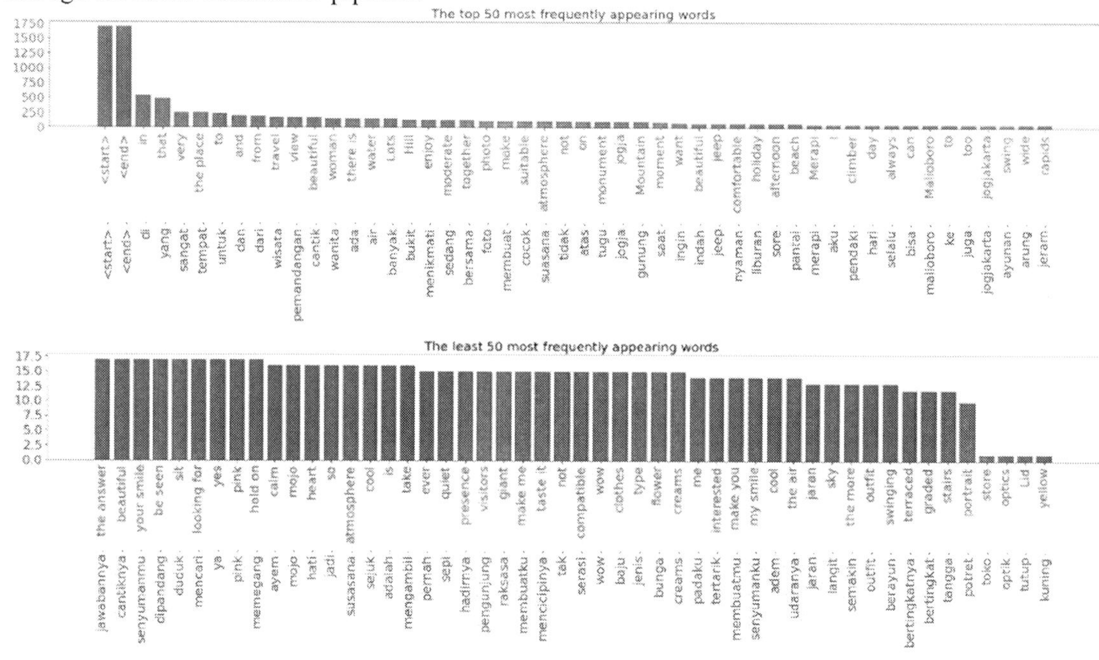

Figure 2. The top and the least 50 most frequently words

3.2. Modeling

Encoder-decoder architecture is implemented during the training process. As for the encoder, we applied pre-trained ImageNet weight transfer learning from two different EfficientNet architectures, namely EfficientNetB0 and EfficientNetB4, and two comparator architectures, namely VGG16 and InceptionV3. Figure 3 illustrates our image captioning network.

ImageNet weight from the four architectures' pre-trained model is used in the image feature extraction part. However, there are differences in the image input size for each architecture. EfficientNetB0 uses 224x224, EfficientNetB4 uses 380x380, InceptionV3 uses 299x299 and VGG16 uses 224x224. At the end of the feature extraction process, the image feature dimensions are also different for each architecture. EfficientNetB0 creates vectors in size of 49x1280, EfficientNetB4 in size of 121x1792, InceptionV3 in size of 64x2048 and VGG16 in size of 49x512. These feature vectors are then passed to a fully connected feedforward network with 256 nodes. The resulting vector enters a loop as much as the maximum length of the training data caption through the Attention layer and then goes to the GRU-based decoder.

Within the Attention layer, we calculate the attention score and generate context vectors. The attention score is calculated from the activated summation of the hidden state value with the previous decoder layer. Softmax is used to finalize the score into probabilities values. The score will be multiplied with the image feature vector to be the context vector in the Attention layer. The context vector will be added with the embedding vector of words before entering the decoder. The concatenation context vector with a target word embedding vector is called the Teacher Forcing method. This technique helps learn the correct sequence or correct statistical property in front of the sequence quickly. The decoder that we use is GRU, with the initial initiation of Uniform Glorot numbers. The GRU layer has 512 hidden units that produce two different outputs: the word prediction matrix and the hidden state vector. The outputs will be used to train the next word. The prediction word vector is then passed to the fully connected layer with an output layer size equal to the existing vocabulary size, 291 units. In the training process, the predicted words will be used to calculate the loss with the target word embedding vector matrix. Finally, the gradient is calculated using an optimizer and runs the backpropagation to adjust the network parameter.

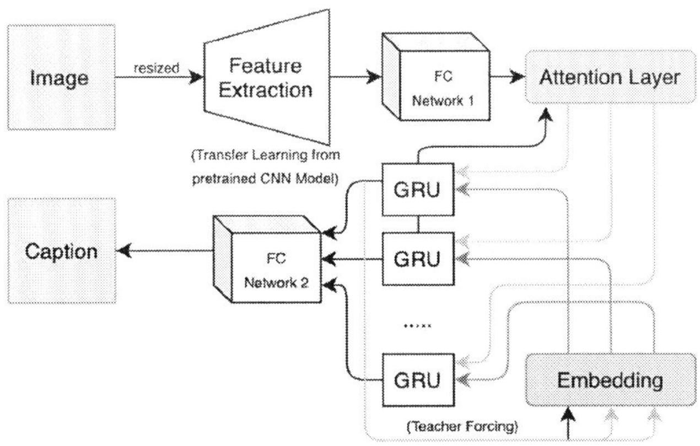

Figure 3. The image captioning network

3.3. Evaluation
Since the image captioning model aims to generate sentences that explain or express the given image, we evaluated the performance of the model's text generation using BLEU metrics. BLEU is a recognized metric for measuring the similarity of one hypothetical sentence to several reference sentences. We are given one hypothetical sentence and several reference sentences, returning a value between 0 and 1. A metric close to 1 means they are very similar [27]. In our evaluation, BLEU metrics take slices of the n-gram in the predicted caption and compare them to the ground truth caption data. We use percentage normalization to represent the BLEU value. The ground truth is obtained from self-image annotation by captioning manually at our custom dataset. In our study, n is equal to 1.

4. Result and Discussion
EfficientNet is the state-of-the-art architecture used as a feature extractor in this research. We take the least and the highest architecture of EfficientNet, namely B0 and B4, in our experiment. We also carry two other well-known models as a comparison (VGG16 and InceptionV3). Figure 4 shows the plot of the four training experiments we have done using four different models. The graphs in Figure 4 show a decline of loss as long as the epoch goes. The four architectures have similarities during the training. Around epoch 8 and 9, the four models started to suffer overfitting about the same loss value. We choose the best checkpoint right about the intersection on evaluation lines, to be our image captioning model.

The four generated models are then evaluated using the BLEU metric. Table 1 shows the detailed average BLEU score of the four models. From the table, we could see that the EfficientNetB0 architecture is the best model for the context of our image captioning model using our collected dataset, with the average BLEU scores of 73.39 and 24.52 for training and validation, respectively. The EfficientNetB4 model, which has a larger number of parameters, has a similar (slightly lower) performance to the B0 version. The result is exciting and fortunate for our local tourism domain. Inference resources can be saved to implement the smaller B0 model. Also, the EfficientNet model is proven to perform better than VGG16 and InceptionV3 in our local tourism domain image captioning. They have lower BLEU scores than the EfficientNet's.

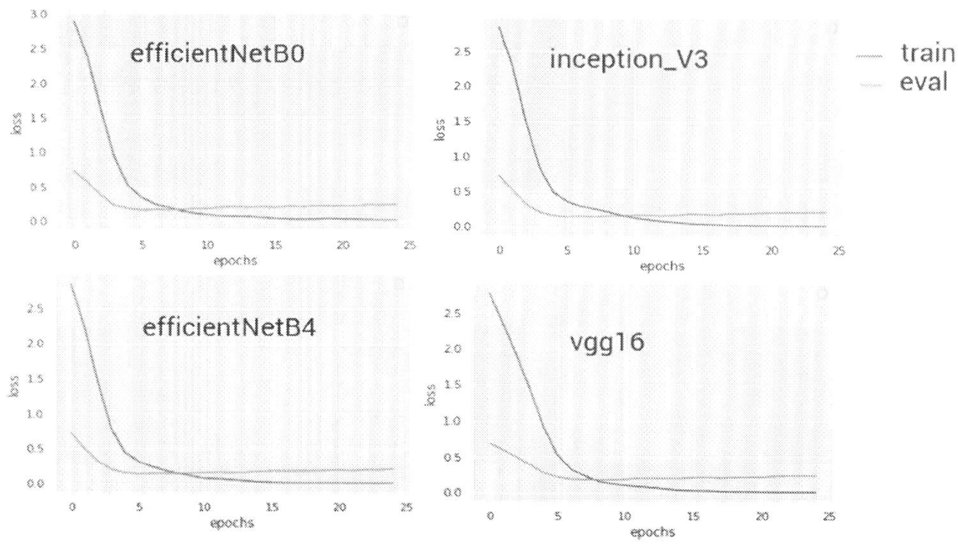

Figure 4. Training and evaluation plot of four different models

Table 1. The Comparison of BLEU Metric Score of Four Different Models.

Architecture	Average of the BLEU_train_score	Average of the BLEU_val_score
EfficientNetB4	72.84	22.24
EfficientNetB0	**73.39**	**24.51**
VGG16	68.67	19.33
InceptionV3	58.67	22.41

To show the inferencing result using our developed model, which is based on EfficientNet, we take the captioning prediction result of three random validation images, as shown in Figure 5, Figure 6, and Figure 7. The first validation image, depicted in Figure 5, has a pretty high BLEU score in the predicted caption (71.42). The ground truth caption is *'menikmati ombak dan angin di pantai parangtritis'* (enjoying the waves and the wind on parangtritis beach) and the predicted caption is *'selalu ada dan angin di pantai parangtritis'* (always there and the wind at parangtritis beach). While the BLEU score is high, it still suffers from slight grammatical errors. This may be improved by feeding more grammatically structured training data. Another thing to highlight is that the predicted caption can point to specific tourism destinations. The second example, shown in Figure 6, is one of the interesting examples. It may not hold a high BLEU score. However, the predicted caption is sensible. The ground truth caption is *'serasa dapat terbang walaupun dengan bantuan ayunan'* (feels like I can fly even with the help of a swing) and the sensible predicted caption is *'banyak tumbuhan di bawahnya'* (many plants below). The third example that holds a pretty average BLEU score (42.85) also gives sensible predicted caption *'wisata merapi kaliadem private jeep tour yogyakarta'* (merapi tours kaliadem private jeep tour yogyakarta) which is close to the ground truth *'wisata jeep gunung merapi sleman'* (mount merapi jeep tour in sleman). Overall, the model has achieved a good sensible image captioning prediction in local tourism, especially in the region of Yogyakarta, Indonesia.

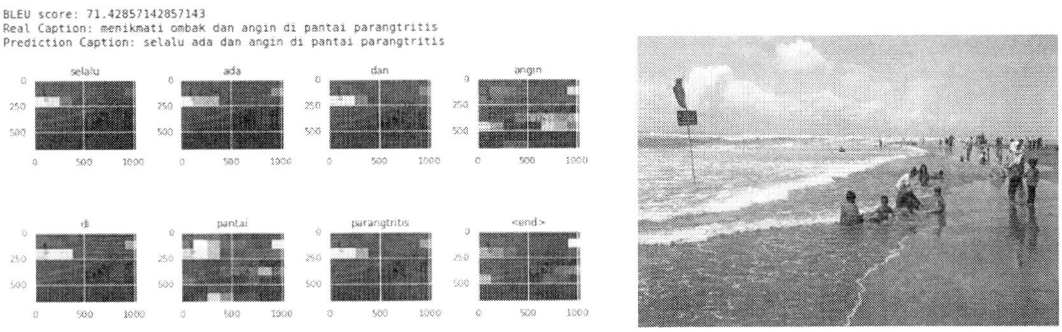

Figure 5. First image captioning prediction result

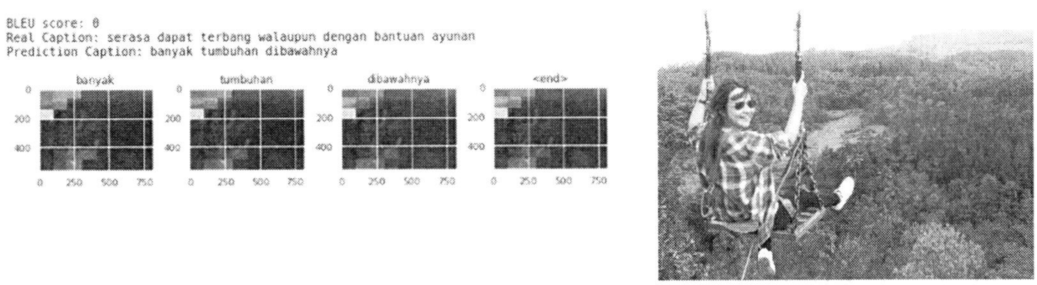

Figure 6. Second image captioning prediction result

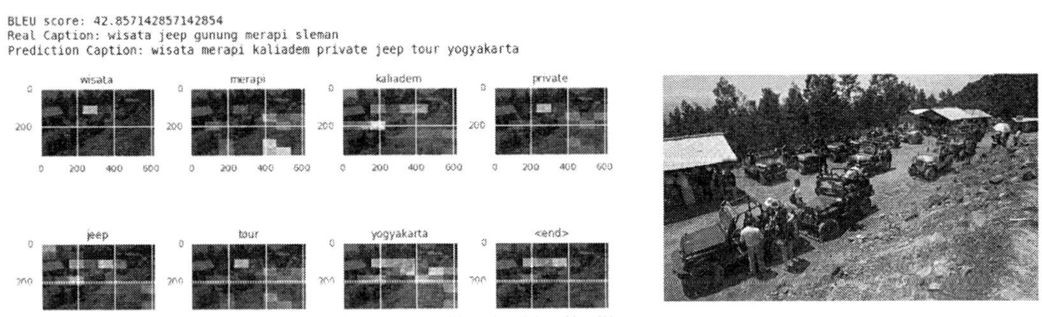

Figure 7. Third image captioning prediction result

5. Conclusion and Future Work

The information dissemination and engagement of smart local tourism systems to broader communities is crucial in tourism development. This work has presented an attention-based image captioning on tourism images using EfficientNet to support a more humane engagement within the local tourism smart system. To evaluate the performance of the model, we compared the EfficientNet-based model to another well-known architecture model. Our experiment showed that our image captioning model using EfficientNetB0 and EfficientNetB4 architectures resulted in similar BLEU scores and got a higher score than another famous trained model, InceptionV3, and VGG16. However, EfficientNetB0 got a higher average train score and the average validation score than EfficientNetB4 with 73.39 and 24.51. The captioning inference has shown that the EfficientNet-based image captioning model can produce a sensible yet detailed caption in the domain of local tourism. In the next development, we will create richer captions with more complex objects or activities in the image to improve the suitability of the attention results in the captions. Besides, we will develop API wrapping for the image caption model that facilitates image captioning features with smart tourism systems, such as customer chatbot.

References

[1] Rif'an A A 2016 Tourism components and tourists characteristic of Prambanan Temple as the world culture heritage site in Yogyakarta, Indonesia *Int. J. of Tourism and Hospitality Study* vol **1** no 1 pp 11–21

[2] Vitasurya V R 2016 Local wisdom for sustainable development of rural tourism, case on Kalibiru and Lopati Village, Province of Daerah Istimewa Yogyakarta *Procedia - Social and Behavioral Sciences* vol **216** pp 97–108

[3] Damayanti M, Scott N and Ruhanen L 2017 Coopetitive behaviours in an informal tourism economy *Annals of Tourism Research* vol **65** pp 25–35

[4] Pujiastuti E E, Nimran U, Suharyono S and Kusumawati A 2017 The antecedents of behavioral intention regarding rural tourism destination *Asia Pacific J. of Tourism Research* vol **22** no 11 pp 1169–81

[5] Pradikta A C 2017 Building partnership across rural tourism destination for sustainability: a case study in special region of Yogyakarta, some challenges and opportunities *J. Bus. Hosp. Tour.* vol **2** no 1 p 87

[6] Manaf A, Purbasari N, Damayanti M, Aprilia N and Astuti W 2018 Community-based rural tourism in inter-organizational collaboration: how does it work sustainably? Lessons learned from Nglanggeran Tourism Village, Gunungkidul Regency, Yogyakarta, Indonesia *Sustain. Sci. Pract. Policy* vol **10** no 7 p 2142

[7] Filipczyk B, Gołuchowski J, Paliszkiewicz J and Janas A 2016 Success and failure in improvement of knowledge delivery to customers using chatbot—result of a case study in a Polish SME *Successes and Failures of Knowledge Management* 2016 pp 175–90

[8] Abashev A, Grigoryev R, Grigorian K and Boyko V 2017 Programming tools for messenger-based chatbot system organization: implication for outpatient and translational medicines *Bionanoscience* vol **7** no 2 pp 403–7

[9] Xu A, Liu Z, Guo Y, Sinha V and Akkiraju R 2017 A new chatbot for customer service on social media *CHI Conf. on Human Factors in Computing Systems* pp 3506–10

[10] Shankar V 2018 How Artificial Intelligence (AI) is reshaping retailing *J. of Retailing* **94** pp vi–xi

[11] Piau A, Crissey R, Brechemier D, Balardy L and Nourhashemi F 2019 A smartphone chatbot application to optimize monitoring of older patients with cancer *Int. J. Med. Inform* vol **128** pp 18–23

[12] Larivière B et al 2017 Service Encounter 2.0: An investigation into the roles of technology, employees and customers *J. Bus. Res* vol **79** pp 238–46

[13] Sugano Y and Bulling A 2016 Seeing with humans: gaze-assisted neural image captioning *Preprint arXiv:1608.05203* [cs.CV]

[14] Wang C, Yang H and Meinel C 2018 Image captioning with deep bidirectional LSTMs *ACM Trans. Multimed. Comput. Commun. Appl.* vol **14** no 2s pp 988–97

[15] Mun J, Cho M and Han B 2017 Text-guided attention model for image captioning *31st AAAI Conf. Artif. Intell.* pp 4233–9

[16] Kilickaya M, Erdem A, N Ikizler-Cinbis and Erdem E 2017 Re-evaluating automatic metrics for image captioning *15th Conf. Eur. Chapter Assoc. Comput. Linguist.* vol **1** pp 199–209

[17] Yao T, Pan Y, Li Y, Qiu Z and Mei T 2017 Boosting Image Captioning with Attributes *Proc. IEEE Int. Conf. Comput. Vis.* pp 4904–12

[18] Chen L, He Y and Fan L 2017 Let the robot tell: Describe car image with natural language via LSTM *Pattern Recognit. Lett.* vol **98** pp 75–82

[19] You Q, Jin H, Wang Z, Fang C and Luo J 2016 Image Captioning with Semantic Attention *IEEE Conf. on Computer Vision and Pattern Recognition (CVPR)* pp 27–30

[20] Pedersoli M, Lucas T, Schmid C and Verbeek J 2017 Areas of attention for image captioning *IEEE Int. Conf. on Computer Vision (ICCV)* pp 1242–50
[21] Guo L, Liu J, Yao P, Li J and Lu H 2019 MSCAP: multi-style image captioning with unpaired stylized text *Proc. IEEE Comput. Soc. Conf. Comput. Vis. Pattern Recognit.* pp 4199–208
[22] Fan C, Zhang Z and Crandall D J 2017 Deepdiary: lifelogging image captioning and summarization *J. Vis. Commun. Image Represent.* vol **55** pp 40–55
[23] Wang W, Chen Z and Hu H 2019 Hierarchical attention network for image captioning *33rd AAAI Conf. Artif. Intell.* pp 8957–64
[24] Adhikari A and Ghimire S 2019 Nepali image captioning *Int. Conf. Artif. Intell. Transform. Bus. Soc.* pp 1–6
[25] Zheng L, Caiming Z and Caixian C 2018 MMDF-LDA: an improved multi-modal latent dirichlet allocation model for social image annotation *Expert Syst. Appl.* **104** pp 168–84
[26] Huang L, Wang W, Chen J and Wei X Y 2019 Attention on attention for image captioning *Proc. IEEE Int. Conf. Comput. Vis.* pp 4633–42
[27] Papineni K, Roukos S, Ward T and Zhu W J 2002 BLEU: a method for automatic evaluation of machine translation *Proc. 40th Annu. Meet. Assoc. Comput. Linguist* pp 311–8

Analysis and Design of UI and UX Web-Based Application in Maiproyek Startup Using User Centered Design Method in Information System Program of Telkom University

C Adhitya[1*], R Andreswari[2], and P F Alam[3]

[1,2,3] Information System Study Program, Faculty of Industrial Engineering, Telkom University, Jl. Telekomunikasi No.1, Terusan Buah Batu, Sukapura, Dayeuhkolot, Bandung, West Java, Indonesia

E-mail: *chandhityaa@gmail.com

Abstract. To develop a good application, the aspect that needs to be considered is to design a User Experience and User Interface that suits the needs of the users themselves. One of the design methods that can be used is User Centered Design method. This method focuses on designing the user's needs where it aims to provide comfort for the user and provide a relevant interface for the user. The study was conducted in Information Systems Program of Telkom University. The design will then be tested on each lecturer and student in Information Systems Program of Telkom University. This test aims to test whether a design is in accordance with the desired objectives or not. The results of the research the author wants to achieve is to design a user interface and user experience design of a web-based application for a Maiproyek application. After conducting research, the results obtained by the author are usability scores obtained from the Maze.design application. The scores obtained are calculated in the medium category and show that the design of the application tested to each respondent already has a good usability value.

1. Introduction

Indonesia population was 261.9 million in 2017 and increased to 265 million in 2018. This fact proves that Indonesia is the fourth most populous country in the world [1]. From 261,9 million of people, Labor force participation rate in 2017 was 66,7%. From those numbers, the open unemployment rate was 5,5% [2] and the highest number is on diploma and university level which is 6.88% and 5.18% each [3]. In the next year, the population number is increased to 265 million and the labor force participation rate also increasing to 67.3%. Nevertheless, the open unemployment rate is decreasing to 5.34% [4] and the unemployment rate in diploma level was 6.02% which is decreasing by 0.86% from the last year. Then, On university level, the open unemployment rate is increased by 0.71% and touching 5.89% [5]. Those numbers shown that there are so many unemployed diploma and university graduates. Fasli Djalal, 2010 Vice of Minister of National Education said "The open unemployment rate in Indonesia based on education completed is quite endangered"[6]. Then, on the article that published by okenews.com, the other reason why the open unemployment rate in university level is still high because the university giving less attention on soft skill development of their students [7]. Another reason why the student quality in Indonesia dropped is because the student had less interest to conducting some research. According to the words of Mochammad Ashari, lecturers who conducting research diligently will have updated insight and knowledge and always new and automatically will affect to the teaching material that taught to their students [8]. This case also happened because lecturer no longer get wages to conducting research and also because the lecturer having some troubles to find the competent student to help their research [9]. Even though, by conducting research, the human resources quality in Indonesia would increase and encourage Indonesia progress level through science and technology aspects and community welfare [10]. Therefore, the conclusion is that one of the factors that reduce the quality of student is the lack of interest in student and lecturer to conducting research.

Based on those facts, survey is conducted to lecturer and student of System Information Program of Industrial Engineering Faculty of Telkom University to figure out the interest of conducting research in Information System's lecturer and student. The survey was conducted on 4 lecturers and 30 students of Information System Program of Telkom University. From 30 students, 63,3% of the students have joined the lecturer project and the rest of the students said never joined and worked on the lecturer project. That 63,3% of students that have joined the lecturer project tells the reason why the students interested to join the lecturer project. The reasons that arise most often is that the students want to have some experience during their college period and train their skills. The students also said that increasing income and knowledge as the reason why the students interested to join and work on the lecturer project. Afterwards, the students tell us about the trouble to join and work on the lecturer project. The results are 20% of the students said that the students find it very difficult to join the lecturer project, then 30% of the students said it's difficult to join the lecturer project and or 46,67% of the students gave neutral answers. 1 respondent also find no difficulties to join the lecturer project. After conducting survey to students, the authors conduct survey to the lecturer. The survey conducted to 4 lecturers of system information program of Telkom University. The first question asked is asking about how the lecturers currently recruiting the student to join their project. The result is that the lecturers asking directly to the student and offering it to the organization or laboratory that the lecturers chose. The second question is asking how difficult for the lecturers to recruit students to join their project. Three of the lecturers felt it's difficult to recruit the students and one of the lecturers is find it very difficult. Then, the authors asked to the lecturer about what is needed to make recruitment process become easier. Most of the lecturer said a good communication and connection is the most important thing to make student join their projects.

Based on the survey results, Maiproyek is developed, the platform that able for connecting lecturer and students so the students and the lecturers able to collaborate together on projects. This platform will help lecturers to recruit students for their projects and help students to join the lecturer project. In addition, Maiproyek giving students some chances to get additional income and giving the lecturers chances to grow their experience during the college period. With this application, the student quality would increase and become a better human resource once the students graduated from the college.

2. Research Method

Research method that used in this research is user centered design [11]. Figure 1 shows how the method will be conducted in this research.

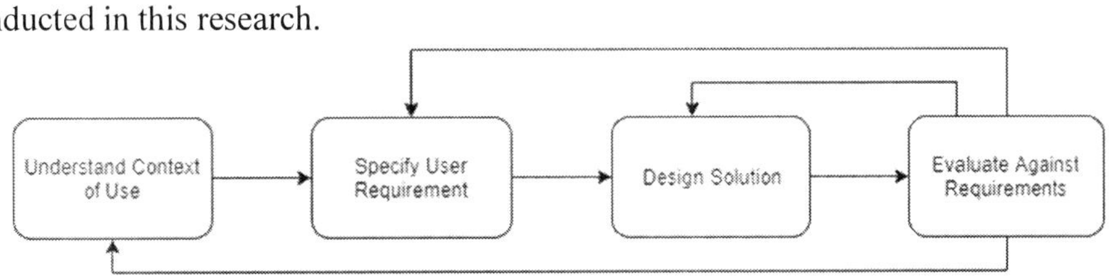

Figure 1. User Centered Design Flow

Based on Figure 1, there are four phases that used in this research. The phases are Understand Context of Use, Specify User Requirements, Design Solution, and Evaluate Against Requirements. Understand Context of Use is the phase where the author identifying the user of the application and conducting research about the user through user requirement survey. The second phase is Specify User Requirements is conducted to figuring out the user requirement based on the result of user requirement survey that already conducted in the understand context of use phase. In this phase, persona and use case diagram will be defined in order to figuring out what user need and expect from thee applications. The third phase is Design Solution, the phase where the author designed the user interface and user experience design of the application. The design solution phase will be carried out into 3 steps, which is wireframe, mock up, and prototyping. The last phase is Evaluate Against Requirement. This phase goal is to assess the design directly to the targeted user. This phase will involve user directly to figuring out if the design is already fulfilling the user need or not. Moreover, the information that gathered in this phase will be used to make to improve the design that have been already made. If the results from this phase shows the target user is not proper for the app, then user identification and user requirement survey will be done all over again then iteration process should start according to the flow. If the results show there are some mistakes that related to the design (misclick, duration, success rate) or the heatmap of the test results shows there is something wrong the design, the process should be back to the third phase where those design mistakes will be fixed and then the second test will be done again. If the results from the last phase is good, then the design process is complete and conclusion can be drawn.

3. Results and Discussion

3.1. Understand Context of Use

In this phase, the users of the application will be identified. There are 2 type of users that will be the user of Maiproyek application. The first one is lecturer, User that providing project for the student by started and advertise their project to the Maiproyek application. Lecturer able to create and manage the ongoing project. The second one is student, User that will help lecturer to work their projects. The student able to find the project and registering themselves and joined to the project that the students chose. After identifying the users, the next step to take is to conducting user requirement survey. This survey will give the author some insight about the users' needs for the application. For the lecturer, the method that used to carried out the survey is conducted by interview and for the student, it's conducted by online form using Google Form to helps the online survey.

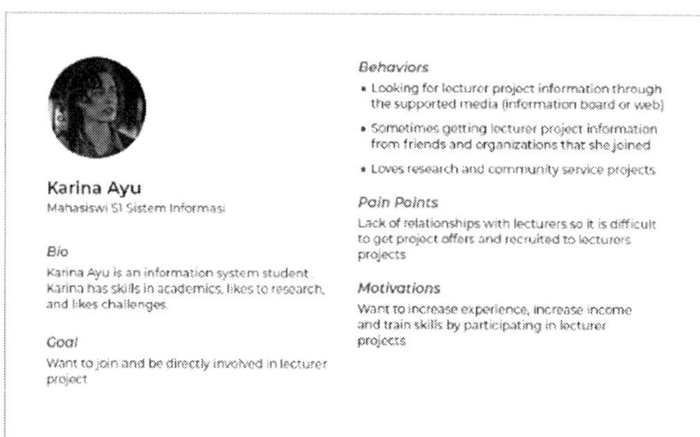

Figure 2. Student Persona

3.2. Specify User Requirement

In this phase, the user requirement survey results will be analyzed and then persona and use case diagram will be defined based on the user requirement survey results. Those are the main reference for the design.

There are 2 personas for each type of user in the Maiproyek application. Persona is defined to finding out the user purpose and user accuracy that obtained from real people observations [12]. One persona for each user category is chosen in order to solve the problem that the user encountered the most.

Furthermore, to make a clear insight about what this research all about and what the biggest problem that the users faced. Figure 2 and 3 are the persona that has been created based on the user requirement survey results.

Figure 2 is the student persona that designed based on the user requirements survey results. Information that will be used in the persona development is how difficult it is to join a lecturer project, how often the students join the lecturer project, the motivation that encourage students to join the lecturer project, what is the problem that the students experience when the students want to join the lecturer project, and what feature that the students need in order to achieve their goal by using the application. After gathering the information, the information of the persona will be defined and persona is designed. The main goal of the persona is to join and directly involved in lecturer project so the subject able to increase experience, income and train the subject skills. Afterwards, the pain points that subject face is the lack of relationships with lecturers so it is difficult for the subject to get projects offers or invitation and recruited to lecturer projects. Therefore, the subject finds the information about the lecturer project through the supported media and getting some information about lecturer project by the subject's friend sometimes. Then, the authors defined another persona for the lecturers. The persona is shown in Figure 3.

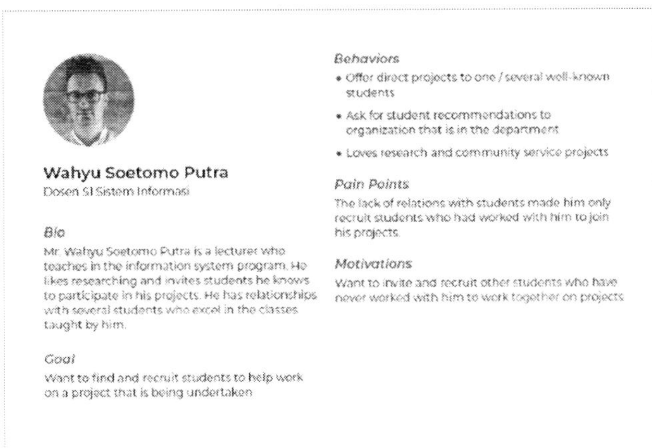

Figure 3. Lecturer Persona

Figure 3 is the lecturer persona. The information used in the persona development are what kind of project that the lecturers like, how the lecturers able to find the student the lecturer needed for the project currently, what is the problem to find the right student for their project, what makes the student search process easier, and what feature the lecturers need to achieve their goal by using this application. After the information has been gathered, the motivation, goal, pain point and user behaviors will be defined and will be the based information for the persona. The main goal of the subject is to find and recruit students to help the subject on a project that is being worked at especially the students who never works with the subject before. The pain point is the lack of relations with students made the subject only offers and recruits students who had worked with the subject to join the projects that the subject had.

After designing the persona, the authors design use case diagram based on the user requirement survey results that was carried out in the previous phase, the use case diagram is defined. Use case diagram is defined to gather the functionalities of the application based on the user observation [13]. Figure 4 is the results of the use case diagram design. Figure 4 show each function that can be accessed to each type of user. Both lecturer and student are able to access authentication and profile function. The function which can only be used by lecturer is project promotion, Registrant Selection, Project Management for Lecturer, Make Projects Payments and Member Review. The function which can only be accessed by student is Project Registration and Project Management for Student

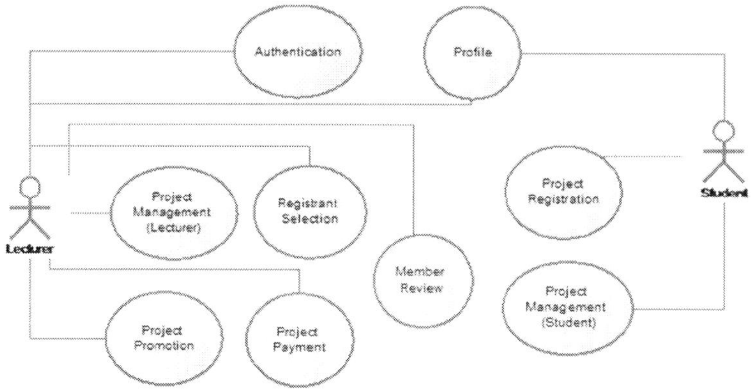

Figure 4. Use Case Diagram of Maiproyek Application

3.3. Design Solution

In this phase, the authors design the user interface and user experience design based on the information that has been obtained in the previous phases. The design is carried out in three stages, wireframe, mockup, and prototype.

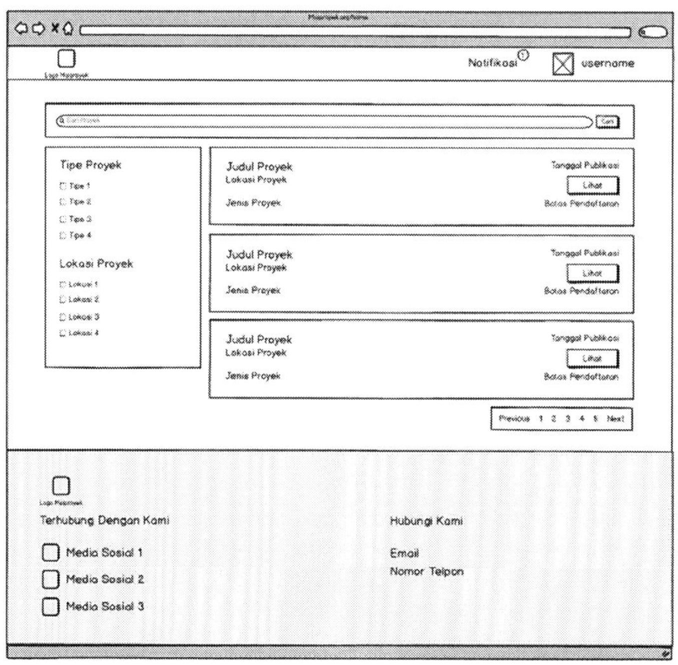

Figure 5. Wireframe of Maiproyek Application

First, the authors design the wireframe in order to design structure of the mock up in the next stage. Wireframe designed to get a brief overview of the design results. Design wireframe designed using the Balsamiq Mockup application. Figure 5 is one of the wireframes already designed

The mockup is designed based on the wireframe that was designed in the previous stage. Before designing a mockup, the author first makes design rules that will be used in designing the mockup. These rules include the colors and fonts used in the application later. After that the authors designed a mockup based on the wireframe that had been used previously. Figure 6 is one of the results of the mockup design that has been designed by the author.

After designing mock-up, the authors designed a prototype using mock-ups that already designed in the previous phase. Prototype is designed using Marvel Application. The prototype will be used in the next phase for the usability testing.

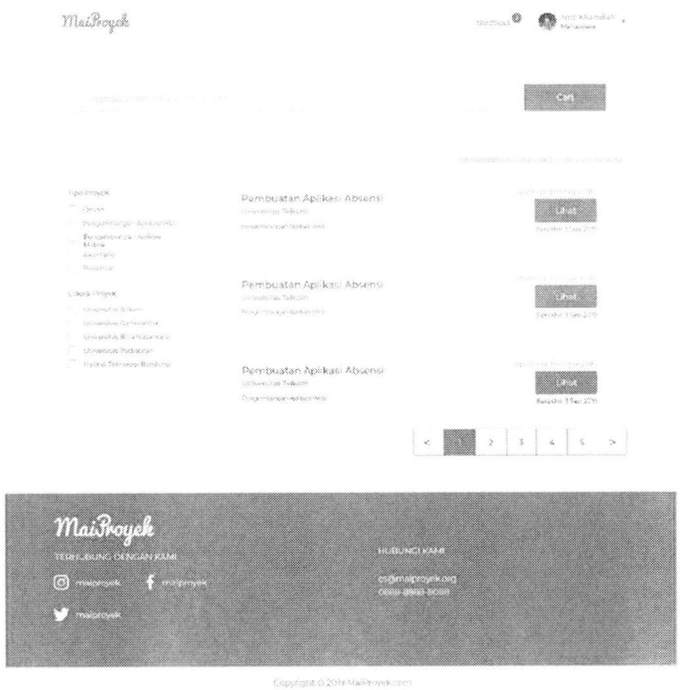

Figure 6. Mock-up of Maiproyek application

3.4. Evaluate Against Requirements.

At this phase, the authors conduct a design assessment that has been designed by performing usability testing to targeted users. Usability testing is conducted using a previously designed prototype. Then the prototype will be linked with the maze.design application to be used in usability testing. In the maze.design application, the author has designed 20 scenarios with details of 10 scenarios for lecturers and 10 scenarios for students. Usability testing will be carried out in two sessions, student session and lecturer session.

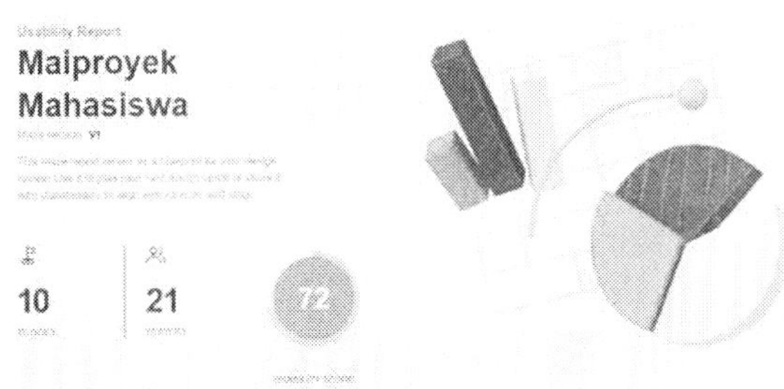

Figure 7. Student Test Results

The final score will be obtained after the test has been completed. The score is calculated using the method that applied by maze.design [14]. The scenarios tested with the lecturer are log in, creating a project, seeing the details of an ongoing project, seeing current project members, viewing details and responding to project registrations made by students, seeing project progress, adding assignments, view task details, project payments, and view payment history. Then, for students, the scenario tested is a list, logging in, viewing personal profiles, editing profiles, joining lecturer projects, viewing ongoing projects, viewing ongoing project members, viewing lecturer profiles, viewing ongoing project progress, and upload the work results from the specified task.

The authors using the accidental sampling to conduct the testing session. This sampling is used because it's difficult to find respondents who are willing to be tester, so the authors took several samples from each type of respondent using accidental sampling method. For sampling, the authors made two sampling periods. Each period lasts 2 weeks. In the first period, there were 11 respondents who were successfully obtained, each consisting of 5 lecturers and 7 students. Then in the second period, the respondents who were successfully obtained were 18 respondents, each consisting of 4 lecturers and 14 students. The total respondents who tested the application design were 30 respondents.

The final score obtained from the testing session for students is 72. Based on the score categories set by maze.design. The score goes into Results of usability testing to students is placed the medium category. This score shows that the application design on the student side has a good usability value.

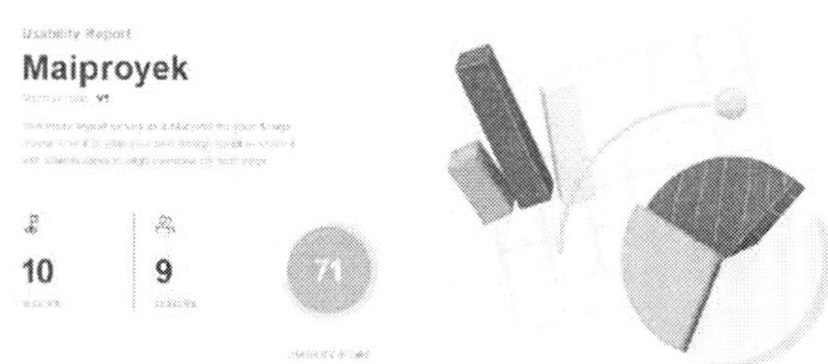

Figure 8. Lecturer Test Results

The final score obtained from the testing session for the lecturer was 71. Based on the score categories set by maze.design. The score categorized into the medium category. This score shows that the application design on the student side has a good usability value.

4. Conclusion

The design is carried out using the User Centered Design method which is chosen because this method focuses more on the users targeted by the application. The design process is carried out through 4 phases of user centered design method, which is Understand Context of Use, where the author identifies application users by idea validation and user requirement survey, Specify User Requirements, where the author analyzes the results of the user needs survey by translating the survey results into a persona and use case diagrams, Design Solution, where the authors design the user interface and user experience by designing wireframes, mock ups, and prototypes based on the analysis results obtained in the previous stage and the last phases which is Evaluate against requirements, where the author tests the design to the respondents who are also the target users of the application. Tests were carried out using the maze.design application.

The method used in this research is effective enough to design a user interface design and user experience for the Maiproyek application. This can be seen from the results of testing the application design which obtained good usability points in each testing session for each respondent. For the lecturer side, the usability score obtained was 71 with 9 respondents which is a medium score and on the student side, the usability score obtained was 72 with 21 respondents which is a medium score.

References

[1] The Atlas 2017 *Top 10 Countries in the World by Population 2017* Retrieved from

https://www.theatlas.com/charts/ry0XD4WHW
[2] Badan Pusat Statistik 2018 *Statistika Indonesia 2018* Retrieved from https://www.bps.go.id/publication/2018/07/03/5a963c1ea9b0fed6497d0845/statistik-indonesia-2018.html
[3] Badan Pusat Statistik 2017 *Agustus 2017: Tingkat Pengangguran Terbuka (TPT) Sebesar 5,50 Persen* Retrieved from https://www.bps.go.id/pressrelease/2019/05/06/1564/februari-2019--tingkat-pengangguran-terbuka--tpt--sebesar-5-01-persen.html [Accessed: 06-Mar-2020]
[4] Badan Pusat Statistik 2019 *Statistika Indonesia 2019* Retrieved from https://www.bps.go.id/publication/2019/07/04/daac1ba18cae1e90706ee58a/statistik-indonesia-2019.html [Accessed: 04-Mar-2020]
[5] Badan Pusat Statistik 2018 *Agustus 2018: Tingkat Pengangguran Terbuka (TPT) Sebesar 5,34 Persen* Retrieved from https://www.bps.go.id/pressrelease/2018/11/05/1485/agustus-2018--tingkat-pengangguran-terbuka--tpt--sebesar-5-34-persen.html [Accessed: 05-Mar-2020]
[6] Gewati M 2016 *Kenapa Lulusan Perguruan Tinggi Makin Susah Mendapat Pekerjaan?* Retrieved from https://edukasi.kompas.com/read/2016/04/23/17424071/Kenapa.Lulusan.Perguruan.Tinggi.Makin.Susah.Mendapat.Pekerjaan?page=2
[7] Okenews 2018 *Kuliah Tak Membantu Kesiapan Kerja Mahasiswa Baru Lulus?* Retrieved from https://news.okezone.com/read/2018/11/04/65/1973150/kuliah-tak-membantu-kesiapan-kerja-mahasiswa-baru-lulus?page=1
[8] Yunelia I 2019 *Dosen Rajin Meneliti Berdampak pada Kualitas Mahasiswa* Retrieved from https://www.medcom.id/pendidikan/news-pendidikan/Rb15pvxb-dosen-rajin-meneliti-berdampak-pada-kualitas-mahasiswa#:~:text=Menggairahkan semangat meneliti dosen sangat,ilmu pengetahuan yang selalu baru
[9] Billy 2016 *Untuk Kamu yang Berprofesi Sebagai Dosen Mungkin Pernah Merasakan Dan Mengalami 7 Tantangan Ini* Retrieved from https://www.duniadosen.com/untuk-kamu-yang-berprofesi-sebagai-dosen-mungkin-pernah-merasakan-dan-mengalami-7-tantangan-ini/.
[10] Devananta A A 2013 *Polemik Pudarnya Gairah Meneliti Mahasiswa* Retrieved from https://news.okezone.com/read/2013/03/18/95/777339/polemik-pudarnya-gairah-meneliti-mahasiswa
[11] Interaction Design Foundation 2018 *User Centered Design* Retrieved from https://www.interaction-design.org/literature/topics/user-centered-design [Accessed: 20-Sep-2010]
[12] Cooper A Reimann R and Cronin D 2007 *About Face 3. The essentials of Interaction Design* (Indianapolis: Wiley Publishing, Inc)
[13] Waykar Y 2015 Role of use case diagram in software development *Int. J. Manag. Econ.* p 3
[14] Maze.design 2020 *How Your Usability Score is Calculated* Retrieved from https://help.maze.design/hc/en-us/articles/360052723353-How-your-Usability-Score-is-calculated

Indonesian Ethnicity Recognition Based on Face Image Using Gray Level Co-occurrence Matrix and Color Histogram

Dinda Mareta Putriany[a], Ema Rachmawati[b], Febryanti Sthevanie[c]

School of Computing, Telkom University, Bandung, Indonesia

E-mail: [a]dindaputriany@student.telkomuniversity.ac.id, [b]emarachmawati@telkomuniversity.ac.id, [c]sthevanie@telkomuniversity.ac.id

Abstract. Facial and ethnic recognition became a popular research topic. Different from face recognition, ethnic recognition classifies faces according to the general features of certain ethnic groups. Ethnic recognition in face image is increasingly becoming a necessity and is used in various fields. This paper proposed the Indonesian ethnic recognition system based on periorbital features on facial images. We use the five largest ethnic groups in Indonesia in this study, namely Banjar, Bugis, Javanese, Malay, and Sundanese. Gray Level Co-occurrence Matrix and Color Histogram were used as methods, and Random Forest was used as a classifier. Based on - cross-validation tests with optimal k values, the model achieved 98.65% accuracy.

1. Introduction

The human face is a biometric key that provides demographic information on gender, age, and ethnicity useful in facial recognition systems [1]. Facial recognition is a popular research topic because it has a wide range of applications [2]. For two decades, this research has been studied to understand how humans perceive and determine a given image's ethnicity because ethnicity plays an important role in applying facial images [1][3]. Based on Potensi Desa (PODES) in 2014, Indonesia has more than 250 ethnicities. As many as 71.78% of Indonesia regions have a heterogeneous composition of citizens and have historically shown that their development is spread heterogeneously with high diversity [4], ethnic recognition is increasingly difficult for humans to do. Ethnic recognition can be used in population data collection activities to assist an agency in making decisions. It can also be used to identify crime suspects to narrow the scope of searches in the database [5] and improve the security of biometric systems in protecting personal and business information by limiting and monitoring access to data [6]. In this study, we propose an Indonesian ethnic identification system based on the periorbital face, namely the periocular region (the area around the eyes including the eyebrows) to the nose as a Region of Interest (ROI), because it is the most dominant and important area on the face [6][7] and is an effective biometric approach. We used ethnic Banjar, Bugis, Javanese, Malay, and Sundanese in this study because these are the five largest ethnic groups in Indonesia. We used a combination of textural and color approaches because they are very important for effective facial information [8]. Specifically, we used the GLCM method to capture texture, Color Histogram to capture colors, and Random Forest as a classifier.

2. Proposed System

This section describes our proposed system. An overview of our system is shown in Figure 1. The proposed Indonesian ethnic recognition system consists of three steps, namely preprocessing, feature

IOP Conf. Series: Materials Science and Engineering 1077 (2021) 012040 doi:10.1088/1757-899X/1077/1/012040

extraction, and classification. First, the RGB image is entered into the system and then preprocessing the image. Second, each preprocessed image is performed feature extraction using GLCM and Color Histogram. Third, the results of feature extraction are classified using Random Forest as the classifier. The classification results are used for system analysis.

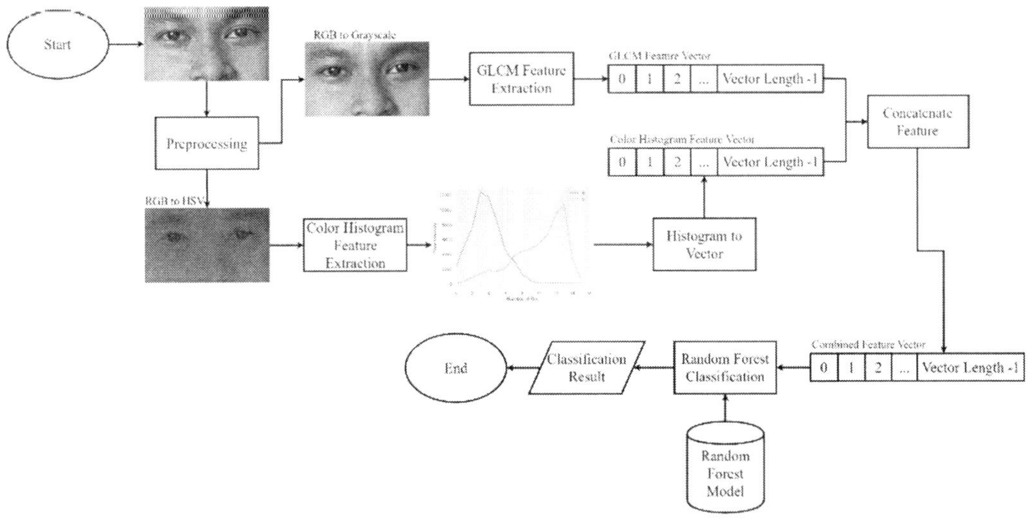

Figure 1. System overview.

2.1. Dataset

The dataset used in this study belongs to the Multimedia Laboratory, Faculty of Informatics, Telkom University [9]. The dataset contains images of the five largest ethnic groups in Indonesia, consisting of Banjar, Bugis, Javanese, Malay, and Sundanese. The image used for the classification is approximately 400 x 200 pixels. The number of each class is 485 for Banjar, 465 for Bugis, 440 for Javanese, 455 for Malay, and 445 for Sundanese, with a total of 2290 periorbital images.

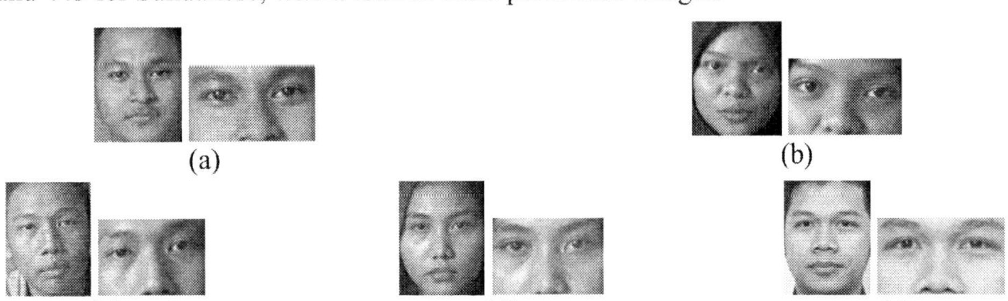

Figure 2. Image samples from ethnic (a) Banjar (b) Bugis (c) Javanese (d) Malay (e) Sundanese.

2.2. Preprocessing

The purpose of preprocessing is to make the data more ideal before the feature extraction process. At this step, the image with the RGB color space is then converted into a grayscale image for GLCM feature extraction and converted into an HSV image for Color Histogram feature extraction. The preprocessing scheme is shown in Figure 3.

2.3. Gray Level Co-occurrence Matrix (GLCM) Feature Extraction.

The GLCM is calculated as a second-order histogram of a gray image. GLCM is a matrix whose dimensions depend on the intensity of the gray level G in an image. The feature extraction scheme using GLCM is shown in Figure 4.

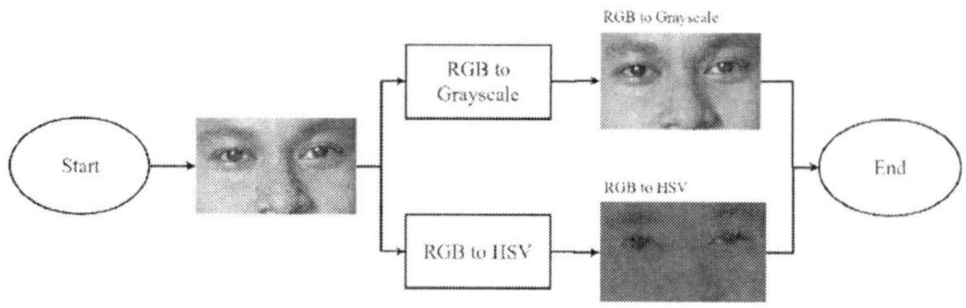

Figure 3. Preprocessing scheme.

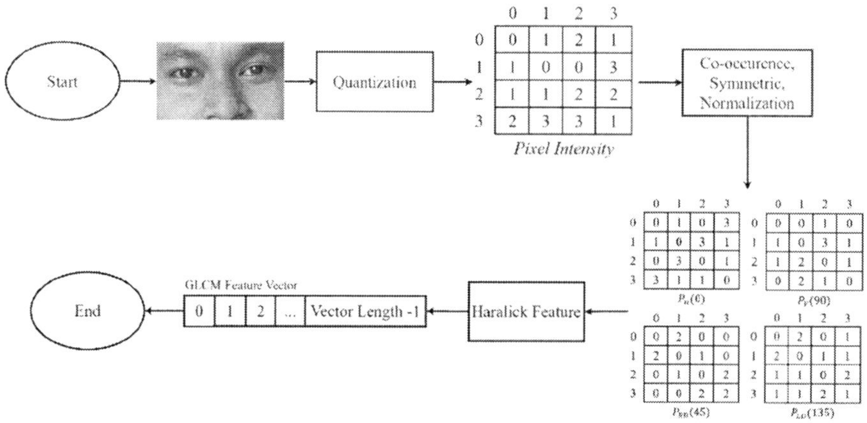

Figure 4. GLCM feature extraction scheme [10][11].

The first step is a quantized grayscale image. The goal is to convert the grayscale value of an image and classify it into $\boldsymbol{G}$ levels (degrees of gray) expressed in integers. In the Co-occurrence stage, the number of occurrences of pixel intensity value level with another pixel intensity is counted within one distance (distance = 1) and the orientation of 0°, 45°, 90°, and 135°. In this step, it will produce 4 (four) co-occurrence matrices. In the Symmetric step, the co-occurrence matrix is added to the transposition matrix, aiming to make the matrix symmetrical on the diagonal. The normalization step divides each set of pixel pairs by the total number of pixel pairs used. After that, a GLCM feature extraction process uses 7 (seven) components measured from the Haralick Feature:

- Contrast
  Used to measure how far the intensity of a pixel differs from its neighboring pixels throughout the image. When the contrast is 0, then the image does not change the pixel intensity (constant). The contrast value increases exponentially. Contrast can be calculated by equation (1) [8]:

$$Contrast = \sum_{i,j=0}^{G-1} (i-j)^2 \, p(i,j) \tag{1}$$

- Energy
  Energy is a value that shows how regular the intensity of the pixels in the image. The more organized an image, the energy value is getting higher, and vice versa. Energy can be calculated by equation (2) [11].

$$Energy = \sqrt{\sum_{i,j=0}^{G-1} p(i,j)^2} \tag{2}$$

- Homogeneity
  Homogeneity is a value that shows how close the intensity of a pixel with neighboring pixels. The higher the homogeneity value, then the image's differences pixel intensity is getting closer with value ranges 0 to 1 [11].

$$Homogeneity = \sum_{i,j=0}^{G-1} \frac{p(i,j)}{1+(i-j)^2} \tag{3}$$

- Correlation
  Correlation is a value that measures how big the relationship between two pixels. Correlation is the value of the gray linear dependence between pixels in two different directions [12].

$$Correlation = \sum_{i,j=0}^{G-1} p(i,j) \left[\frac{(i-\mu_i)(j-\mu_j)}{\sigma_i{}^2 \sigma_j{}^2}\right]$$

With $\mu_i = \sum_{i,j=0}^{G-1} i \times p(i,j)$, $\mu_j = \sum_{i,j=0}^{G-1} j \times p(i,j)$, (4)

$\sigma_i{}^2 = \sum_{i,j=0}^{G-1} p(i,j)(i-\mu_i)^2$, dan $\sigma_j{}^2 = \sum_{i,j=0}^{G-1} p(i,j)(j-\mu_j)^2$

- Dissimilarity
  Dissimilarity is a measure of the dissimilarity of an image. The value will be high if the image has random pixels, and vice versa will be low if the image pixels are uniform. The value increases linearly. Dissimilarity can be calculated by equation (5) [11].

$$Dissimilarity = \sum_{i,j=0}^{G-1} |i-j| p(i,j) \tag{5}$$

- Angular Second Moment (ASM)
  ASM is a measure of the homogeneity of an image. ASM has a high value when the image has great homogeneity and vice versa. It can be calculated by equation (6) [11].

$$ASM = \sum_{i,j=0}^{G-1} p(i,j)^2 \tag{6}$$

- Entropy
  Entropy is used to measure the complexity or abnormality of an image. Entropy can be calculated by equation (7) [11].

$$Entrophy = \sum_{i,j=0}^{G} p(i,j)\ log_2\, p(i,j) \tag{7}$$

In the equation, $\boldsymbol{i}$ and $\boldsymbol{j}$ is the appearance of a pixel pair, $\boldsymbol{p(i,j)}$ is the GLCM normalized value, and $\boldsymbol{G}$ is the gray level.

2.4. Color Histogram Feature Extraction

This extraction uses the HSV color space, commonly used and effective in extracting color features for facial recognition [8][13]. HSV also has a better performance in distinguishing colors when compared to RGB. There are 3 (three) steps: color quantization, normalization, and color histogram construction. Color quantization aims to reduce the possible number of colors to reduce the computation process, and the required process will be easier. Quantization divides each color pixel by a specific range of values.

Pixels are grouped into n groups of color pixel value ranges, called bins. Each bin in the color histogram will contain a statistical value in the form of a percentage of the number of pixels. Then, each group was normalized to simplify the color distribution values on the histogram. Besides, normalization also aims to make the histogram value remain the same in images with the same color distribution with different image sizes.

2.5. Classification

Random Forest (RF) algorithm was used as a classifier in this study. The RF classification scheme is shown in Figure 5. This classification process aims to determine an image's class into 5 (five) ethnic classes, namely Banjar, Bugis, Javanese, Malay, and Sundanese. There are 2 (two) processes, namely training and testing. RF conducts a training process on several random and different data samples (bootstrap). As much as one-third of the sample data from the training set is used as Out-of-Bag (OBB) data, namely, data that is not selected as a bootstrap to calculate errors and determine important variables. The rest is used as In-Bag data to form trees. Each tree can provide predictions with the same or different results. The final prediction is selected based on majority voting.

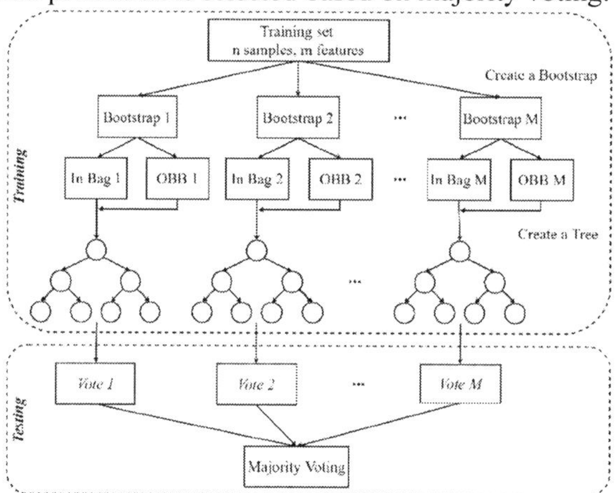

Figure 5. Random forest classification scheme [18].

3. Experimental Result

This evaluation step uses stratified k-fold cross-validation, a k-fold variation that divides data into k partitions by ensuring that each partition has the same class proportions [14]. There are k repetitions, and the data divided into k partitions by making one part of the training data while the other becomes test data. Each fold has at least been validated once. The final result is the accuracy of the model performance obtained from the average of each iteration. In this study, there are 4 (four) parameter tuning scenarios: 1) The Gray Level Co-occurrence Matrix (GLCM) parameter tuning; 2) Color Histogram parameter tuning; 3) Combining the GLCM and Color Histogram methods; 4) Comparison of classification and cross-validation methods. In the scenario of tuning parameters 1, 2, and 3 using the Random Forest classification method with 100 trees and $k = 10$ for cross-validation.

3.1. Parameter Tuning Results

3.1.1. Gray Level Co-occurrence Matrix (GLCM) Parameter Tuning Results. There are 7 (seven) global features of the Haralick Feature, namely contrast, energy, homogeneity, correlation, dissimilarity, ASM, and entropy, used in this parameter tuning. These seven features are features commonly used by researchers in facial recognition research. The features that produce good accuracy will be selected in this Indonesian ethnic recognition system. The scenario of using the Haralick Feature and the results is given in Table 1.

Table 1. Results of GLCM parameter tunning.

Scenario	Haralick Feature	Accuracy (%)
Scenario-1	Contrast	57.90
Scenario-2	Energy	46.90
Scenario-3	Homogeneity	47.03
Scenario-4	Correlation	56.64
Scenario-5	Dissimilarity	54.72
Scenario-6	ASM	46.68
Scenario-7	Entropy	56.46
Scenario-8	Contrast, energy, homogeneity, correlation [15]	78.16
Scenario-9	Contrast, energy, correlation, entropy [16][17]	82.05
Scenario-10	Energy, homogeneity, correlation, entropy [13]	79.43
Scenario-11	Contrast, energy, homogeneity, correlation, entropy [12]	82.71
Scenario-12	Contrast, homogeneity, correlation, ASM, entropy	82.80
Scenario-13	Contrast, energy, homogeneity, correlation, dissimilarity, entropy	82.53
Scenario-14	Contrast, homogeneity, correlation, dissimilarity, ASM, entropy	82.53
Scenario-15	Contrast, energy, homogeneity, correlation, dissimilarity, ASM, entropy	81.57

This scenario aims to find the Haralick feature that gets the best accuracy results. Scenario-1 to Scenario-7 is a scenario to get the accuracy of each feature. Scenario 8 to scenario 11 is based on previous research. Scenario-12 is a combination of 4 (four) features that has the highest accuracy. Scenario-13 and Scenario-14 are based on the highest accuracy of a combination of 6 (six) features. Scenario-15 is a combination of 7 (seven) features. The Haralick feature tuning parameter results show that the highest accuracy is 82.80%, obtained by Scenario-12 with a combination of contrast, energy, homogeneity, correlation, ASM, and entropy features. It shows that 5 (five) features can produce good accuracy than 7 (seven) features.

3.1.2. Color Histogram Parameter Tuning Results. There are 2 (two) Color Histogram parameters used, namely the channel and the number of bins. The color space used is HSV with 4, 8, and 16 bins. The results of the tuning color histogram parameters are given in Table 2.

Table 2. Results of Color Histogram parameter tunning.

Channel	Number of Bins	Accuracy (%)
H	4	64.72
	8	72.93
	16	80.17
S	4	80.66
	8	91.00
	16	93.41
V	4	85.98
	8	93.62
	16	95.41
H, S	16;16	95.59
H, V	16;16	97.16
S, V	16;16	98.12
H, S, V	16;16;16	98.08

The results show that each parameter of the H, S, and V channels with 16 bins gives better accuracy than using 4 and 8 bins. The tuning parameters' results with the best accuracy, with 98.12%, on 16 bins and a combination of S-V channels. Channel S is the percentage of white light added to the base color.

Therefore, light base colors have high saturation intensity, pastel colors have low saturation intensity, and monochrome colors (black and white) have no saturation. Channel V refers to the light intensity in the image. It can be said that the S-V channel has a significant impact because the human face image has a slight color variation so that information can be taken based on the level of color saturation (light-dark) and retrieve information based on the level of light intensity in the image. It was also stated in [18] that S and V channels have greater power to differentiate ethnicity.

3.1.3. Results of Combining GLCM Method and Color Histogram. In this study, the GLCM and Color Histogram methods are combined. This combination determines how much the performance increases if the two methods are used in the Indonesian ethnic recognition system. Besides, it also compares the use of the holistic features and the periorbital features of the two methods. Holistic facial features are features that are contained in the face as a whole. Periorbital features are features of the periocular region (the area around the eyes, including the eyebrows) to the nose. The best parameters from GLCM with Scenario-12 and Color Histogram with 16 bin and S-V channels were used. The result of combining feature extraction methods are given in Table 3.

Table 3. Result of combining feature extraction methods.

Features	Accuracy (%)	Vector Length
Holistic	96.12	52
Periorbital	98.73	52

Combining the GLCM method and the Color Histogram with a vector length of 20 + 32 = 52 obtained 96.12% accuracy for holistic facial features and 98.73% for periorbital features. The best performance is in the periorbital features. Thus, periorbital features are better than the holistic features of the face in ethnic recognition [6][7][19]. Table 5 shows that combining texture and color features can improve the ethnic recognition system's performance rather than only its texture or color features.

Table 4. Performance of the feature extraction method on periorbital features.

Methods	Accuracy (%)
GLCM	82.80
Color Histogram	98.12
Combined	98.73

3.1.4. Results of Comparison of Classification Methods and Parameter Tuning Cross-Validation. In this case, we compare the classification method and cross-validation parameters. The proposed classifier is compared with the Support Vector Machine (SVM) [20]. Using 200 trees (n-estimators = 200) as Random Forest Parameters. Uses the values 10^{-1} for regularization, 1 for gamma, and the Polynomial kernel as SVM parameters. These parameters are obtained from hyperparameter tuning using Grid Search. Also, using $k = 10$ for the cross-validation parameter. The results of the comparison of classification methods are given in Table 5.

Table 5. Results of comparison of classification methods.

Classifiers	Accuracy (%)
Random Forest	98.60
SVM	92.27

The results show that the Random Forest classification method's accuracy gets the highest accuracy with 98.60%. It shows that Random Forest has better than SVM [20]. Furthermore, looking for the optimal k value in cross validation using the optimization method, namely the elbow method. This method uses an elbow as a cut-off point to select a point where the curve appears to be bent, particularly from high to low slopes, flat or close to flat, or in another direction. The elbow point is the optimal point of several decisions whose value is no longer increasing rapidly and is no longer proportional to further (stable) increase [21]. Figure 6 shows that the k value in cross-validation is in the range of 2 to 20 with each accuracy result. In the search for values using Random Forest and SVM, both elbows are at a point 6 (six) with an accuracy of 98.65% and 91.75%, respectively.

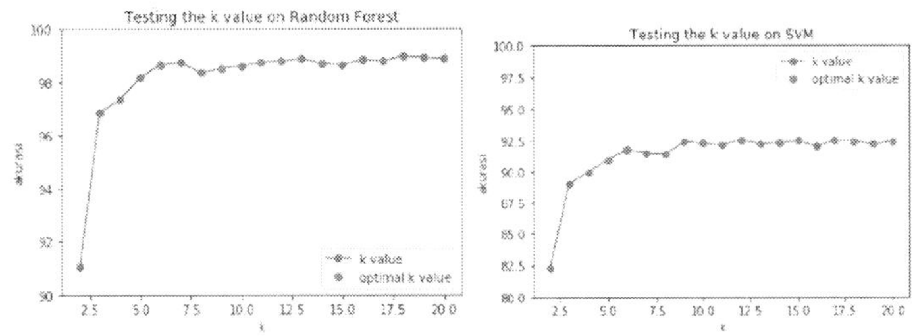

Figure 6. Graph of optimal k values in cross-validation.

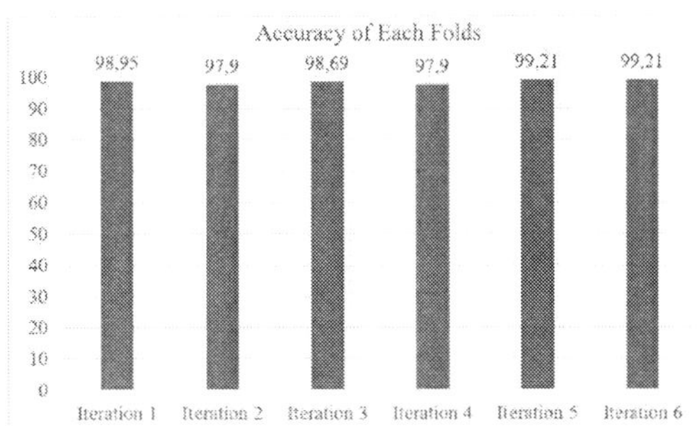

Figure 7. Accuracy of each fold.

3.2. Analysis of Test Results.

We conduct the testing using cross-validation on 2290 periorbital image by incorporating the optimal color and texture feature parameter. We use GLCM parameters, namely contrast, homogeneity, correlation, ASM, and entropy; S -V channel and 16 bins for Color Histogram; 200 trees for Random Forest. Based on optimal k at point 6, the accuracy is 98.65%. This value is obtained from the calculation of the average accuracy of each iteration of the fold. The accuracy details for each fold are shown in Figure 7. The evaluation results for the overall performance are obtained from the average score for each iteration. These results were measured using accuracy, precision, recall, and f1-score [22], shown in Table 6.

Table 6. Results of the overall system performance evaluation.

	Accuracy	Precision	Recall	F1-Score
Average (%)	98.65%	98.66%	98.65%	98.65%

Table 7. Euclidean distance feature comparison between misclassified data with actual and predictive classes.

Class	Distance Average of		
	GLCM Feature	Histogram Feature	Combined Feature
Actual	90.35	37.636,24	37.636,52
Predicted	65.88	34.361,40	34.361,57

Misclassification in the recognition system is common. A misclassified image has features similar to the image features in the prediction class. To determine the feature similarity, we calculate the distance of the actual class and the predicted class using Euclidean Distance. For example, there is an image of the Javanese class, classified into the Sundanese class. Table 7 compares the distance of feature, the actual class feature, and the predicted class feature using the Euclidean Distance.

Table 7 shows that the Java class image classified as a Sundanese class has an average distance of the GLCM and Color Histogram features in the prediction class, which is smaller than the average distance of the GLCM and Color Histogram features in the actual class. Thus, the Javanese class is classified as a Sundanese class because the feature distance is closer to the Sundanese class. Also, the combined feature's mean distance is similar to the average distance of the Color Histogram feature. Thus, the Color Histogram feature is more dominant and affects the classification process.

4. Conclusion

Based on the results of testing and analysis, this study obtained an optimal accuracy of 98.65% with a precision of 98.66%, recall, and f1-score of 98.65%, respectively, in $k = 6$. The best parameters on GLCM are 5 (five) Haralick Features, namely contrast, homogeneity, correlation, ASM, and entropy. The best parameter on the Color Histogram is the S-V channel with 16 bins. The best parameter of Random Forest is 200 trees. The suggestion for further research is to expand the class of Indonesian ethnicity to several other classes.

References

[1] Momin H and Tapamo J R 2016 A comparative study of a face components based model of ethnic classification using gabor filters *Appl. Math. Inf. Sci.* **10** 2255–65

[2] Li Z, Zhang Q, Duan X and Wang Y 2018 A novel semantic approach for multi-ethnic face recognition *Int. J. Pattern Recognit. Artif. Intell.* **32**

[3] Huri K, Omid David E and Netanyahu N S 2018 DeepEthnic: multi-label ethnic classification from face images *Lecture Notes in Computer Sci. (including subseries Lecture Notes in Artificial Intelligence and Lecture Notes in Bioinformatics)* vol **11141** LNCS (Springer International Publishing) pp 604–12

[4] Dokhi M S, Theodora H S, Wulansari I Y H and Dwi W S N 2016 Analisis kearifan lokal ditinjau dari keragaman budaya *PDSPK Kemendikbud RI* 1–67

[5] Prasad P and Gupta A 2018 Moving object tracking and detection based on kalman filter and saliency mapping *Advances in Intelligent Systems and Computing* vol **542** pp 639–46

[6] Mohammad A S and Al-Ani J A 2017 Towards Ethnicity Detection Using Learning Based Classifiers *2017 9th Comput. Sci. Electron. Eng. Conf. CEEC 2017 - Proc.* pp 219–24

[7] Rasyid H F, Ramadhani K N and Sthevanie F 2018 Mongoloid and non-mongoloid race classification from face image using local binary pattern feature extractions *2018 6th Int. Conf. on Information and Communication Technol., ICoICT 2018* (IEEE) pp 329–32

[8] Rehman A, Khan G, Siddiqi A, Khan A and Khan U G 2018 Modified texture features from histogram and gray level co-occurence matrix of facial data for ethnicity detection *5th Int. Multi-Topic ICT Conf.: Technologies For Future Generations, IMTIC 2018 - Proc.* (IEEE) pp 1–6

[9] Suyanto S, Arifianto A, Sirwan A and Rizaendra A P 2020 End-to-End speech recognition models for a low-resourced Indonesian Language *2020 8th Int. Conf. on Information and Communication Technol., ICoICT 2020*

[10] Al-Kilidar S H S and George L E 2017 Texture recognition using co-occurrence matrix features and neural network *J. Theor. Appl. Inf. Technol.* **95** 5949–61

[11] Abbas Z, Rehman M U, Najam S and Danish Rizvi S M 2019 An efficient gray-level co-occurrence matrix (GLCM) based approach towards classification of skin lesion *Proc. - 2019 Amity Int. Conf. on Artificial Intelligence, AICAI 2019* (IEEE) pp 317–20

[12] Bianconi F, Smeraldi F, Abdollahyan M and Xiao P 2017 On the use of skin texture features for gender recognition: An experimental evaluation *2016 6th Int. Conf. on Image Processing Theory, Tools and Applications, IPTA 2016*

[13] Nazir A, Ashraf R, Hamdani T and Ali N 2018 Content based image retrieval system by using HSV color histogram, discrete wavelet transform and edge histogram descriptor *2018 Int. Conf. on Computing, Mathematics and Engineering Technologies* pp 1–6
[14] Yadav S and Shukla S 2016 Analysis of k-fold cross-validation over hold-out validation on colossal datasets for quality classification *Proc. - 6th Int. Adv. Comput. Conf. IACC 2016* pp 78–83
[15] Sukiman T S A, Suwilo S and Zarlis M 2019 Feature extraction method GLCM and LVQ in digital image-based face recognition *Sink. J. dan Penelit. Tek. Inform.* **4** pp 1–4
[16] Yahia S and Salem Y Ben 2016 3D face recognition using local binary pattern and grey level co-occurrence matrix *2016 17th Int. Conf. Sci. Tech. Autom. Control Comput. Eng.* **1** pp 328–38
[17] Dixit U D and Shirdhonkar M S 2018 Face-based document image retrieval system *Procedia Comput. Sci.* **132** pp 659–68
[18] Du H, Salah S H and Ahmed H O 2014 A color and texture based multi-level fusion scheme for ethnicity identification *Mobile Multimedia/Image Processing, Security, and Applications 2014* vol **9120** p 91200B
[19] Wang C, Zhang Q, Liu W, Liu Y and Miao L 2019 Facial feature discovery for ethnicity recognition *Wiley Interdiscip. Rev. Data Min. Knowl. Discov.* **9**
[20] Kremic E and Subasi A 2016 Performance of random forest and SVM in face recognition *Int. Arab J. Inf. Technol.* **13** pp 287–93
[21] Syakur M A, Khotimah B K, Rochman E M S and Satoto B D 2018 Integration k-means clustering method and elbow method for identification of the best customer profile cluster *IOP Conf. Ser. Mater. Sci. Eng.* **336**
[22] Sokolova M, Japkowicz N and Szpakowicz S 2006 Beyond accuracy, F-score and ROC: a family of discriminant measures for performance evaluation *AI 2006 Adv. Artif. Intell.* **4304**

Canvas Learning Management System Feature Analysis Using Feature-Oriented Domain Analysis (FODA)

Galih Wasis Wicaksono[1], Pamula Brian Nawisworo[2], Evi Dwi Wahyuni[3], Yus Muhammad Cholily[4]

Informatics Study Program, Muhammadiyah University Malang, Jalan Raya Tlogomas No. 246, Malang, East Java

E-mail: [1]galih.w.w@umm.ac.id, [2]pamulabriannawisworo@gmail.com, [3]evidwi@umm.ac.id, [4]yus@umm.ac.id

Abstract. The impact of Covid-19 on the world of education requires the implementation of online learning. Information and computer technology can provide solutions in education through the Learning Management System (LMS). Canvas LMS is a form of LMS that can be utilized by lecturers and students in online learning during the Covid-19 pandemic. In this study, researchers analyzed the features of Canvas LMS using the Feature-Oriented Domain Analysis (FODA) method. Researchers also conducted a mapping of canvas LMS features with the National Standard of Higher Education (SN Dikti). This study showed two stages of validation by Education and LMS experts to verify the results of the analysis. At the end of the course, we recommended a new LMS feature that complies with SN Dikti for Lecturer and Student users.

1. Introduction

The development of information and computer technology can be a solution in several areas during the Covid-19 pandemic. One of the affected areas is in the education sector; distance learning activities require educators and students to be able to utilize technology massive. As well as the higher education teaching system, the need to use the Learning Management System becomes one of the supporters of teaching and learning activities conducted online. Information technology will be used as a learning medium to facilitate the learning process through the Learning Management System (LMS). LMS has the primary function as a data center for all kinds of data needs in the learning process. The main features that LMS developers have always accommodated are 1) curriculum design, 2) instant evaluation, 3) student interest, and 4) content management [1]. LMS is generally a software application and web technology-based that comes with various functions such as creating learning plans, implementing learning processes, and assessing or evaluating learning processes. LMS presents instruction through content creation and storage, sees student participation, and assesses student performance online [2].

In some Universities around the world, LMS is used to support the implementation of mixed learning, or in other meaning of distance learning. LMS provides facilities that facilitate the interaction of lecturers and students, even through online learning. Some examples of the use of LMS include video conferences where lecturers can convey learning materials to students using live streaming. Another example of LMS function is that lecturers can give tasks online to students, so students can then complete and collect assignments through LMS. Lecturers can also provide information related to lectures to their students so that lecturers and students can connect and communicate anytime and anywhere.

Canvas is one of the LMS that serves to manage the learning process online. Canvas was launched in 2011, and the user of the Canvas network was established in 2012. Canvas, as a good enough LMS, is already widely used by 3,000 universities and Institutions around the world. Canvas LMS was developed to support highly user-friendly and reliable online learning with intuitive interface features designed to save time. Canvas is adopted faster and broader than other LMS. Canvas LMS is also known for its ease of use, accessibility, Cloud-Native architecture, Reliability, and openness [3]. Canvas LMS is claimed to help educators to improve the effectiveness of learning processes. Canvas LMS is also straightforward to use by educators who have recently switched from other LMS before [4].

The National Standard of Higher Education (SN Dikti) is the standard of higher education in Indonesia, including the National Standard of Education, The National Standard of Research, and the National Standard of Community Service [5]. The National Standard of Education is composed of 8 Standards, namely: 1) Graduate Competency Standards; 2) Standard content learning; 3) Learning Process Standards; 4) Learning Assessment Standards; 5) Lecturer and Education Standards; 6) Standard Learning Facilities and Infrastructure; 7) Learning Management Standards, and 8) Learning Financing Standards.

In this study, researchers analyzed the features on canvas LMS using the Feature-0riented Domain Analysis (FODA) method. Canvas LMS feature analyst using the FODA method focuses more on identifying features related to learning and assessment processes in Canvas LMS. The researcher chooses Canvas LMS because of its ability to integrate learning content, and there is no research before that make mapping between Canvas LMS and SN Dikti standardization. It relevant to the purpose of researchers analyzing the suitability of Canvas LMS features based on SN Dikti. The results of the analysis will be used to formulate features on the recommended new LMS. The reason for choosing FODA because this method can provide several ways of applying domain analysis results to support software development [6]. FODA supports the reuse of features from a pre-existing system. FODA method has necessary activities: context analysis, domain modeling, and architecture modeling [7].

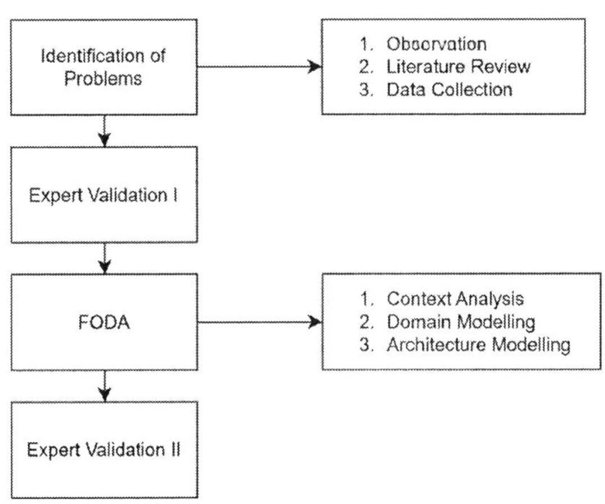

Figure 1. Research Methods

Previous research has analyzed 3 LMS features, namely Edmodo, Google Classroom, and Moodle. Analysis of LMS features in the study using the FODA method by adding expert validation stages after the expert-recommended LMS feature was successfully formulated. The study also only mapped 3 LMS features with assessment standards in SN Dikti [6]. Firstly, the fundamental difference with this study is that the analysis of the LMS features that are referenced is the features of Canvas LMS.

Second, this study also developed feature mapping that is not limited to assessment standards only but also maps features to learning process standards. Third, this study implemented two stages of validation before and after the FODA analysis was conducted.

Researchers choose Canvas LMS as the object of the study because Canvas LMS can help lecturers improve the learning process of students. The purpose of using SN Dikti on the criteria in the standard of the learning process and assessment standards because it has not been formulated LMS feature that refers to the criteria of SN Dikti. This research result may become the guideline and benchmark of the implementation of Higher Education, especially for online learning media in universities in Indonesia.

2. Research Methods

This research was conducted through 4 main stages, as outlined in Figure 1.

2.1. Problem Identification

2.1.1. Observation. At the observation stage, researchers observed the functionality of Canvas LMS features as research objects and how to use Canvas LMS directly, both as a Teacher and Student user.

2.1.2. Literature Study. At the study stage of literature, researchers gather and study various literature such as books, laws, and regulations and articles in journals related to the topics Canvas LMS, Learning Management System, FODA Method, and SN Dikti.

2.1.3. Data Collection. At the data collection stage, the author registers the features on canvas LMS, obtained the number of features in Canvas LMS as many as 71 features consisting of elements in lecturers and student users, as exemplified in Table 1.

Researchers also grouped the SN Dikti criteria into two categories. The category of the learning process that refers to the standard of learning process SN Dikti consists of criteria: 1) Interactive; 2) Holistic; 3) Integrative; 4) Scientific; 5) Contextual; 6) Thematic; 7) Effective; 8) Collaborative; 9) Student-Centered. While the assessment category that refers to the SN Dikti assessment standard consists of 1) Educational; 2) Authentic; 3) Objective; 4) Accountable; 5) Transparent; 5) Integrative. Furthermore, researchers grouped features based on activities that support the learning and assessment process. The result of the grouping will be the new LMS modeling analysis material.

2.2. FODA

FODA analysis in this study is used to analyze LMS features. This method provides several ways to apply domain analysis results from canvas LMS, to be developed on the new LMS [6]. So application developers do not take a long time and a lot of analytics tools. FODA consists of 3 stages as follows:

2.2.1. Context Analysis. At this stage, the author creates a context diagram for visualization of processes running on the LMS. The result at this stage is a model context that is a context diagram

2.2.2. Domain Modeling. At this stage, domains that have been restricted to context analysis will be analyzed for similarities and differences. The domain modeling phase consists of entity-relationship modeling feature analysis and functional analysis of Canvas LMS [7].

Feature analysis aims to find feature needs on LMS Recommendations. The process of finding these needs was through a feature analysis diagram. Feature analysis diagrams can also be called hierarchy diagrams that have tree shapes that will connect between features with sub-features [7]. In the analysis diagram, the feature was set to symbolize the relationship between the features and sub-features located in the hierarchy below them.

Entity-relationship modeling is compiled to find relationships between entities in the LMS model domain. Researchers present entity-relationship diagrams that serve as guidance in creating database models implemented in the Recommended LMS.

Table 1. List of Canvas LMS Features

Feature Code	Feature
FB001	In the account feature, lecturers and students display self-profiles.
FD002	On the dashboard feature, the lecturer has other functions to make skinny or classroom.
FM003	In the dashboard feature, students can see the courses or classes that followed.
FB004	In the course features, lecturers and students have a function as a means of learning.
FB005	In the calendar feature, lecturers and students can determine the date of the event or schedule
FB006	In the inbox feature, lecturers and students can send and receive messages from other students or lecturers.
....	....

2.3. *Expert Validation Phase I.*

At this stage, the researchers validate features by experts, as expert validation is needed to verify the decisions made [8]. Verification has proceeded by interviewing four experts, representing expert competencies of one expert in Canvas LMS and three experts in education. They master the SN Dikti sub learning and assessment process. The author asks the question openly by using the Canvas LMS feature mapping table with SN Dikti. Experts will give approval or correction to the mapping results. Expert validation at this stage aims to align SN Dikti criteria with Canvas LMS.

Functional analysis is created to determine functional features on LMS Recommendations. At this stage, the researchers compiled a use case diagram to describe the functionality of the recommended features.

2.3.1. Architectural Modeling. At the architecture modeling, researchers designed a recommended LMS architecture model. The new architecture will develop web-based LMS, whose main focus is the addition of features according to domain needs [9].

2.3.2. Expert Validation Phase II. Canvas LMS experts conduct phase 2 expert validation. This validation aims to verify the recommendations of LMS functional features to be built.

Table 2. LMS Canvas Feature Grouping and Mapping Results on SN Dikti

No	Standard	SN Dikti Criteria	Number of Canvas Features
1	Learning Process	Interactive	14
2		Holistic	13
3		Integrative	18
4		Scientific	21
5		Contextual	11
6		Thematic	12
7		Effective	21
8		Collaborative	6
9		Student-Centered	6
10	Assessment	Educational	13
11		Authentic	26
12		Lens	8
13		Accountable	11
14		Transparent	12
15		Integrative	6

3. Research and Discussion Results

The results of a series of research stages that have been conducted find various facts that support the formulation of LMS recommendations.

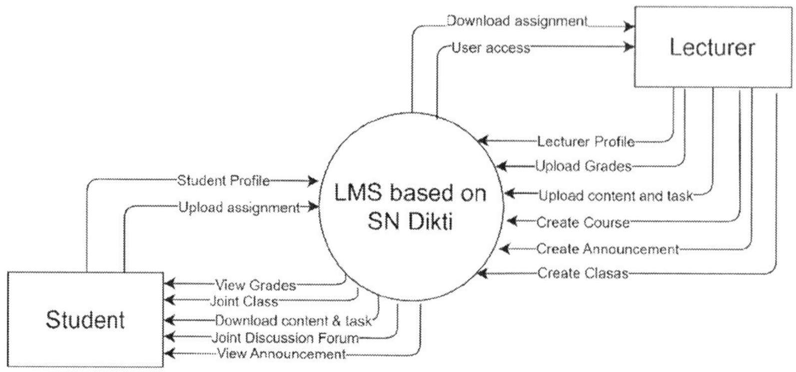

Figure 2. Context Diagram

Table 3. List of Recommended LMS Features Recommendations on Lecturer users

No.	Feature Code	Feature
1.	FR001	Lecturers can manage lecturer profiles
2.	FR002	Lecturers can manage calendars
3.	FR003	Lecturers can communicate using Messages
4.	FR004	Lecturers can create classes
5.	FR005	Lecturers can create syllabuses
6.	FR006	Lecturers can see a list of active students
7.	FR007	Lecturers can Make Announcements
8.	FR008	Lecturers can Attach Announcement Document Files
9.	FR009	Lecturers can add new students.
10.	FR010	Lecturers can Display all menus in the classroom.
11.	FR011	Lecturers can Comment
12.	FR012	Lecturers can view class information.
13.	FR013	Lecturers can create tasks.
14.	FR014	Lecturers can create groups.
15.	FR015	Lecturers can Determine the weight and point of the task.
16.	FR016	Lecturers can Specify the format of the task.
17.	FR017	Lecturers get a Deadline for Collecting Tasks.
18.	FR018	Lecturers can Manage all tasks.
19.	FR019	Lecturers can Upload material learning files.
20.	FR020	Lecturers can Store material files.
21.	FR021	Lecturers can search for files.
22.	FR022	Lecturers can create discussion forums.
23.	FR023	Lecturers can Comment on the discussion.
24.	FR024	Lecturers can Determine the duration of the discussion.
25.	FR025	Lecturers can include discussion attachments.
26.	FR026	Lecturers can Define discussion groups.
27.	FR027	Lecturers can Manage discussions.
28.	FR028	Lecturers can create quizzes.
29.	FR029	Lecturers can Specify points and rubrics on quizzes.
30.	FR030	Lecturers can Determine the type of Quiz.
31.	FR031	Lecturers can Define quiz groups.
32.	FR032	Lecturers can Specify the duration of the Quiz.
33.	FR033	Lecturers can create a list of questions.
34.	FR034	Lecturers can Give and display the value.
35.	FR035	Lecturers can Display the progress of students.
36.	FR036	Lecturers can share the value of results.

3.1. Feature Grouping and Mapping Results

Based on the list of features presented in Table 1, the grouping and mapping of the Canvas LMS feature against SN Dikti obtained results as outlined in Table 2. Each feature group is then detailed and documented to facilitate the next stage.

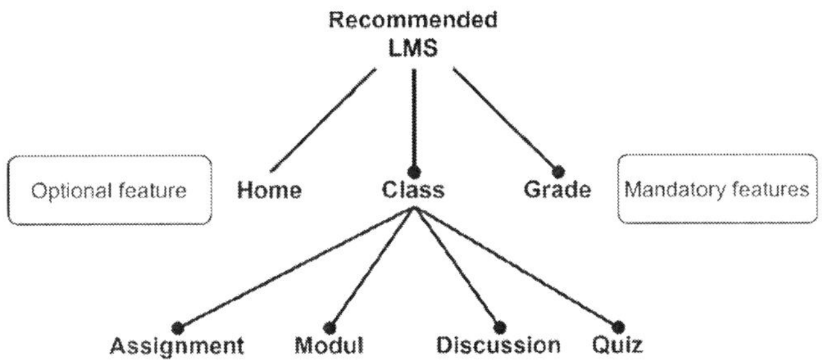

Figure 3. LMS Feature Analysis

Table 4. List of Recommended LMS Features on Student Users

No.	Feature Code	Feature
1.	FR037	Students can manage their Profile.
2.	FR038	Students can see the Classes they are in.
3.	FR039	Students can see class activity notifications
4.	FR040	Students can view calendars
5.	FR041	Students can communicate with Messages
6.	FR042	Students can see the Syllabus
7.	FR043	Students can see announcements
8.	FR044	Students can download the announcement file
9.	FR045	Students can comment on announcements
10.	FR046	Students can See class members.
11.	FR047	Students can Display menus in the classroom.
12.	FR048	Students can see Assignment Notifications.
13.	FR049	Students can Do Chores
14.	FR050	Students can attach assignment documents.
15.	FR051	Students can Comment on assignments.
16.	FR052	Students can search for materials
17.	FR053	Students can Download Materials
18.	FR054	Students can join and view information on discussion forums
19.	FR055	Students can Download discussion files.
20.	FR056	Students can comment on the discussion
21.	FR057	Students can create groups
22.	FR058	Students can see quiz notifications
23.	FR059	Students can Do Quizzes
24.	FR060	Students can see the results of the Quiz.
25.	FR061	Students can see the results
26.	FR062	Students can score

3.2. FODA

3.2.1. Context Analysis Results. Based on the results of context analysis, a context diagram illustrates the relationship between Lecturer and Student users, including the interaction of data streams from users to LMS systems and vice versa.

3.2.2. Domain Modeling: Feature Analysis Results. At the analysis stage, the recommended LMS results will have optional features, namely home and mandatory features in the first level of Class and Grade. While in the second level, the recommended mandatory features are assignment, module, discussion, and Quiz features.

Furthermore, based on the validation results of phase I and II, experts obtained the results of functional feature recommendations that are following the results of the alignment to the criteria of SN Dikti in the category of learning and assessment process. Table 3 shows a list of functional feature recommendations for Lecturer users, consisting of 36 features.

Table 4 shows a list of functional feature recommendations for Student users, consisting of 26 features.

Based on the formula of new LMS features, the features validated by experts and by the criteria of SN Dikti, in the category of learning and assessment processes. We obtained validation results that the recommendation LMS features have been 100% appropriate. It because experts have validated the recommendations and based on the previous LMS reference feature. Otherwise, the recommendation feature can be applied to the software development environment.

4. Conclusion

The study successfully selected 62 features out of a total of 71 mapped features. The number of features selected is not much reduced because, in this research, we use two standards as a mapping base, namely learning standards and assessment standards, unlike previous studies that used only assessment standards. The result of the analysis is obtained 36 functional features for lecturers and 26 functional features for students under the learning process and assessment process based on the Learning Standards and SN Dikti Assessment Standards.

Acknowledgment

The data in this study is sourced from the use of LMS based canvas LMS at Universitas Muhammadiyah Malang.

References

[1] Kulshrestha T 2013 Benefits of learning management system (LMS) in Indian education *Int. J. Comput. Sci. Eng. Technol.* **4** pp 153–64

[2] Alias N and Zainuddin A 2005 Innovation for better teaching and learning: adopting the learning management system *Malaysian online J. Instr. Technol.* **2** pp 27–40

[3] Anon 2020 *About Canvas | Edtech Learning Platform | Instructure* Retrieved from https://www.instructure.com/canvas/about

[4] Rubio F A, Santamaría G F, Merino P J M and Delgado K C 2017 A Data collection experience with canvas LMS as a learning platform *Proc. of the Learning Analytics Summer Institute Spain 2017: Advances in Learning Analytics*

[5] Kemenristek 2016 *Sistem Penjaminan Mutu Pendidikan Tinggi dan Standar Nasional Pendidikan* Retrieved from https://www.lldikti4.or.id/wp-content/uploads/2019/09/A.-Kebijakan-Nasional-SPM-Dikti-300119-k.pdf

[6] Wicaksono G W, Juliani G A, Wahyuni E D, Cholily Y M, Asrini H W and Budiono 2020 Analysis of learning management system features based on indonesian higher education national standards using the feature-oriented domain analysis *2020 8th Int. Conf. on Information and Communication Technol. (ICoICT)* (IEEE) pp 1–6

[7] Kang K C, Cohen S G, Hess J a, Novak W E and Peterson a S 1990 Feature-Oriented Domain Analysis (FODA) feasibility study *Distribution* **17** 161

[8] Meyer M A and Booker J M 2001 *Eliciting and Analyzing Expert Judgement* (London: Academic Press Limited)

[9] Iqbal M, Faisal M R and Budiman I 2016 Penentuan fitur website bidang pariwisata dan kebudayaan dengan metode feature-oriented domain analysis (FODA) *Kumpulan Jurnal*

Ilmu Komputer vol **03** no 2 pp 172–81

Does it make you sad? A lexicon-based sentiment analysis on COVID-19 news tweets

Dedy Suryadi

Industrial Engineering Department, Parahyangan Catholic University
Jl. Ciumbuleuit No. 94, Bandung 40141, Indonesia

E-mail: dedy@unpar.ac.id

Abstract. This research utilizes a lexicon-based sentiment analysis to reveal the emotions conveyed by various news media and analyze the differences among the media. NRC Affect Intensity Lexicon is utilized; it provides the emotions of words in 8 categories of emotions. The lexicon is modified to reflect the context of the news, i.e., COVID-19 pandemic. Tweets are collected from four Indonesia's news media and three international English-language news media. Each tweet from a news media is assigned a total emotion score for each of the 8 emotions. The scores are averaged for each day to obtain the Daily Emotion. Based on the Daily Emotion, the Dominant Daily Emotion may be identified, i.e., the emotion with the highest average score in a particular date. The Dominant Daily Emotion visualization shows that the dominant emotions in Indonesian-language media are Sadness and Trust, while Fear and Trust are considerably more pronounced in English-language media. Furthermore, among Indonesian-language media, kompascom is significantly more intense than the others in Joy, Sadness, Fear, and Trust. Among English-language media, XHNews is more intense than others in conveying Fear, Sadness, and Trust; while XHNews and timesofindia are both the most intense in Joy. The sentiment analysis in this research concludes that the Emotion Mix of news media are significantly different. Furthermore, news media also convey significantly different levels of emotion intensity in reporting COVID-19 news.

1. Introduction

Twitter has been one of the main sources of information nowadays. News media have been actively utilizing Twiter as well to broadcast information, including the news about the COVID-19 pandemic. Due to the limitation of number of characters, news media typically tweet concise news or striking headlines.

Previous research has shown that the news content triggers different types of emotion. Lin et al. [1] predict the emotions of Useful, Bored, Happy, Angry from the news in Yahoo Chinese with higher than 73% accuracy. Furthermore, Zhu et al. [2] model emotions with respect to time, such that the evolution of the emotions may be tracked.

More specifically, Kim and Cameron [3] reveal that news frames (e.g., sadness-inducing) induce distinct emotions. In other words, different framing of the same news may affect the degree of message processing by the readers [4], as well as the readers' emotional response, i.e., media influence what people feel. Boukes and Vliegenthart [5] conclude that the consumption of news that focus on misery, powerlessness, and negativity negatively affect the mental well-being. Therefore, it is clear that the emotions in the news, including news tweets, do affect the readers. Thus, the first objective of this research is to analyze the emotions that are conveyed by the news tweets from various news media. The emotion analysis in this paper takes the approach of a lexicon-based sentiment analysis.

From the perspective of news media, emotional communication is indeed one of journalists' main strategies in writing news, according to a research by Saxer and Märki-Koepp [6]. Negativity, in particular, is one of journalists' key news values [7]. Therefore, this research's second objective is to analyze the differences in the emotional communication among news media, e.g., whether or not some news media conveys significantly higher sadness than others.

2. Literature review: Sentiment analysis

Sentiment analysis is defined as a computational study of sentiments, emotions, opinions, as well as attitude, towards a particular aspect in texts [8]. A typical sentiment analysis task is determining whether a sentence expresses a positive or negative sentiment towards an aspect [9]. According to Ravi and Ravi [9], sentiment analysis may be performed using machine learning, lexicon-based, or hybrid approach. A machine learning approach generally requires annotated sentences to train the machine learning model and the model is subsequently applied to classify the unlabeled sentences. A lexicon-based approach, on the other hand, relies on the sentiment lexicon that has been developed from crowdsourcing or similar method.

Sentiment lexicons may have different number of categories of emotion. In NRC Affect Intensity Lexicon [10], there are 8 emotion categories. The categories correspond to the basic emotions that are defined by Plutchik [12]. Plutchik's basic emotions are conceptualized in terms of pairs of opposites, i.e., Joy-Sadness, Trust-Disgust, Fear-Anger, and Surprise-Anticipation.

The NRC Affect Intensity Lexicon is chosen in this research because it provides a wider range of emotions to analyze the news tweets, compared to using binary sentiment lexicon, i.e., positive and negative sentiments only. Furthermore, the lexicon also provides real-valued scores for rating the intensity of the 8 basic emotions. The lexicon is built using Best Worst Scaling technique, in which annotators pick the best and the worst items among *n* items regarding to the property of interest [11], e.g., a particular emotion. Based on the selections, the real-valued emotion intensity score of a word may be inferred.

3. Methodology, data and results

3.1. Data collection and preprocessing

The data in this paper is obtained from Twitter. In particular, the tweets are collected from the official Twitter accounts of several news media in several countries.

The news media from Indonesia and their respective verified Twitter accounts are as follows: The Jakarta Post @jakpost, CNN Indonesia @CNNIndonesia, Republika.co.id @republikaonline, Kompas.com @kompascom. Among those four news media, The Jakarta Post is the only news media that uses English, while the others use Indonesian. The other three news media whose tweets are collected are China's China Xinhua News @XHNews, India's The Times of India @timesofindia, and Singapore-based CNA @ChannelNewsAsia. All three news media use English. The Twitter account name will subsequently be used to refer to each news media.

The tweets are collected using a Python package called GetOldTweets3. The collection period is from January 1, 2020 to May 31, 2020. Tweets are collected if they contain either "corona" or "covid" (case insensitive). The highest number of tweets is obtained from kompascom (19,357) and the lowest is jakpost (2,032). An example of jakpost's five last tweets during the data collection period are shown in Table 1. The tags and HTML links are subsequently removed from the tweets for further text processing.

The tweets are preprocessed as follows: replacing every capital letter with a lowercase, removing words that start with "http" and a hashtag ("#"), and removing any non-alphanumeric character at the beginning or end of a word. For the tweets in Indonesian language, a highly frequent bigram "rumah sakit" (hospital) is replaced by "rumah_sakit" and the abbreviations that refer to a hospital (i.e., "rs", "rsud", and "rsup") are all replaced by "rumah_sakit".

3.2. NRC Affect Intensity Lexicon Modification

The NRC Affect Intensity Lexicon [10] is utilized to score each tweet on each of the 8 emotions. The lexicon is also available for Indonesian language. Therefore, the tweets from Indonesian news media may be scored using the same lexicon.

Table 1. Example of jakpost's tweets

Timestamp	Tweet
2020-05-31 13:10:44	UN to use Liberian president Weah's COVID-19 awareness song #jakpost
2020-05-31 12:21:08	Via Vallen explained the story about her younger brother who has been tested positive for COVID-19. #jakpost
2020-05-31 11:34:05	Canada to promote holidays at home because of COVID-19 border closures #jakpost
2020-05-31 11:04:47	The Health Ministry announced 700 new confirmed COVID-19 cases on Sunday, bringing the total number of infections nationwide to 26,473. #jakpost
2020-05-31 10:10:47	COVID-19: Surabaya mayor feuds with East Java governor over mobile PCR labs #jakpost

Nevertheless, several modifications must be made to the lexicon such that the emotion scores reflect the context of the tweets, i.e., COVID-19 pandemic. The modifications are subjective, yet deemed reasonable.

For the English lexicon, two modifications are made as follows:

1. Swapping: the emotions of "positive" and "negative" are swapped because, in the pandemic context, "positive" has a negative meaning.
2. Removing the words that do not convey emotions in the pandemic context, such as the names of days in a week; as well as some words that are considered containing no emotions when they appear in a phrase, i.e., "health" in "health ministry", and "dream" in "dream world".
3. Flipping: the emotions of "highest" are flipped to their corresponding opposites when it appears together with "case(s)" and "death(s)".

For the Indonesian lexicon, the modifications are made as follows:

1. Swapping: the emotions of "positif" (positive) an.d "negatif" (negative) are swapped for the same reason with the English lexicon.
2. Removing the words that, in the pandemic context, do not match the original English terms from which they got translated; such that these words are assigned zero score for all emotions. Those words are as follows:
    a. "sangat" (very; adverb) which is translated from "wonderfully" in the English lexicon,
    b. "sedang" (be; verb) is translated from "temperate",
    c. "tetap" (still; adverb / remain; verb) is translated from "fixed",
    d. "jantung" (heart) –mostly refers to heart disease; is translated from "heart" with joy as one of the emotions,
    e. "khusus" (special) –mostly refers to the special spokeperson of COVID-19; is translated from "special" with joy as as one of the emotions;
    f. "kesehatan" (health) –mostly refers to the health ministry; is translated from "health" with joy as one of the emotions.
3. Adding: an entry "rumah_sakit" (hospital) is added to the lexicon and the emotions are copied from the English lexicon for the word "hospital".

4. Flipping: the emotions of the following words, i.e., "meningkat" (increase), "mencapai" (reach), "melonjak" (soar), "jiwa" (lives), "ledakan" (explosion); are flipped when they appear together with "kasus" (case), "kematian" (death), "meninggal" (die).

The modifications are made based on the preliminary observations that some tweets were assigned high scores for joy, despite those tweets clearly being bad news. For example, before the modifications, tweets that mention "highest deaths" score high on joy. These modifications are not claimed to be exhaustive but they certainly help in eliminating the extreme cases, such as the aforementioned case of "highest deaths".

3.3. *Emotion scoring for tweets: total emotion, daily emotion, dominant daily emotion*

Each tweet is split into words. If a word appears in the NRC Affect Intensity Lexicon, then its corresponding emotion(s) are collected. The total score for each emotion in a tweet is then obtained by summing the scores over all words related to the emotion. The procedure is illustrated in Table 2 with a tweet from jakpost on May 8, 2020. The tweet content is "Under normal circumstances, Buddhists would have gathered in temples across the country on Thursday to celebrate Waisak. But this year, with COVID-19 measures preventing most gatherings, the temples have gone quiet. #jakpost".

Table 2. An example of obtaining the total emotion scores of a tweet

Word in Lexicon	Score (joy)	Score (trust)	Score (sadness)
circumstances	0.125	0.359	0
celebrate	0.844	0	0
gone	0	0	0.453
quiet	0.188	0	0.094
Total Emotion	1.157	0.359	0.547

Each day, a news media may broadcast several tweets. To reflect the daily emotion, the total emotions for every tweets on a particular date are averaged. An example of the daily emotion in jakpost on May 31, 2020, over 9 tweets on that particular date, is shown in Table 3. Based on the daily emotion, the dominant daily emotion may be obtained, i.e., the emotion with the highest score in the daily emotion. Using the example in the Table 3, the dominant daily emotion for jakpost on May 31, 2020 is Trust.

Table 3. Daily Emotion of jakpost's Tweets on May 31, 2020

Anger	Anticipation	Disgust	Fear	Joy	Sadness	Surprise	Trust
0	0.038222	0	0.073778	0.142222	0.152778	0	**0.506778 (dominant)**

The plots of daily emotion for different news media indicate the differences between media. For example, the plots in Figure 1 display the difference in daily Joy and Sadness emotions between kompascom and jakpost. Sadness (purple line) is obviously above Joy (orange line) in kompascom, while such dominance does not exist in jakpost. This observation motivates the further analysis whose results are presented in the next sections.

3.4. *Emotion Mix of News Media*

First of all, the emotion mix, i.e., the average proportions of the emotions in the daily emotion; of each news media is shown in Figure 2. The visualizations in this and subsequent subsections are produced by the Python package called Matplotlib [13].

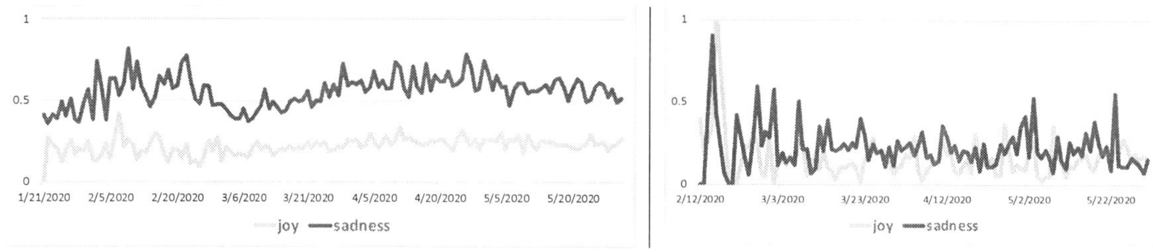

Figure 1. Daily Emotions in kompascom (Left) and jakpost (Right).

On the left part of Figure 2, it can be observed that the emotion mix of Indonesian-language news media contains a significantly higher portion of Sadness (i.e., at least 29.5%) than jakpost. Similar to jakpost, the international English-language news media also contain just around 20% of Sadness in their emotion mixes.

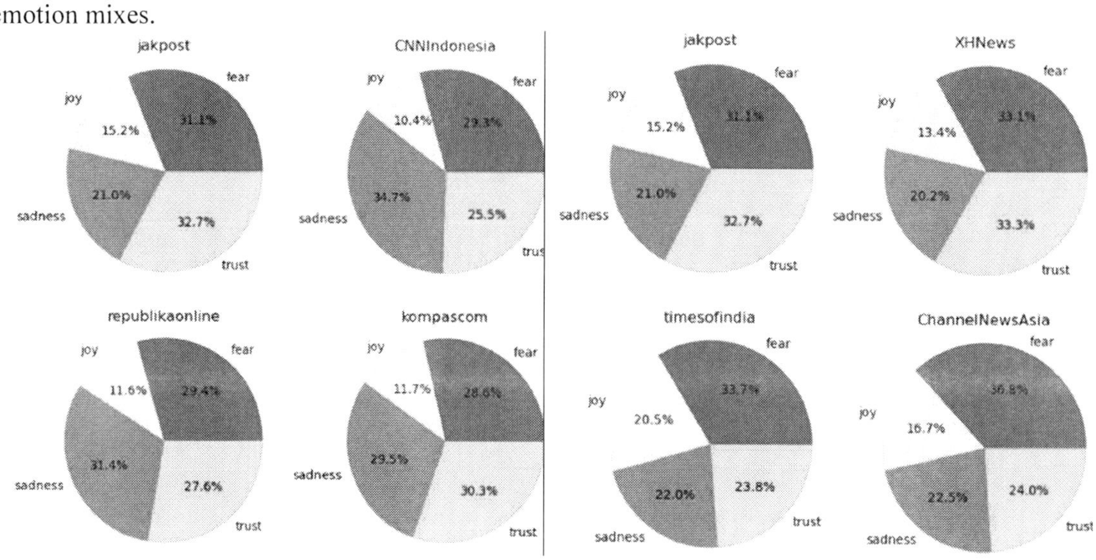

Figure 2. Emotion Mix in Indonesia's News Media (Left) and English-language News Media (Right).

On the other hand, the portion of Fear appears slightly higher in English-language news media (i.e., at least 31.1%) than their Indonesian-language counterparts, with ChannelNewsAsia as the highest with 36.8%. The portion of Sadness is just around 20% in these news media.

3.5. Dominant Daily Emotion of News Media

The four emotions are selected based on the most common dominant daily emotions, which are displayed in Figure 3 for Indonesia's news media and Figure 4 for English-language news media. It is apparent that Indonesian-language news media's preferred dominant emotion is Sadness, with CNNIndonesia as the leader with 72.6%; while kompascom's dominant emotions are Trust and Sadness.

On the other hand, the English-language news media choose Fear and Trust. In particular, Fear becomes ChannelNewsAsia's clear dominant daily emotion with 81.1%. The dominant daily emotion amplifies the difference among news media, as it can be seen from how the plots in Figure 3 and 4 differ greatly from the plots in Figure 2.

3.6. Emotion Intensity Comparison (Indonesia's News Media)

In the previous section, it has been shown that the emotion mix and dominant daily emotion in each news media may greatly differ. In this section, it is shown that the intensity of the emotions may be significantly different among news media, as well. In other words, even if two media have similar emotion mix or similary dominant daily emotion, one media may be more emotionally intense than the other.

A Z-test is performed using a Python package called statsmodels [14] to compare the average daily emotion between pairs of news media with respect to a particular emotion. The result for Sadness comparison among Indonesia's news media is shown in Table 4. The column H1 denotes the sign in the alternate hypothesis and the p-value that is less than 0.05 is considered as statistically significant. Table 4 indicates that kompascom is the most intense in Sadness, followed by republikaonline, CNNIndonesia, and jakpost. Interestingly, despite CNNIndonesia having the highest portion of Sadness in its dominant daily emotion, its intensity is significantly less than two other media.

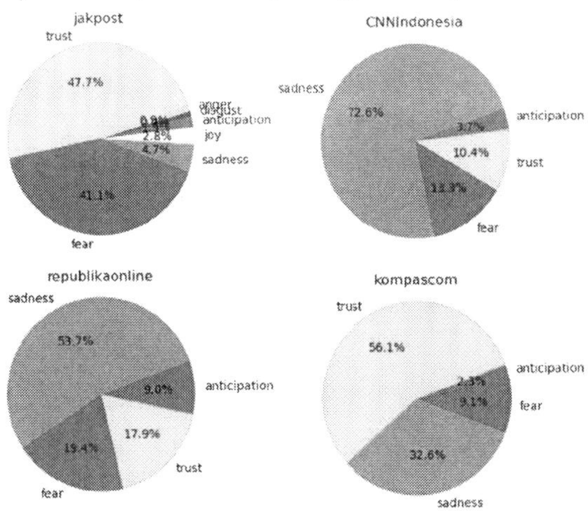

Figure 3. Dominant Daily Emotion in Indonesia's News Media

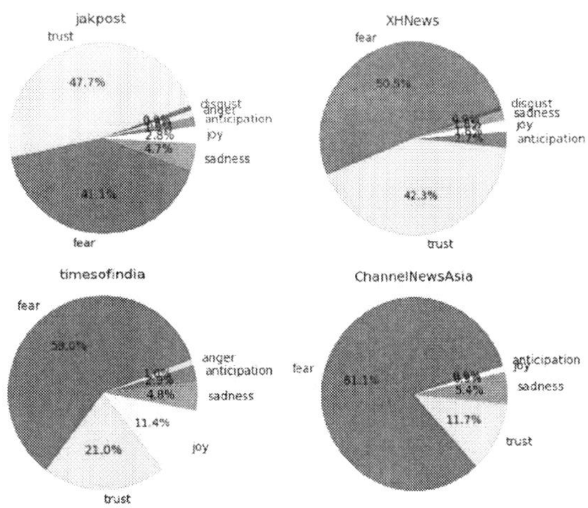

Figure 4. Dominant Daily Emotion in English-language News Media

Table 4. Z-test for comparing the average daily emotion of Sadness among Indonesia's news media

Media 1	H1	Media 2	Z-value	p-value
jakpost	<	CNNIndonesia	-8.92	2.27E-19
jakpost	<	republikaonline	-11.58	2.74E-31
jakpost	<	kompascom	-22.42	1.14E-111
CNNIndonesia	<	republikaonline	-3.96	3.72E-05
CNNIndonesia	<	kompascom	-17.83	2.17E-71
republikaonline	<	kompascom	-12.94	1.41E-38

The result in Table 4 is confirmed by the boxplot visualization of month-by-month daily emotion of Sadness of each news media in Figure 5. The Y-axis indicates the Sadness emotion score, while the labels on X-axis indicate the media name and the month, e.g., "J2" means jakpost in month 2. Meanwhile, "C" represents CNNIndonesia, "R" represents republikaonline, and "K" represents kompascom. There is no boxplot in J1 because the earliest tweets for jakpost was from February. It may be observed that kompascom's scores for Sadness soar above other news media in all months.

Interestingly, kompascom is also the most intense in Joy, followed by jakpost and republikaonline which are not significantly different, and CNNIndonesia is the last. This matches the fact that the dominant daily emotion of CNNIndonesia is Sadness. For the other two emotions, i.e., Fear and Trust, kompascom is also significantly more intense than the others. Due to page limitation, the Z-test results and boxplots are not displayed here.

Table 5. Z-test for comparing the average daily emotion of Fear among English-language news media

Media 1	H1	Media 2	Z-value	p-value
jakpost	<	XHNews	-11.97	2.47E-33
jakpost	<	timesofindia	-3.40	3.42E-04
jakpost	>	ChannelNewsAsia	1.48	6.94E-02
XHNews	>	timesofindia	6.18	3.15E-10
XHNews	>	ChannelNewsAsia	18.78	5.22E-79
timesofindia	>	ChannelNewsAsia	5.27	6.69E-08

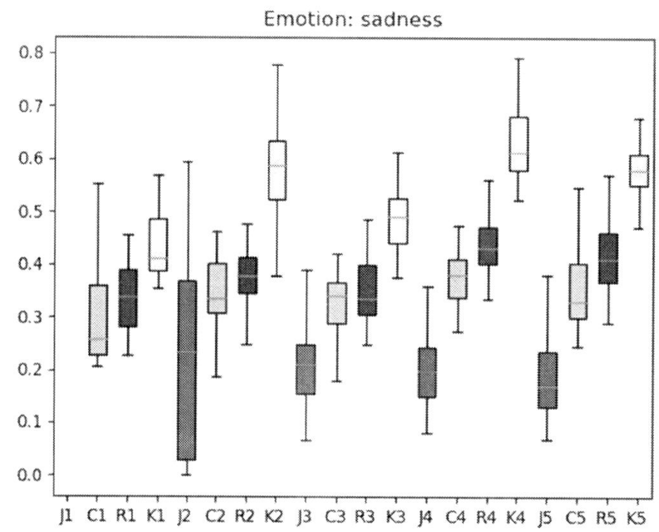

Figure 5. Comparison of Sadness in Indonesia's News Media

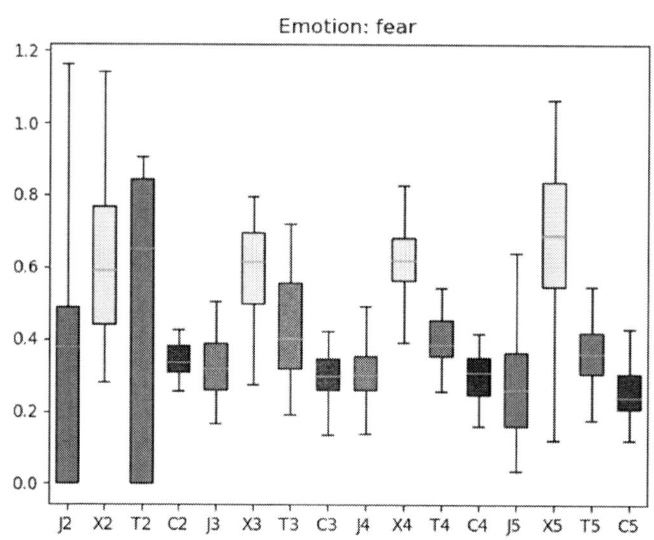

Figure 6. Comparison of Fear in English-language News Media

3.7. Emotion Intensity Comparison (English-language News Media)

Among English-language news media, Table 5 indicates that XHNews has the most intense emotion in Fear, followed by timesofindia, and finally both jakpost and ChannelNewsAsia do not differ significantly. Correspondingly, Figure 6 shows that XHNews' scores for Fear are above others in most of the months. The "J", "X", "T", "C" labels in the X-axis of Figure 6 represent jakpost, XHNews, timesofindia, and ChannelNewsAsia, respectively; followed by a number that indicate a particular month. It is worth noting here that ChannelNewsAsia's daily emotion of Fear is relatively stable in all months. Despite its dominant daily emotion is Fear, ChannelNewsAsia delivers relatively low intensity of Fear compared to other news media. Furthermore, the intensity neither increases nor decreases significantly along time.

Interestingly, ChannelNewsAsia also becomes the news media that is the least intense in the other three emotions, i.e., Joy, Sadness, and Trust. XHNews is the most intense in both Sadness and Trust. XHNews and timesofindia do not differ in the intensity of Joy.

4. Conclusion and future work

This research uses the NRC Affect Intensity Lexicon to analyze the emotions within the news tweets from various news media. Using statistical testing and data visualization, it may be concluded that news media have different emotional communication strategies. It is shown by the differences in Emotion Mix, Dominant Daily Emotion, and Emotion Intensity.

The contribution of this research is that readers may be aware of the differences in emotion mix and intensity among news media, such that they may make an informed decision on reading the news tweets from a particular media. For the news media, the result in this research may be used to reflect their emotional communication strategy.

For the future work, to address the limitation of the subjective modifications that are proposed in this paper, a more objective and automatic approach may be used to modify a sentiment lexicon according to the context of the corpus. For example, word embedding technique, such as word2vec [15], may be useful to automatically identify that the word "positive" in the COVID-19 corpus generally has a negative meaning. Adding the data such as time of day, the number of likes and retweets; this research may be the basis of the emotion mix optimization to achieve the objective such as maximizing the number of readers. This research may also become the initial step in linking the emotions conveyed by the news media to the other issues, such as cultural background, emotion in different language, etc. In addition, analogous to the research about the changes in emotion expression in the developmental context [16], a time series analysis may be conducted to reveal the trend of COVID-19 news tweets along with the development of the pandemic.

References

[1] Lin K H-Y, Yang C and Chen H-H 2007 What emotions do news articles trigger in their readers? *SIGIR 2007 Proc.* Amsterdam, The Netherlands 23–27 July pp 733–4
[2] Zhu C, Zhu H, Ge Y, Chen E and Liu Q 2014 Tracking the evolution of social emotions: A time-aware topic modeling perspective *2014 IEEE Int. Conf. on Data Mining* Shenzhen, China 14-17 December pp 697–706
[3] Kim H J and Cameron G T 2011 Emotions matter in crisis: The role of anger and sadness in the publics' response to crisis news framing and corporate crisis response *Communication Research* **38** pp 826–55
[4] Nabi R L 2003 Exploring the framing effects of emotion: Do discrete emotions differentially influence information accessibility, information seeking, and policy preference? *Communication Research* **30** pp 224–47
[5] Boukes M and Vliegenthart R 2017 News consumption and its unpleasant side effect *J. of Media Psychology* **29(3)** pp 137–47
[6] Wirth W and Schramm H 2005 Media and Emotions *Communication Research Trends* **24(3)** pp 3–43

[7] Harcup T and O'Neill D 2001 What is news? Galtung and Ruge revisited *Journalism Studies* **2(2)** pp 261–80
[8] Medhat W, Hassan A, Korashy H 2014 Sentiment analysis algorithms and applications: A survey *Ain Shams Eng. J.* **5(4)** pp 1093–113
[9] Ravi K and Ravi V 2015 A survey on opinion mining and sentiment analysis: Tasks, approaches and applications *Knowledge-Based Systems* **89** pp 14–46
[10] Mohammad S M 2018 Word affect intensities *Proc. of the Eleventh Int. Conf. on Languate Resources and Evaluation (LREC 2018)* Miyazaki, Japan 7-12 May pp 174–83
[11] Mohammad S M, Bravo-Marquez F, Salameh M and Kiritchenko S 2018 SemEval-2018 Task 1: affect in tweets *Proc. of the 12th Int. Workshop on Semantic Evaluation (SemEval-2018)* New Orleans Louisiana 5-6 June pp 1–17
[12] Plutchik R 1980 A general psychoevolutionary theory of emotion *Theories of Emotion* pp 3–33
[13] Hunter J D 2007 Matplotlib: A 2D graphics environment *Computing in Sci. & Eng.* **9(3)** pp 90–5
[14] Seabold S and Perktold J 2010 Statsmodels: econometric and statistical modeling with Python *Proc. of the 9th Python in Sci. Conf.* Austin, USA 28 June – 3 July pp 92–6
[15] Řehůřek R and Sojka P 2010 Software framework for topic modelling with large corpora *LREC 2010 Workshop on New Challenges for NLP Frameworks* Valletta, Malta 17-23 May pp 45–50
[16] Hipson W E 2019 Using sentiment analysis to detect affect in children's and adolescents' poetry *Int. J. of Behavioral Development* pp 1-8

Developing a Framework on Predicting Evacuee Needs for Disaster Risk Reduction Management in the Province of Albay

R N Monreal[1] and T D Palaoag[2]

[1]Divine Word College of Legazpi, Legazpi City, Philippines
[2]University of the Cordilleras, Baguio City, Philippines

E-mail: [1]richmonreal83@gmail.com and [2]tpalaoag@gmail.com

Abstract. An evacuee needs allocation framework wants to ease the predicting process on disaster preparedness in the province of Albay. The evacuee needs allocation framework will help the provincial government to identify and solve the problem of the current and manual process of the disaster management system in preparation down to the allocation of evacuee needs. An evacuation centre is a place that the Provincial Government provides basic human needs including food, water as well as accommodation during or before the disasters to the affected people. Detailed assessment was conducted by authorized personnel in the government organization to look at how to improve the chances of disaster victims with their stay in an evacuation centre. In this paper, the study aims to develop a framework that predicts the evacuee needs preparedness as well as the allocation needs on before and during the calamity to help and enhance the current process in the province of Albay. The researcher includes information technology tools such as RFID and GPS to the framework to enhance the efficiency and accuracy in the implementation of the needs allocation during the calamity in the evacuation centres. The paper will lead the provincial government of Albay in predicting the evacuee needs to insure a sufficient and avoid the shortage and excess of goods from the preparation down to the allocation in the evacuation centres and to help the Provincial Disaster Risk Reduction Management Council in allocating and monitoring of budgets for transparency purposes.

1. Introduction

Developing a framework on Predicting Evacuee needs wants to address that disaster has evolved through time into a complex policy subsystem, and disaster policy is enforced through a collection of functions referred as emergency management and response. In the modern technology, approaches in the disaster management involves its multidimensional efforts to reduce the vulnerability to hazards; to limit the effects of disasters; and to arrange for, respond to, and recover from the disaster occur.

Disaster readiness refers to do things or accomplish a purpose to get ready and lessen the effects of disasters. In the 2018 information of World Risk Report, the Philippines rank third among all of the nations worldwide which has very high in disaster risk, with index value of 25.14% [1]. The nation's total area of at least 60% is vulnerable to multiple threats or dangers, and 74% of the population is vulnerable to their impact [2]. The risk including coastal threats such as typhoons, tsunami and rising sea level is due to the location and geographical context of the country. Also, since the island is located within the "Ring of Fire" the people are seriously exposed to earthquakes and volcanic eruptions, flooding, landslides, droughts and tsunamis that further contribute to the exposure of natural disasters [3]. Hydro-meteorological events together with typhoons and floods, accounted for over 80% of the natural disasters within the country throughout the year [4]. Some of the region's most affected by natural disasters in the Philippines are Bohol, Eastern Visayas, Bicol and Ilocos [5][6].

Albay is a Philippine region situated in Bicol about 500 kilometres south of Manila. In 2015, the population of the province pegged at 1,314,826, with the weight of 481 people for each square kilometre. The province is considered as one of the most disaster-prone areas in the country because of its area, geography and land. However, the provincial government mostly managed to stay its zero-casualty goals throughout disasters. The Provincial Government of Albay hailed by many International Organizations because of its highly effective Disaster Risk Response and Management.

The National Risk Reduction and Management Plan (NDRRMP) accomplish the necessity of RA No. 10121 of 2010, which come up of the legitimate reason for policies, plans and projects to manage calamities. The NDRRMP incorporates four topical zones like as Disaster Interference and Mitigation; Disaster Preparedness; Disaster reaction; and Disaster restoration and Recovery that adjust to the structure of the National Disaster Risk Reduction and Management Council (NDRRMC) [7]. Evacuation management is one among the previous plans in disaster interference and preparation to avoid wasting life in every disaster.

Evacuation is a risk management technique that might be utilized to decrease death toll or reduce the impacts of a crisis on a network, before the beginning of, or during, a crisis. It includes the development of individuals compromised by a risk to a more secure area and, normally, their possible sheltered and timely return. For an evacuation to be as effective as could be expected under the circumstances, it must be appropriately arranged and implemented. Depending on the risk and its effect on the network, the clearing procedure, including withdrawal and return, may take days, weeks or months to finish. Some evacuations might be completed rapidly and over short distances [8].

Disaster Risk Reduction Management is an investment, not an expense and it increases the business returns. The province of Albay has seen an expansion of ventures, even when there are calamities such as typhoons and volcanic eruptions. Climate change transformation and risk reduction management permit advancement to continue in the midst of disasters, since the provincial government takes responsibility of the disaster and they don't disrupt the lives of their subordinates [9][10][11][12].

Thus, the study aims to develop a Framework on Predicting Evacuee Needs Allocation for Disaster Preparedness in the Province of Albay to allow and understand the situation, make a connection and interact with the government agencies in the preparation and implementation of needs allocation in every disaster or calamity. Also, the framework wants to lessen the time in the implementation of needs allocation during the calamity and to lessen the time in preparing the report for the annual DRRM Evacuation needs budget. Making predictions is a valuable strategy to improve the decisions for DRRM Planning and implementation.

2. Review of related literature

2.1. *The 2011-2028 Philippines National Disaster Risk Reduction and Management Plan (PNDRRMP)*

The Philippine National DRRM Framework fills in as the principal manual to the disaster risk reduction and management endeavours to the nation. The Framework envisions a country of a safer, flexible and disaster versatile Filipino people group toward maintainable turn of events. It grants a paradigm change from responsive to proactive DRRM whereby people have expanded their mindfulness and comprehension of DRRM. The NDRRMP sets the sets the normal results, outputs, key activities, markers, lead offices, executing accomplices and timelines under each of the four particular yet commonly strengthening topical regions. The goals of every thematic space result in the fulfilment of the nation's general DRRM vision [13].

2.2. *International Frameworks for Disaster Risk Reduction (DRR)*

The National Disaster Risk Management Framework (NDRRMF) of Palau plots an institutional structure to ensure successful coordination and cooperation in preparing for and reacting to, property moderation against and recuperation from the effect of hazards that proportion of high danger to Palau. The Framework likewise accommodates powerful coordination and direction for leaders in the improvement of crisis reaction plans and techniques and calamity hazard decrease systems by government and non-government offices that have commanded obligations regarding such exercises. [14][21].

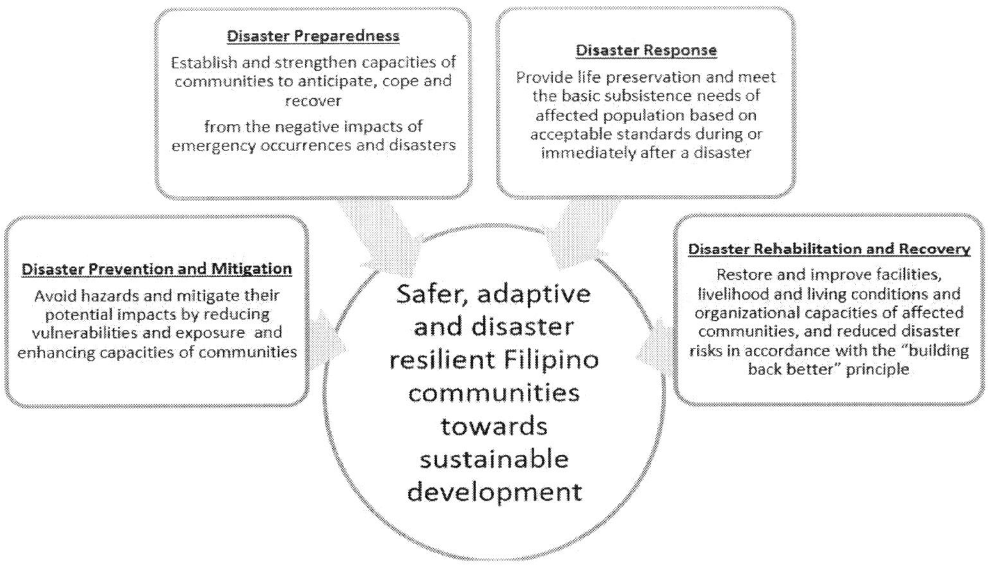

Figure 1. Philippines National Disaster Risk Reduction and Management Plan

Another Disaster Risk Reduction (DRR) framework is the Sendai Framework (2015-2030), it plots the seven clear targets and four requirements for activity to prevent new and decrease existing calamity dangers, for example, understanding calamity hazard; fortifying calamity hazard administration to manage disaster hazard; Investing in disaster reduction for flexibility and; Enhancing disaster readiness for practical response, and to Build Back Better in recovery and reproduction. It intends to achieve the generous decrease of calamity threat and losses in lives, livelihoods and well-being and in the budgetary, physical, social, social and ecological assets of people, associations, organizations and countries throughout the following 15 years. The Framework was embraced at the Third United Nation World Conference on Disaster Risk Reduction in Sendai, Japan, on March 18, 2015 [11][12][15].

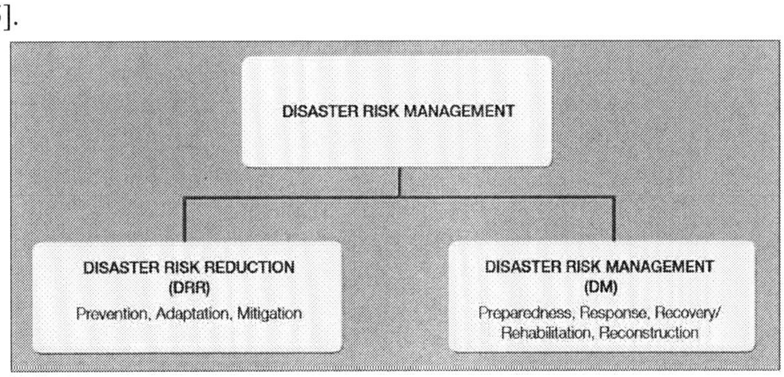

Figure 2. Palau Disaster Risk Reduction and Management Model

IOP Conf. Series: Materials Science and Engineering 1077 (2021) 012043 doi:10.1088/1757-899X/1077/1/012043

2.3. *Information Technology tools Framework*

Radio frequency identification (RFID) is a remote correspondence technology that facilitates automatic identification and data capture without human intervention. It has the advantage of having the option to catch data without the requirement for the distinguishing proof module and the probe to be in the line of sight. (Gupta et .al, 2015), Automated care and safety design is accomplished utilizing RFID. From the perspective of workflow management, several manual works like the determination of patient discharge time, reservation for regular check-up can be achievable through automation. RFID tags guarantee exact information collection with minimal human intervention [16].

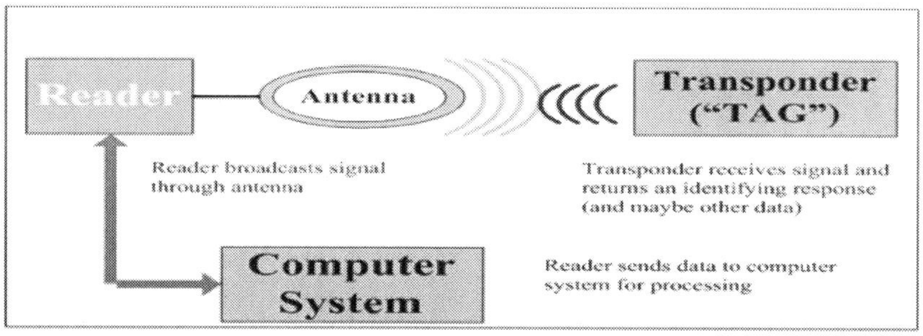

Figure 3. Function of RFID Technology

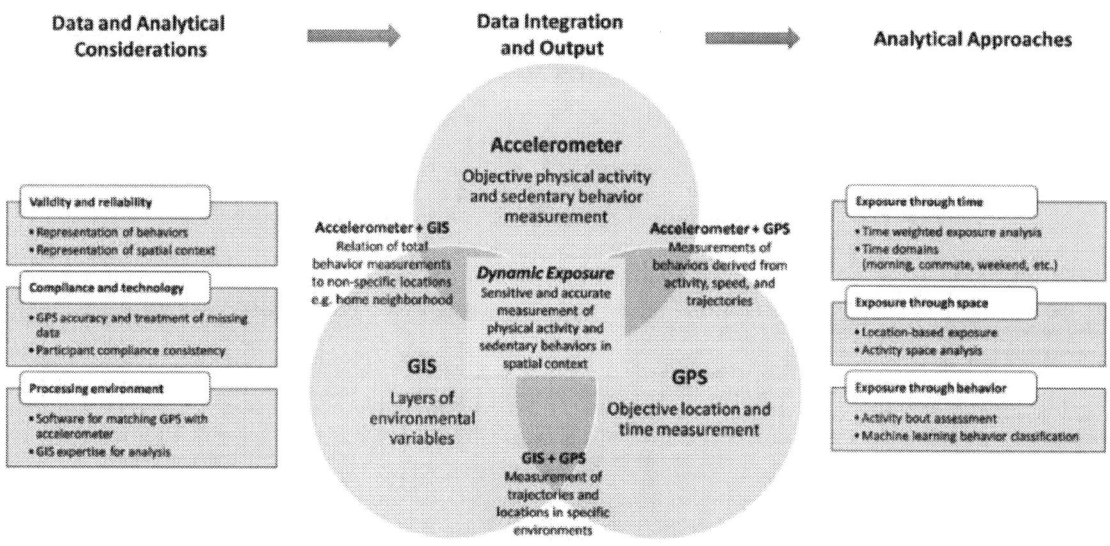

Figure 4. Framework for integration of GPS, accelerometer, and GIS technology

GPS data provides latitude, longitude, and time – three variables that enable us to assess when an individual is found wherever. Data are collected at varied intervals ranging from each few hours. In the study of Jankowska et. al, they conferred a framework for integration of GPS, GIS, and accelerometer data, as well as information collection concerns, attainable data outputs from joining specific technologies, and analytical methods when using all three measurement ways along. The framework is supposed to guide researchers through abstract development of GPS based mostly PA and SB comes by deepening the understanding of what info represents, the categories of research questions that are and aren't attainable to answer with the data, how to combine datasets to create better representations of human-environment interactions, and therefore the details for information assortment, handling, and analysis that has to be considered to make sure successful productive project completion [17].

2.4. Predictive Algorithm

Predictive Analytics (PA) tools are presented by few distinctive pattern or paradigm and algorithms which is able to apply in a wide scope of utilization instances. The top 5 predictive Analytics models are classification model, clustering model, forecast model, outliers model, and the time series model. The key in obtaining the best out of a predictive analytics answer and support information to create perspective choice is to determine what predictive model are fitted for the organization. The PA algorithms are mostly divided into two groups, the Machine Learning (ML) and the Deep Learning (DL). The ML predictive modelling has several different algorithms that can be enforced. Majority of the frequent algorithms that are being utilized to influence the predictive analytics such as the Random Forest (RF) , the Generalized Linear Model (GLM), the K-means and the Prophet algorithm. The first algorithm is the RF algorithm, it is the most utilized and preferred classification rule that is capable of classification and regression. An RF accurately classifies a huge volume of data. Second, the GLM may be an additional complicated version of the General Linear Model. It takes the latter model's comparison of the results of multiple variables on continuous variables before drawing from array of varied distributions to look out the best fit model. Another algorithm is the K-means algorithm; it may be a widespread and fast algorithm that includes putting unlabelled data points in different groups based on correspondence. The algorithm is also employed for the clustering model. And last the Prophet algorithm; it is an open-source algorithm developed by Facebook Company to utilize in the time series and forecasting model. It is utilized internally for forecasting an open-source algorithm developed by Facebook, used internally by the company for forecasting [18].

3. Methodology

A combination of data analysis and interview were applied for data collection. Mr. Andes, the former Local Disaster Risk Reduction Management Council (LDRRMC) officer and Ms. Siwa the coordinator of Municipal Social Welfare and Development Office were interviewed about the policies and procedure of needs assessment and allocation for pre and during the disaster. They provided an adequate amount of information on how the framework can be develop by giving some requirements and the possible agencies that can be use its benefits. The researchers used a Document Analysis in able to know the duties and responsibilities of government sectors that are connected in the evacuee needs management of every municipality in the province of Albay. Public government files were also used in knowing the responsibilities of each sector related to evacuation and its importance to the community.

The conceptual Framework was designed by considering the issues encountered by the Local Government Unit, Barangay, MSWD, Evacuee and the DRRM council. GPS and RFID were also considered in the design of the evacuee needs allocation framework. Also, it considers the algorithm and predicting needs preparation and allocation.

4. Results and Findings

4.1. Standard allocated evacuee needs per household

Table 1 show the standard relief package distributed by the Department of Social Welfare and Development (DSWD). The average family food pack provided by the DSWD to the victims of calamity contains six (6) kilos of rice, eight sachets of coffee, eight (8) packs of instant noodles, and six (6) canned goods (three (3) sardines, and three (3) corned beef). It is meant to last a family of five (5) for two (2) to three (3) days. The DSWD has introduced new relief packages being distributed to affected families. The said relief goods is good for a family of five (5) for two (2) days. It should be sealed when distributed [19].

Table 1. Evacuee Relief Goods Allocation.

Needs Description	Amount/Qty
Rice	6 kg
Coffee/ Cereal Drink	6 sachet
Corned Beef	4 tin
Sardines	4 tin

4.2. Pre-emptive Evacuation Management

The Barangay person in-charge identifies the families who are in need during the hazard/ calamity and create a Risk Analysis of Population report according to the type of hazard/calamity and submit it to the Municipal DRRM. The Municipal DRRM person in-charge identifies the pre-needs of the reported resident for evacuee planning. It also prepares a pre-emptive evacuation report based on the information submitted by the Barangay DRRM. The MSWD person in-charge receive the pre-emptive evacuation report contains of the total family members by type of hazard. The Barangay person in-charge provides the total list of resident at risk in every Barangay according to hazard. The Municipal DRRM person in-charge summarizes the total evacuee according to age and gender per hazard or calamity. The MSWD person in-charge will identify the list family risk to calamity or hazard and target evacuation center [20].

4.3. Factors influencing in the design of digital evacuee needs management

In mid 2013, the Department of Social Welfare and Development (DSWD) wound up trapped in an extremely twirl of debates identifying with the repacking and conveyance of alleviation stock. A few news organizations supposed that help stock from land were repacked utilizing DSWD's legitimate packs. There was conjointly a Commission on Audit (COA) report expressing that 7,527 family food bundles estimation of Php 2,784,900,19, 172 canned merchandise, 81 packs of noodles, 21 sacks of rice, in addition to the 76,300 different canned products were not reasonable for human utilization because of "inappropriate handling" [22].

4.4. The Proposed Evacuee Needs Management Framework

The figure shows the proposed Evacuee needs allocation framework for the province of Albay. This framework has direct linkages to the Local Government Agencies that concerns the Evacuations Needs Management. The Evacuee needs allocation system has 3 major key roles such as the BDRRMC, MDRRMC and the MSWD. It also shows the function of an RFID and GPS devices that helps the local government in monitoring the evacuee needs distribution during the calamity.

The role of the Barangay Disaster Risk Reduction Management Council (BDRRMC) is to collect the information from the residents who are at risk in calamity/disaster such as flood, typhoon, earthquake, landslide and tsunami. The BDRRMC are the in-charge in identifying who are at risk in every Barangay. The officer in-charge will encode the information conducted from every Barangay to the computerized system.

The role of the Municipal Disaster Risk Reduction Management Council is to consolidate all the information (residents who are at risk) taken from the Barangay covered by the municipality. The municipal office will generate a pre-emptive report based on the submitted information by the BDRRMC. The council may use the computerized system in a decision making in generating the Annual Disaster Risk Reduction Management budget for the needs allocation in the evacuation centers.

The role of Provincial Risk Reduction Management Council is to monitor and support the needs of every municipality. The PDRRMC office will monitor the distribution of needs in all the evacuations centers. The council may generate a statistical graph of the needs allocation using the proposed computerized system.

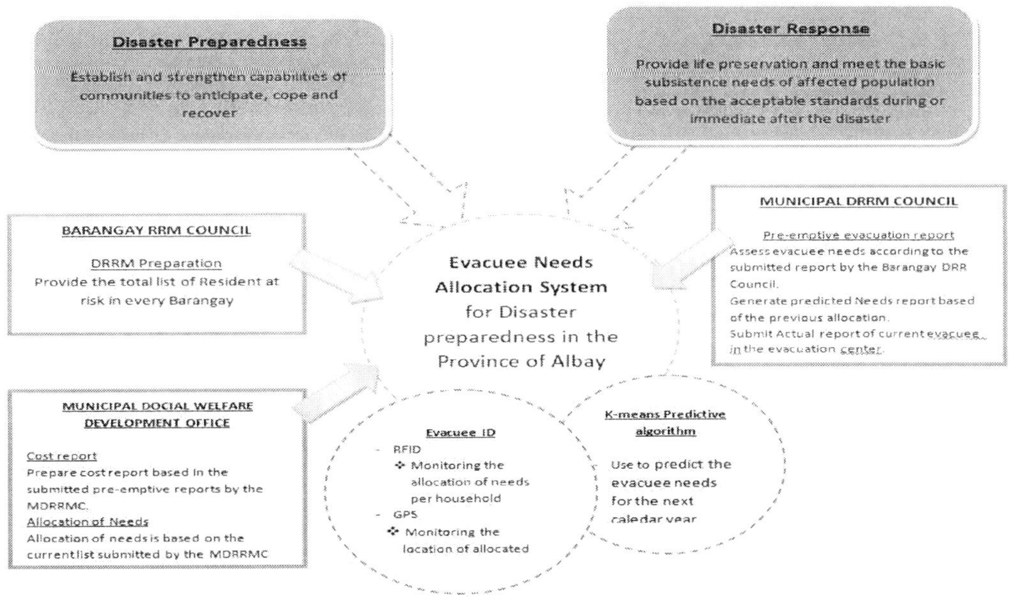

Figure 5. Framework for predicting Evacuee needs allocation for Disaster Preparedness

The role of Municipal Social Welfare Development is to generate a generated report submitted by the MDRRM. The report will be posted electronically and will access in the proposed computerized system. During the calamity, the officer in-charge in every evacuation centre will verify the information of evacuee through the use of Evacuee ID (integrated RFID) for a fast allocation of needs.

The GPS device refers to automatic monitoring of activities in the evacuation center. The GPS will be attached in the computing machine (Computer or laptop) of the officer in charge in the allocation of needs in the center. The MDRRMC can verify the actual GPS location of the machine to the data encoded to the computerized system. All the information will be send to the office of MSWD and MDRRC that the needs are allocated properly.

The RFID device refers to minimizing or lessening the multiple inputs of evacuees in monitoring the casualties during the disaster. Every Evacuee Family head will be given an Evacuee ID with integrated RFID. The Evacuee ID will be use in monitoring the attendance of each family and family member in the evacuation centers. Also, the ID will use in an accurate monitoring and allocating of needs in the centers.

The predictive k-means algorithm is a machine learning technique that will be applied in the proposed system. It will give an automatic prediction of needs allocation for the next planning of Disaster Risk Reduction Management budget in every municipality for the next year and it will lessen the time in preparing of reports.

4.5. The Findings

On the framework and data presented, the GPS and RFID digitals tools will help the monitoring of evacuees and needs allocated during the calamity. The use Evacuee ID with integrated RFID will help the DRRM to monitor the accurate distribution of needs in the evacuation centers and by the use of GPS technology the local government unit can easily monitor the location of distributed goods. The k-means predictive algorithm will help the Provincial DRRM to easily predict the evacuee needs for the pre-emptive report of DRRM in the DRRM planning for the next Calendar year.

5. Conclusion

The researchers have presented the Evacuee needs management framework of the study. The Framework will guide to the Provincial Disaster Risk Reduction Management (PDRRM) to enhance the process of the preparation and allocation of needs in the province of Albay. Using the RFID attached in the Evacuee ID, the LGU will easily monitor the distribution and Evacuee information of the Province.

The researchers recommend developing the evacuee needs management system to implement and lessen the process of allocation of evacuee needs in every evacuation centers in every calamity. Also, recommends testing of the usability to check its accuracy and efficiency of the system with algorithm used in the prediction.

References

[1] World Economic Forum 2018 *The Global Risks Report 2018: 3rd Edition* Retrieved on 30 May 2020 from http://www3.weforum.org/docs/WEF_GRR18_Report.pdf
[2] GFDRR 2017 *GFDRR: Philippines* Retrieved on 25 May 2020 from https://www.gfdrr.org/en/philippines
[3] CFE-DM 2018 *Philippines: Disaster Management Reference Handbook* Retrieved on 25 May 2020 from https://reliefweb.int/sites/reliefweb.int/files/resources/Philippines_2018-0318.pdf
[4] Jha S, Quising P F, Ardaniel Z, Martinez A J and Wang L W 2018 *Natural Disasters, Public Spending, and Creative Destruction: A Case Study of the Philippines* Retrieved on 25 May 2020 from https://www.adb.org/sites/default/files/publication/408351/adbi-wp817.pdf
[5] Gonzales J 2018 *Areas in the Philippines Most Affected by Natural Disasters* Retrieved on 25 May 2020 from https://daydreaminginparadise.com/areas-in-the-philippines-most-affected-by-natural-disasters/
[6] Guleria S 2020 Sustainable development goals (SDGs) and risks to coastal communities *Development in Coastal Zones and Disaster Management - Disaster Research and Management Series on the Global South* ed A Singh *et al* (Singapore: Palgrave Macmillan) DOI: 10.1007/978-981-15-4294-7_11
[7] Naz G A, Malonzo A G and Salvador J B 2019 Zero casualty: the disaster risk reduction and management strategy of Albay Province, Philippines *Advance Social Science and Humanities* DOI: 10.31124/advance.8080748.v1
[8] NSW health organization 2018 *Major Evacuation Centres: Public Health Considerations* Retrieved on 20 April 2020 from https://www1.health.nsw.gov.au/pds/ActivePDSDocuments/GL2018_002.pdf
[9] Philippine Information Agency 2010 *Philippines: Albay Governor Gathers Stakeholders to Mainstream RP's Climate Change Programme* Retrieved on 20 April 2020 from https://www.preventionweb.net/news/view/16339
[10] Madu C N and Chu-Hua Kuei 2017 *Handbook of Disaster Risk Reduction & Management World* (New Jersey: World Scientific Pub Co Pte Lt) DOI: 10.1142/10392
[11] Cayamanda K and Paunlagui M 2020 A participatory management model for flood-risk communication (PMMFRC) in Davao City, Philippines *European J. of Business and Management Research* vol **5** no 4 DOI: 10.24018/ejbmr.2020.5.4.451
[12] Shiwaku K and Shaw R 2016 Introduction: disaster risk reduction and education system *Disaster Risk Reduction (Methods, Approaches and Practices)* ed K Shiwaku *et al* (Tokyo: Springer) DOI: 10.1007/978-4-431-55982-5_1
[13] UN Office for Disaster Risk Reduction 2019 *Disaster Risk Reduction in the Philippines* Retrieved on 25 May 2020 from https://www.unisdr.org/files/68265_682308philippines drmstatusreport.pdf
[14] Republic of Palau 2016 *National Disaster Risk Management Framework 2010 (amended 2016)* Retrieved on 25 May 2020 from https://bsrp.gsd.spc.int/wp-content/uploads/2016/09/palau-national-disaster-rsk-framework-for-web.pdf

[15] United Nations 2015 *Sendai Framework for Disaster Risk Reduction 2015-2030* Retrieved on 25-05-2020 from https://www.preventionweb.net/files/43291_sendaiframeworkfordrren.pdf

[16] Jankowska M, Schipperijn J and Kerr J 2015 A Framework for using GPS data in physical activity and sedentary behavior studies *Exerc. Sport Sci.* Rev. Jan 2015 **43(1)** pp 48–56

[17] Gupta G, Kundu T and Appanna A 2015 An integrated conceptual framework for RFID enabled healthcare *Independent J. of Management & Production* vol **6** no 4 issue October-December 2015 pp 946–61 https://www.redalyc.org/pdf/4495/449544332007.pdf

[18] PAT Research 2020 *Predictive Analytics Tools* Retrieved on 25 May 2020 from https://www.predictiveanalyticstoday.com/predictive-analytics-tools/

[19] Department of Social Welfare and Development 2013 *Average contents of a DSWD Family Food Pack* Retrieved on 25 May 2020 from https://www.officialgazette.gov.ph/2013/11/12/average-contents-of-a-dswd-family-food-pack/pdf

[20] Zimmermann M and Keiler M 2015 International frameworks for disaster risk reduction: useful guidance for sustainable mountain development *Mountain Research and Development* **35(2)** pp 195–202

[21] Monreal R and Palaoag T 2019 Evacuee pre-needs assessment using decision tree algorithm of MDRRMC-Daraga records *Intelligent Systems and Applications - IntelliSys 2018 - Advances in Intelligent Systems and Computing* vol 869 ed K Arai *et al* (Cham: Springer) DOI: 10.1007/978-3-030-01057-7_5

[22] Cabigao F 2015 *Relief Protocols and Rules* Retrieved on 25 May 2020 from https://pcij.org/article/1686/relief-protocols-rules

Bilingual Lexicon Approach to English-Filipino Sentiment Analysis of Teaching Performance

Caren Pacol[1] **and Thelma Palaoag**[2]

[1]Department of Information Technology, Pangasinan State University, Urdaneta City, 2428, Philippines
[2]Department of Computer Science, University of the Cordilleras, Baguio City, 2600, Philippines

E-mail: [1]ambat_caren@yahoo.com

Abstract. The aim of this study is to formulate a strategy that can possibly calculate teacher performance by analyzing textual feedback. Expressing textual responses in quantitative form like average sentiment rating can actually provide opportunities for administrators to see if the numerical ratings given complement that of the comments. Our approach was designed to enable processing bilingual textual data. Findings of this study shows that there is strong correlation between teaching performance actual mean rating and average sentiment rating. Furthermore, the approach employed obtained 86% accuracy indicating that it is an encouraging technique, capable of analyzing the students' textual responses. In future work, the use of POS tagging can be explored to improve sentiment analysis accuracy. Employing machine learning methods may also be considered to discover techniques and alternative approaches to sentiment classification.

1. Introduction

Teaching performance assessment is important to ensure student learning and provide basis for administrative decision making. A common practice to find out the level of teaching competency is the assessment by students. Assessment is conventionally done using survey questionnaire in each course and is composed of queries about learning undertakings.

The assessment is usually conducted by letting the student answer pool of questions using a Likert scale. The aggregated numerical ratings are used as basis to determine the teaching performance of the concerned faculty member [1]. However, assessment forms include sections for comments allowing students to freely write their observations about their teachers. The observations written on this section are usually not processed in the performance assessment due to the absence of automatic text analyzers [1, 2]. These textual feedbacks could possibly hold meaningful perceptions regarding the course, teacher proficiency, punctuality and classroom management. When analyzed effectively, these data can help administrators and the teacher reflect on what should be done to address students' issues on their teachers. This is something that is difficult to quickly derive from the Likert scale-based feedback.

Metric based results of teaching performance evaluation are highly used to represent the quality of teaching, but it is difficult to draw interpretations from them. Though text feedbacks usually do not have weight in the calculation of teaching performance rating, they can make it easier for the administrator to identify teachers' strengths and weaknesses. These insights can be utilized to plan for faculty development program through trainings and seminars or perform appropriate intermediary action such as coaching and mentoring that is necessary to address crucial issues in teaching performance. Since textual responses are more informative than numerical ratings, then expressing them in quantitative form like average sentiment rating can actually provide opportunities for administrators to see if the numerical ratings given complement that of the comments. Hence, our aim is to formulate a strategy that can possibly calculate teacher performance rating based on textual responses. This is where natural language processing come into play. Existing data mining technologies with text analytics tools can be used for natural language processing. One of the applications of natural language processing is sentiment analysis.

A person's viewpoint conveying thoughts, attitudes and emotion is referred to as sentiment. Analyzing a group of words purposively to grasp its impression is the main focus of sentiment analysis. Sentiments are represented as either positive or negative, referred to as polarity. Looking at the sign of the polarity score, the overall sentiment is concluded as positive if it has plus (+) sign, negative if it has minus (-) sign. Students' textual feedback in the student-faculty evaluation can be categorized as positive, negative or neutral. Thus, sentiment analysis is the appropriate approach to use in this study.

Lexicon-based sentiment analysis extract all sentiment terms for any given text and assign their sentiment value using sentiment lexicon [3]. Development of a broader scope lexicon and efficient handling of context-sensitive terms, negations and intensifiers can lead to an effective sentiment analyzer. [4]. Many sentiment analyzers that have been created were designed to evaluate single language sentences. Nevertheless, not in all cases the documents that need to be processed are expressed in single language only, especially in areas where people are multilingual. It is a common practice in sentiment analysis that sentences expressed in a different language are translated. However, translation requires maintaining the exact meaning of a sentence after translation, which is a difficult task for an automatic translator and even more difficult and time consuming when done manually. Relevant to this, we devised a lexicon of English and Filipino terms that is appropriate in the context of evaluating teaching performance in a bilingual setting. The basis of this is the similarity between the two languages such as both are written from left to right and they have similar parts of speech. This lexicon can be used to analyze students' comments based on polarity and sentiment scores.

Analyzing students' feedback through sentiment scoring allows measuring how intense a sentiment is instead of classifying based on positive or negative labels only. In addition, this can efficiently handle combination of positive and negative phrases in sentences. The sentiment score could either have plus (+) or minus (-) sign depending on which of the positive or negative terms has greater weight.

Specific objectives of this study are: (a) automatically calculate the mean sentiment score per faculty member (b) determine if there is correlation between the mean sentiment score in a dataset (pertaining to the comments given on one faculty member) and the actual mean rating, and (c) validate sentiment scores.

2. Related Literature

To further establish the relevance of our study, we decided to present some of lexicon-based sentiment analysis research conducted that we have reviewed.

Haider, Ghani and Rajput [1] studied on a sentiment analysis approach identifying comments with positive, negative or neutral polarity. A dictionary of sentiment terms and their labels was used. An accuracy of 91.2% was achieved in their approach. Analyzing their calculations, they found that there is correlation between sentiment scores and likert scale-based scores. Similar to [1], we used lexicon-based approach in our study. However, the difference of our strategy is that we designed our approach to process bilingual sentences incorporating valence shifters in sentiment analysis. In another study, Aung [5] designed a lexicon-based sentiment analysis framework predicting the proficiency of the teacher based on students' text feedback. Balahadia, Juanatas and Fernando [6] also utilized a database of words in sentiment analysis in teacher evaluation. In the study by Rajput et.al, there were few misclassified feedbacks which can probably be attributed to the presence of negations in the sentences. In [5] and [6], their study included checking of negation wherein certain amount is added to or deducted from the sentiment score. Both [5] and [6] handled negations in their approaches but did not mention handling of amplifiers, de-amplifiers and adversative conjunctions. In our study, we used function that automatically calculates sentiment scoring based on a modified lexicon of sentiment words and valence shifters.

Shaikh, Meghji and Taj [7], worked on news articles and demonstrated a lexicon-based sentiment analysis approach. When news articles have a score of +1, they were classified as positive while those yielding a score of -1 were considered negative. News articles having score of 0 were classified as neutral. An operator named "extract sentiment" was utilized to compute the sentiment rating of a full news article. This operator utilizes a SentiWordNet 3.0.0 lexicon extension which is a wordNet lexicon extension. Moreover, calculation of total sentiment rating of news article was generated using Score sentiment function.

In the paper written by Fazal Masud Kundi et.al [8], they presented a lexicon-based sentiment classification framework distinguishing and scoring slangs found in tweets. Polarity of tweets were identified as neutral, positive or negative. Comparing its results with current systems, their framework achieved higher accuracy of 92% in two-class cases and 87% in multi-class cases.

Reddy et.al [9] used the available Telugu SentiWordNet to perform sentiment analysis for Telugu eNewspapers sentences. They evaluated their proposed approach on sentiment analysis and obtained 74% accuracy on subjectivity classification and 81% on sentiment classification in news data.

The similarity of our study with Taj et.al [7], Kundi et.al [8] and Reddy et.al [9] is that it presents a lexicon-based classifier of sentences to neutral, positive or negative. However, instead of just classifying through labels, it calculates sentiment scores and aggregates them to produce average sentiment ratings. Unlike their dataset which are news articles and tweets, the dataset in this study is textual feedback of students on teaching performance. Moreover, it utilized a customized lexicon consisting of English and Filipino terms frequently used in the context of teaching performance assessment.

In [10], a method of determining polarity and calculation for lexicon-based sentiment analysis was devised and evaluated on Turkish tweets utilizing a Twitter API. An accuracy of 88.2% was achieved. To find the correct meaning of each tweet, they analyzed the data in the following levels: first is by word, second is by word group and third is by idiom/proverb.

The study in [10] employed lexicon-based approach on purely Turkish tweets. On the other hand, our study applied lexicon-based approach on English, Filipino and mixed English-Filipino sentences.

3. Methodology

The methodology is divided into the following steps.

3.1. Data Gathering

Teaching performance data were sourced from nine (9) campuses of Pangasinan State University (PSU). The data were taken from summary results of faculty teaching performance evaluation questionnaires of 367 faculty members. At present, PSU utilizes a Faculty Teaching Performance Evaluation composed of four (4) areas of evaluation namely commitment, knowledge of subject matter, teaching for independent learning and management of learning. Students rate their faculty from 1 to 5 (5 is the maximum and 1 is the minimum) based on main criteria in each area. A section for narrative comments labelled Strengths and Weaknesses is in the latter portion of the evaluation form. The results are consolidated calculating ratings in each area and the overall mean rating of each faculty member. Comments on strengths and weaknesses are also listed for individual teachers.

3.2. Data Preparation

After data were gathered, we summarized the average rating and comments corresponding to every faculty ID. Faculty members' data were randomly selected based on the teaching performance actual mean rating. Then, comments were segregated by faculty ID to generate dataset for our experiment. The data were manually cleaned removing sentences with no clear meaning and correcting misspelled words.

Some comments were written in pure English while others were written in pure Filipino. There were also sentences that were written in mixed English-Filipino. Sample comments of students from the dataset were shown in Table 1.

The function utilized in the experiments uses a generic English-language based sentiment lexicon. Thus, we modified the sentiment lexicon to suit the context of evaluating teaching performance. More English sentiment terms had to be added and sentiment scores were modified to suit the setting. For instance, the words 'soft' and 'fast' have positive sentiment scores in the generic lexicon but have negative connotation in the context of teaching performance. We initially devised a separate Filipino sentiment lexicon using a generic dictionary. Then we added more Filipino terms that commonly occurred in students' comments in the dataset. The function we used performs sentiment analysis at the sentence level, hence, we combined the Filipino and English sentiment lexicons. This enabled the function to analyze even the sentences written in pure Filipino as well as those written in mixed English-Filipino. The valence shifters lexicon was also modified to include Filipino terms.

Basic sentiment analysis algorithms calculate a final aggregated score by counting how many times positive and negative words occur and getting their weight or label if they occur in the sentiment dictionary. This method leads to numerous false positives when negations are not handled properly. In addition, adversative conjunctions, amplifiers and de-amplifiers have significant impact in the final sentiment score but are not accounted accordingly. Our approach simplified the procedure by using pre-defined functions that integrate English and Filipino valence shifters within a few lines of code.

To prepare our dataset for the validation of the sentiment scores, we labelled the data with positive (1), negative (-1) and neutral (0). Also, we asked two teachers to label the sentences separately. Then, we finalized the labels based on majority rule.

Table 1. Samples of Students' Comments.

	Sample students' comments	Language used
1	He helps the students to understand the lessons.	English
2	He explain but very few explanations.	English
3	Lack of patience sometimes and very strict in our quizzes and exams.	English
4	Magaling magturo, maunawain.	Filipino
5	Magaling siya sa explanation and also she has a good attitude.	Mixed
6	Nagtuturo ng mabuti and explains the lesson.	Mixed
7	If the student is not learning the topic, medyo nag-iinit yung ulo niya.	Mixed
8	Lagi siyang absent at kapag pumapasok hindi siya dumadating sa tamang oras.	Mixed
9	Masyadong maraming requirements na ibinibigay sa amin.	Mixed
10	Binibigay niya ang best niya matutunan talaga namin iyong topic na tinututro niya.	Mixed

3.3. Sentence Level Sentiment Analysis Using Sentimentr

R was used for the sentiment analysis process. First, *sentimentr* package was loaded in R. Then, data were extracted from csv file and loaded into R. The *get_sentences()* function partitions the data to individual sentences.

Throughout our experiments, we have seen that there were terms that were not listed in the lexicon which caused misclassification of the sentences. Thus, these terms were included.

Next, we utilized the *sentiment_by()* function to get mean sentiment score for a given sentence. By default, sentiment terms and valence shifters lexicons that are pre-defined in *sentimentr* package are the parameters passed to *sentiment_by()* function. A challenge we faced in the experiment is enabling the function to accept parameters that are sourced from outside the package. Both the customized sentiment terms and valence shifters lexicons were saved as csv files. A data frame with four attributes is returned by the function. An attribute labelled *ave_sentiment* is most interesting as this represents the sentiment of the sentences in one row. The resulting value could be negative or positive expressing the intensity and the polarity of the sentiment. In some instance the result can be 0 if the sentence contained words not found in the lexicons.

Sentiment scores are calculated by searching and comparing the words in each sentence to a lexicon of polarized words. Positive ($w_{i,j,k}^{+}$) words are marked with a +1 and negative ($w_{i,j,k}^{-}$) words, with a −1 (or other positive/negative weighting).

Words that appear four (4) words prior-to and two (2) words following polarized words are considered valence shifters. When polarized word (pw) is found, words nearby it is examined to check for the presence of valence shifters and the words' polarized context cluster ($c_{i,j,l}$) is drawn. The cluster can be represented as equation 1:

$$(c_{i,j,1} = \{pwi, j, k - nb, \ldots, pwi, j, k, \ldots, p^{w}{}_{i,j,k-na}\}) \quad (1)$$

In equation 1, nb and *n**a* variables correspond to the n.before and n.after that the user assigned. The words in this polarized context cluster are marked neutral ($w_{i,j,k}^{0}$), negator ($w_{i,j,k}^{n}$), amplifier ($w_{i,j,k}^{a}$), or de-amplifier ($w_{i,j,k}^{d}$). Neutral words are not given value in the equation. When polarized word is found, it is weighted (w) by getting its corresponding weight in the *polarity_dt* argument. Then, the function further weights it based on the count of valence shifters that directly surround the polarized word *(pw)*. Punctuation locations including colons, semicolons and commas are included in the calculation of the polarized context cluster's lower and upper limit. The reason for this is pause punctuations specify a shift in thought and words that appear before these marks are not necessarily related with words that appear after these punctuations.

The polarized word is the main value in the cluster and is altered depending on the presence of valence shifters. When amplifiers are found, the polarity is increased by 1.8. Amplifiers are treated de-amplifiers when an odd number of negators is found in the context cluster. When a de-amplifier is found, it decreases the polarity. When negations are found with amplifiers or de-amplifiers, the sign of the polarized word is reversed. To calculate negation, -1 is raised to the power of the number of negators ($w_{i,j,k}^{n}$) and a value of 2 is added. The basis of this is the notion that the presence of even numbered negatives is equivalent to a positive and odd number negatives is equivalent to a negative.

Adversative conjunctions such as 'nevertheless', 'though' and 'but' likewise changes the context cluster. When found before the polarized word, it up-weights the cluster by equation 2:

$$c_{i,j,1}\ \text{up-weight} = 1 + z_2 * \{|^{w}{}_{\text{adversative conjunction}}|, \ldots, {}^{w}{}_{i,j,k}{}^{p}\} \quad (2)$$

where the default weight (z_2) is 0.85 and $|^{w}{}_{\text{adversative conjunction}}|$ are the count of adversative conjunctions that appear before the polarized word. When found after a polarized word, adversative conjunction down-weights the cluster by equation 3:

$$c_{i,j,1}\text{down-weight} = 1 + \{w_{i,j,k}{}^{p}\ , \ldots, |^{w}{}_{\text{adversative conjunction}}| * - 1\} * z_2 \quad (3)$$

The basis of this is the idea that when an adversative conjunction is found, greater values is placed on the succeeding phrases while the value given on previous phrases are lowered.

Amplifiers and de-amplifiers were set to 0.8. Lower limit of weight for de- amplifiers is -1. The sum ($c'_{i,j}$) of these weighted context clusters ($c_{i,j,1}$) is computed. Next, the square root of the word count is calculated. Finally, the sum ($c'_{i,j}$) is divided by the square root as shown in equation 4:

$$\delta i, j* = (c'_{i,j}) / (\sqrt{}_{wi,*j**n}) \quad (4)$$

This yields an unbounded polarity score ($\delta i, j*$) for each sentence. To get the mean of all sentences ($s_{i,j}$) within a paragraph (p_i), simply take the average sentiment score in equation 5:

$$p_i, \delta_{i,j} = 1/n \cdot \sum \delta_{i,j} \quad (5)$$

or utilize the default *average_weighted_mixed_sentiment* upweighting the negative values in a vector while also downweighting the zeros in a vector or *average_downweighted_zero* which simply downweights the zero polarity scores [11]. Summary statistics of the calculated sentiment scores were also observed.

One more function that was used is *highlight()*, combined with *sentiment_by()*. Combining these two functions generated an output in html highlighting sentences with green and red color. A green color denotes positive polarity while a red color denotes negative polarity. Sentences which were not highlighted are analyzed as neutral by sentimentr. We used these functions to easily observe misclassified sentences in the dataset.

3.4. Determining Correlation of Mean Sentiment Score and Actual Mean Rating

The actual mean ratings and calculated mean sentiment scores were consolidated in a csv file. We used *ggscatter()* function under *ggpubr* library in R to compute the Pearson's correlation coefficient and the significance level(or p-value) of the correlation. Pearson's correlation coefficient evaluates the statistical relationship of two continuous variables [12]. The function also generated a scatter plot illustrating the correlation of the two variables.

3.5. Validating the Sentiment Scores

Cross-tabulation of actual and predicted classes were generated using *confusionMatrix()* function. The overall accuracy and unweighted Kappa measurement were determined. Kappa measurement describes the true and predicted values' level of agreement [13]. Utilizing *mcnemar.test*, a p-value is calculated. Calculation of the overall accuracy rate was done using confidence interval of 95% for this rate and a one-sided test to check whether the accuracy is superior compared to the "no information rate", the percentage of the largest class in the dataset.

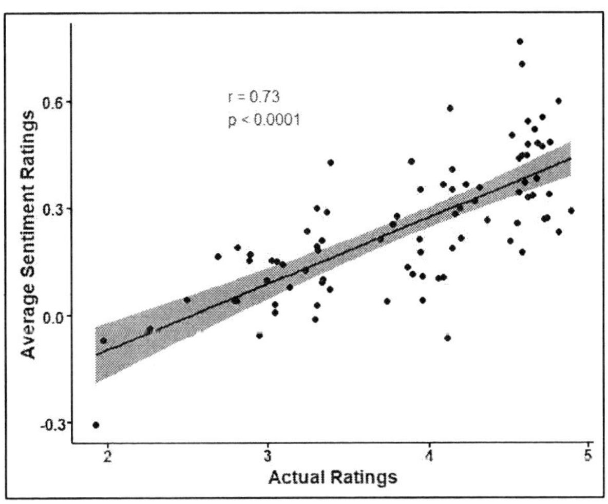

Figure 1. Scatterplot graph of actual mean ratings and average sentiment ratings.

4. Results and Discussions

This section presents and discusses the findings in our experiments. Average sentiment ratings were automatically computed for each faculty. Utilizing the data of 85 teachers, the Pearson's correlation coefficient that came out from the computation is 0.73.

To determine the degree of correlation, when the coefficient value is close ± 1, at that point the correlation is considered perfect. This indicates that as one variable increments, the other variable will in general additionally increment (if positive) or decrease (if negative). A coefficient value within ± 0.50 and ± 1 means the correlation is strong. A coefficient value within ± 0.30 and ± 0.49 means the correlation is medium. A coefficient value underneath ± 0.29 means the correlation is small. If the value equals zero, then, there is no correlation [12].

This signifies that obtaining 0.73 of correlation coefficient, which is within ± 0.50 and ± 1, there is strong correlation between actual mean rating and average sentiment rating. These findings support the results found by Rajput, Haider, and Ghani [1] that there is positive correlation between actual mean rating and sentiment rating. The linear relationship between the two variables is illustrated in Figure 1.

In this case, there are three classes used to classify the sentences namely positive (1), negative (-1) and neutral (0). A total of 1669 sentences were analyzed in this study. Based on the analysis of the confusion matrix, 900 sentences were correctly classified as positive sentiments out of 947 True Positive sentences. Meanwhile, 518 sentences were correctly classified as negative comments out of 677 True Negative comments. Concurrently, among 45 True Neutral sentiments only 20 sentences were correctly classified.

The approach presented in this paper obtained 0.86 accuracy indicating that it is approximately 86% accurate using a confidence interval of 95% ranging from 0.8441 to 0.8778. The no-information rate of 0.5674 means approximately 57% accuracy can be achieved by always choosing the majority class label. The accuracy obtained is far higher than the no information rate indicating that our classifier is significantly better in terms of performance compared to the no-information rate. This is also showed by the p-value. The *Mcnemar's test* P-value describes if there is significant difference between true and predicted values [14].

Table 2. Overall Statistics.

Accuracy	0.8616		
95% CI	(0.8441, 0.8778)		
No Information Rate :	0.5674		
P-Value [Acc > NIR]	< 2.2e-16		
Kappa	0.7313		
Mcnemar's Test P-Value	< 2.2e-16		
Statistics by Class:			
	Class: -1	Class: 0	Class: 1
Sensitivity	0.7651	0.44444	0.9504
Specificity	0.9708	0.95628	0.8186
Pos Pred Value	0.9470	0.21978	0.8729
Neg Pred Value	0.8583	0.98416	0.9263
Prevalence	0.4056	0.02696	0.5674
Detection Rate	0.3104	0.01198	0.5392
Detection Prevalence	0.3277	0.05452	0.6177
Balanced Accuracy	0.8680	0.70036	0.8845

A Kappa score within 0 and 0.20 is considered slight agreement. If within 0.21 and 0.40, it is fair agreement. A value within 0.41 and 0.60 is considered moderate. If within 0.61 and 0.80 it is substantial.

A value within 0.81 and 1.0 is perfect agreement [13]. Since the kappa score of 0.73 is within 0.61 and 0.80, this indicates that there is a substantial level of agreement between the two values.

Predicting positive sentiments is necessary to see best practices of teachers. More so, predicting negative sentiments is crucial in this case since this require intervention measures to address the concern of students. Thus, higher value of sensitivity in class -1 (pertaining to negative sentences) is desired. Sensitivity in this context is the actual negative sentences that got predicted correctly. The overall statistics shown in Table 2 indicates that there is lower sensitivity of 0.77 for class -1 (negative sentences) as compared to 0.95 of class 1 (positive sentences). Similarly, higher positive predictive value (PPV) in class -1 is desired. Positive predictive value in this case is the confidence in the classifier that when it predicts a positive result for a class, it has the capability to distinguish successfully which belong to the class and which do not.

Specificity of class -1 (Negative) in this context is the actual not negative sentences that got predicted correctly. Table 2 indicates a higher specificity of 0.97 in class -1 (Negative) as compared to 0.82 in class 1 (Positive). This means more not negative sentences were predicted correctly compared to not positive sentences that were predicted correctly.

Furthermore, Table 3 shows the breakdown of samples for English, Filipino and mixed sentences. Classification accuracy for English sentences got 87% and mixed English-Filipino followed at 85%. Though there is uneven distribution of samples, it can be implied that the approach has greater difficulty in analyzing Filipino sentences.

It was observed that most of the incorrectly classified English statements has words that belong to more than one parts of speech. For instance, the word 'needs' can be a noun or verb. Depending on how it is used, this alter the polarity of a sentence that has the word. Thus, in future research, integration of parts of speech tagging can be explored to see if it can remedy this problem. More terms, which are often used in academic environment, can also be added to the sentiment lexicon to improve the accuracy.

Table 3. Breakdown of Samples for English, Filipino and Mixed Sentences.

Language	Total Number of Sentences	Correctly Classified
English	1340	1167
Filipino	282	231
Mixed	47	40

5. Conclusion

In this paper, we presented a bilingual lexicon-based strategy for analyzing students' textual responses utilizing sentence level sentiment analysis in R. Automatic computation of sentiment polarity scores was done utilizing *sentimentr*. Results show that the Pearson coefficient is 0.73 which implies that there is strong correlation between real mean rating and average sentiment rating. From these discoveries, it very well may be said that as teaching performance mean rating increments, the average sentiment rating likewise increments. This demonstrates analysis of students' comments utilizing the lexicon-based sentiment analysis technique can possibly be applied to evaluate teaching performance. Our approach got 86% accuracy. From these, it very well may be said that the methodology is encouraging as it is time-efficient and capable of analyzing the students' textual responses as compared to manual listing. With this methodology, textual responses of students can be expressed to quantitative information.

In future work, the use of POS tagging can be explored to improve sentiment analysis accuracy. Employing machine learning methods may also be considered to discover techniques and alternative approaches to sentiment classification.

References

[1] Rajput Q, Haider S and Ghani S 2016 Lexicon-based sentiment analysis of teachers' evaluation *Applied Computational Intelligence and Soft Computing* Volume **2016** (London: Hindawi Publishing Corporation)

[2] Wook M, Razali N, Ramli S, Wahab N, Hasbullah N, Zainudin N and Talib M 2019 Opinion mining technique for developing student feedback analysis system using lexicon-based approach *Educ. Inf. Technol.* **25** pp 2549–60

[3] Rintyarna B. S, Sarno R and Fatichah, C 2019 Evaluating the performance of sentence level features and domain sensitive features of product reviews on supervised sentiment analysis tasks *J. Big Data* **6** p 84

[4] Mukhtar N and Khan M. A 2019 Effective lexicon-based approach for Urdu sentiment analysis *Artif. Intell. Rev.* **53** pp 2521–48

[5] Aung K Z 2016 A lexicon based sentiment analyzer framework for student-teacher textual comments *Int. J. of Scientific and Research Publications* vol **6** issue 2 February 2016 ISSN 2250-3153 pp 277–80

[6] Balahadia F, Juanatas I and Fernando C 2016 Teacher's performance evaluation tool using opinion mining with sentiment analysis *2016 IEEE Region 10 Symp. (TENSYMP) Bali* pp 95–8 DOI:

10.1109/TENCONSpring.2016.7519384

[7] Taj S, Shaikh B and Meghji A 2019 Sentiment analysis of news articles: a lexicon based approach *2019 Int. Conf. on Computing, Mathematics and Eng. Technologies – iCoMET* pp 1-5

[8] Kundi F, Khan A, Ahmad S and Asghar M 2014 Lexicon-based sentiment analysis in the social web *J. Basic. Appl. Sci. Res.* **4(6)** pp 238–48

[9] Reddy N, Kumar B.S, Korra B, Mohapatra S and Kumar R 2017. Sentiment analysis using Telugu sentiWordNet *Int. Conf. on Wireless Communications Signal Processing and Networking (WiSPNET) - SSN College of Engineering, Chennai, India - 22-24 March 2017*

[10] Yurtalan G, Koyuncu M and Turhan C 2019 A polarity calculation approach for lexicon-based Turkish sentiment analysis *Turk. J. Elec. Eng. and Comp. Sci.* **27** pp 1325–39 DOI: 10.3906/elk-1803-92

[11] github.com 2020 *Trinker/sentimentr* Retrieved on 14 March 2020 from https://github.com/trinker/sentimentr

[12] StatisticsSolutions 2020 *Pearson's Correlation Coefficient* Retrieved on 13 March 2020 from https://www.statisticssolutions.com/pearsons-correlation-coefficient/

[13] Chen Y S 2019 *Interpretation of Kappa Values* Retrieved from https://towardsdatascience.com/ interpretation-of-kappa-values-2acd1ca7b18f

[14] Ashcroft M 2016 *The Essentials of Data Analytics and Machine Learning* Retrieved from https://courses.edsa-project.eu/pluginfile.php/1330/mod_resource/content/0/ Module%2010%20- %20Analysis%20of%20Classification%20Models_V1.pdf

IOP Conf. Series: Materials Science and Engineering 1077 (2021) 012045 doi:10.1088/1757-899X/1077/1/012045

Comparison of Main Algorithms in Big Data Analysis of Telecom Customer Retention

Yuanhu Gu[1], Thelma Domingo Palaoag[2] and Josephine S. Dela Cruz[3]

University of the Cordilleras, Baguio City, Philippines

E-mail: [1]tiger18921884431@gmail.com, [2]tpalaoag@gmail.com, [3]delacruzpen@gmail.com

Abstract. In today's fierce telecommunications market competition, customer churn is very severe. In order to retain customers, telecommunications companies have made various attempts from various data and consumption characteristics analysis to big data analysis. However, since the actual situation of customer churn is very complicated, how to predict customer churn accurately and quickly is a difficult problem. After the researchers successfully conducted big data analysis of customer churn and successfully retained customers, in this article, the researchers mainly compared several commonly used algorithms in order to find a better algorithm for big data analysis of telecommunications customer churn. Compare and analyze the accuracy and efficiency of these several algorithms and suggest that the business support staffs of telecommunications companies adopt major methods for big data analysis. The researcher found that the Decision Tree (CART) algorithm is better for the prediction of customer churn and guided other branch staffs to predict customer churn and retain customers in a timely manner. This kind of big data analysis can be used to retain customers in the telecommunications industry.

1. Introduction

In the 21st century, the telecom industry has developed rapidly. It took 8 years for China to go from 3G to 4G and only 4 years to 5G. But too fast development has also brought strong competition. When all telecom operators provided similar services, the competition became fierce. Especially in recent years, some telecom companies have found that the number of customers has even shown a negative growth. Obviously, many customers are churned. Customer churn means the loss of customers from their company [1]. It is often difficult to recover the churn customers because these they have already used the services of competitors. With the development of big data analysis, researchers have found that using big data analysis can better judge customer churn called Customer Churn Prediction [2].

Big data analysis has a significant impact on the development of modern society [3]. It is increasingly difficult for telecom companies to get new customers, so maintaining existing customers has become particularly important [4]. Using big data analysis to predict customer churn can help telecom companies know in advance which customers may be churn [5] and take certain preventive measures. At present, churn prediction has become one of the main challenges for telecom companies. [6] It is difficulty in modeling customer churn prediction. Dimensionality reduction is an important aspect of considering algorithm efficiency [7]. However, it is also more complicated to determine the reduced dimension.

Although there are many difficulties in predicting customer churn, we can study what algorithm is more suitable for Customer Churn Prediction and train the research results to more data support personnel to perform daily big data analysis. Customer Churn Prediction has been performed in the literature using various techniques including machine learning, data mining, and so on. First of all, let us review the top 10 data mining algorithms identified by the IEEE International Conference on Data Mining (ICDM) in December 2006: C4.5, k-Means, SVM, Apriori, EM, PageRank, AdaBoost, KNN, Naive Bayes, and CART. These 10 algorithms are one of the most influential data mining algorithms in recent years [8]. One of the commonly used tools in data mining is C4.5, C5.0 and CART. The decision tree algorithm is cost-sensitive, while improving the robustness [9], [10]. According to the literature, these algorithms can be used for big data analysis and decision-making of our telecom customer churn. It strikes a good balance between accuracy and cost. In machine learning, support vector machines (SVM) must be tried [8]. K-nearest neighbor (KNN) classification is a more complex method [11]. KNN is suitable for low-dimensional data, and it can help decrease computational costs [8]. Naive Bayes method is a very important method. A general discussion of the Naive Bayes method and its advantages is given in [12], [13]. The Naive Bayes model is simple, elegant and robust [14], [15]. After referring to many documents, the researchers know that many algorithms need to be studied and will focus on the impact of these algorithms on the performance and accuracy of telecommunication customer churn analysis. The researchers will mainly compare C4.5, K-means, SVM, KNN, Naive Bayes and Cart algorithms, and find more practical algorithms, so that daily analysis becomes fast and accurate. These algorithms should be simple and easy to use so that more colleagues can analyze various telecommunication services.

The researcher is engaged in data analysis in a branch of China Telecom. Our company is mainly engaged in telecommunications-related services, including fixed-line, mobile, broadband and various value-added services. Among them, the mobile phone business has a high churn rate and fierce competition, which is also our key business. Therefore, the analysis of mobile phone customer churn must be timely and accurate to effectively retain customers. There are many algorithms in big data analysis. What algorithm is better to use? This is a problem that puzzles many data analysts. In order to obtain the best algorithm for a certain problem, so as to train and guide the support staffs at all levels of the company, the researchers analyzed and compared several main algorithms, in order to find a more accurate and faster algorithm to guide various business support staffs conduct big data analysis. The researchers collected 221,770 user data through the support of the superior company, of which 11,282 are churn customers, involving 37 features. This article uses the same data to compare the accuracy and operating efficiency of the following algorithms: DecisionTree (C4.5, CART), KNN,

Linear, SVM (SVC, SVR), Naive Bayes (GaussianNB, BernoulliNB), RandomForest, Neural Network. The research analyzed train accuracy, test accuracy, AUC and runtime of each algorithm. Finally, it is found that it is best to use Decision Tree (CART) to predict customer churn. The researcher shared the research results with colleagues in each branch and trained and guided other staffs to predict customer churn and retain customers in a timely manner.

2. Methodology

For the data of telecom customer churn, what algorithm can improve the prediction accuracy and operating efficiency? The researcher used the same data to evaluate the accuracy and efficiency of different algorithms. Several commonly used algorithms hope to be found to guide various business support personnel in big data analysis. This article mainly studies the accuracy and efficiency of several main algorithms. Figure 1 shows the research method.

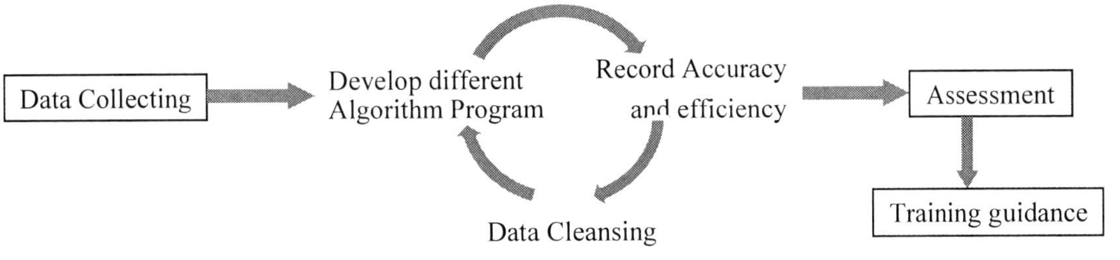

Figure 1. Methodology

First of all, the telecommunications business is very large, with all kinds of data. Data needs to be collected from different departments and platforms. Provincial big data is stored on the Hadoop platform and downloaded to the local ORACLE database every month using ETL tools. The data from the local external department is extracted from the relevant platform through DB link and the program written by JAVA. The data mainly includes user attributes, consumption data, and consumption characteristic data, such as customer age, consumption, online time, online time period, online traffic and other 37 kinds of data.

Secondly, the researcher use python to develop big data analysis program on each algorithm before and after cleaning up abnormal data. The researcher mainly compares the 7 algorithms mentioned in introduction. Record the training accuracy and test accuracy, AUC and running time of each algorithm, and then assess. The efficiency and accuracy of predictions through different algorithms are tested. After repeated predictions with different parameter, evaluations and optimizations, the most accurate and fast algorithm will be found.

The researcher used the data before and after data cleaning to test each algorithm separately. The original data has a lot of null values, and there are data with extremely large values, such as Internet traffic and deposits, which need to be cleaned up. Data cleaning mainly includes cleaning up null values and abnormal values. The purpose of data cleaning is to enhance and maintain the quality of the data [16]. This article also discusses a quick method of clearing abnormal values; simply and easily adjust abnormal data (outliers) to the non-abnormal range.

Finally, the researcher evaluates the prediction results by evaluating the accuracy and efficiency of each algorithm, AUC and running time. In order to improve execution efficiency, the expected test accuracy rate should be at least 80%, the higher the better, the AUC is greater than 0.8, and the shorter the running time, the better. Of course, it also need to consider the complexity of the algorithm and the simplicity of training and use. Through EXECL comparison, after finding the best algorithm, the researcher will train and guide support personnel at all levels to conduct big data analysis in order to react quickly to retain customers.

3. Result and Discussion

3.1. Background

In order to compare the pros and cons of different algorithms for big data analysis of telecom customer churn, the researchers prepared a period of data of churn users with a data volume of 221,770 pieces, which were divided into 161,770 pieces of training data and 60,000 pieces of test data. When the data dimensions are very large, some algorithms will be very difficult to establish an effective model. This is called the "curse of dimensionality" [17]. Feature models are often used to model variability [18]. In this study, the researcher has done a lot of feature analysis and multi-feature analysis based on experience and selects 37 features.

3.2. Data Cleansing

The researcher used the data before and after cleansing to compare the accuracy of each algorithm. There are many out-of-range data in the consumption data of telecom users. These data are called outliers. In order to better analyze big data, outliers should be replaced. In order to better represent the relationship between data sizes, the researchers linearly narrowed the abnormal data to the normal range. The following is batch processing method.

The IQR is from 1 as follows:

$$IQR = Q3 - Q1 \tag{1}$$

Q1 is the first quartile, Q3 is the third quartile, Q3 - Q1 is called the Inter-Quartile Ranger (IQR).

The 1*IQR+Q3 is from 2 as follows:

$$Q5 = IQR + Q3 = 2Q3 - Q1 \tag{2}$$

Q5 is 1*IQR+Q3. The researchers treat data larger than Q5 as outliers and narrow down this type of data to make it within the inner fence. The formula is from 3 as follows:

$$Q = (Q - Q5)/(Qmax - Q5) * IQR/2 + Q5 \tag{3}$$

Qmax is the maximum value of the column. The researchers first replace the null value with an appropriate value, and then convert data beyond Q5 to within 1.5* IQR+Q3. As we all know, 1.5* IQR+Q3 is inner fence.

This is a simple, batch, and fast way to handle outliers. After the researchers' test, it can better improve the prediction accuracy of big data.

3.3. Customer Churn Prediction of Each Algorithm

The researcher used the data before and after data cleansing to test each algorithm separately, and carefully adjust the parameters of some algorithms in order to obtain a better test accuracy and recall rate. Each algorithm is tested and evaluated as below.

3.3.1. Decision Tree C4.5. The researcher first used the simpler C4.5 algorithm, and the training accuracy was 85%, the test accuracy was 83%, the AUC (AUC stands for "Area under the ROC Curve.") was 0.78, and the ROC (Receiver Operating Characteristic) curve is shown in Figure 2.

3.3.2. Decision Tree CART. The researcher again uses the CART algorithm in the decision tree, and the training accuracy is 86%, the test accuracy is 82%, the AUC is 0.81, and the ROC curve is shown in Figure 3. This shows that the C4.5 and CART algorithms of decision tree are relatively similar, and the difference is not very big.

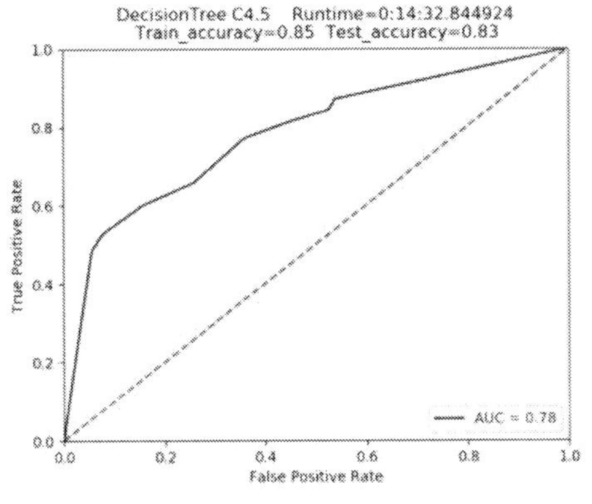

Figure 2. Decision Tree C4.5

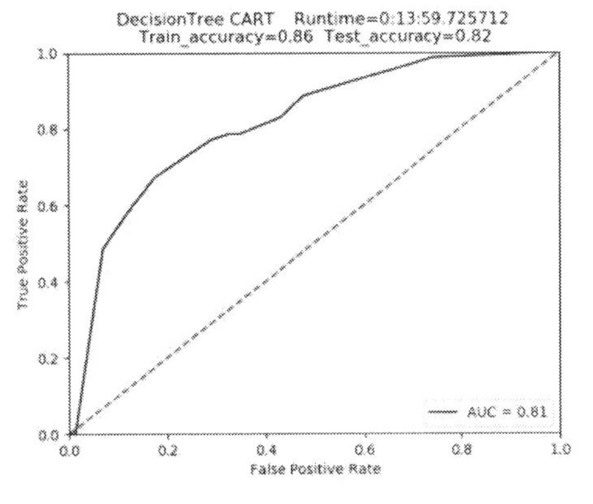

Figure 3. Decision Tree CART

3.3.3. KNN. The researcher continued to use the same data and used KNN to make predictions. The training accuracy rate was 89%, the test accuracy rate was 80%, and the AUC was only 0.72, as shown in Figure 4. This shows that the training situation is good, but the test situation is not ideal. And the AUC is low, indicating that there is over fitting.

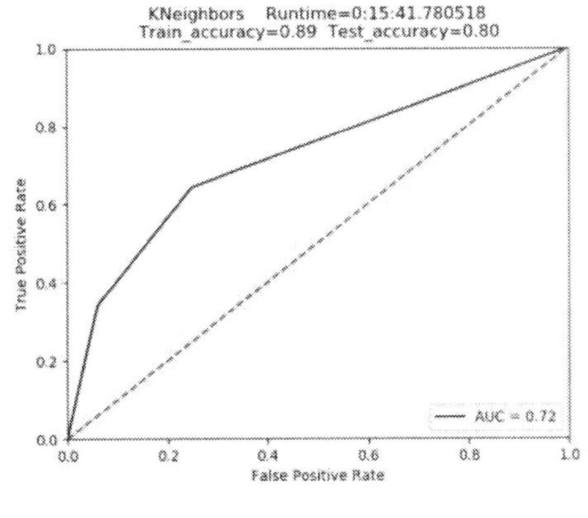

Figure 4. KNN

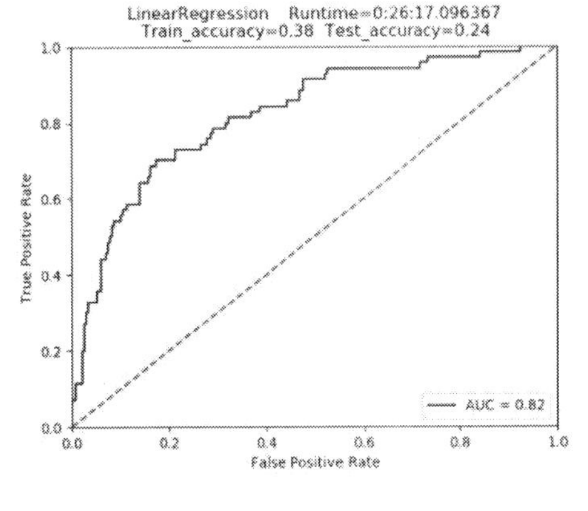

Figure 5. Linear Regression

3.3.4. Linear Regression. The researcher also tested the linear regression algorithm. The training accuracy and test accuracy are both low, as shown in Figure 5. Such prediction results cannot be used. This may indicate that the complexity of telecommunications customer churn data is very large, and it is difficult to use linear regression algorithms to have better prediction results.

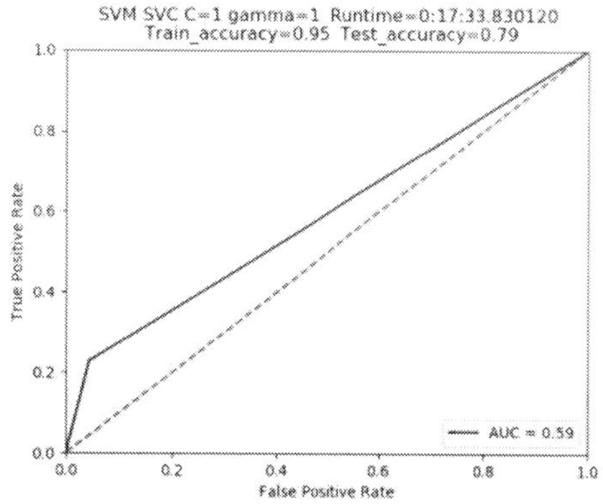

Figure 6. SVM SVC (C=1 gamma=1)

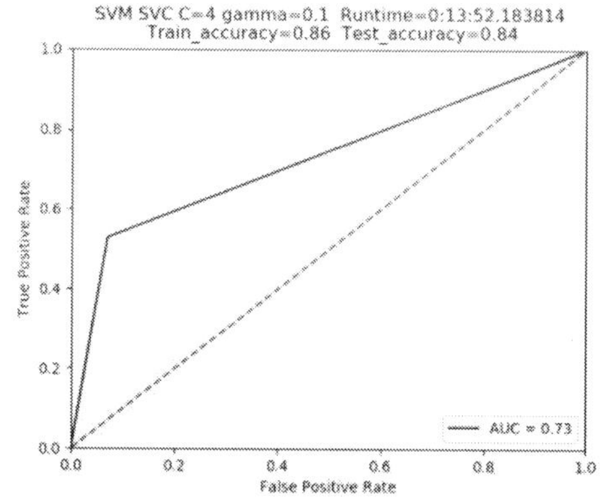

Figure 7. SVM SVC (C=4 gamma=0.1)

3.3.5. SVM. The researcher used SVM to predict the telecom customer churn data. When taking the default value C=1 gamma=1, as shown in Figure 6, the training accuracy rate is 95%, the test accuracy rate is 79%, but the AUC is only 0.59. After trying to adjust the parameters, the best result is shown in Figure 7, where C=4 and gamma=0.1. The training accuracy rate is 86%, the test accuracy rate is 84%, and the AUC is 0.73. When using the SVR algorithm, the best training accuracy rate is only 36%. Although after a lot of adjustments, the test effect is very poor, as shown in Figure 8.

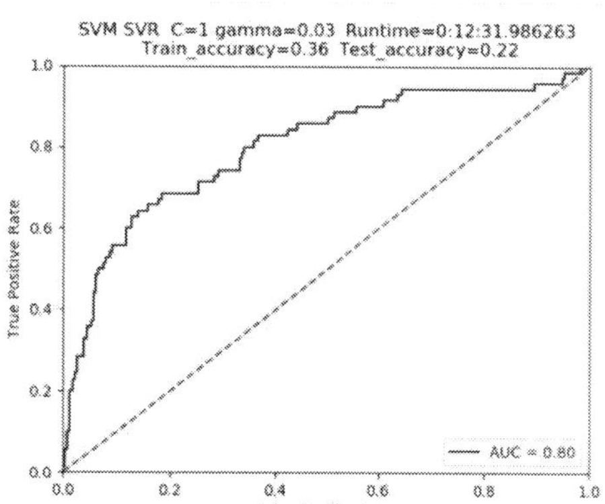

Figure 8. SVM SVR (C=1 gamma=0.03)

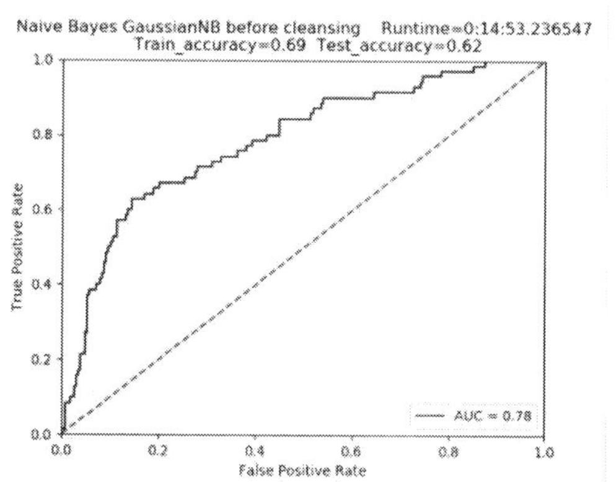

Figure 9. Naive Bayes (before cleansing)

3.3.6. Navie Bayes Then, the researchers studied the Naive Bayes algorithm. Using the GaussianNB algorithm, the effect is shown in Figure 9. The training accuracy rate is 69%, the test accuracy rate is 62%, and the AUC=0.78, the accuracy rate is relatively low. After cleansing the outlier data, the researchers compared the GaussianNB and BernoulliNB algorithms. The results are almost same only

GaussianNB shown in Figures 10 and Figures 11. The test accuracy has increased by 10%, indicating that the cleaning of abnormal data is very important. The effect of these two algorithms is not much different, but the accuracy is relatively low. The comparison shows that data cleansing is very important to Naive Bayes algorithm. In addition, using this algorithm must first deal with null values.

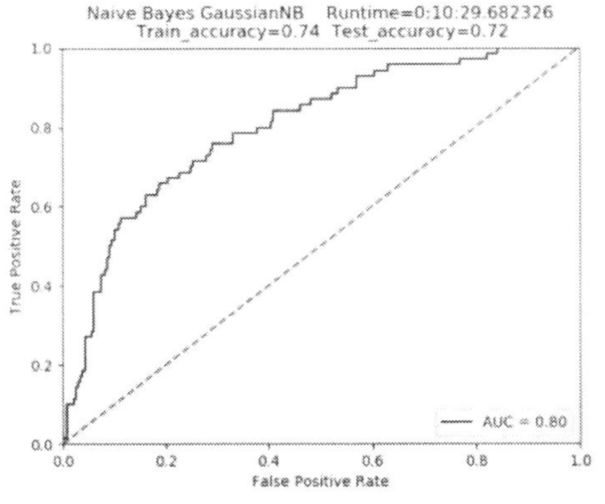

Figure 10. Naive Bayes GaussianNB

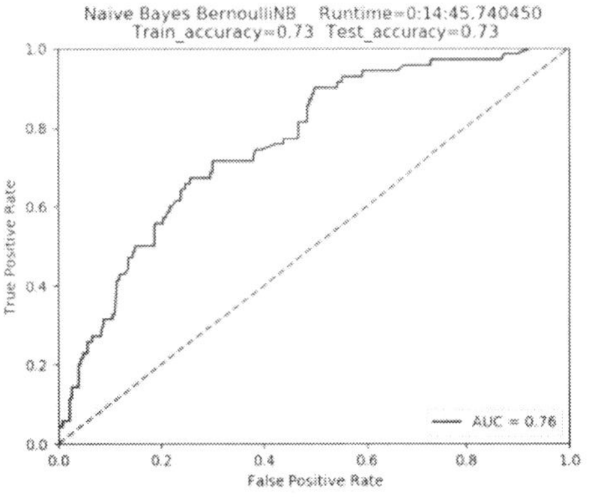

Figure 11. Naive Bayes BernoulliNB

3.3.7. Random Forest. The researcher used RandomForest to perform big data analysis. The training accuracy rate is very high, reaching 99%, and the test accuracy rate is also 80%, with AUC=0.76, as shown in Figure 12, indicating that the RandomForest algorithm has a high over fitting phenomenon. After trying several times, AUC from 0.76 to 0.80 has a relatively large range of variation, indicating that the results of each operation of the RandomForest are significant different, which proves that it has a random result, as shown in Figure 13. Maybe this algorithm is suitable for scientific research, but not very suitable for enterprise production.

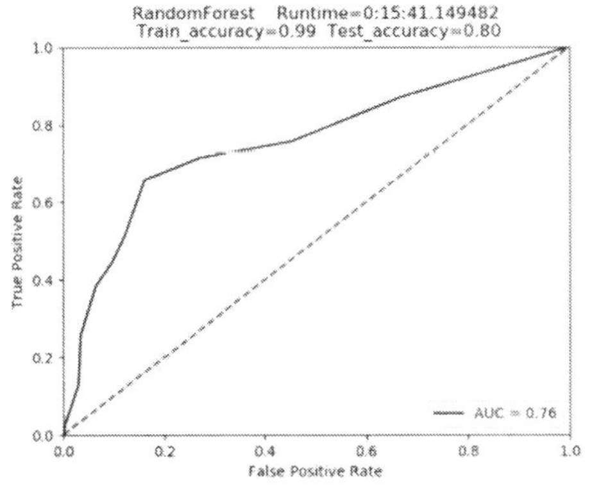

Figure 12. Random Forest (1)

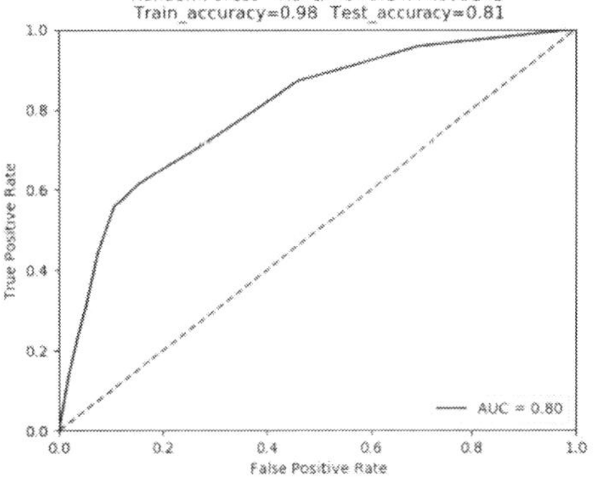

Figure 13. Random Forest (2)

3.3.8. Neural Network. The researcher finally tried the Neural Network algorithm. As shown in Figure 14, the training accuracy rate is 83%, the test accuracy rate is 80%, and AUC=0.78. This is closer to the decision tree algorithm, but the execution time is much longer.

3.4. *Algorithms Comparison*

The researcher compared these algorithms, as shown in Table 1. Researcher believes that it is better to use the decision number (CART) method for big data analysis such as telecom customer churn. This method is simple and practical, and the efficiency and accuracy are high. Although the accuracy of SVM is high, the AUC is not high, and it takes too long to adjust the parameters for repeated training. RandomForest has also obtained good accuracy and AUC, but the outliers need to be sorted in advance. The results of each operation of RandomForest algorithm are quite different, and it is necessary to try to find better prediction results many times. The neural network also has good accuracy and AUC, but it takes too long.

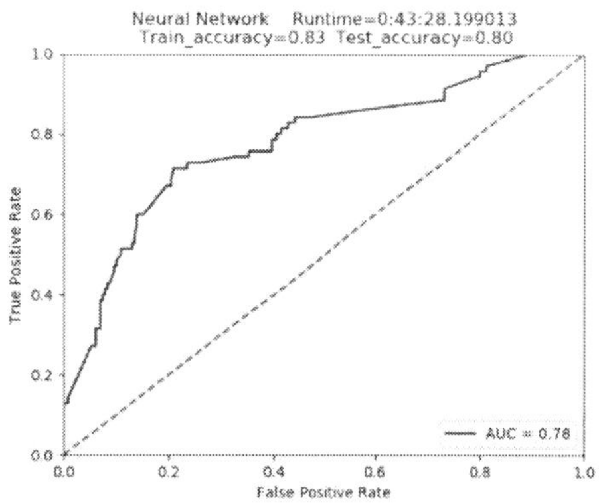

Figure 14. Neural Network

Table 1. Algorithms comparison

Algorithm	Train_Accuracy	Test_Accuracy	AUC	RunTime
DecisionTree C4.5	85%	83%	0.78	14:33
DecisionTree CART	86%	82%	0.81	14:00
KNN	89%	80%	0.72	15:42
LinearRegression	38%	24%	0.82	26:17
SVM SVC(C=1, gamma=1)	95%	79%	0.59	17:34
SVM SVC(C=4, gamma=0.1)	86%	84%	0.73	13:52
SVM SVR(C=1, gamma=0.03)	36%	22%	0.8	12:32
Naive Bayes(GaussianNB)(outlier)	69%	62%	0.78	14:53
Naive Bayes(GaussianNB)	74%	72%	0.8	10:30
Naive Bayes(BernoulliNB)	73%	73%	0.76	14:46
RandomForest(1)	99%	80%	0.76	15:41
RandomForest(2)	98%	81%	0.8	14:45
Neural Network	83%	80%	0.78	43:28

In short, through these experiments and analysis, the researchers found that the CART algorithm is most suitable for big data analysis of telecommunications companies retaining customers, because it has better performance in accuracy and running speed. In addition, the algorithm is simple, does not require complex parameter adjustments, and does not require too high technology. It is suitable for data support personnel at all levels to perform big data analysis.

Finally, telecommunications companies can use the K-means clustering method [19] to classify churn customer and introduce corresponding policies to retain customers.

4. Conclusion

The analysis of customer churn is a particularly difficult problem faced by telecom companies in the fierce market competition. When data acquisition, outlier processing, and analysis using various big data algorithms become daily tasks, it is necessary to further study how to conduct big data analysis accurately and quickly. In this paper, the researcher first discussed a quick and practical method of data cleaning, and then compared and analyzed several major algorithms in order to obtain better prediction results, and further train and guide analysts in our company. After comparing the main algorithms, the researcher found that the Decision Tree (CART) algorithm is most suitable for predicting customer churn. The algorithm is fast and accurate. The accuracy of the model on the test set reached 82%. In addition, the CART algorithm is simple and does not require too high technology, so it is suitable for data support personnel at all levels to quickly analyze big data for customer churn. For telecommunications companies, professional personnel can be used to find suitable algorithms for specific problems to train and guide support personnel at all levels to perform big data analysis. Through this process it can save time, standardize operations, and respond quickly to the market to retain customers.

References

[1] Maji G, Mandal S, Bhattacharya S and Sen S 2017 Designing combo recharge plans for telecom subscribers using itemset mining technique *2017 IEEE Int. Conf. on Industrial Technol. (ICIT)* DOI: https://doi.org/10.1109/ICIT.2017.7915539

[2] Amin A, Anwar S, Adnan A, Nawaz M and Huang K 2017 Customer churn prediction in the telecommunication sector using a rough set approach *Neurocomputing* vol **237** no 10 pp 242–54 DOI: https://doi.org/10.1016/j.neucom.2016.12.009

[3] Iqbal R, Doctor F, More B, Mahmud S and Yousuf U April 2018 Big data analytics: computational intelligence techniques and application areas *Technological Forecasting and Social Change* **153** p 119253 DOI: https://doi.org/10.1016/j.techfore.2018.03.024

[4] Abd-Allah M N, El-Beltagy S R and Salah A 2017 DyadChurn: Customer Churn Prediction using Strong Social Ties *IDEAS 2017: Proc. of the 21st Int. Database Engineering & Applications Symp.* pp 253–63 DOI: https://doi.org/10.1145/3105831.3105832

[5] Tiwari A, Sam R, Shaikh S Analysis and prediction of churn customers for telecommunication industry *2017 Int. Conf. on I-SMAC (IoT in Social, Mobile, Analytics and Cloud) (I-SMAC)* pp 218–22 DOI: https://doi.org/10.1109/I-SMAC.2017.8058343

[6] Abd-Allah M N, Salah A and El-Beltagy S R 2014 Enhanced customer churn prediction using social network analysis *DUBMOD '14: Proc. of the 3rd Workshop on Data-Driven User Behavioral Modeling and Mining from Social Media* pp 11–2 DOI: https://doi.org/10.1145/2665994.2665997

[7] Li H, Yang D, Yang L, Lu Y and Lin X Oct 2016 Supervised massive data analysis for telecommunication customer churn prediction *2016 IEEE Int. Conf. on Big Data and Cloud Computing (BDCloud)* pp 163–9 DOI: https://doi.org/10.1109/BDCloud-SocialCom-SustainCom.2016.35

[8] Wu X *et al* 2008 Top 10 algorithms in data mining *Knowl. Inf. Syst.* **14** pp 1–37 DOI: https://doi.org/10.1007/s10115-007-0114-2

[9] Han L, Li W and Su Z 2019 An assertive reasoning method for emergency response management based on knowledge elements C4.5 decision tree *Expert Systems With Applications* **122** pp 65–74 DOI: https://doi.org/10.1016/j.eswa.2018.12.042

[10] Y Chen L, C Wu C and Tang K 2016 Time-constrained cost-sensitive decision tree induction *Information Sciences* vol **354(1)** pp 140–52 DOI: https://doi.org/10.1016/j.ins.2016.03.022

[11] Tao D, Li X, Wu X, Hu W and Maybank S J 2007 Supervised tensor learning *Knowl. Inf. Syst.* **13(1)** pp 1–42

[12] Domingos P and Pazzani M 1997 On the optimality of the simple Bayesian classifier under zero-one loss *Mach. Learn.* **29** pp 103–30

[13] Hand D J and Yu K 2001 Idiot's Bayes—not so stupid after all? *Int. Stat. Rev.* **69** pp 385–98

[14] Friedman N, Geiger D and Goldszmidt M 1997 Bayesian network classifiers *Mach. Learn.* **29** pp 131–63

[15] Ridgeway G, Madigan D and Richardson T 1998 Interpretable boosted naive Bayes classification *Proc. of the Fourth Int. Conf. on Knowledge Discovery and Data Mining* ed R Agrawal *et al* (California: AAAI Press) pp 101–4

[16] Alotaibi S B 2017 ETDC: An efficient technique to cleanse data in the data warehouse *ICAIP 2017: Proc. of the Int. Conf. on Advances in Image Processing* pp 135–8 DOI: https://doi.org/10.1145/3133264.3133296

[17] Elitedatascience 2017 *Dimensionality Reduction Algorithms: Strengths and Weaknesses* Retrieved from https://elitedatascience.com/dimensionality-reduction-algorithms

[18] L A Zaid, F Kleinermann and O D Troyer 2011 Feature Assembly Framework: towards scalable and reusable feature models *VaMoS '11: Proc. of the 5th Workshop on Variability Modeling of Software-Intensive Systems* pp 1–9 DOI: https://doi.org/10.1145/1944892.1944893

[19] W Qadadeh and S Abdallah 2018 Customers segmentation in the insurance company (TIC) dataset *Procedia Computer Sci.* vol **144** pp 277–90 DOI: https://doi.org/10.1016/j.procs.2018.10.529

Lightweight Residual Network for Person Re-identification

Reza Fuad Rachmadi, Supeno Mardi Susiki Nugroho, and I Ketut Eddy Purnama

Dept. of Computer Engineering, Institut Teknologi Sepuluh Nopember, Surabaya, Indonesia 60111

E-mail: {fuad,mardi}@its.ac.id, ketut@te.its.ac.id

Abstract. In this paper, we investigated several lightweight convolutional neural network (CNN) classifiers for person re-identification problems. Person re-identification can be described as finding a person by performing a query with a specific person image on some defined gallery of people images. We construct the lightweight CNN classifier by utilizing the lightweight residual network architectures originally used for the CIFAR dataset. The goal is to design a lightweight CNN classifier with a maximum of 3 million number of parameters that produce good accuracy on person re-identification problems. The last layer of the residual network is detached and changed by two fully-connected layers, one for learning the discriminant features, and the other fully-connected layer for calculated the final classification score in the training process. Experiment results show that the ensemble of lightweight residual networks achieved a good performance on Market-1501 (rank-1 accuracy of 89.46% with mAP of 84.48% for single-query and rank-1 accuracy of 92.23% with mAP of 88.07% for multi-query). Although the lightweight residual network does not achieve state-of-the-art performance, our analysis shows that the lightweight residual network has higher information density comparing with other state-of-the-art models. More information density means that the classifier is more efficient in terms of performance with the number of parameters in the classifier. Implementation of these experiments are available at https://github.com/rezafuad/person-reid-lightweight-residualnet.

1. Introduction

The development of a smart security system is one of the active research areas that accelerate very fast after the introduction of deep learning technology. One of the most implemented technologies of security system is a network of CCTV cameras placed in several areas of the city or public places for monitoring purposes. Searching specific objects, such as a person or car, in the CCTV video data is one of the tasks for a smart security system. Person re-identification problems can be described as finding a specific person on the entire people's data extracted from the CCTV video in the same area and timestamp. In the last five years, the development of person re-identification models is in a very active state with several datasets [1, 2, 3, 4] and models [5, 6, 7, 8] had been proposed by the researchers. Person re-identification problems are equivalent to the image retrieval problem that specific for person images. Given a query of a specific image person, the person re-identification model returns all of the images in the gallery that match the query image. Although the development of the person re-identification models has been produced several state-of-the-art models, the investigating of lightweight CNN

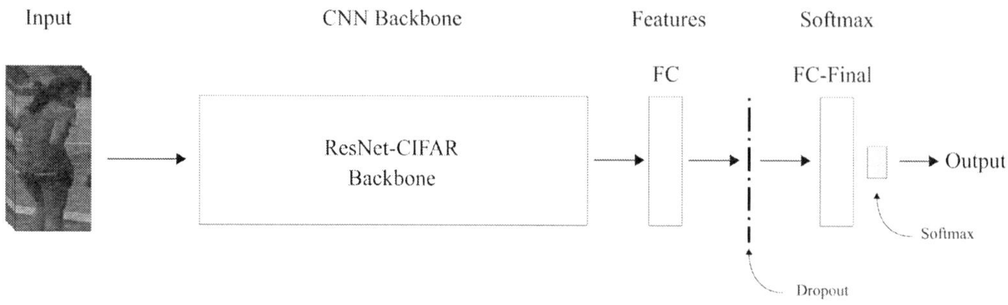

Figure 1. Simple diagram of our proposed classifier, with two fully-connected layers at the end of the classifier.

models for person re-identification problems are not yet addressed. A lightweight CNN classifier for person re-identification problems is useful for deploying the models in the edge computing scheme which have limited hardware specification choices.

In this paper, we investigated several lightweight residual CNN architecture as a solution to person re-identification problems. We constructed the lightweight residual CNN classifier based on the residual network [9] that used to solve CIFAR image classification problems. We removing the fully-connected layers of the original classifier and added two new fully-connected layers with different configuration from the original classifier. The contributions we make can be listed as follows.

- We investigated several lightweight residual network classifiers for person reidentification problems. The classifier constructed using residual network architecture that was originally used for the CIFAR dataset. Five different residual network architectures are used for the experiments, including ResNet20, ResNet32, ResNet44, ResNet56, and ResNet110.
- From the experiments, we proved that lightweight residual network classifiers with ensemble configuration can also produce very good performance comparing with other state-of-the-art methods on the Market-1501 dataset. Although the lightweight is not achieved state-of-the-art performance, the information density shows that the lightweight residual network has more information density comparing with other state-of-the-art models which proved that the classifier is more efficient in term of performance with the number of parameters in the classifier.
- We perform an ablation study for several lightweight residual networks to see the performance impact of the classifier if some of the features are removed or added to the classifier. Experiments show that reducing the number of neurons on the fully-connected layer can reduce the overfitting problem and improve the performance of the classifier.

The rest of the paper is organized as follows. The setting of the experiments of lightweight CNN classifiers is discussed in section two. The ablation study on the Market-1501 dataset is discussed in section three. Additional experiments by adjusting several parameters/method, including dropout rate, re-ranking, and multi-scale, is discussed in section four. Finally, we conclude the experiments in section five.

2. Experiments Setup

In this section, we describe the experiment setting, including the classifier we use for the experiments, the dataset used in the experiments, the training, and the testing process. All of the experiments are done using PyTorch framework and NVIDIA RTX 2080 TI with 11 GB memory.

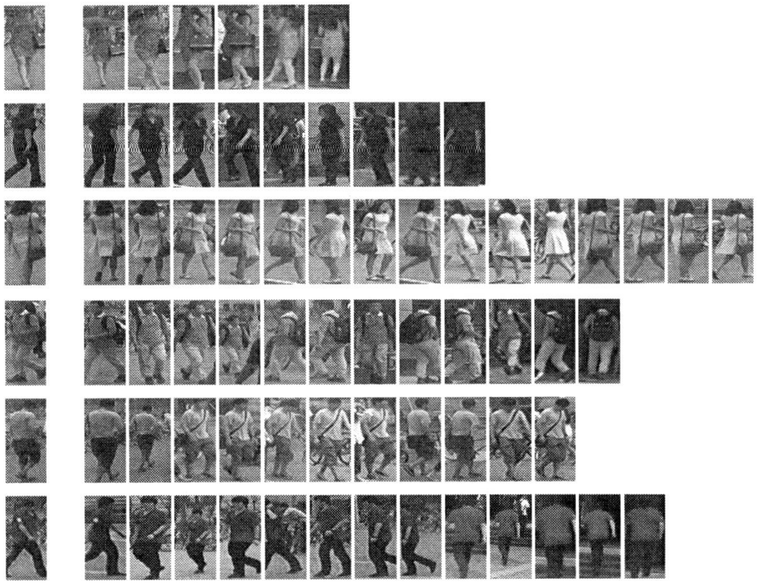

Figure 2. Some person images from Market-1501 person re-identification datasets.

2.1. Lightweight CNN
Figure 1 shows the simple diagram of our proposed lightweight CNN classifier. We construct the lightweight CNN classifier based on residual network architecture that used to solve CIFAR image classification problems. The final fully-connected of the residual network is removed and two new fully-connected layers is added at the end of the classifier. The first fully-connected is used to learn discriminant features while the last fully-connected layer is used to classify the identity of the person in the training data. To ensure that the first fully-connected layer learns good discriminant features, no activation function is attached after the first fully-connected layer. In the testing phase, the last final fully-connected layer is removed and the features used for the retrieval process are extracted from the first fully-connected layer. Unlike the original residual network, our lightweight CNN classifier use an input resolution of $32\times64\times3$ instead of $32\times32\times3$. We use five different residual network, including ResNet20, ResNet32, ResNet44, ResNet56, and ResNet110. Although the number of layers is very deep (e.g. 110 layers), the number of parameters of the deepest classifier is still under 3 million parameters.

Is there any other lightweight CNN architectures that can be used for solving the person re-identification problem? Based on our study, some other lightweight CNN architectures can be used for solving the person re-identification problems but in this paper, we focus the experiments for residual-based lightweight CNN architecture.

2.2. Person Re-identification Dataset
To evaluating the performance of the lightweight classifier, we mainly use the Market-1501 dataset which consists of 1,501 unique identities captured using six different cameras. The dataset was divided into 750 identities for the training process and 751 identities for the testing process with a total number of 32,668 bounding boxes. Each identity is unique and no overlapping between the training and testing process. Each person image is cropped and normalize to 128×64 using a combination of DPM (Deformable Parts Model) and handcrafted method. Figure 2 shows several samples of person images in Market-1501 person re-identification dataset.

Figure 3. Examples of data augmentation images using random crop and random erasing method.

In this paper, we use a different resolution for our classifier input. We use resolution of 64×32 for the training and testing process which half of the original Market-1501 resolution. The training dataset of Market-1501 is reorganized such that the data can be used for the training process using classification loss function with one person image per identity is used for the validation process. We choose the classifier with the highest accuracy based on the validation set for further evaluation.

2.3. Training and Testing Process
To ensure the performance analysis is reliable, we perform the training and testing for five iterations and averaging the evaluation metric. Three evaluation metric is used for evaluating the performance of the classifier, including rank, mAP, and information density.

The training process is done for 100 epochs with the learning rate initialize at 0.1 and decreased by a factor of 0.1 when reaching 40 and 80 epoch. All of the convolutional layer weights were initialize using weights from the same CNN architecture trained using CIFAR datasets [10]. The different learning rate is used for the convolutional layers with an initial value of 0.01 with the same policy as used in the global learning rate. We use two different data augmentation process, random crop and random erasing data augmentation follow by subtracting using the mean of the training data. Figure 3 shows some examples of data augmentation images using those two methods. In the first fully-connected layer we added a dropout layer with several different dropout ratios. The cross-entropy loss function is used to train the classifier with softmax activation at the end of the classifier.

In the testing process, the features for the retrieval process are taken from the first fully-connected layer. Same as in the training process, the input image is resized to 64×32 and subtracted by the mean of the training data. The similarity score is computed by correlating two extracted features and sorted the results from low to high. A low score means that the two features are correlated and vice-versa.

3. Ablation Study
To evaluate the ResNet-CIFAR lightweight classifier for person re-identification problems, we perform an ablation study using three different protocols, by changing the random erasing

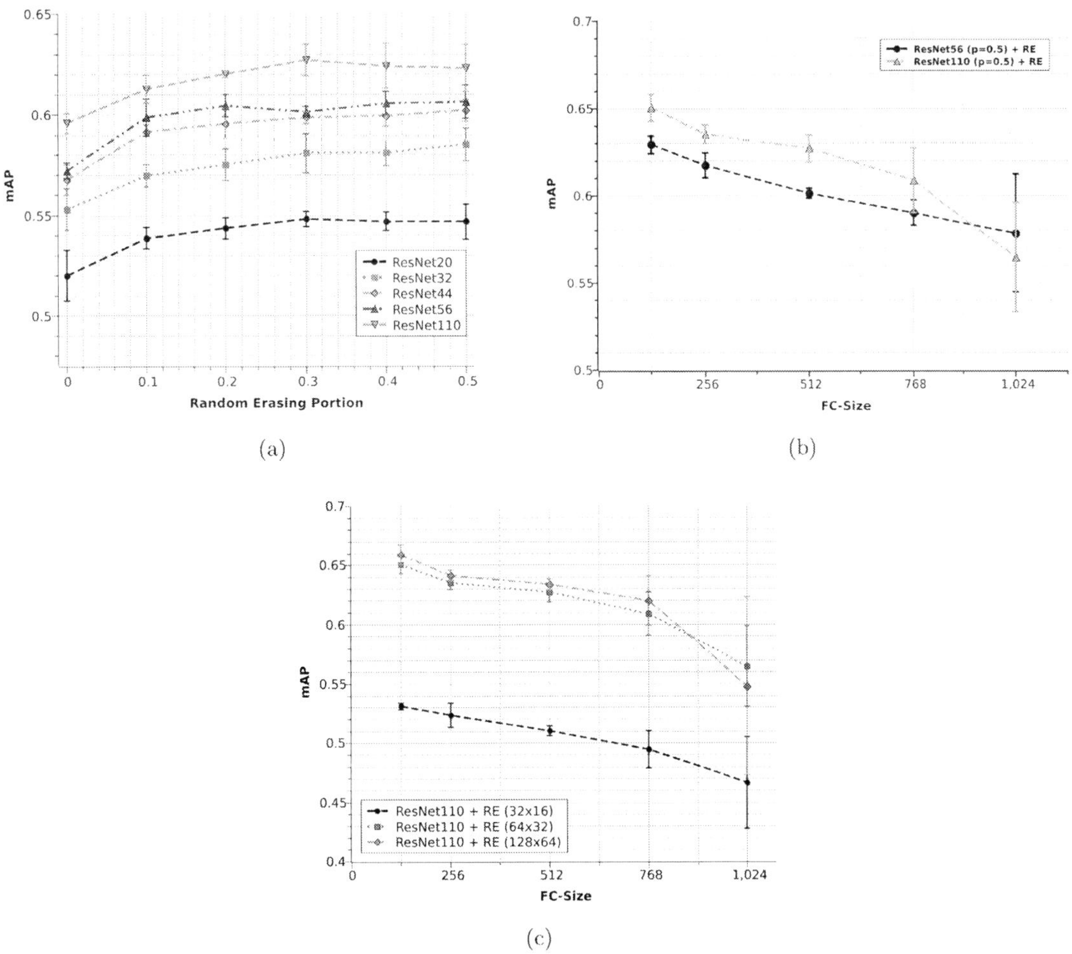

Figure 4. Results from the ablation study of the lightweight CNN classifier for person re-identification problem using Market-1501, (a) random erasing portion, (b) size of FC (or size of features), and (c) impact of input resolution.

portion, by changing the size of the first fully-connected layer, and by changing the input resolution of the classifier. We use mAP (mean average precision) as the main metric in our ablation study. Figure 4 shows the ablation study results in graphic form.

In the first ablation study, we change the random erasing portion from 0 to 50% with an increment of 10%. As shown in Figure 4-(a), the random erasing data augmentation proved can improve the classifier with appropriate portion value. After the random erasing portion of 30%, the performance is not consistent for all classifiers due to the large erasing portion of the input image which effected the classification performance.

Several different sizes of the first fully-connected layer (FC-size) in the classifier is our next ablation study. In this and next experiments, we only use ResNet56 and ResNet110 which in the last experiments achieves first and second highest mAP. Figure 4-(b) shows the plot of FC-size with the performance of the classifier (in mAP). As shown in Figure 4-(b), the best mAP is achieved using an FC-size of 128 which consider the lowest FC-size configuration in our experiments. After analyzing the training and validation process, we concluded that bigger FC-size lead to overfitting problems which effected the testing performance of the classifier.

Table 1. Summary of lightweight residual network classifier experiments on Market-1501 datasets (averaging from five different runs).

Model	Single Query (%)			Multi Query (%)		
	rank-1	rank-5	mAP	rank-1	rank-5	mAP
Res56-RE	82.56±0.27	92.99±0.71	62.03±0.51	88.51±0.22	95.95±0.58	72.66±0.76
Res56-SPP-RE	83.31±0.59	93.67±0.66	63.76±0.26	89.09±0.66	96.13±0.25	73.05±0.22
Res56-RE-Rerank	85.93±0.62	92.01±0.63	80.03±0.44	90.13±0.25	94.92±0.18	84.84±0.24
Res56-SPP-RE-Rerank	86.83±0.34	92.68±0.25	80.89±0.17	90.66±0.39	95.26±0.25	85.39±0.13
Res110-RE	84.28±1.05	93.77±0.52	65.05±0.76	90.02±0.79	**96.49±0.23**	74.64±0.91
Res110-SPP-RE	84.81±0.60	**94.01±0.37**	65.87±0.63	90.15±0.63	96.32±0.15	74.89±0.45
Res110-RE-Rerank	87.12±0.94	92.77±0.49	81.45±1.09	91.17±0.54	95.47±0.28	86.04±0.90
Res110-SPP-RE-Rerank	**87.67±0.47**	92.98±0.28	**82.04±0.35**	**91.48±0.83**	95.54±0.48	**86.27±0.31**

In the last ablation study, we try several different input resolutions for the classifier and measure the mAP metric. Figure 4-(c) shows the plot of FC-size with the performance of the classifier (in mAP) using three different input resolutions. As shown in Figure 4-(c), input resolutions of 128×64 and 64×32 have a similar performance with a very narrow gap. The opposite results happen when the input resolution reduces to 32×16 which produces around 50% drop of mAP comparing with other input resolutions.

4. Results

The Market-1501 dataset testing process is split into two different protocols, single-query testing protocol, and multi-query testing protocol. The single-query testing protocol done by given only one query image to the classifier while the multi-query is given more than one query image. Naturally, the multi-query results will be higher than the single query due to the more query samples exposes to the classifier.

Based on the results in our ablation study, we decided to use an input resolution of 64×32, a droprate of 0.5, a random erasing portion of 0.3, and an FC-size of 128. We only choose ResNet56 and ResNet110 CNN architecture for further experiments. The hyperparameter used in the training process is also the same as in the ablation study. Additional re-ranking method [11] and SPP (Spatial Pyramid Pooling) [12] is used in the experiments which proved can improve retrieval accuracy (especially rank-1 and mAP). Table 1 shows the summary of our experiments using ResNet56 and ResNet110 with additional RE (Random Erasing) and re-ranking method.

As shown in Table 1, the additional methods (SPP and Reranking) can increase the performance of the classifier by around 1% for SPP and around 3% for reranking. The best performance is achieved using a combination of ResNet110, random erasing, SPP, and re-ranking method. The different results for rank-5 accuracy in which the best performance is not achieved by the classifier with all additional methods. Based on Table 1 results, the phenomena also appear for other classifiers when the testing processes using re-ranking method.

4.1. Ensemble

To increase the performance of the classifier, we also conducted testing process using ensemble of ResNet56 and ResNet110 classifiers. Ensemble of two or more CNN classifiers proved can increase the performance of the classifier by around 2%. We use a simple average ensemble

Table 2. Experiment using an ensemble of ResNet56 and ResNet110 on Market-1501 datasets (averaging from five iterations).

Model	Single Query (%)			Multi Query (%)		
	rank-1	rank-5	mAP	rank-1	rank-5	mAP
Ensemble-RE	86.95±0.61	**94.99±0.20**	69.94±0.46	91.52±0.37	**96.97±0.14**	78.01±0.42
Ensemble-SPP-RE	86.97±0.49	94.98±0.18	69.87±0.19	91.16±0.38	96.79±0.18	77.60±0.26
Ensemble-RE-Rerank	89.23±0.28	93.96±0.44	84.45±0.75	**92.39±0.27**	96.09±0.31	**88.24±0.38**
Ensemble-SPP-RE-Rerank	**89.46±0.44**	94.08±0.47	**84.48±0.24**	92.23±0.37	96.01±0.27	88.07±0.07

Table 3. Comparison of our lightweight classifier with several state-of-the-art models.

Years	Method	#Params (million)	Market-1501 Single (%)			Market-1501 Multi (%)		
			r = 1	mAP	Inf. Den.	r = 1	mAP	Inf. Den.
2015	BoW + KISSME [2]	n/a	44.42	20.76	-	-	-	-
2016	SL [13]	n/a	51.90	26.35	-	-	-	-
2016	DNS [14]	n/a	61.02	35.68	-	71.56	46.03	-
2016	SSDAL [15]	~61	39.4	19.6	0.321	49.0	25.8	0.422
2017	CAN [16]	~50	60.3	35.9	0.718	72.1	47.9	0.958
2017	Fisher Net [17]	5.2	48.15	29.94	5.757	-	-	-
2017	Verif-Iden [18]	~23	79.51	59.87	2.603	85.84	70.33	3.057
2018	PAN [19]	~45	88.57	81.53	1.811	91.45	87.44	1.943
2019	Spatial-Temporal [20]	~30	98.0	95.5	3.183	-	-	-
2019	PyrNet [21]	35.9	94.6	91.4	2.545	96.1	94.0	2..618
2019	ARP [22]	~23	87.04	66.89	2.908	-	-	-
2020	VA-ReID [23]	52.64	96.79	95.43	1.812	-	-	-
2020	SARL [24]	~32	96.1	88.0	2.750	-	-	-
2020	RGA-SC [25]	56.4	96.1	88.4	1.567	-	-	-
	Our (Single)	1.8	87.67	82.04	45.57	91.48	86.27	47.93
	Our (Ensemble)	2.7	89.46	84.48	31.29	92.23	88.07	32.62

method between those two models. Table 2 shows the results of ensemble classifiers using the Market-1501 person re-identification dataset. In single-query testing, the best performance is achieved using Ensemble of ResNet56 and ResNet110 with additional SPP method, random erasing, and reranking. In multi-query testing, the best performance is achieved using Ensemble of ResNet56 and ResNet110 with additional random erasing and reranking. The same phenomena appear in previous experiments is also appear in this experiment in which the ensemble classifier with an additional reranking method produces a lower rank-5 accuracy comparing with the ensemble classifier that does not use the re-ranking method.

4.2. Comparison

We added information density metrics which measure the effectiveness of the classifier in term of performance (in mAP) and the number of parameters in the classifier. Let f_{mAP} is measured performance of the classifier (in mAP) and f_p is number of parameters (in million) in the classifier, the information density can be calculated as follows.

$$I_{den} = \frac{f_{mAP}}{f_p} \tag{1}$$

Table 3 shows the performance of our lightweight residual network classifier comparing with several other methods. As shown in Table 3, although our lightweight residual network classifier is not achieved a state-of-the-art performance but after we computer the information density, our lightweight residual residual network classifier have very high with around 10 to 20 times higher than other methods. In terms of single query rank-1 and mAP metric, the lightweight residual network classifier is around 10% lower than the state-of-the-art method while in terms of multi-query rank-1 and mAP metrics, the lightweight residual network classifier is 4-8% lower than the state-of-the-art classifier.

5. Conclusion

We have investigated several lightweight CNN architectures for person re-identification problems. The lightweight CNN architecture formed using ResNet-CIFAR CNN architecture by removing the last fully-connected layer and adding two fully-connected layers. The first fully-connected layer is used for learning the discriminant features and the second fully-connected layer for the final classification score in the training process. To avoid dead neurons in the first fully-connected, no activation function is added in the first fully-connected layer. In the testing process, we removed the second fully-connected layer and extracted the features for the retrieval process from the first fully-connected layer. Based on the experiments using Market-1501 person re-identification, the best performance achieved using Ensemble of ResNet56 and ResNet110 with rank-1 accuracy of 89.46%, mAP of 84.48% in single query testing and rank-1 accuracy of 92.23%, mAP of 88.07% in multi-query testing. Although the lightweight ResNet-CIFAR CNN classifier not achieved state-of-the-art performance, the information density of the lightweight ResNet-CIFAR CNN classifier is very high which proved that the classifier is more efficient comparing with other methods.

Deeper investigations of the lightweight CNN classifier on other person re-identification datasets is one of our concerns for future work. A combination of the lightweight CNN classifier with deep metric loss also worthed to investigate because the deep metric learning, such as sphereface or center-loss, proved can increase the performance of the classifier on face recognition problems.

Acknowledgments

This research is partially supported by Indonesia Ministry of Education and Culture Research Grant (under PDUPT Scheme) No. 1206/PKS/ITS/2020.

References

[1] Li W, Zhao R, Xiao T and Wang X 2014 Deepreid: deep filter pairing neural network for person re-identification *Proc. of the IEEE Conf. on Computer Vision and Pattern Recognition* pp 152–9

[2] Zheng L, Shen L, Tian L, Wang S, Wang J and Tian Q 2015 Scalable person re-identification: a benchmark *Proc. of the IEEE Int. Conf. on Computer Vision* pp 1116–24

[3] Ristani E, Solera F, Zou R, Cucchiara R and Tomasi C 2016 Performance measures and a data set for multi-target, multi-camera tracking *European Conf. on Computer Vision workshop on Benchmarking Multi-Target Tracking* pp 17–35

[4] Zheng Z, Zheng L and Yang Y 2017 Unlabeled samples generated by gan improve the person re-identification baseline in vitro *Proc. of the IEEE Int. Conf. on Computer Vision* pp 3774–82
[5] Sun Y, Zheng L, Yang Y, Tian Q and Wang S 2018 Beyond part models: person retrieval with refined part pooling (and a strong convolutional baseline) *Proc. of the European Conf. on Computer Vision (ECCV)* pp 480–96
[6] Zheng Z, Yang X, Yu Z, Zheng L, Yang Y and Kautz J 2019 Joint discriminative and generative learning for person re-identification *Proc. of the IEEE Conf. on Computer Vision and Pattern Recognition* pp 2138–47
[7] Fu Y, Wei Y, Wang G, Zhou Y, Shi H and Huang T S 2019 Self-similarity grouping: a simple unsupervised cross domain adaptation approach for person re-identification *Proc. of the IEEE Int. Conf. on Computer Vision* pp 6112–21
[8] Zhang X, Cao J, Shen C and You M 2019 Self-training with progressive augmentation for unsupervised cross-domain person re-identification *Proc. of the IEEE Int. Conf. on Computer Vision* pp 8222–31
[9] He K, Zhang X, Ren S and Sun J 2016 Deep residual learning for image recognition *Proc. of the IEEE Conf. on Computer Vision and Pattern Recognition* pp 770–8
[10] Idelbayev Y *Proper ResNet Implementation for CIFAR10/CIFAR100 in PyTorch* retrieved on 31 March 2020 from https://github.com/akamaster/pytorch_resnet_cifar10
[11] Yu R, Zhou Z, Bai S and Bai X 2017 Divide and fuse: A re-ranking approach for person re-identification *Proc. of the British Machine Vision Conf. (BMVC)* ed Kim T K, Zafeiriou S, Brostow G and Mikolajczyk K (BMVA Press) pp 135.1–13 ISBN 1-901725-60-X dOI: 10.5244/C.31.135
[12] He K, Zhang X, Ren S and Sun J 2015 Spatial pyramid pooling in deep convolutional networks for visual recognition *IEEE Transactions on Pattern Analysis and Machine Intelligence* **37** pp 1904–16
[13] Chen D, Yuan Z, Chen B and Zheng N 2016 Similarity learning with spatial constraints for person re-identification *Proc. of the IEEE Conf. on Computer Vision and Pattern Recognition* pp 1268–77
[14] Zhang L, Xiang T and Gong S 2016 Learning a discriminative null space for person re-identification *Proc. of the IEEE Conf. on Computer Vision and Pattern Recognition* pp 1239–48
[15] Su C, Zhang S, Xing J, Gao W and Tian Q 2016 Deep attributes driven multi-camera person re-identification *European Conf. on Computer Vision* (New York: Springer) pp 475–91
[16] Liu H, Feng J, Qi M, Jiang J and Yan S 2017 End-to-end comparative attention networks for person re-identification *IEEE Transactions on Image Processing* **26** pp 3492–506
[17] Wu L, Shen C and Van Den Hengel A 2017 Deep linear discriminant analysis on fisher networks: A hybrid architecture for person re-identification *Pattern Recognition* **65** pp 238–50
[18] Zheng Z, Zheng L and Yang Y 2017 A discriminatively learned cnn embedding for person reidentification *ACM Transactions on Multimedia Computing, Communications, and Applications (TOMM)* **14** pp 1–20
[19] Zheng Z, Zheng L and Yang Y 2018 Pedestrian alignment network for large-scale person re-identification *IEEE Transactions on Circuits and Systems for Video Technology* **29** pp 3037–45
[20] Wang G, Lai J, Huang P and Xie X 2019 Spatial-temporal person re-identification *Proc. of the AAAI Conf. on Artificial Intelligence* vol 33 pp 8933–40
[21] Martinel N, Luca Foresti G and Micheloni C 2019 Aggregating deep pyramidal representations for person re-identification *Proc. of the IEEE Conf. on Computer Vision and Pattern Recognition Workshops* pp 0–0
[22] Lin Y, Zheng L, Zheng Z, Wu Y, Hu Z, Yan C and Yang Y 2019 Improving person re-identification by attribute and identity learning *Pattern Recognition* **95** pp 151–61
[23] Zhu Z, Jiang X, Zheng F, Guo X, Huang F, Sun X and Zheng W 2020 Viewpoint-aware loss with angular regularization for person re-identification. *AAAI* pp 13114–21
[24] Jin X, Lan C, Zeng W, Wei G and Chen Z 2020 Semantics-aligned representation learning for person re-identification. *AAAI* pp 11173–80
[25] Zhang Z, Lan C, Zeng W, Jin X and Chen Z 2020 Relation-aware global attention for person re-identification *Proc. of the IEEE/CVF Conf. on Computer Vision and Pattern Recognition* pp 3186–95

IOP Conf. Series: Materials Science and Engineering 1077 (2021) 012047 doi:10.1088/1757-899X/1077/1/012047

Usability assessment of virtual reality as a training tool for oral presentation

M M Daniels

Ifugao State University, Ifugao, Philippines

E-mail: mylemelodydaniels@gmail.com

Abstract. One of the widely used teaching-learning strategies in the educational sector is the delivery of oral presentation by students. They are tasked to present their outputs–such as research paper, project, case study findings—to a group of audience. The problem experienced by most students in this type of task is the fear of public speaking combined with the pressure of relaying information effectively. In practice, this problem is resolved by giving more opportunity to students to rehearse repeatedly in front of peers. Organizing such setup, however, becomes almost impossible during times of pandemic when face-to-face interaction is limited, if not prohibited. Hence, this study ventured on assessing the usability of an alternative solution. This is with the use of virtual reality technology that can simulate the traditional training scenario and render it in a virtual environment. Three categories of participants were involved in the rating of its usability – IT students, non-IT students and Trainers. They were given ample time to navigate a public speaking app using a VR headset. Afterwards, they accomplished a survey questionnaire where they assessed the usability of the VR tool as a mode of training for oral presentation. Findings show that the usability rating of virtual reality as a training tool for oral presentation can vary depending on the technological background of users. This is due to the lack of on-screen assistant that would guide users in navigating the system. Despite these minor problems, the use of virtual reality as a training tool for oral presentation is highly recommended. This is supported by the unanimous positive response of participants in the SUS scale measuring their interest in using the VR tool for oral presentation.

1. Introduction

One of the critical skills focused on by 21st Century education is oral communications due to its high demand in the industry. The ability to present effective business proposals and sales pitches is vital for business growth. This is why language curriculums have integrated diverse forms of communication activities to reinforce this skill among students. A very popular activity performed inside the classroom is the delivery of oral presentations. The goal of this activity is to train students to prepare, organize and deliver effective oral presentations for future business meetings and other social professional engagements [1]. In the classroom, students are tasked to present their output, either an assignment, project, research paper, thesis, or capstone project to a group of audience [2]. A more formal structure involves a panel of evaluators who closely examine the procedures, processes and output of the project, as in the case of thesis and capstone projects.

One of the main difficulties experienced by students in oral presentation is public speaking anxiety or the fear of speaking to a group of audience. As research reveals, anxiety can impede students' production and achievement [3] resulting to their poor delivery. A practical solution to this is to provide more opportunity for students to deliver oral presentation by organizing public speaking activities. This would involve setting up a conference room and gathering of audience. Yet another problem arises when face-to-face interaction becomes limited due to pandemic. Hence, this study endeavors to look into a technological tool that can help resolve these issues.

In digital technology, a 3-dimensional (3D) environment called virtual reality (VR) is rapidly proliferating the market especially in the gaming industry. This is due to its highly immersive design that attract consumers [4]. It uses sensor devices to capture user movements and facilitate interaction with the virtual environment [5]. The display is rendered using high-definition graphics that can be navigated 360 degrees. With this technology, a virtual conference room can be generated to serve as an alternative training environment for students. In fact, there are a few VR apps commercially available for public speaking. In these apps are virtual conference rooms with virtual humans as audience. The user is positioned at the stage, recreating the traditional scenario of public speaking.

Another significant facet of virtual reality is its ability to facilitate an environment in treating general phobia [6], credit to the realistic sense of presence experienced within the virtual environment [7]. Several studies demonstrated that the use of virtual reality systems in the therapeutic setting yields a promising outcome [8] [9]. Authors refer to the treatment as virtual reality exposure therapy (VRET). Instead of exposing the subject to the real object which they fear, they are exposed to a virtual version of it. The sight of the virtual object stimulates similar emotions to that of the real scenario, thereby achieving the goal of the treatment. This capability of VRET encouraged authors to look into its application over fear of public speaking. Results collated in a systematic review showed consistent findings from different authors, demonstrating the efficacy of this treatment in helping subjects overcome fear of public speaking [10].

The challenge faced by teachers and students in organizing public speaking engagement opportunities can be resolved with the use of virtual reality. It is then the goal of this study to assess the usability of this tool as perceived by its intended users. System usability is a test to investigate whether a system successfully delivers its purpose. It measures the extent to which a product can be used by intended users to achieve specified goals with effectiveness, efficiency, and satisfaction in a specified context of use [11]. There are several usability scales crafted by various authors over the years, but the most popular and widely used instrument is the System Usability Scale or SUS [12]. This scale covers a variety of aspects of system usability such as the need for support, training, and complexity, and thus have a high level of face validity for measuring usability of a system [13]. The SUS instrument is generally used after the respondent has had an opportunity to use the system being evaluated, but before any debriefing or discussion takes place [14]. This instrument has been used for a variety of evaluations such as Learning Management System [15], and was modified to match the context of what is being evaluated.

Findings of this study will contribute greatly to the improvement of this technology in its capacity to provide skills training which can benefit both students and teachers. In a classroom with a teacher-to-student ratio of 1 is to 40, there is limited time for teachers to monitor and coach each student especially when targeting specific skills. Hence, the introduction of such technological tool will provide additional option for teachers and students to utilize to supplement skills development.

2. Methodology

An overview of the undertaking of the study is illustrated in Figure 1. Participants were recruited from a state university in a province of Cordillera Administrative Region, Philippines. Participant categories are IT and non-IT students. This would allow comparison of usability rating between students who has strong background in computer technology and those with lesser computer literacy.

For the IT program category, participants were those recruited from Capstone Project class. The final requirement of this class is to present their project output (system prototype) to their teacher and classmates for evaluation. For the non-IT program category, participants were recruited from students of Thesis course—a major curriculum requirement that mandates students to conduct a study and present their research output to a panel of evaluators. They were randomly selected to match the population in the IT program category. The joining of participants was voluntary such that anyone who opted out were given the opportunity to do so. There was a total of 20 IT students and 20 non-IT students. Furthermore, English instructors (N=5) were also recruited to participate in the usability assessment. This is to capture the VR tool's usability from trainer's point of view.

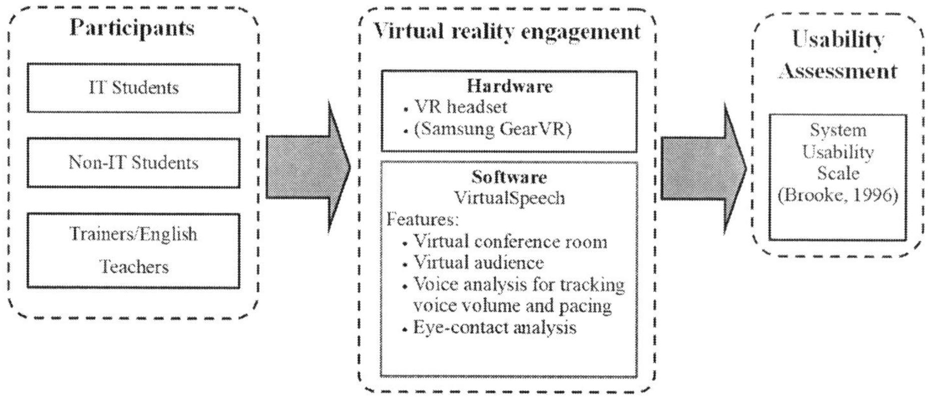

Figure 1. Conceptual Framework

Participants were first oriented as a group regarding the procedures of the study and were instructed to report to the training venue during their given schedule. During the VR training period, each participant had the chance to wear a VR headset (Samsung Gear VR) and explore a public speaking app called VirtualSpeech. Figure 2 is a screenshot of the VirtualSpeech app. The virtual humans look very similar to human appearance. They seem to be listening intently with slight nodding gestures, just like in a conference meeting.

Figure 2. Screenshot of VirtualSpeech

Additionally, the app has a voice and eye movement analysis feature that evaluates user's voice volume, pacing and eye-contact and gives scores at the end of the speech. This is activated by clicking on the Start Analyze button at the beginning of the presentation. They system displays a performance score after the delivery of the presentation. A sample performance score is shown in Figure 3. They app also provides live feedback during speech delivery to remind the speaker to look either to the left or to the right side of the audience, if neglected during the presentation.

The participants navigated the app and rehearsed their oral presentation. They were able to view their performance scores after each rehearsal. The average time spent by all participants in rehearsing in the VR tool is 20 minutes. After using the VR tool, they answered the System Usability Scale (SUS).

The System Usability Scale (SUS) is composed of 10 statements rated basing on respondent's agreement, using a Likert-type scale: strongly agree, agree, neutral, disagree, and strongly disagree [12]. The usability score using SUS is computed using the formula:

$$\text{SUS Score} = (X + Y) * 2.5 \tag{1}$$

where:

X = Sum of the points for all odd-numbered questions – 5

Y = 25 – Sum of the points for all even-numbered questions

Resulting score is interpreted with the following guideline: greater than 80.3 = Grade A or Excellent, 68 to 80.3 = Grade B or Good, 68 = Grade C, or Okay, 51 – 68 = Grade D or Poor; and less than 51 = Grade F or Awful [16].

The usability rating from the three participant categories were tabulated separately. The mean SUS scores and their corresponding interpretation are summarized in tables.

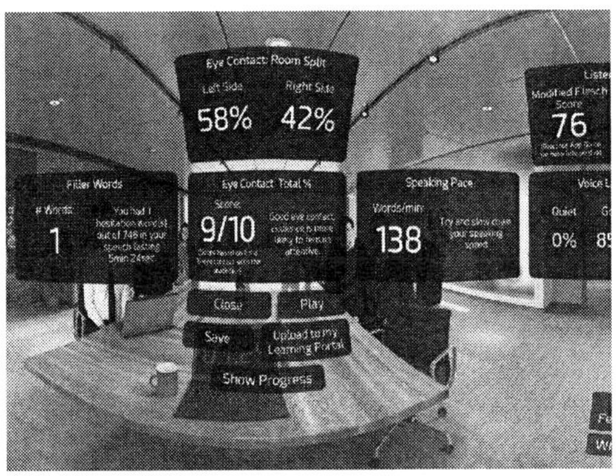

Figure 3. Sample performance score

3. Findings

Using the SUS equation, the usability scores given by IT, non-IT, and trainer respondents were extracted. They are presented in Table 1.

Table 1. SUS Scores of the VR tool

Respondents	SUS Score	Grade	Interpretation
Trainers	69.7	B	Good
IT students	70	B	Good
Non-IT students	66.3	D	Poor

According to the assessment of Trainers and IT students, the usability of the VR tool is GOOD, or acceptable; but for the non-IT students, the resulting SUS score is POOR. Factors for such ratings can be inferred from the individual statements in the assessment scale. The statements in the scale are equally divided into positive and negative constructs. Ratings for the positive statements (odd-numbered items) are presented in Table 2, while for negative statements (even-numbered items), Table 3.

Notably, the mean rating across all groups of respondents in Table 2 indicate a unanimous agreement on the positive statements of the scale. This denotes that majority would like to use the VR tool, and they felt that it was easy to use due to the organized integration of its functions. Despite the different levels of technical adeptness of the respondents, they managed to navigate the tool confidently after a short orientation. As first-time users of such technology, their impression was that the system is quite complex as reflected in Table 3. During the orientation, they were guided step-by-step in navigating the virtual environment. The IT students were able to learn to operate the VR tool more quickly than the non-IT respondents. The latter felt the need to be assisted for a longer time before they can comfortably operate the tool on their own.

Table 2. Mean score of SUS Positive statements

Statement	**Mean**		
	English teachers	IT Students	Non-IT students
1. I think I would like to use this system to train for oral presentation.	4.4	4.4	4.5
3. I thought the system was easy to use.	4.1	4.1	4.1
5. I think the various functions in the system were well integrated.	4.3	4.1	4.3
7. I would imagine that most people would learn to use this system very quickly.	4.4	4.1	4.2
9. I felt very confident using the system.	4.1	3.6	4.0
Mean	**4.3**	**4.0**	**4.2**

Legend: 1 – Strongly Disagree, 2 – Disagree, 3 – Neutral, 4 – Agree, 5 – Strongly Agree

Most of them had difficulty selecting buttons since there was no visible cursor on screen. The only way to determine whether the button is pointed at is when it is highlighted. The non-IT respondents can hardly recognize such indication, making it difficult for them to select the right buttons. As a result, the adjacent button is often selected and they felt it was an inconsistency in the functions of the VR tool. Aside from this technical challenge, the respondents are unfamiliar with the use of a headgear that's why they felt it was cumbersome to use and tiresome to carry. Another problem observed as the respondents used the VR tool is that during their initial use, they felt the need to remove the head gear within the first 5 minutes because they felt nauseous. This condition was slowly reduced as they had longer exposure to the virtual environment.

Table 3. Mean score of SUS Negative statements

Statements	**Mean**		
	English teachers	IT Students	Non-IT Students
2. I found the system unnecessarily complex.	2.6	2.2	2.6
4. I think I would need the support of a technical person to be able to use the system.	3.0	2.7	3.1
6. I thought there was too much inconsistency in this system.	2.1	2.4	2.6
8. I found the system very cumbersome to use.	2.8	2.3	3.0
10. I needed to learn a lot of things before I could get going with this system.	3.1	2.8	3.3
Mean	2.7	2.4	2.9

Legend: 1 – Strongly Disagree, 2 – Disagree, 3 – Neutral, 4 – Agree, 5 – Strongly Agree

In general, the acceptable usability rating given by teachers and IT students is mostly credited to the ease of use of the VR tool. On the other hand, the poor usability rating by non-IT respondents is due to their need for technical assistance and longer period of time to get familiarized with the tool. While it is true that a technical person was present, the assistance relayed was limited because only the user can view the virtual scene. There is then a need to improve the user-friendliness of the system. It would make it easier for new users if there was an on-screen assistance to guide them on how to navigate through the virtual environment, or to instruct them on what to do next.

4. Conclusion

As reflected in the findings, the usability rating of virtual reality as a training tool for oral presentation can vary depending on the technological background of users. For IT students and Trainers, the usability rating is Good, or acceptable; while for non-IT students, the usability rating is Poor. The low rating from non-IT students was due to the need of a technical person, but only during the initial use. This is expected from any other new technology. This study recommends an alternative solution which is the inclusion of an on-screen assistant in the virtual scene that would guide users on how to navigate through the system. Despite these minor problems, the use of virtual reality as a training tool for oral presentation is highly recommended. This is supported by the unanimous positive response of participants in the SUS scale measuring their interest in using the VR tool for oral presentation. While it is true that first-time users face the challenge of navigating through the system, the VR tool is generally very easy to use and its benefits supersede the challenge.

References

[1] Živković S 2014 The importance of oral presentations for university students Mediterranean *J. of Social Sciences* p 5:19
[2] Daud W, Aziz K and Sakib E 2011 An evaluation of students' performance *Proc. of Universiti Malaysia Terengganu 10th Int. Annual Symp.*
[3] Al-Nouh A, Taqi H and Abdul-Kareem M 2014 EFL primary school teachers' attitudes, knowledge and skills in alternative assessment *Int. Education Studies* pp 68–84
[4] Freeman D *et al* 2017 Virtual reality in the assessment, understanding, and treatment of mental health disorders *Psychol Med* p 47
[5] Lenovo 2019 *What is VR Gaming? What accessories are used?* Retrieved on 1 April 2019 from www.lenovo.com/us/en/faqs/pc-life-faqs/what-is-virtual-reailty-gaming
[6] Bouchard S *et al* 2017 Virtual reality compared with in vivo exposure in the treatment of social anxiety disorder: a three-arm randomised controlled trial *The British J. of Psychiatry* vol **210** no 4 pp 276–83
[7] Daniels M M, Palaoag T and Dela Cruz J 2020 Factors influencing user behavior in virtual reality: A systematic review *J. of Adv. Research in Dynamical & Control Systems* vol **12** no 01-Special Issue
[8] L Valmaggia, L Latif, M Kempton and M Rus-Calafell 2016 Virtual reality in the psychological treatment for mental health problems: a systematic review of recent evidence *Psychiatry Tres* vol **236** pp 189–95
[9] Carl E *et al* 2019 Virtual reality exposure therapy for anxiety and related disorders: A meta-analysis of randomized controlled trials *Anxiety Disord* vol **61** pp 27-36
[10] Daniels M M, Palaoag T and Daniels M 2020 Efficacy of virtual reality in reducing fear of public speaking: a systematic *Int. Conf. on Information Technol. and Digital Applications 2019*
[11] ISO 2018 *Ergonomics of Human-System Interaction — Part 11: Usability: Definitions and concepts* Retrieved from https://www.iso.org/obp/ui/#iso:std:iso:9241:-11:ed-2:v1:en
[12] Brooke J 1996 SUS: A "quick and dirty" usability scale *Usability Evaluation in Industry* **189(194)** pp 4–7
[13] Peres S, Pham T and Phillips R 2013 Validation of the system usability scale (SUS) *Proc. of the Human Factors and Ergonomics Society Annual Meeting* vol **57** pp 192–6
[14] Ng W Y, Lo H and Chan A 2011 Measuring the usability of safety signs: a use of system usability scale (SUS) *Proc. of the Int. Multi-conference of Engineers and Computer Scientists*
[15] Orfanou K, Tselios N and Katsanos C 2015 Perceived usability evaluation of learning management systems: empirical evaluation of the system usability scale *Int. Review of Research in Open and Distributed Learning* vol **16** no 2
[16] UIUX Trend 2019 *Measuring and Interpreting System Usability Scale (SUS)* Retrieved from https://uiuxtrend.com/measuring-system-usability-scale-sus/
[17] Eastwick P and Gardner W 2009 Is it a game? Evidence for social influence in the virtual world *Social Influence* vol **4** no 1 pp 18–32

IOP Conf. Series: Materials Science and Engineering 1077 (2021) 012048 doi:10.1088/1757-899X/1077/1/012048

An Integrated Metamodel for Enterprise Architecture using Open Government Data Approach: A Conceptual Framework

Intan Maizura Arzimi[1], Nur Azaliah Abu Bakar[2], Mohd Naz'ri Mahrin[3], Mohd Nazri Kama[4], Surya Sumarni Hussein[5]

[1,2,3,4]Universiti Teknologi Malaysia, Kuala Lumpur, Malaysia,
[5]Malaysian Administrative Modernisation and Management Unit

E-mail: [1]intanmaizura@graduate.utm.my, [2]azaliah@utm.my, [3]mdnazrim@utm.my, [4]mdnazri@utm.my, [5]surya@mampu.gov.my

Abstract. The government is always looking forward to providing citizens with excellent service. One of the government's significant agenda is to identify ways to improve the quality of service and deliver the best service to their citizens. Both Malaysia Open Government Data (MOGD) and Malaysia Government Enterprise Architecture (MyGovEA) initiative play a vital role in assisting the government in providing excellent service deliverables to the citizens. These domains must integrate each other towards the country's economic and social growth. In this paper, we propose the new conceptual model on a new metamodel for the OGD using the Government Enterprise Architecture (GEA) modeling process. Based on the OGD problem and challenges define map with the Archimate metamodel notation to classify the OGD context with the EA metamodel. The study becomes an opportunity for the MyGovEA team and the researchers to highlight EA integration across industry and agencies.

1. Introduction

The most significant of the government's agenda is always looking forward to providing an excellent service deliverable to the citizens. Recently, the innovation on re-strategized their National Information, Communication, and Technology (ICT) Agenda (NITA) by incorporating the EA initiative, strengthening the policy, standards, and practices in the context of ICT plays a vital role in empowering this agenda [1]. Government Enterprise Architecture (GEA) initiatives are the strategic management tool that helps to consolidate the development of business strategies [2]. Malaysian digital transformation and e-government programs require the integration between the agencies in the context of business and ICT towards achieving the agency's functionalities preparing the useful data and information for the decision-making to provide seamless services to the citizens and help the country economic and social growth.

Open Government Data (OGD) was part of Big Data initiatives that emerged globally with the intention of getting public data available to anyone as free and without restrictions [3][4]. Initially, OGD promoting government transparency and accountability but deep interest from the other sectors creates an opportunity for the country's economic growth [5]. The government now days are the biggest social entities with high data complexity. The rising data demand creates rapid data growth and contributes to the massive data management process [6]. Listed 'massive data collection effort and poor quality EA model data as the most frequently appeared [7].

In Malaysia, both GEA and OGD initiatives were introduced in 2014 and become parts of the Malaysian Public Sector ICT Strategic Plan (2016-2020) main agenda to support the 11th Malaysia plan. GEA presented as the holistic approach that aligns the Malaysian government IT resources with the agencies' business processes and strategies through MyGovEA [8]. This is due to the increasing number of IT applications that exist in silos at every government agencies but then it serves the same functionalities. Hence, these circumstances cause redundancy in cost, technology, and even IT experts that develop those applications. This has urged MAMPU, the IT centralization agency in MPS to encourage the usage of the shared resources which can be achieved through MyGovEA [9].

On the other hand, OGD initiative in the Malaysian Public Sector (MPS) aims to publicly and freely share government data using the online platform [10]. However, the implementation of OGD in MPS faced a quality challenge such as lack of data integration between agencies and the nonexistence of OGD metadata [11] cause the data owner is required to update data repeatedly. Furthermore, the government data policy which is not gazettes yet, also one of the hindering factors for the successful OGD implementation in MPS [12].

GEA and OGD should have a bilateral relationship to support new data management techniques [13] to enable industries and businesses the fastest growth. Looking at these issues exists, this study aims to investigate the possibility to integrate the OGD metadata structure in the context of problems and challenges with EA metamodel on top of data layers. Currently, there is no yet exist a specific metamodel GEA data layer that utilizes OGD. This paper will propose the new conceptual framework on a metamodel for GEA based on OGD.

2. Literature Review

2.1. Enterprise Architecture Overview

Enterprise Architecture (EA) defines as the business process management in a large-scale organization with an understanding of the information system supporting the business function. Basically, it consists of the functional description as "input-process-output" to develop the business strategy to be linked with the information system [14]. It also helps the organization in designing its various organization facets such as business, data, application, and information to achieve the organizational goal. EA fundamentally starts from the introduction at the first stage is to familiarize the concepts and functionalities towards understanding, planning, and implementation. In a wide, the discipline EA visualizes clear artifacts that help the organization to identify its various organization facets among business, data, application, and information layer to achieve the organization's goals and objectives [15].

2.2. Enterprise Architecture Modeling

Enterprise Architecture modeling is defined as a model as a simplification of reality in order to get a better understanding of the flow, system structure [16]. EA Modeling encapsulated the illustration of the complexity of the organization. The concept interrelated to operational aspects of the organizational weather on the holistic view or in-depth [17]. The first essential consist of the methodology and language use to develop the model [18]. The modeling approach usually used at the planning and implementation stage to identify the quality and satisfaction of information and system towards the organizational impact [15][19]. A study by Sudarsono et al [19] shows the emergence of the design comes from a series of interrelated concepts. It highlights the architectural elements to contribute to implementing the organizational strategies. The modeling procedures are based on the use of the information described in the models to support various enterprise tasks [20]. The integration is possible in the future long-term challenge is to involve the business people in the adaptation of EA and the implementation and adaptation of enterprise information systems.

The EA model proposed as the holistic view was at the strategic and business level. The modeling process is a crucial enabler for the solution to many business problems. There are different types of modeling tools used to model EA such as Open Source ArchiMate modeling and UML modeling. This paper will be using ArchiMate Modelling as the most closely related to research modeling [16]. Figure 1 shows the modeling approach as a solid basis for this future challenge.

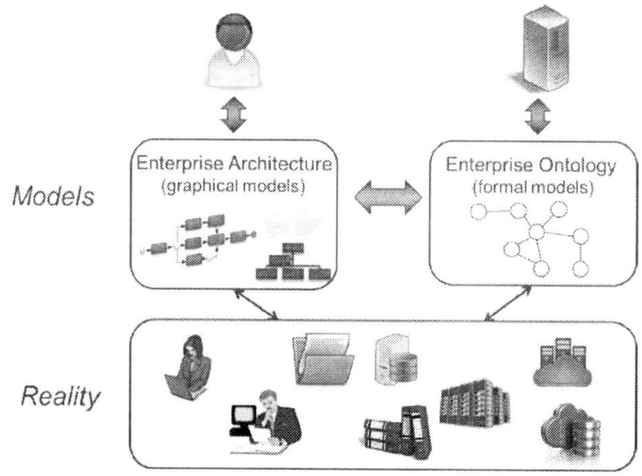

Figure 1. The modelling approach for continuous business-IT alignment. (Hinkelmann, Knut, et al. 2016)

TOGAF Architectural Development Method (ADM) combines with Archimate Modelling Approaches consists of designing and implementing the component Strategy and Motivation, Business Layer, Application Layer, Software Layer, and Implementation and Migration. It serves not only as an EA framework but also tools to promote architectural development [20]. Figure 2 depicts the ADM as discussed.

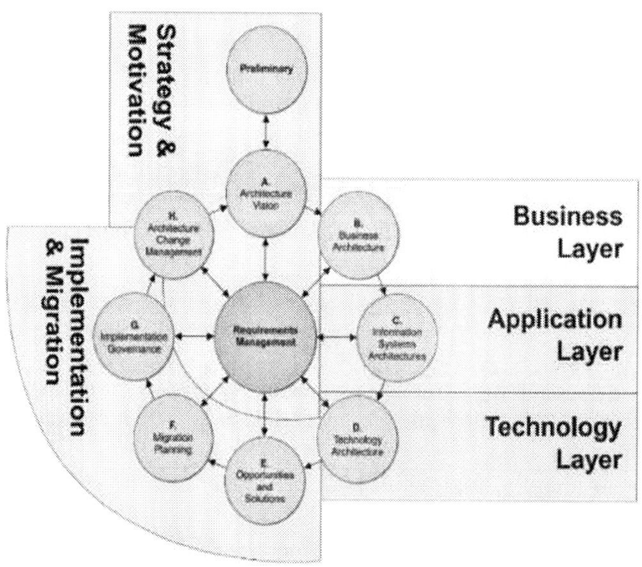

Figure 2. Archimate Modelling (TOGAF www.opengroup.org)

2.3. *Enterprise Architecture Metamodel*

Metamodel describes a set of artifact models develop from the process of defining the problem statement, analysis, and construct the problem into the useful and applicable model. Meanwhile, EA metamodel is a clear common language and structure view that shows the relevant dependencies and relationship between business, data, application, and technology. The metamodel is located between formal and informal expressions [8].

It is a formal basis that ensures that models can be interpreted unambiguously and are ideal for automated analysis [21]. The fundamental organization of a system is often represented by models of the 'as-is' state or the 'to-be' state of a system. Korean government's EA describes the EA metamodel as the compilation of EA deliverables standard that organizations need to establish and disclose towards the EA government-wide Information Technologies interoperability among the agencies. It is also a foundation model used to determine the EA business by specifying the architectural information required and their relationships [22].

EA modeling plays an essential role in the practice, as shown in Figure 3 shows the ArchiMate metamodel, which is now adopted by the Open Group interconnecting the business, application, and technology in a top-down view of the actors. It built the easy and understand model which will support the existing and develop the new form flows of integration organizational functional process [23]. This generic metamodel will be used to map on top of the OGD problem and challenges discussed to classify on OGD issues and which layer should take the responsibility to expend the proper metamodel on problem and challenges solution.

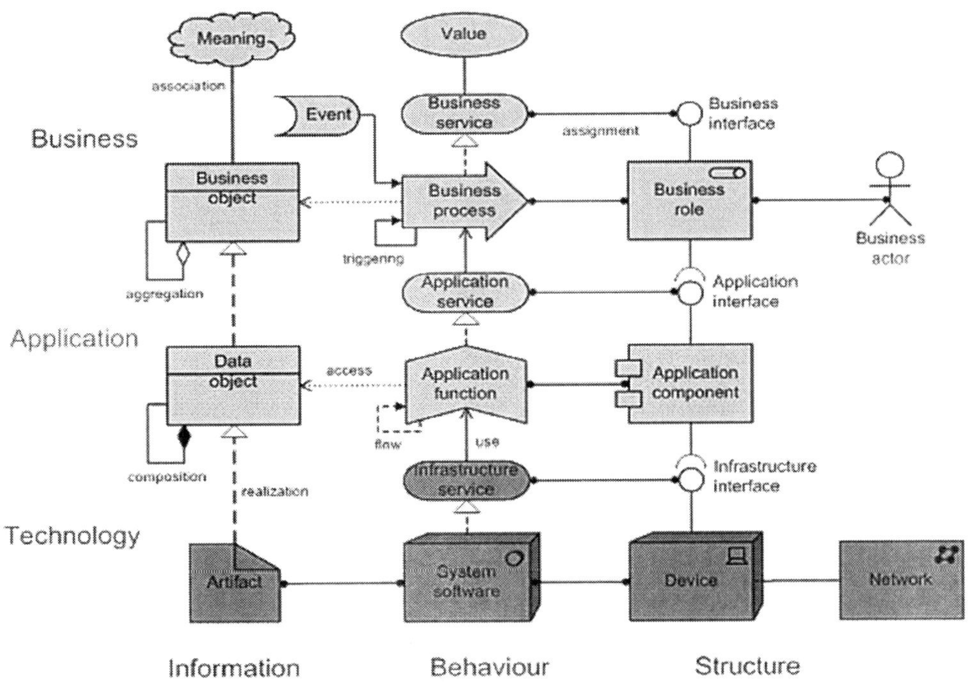

Figure 3. The ArchiMate Meta-Model (http://www.archimate.org)

2.4. Open Government Data

Open Government Data (OGD) is a government set of data sharing publically to their citizens. It helps the citizens to understand their countries better with facts and figures as the information can be openly shared and viewed by the citizens [24]. Three different approaches contribute to OGD are the government data approaches, open data approaches, and open government approaches [25]. All the approaches interconnected each other as illustrated in Figure 4.

The main features of OGD are transparency, participation, and collaboration. Transparency means the acquired information collected by the government's institutions are available to the public with the certain data policies provided. Meanwhile, participation refers to allow the citizen to participate in the decision making as well as contributing idea. Finally, collaboration refers to both private and public sectors to increase government decision-making and improve the effectiveness of government services deliverables [26].

Overall, OGD is still in a context of implementation that requires personnel with the necessary skills to manage data from multiple perspectives and views [12]. Since 2000, the OGD portal is developed widely to promote transparency, accountability, and value creation by making government data available to their citizens [10] [27]. OGD was expected to promote democratic governance and political engagement to foster improvements in service and industry and social innovation. According to McKinsey, OGD could potentially boost global economic growth worth \$3 trillion by delivering better services, and greater transparency and accountability [28].

Open government data can bring many benefits for both government and society [27] but only five studies were found addressing various challenges related to policy, technology, financing, organization, culture, and legal frameworks is important to create an ecosystem and sustainable business models for OGD initiatives that can deliver the desired benefits [24]. In the context of data management, in the study conducted by Mourino et al. on top of Italian OGD, only 40% of the state and municipality portal were completely updated [11] in the context of OGD data quality observation. Furthermore, not many publications address this data quality challenge as most of the study more focuses on organizational OGD purposes [27].

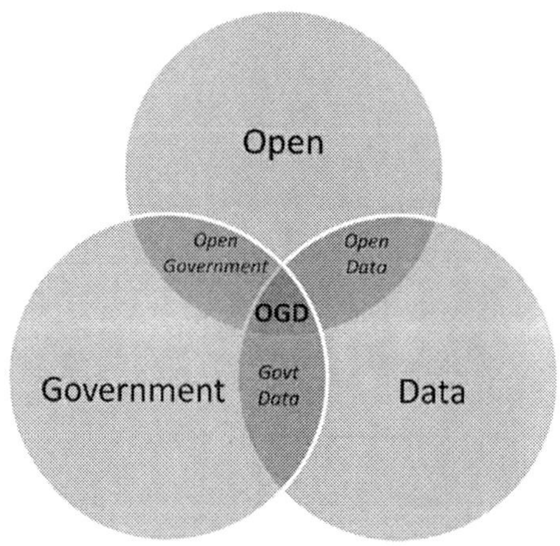

Figure 4. Foundations of Open Government Data (Adapted from Gonzalez-Zapata et al. 2015)

2.5. Malaysia Government Enterprise Architecture

The Malaysian government launched the Malaysian Government Enterprise Architecture (myGovEA) in 2014 [8]. The principles being developed to initiate the EA practice in Malaysia. The principles are a fundamental set of overarching architectural standards that support the development of standardized and consistent architecture across the organization. They are often used to guide an organization to implement future initiatives in public sector agencies. They should be adopted when agencies begin to develop their business and technical architectures [29]. Figure 5 shows the MyGovEA Principles consist of the general, business, data, application, and technology. Another set of document develop are the framework and methodology for guiding the public sector agencies in Malaysia in building EA practices. It is the other new way of approaching the business and technology to be aligned together and establish a single view of the current business and technical environment for the particular public sector organization. However, some Malaysian government agencies are still hesitating to carry out this EA in their organizational structure due to organizational complexity, public values issues, and the political climate [30].

2.6. Malaysia Public Sector Open Government Data

Malaysia Public Sector OGD(MPSOGD) initiatives were launched in 2014 to encourage data sharing to support a reliable, fast, and transparent delivery system. The implementation was expected to have a significant impact on the economic growth, social well-being, and the society of the country to increase the competitiveness of a country in the digital initiatives [31]. Both MyGovEA and MPSOGD represent different functionalities in the government sectors but the most similar in its goals, which is towards government digital transformation. Most of the EA study focused on EA understanding, implementation, and management but not many on EA modeling study [32]. Until now there is a lack of an EA model that entirely enables integrated enterprise modeling. The architects keep on using their modeling techniques and principles, tool support, visualization techniques for each architectural domain [21]. Even though the data layer where stated closed to the business layer, there is yet a proper interlink between this business layer and the data layer [33]. In this point, highlighted Malaysia EA expert and practitioners also plays a vital role in EA data management [34].

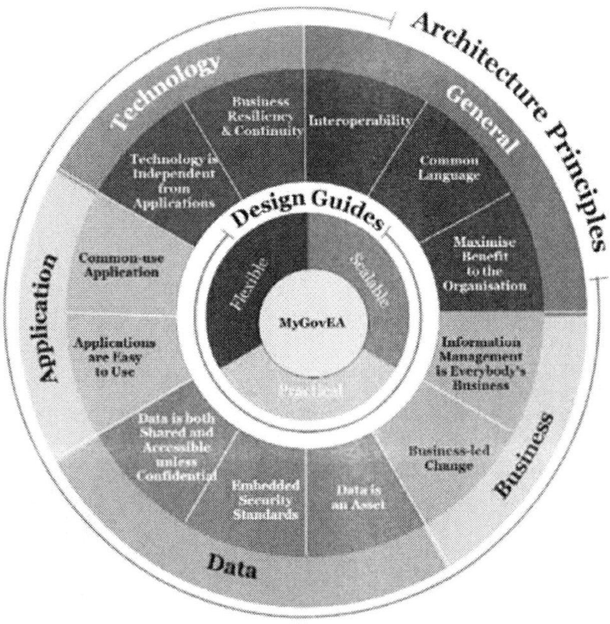

Figure 5. MyGovEA Principles (MAMPU 2016)

3. Methodology

This study adapting the Design Science Research Methodology (DSRM) by Hevner [35] to design the conceptual framework on the metamodel for the MyGovEA data layer using the OGD. Figure 7 shows the Design Science Research Methodology used in this study. The problem identifies shows that both MyGovEA and OGD are having the same responsibility in obtaining data objects, data set to be interpreted, and providing valuable data to the citizens.

Both of the domains MyGovEA and OGD in a context of technical perspective holding a different role with similar function supplying the end product or valuable data to support the decision-making process and contribute to the country's economic growth. The next step is keyword searching which is related to the topic of integration and domain public sector, MyGovEA, and OGD.

The next step is keyword searching which is related to the topic of integration and domain public sector, MyGovEA, and OGD. A literature review is done to validate knowledge and getting the essential information to fill the novelty for the current study [36]. Effective literature searching defines the related topic to analyze and contribute to the writing process [37]. Four major activities in Figure 7 such as searching gathering, screening categorizing, analyzing, structuring, and writing have been done accordingly to fit the purpose and ready to put in into writing.

The objective of the solution is done through the literature review on GEA and Open Data related to the government sector. Finally, at the design and development, we choose the Archimate metamodel as the base model to be a map on top of the OGD issues founded during the OGD implementation [27] and leverage the new conceptual model to be integrated with the technical design OGD initiatives [12]. The next study will cover the context of demonstration, evaluation, and validation that will be using some cases to show the usefulness for achieving the purposes.

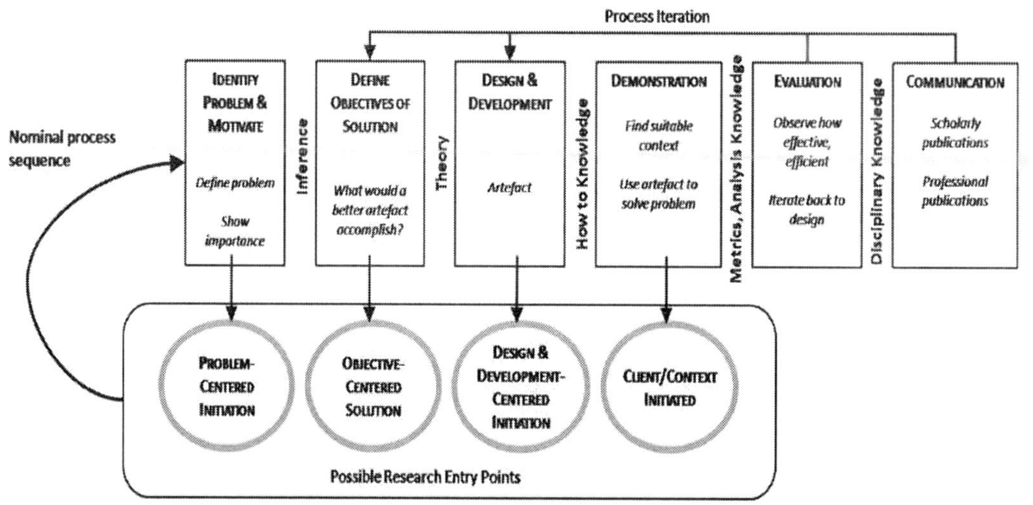

Figure 6. Design Science Research Methodology (Peffers et al. 2008)

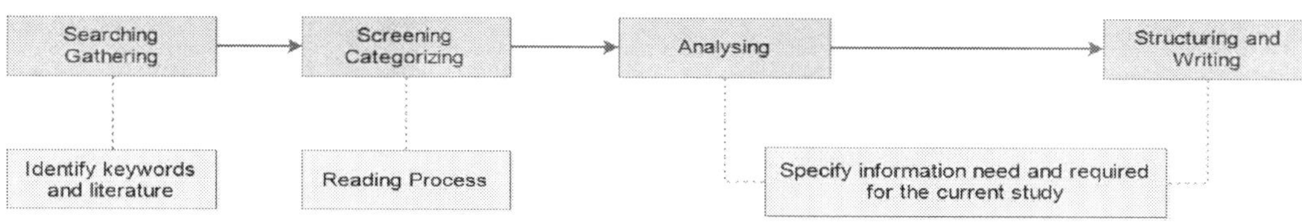

Figure 7. Basic flow on the literature review process

3.1. Data Collection

Five databases were used to conduct the literature are the Scopus, Science Direct, Web of Science, IEEE, and Google Scholar. The keyword used for this study is ("enterprise architecture"), ("open government data"), ("integrate*") then the boolean use is "AND". The most essential information requires is the integration between both of these GEA and OGD. Most of the study on OGD highlighted the main problem and challenges affecting the OGD initiatives. This problem and challenges will be defined and map with the Archimate metamodel. Two studies highlighted the most similar OGD problem appear are the organizational, policy and legal, economic, and financial posses of a similar perspective of the problem.

3.2. Data Analysis

To extract the required OGD problem and challenges qualitative content analysis is taken into action. Inductive and deductive analysis help to build the categorizing from the bottom technical content towards the organizational content [39]. Table 1 shows the findings from the data analysis process. Based on the researcher's perspective conclude having a similar understanding of OGD problem and challenges categorize such as Organizational, Policy and Legal, Economic, and Financial. Citizens/Cultural were combined as one category as some of the perspectives of the study are similar. Data Quality and Technology as category problems and challenges related to service deliverables. Each of these categorizes problems and challenges still arise, will be aligned to the Archimate metamodel as the guideline for the new study in the theme of integration between the GEA and OGD.

4. Discussion

The integration between EA and OGD is still new in any government initiatives. EA in the public sector gains attention as an approach to strengthening the e-government function. This hierarchical approach helps the government a lot strategically aligning the business by integrating the information structures, procedures, organizational units, and staff in the enterprise. In the sense of Malaysia's public sector perspective, OGD, as an initiative connected to the Digital Economy Transformation Program, launched to embed with the Malaysian Government's National Agenda the sharing of sets of data that civilians can freely access.

Table 1. Main Problem and challenges affecting OGD initiatives

Categorize	In the perspective of [27]	In the perspective of [38]
Organizational	Authority involvement Awareness Benefit metrics Competition Institutionalization Inter-organization collaboration Motivation Policy	Institutionalization Overlapping Scope Technical Support
Policy and legal	Copyright & licensing Liability Regulations Privacy & data protection	Copyright/Licensing Liability Regulations Privacy/Data Protection Conflicting
Economic and financial	Budget provision Cost metrics	Budget Provision
Citizens / Cultural	Public participation User skills	Public Participation Awareness Motivation Competition
Data Quality	Data accuracy Data ambiguity Data completeness Data formats Data Data misuse Data Data quality Data representation Data scarcity	
Technical	Data access Data interoperability Storage capacity Technical knowledge Technical support	Formats Ambiguity Discoverability Representation Capacity

MyGovEA consists of the blueprint which includes the EA framework focus on business process solutions and technology, methodology, and implementation plan [2]. The people factors with significant EA talent and experts also play a vital role to ensure this integration can be understood and develop in the future [38]. EA modeling and metamodel still an ongoing topic in EA publication but significantly fewer than other EA topics [32] contribute to a lack of understanding deeply to integrate between the GEA to other domains. By the time EA should become the solution in a holistic view.

Eventually, since 2014, OGD introduces in the Malaysian public sector, the dataset is growing from 115 to 12,955 in the MOGD portal [10] shows the effort of the practitioner towards data sharing available to the citizens are actively ongoing promoted and run. The problem and issues that arise at the beginning of the OGD implementation still exist with unclear perspectives. Some other how, both the

MyGovEA and MOGD enroll similar roles must produce valuable data for decision making towards social and economic growth.

OGD problems and challenges are identified to be placed accordingly on top of the generic Archimate metamodel that highlights a similar problem and challenge category map on the EA aspect as in Figure 8. The categorize of Organizational, Policy and Legal, Economic, and Financial, suitable to be located at the Business level in the context of EA. The perspective in this problem and challenge categorize were suitable to be solved at this level as the categorize are still stated at the strategic view.

The Citizens/Cultural are closely related, they are the people and theme accessing the data through the application, the problem and challenges related to this category should be placed at this level as a total solution to data usability. Only two studies touched on the problem and challenges of Data Quality as poor data quality that may be ubiquitous and can impede the successful reuse of open data. The data collection process now forces data owners facing of updating data repeatedly [41]. This requires the new data management technique to be created at this level. Finally, technical problems and challenges related to infrastructure, hardware, and software readiness are placed at the EA technology level.

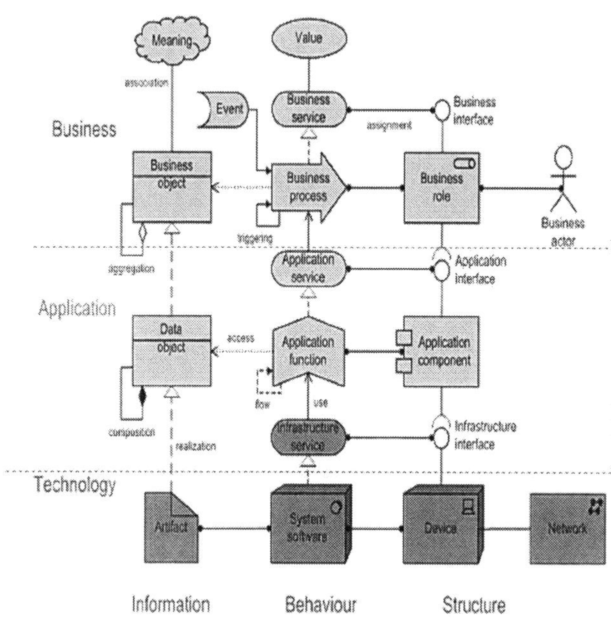

Categorize	In the perspective of [28]	In the perspective of [38]
Organizational	Authority involvement Awareness Benefit metrics Competition Institutionalization Inter-organization collaboration Motivation Policy	Institutionalisation Overlapping Scope Technical Support
Policy and legal	Copyright & licensing Liability Regulations Privacy & data protection	Copyright/Licensing Liability Regulations Privacy/Data Protection Conflicting
Economic and financial	Budget provision Cost metrics	Budget Provision

Categorize	In the perspective of [28]	In the perspective of [38]
Citizens / Cultural	Public participation User skills	Public Participation Awareness Motivation Competition
Data Quality	Data accuracy Data ambiguity Data completeness Data formats Data Data misuse Data Data quality Data representation Data scarcity	

Categorize	In the perspective of [28]	In the perspective of [38]
Technical	Data access Data interoperability Storage capacity Technical knowledge Technical support	Formats Ambiguity Discoverability Representation Capacity

Figure 8. Conceptual Framework metamodel for GEA map on OGD problem and challenge

Data Quality problems become a significant topic highlight and suitable to be located at this application level should contribute more publication, awareness, competition, and motivation to be isolated at the EA application level. This level plays a vital role to ensure fully design and develop the data architecture or data layer to ensure the data structure interoperability towards the decision-making process [38]. The set of the new conceptual framework is proposing to bridge both EA and OGD to illustrate the new collaboration between both domains. Predefining and analyzing the clear problem statement by referring the problem and challenge categorize map towards each level as the guideline that will help the researchers and practitioners to develop the appropriate detail metamodel and metadata to reduce and overcome each category of problem and challenge. Finally, the future phase defines and validates the integrated metamodel for OGD according to the EA Archimate metamodel map with the problem and challenges of OGD perspectives as the next study contribution. Lack of research and publication on this topic contribute as an opportunity for academic and expert to further research.

5. Conclusion and Future Works

The innovation of ICT technology plays a vital role in empowering the service deliverables. Transitioning from e-government to digitalization requires to remodel and change management process to be implemented towards moving the digital transformation or digitalization. Integration among ICT tools and infrastructure across the EA domain and others become more significant topics to be the highlight. OGD problems and challenges are highlighted to enable the opportunity to detail research on the integration process between the EA and other domains. Public Sector OGD is still new approaches and only practicing the data sharing internally among the government agencies. The conceptual model should assist the practitioner and academician to initiate a detailed study on the integration process between GEA and OGD in a public sector to reduce redundancy on data management and increasing the services deliverables performance to the citizens and across the public and private sectors.

The new conceptual model integrating the OGD using EA modeling proposed and more new study should be discovered. The impact of this further study will contribute to a beneficial economy and a more systematic business solution. The future study will take into action in the context of implementation and modeling the specific metamodel on each level the integration between the GEA and OGD.

Acknowledgments

The research is financially supported by Ministry of Higher Education Malaysia and Universiti Teknologi Malaysia under grant vot no R.K130000.7856.5F072

References

[1] Bakar N A A, Harihodin S and Kama N 2013 Influence factors in government enterprise architecture establishment process: a preliminary findings *23rd IBIMA Conf. on Vision 2020: Sustainable Growth, Economic Development, and Global Competitiveness* no June 1-6

[2] Bakar N A A, Kama N and Harihodin S 2016 Enterprise architecture development and implementation in public sector: The Malaysia Perspective *J. of Theoretical & Applied Information Technol.* **88(1)** pp 176–88

[3] Kalampokis E, Tambouris E and Tarabanis K 2011 Open government data: a stage model *Int. Conf. on Electronic Government* pp 235-246 (Berlin, Heidelberg: Springer)

[4] Zainal N Z, Hussin H and Nazri M N M 2018 A trust-based conceptual framework on open government data potential use *2018 Int. Conf. on Information and Communication Technology for the Muslim World (ICT4M)* (New Jersey: IEEE) pp 156–61

[5] Alexopoulos C, Loukis E, Mouzakitis S, Petychakis M and Charalabidis Y 2018 Analysing the characteristics of open government data sources in Greece *J. of the Knowledge Economy* 9(3) pp 721–53

[6] Saha P 2010 Advancing the whole-of-government enterprise architecture adoption with strategic (systems) thinking *Enterprise Architecture as platform for connected government* (Singapore: National University of Singapore)

[7] Banaeianjahromi N and Smolander K 2019 Lack of communication and collaboration in enterprise architecture development *Information Systems Frontiers* **21(4)** pp 877–908

[8] Bakar N A A, Yaacob S, Hussein S S, Nordin A and Sallehuddin H 2019 Dynamic metamodel approach for government enterprise architecture model management *Procedia Computer Sci.* **161** pp 894–902

[9] Chung H M and McLeod G 2002 Enterprise architecture, implementation, and infrastructure management *Proc. of the 35th Annual Hawaii Int. Conf. on System Sciences* (New Jersey: IEEE) pp 1256–7

[10] Zainal N Z, Hussin H, Abd Rahim N H, Nazri M N M and Suhaimi M A 2019 Open government data use by Malaysian researchers some empirical evidence *2019 6th Int. Conf. on Research and Innovation in Information Systems (ICRIIS)* (New Jersey: IEEE) pp 1–6

[11] Vetrò A, Canova L, Torchiano M, Minotas C O, Iemma R and Morando F 2016 Open data quality

measurement framework: definition and application to open government data *Government Information Quarterly* **33(2)** pp 325–37
[12] Mustapa M N, Hamid S and Nasaruddin F H M 2019 Exploring the issues of open government data implementation in Malaysian public sectors *Int. J. on Advanced Sci., Eng. and Information Technol.* **9(4)** pp 1466–73
[13] Kehrer S, Jugel D and Zimmermann A 2016 Categorizing requirements for enterprise architecture management in big data literature *2016 IEEE 20th Int. Enterprise Distributed Object Computing Workshop (EDOCW)* (New Jersey: IEEE) pp 1–8
[14] Zachman J A 1987 A framework for information systems architecture *IBM Systems J.* **26(3)** pp 276–92
[15] Ahmad N A, Drus S M, Abu Bakar N A and Othman M M 2018 A systematic review of enterprise architecture adoption models *Int. J. of Eng. & Technol.* pp 369–74
[16] Holt J 2010 *Modelling enterprise architectures* (London: The Institution of Engineering and Technology)
[17] Jonkers H, Lankhorst M, Van Buuren R, Hoppenbrouwers S, Bonsangue M and Van Der Torre L 2004 Concepts for modeling enterprise architectures *Int. J. of Cooperative Information Systems* **13(03)** pp 257–87
[18] Hinkelmann K, Maise M and Thönssen B 2013 Connecting enterprise architecture and information objects using an enterprise ontology *Proc. of the First Int. Conf. on Enterprise Systems: ES 2013* (New Jersey: IEEE) pp 1–11
[19] Sudarsono B G, Andry J F and Nirwan 2020 Design information system order fulfillment using archimate modelling *Int. J. of Advanced Trends in Computer Sci. and Eng.* **9(02)** pp 1303–10
[20] Gampfer F, Jürgens A, Müller M and Buchkremer R 2018 Past, current and future trends in enterprise architecture—a view beyond the horizon *Computers in Industry* **100** pp 70–84
[21] Hinkelmann K, Gerber A, Karagiannis D, Thoenssen B, Van der Merwe A and Woitsch R 2016 A new paradigm for the continuous alignment of business and IT: combining enterprise architecture modelling and enterprise ontology *Computers in Industry* **79** pp 77–86
[22] Lankhorst M M 2004 Enterprise architecture modelling—the issue of integration *Advanced Eng. Informatics* **18(4)** pp 205–16
[23] Du Lee J and Kwon Y I 2013 A study on strategy planning and outcome of EA in Korea *2013 15th Int. Conf. on Advanced Communications Technol. (ICACT)* (New Jersey: IEEE) pp 873–9
[24] Ricca M 2011 Bridging the gap between enterprise architecture and solution architecture for maximum benefit *Technical Article* Retrieved on 31 August 2020 from https://www.oracle.com/technical-resources/articles/enterprise-architecture/ricca-bridging-ea-sa.html
[25] Ubaldi B 2013 Open government data: towards empirical analysis of open government data initiatives *OECD Working Papers on Public Governance* no 22 (Paris: OECD Publishing)
[26] Gonzalez-Zapata F and Heeks R 2015 The multiple meanings of open government data: understanding different stakeholders and their perspectives *Government Information Quarterly* **32(4)** pp 441–52
[27] Petrušić D, Segedinac M and Konjović Z 2016 Semantic modelling and ontology integration of the open government systems *Tehnički vjesnik* **23(6)** pp 1631–41
[28] Roa H N, Loza-Aguirre E and Flores P 2019 A survey on the problems affecting the development of open government data initiatives *2019 Sixth Int. Conf. on eDemocracy & eGovernment (ICEDEG)* (New Jersey: IEEE) pp 157–63
[29] Dawes S S, Vidiasova L and Parkhimovich O 2016 Planning and designing open government data programs: an ecosystem approach *Government Information Quarterly* **33(1)** pp 15–27
[30] Malaysian Modernisation And Management Planning Unit (MAMPU) 2016 *Background of MyGovEA* Retrieved on 31 August 2020 from http://mygovea.mampu.gov.my/en/enterprise-architecture/background-mygovea

[31] Bouwman H, van Houtum H, Janssen M and Versteeg G 2011 Business architectures in the public sector: experiences from practice *Communications of the Association for Information Systems* **29(1)** p 23
[32] Modernisation and Management Planning Unit (MAMPU) 2019 *Open Government Data* Retrieved on 31 August 2020 from https://www.malaysia.gov.my/portal/content/30589
[33] Dang D D and Pekkola S 2016 Root causes of enterprise architecture problems in the public sector *PACIS 2016 Proc.* vol **287**
[34] Bakar N A A, Harihodin S and Kama N 2016 Enterprise architecture implementation model: Measurement from experts and practitioner perspectives *2016 4th IEEE Int. Colloquium on Information Sci. and Technol. (CiSt)* (New Jersey: IEEE) pp 1–6
[35] Hevner A R 2007 A three cycle view of design science research *Scandinavian J. of Information Systems* **19(2)** 4
[36] Rowley J and Slack F 2004 Conducting a literature review *Management Research News* vol **27** no 6 pp 31–9
[37] Timmins F and McCabe C 2005 How to conduct an effective literature search *Nursing Standard* **20(11)** pp 41–7
[38] Creswell J W and Clark V L P 2017 *Designing and conducting mixed methods research* (California: Sage Publications)
[39] Attard J, Orlandi F, Scerri S and Auer S 2015 A systematic review of open government data initiatives *Government Information Quarterly* **32(4)** 399–418
[40] Bakar N A A and Hussien S S 2018 Association of people factors with successful enterprise architecture implementation *Int. J. of Eng. & Technol.* **7(4.31)** pp 52–7
[41] Zainal N Z, Hussin H, Abd Rahim N H, Nazri M N M and Suhaimi M A 2019 open government data use by Malaysian researchers: some empirical evidence *2019 6th Int. Conf. on Research and Innovation in Information Systems (ICRIIS)* (New Jersey: IEEE) pp 1–6

Fingerprint Salt & Pepper Noisy Image Enhancement with Threshold Method

Mahmut Ince[1], Songfeng Lu[2], Saad Ali Alfadhli[3], Ismail Maulood[4], Shayem Saleh Alresheedi[5]

[1,3]School of Computer Science and Technology, Huazhong University of Science and Technology,Wuhan 430074, China
[2]School of Cyber Science and Engineering, Huazhong University of Science and Technology, Wuhan 430074, China
[4]Department of Research and Development, Directorate of Scientific Affairs, Ministry of Higher Education and Scientific Research Erbil, Kurdistan Regional Government, Iraq
[5]War College, National Defense University, Riyadh, KSA

E-mail: [2]lusongfeng@hust.edu.cn, [1]mahmutince@hust.edu.cn

Abstract. We present an adaptive threshold method for salt and pepper noisy fingerprint images. At first, we applied a threshold algorithm to detect noisy and corrupt pixels. Fingerprint image optimized with window filter produced pixels values. We measured the quality of the fingerprint images with peak signal to noise ratio and other efficient noise removing techniques were compared with it. The result shows improving performance of our method with remove salt and pepper noise from a fingerprint image. Our present work is useful to reduce or minimize medium level noise in fingerprint image. In future, we will work to reduce more than 80 percent noise level in fingerprint image.

Keywords: Salt-pepper noise, Image denoising, Adaptive threshold

1. Introduction

A fingerprint has become a more popular biometric recognition system due to permanency, easy implementation, small size sensor and affordable cost. Despite of these advantages, fingerprint identification and verification system have a challenge with low-quality fingerprint image and noise. The fingerprint images may have noise cause of low-quality images due to sensors, dust particles, oil, dryness, non-cooperated user [1-5].

Consequently, noise detection and noise filtering is still challenging in fingerprint image processing. Noise filtering techniques are important in the enhancement process effect fingerprint image quality, image quality effect fingerprint features, and features affect biometric system accuracy. Previous literature proposed several nonlinear filters to remove unwanted effect as salt -pepper noise from fingerprint images [6-8]. As the median filter shows simple and efficient computational performance [9-10] but disadvantage of this filter is that they are efficient only in low noise density if fingerprint image includes high-density noise than some patches are visible and edge details will disappear [11]. Centre weighted median filter, Dual window selective median switching filter, adaptive median filter are proposed to protect edge pixel [12-13].

Filters are applied on the previous pixel and if current pixels are damaged or include noise it can degrade image quality [14]. A low-quality image can lost useful image details. To solve this shortcoming, two-stage methods have proposed. The main idea of this method helps to detect noisy or corrupted pixel which has separated from other pixels [15].

Most of the previously proposed methods use two stages to improve image quality such as an improved Progressive Switching Median Filter [16], Recursive Weighted Median Filter [17], Dual Windows Selective Median Switching Filter [18], Adaptive Image Denoising Method [19], etc. We used 3 x 3 windows for noise detection, to detect noisy, clear and corrupted pixels. We describe test window, a central pixel of window. In this study we determine the threshold for detecting noise. This noise will change with the algorithm result value. The outline of the work is organized as follows, after Section 1; Section 2 shows the proposed algorithm method. Section 3 shows the proposed method, experiments, comparison of results from different efficient filters and discussion. Section IV consist of conclusions.

2. Proposed Algorithm

2.1. Modified Impulse Noise Model

In this model image inspected in pixel level, pixels are effected with [0, 255] peak values. Unchanged pixels stay the same as before (genuine) [20-22].

The method is shown at the below:

$$\text{S(X)} = \begin{cases} \frac{q}{2} \; X_{ij} = 0 \\ \left(1 - \frac{q}{2}\right) X_{ij} = W_{ij} \\ \frac{q}{2} \; X_{ij} = 255 \end{cases} \tag{1}$$

Where q is noisy pixel density, S(X) is possibility function, W(i,j) is the original image pixel intensity, and X(i,j) is the intensity of the noisy pixel.

2.2. Modified Decision-Based Algorithm

In [23] presented algorithm accepts all pixel intensity in 0and 255 noisy pixel detection window is 3 x 3. If the calculated value of a pixel in the range [0, 255] then it is, a clear pixel and the algorithm will not change pixel. If value is not in range then the pixel is affected by noise and it changes with the median value of the window method.

3. Proposed Method

In this study, our proposed method convolutes decision-based method and adaptive threshold methods. We used a threshold to determine a noisy pixel. In this process, the central pixel value of (3 x 3) window plays a vital role. For threshold, central pixel values used to decide that pixel are corrupted or not, if pixel values are higher or less than the central pixel values of the range then those pixels accept the noise.

The decision-based algorithm used for change corrupted or noisy pixels after the detection phase. In another word, let the grey level Image S of size [M x M], where M $\in R$ & $0 \leq M \leq 255$. X (i , j) pixel of S filtered by [3 x 3] filtering window "W", formed by dividing the image. Also, we propose the minimum level threshold (Thr_{min}) and maximum level threshold (Thr_{max}) that computed in Eq. (3) and (4) as a threshold which we used to detect corrupted pixels.

Pixels are filtering by window filter can be shown as below:

$$M_w = \frac{1}{W \, x \, W} \sum_{i=0}^{2} \sum_{j=0}^{2} X_w(i,j) \tag{2}$$

Thus, the Eq.2 can efficiently give the mean of window value (M_W) for every filtering window in the image.

$$Thr_{min} = M_w * (1 - \theta) \tag{3}$$

$$Thr_{max} = M_w + \{255 - M_w\} * \theta \tag{4}$$

Where $\theta \in R$ & $0 \leq \theta \leq 1$ is a constant value calculated by the image [24] as shown in Eq. 5

$$\theta = \sqrt{\frac{2\sigma^2 log\,(m)}{m}} \tag{5}$$

We compare the Thr_{min} and Thr_{max} with central pixel values to detect noisy pixels. So, if the values are within the range of M, the system accepts them as a noise-free as shown in the below formula (ClP_W: Noise free pixel):

$$ClP_w \begin{cases} Thr_{min} \leq ClP_w \leq Thr_{max} \ (clean\ pixel) \\ Else \quad Corrupted \end{cases} \tag{6}$$

Corrupted pixel detected then the DbA used to replace a corrupted pixel with calculated window values. Proposed modified Denoising algorithm follows the below steps:

Input: The grayscale fingerprint image F(i, j)

Output: Denoised, repaired fingerprint image

Step 1: 3 x 3 windows select

Step 2: Mean, M_w of F(i, j) in filtering window estimates

Step 3: θ calculated by Eq.5

Step 4: Thresholds Thr_{min} and Thr_{max} calculated from Eq.3 and Eq.4

Step 5: If $Thr_{min} \leq ClP_w \leq Thr_{max}$ then ClP_w is clear pixel; else ClP_w will be replaced by DbA value.

Step 6: Optimal values obtained

Step 7: Exit

The result obtains in the following platform as follows: The hardware platform: CPU Intel i5 3.00 GHz, Software platform: Matlab 2019a. The new algorithm verified with the following methods. Standard median filter (SmF), Window median filter (WmF), Progressive switching median filter (PsMF) and Recursive enhanced median filter (ReMF) [15-18].

Fingerprint verification competition (FVC 2002) DB1 and DB2 used for validating the proposed algorithm. Fingerprint images set to 256 x 256 pixels. The quality and performance measurement is the peak-signal-to-noise ratio (PsNR) used for renovation. Renovated image (F_R) of size M x M concerning the original image original image (Fo) as:

$$\text{PsNR}\ (F_0, F_R) = 10log_{10}\frac{255^2}{Mse} \tag{7}$$

$$\text{MsE}\ ((F_0, F_R) = \frac{1}{MN}\sum_{i=0}^{M-1}\sum_{j=0}^{N-1}(F_R(i,j) - F_0(i,j))^2 \tag{8}$$

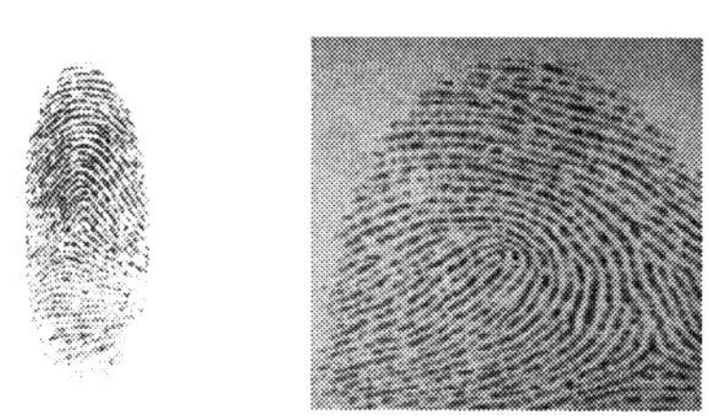

i ii

Figure 1. The resized original grey images: from FVC2002 (i) DB1, (ii) DB2

Table 1. Fingerprint image in FVC 2002 DB1 with PsNR results for salt and pepper noise

Method	Noise Ratio (PsNR)			
	50%	60%	70%	80%
SmF	9.6	7.3	6.46	4.43
CwMF	14.52	11.43	7.86	4.96
PsMF	15.51	13.92	12.36	8.75
ReMF	19.83	14.72	11.92	7.84
Proposed	19.95	19.8	16.98	15.62

Table 2. Fingerprint image in FVC 2002 DB 2 with PsNR results for salt and pepper noise

Method	Noise Ratio (PsNR)			
	50%	60%	70%	80%
SmF	10.6	8.3	6.46	5.43
CwMF	16.52	11.48	7.86	4.96
PsMF	18.53	15.40	13.92	6.14
ReMF	21.83	20.78	14.72	10.61
Proposed	23.68	22.53	19.80	17.76

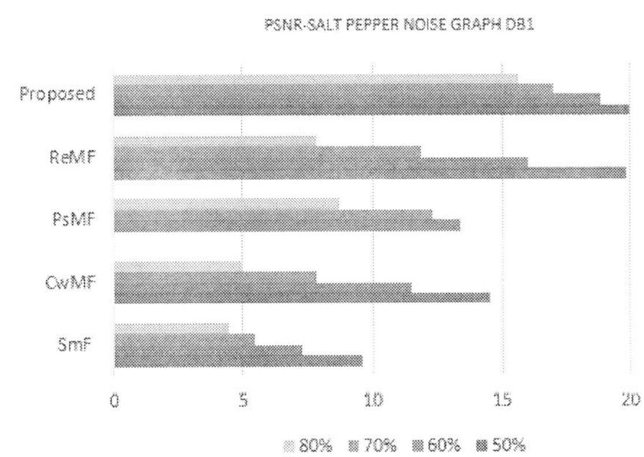

Figure 2. PsNR result for different de-noising methods for DB1 (CwMF is Curved weighted median filter)

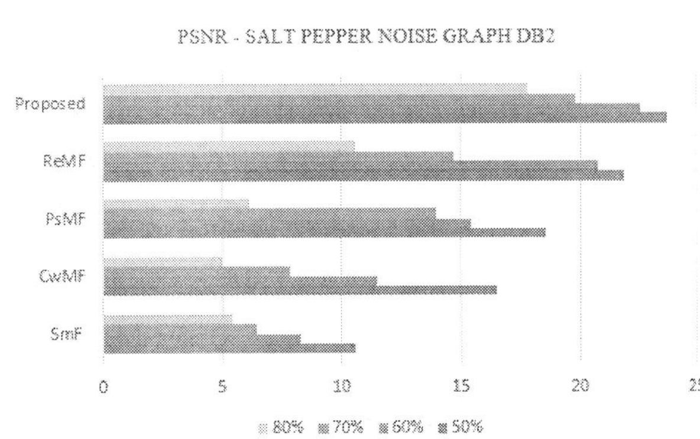

Figure 3. PsNR result for different denoising method graph for FVC 2002 DB2

4. Results and Discussion

Fingerprint image noise density is chosen from 50% to 80% and result value of different denoising method and our proposed method are given in Table 1 and Table 2. Quality performance measured by PsNR. The graphical result is given for DB1 and DB2 in Figure 2 and Figure 3. Our proposed algorithm shows high performance in more noise fingerprint images. PsMF also has showed good performance.

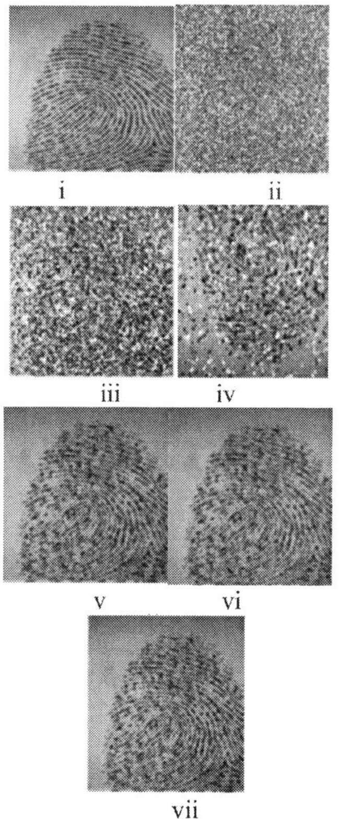

Figure 4. Original fingerprint image, noisy fingerprint image and filtered fingerprint image shown (i) Original image; (ii) Noise image; (iii) SmF; (iv) CwMF; (v) PsMF; (vi) ReMF; (vii) Proposed Algorithm.

Our present work is useful to reduce or minimize medium level noise in fingerprint image. In future, we will work to reduce more than 80 percent noise level in fingerprint image

5. Conclusion

We introduced an adaptive threshold-based algorithm, used salt pepper noise for increase performance of the biometric system. Our method uses a PsNR measurement scale to compare with other methods. Noise level is chosen in between 50% to 80%. In each selected scale noise level fingerprint image, our proposed method had a better result than other methods.

References

[1] Batool A and Tariq A 2011 Computerized system for fingerprint identification for biometric security *IEEE 14th Int. Multitopic Conf.* 102 DOI: 10.1109/INMIC.2011.6151452

[2] Singh S P, Ayub S and Saini J P 2016 Literature survey on different type of fingerprint recognition *3rd Int. Conf. on Computing for Sustainable Global Development (INDIACom) New Delhi* pp 3748–55

[3] Wang Q and You S 2008 Feature selection for real-time image matching systems *19th Int. Conf. on Pattern Recognition Tampa* pp 1–4 DOI: 10.1109/ICPR.2008.4761164
[4] Kolsch T, Keysers D, Ney H and Paredes R 2004 Enhancements for local feature based image classification *Proc. of the 17th Int. Conf. on Pattern Recognition Cambridge* **1** pp 248–51 DOI: 10.1109/ICPR.2004.1334070
[5] Gottschlich C 2012 Curved-region-based ridge frequency estimation and curved gabor filters for fingerprint image enhancement *IEEE Transactions on Image Processing* **21** pp 2220–7 DOI: 10.1109/TIP.2011.2170696
[6] Zhang P and Li F A 2014 New adaptive weighted mean filter for removing salt-and-pepper noise *IEEE Signal Processing Letters* **21** pp 1280–3 DOI: 10.1109/LSP.2014.2333012
[7] Soni A and Shrivastava R 2017 Removal of high density salt and pepper noise removal by modified median filter *Inventive Communication and Computational Technologies (ICICCT)* pp 282–5 DOI: 10.1109/ICICCT.2017.7975204
[8] Fareed S B S and Khader S S 2018 Fast adaptive and selective mean filter for the removal of high-density salt and pepper noise *IET Image Processing* **12** pp 1378–87 DOI: 10.1049/iet-ipr.2017.0199
[9] Han K, Wang Z and Chen Z 2018 Fingerprint image enhancement method based on adaptive median filter *24th Asia-Pacific Conf. on Communications (APCC)* pp 40–4 DOI: 10.1109/APCC.2018.8633498
[10] Makhija S, Khatwani A and Roja M M 2017 Performance analysis of latent fingerprint enhancement techniques *Int. Conf. on Innovative Mechanisms for Industry Applications (ICIMIA)* pp 96–100 DOI: 10.1109/ICIMIA.2017.7975580
[11] Zhang M M, Kouri D J and Zhang D S 2017 Dual window selective median switching filter *11th Int. Conf. on Signal Processing and Communication Systems (ICSPCS)* pp 1–3 DOI: 10.1109/ICSPCS.2017.8270512
[12] George G, Oommen R M, Shelly S, S Philipose S and Varghese A M 2018 A survey on various median filtering techniques for removal of impulse noise from digital image *Conf. on Emerging Devices and Smart Systems (ICEDSS) Tiruchengode* pp 235–8 DOI: 10.1109/ICEDSS.2018.8544273
[13] Hou L, Kuo C, Lin Z and Yu C 2015 Efficient image denoising scheme for removal of impulse noise*2015 IEEE 4th Global Conf. on Consumer Electronics (GCCE)* pp 298–9 DOI: 10.1109/GCCE.2015.7398612
[14] Raza M T and Sawant S 2012 High density salt and pepper noise removal through decision based partial trimmed global mean filter *2012 Nirma University Int. Conf. on Eng. (NUiCONE)* pp 1–5 DOI: 10.1109/NUICONE.2012.6493236
[15] Qian Y 2019 Removing of Salt-and-pepper Noise in images based on adaptive median filtering and improved threshold function *2019 Chinese Control And Decision Conf. (CCDC)* pp 1431–6 DOI: 10.1109/CCDC.2019.8832612
[16] Wang Z and Zhang D 1999 Progressive switching median filter for the removal of impulse noise from highly corrupted images *IEEE Transactions on Circuits and Systems II: Analog and Digital Signal Processing* **46** pp 78–80 DOI: 10.1109/82.749102
[17] Arce G R and Paredes J L 2000 Recursive weighted median filters admitting negative weights and their optimization *IEEE Transactions on Signal Processing* **48** pp 768–79 DOI: 10.1109/78.824671
[18] Zhang M M, Kouri D J and Zhang D S 2017 Dual window selective median switching filter *2017 11th Int. Conf. on Signal Processing and Communication Systems (ICSPCS)* Gold Coast, QLD pp 1–3 DOI: 10.1109/ICSPCS.2017.8270512
[19] Zhang X, Xiong R, Ma S and Gao W 2014 Artifact reduction of compressed video via three-dimensional adaptive estimation of transform coefficients *IEEE Int. Conf. on Image Processing (ICIP)* pp 4567–71 DOI: 10.1109/ICIP.2014.7025926
[20] Soni A and Shrivastava R 2017 Removal of high density salt and pepper noise removal by

modified median filter *Int. Conf. on Inventive Communication and Computational Technologies (ICICCT) Coimbatore* pp 282–5 DOI: 10.1109/ICICCT.2017.7975204

[21] Cumpim C and Punchalard R 2017 Sliding block shepard interpolation method for removing high-density impulse noise *Int. Electrical Eng. Congress (iEECON) Pattaya* pp 1–4 DOI: 10.1109/IEECON.2017.8075884

[22] Cherrat E M, Alaoui R, Bouzahir H and Jenkal W 2017 High density salt-and-pepper noise suppression using adaptive dual threshold decision based algorithm in fingerprint images *Intelligent Systems and Computer Vision (ISCV)* pp 1–4 DOI: 10.1109/ISACV.2017.8054913

Designing and Developing an Agricultural Product Sales Application Catalog with a Hybrid Application Development Framework

Argo Wibowo[1], Rosa Delima[2], Antonius Rachmat C[3]

[1] Information System Department, Faculty of Information Technology, Duta Wacana Christian University, Yogyakarta

[2,3] Informatics Department, Faculty of Information Technology, Duta Wacana Christian University, Yogyakarta

Abstract. The use of e-commerce systems continues to increase in Indonesia, but the field facts show that e-commerce in agriculture is still relatively low compared to other fields. The problems faced are limited e-commerce infrastructure, lack of the farmers' knowledge on the benefits of e-commerce, and application design, which is challenging to use. Those problems cause a lack of e-commerce use in this field. Besides, the characteristics of technology users in agriculture also have a considerable influence. In this paper, a mobile-based application is developed to facilitate farmers in promoting agricultural products sales. One of the problems faced in developing mobile applications is the diversity of mobile device platforms. A mobile application can ideally run on multiple platforms. This paper contributes to building an application by using a hybrid method for application development to be used on the Android and IOS platforms. The application built aims to display agricultural products of farmers. The filter feature also makes it easier to find agricultural products in this application. The application is built using the Flutter framework to use one source of the same code on several mobile operating systems at once. There are some slight differences between the output versions of the Android and iOS platforms. Although there are some subtle differences, the similarity rate reaches 90%.

Keywords: agricultural commerce application, hybrid application development, platform, comparison

1. Introduction

Agriculture is one of the sectors producing natural resources and plantations, fisheries, forests, and livestock. Natural resources are the economic backbone of an area that contributes to the welfare of society [1]. In Java Island, Indonesia, the agricultural land population around 41% [2], but the marketing of agricultural products, in general, is still traditionally constrained by information and computer technology [3]. A solution that can help in the marketing process of agricultural products is needed. One solution that can be taken is to build an application that allows farmers to display and market their products. The system uses the development of a catalog system for agricultural products that can be easily accessed, supports high mobility, can be used by many parties, both sellers and buyers, and can be used on platforms that are most widely used by public in general.

In this research conducted, a catalog system was developed for the sale of agricultural products. This catalog was developed on a mobile platform. Mobile technology was chosen because of the ease of use of access to technology. Based on the results of this research conducted, it is known that most of the potential users of the system have mobile devices in the form of cellphones or tablets. Based on research conducted in [4], it is known that there are two main issues in the development of mobile devices. First, the issue of various mobile device screen sizes and second is the issue of differences in system platforms on mobile devices. One way to overcome the differences in platforms on mobile devices, in this study, a mobile application development was carried out using a hybrid application development method. Applications can be used on two types of operating systems: Android and IOS platforms.

This research has contributed to building the application of the hybrid method for mobile application development. The application also contributes to a system that supports the expansion of the agricultural product market through a mobile catalog. The implementation of this application will impact increasing the sale of products directly by farmers and make it easier for buyers to select and order the agricultural products needed.

This paper is organized into five sections. The first section contains an introduction, followed by a literature review in the second section. The third section discusses the application of the method followed by analysis and discussion in the fourth section and conclusions in the last section.

2. Literature Review

2.1. Catalog

Catalogs are one of the media for publication to promote the products offered. The product catalogs are known to be practical in conveying all types of information as generally, catalogs can also be said as a presentation tool of product features. In the product catalogs, the content includes how the seller's relationship with the buyers, the seller's profile, and the most important part is such as to explain the quality of the product or service.

Not only that, product catalogs also contain the required seller data such as telephone numbers, address information, email, website (if any), operational hours, how to order, and payment methods.

In general, product catalogs are often found in certain places such as libraries, shopping centers, restaurants, and many other places. The more attractive the design of a catalog, the more potential it will be to increase the sale value of the products being marketed.

Regarding the positive side of the catalogs, this paper will also build a sales catalog application for agricultural products. Using catalogs and attractive designs will potentially increase the sale value of the agricultural products being marketed. The modern marketing process has a very big role as direct support for increasing a business [5]. The purpose of marketing is to bridge what producers and buyers want in completing the production process [6]. The publication in catalogs is not only for agricultural products but also means of agricultural production.

2.2. Mobile Application

Mobile applications are quite widely used by current users. Smartphone users in Indonesia continue to grow at an annual compound growth rate of 33% from 2013-2017 [7]. This can be seen as an opportunity to create marketing media for agricultural products in mobile applications.

Mobile applications are one of the results of technological developments. The applications that used to be available only on PC, but nowadays can also be used on a handheld device. The various mobile operating systems that can be found on smartphones are Symbian OS, iOS, RIM BlackBerry, Windows Mobile, Linux, Palm, WebOS, and Android [8]. The majority of operating systems used by users in various countries, particularly Indonesia, are the Android and iOS operating systems. Based on data in July 2018, it showed that the Android operating system is 90.64%, iOS has increased to 5.34%, Blackberry is 0.38%, Series 40 is 0.37%, Nokia is 0.33%, and others are up to 2.31% [9] .

No matter what the mobile operating system is, but in essence, the mobile application must meet the following things: 1. Improve the quality of human life, 2. Improve communication, 3. Improve existing business processes, 4. Increase customer satisfaction. The ideal mobile application is user-friendly, especially for customers. User experience is one of the most important things which is then related to customer satisfaction.

2.3. Hybrid Framework

A framework is a tool used in developing an application. Frameworks can be functions, plugins, or concepts so that it forms a system. By using a framework, an application can be arranged and structured neatly.

However, using a framework does not mean it is free from coding. In this paper, we will use a hybrid framework. The goal is to be able to build applications that can run on various operating systems using one program code source. The operating system that is targeted is the operating system that is mostly used by users in multiple countries, such as Android and iOS [9].

Different from native programming, of course, hybrid also has its advantages and disadvantages. Here are the advantages and disadvantages of hybrid and native frameworks:

1. Native Programming has a more challenging development process, while the hybrid process is much easier and cheaper.
2. Applications can run on several operating systems at once if you use Hybrid, while native can only be on one operating system
3. Hybrid performance cannot be as good as native
4. Native program features do not fully exist in hybrid; therefore, it will be difficult to make any special features.

Seeing the facts above, the performance problem can be a serious problem. To overcome this problem, this paper uses a flutter framework which is claimed to have a fairly good performance of up to 60 FPS [10]. The hot reload feature is also provided by this framework, which allows programmers to see the results of code changes more quickly. Flutter is a hybrid framework made by Google, which is still developing. Although it is still developing, many companies have used this framework because it can run on several operating systems and have good performance [11].

2.4. Agricultural Product Sales Application

Agricultural products can be sold in several ways. In 2014 there was an agribusiness application that introduced bananas [12]. The method used is Rapid Application Development (RAD). This application supports the management of user data, customer data, product data, and sales data. In 2017 there was research on E-farming application [13]. The application made was web-based using the Software Development Life Cycle (SDLC) method. In 2018, research in the field of agricultural e-commerce was conducted by Fachriyan by creating an agribusiness marketplace [14]. Unlike previous research, this research will be designed and built a multi-platform mobile application. Another multiplatform application is tanihub, which also connects directly between farmers and buyers [15]. However, one of the features developed in this study is the pre-order feature. The data managed is not only product data, but also seller data and buyer data. The method used in making the application in this study is Scrum. Scrum is an additional framework for software or product development and can increase the speed and flexibility in development [16], this is the basis for implementing Scrum for the development of this agricultural application.

3. Research Methodology

In this research, we use The Scrum methodology. This system development is carried out using the Scrum method which consists of planning, staging, development and release. The first stage is planning, at this stage the developer sets the right goals and funding. the requirements and iterations that will be prioritized will be identified. At the development stage, the system is implemented and the release stage is the final stage of deploying. Details are explained in the paragraphs below.

In this research, the data on agricultural products used is not limited to natural agricultural products and applies to agricultural production facilities. From the requirements analysis stage, the sales system developed in this study aims to display the farmer's agricultural products. If there are interested buyers, they can interact with the seller through the address and telephone number provided by the system. The system can also direct the seller's locations using maps. Another feature provided by this system is the availability of filters, which makes it easy to find a product. Based on these filters, potential buyers can sort products based on specific criteria.

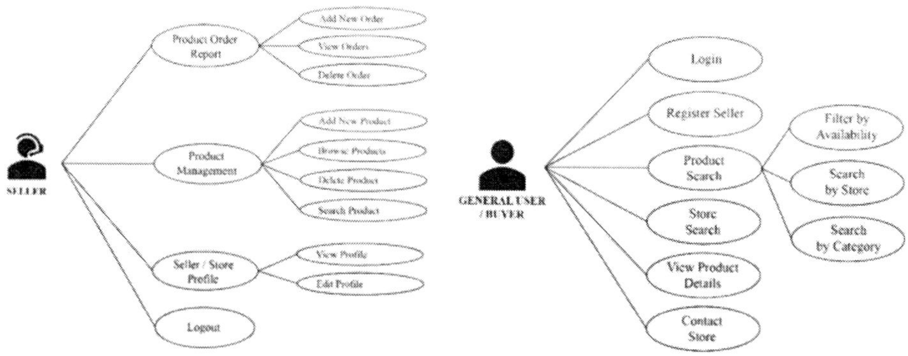

Fig. 1. Use Case Application Diagram

At the development stage, this study uses a multi-platform approach. The purpose of using this multiplatform is to support the scalability, availability, and reusability aspects of the program. The scalability aspect is used to support developing applications with other system integration. The availability aspect supports applications available on various platforms. The reusability aspect supports speed in making systems available on various platforms in a short amount of time. The code made is only one code, but it can be applied and used on many platforms. The framework used is Flutter.

3.1. Business Process Application Desain

The application development process is carried out by the Scrum method, involving a team of 4 persons, with their respective assignments as the product owner, scrum master, front-end developer, and back-end developer. The product owner manages how the business process of the application runs and provides direction when one or more features change.

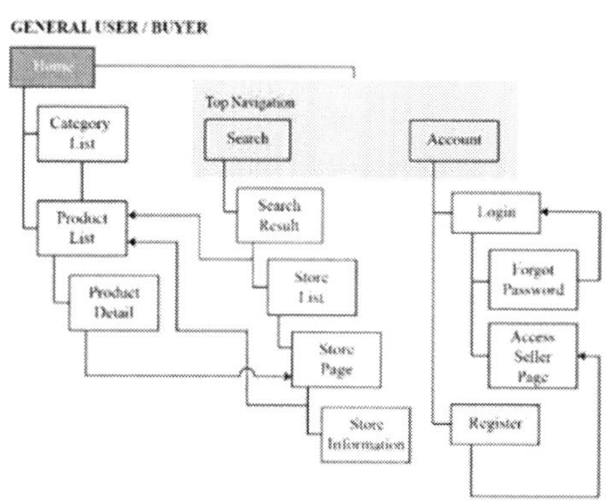

Fig. 2. Sitemap application for buyers

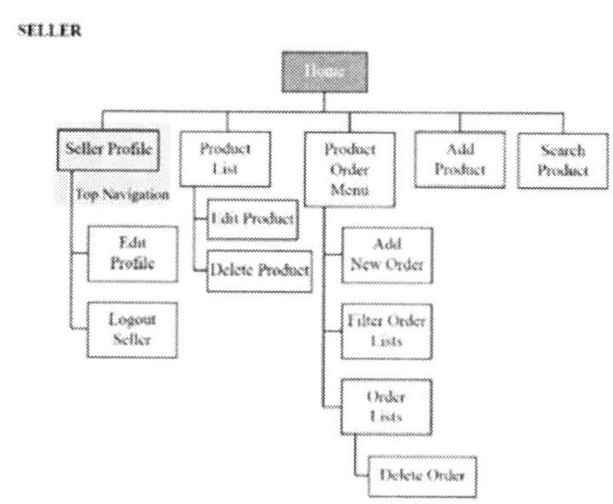

Fig. 3. Sitemap application for sellers

In Fig. 1. there is a use case diagram application. The two main roles of the application are buyers and sellers. Buyers can search for products (by category, keywords, stores, and the status of products sold), view detailed information about products sold, information about sellers, as well as view seller contacts. Buyers can also register as sellers through the registration process available in the application, after which they log in to enter the store admin page.

Activity diagrams and sitemaps explain the application features and overall usage process. The application sitemap is different for each user role. Fig. 2 is a sitemap application for buyer pages, while Fig. 3 is a sitemap application for admin/seller pages. When the user has logged in as a seller, the application will automatically enter the admin page when they reopen the application again. Therefore, the seller does not need to log in every time he/she wants to use it.

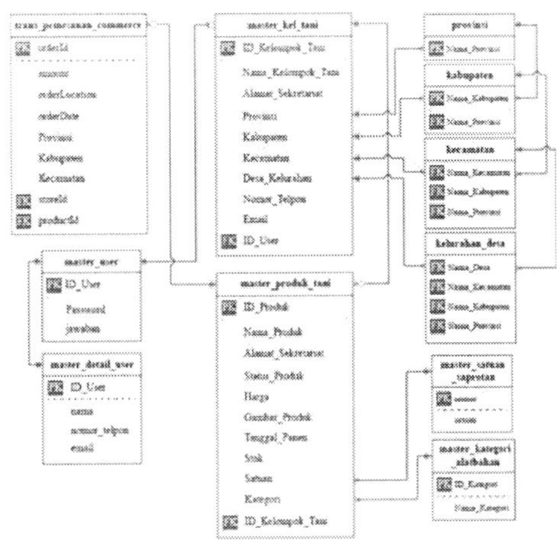

Fig. 4. Entity Relationship Diagram of Dutatani m-commerce database

3.2. Database Application Design

The database used in this application is the Dutatani database. Dutatani is a system developed specifically to support the business process in agriculture toward precision farming. This system has been in development since 2016 [17]. The system is named Integrated Agricultural Information System (IAIS). Several applications in agriculture have been carried out [18]. After that, an integrated SIP blueprint was developed by developing Architecture Vision [19], Business Architecture [20], and defining IAIS stakeholders [21]. Besides, there are several systems developed, such as the Agricultural Portal [22], Farmer and Farmer Groups Information System [23], Farmers Activity Information System [24], Information System of Agricultural Products' Purchases and Sales [25] and Agriculture Mapping System [26]. Three systems that have been developed can be accessed via the website at http://dutatani.id. Dutatani is an integrated database for the whole system in IAIS. The database entity relationship diagram and details of the data variables used special for mobile commerce needs are explained in Fig. 4.

The Dutatani database uses MySQL, which is accessed by the REST API in php programming language. There are 11 tables that are used, such as product table (master_produk_tani), user (master_user), user details (master_detail_user), ordering (trans_pemesanan_commerce), stores (master_kel_tani), product units (master_satuan_saprotan), product categories (master_kategori_alatbahan), province, district, sub-district, and kelurahan_desa.

3.3. Application Interface Design

The design of the application interface is made to cover all pages or activities of the application. The login and register page on Fig. 5 and Fig. 6 will appear first for new users. However, if the user has registered or entered in the application, the page will not occur again. The main page in Fig. 7 contains the categories of products sold that are displayed with 2 types of test displays, which aimed to enable users to make it easier to search for the products they want directly. On the main page, there is also navigation to the search page or shopping cart page. The shop detail page will be displayed like Fig. 8 for display with card view, and Fig. 9 to display the list view. The selected product details will be displayed with a model view like Fig. 10.

Fig. 5. New user registration interface activity

Fig. 6. User login interface activity

Fig. 7. Home interface activity with list view

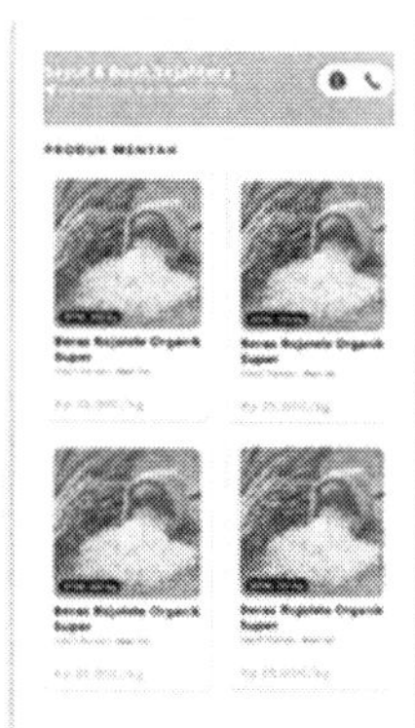

Fig. 8. Shop detail interface activity

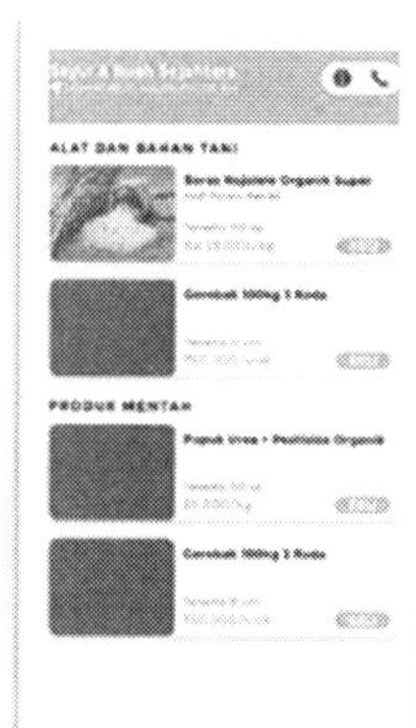

Fig. 9. Shop detail interface activity

Fig. 10. Interface for product details

The design of the application interface is made to cover all pages or actions of the application. The login and register page on Fig. 5 and Fig. 6 will appear first for new users. However, if the user has registered or entered in the application, the page will not appear again. The main page in Fig. 7 contains the categories of products sold that are displayed with two types of test displays, which aimed to enable users to make it easier to directly search for the products they want. On the main page, there is also navigation to the search page or shopping cart page. The shop detail page will be displayed like Fig. 8 for display with card view, and Fig. 9 to display the list view. The selected product details will be displayed with a model view like Fig. 10.

4. Results and Discussion

This application has been successfully developed with a display in accordance with the design. This application has been developed with Flutter which means it can be installed on Android or iOS Operating System (OS). In this paper, each figure contains 2 versions of OS. The left side is the Android version, and the right side is iOS version as in Fig. 11, which is the first-page image displayed when the application is opened. The display of Fig. 11 contains product categories as well as recommended products.

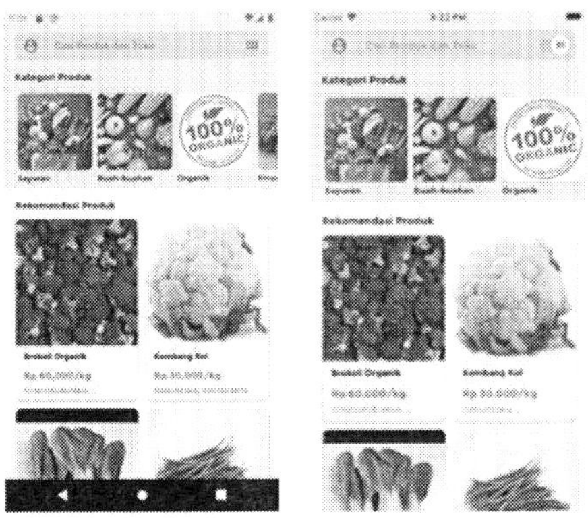

Fig. 11. The Front Page of the application

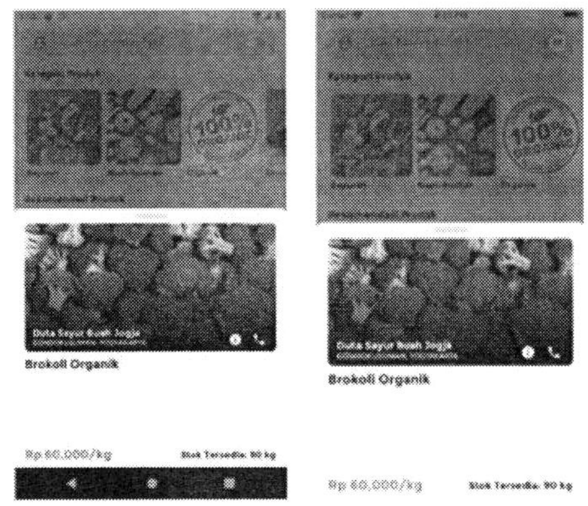

Fig. 12. Product Detail Page

The product detail page is shown in Fig. 12. This page is equipped with a product image, a shortcut button for the seller's telephone number, as well as the seller's location shortcut. The seller location button will be connected to the Google Map, while the phone number button will be connected to the mobile call feature. Application users do not need to log in because this application uses a model to meet directly between the seller and buyer. This design makes it easy for users and aims to increase prospective buyer interest because they can see the product without logging in first. If prospective buyers find it easier to see the product catalog, of course, it will provide added value for the sale of these products. The possibility of prospective buyers to buy the product will be greater.

Fig. 13. Seller Enter Menu

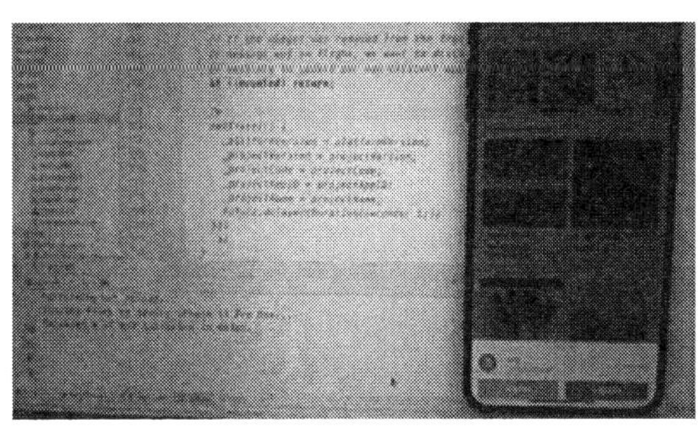

Fig. 14. Error Code on Seller Enter Menu

The login page is intended for sellers who will publish their products. This page is shown in Fig. 13. There is a problem with the Seller Enter Menu. The problem happened on iOS OS. The code is searched for Application ID and can not found the ID, shown in Fig. 14. The problem can be solved when the code is commented on, or the text field is manual entry. The application can be run normally again after the error fixed.

In this paper, we can see a comparison between 2 OS outputs, as seen on Table 1. There are 20 pages that are compared, and there are 2 differences in them. There are some colors that are darker on iOS when compared to Android in Shop Product Detail. There is an error page in Seller Enter Page when retrieving the application ID on iOS. So the calculation of the similarity of output obtained by using Hybrid Programming in both OS versions is 18/20 = 90%.

Table 1. Comparison Result

Page	**Similarity**		**Page**	**Similarity**	
	Android	**Ios**		**Android**	**Ios**
Front Page	v	v	Seller List	v	v
Product Detail	v	v	Disclaimer	v	v
Seller Enter	v		Personal Data Completion	v	v
Product Category	v	v	Seller's Shop	v	v
Product Search	v	v	Profile	v	v
Search Shop	v	v	Profile Update	v	v
Shop Product Detail	v		Manage Product	v	v
Shop Information	v	v	Add new Product	v	v
Seller Login	v	v	Purchase Recording	v	v
Forget Password	v	v	Purchase Reporting	v	v
			Passed	**20**	**18**

5. Conclusion

In this paper, the authors have succeeded in designing and developing an application for selling agricultural products. This application is developed by designing interfaces and databases, and the results made are also in accordance with the design. The application is developed using the Flutter Framework. The Flutter Framework is a hybrid framework that can support application development for various operating systems (such as Android, iOS, etc) on mobile devices. The application can be compiled into several applications on Android and iOS with just the same source code.

Testing were carried out on twenty features on the system using two operating systems (Android and iOS). The testing results show that there is a slight difference between the output versions of Android and iOS, as there are some colors that are darker on iOS when compared to Android. Besides that, there is an error page when retrieving the application ID on iOS, even though it comes from the same source code. Although there are two differences, the similarity rate reaches 90%. This research shows that hybrid development in agriculture mobile application using Flutter Framework can make development system easier and also faster.

Acknowledgment

The authors thank the Institute of Research and Community Empowerment (Lembaga Penelitian dan Pengabdian kepada Masyarakat / LPPM) Universitas Kristen Duta Wacana which has funded this research with contract numbers: 047/D.01/LPPM/2020 and the Faculty of Information Technology Universitas Kristen Duta Wacana.

References

[1] Heryawan A, Fauzi A and Hidayat A 2014 Analisis ekonomi dan kebijakan sumber daya alam

Provinsi Jawa Barat *J. of Agriculture, Resource, and Environmental Economics* vol **2** no 1 pp 1–11

[2] Wahyunto 2009 Lahan Sawah di Indonesia sebagai pendukung ketahanan pangan nasional *Informatika Pertanian* vol **18** no 2 pp 133–52

[3] Riyadh M I 2018 Analisis saluran pemasaran lima pangan pokok dan penting di lima Kabupaten Sumatera Utara *Jurnal Ekonomi & Kebijakan Publik* vol **9** no 2 pp 161–71

[4] Mushtaq Z and Wahid A 2018 Mobile application learning: the next generation e-learning *Proc. of the 3rd Int. Conf. on Inventive Computation Technologies ICICT 2018* pp 826–9

[5] Asrina and Martina 2017 Strategi pemasaran usaha kerupuk tempe di desa Blang Geulanggang Kecamatan Peusangan Kabupaten Bireuen (studi kasus: usaha kerupuk tempe Ibu Yusnita) *Jurnal S. Pertanian* vol **1** no 1 pp 48–62

[6] Asmarantaka R W, Atmakusuma J, Muflikh Y N and Rosiana N 2017 Konsep pemasaran agribisnis: pendekatan ekonomi dan manajemen *Jurnal Agribisnis Indonesia* vol **5** no 2 pp 151–72

[7] Putra D R and Nugroho M A 2016 Pengembangan game edukatif berbasis android sebagai media pembelajaran akuntansi pada materi jurnal penyesuaian perusahaan jasa *Jurnal Pendidikan Akuntansi Indonesia* vol **14** no 1 pp 25–34

[8] Daeng I T M, Mewengkang N and Kalesaran E R 2017 Penggunaan smartphone dalam menunjang aktivitas perkuliahan oleh mahasiswa Fispol Unsrat Manado *Jurnal Acta Diurnae* vol **6** no 1 p 15

[9] Zaini M and Soenarto 2019 Persepsi orangtua terhadap hadirnya era teknologi digital di kalangan anak usia dini *Jurnal Obsesi: Jurnal Pendidikan Anak Usia Dini* vol **3** no 1 pp 254–64

[10] Dev F 2020 *Flutter UI Performance* Retrieved on 4 June 2020 from https://flutter.dev/docs/perf/rendering/ui-performance

[11] Dev F 2020 *Flutter Showcase* Retrieved on 4 June 2020 from https://flutter.dev/showcase

[12] Sugiarti Y 2014 Perancangan sistem informasi agribisnis ecommerce buah pisang *Jurnal Agribisnis* vol **8** no 1 pp 71–82

[13] Suweantara E, Martana I G S A, Suarya A and Widiartha I M 2017 Aplikasi e-tani sebagai strategi peningkatan kualitas dan kesejahteraan para petani *Seminar Nasional Teknologi Informasi & Aplikasinya - Bali*

[14] Fachriyan H A and I Wijaya P E 2018 Apliaksi model e-marketplace dalam e-agribusiness *Jurnal Ilmu-Ilmu Pertanian* vol **14** no 1 pp 12–24

[15] Junaidi M A and Maghdahfanti E P 2020 Dampak pola kemitraan melalui e-commerce pertanian (Kasus pada petani jeruk dengan PT TaniHub Indonesia di Kabupaten Malang) *Magister Agribisnis* vol **20** no 2 pp 88–93

[16] Permana P A G 2015 Scrum method implementation in a software development project management *(IJACSA) Int. J. of Advanced Computer Sci. and Applications* vol **6** no 9 pp 198–204

[17] Delima R 2016 Analisis kondisi dan kesiapan masyarakat tani di Daerah Istimewa Yogyakarta untuk memanfaatkan TIK di bidang pertanian *National Conf. on Information and Communication Technol. Application - Yogyakarta*

[18] Delima R, Santoso H B and Purwadi J 2016 Kajian aplikasi pertanian yang dikembangkan di beberapa negara Asia dan Afrika *Seminar Nasional Aplikasi Teknologi Informasi 2016 - Yogyakarta*

[19] Delima R, Santoso H B and Purwadi J 2016 Architecture vision for Indonesian integrated agriculture information systems using TOGAF framework *Int. Conf. on Informatics and*

Computing - Lombok

[20] Delima R, Santoso H B and Purwadi J 2017 Business architecture development for integrated agriculture information system (IAIS) Using Togaf Framework *Researchers World* vol **8** no 2 pp 1–13

[21] Santoso H B and Delima R 2016 Stakeholder definition for Indonesian integrated agriculture information system (IAIS) *The Int. Conf. on Information Technol. and Digital Applications 2016 - Yogyakarta*

[22] Delima R, Santoso H B and Purwadi J 2017 Development of Dutatani website using rapid application development *Int. J. of Information Technol. and Electrical Engineering* vol **1** no 2 pp 36–44

[23] Santoso H B, Malvin C and Delima R 2017 Pengembangan sistem informasi pendataan petani dan kelompok tani *Seminar Nasional Sistem Informasi Indonesia - Sanur*

[24] Delima R, Galih F and Wibowo A 2017 Development of crop and farmer activity information system *Researchers World* vol **8** no 4 pp 180–9

[25] Delima R, Santoso H B, Andriyanto N and Wibowo A 2018 Development of purchasing module for agriculture E-commerce using dynamic system Development *Int. J. of Advanced Computer Sci. and Application* pp 86–96

[26] Rachmat A, Wibowo A, Santoso H B and Delima R 2019 Developing agriculture land mapping using rapid application development (RAD): a case study from Indonesia *Int. J. of Advanced Computer Sci. and Applications* pp 232–41

KaByahe: a mobile application focusing on the historical and cultural places within the city of manila

G M Cabrera, L P Caringal, J P Manocan, J G Senga and N Estrella

Institute of Information and Computing Sciences, University of Santo Tomas, España Boulevard, Sampaloc, Manila, Philippines

Abstract. With the Philippines' rich historical and cultural heritage, tourism remains a vital sector and is one of the greatest contributors to its rising economy. Manila, its capital, has a vast list of historical and cultural places that mirrors the country's rich historical and cultural heritage. The developers were inspired to develop a mobile application called KaByahe with the purpose to determine the different historical and cultural places within Manila and serve as its users' own guide in exploring more of its historical and cultural roots. The developers strongly believe that this project is important not only for the benefit of the tourists in exploring Manila, but also, for its local people, to aid them in knowing what has transpired beyond these places and promote appreciation and understanding of the city's origins. The developers used Agile Development Methodology since the strategy is to develop using short development cycles to focus more on addressing potential issues and produce quality output per cycle. With this, the developers were able to work more efficiently and collaboratively. The developers were able to gather substantial information through various testing procedures conducted to address both functional and non-functional requirements. Based on the results of these, KaByahe was deemed to successfully meet the objectives set by the developers.

1. Introduction

Tourism remains one of the greatest contributors to the rising economic status of the Philippines. In 2017, the tourism industry contributed approximately 3.35 trillion to the country's economy. [1] The country's capital, Manila, has a tremendous rundown of historical and cultural places reflecting the nation's past. One of Manila's oldest and most popular historical and cultural places is Intramuros, which was once the Spaniard's political and military base in Asia. [2] In 2019, at least 3.7 million tourists visited Intramuros, specifically Fort Santiago, according to an article by Faye Almazan of The Manila Times. Intramuros Administration (IA) head, Guiller Asido, detailed that there was an evident increase from the usual 400 to 500 daily to an average of 1,200 to 1,300 tourists who visited the historical site adding that they were able to generate P100 million in revenue versus the P94 million they made in 2018. Demographics showed that 70% were local tourists and 30% were foreign tourists. [3] Based on the volume of tourists who have been visiting the country, the number of accredited tour-guides available cannot accommodate the rapid growing count. For instance, in an article written by JM Agreda, which was published last 2013 by Sunstar, Baguio City, one of the top tourist destinations in the country, ironically has only three accredited tour guides.

Based on the information gathered and to address the stated problem of lack of tour guides, the developers were able to develop KaByahe which consisted of both web and mobile applications. KaByahe mobile is intended to help its target users to explore more of Manila's historical and cultural places by functionalities such as generation of list of places either based on user preferences or user's current location, the capability to post a comment/review or photo on a thread which will serve as the user's moment on a certain place, the capability to show a pop-up message box whenever a user passes by a historical or cultural place, show weather information, and suggest the shortest possible route from the user's current location to the desired destination. KaByahe web, on the other hand, was created for the administrator's use in managing and maintaining the contents of KaByahe mobile.

2. Objectives of the Study

The project's main goal was to develop a mobile application to determine Manila's rich historical and cultural places and aid local or foreign tourists in exploring more about these places. One of the main features of the application is to generate a list of places based on the preference of the users. Another is to allow them to openly share their experiences by posting a comment/review or uploading photos. It also provided the shortest possible route from the user's current location to the desired destination and while on the road, provided a brief description of each historic or cultural place a user passed by through a pop-up. The developers also created a web application to manage the contents of the mobile application.

The project provided a web application for the administrator and a mobile application for its users. For KaByahe web, the administrator had the capability to view the total number of registered users, the total count who downloaded the application, the total number of places, and the total number of preferences. The administrator was also able to modify new and existing historical and cultural places by adding, editing, and deleting. Hence, new and existing preferences can also be managed. The initial categories of preferences were churches, monuments, museums, parks, institutions, libraries, and theatres. These were considered initial because it can be modified through the administrator side.

For KaByahe mobile, the application required the user to create an account if the user wanted to post a moment (photo, comment/review) on a certain place. It also asked the user regarding preferences upon initial login. These preferences can be changed by the user later on. It displayed a list of places according to the user's preferences and a list of nearby places based on the user's current location. Hence, a map was also displayed. Lastly, it was also able to provide weather information to the places where historical or cultural landmarks were present.

KaByahe mobile was limited to run on the Android platform and with the presence of Internet connection. The user may use the application first-hand. However, the user was not able to share a moment on a specific place unless he is already logged in and his account is already verified. The generated list of places was limited only to historical or cultural places within the City of Manila. Photos to be uploaded on a thread was limited to one photo per upload. Uploading of videos was also out-of-scope. Lastly, in the generation of the shortest possible route, the default mode of transportation was by car and traffic was not a factor in determining the route.

3. Review of related literature, systems, and technologies

Most people, either at a young or old age, have smartphones today and it became an integral part of every individual's life due to useful mobile applications which were relatively easy to access and simply make one's life better as a result. [4]. The developers decided to come up with a mobile application because smartphones are more ubiquitous than laptops or personal computers. [5] Today, travel applications are the 7th most downloaded in the application categories and as technology continues to bloom even more, new trends are continuously rising.

Among the vast list of tourist destinations throughout the world is the Philippines, which was considered as one of the top tourist destinations by people all over the world. Not only does this country have beautiful beaches and scenic spots for people to enjoy, but also a rich historical and cultural background that tourists find interesting. The Philippines became a melting pot of various cultures like Spanish, Chinese, Japanese, and American which had greatly affected the Filipino customs, traditions, and norms. With these influences, its historic landmarks have played a major part in its tourism industry. [6] Among the destinations that tourists can explore in the Philippines is Manila, its capital. The complex past of Manila still stands through due to its historical and cultural landmarks evident until now. In fact, according to an article written by Stephanie Mayo, Intramuros, the oldest administrative district in Manila, and the ever-famous Rizal Park were just one of the top ten historical landmarks to be seen and visited in Manila. [7] These landmarks speak beyond what is seen by the naked eye, Manila's historical and cultural roots.

Within the sixteen administrative districts of Manila namely Binondo, Ermita, Intramuros, Malate, Paco, Pandacan, Port Area, Quiapo, Sampaloc, San Andres, San Miguel, San Nicolas, Santa Ana, Santa Cruz, Santa Mesa, and Tondo lies different historical and cultural structures, even places, which serves as an instrument to look back on the story of the past. Adhering to KaByahe's vision, to determine these places and aid local or foreign tourists know how rich the history of Manila is, below are some features of KaByahe along with its references to other existing applications in the market:

Table 1. Some of KaByahe features and comparison to other existing mobile applications.

KaByahe features	Comparison to other applications
1. To provide a list of places based on gathered user preferences where distance and preferences were the basis of generating a list of places for the user. Based on the user's preference selection and distance approximation, the application generated a list of places for the user.	Comparable to Sygic, a travel application, which generates itineraries for the user, except that it does not include user preferences. [8]
2. To automatically pin and display the user's current location, given that permission was enabled, using the coordinates returned by the API.	This is similar to the study titled "PWDirectory - Mobile Application", [9] which also utilized the Google Maps API.
3. To see nearby historical and cultural places based on the user's current location and share moments by uploading a photo or a review which will be included in a place's thread upon successful upload.	Comparable to the "Near Me Now" and "Travel Advice" features of the application TripAdvisor. [10].
4. To provide accurate weather information for a certain historical or cultural place which is vital most especially in planning activities for a certain day.	Similar to Packpoint's functionality to produce weather information to help aid the user in creating his pack list. [11].
5. To generate the shortest possible route from the user's current location to the destination, assuming that the default mode of transportation was car.	Similar to Waze, instead, the route was automatically set to shortest, not an optional fastest or shortest route. [12].

4. Methodology

The totality of the Agile Development Methodology comprised of five phases:

a) Brainstorming, wherein the project was envisioned and prioritized. In the Philippines, educational tours were one of the most unforgettable moments of a student's academic journey. It happens every year and itineraries are mostly focused on the historic and cultural places around the reach of the academic institution a student is enrolled in. The developers actually incorporated their own experiences that led to the creation of KaByahe specifically the integration of the pop-up feature where a pop-up containing a brief description, fun fact, or a trivia appears whenever a user passes by a historical or cultural place. This resembled the developers' experience of their own tourist guides giving trivia whenever their bus passed by a historic or cultural place back then. Apart from these experiences, the developers also considered the rapid growth of technology and its impact to educational tours. Now, more than ever, educational tours started to utilize the use of travel applications to further amplify the purpose of the trip. The developers thought that it would be a great opportunity not just for the students, but also for the local and foreign tourists to have a deeper understanding of the different historic and cultural places within Manila.

b) Designing, wherein the developers decided on how the application would look like to its target users keeping in mind to make it easy to use, easy to understand, and easy to learn. The developers designed the application to overall make it simple and provide the user ease in usage and seamless user experience.
c) Development wherein the actual implementation and coding took place. The developers used short development cycles when developing each functionality identified. At the end of each cycle, a test was conducted to perform changes whenever there were issues to further build a quality application.
d) Quality Assurance to test whether the functionalities developed are working as expected. Apart from the tests conducted per cycle, the developers conducted a test not just for a specific functionality but for the entirety of the application itself to make sure that the integration of functionalities were all working as expected. The developers were able to conduct unit, integration, and system integration testing to test whether both major and minor functional requirements were aligned to both KaByahe web and mobile applications.
e) Deployment wherein the application is subject for release. Below are some screenshots of KaByahe Web and Mobile application:

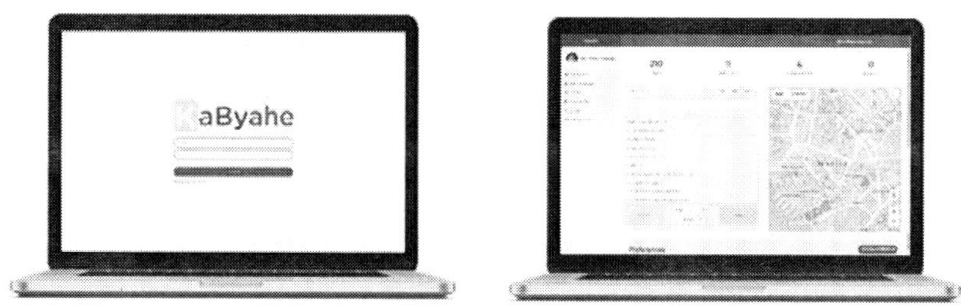

Figure 1. KaByahe Web Application

Figure 2. KaByahe Mobile Application - Selecting Preferences and Nearby Places

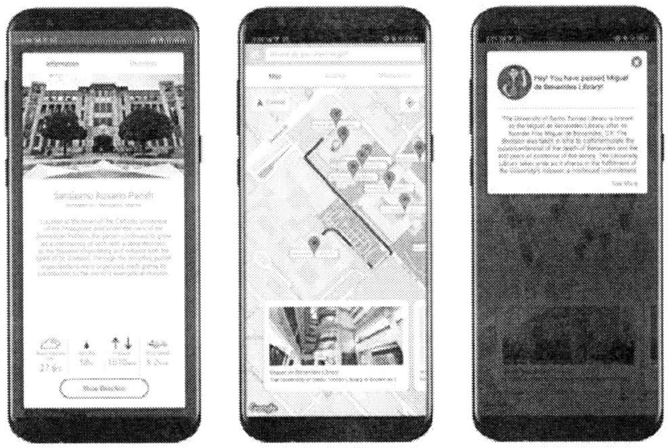

Figure 3. KaByahe Mobile Application - Shortest Possible Route and Pop-up Functionality

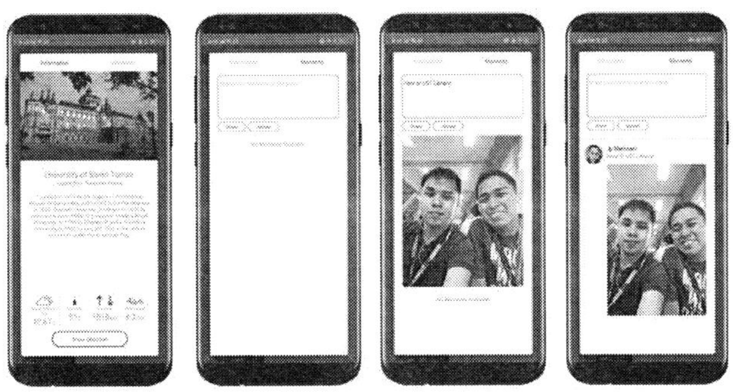

Figure 4. KaByahe Mobile Application - Moments

The developers decided to utilize the Agile Development Methodology instead of the traditional Waterfall model, which is sequential, for the strategy is to develop the application through the use of incremental and iterative work [13] to focus on the improvements in the development phase. Hence, this methodology was able to help the team work more efficiently and collaboratively.

5. Results and discussion

The developers were able to develop the defined major and minor functional requirements and non-functional requirements accompanied with corresponding test cases for each requirement. For the functional requirements and reliability, the developers conducted unit, integration, and system integration testing to test whether both major and minor functional requirements align to both KaByahe web and mobile applications.

For the unit testing, in KaByahe web, the developers conducted two versions of unit testing. Errors encountered were minor in nature such as highlighting of fields in red to denote error and presence of validation messages. Failure marks were immediately resolved after debugging. On the other hand, for KaByahe mobile, there were three versions of unit testing conducted. Again, errors were minor in nature and were immediately resolved after debugging.

For integration testing, one version was conducted for KaByahe web since when tested, all test cases planned obtained pass marks when executed. For KaByahe mobile, there were two versions of integration testing. In the first version, 8 out of 10 test cases obtained pass marks. Again, errors were minor in nature and were resolved after debugging leading to a 10 out of 10 pass marks in the second version.

Lastly, for system integration, both KaByahe web and mobile were enclosed in one version of system integration testing since there were no failures documented when the testing was done.

For the non-functional requirements, the developers conducted SQL injection testing, XSS testing, scalability, and portability testing. Pass marks were obtained after every test case for every requirement was done. For usability, the developers conducted User Acceptance Testing (UAT) at its target sites Intramuros and Rizal Park. They were able to get 20 respondents. In evaluating the responses, the System Usability Scale (SUS) metric was used. The developers computed the average of answers per question by getting the total sum of each question and divided it into the number of respondents. [14] Table 2 below shows how the usability score was computed:

Table 2. Calculating the usability score.

Questions	Avg. score of 20 respondents	Odd numbers	Even numbers
1. I think that I would like to use this system frequently.	4.85	3.85	
2. I found the system unnecessarily complex.	1.5		3.5
3. I thought the system was easy to use.	4.8	3.8	
4. I think that I would need the support of a technical person to be able to use this system.	1.25		3.75
5. I found the various functions in this system were well integrated.	4.85	3.85	
6. I thought that there was too much inconsistency in this system.	1.2		3.8
7. I would imagine that most people would learn to use this system very quickly.	4.85	3.85	
8. I found the system very cumbersome to use.	2.5		2.5
9. I felt very confident using the system.	4.6	3.6	
10. I needed to learn a lot of things before I could get going with this system.	1.25		3.75
	Total	18.95	17.3
	Final Score	[(18.95 + 17.3) * 2.5] = ***90.625***	

Table 2 shows that the total final score of the system was 90.625 out of 100. This means that in accordance with the scoring scale shown in Figure 5, the mobile application gained an excellent in adjective ratings and B+ to A in grade scale. Therefore, the results of the testing show that the KaByahe mobile application satisfies the needs of its target users.

6. Conclusion and recommendations

6.1. Conclusion

The developers were able to gather substantial information through various testing procedures they have conducted to address the functionality and reliability requirements on both web and mobile applications. For usability, the end users were also given the chance to provide the strong points of the mobile application and point out possible areas for improvement and their suggestions to further improve the application.

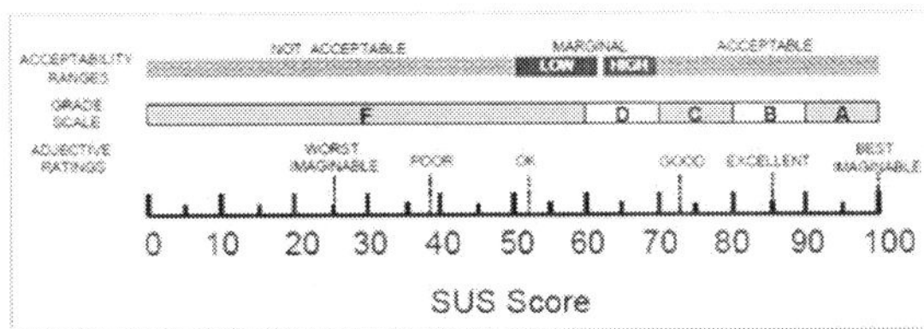

Figure 5. SUS Scoring Scale [15]

The development of this project provided the developers new knowledge and skills that may aid them in the future. Based on the gathered results, the developers therefore came up with the conclusion that the KaByahe web and mobile application have successfully met the defined objectives, generally, to determine historical and cultural places within the City of Manila and specifically, provide a web application for content management; generate list of places either based on user preferences or user's current location; provide a thread of comments/reviews or photos to serve as moments and lastly, generation of shortest possible route. All these were met not exceeding the scope and limitations of the project itself.

6.2. Recommendations

Based on the gathered subjective feedbacks and suggestions from the users, the following are the recommendations for KaByahe mobile: improve user experience when uploading photos in the moments tab of a specific place by letting the user upload multiple photos and customize these photos through adding stickers, texts, and effects, integrate modification, deletion, and filtering of bad words placed by a certain user, provide text-to-speech functionality similar to Google Maps for enhanced accessibility and lastly, develop KaByahe to run in the iOS platform.

Acknowledgement

The developers would like to express their deepest gratitude to the following people who helped, guided, and supported them in the whole duration of developing this Capstone project:

To Asst. Prof. Noel E. Estrella, their technical adviser, for his unending support and motivating words that really helped and inspired the developers to continue amidst all circumstances they have encountered during development and even in his utmost support to their title defense down to their final defense.

To Mr. Johnrebb Ortal, Mr. Roy Martin Cerrado, RMT, and Mr. Jan Edzekiel Cabigao, for their helpful insights and tips to further improve KaByahe.

To Ms. Naomi M. Jemera, M.A, the developers' Philippine History professor, for her time and help in verifying the historical and cultural places the developers have included in their database.

To their family and friends, for their unending love, understanding, and support.

And lastly, to God, for giving the developers the wisdom, strength, knowledge, patience, and most importantly, determination to finish this Capstone project and successfully defend it. To God be all the glory.

References

[1] DOT 2018 *DOT Statement on July 2018 Visitor Arrivals* Retrieved on 23 August 2018 from http://www.tourism.gov.ph/news_features/visitor_arrivals_july2018.aspx

[2] Conde M 2018 The history of Manila's walled city of Intramuros *Culture Trip* Retrieved on 6 July 2018 from https://theculturetrip.com/asia/philippines/articles/the-history-of-manilas-walled-city-of-intramuros/
[3] Almazan F 2020 3.7M tourists visited Intramuros in 2019 *The Manila Times* Retrieved on 27 January 2020 from https://www.manilatimes.net/2020/01/27/news/top-stories/3-7m-tourists-visited-intramuros-in-2019/677587/
[4] Oza H 2017 *The Importance of Mobile Applications in Everyday Life!* Retrieved on 28 March 2017 from https://www.hyperlinkinfosystem.com/blog/the-importance-of-mobile-applications-in-everyday-life
[5] Laszlo J, Daniels D and Einstein N 2018 The connected companion: the implications of mobile and smartphone ubiquity – 2018 *The Relevancy Group* Retrieved on February from http://www.relevancygroup.com/shop/the-connected-companion-the-implications-of-mobile-and-smartphone-ubiquity-2018
[6] Balinski A 2018 Spanish influence on the Philippines *Lifey* Retrieved on 1 August 2018 from http://lifey.org/spanish-influence-on-the-philippines/
[7] Mayo S 2016 *Top 10 Historical Tourist Attractions in Manila* Retrieved on 14 June 2016 from https://www.madmonkeyhostels.com/h20_blog/top-10-historical-tourist-attractions-manila/
[8] Patkar M 2018 *5 Smart Travel Planning Apps for Easier Trip Itineraries* Retrieved on 17 March 2018 from https://www.makeuseof.com/tag/smart-travel-planning-apps/
[9] Dacanay J, Co J, Cruz V, Santos J and Eleazar M 2017 *PWDirectory - Mobile Application*
[10] TripAdvisor 2018 TripAdvisor hotels restaurants (Version 25.8.1) *Google Playstore* Retrieved from https://play.google.com
[11] Price E 2017 15 best travel apps for your next trip *CNN* Retrieved on 20 November 2017 from https://edition.cnn.com/travel/article/best-travel-apps/index.html
[12] Becker T 2015 How to turn waze into the ultimate navigation app *Social Media week* Retrieved on 11 November 2015 from https://socialmediaweek.org/blog/2015/11/how-to-turn-waze-into-the-ultimate-navigation-app/
[13] Drasnar J 2013 *Agile Methodology – Scrum* Retrieved on 04 March 2013 from https://jandra.wordpress.com/2013/03/04/agile-methodology-scrum/
[14] Thomas N 2015 How to use the system usability scale (SUS) to evaluate the usability of your website *Usability Geek* Retrieved on 13 July 2015 from https://usabilitygeek.com/how-to-use-the-system-usability-scale-sus-to-evaluate-the-usability-of-your-website/
[15] Userlytics 2020 *System Usability Scale (SUS)* Retrieved from https://www.userlytics.com/system-usability-scale-sus

Random Effect Meta-Analytic Structural Equation Modeling (MASEM) Estimation Using The Method Of Moment: Case Study On The Poverty In The Island Of Java

R E Standsyah[1], **B W Otok**[2*], **and A Suharsono**[2]

[1]Doctoral Student Of Department Of Statistic, Institut Teknologi Sepuluh Nopember, 60111 Surabaya, Indonesia
[2]Department Of Statistic, Institut Teknologi Sepuluh Nopember, 60111 Surabaya, Indonesia

E-mail: *dr.otok.bw@gmail.com

Abstract. Meta-Analytic Structural Equation Modeling (MASEM) is divided into two types according to the homogeneity of the effect size, the fixed-effects MASEM, and the random-effects MASEM. The random effects of MASEM are heterogeneous or there are variations in the effects of the study population, thus implying both between-study variance and within-study variance. Accordingly, the process of estimating the MASEM random-effects parameters involves an additional process, the estimation of the variance between studies. The initial process of MASEM parameters was estimated with generalized least squares (GLS) and then estimated the variance between studies. The iterative EM algorithm is used to estimate the variances between studies of random effects MASEM GLS but this method may not achieve convergence. So, a non-iterative procedure was developed to estimate the variance between studies using the moment method. Data on poverty on the island of Java are likely to differ. Poverty data, if collected from a different population in between province, will be more accurate and in line with targets. The poverty model with the economy, human resources, and health as the right exogenous variables on the island of Java. The results of applied random-effect MASEM GLS with the method of moment approach to estimate the variance between studies on a poverty model show that the coefficient of each economic, health, and human resources variable on poverty is 0.4942, -0.1106, and -0.2843. When human resources increase along with health, people's creativity and productivity also increase, it affects the decline in the value of economic indicators or is equivalent to the increase in the economy of the community so that poverty decreases.

Keyword: MASEM, Random Effect Model, Method Of Moment, Poverty

1. Introduction

The meta-analysis was first proposed by Glass (1976) which can be defined as a statistical method to integrate the result analysis from previous studies which has similarities [1]. Meta-analysis involves the process of identifying, collecting, reviewing, coding, and interpreting various research studies. The main goal of the meta-analysis is to find the effect size. Effect size is a measure of the size of the effect, the difference, and the relationship between a variable and other variables [2].

Structural Equation Modeling (SEM) is a multivariate statistical analysis technique. This structural equation modeling is a combination of factor analysis and multiple regression analysis making it possible to analyze the structural relationship between the measured variable and the latent construct [3]. Research using structural equation modeling (SEM) [4–8] can be integrated into the meta-analysis method based on effect size correlation. The combination of SEM and meta-analysis methods is called Meta-Analytic Structural Equation Modeling (MASEM).

The meta-analytic structural equations modeling (MASEM) involves 2 stages of analysis, synthesizes correlation or covariance matrices from several studies or results, then applies SEM techniques to test the fit or explain the relationship between the variables to using a combined correlation or covariance matrix [9]. MASEM is a correlation-based meta-analysis so that the MASEM data type is a correlation matrix of the research studied.

In general, according to the homogeneity of the effect size, the MASEM model is divided into two types, the fixed-effects MASEM and the random-effects MASEM [2, 10]. The fixed-effects model assumes that the effect on the population is homogeneous across studies, while the random-effects model is heterogeneous or that there are variations in the effects of the study population, thus implying between studies variances and within studies variances. Accordingly, the process of estimating random effect MASEM parameters involves an additional process, estimating the variance between studies. This type of search effect size is rarely homogeneous because it is difficult to find studies that give the same results.

The first MASEM estimation model was developed by Becker in 1992 using the generalized least squares (GLS) estimation method. The initial process of the MASEM parameters was estimated with GLS, the next test of homogeneity if the test of homogeneity of the results is heterogeneous then estimated the variance between studies for the random-effects model and the final step is a fitting model [11, 12]. MASEM estimator In addition to the GLS method, there are other methods, namely the TSSEM method [13–15]. However, this method was developed in a different simulation with MASEM GLS which was analyzed analytically.

Random effect MASEM GLS using the iteration of the EM algorithm to estimate the variance between studies. In addition, the estimation variance between studies random effect MASEM has also been developed using the iterative restricted maximum likelihood [16, 17]. However, this method may not achieve convergence. Thus, a non-iterative procedure was developed to estimate the variance between studies using the method of moment. Estimation of the variance between studies random effect MASEM GLS with the moment method gave a better statistical Wald value compared to the iteration of the EM algorithm [18].

The implementation of MASEM on socio-economic issues in Indonesia has been completed, including health status issues in East Java Province [19], and poverty [20, 21]. Poverty data on the island of Java between provinces are likely to differ. Poverty data, if collected from a different population in between province, will be more accurate and in line with targets. The model of poverty with the economy, human resources, and health as exogenous variables is appropriate on the island of Java [22, 23].

In this paper, the novelty random effect MASEM GLS with the method of moment [18] is used to analyze the poverty model in Java. In addition, the results obtained are compared to the analysis of poverty on paper [21]. This research aims to provide the best results of the poverty model so that it can become a policy proposal for the Indonesian government to overcome poverty.

2. Method

2.1. MASEM with GLS approximation

Correlation matrix as many as k studies and p predictor variables will exist $p^* = \frac{p(p+1)}{2}$ the correlation will be estimated. The MASEM model, estimated using the generalized least squares (GLS) method, has 4 variables. There is a correlation matrix, an identity matrix, a correlation matrix estimator, and an error matrix which is approximated by the variant covariance matrix Σ. It is supposed to complete the equation: [18, 21]

$$b_{kp^{*}\times 1} = W_{kp^{*}\times p^{*}}\,\beta_{p^{*}\times 1} + e_{kp^{*}\times kp^{*}} \tag{1}$$

$$\Sigma = \begin{bmatrix} \Sigma_1 & 0 & \cdots & 0 \\ 0 & \Sigma_2 & & 0 \\ \vdots & & \ddots & \\ 0 & 0 & \cdots & \Sigma_k \end{bmatrix} \tag{2}$$

Equation (1) is estimated using the GLS method, the following forms of the estimator are obtained:

- Fixed Effect Model

$$\hat{\beta}_F = \left(\mathrm{W}^{\mathrm{T}}\Sigma^{-1}W\right)^{-1}\mathrm{W}^{\mathrm{T}}\Sigma^{-1}b \tag{3}$$

$$Cov\,(\hat{\beta}_F) = \left(\mathrm{W}^{\mathrm{T}}\Sigma^{-1}W\right)^{-1} \tag{4}$$

- Random Effect Model
  Random effect model is assumed that

$$b = W\beta + \delta + e \tag{5}$$

where δ_i $(1 \le i \le k)$ follows $\delta \sim N(0, T)$

$$\Omega = \begin{bmatrix} \Sigma_1 + T & 0 & \cdots & 0 \\ 0 & \Sigma_2 + T & & 0 \\ \vdots & & \ddots & \\ 0 & 0 & \cdots & \Sigma_k + T \end{bmatrix} \tag{6}$$

so,

$$\hat{\beta}_R = \left(\mathrm{W}^{\mathrm{T}}\Omega^{-1}W\right)^{-1}\mathrm{W}^{\mathrm{T}}\Omega^{-1}b \tag{7}$$

$$Cov\,(\hat{\beta}_R) = \left(\mathrm{W}^{\mathrm{T}}\Omega^{-1}W\right)^{-1} \tag{8}$$

The heterogeneity test used to test the effect size model type conforms to the following equation [24]:

$$Q_i = d_i'\widehat{M_i}d_i \tag{9}$$

with $\boldsymbol{d}$ are defined by $M_i = \Lambda^{(i)} - \Lambda^{(i)}ee'^{\Lambda^{(i)}} / \, e'\Lambda^{(i)}e$ and $\Lambda^{(i)} = \boldsymbol{\Sigma^{-1}}^{(i)}$. So that the homogeneity test with the GLS method is defined by:

$$Q = b^T\left[\Sigma^{-1} - \Sigma^{-1}W\left(\mathrm{W}^{\mathrm{T}}\Sigma^{-1}W\right)^{-1}W^T\Sigma^{-1}\right]b \tag{10}$$

test statistics:

If $Q < \chi^2_{(k-1)p^*}$ then the Fixed Effect Model or

If $Q > \chi^2_{(k-1)p^*}$ then the Random Effect Model.

The second step of this method is a fitting model. Estimating coefficient and the suitability of the model based on [21], pooled correlation matrix $\boldsymbol{R}$ is partitioned into: (assuming consist of 4 variables)

$$R = \begin{bmatrix} 1 & \beta_1 & \beta_2 & \beta_3 \\ \beta_1 & 1 & \beta_4 & \beta_5 \\ \beta_2 & \beta_4 & 1 & \beta_6 \\ \beta_3 & \beta_5 & \beta_6 & 1 \end{bmatrix} = \begin{bmatrix} 1 & R_{10} \\ R_{01} & R_{11} \end{bmatrix} \tag{11}$$

where

$$R_{11} = \begin{bmatrix} 1 & \beta_4 & \beta_5 \\ \beta_4 & 1 & \beta_6 \\ \beta_5 & \beta_6 & 1 \end{bmatrix}, R_{01} = \begin{bmatrix} \beta_1 \\ \beta_2 \\ \beta_3 \end{bmatrix} \quad (12)$$

The value of the estimated Coefficients latent variable is defined as follow:

$$b^* = R_{11}^{-1} R_{01} \quad (13)$$

then the partial test as follows:

$$T_i^* = \frac{b_i^*}{\sqrt{s_{ii}}} \quad (14)$$

test statistics:

If $|T_i^*| < Z_{\alpha/2}$ then the effect is not significant

If $|T_i^*| > Z_{\alpha/2}$ then the effect is significant.

The RSMEA assessment is used to fitting models. According to Cheung [25]

$$RMSEA = \sqrt{\frac{\max\left(\frac{\chi^2 - df}{N-1}, 0\right)}{df}} \quad (15)$$

2.2. Random effect MASEM With Moment Method

The main step of the MASEM random-effects model is to estimate the between-study variance-covariance matrix T. The result of the moments method for the estimate T based on [18], let

$$\Psi = Cov\,(\hat{\beta}_F) = (W^T \Omega^{-1} W)^{-1} = \left(\sum_{j=1}^{k} \Sigma_j^{-1}\right)^{-1} \quad (16)$$

$$\Phi = \Psi^{-1} - \sum_{i=1}^{k} \Sigma_i^{-1} \Psi \Sigma_i^{-1} = \sum_{i=1}^{k} \left(\Sigma_i^{-1} - \Sigma_i^{-1} \left[\sum_{j=1}^{k} \Sigma_j^{-1}\right]^{-1} \Sigma_i^{-1}\right) \quad (17)$$

$$A = \sum_{j=1}^{k} \Sigma_j^{-1} \left(b_j - \hat{\beta}_F\right)\left(b_j - \hat{\beta}_F\right)^T - (k-1) I_{p^* x\, p^*} \quad (18)$$

where $\Sigma_j, b_j, \hat{\beta}_F$ are defined in equation (1)-(4), then

$$E(A) = \Phi T \quad (19)$$

so the estimator T with a method of moment is

$$\hat{T} = \frac{\Phi^{-1} A + A^T \Phi^{-1}}{2} \quad (20)$$

2.3. Research Data Sources and Variables

The main goal of Meta-Analytical Structural Equation Modeling (MASEM) is to summarize various studies on the same topic to come to a conclusion. This is why the poverty model of this study is a slice of research on the factors of poverty on the island of Java that has been conducted. The poverty model is consistent with Figure 1.

Indicators of each latent variable are according to previous studies [21]. In addition, this research data is secondary data from the results of the Statistics Indonesia (SUSENAS) survey in 2017.

3. Result and Discussion

3.1. Random Effect MASEM Algorithm With Method Of Moment

The first step of MASEM analysis is to synthesize a correlation matrix which is a process of meta-analysis method. At this point, there is a test of homogeneity of the effect size model, and the results of the MASEM parameter estimation will be obtained. So, there is a difference of steps between the random effect MASEM GLS with the iteration of the EM algorithm and the method of moment, namely how to obtain the variance matrix between the studies.

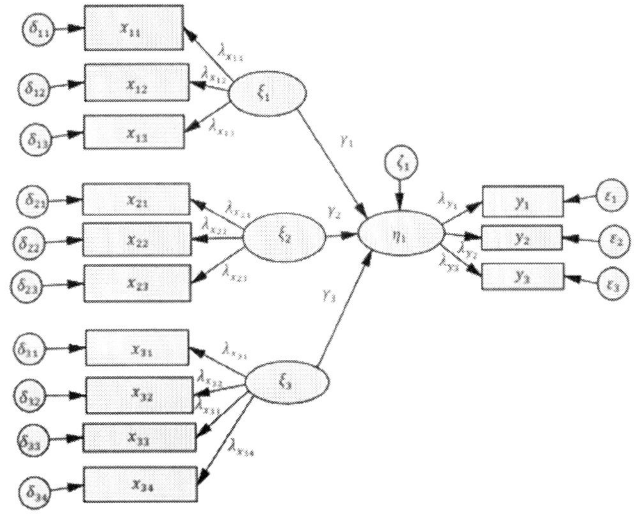

Figure 1. Java Island Poverty Model

The random effect MASEM variant matrix is the sum of variance matrix the within-study and the between-study. In a previous study [21], it was described to obtain a variance matrix between MASEM GLS studies with the EM Algorithm iteration as follows:

$$\hat{\tau}_{\alpha\gamma} = \frac{S_{\alpha\gamma}}{k-1} - \left(\frac{1}{k}\right)\sum_{i=1}^{k}\hat{\sigma}_{i\alpha\gamma},$$

with $\hat{\sigma}_{i\alpha\gamma}$ is a matrix element $\widehat{\Sigma}_i$ like Equation (2) and $S_{\alpha\gamma} = \sum_{i=1}^{k}(b_{i\alpha} - \bar{b}_\alpha)(b_{i\gamma} - \bar{b}_\gamma) = b'_\alpha B b_\gamma, \bar{b}_\alpha = \sum_{i=1}^{k} b_{i\alpha} / k$ and matrix B is defined as follows [11, 12]:

$$B = \begin{bmatrix} \frac{k-1}{k} & -\frac{1}{k} & \cdots & -\frac{1}{k} \\ -\frac{1}{k} & \frac{k-1}{k} & & -\frac{1}{k} \\ \vdots & & \ddots & \\ -\frac{1}{k} & -\frac{1}{k} & \cdots & \frac{k-1}{k} \end{bmatrix}$$

while a variance matrix of MASEM GLS with the moment method conforms to equations (16) - (20).

The second step of MASEM consists of fitting the model and at this stage, there is no difference of steps between the MASEM GLS with the EM algorithm and the method of moment. Therefore, the random effect MASEM GLS step or algorithm with the method of moment approach on the variance between studies can be summarized as follows:

Step Or Algorithm: Random effect MASEM GLS with a method of moment

1 : Collect the correlation matrices for each study

$$b = \begin{bmatrix} b_1 \\ b_2 \\ \vdots \\ b_k \end{bmatrix}$$

with

$$b_i = \begin{bmatrix} b_{i1} \\ b_{i2} \\ \vdots \\ b_{ip^*} \end{bmatrix}$$

2 : Construct a variant matrix $\Sigma_{kp^* \times kp^*}$ is defined as follows:

$$\sigma_{i\alpha} = \mathrm{Var}(r_{i\alpha}) = \frac{\left(1 - \rho_{i\alpha}^2\right)^2}{n_i}$$

$$\sigma_{i\alpha} = \mathrm{Cov}(b_{i\alpha}, b_{i\gamma})$$

$$\mathrm{Cov}(b_{ist}, b_{iuv}) = \big[0.5\rho_{ist}\rho_{iuv}\left(\rho_{isu}^2 + \rho_{isv}^2 + \rho_{itu}^2 + \rho_{itv}^2\right) + \rho_{isu}\rho_{itv} + \rho_{isv}\rho_{itu} - \left(\boldsymbol{\rho}_{ist}\boldsymbol{\rho}_{isu}\boldsymbol{\rho}_{isv} + \boldsymbol{\rho}_{its}\boldsymbol{\rho}_{itu}\boldsymbol{\rho}_{itv} + \boldsymbol{\rho}_{ius}\boldsymbol{\rho}_{iut}\boldsymbol{\rho}_{iuv} + \boldsymbol{\rho}_{ivs}\boldsymbol{\rho}_{ivt}\boldsymbol{\rho}_{ivu}\right)\big]/n_i$$

3 : Homogeneity test effect size model using Equation (10)

4 : Arrange a variance matrix of the random effect model

$$\Omega = \begin{bmatrix} \Sigma_1 + T & 0 & \cdots & 0 \\ 0 & \Sigma_2 + T & & 0 \\ \vdots & & \ddots & \\ 0 & 0 & \cdots & \Sigma_k + T \end{bmatrix}$$

with Equation (16)-(20)

5 : Calculates the estimate of the MASEM parameter
1. Fixed effect Equation (3) and (4)
2. Random Effect Equation (7) and (8)

6 : Arrange correlation matrix ***R*** corresponds to equation (11) of the element

$$\hat{\beta} = \begin{bmatrix} \beta_1 \\ \beta_2 \\ \vdots \\ \beta_{p^*} \end{bmatrix}$$

7 : Calculates of the coefficient between the latent variables according to equation (13) and the statistical test of equation (14)

8 : The fit of the model with the value Root Mean Square Error Approximation (RMSEA) according to equation (15).

Table 1. Result of Random Effect MASEM With Moments Method

Result Of Random Effect MASEM With Moments Method		
Coefficient Beta	Coeff. Economy → Poverty (b_1^*)	0.4942
	Coeff. Health → Poverty (b_2^*)	-0.1106
	Coeff. HR → Poverty (b_3^*)	-0.2843
Significant Test $Z_{0.025} = 1.96$	$\lvert T_{b1}\rvert = \lvert b_1/\sqrt{s_{11}}\rvert$	3.7309 (Significant)
	$\lvert T_{b2}\rvert = \lvert b_2/\sqrt{s_{22}}\rvert$	0.9838 (Not Significant)
	$\lvert T_{b3}\rvert = \lvert b_3/\sqrt{s_{33}}\rvert$	2.3543 (Significant)
Coefficient of Determination	R^2	56.51%
Fitting Models	RMSEA	0.000

Source: Matlab R2018b Processed Data Output.

3.2. Data analysis and processing

MASEM is a technique for synthesizing a correlation or a covariance matrix and then testing the fit of the model from a combined matrix correlation or covariance with the fitting SEM method. Therefore, in the first step, a correlation matrix of several studies on poverty is needed. In this article, the results of the SEM modeling with the conceptual framework according to Figure 1 in each province become the source of the synthesized research. Based on the research of B. W. Otok, R. E. Standsyah, A. Suharsono, and Purhadi (2019), the correlation matrix can be organized into:

$$R = \begin{bmatrix} 1 & 0.59 & -0.476 & -0.612 \\ 0.59 & 1 & -0.311 & -0.563 \\ -0.476 & -0.311 & 1 & 0.666 \\ -0.612 & -0.563 & 0.666 & 1 \\ 1 & 0.591 & -0.495 & 0.326 \\ 0.591 & 1 & -0.591 & 0.639 \\ -0.495 & -0.591 & 1 & -0.378 \\ 0.326 & 0.639 & -0.378 & 1 \\ 1 & 0.827 & -0.604 & -0.753 \\ 0.827 & 1 & -0.529 & -0.738 \\ -0.604 & -0.529 & 1 & 0.898 \\ -0.753 & -0.738 & 0.898 & 1 \\ 1 & 0.698 & -0.399 & -0.387 \\ 0.698 & 1 & -0.747 & -0.251 \\ -0.399 & -0.747 & 1 & 0.231 \\ -0.387 & -0.251 & 0.231 & 1 \\ 1 & 0.963 & -0.946 & -0.93 \\ 0.963 & 1 & -0.939 & -0.985 \\ -0.946 & -0.939 & 1 & 0.888 \\ -0.93 & -0.985 & 0.888 & 1 \\ 1 & 0.829 & -0.848 & -0.447 \\ 0.829 & 1 & -0.888 & -0.742 \\ -0.848 & -0.888 & 1 & 0.826 \\ -0.447 & -0.742 & 0.826 & 1 \end{bmatrix} \quad (21)$$

The next step is to process the matrix $\boldsymbol{R}$ equation (21) with the random effects MASEM according to step or algorithm of Random effect MASEM GLS with a method of moment. The author implements this algorithm in the Matlab R2018b software with reference to Becker [11, 12]. The results of the implementation of the Matlab R2018b software show that the results of the homogeneity of $Q = 425.5881 > (X^2_{0.05;30})$ so that the effect size of this model is a random-effects model. as for the combined correlation matrix is:

$$R_{pooled} = \begin{bmatrix} 1 & 0.6964 & -0.5995 & -0.5442 \\ 0.6964 & 1 & -0.7332 & -0.4262 \\ -0.5995 & -0.7332 & 1 & 0.4453 \\ -0.5442 & -0.4262 & 0.4453 & 1 \end{bmatrix}$$

and the value of the estimated Coefficients latent variable is shown in Table 1.

3.3. Results Interpretation

The results of the random effect MASEM, where the variance between studies was estimated using the method of moment, showed an RMSEA value of 0.000. The RMSEA value is the same as the fit criteria of the SEM model, which is less than 0.08, so data can be explained well by the model. The interpretation of this MASEM result shows that the economy has a significant effect on poverty at alpha 0.5 with a coefficient of 0.4942, health has no significant effect on poverty at alpha 0.5 with a coefficient of -0.1106 and human resources have a significant effect on poverty at alpha 0.5 with a coefficient of -0.2843.

The economy has a significant positive effect because the economic variable indicator is directly proportional to the poverty variable indicator. If the values of the economic indicators decrease, it means that the economy of the community has increased, then the community has been able to meet its needs independently, which has resulted in a decrease in poverty. This is different from human resources and health which negatively affect poverty. If the HR and Health indicators increase, the HR and Health variables will increase. When human resources increase along with health, people's creativity and productivity for their lives will increase, which will improve their standard of living and poverty will decrease.

The comparison of the random effect MASEM GLS which estimates the variance between studies with the method of moment and the EM algorithm shows that the RMSEA value of the approach method of moment is smaller than that of the EM algorithm according to the results of the article [21]. Moreover, based on the value of R^2, it also shows that the method of moment (56.51%) is larger than the EM algorithm (50.60%). This explains that the performance and results interpretation of the random effect MASEM GLS with the current method is better.

4. Conclusion

This study can conclude that the economy has a significant positive effect with a coefficient of 0.4942 because the indicators of economic variables are directly proportional to the indicators of poverty variables. However, health and human resources have a negative effect with coefficients of -0.1106 and -0.2843 respectively on poverty. When human resources increase at the same time as people's health, creativity, and productivity increase, it also affects the decline in the value of economic indicators or the same as the increase in the economy of the community. When these 3 variables show an increase, it can lead to a decrease in the poverty rate in Java.

The results of the random effect MASEM application which estimates the variance between studies using the method of moment on the poverty model show better performance and interpretation results compared to the EM algorithm approach. The results of this poverty model may improve on the previous poverty model which used the EM algorithm approach.

Acknowledgments

This study is the original idea of the author and the author is a Doctoral Student Of the Department Of Statistic, Institut Teknologi Sepuluh Nopember and also the lecturer in the Faculty Of Teacher Education And Educational Sciences The University of Dr. Soetomo Surabaya.

References

[1] Glass G V 1976 Primary, secondary and meta-analysis of research *Am. Educ. Res. Assoc.* **5** pp 3–8

[2] Borenstein M, Hedges L V, Higgins J P T and Rothstein H R 2009 *Introduction to Meta-Analysis* (United Kingdom: John Wiley & Sons, Ltd.)

[3] Bollen K A 1989 *Structural Equations with Latent Variables Canada* (United Kingdom: John Wiley & Sons, Ltd.)

[4] Anekawati A, Otok B W, Purhadi and Sutikno 2017 Structural equation modelling with three schemes estimation of score factors on partial least square (case study: the quality of education level SMA/MA in Sumenep Regency) *J. Phys. Conf. Ser.* **855** 1 DOI: 10.1088/1742-6596/855/1/012006

[5] Anuraga G and Otok B W 2013 Pemodelan kemiskinan di Jawa Timur dengan structural equation modeling-partial least square *Statistika* **1** 2

[6] Otok B W, Sustrami D, Hastuti P, Purhadi, Sutikno, Purnami S W and Suharsono A 2018 Structural equation modeling the environment, psychology, social relationships against physical health in determination quality of elderly community surabaya *Int. J. Civ. Eng. Technol.* **9** pp 926–38

[7] Suyanto S, Purnomo B and Standsyah R E 2019 The influence of education on the depressing of unemployment and the increasing of the society economy in East Java Surabaya – Indonesia *Proc. of 1st Asian Conf. Humanit. Ind. Technol. Soc. Universitas Dr. Soetomo Surabaya* pp 292–8

[8] Suyanto S, Purnomo B and Standsyah R E 2019 Impact of education and health on the unemployment rate and economy of East Java *Sinergi Jurnal Ilmiah Ilmu Manajemen* **9** 1

[9] Viswesvaran C and Ones D S 1995 Theory testing: combining psychometric meta-analysis and

structural equations modeling *Pers. Psychol.* **48** pp 865–85
[10] Hedges L V and Vevea J L 1998 Fixed- and random-effects models in meta-analysis *Psychol. Methods* **3** pp 486–504
[11] Becker B J 1992 Using results from replicated studies to estimate linear models *J. of Educational Statistics* **17** pp 341–62
[12] Becker B J 1995 Corrections to "using results from replicated studies to estimate linear models" *J. of Educational and Behavioral Statistics* **20** pp 100–2
[13] Cheung M W and Chan W 2005 Meta-analytic structural equation modeling: a two-stage approach *Psychological Methods* **10** pp 40–64
[14] Cheung M W L 2014 Fixed- and random-effects meta-analytic structural equation modeling: Examples and analyses in R *Behav. Res. Methods* **46** pp 29–40
[15] Cheung M W and Fai S 2016 Random-effects models for meta- analytic structural equation modeling: review, issues, and illustrations *Research Synthesis Methods* **7** pp 140–55
[16] Jackson D, White I R and Thompson S G 2010 Extending DerSimonian and Laird's methodology to perform multivariate random effects meta-analyses *Stat. Med.* **29** pp 1282–97
[17] Cheung M W L 2013 Implementing restricted maximum likelihood estimation in structural equation models *Struct. Equ. Model* **20** pp 157–67
[18] Chen H, Manning A K and Dupuis J 2012 A method of moments estimator for random effect multivariate meta-analysis *Biometrics* **68** pp 1278–84
[19] Tohari A and Widjanarko B 2014 Meta Analytic Structural Equation Modeling (MASEM) pada faktor-faktor yang mempengaruhi derajat kesehatan di Jawa Timur *Prosiding Seminar Nasional Matematika, Universitas Jember* pp 35–50
[20] Nur A and Widjanarko B 2014 Meta-Analityc Structural Equation Modeling (MASEM) pada faktor-faktor yang mempengaruhi kemiskinan di Pulau Jawa *Prosiding Seminar Nasional Matematika, Universitas Jember* pp 51–62
[21] Otok B W, Standsyah R E, Suharsono A and Purhadi 2019 Development of model poverty in Java using Meta-Analysis Structural Equation Modeling (MASEM) *AIP Conf. Proc.* pp 1-8
[22] Suryati C 2005 Memahami kemiskinan secara multidimensional understanding *JMPK* **08** pp 121–9
[23] BPS dan Bappeda 2017 *Kerjasama balai statistik daerah Bappeda DIY dengan Badan Pusat Statistik (BPS) Provinsi DIY 2017* (Yogyakarta: Badan Pusat Statistik)
[24] Hedges L V and Olkin I 1985 *Statistical Methods for Meta-Analysis* (Orlando, Florida: Academic Press,Inc)
[25] Cheung M W-L 2015 *Meta-Analysis a Structural Equation Modeling Approach* (United Kingdom: John Wiley & Sons, Ltd.)

Ensemble convolutional neural network for robust batik classification

Yufis Azhar, Moch. Chamdani Mustaqim[1], Agus Eko Minarno

Universitas Muhammadiyah Malang
Jl. Raya Tlogomas 246, Malang, East Java, Indonesia

E-mail: [1]danicham14@gmail.com

Abstract. Some researchers propose using the Convolutional Neural Network (CNN) method to classified batik images. It can extract features automatically without the need to define feature manually from the image. However, CNN's weakness is that its accuracy is quite low, especially for small-sized datasets, compared to machine learning methods that use hand-crafted feature extraction. In this research, an ensemble CNN method is proposed to improve the accuracy of the CNN method in classifying batik images. This method will train several CNN models at once, and then by voting and averaging techniques, the output label will be determined. Test results for two different datasets show this method can improve the accuracy of the CNN method and get an accuracy value of 100%. This method is also proven to extract features faster than the state-of-the-art method like MTCD+SVM, which is included in the hand-crafted feature extraction category.

1. Introduction

Batik is one of Indonesia's cultural heritage. Research on batik has started in 1974 [1], and since then, the development of research on batik is multiplying. In computer vision, batik is often used as a research object about pattern recognition, image retrieval, object detection, and so on.

Machine learning is an approach that is often chosen for the classification of batik images. Handayani et al. managed to get a maximum accuracy value of 99% after using the voting features interval method to classify 163 batik images into five categories [2]. In machine learning, feature extraction plays a significant role. As evidenced by Suciati et al. when using a fast discrete curvelet transform on HSV color features. The accuracy results showed that the use of HSV is better than RGB color features [3]. Rangkuti also mentioned color features as a critical feature to recognize batik patterns in his research. In the study, Rangkuti used treeval and treefit as decision tree function in optimizing content-based batik image retrieval. The results showed that the decision tree approaches properly-recognized batik patterns with 90% accuracy [4]. Batik pattern recognition was also done by Nurhaida et al. in 2015. The SIFT method that used in this research was able to recognize a batik pattern with an error rate of 0.08 [5]. In 2018, Caraka et al. used geometric invariant moments to detect batik parang rusak patterns. This method is proven to perform exceptionally well when used on binary classification datasets with 92% accuracy [6]. While in 2015, Aditya et al. proposed Grey Level Co-occurrence Descriptor (GLCM) and Neural Network to recognize batik patterns. The accuracy obtained by this research was 91% [7]. Both studies prove that in addition to color features, edge detection and texture detection can also be used to recognize batik patterns. GLCM's success in improving models' accuracy to recognize batik patterns makes this method widely used in other studies [7–11]. In 2020, Azhar et al. also proposed a combined approach of the Multi Texton Co-occurrence Descriptor, which is based on GLCM, with the TF-IDF weighting method commonly used in text retrieval. This method is called Texton Frequency - Inverse Image Frequency (TF-IIF). The results obtained show an increase in 7% precision compared to the previous process [12]. In addition to GLCM, the SIFT approach is also widely used in research on batik pattern recognition. Azhar et al. and Setyawan combine SIFT and KNN [13,14]. Pradnyana et al. also used KNN as a Balinese fabric classifier because of its simplicity [15].

Besides of machine learning approach, deep learning is also often used for the classification of batik images. Among the many proposed methods, the Convolutional Neural Network (CNN) is one of the most widely used methods [16–20]. In 2017, Handayani et al. used CNN and the Deep Belief Network (DBN) to classify batik datasets. However, both methods provide poor accuracy results, with 63% on CNN and 44% on DBN [21]. In 2018, Agastya et al. used augmentation data techniques to increase the number of datasets before they were classified using CNN and its derivative variants (VGG16 and VGG19). Although the accuracy results reached 90%, its accuracy dropped to 56% when the augmentation data was applied [22]. Based on research that has been proposed by Arsa et al., classification using deep learning also has not been able to provide maximum results [23]. It depends on the dataset used. The amount of data in each class dramatically affects the learning process in the deep learning algorithm. Also, the determination of convolution layer models, hidden layers, and other CNN parameters significantly affect the accuracy results.

In the previous research, Minarno et al. classified batik images by utilizing the sub-band image [11]. The wavelet and GLCM methods are combined with the Probabilistic Neural Network (PNN) to obtain 72% accuracy. In the next research, Minarno et al. proposed Multi Texton Histogram (MTH) method with six textons as a feature extractor for batik combined with PNN, KNN, and SVM methods solve the same problem. The results obtained are accuracy scores of 92%, 82%, and 76% respectively [24,25]. In 2019, Minarno et al. using the GLCM method combined with KNN and SVM. The results obtained are 78% for KNN and 92% for SVM [26]. Based on the previous research, both GLCM and MTH could be used to classify batik images quite well. In 2020, Minarno et al. combined MTH and GLCM in a new method called Multi Texton Co-occurrence Descriptor (MTCD). This method used six filters called textons to generate a color histogram and edge on an image by convolution. The two histograms are then combined with the entropy, correlation, energy, and contrast values obtained through the GLCM method. In this research, MTCD was paired with KNN and SVM for the classification of batik images. The results obtained were proven to be able to outperform the previous studies. Both KNN and SVM were able to get an accuracy value of 96% [27]. From these studies, it can be seen that the accuracy value of the classification model is largely determined by the features selected to be extracted. And these features must be determined manually or better known as hand-crafted feature extraction. This is one of the weaknesses of the machine learning method

In terms of practicality of use, deep learning is still better than machine learning because it did not need to define features that must be extracted first, as it does in machine learning methods. Deep learning was used to cover up the deep learning method's weaknesses that are still getting low accuracy on small datasets. An ensemble CNN method was proposed to improve the accuracy of the CNN method in classifying batik images. This method will train several CNN models at once, and then by voting and averaging techniques, the output label will be determined. Test results for two different datasets show this method can improve the accuracy of the CNN method and get an accuracy value of 100%. This method is also proven to extract features faster than the MTCD + SVM method, which is included in the category of hand-crafted feature extraction.

2. Research method

This research's primary focus is to create a deep learning architecture that can classify batik images well. The primary method used is the Convolutional Neural Network (CNN). A dropout layer was added to the CNN architecture to produce better accuracy. Figure 1 shows the architecture used in this research.

As we know, overfit is one of the problems that often arise in implementing deep learning in image classification. And the way to handle it is by using the concept of regularization. Adding layer dropout to the network is one form of regularization that can be applied. How it works from the dropout layer was to break the input relationship from the previous layer to the next layer with a certain probability. In this research, the probability value was set to 0.5, so it was expected that there would be half of the interrupted relationship between the extraction and the fully connected layers. Thus, overfit is expected to be prevented. From Figure 1, it can be seen that there are two dropout layers installed on the network. The first layer is before the Fully Connected (FC) layer (after the feature was extracted), and the second is between the dense layer on the FC Layer.

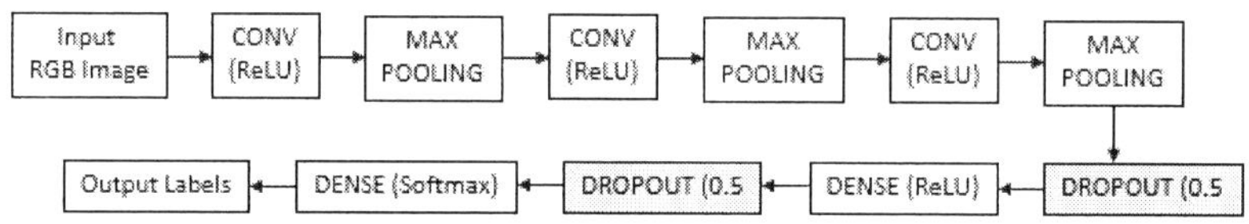

Figure 1. Network architecture with the addition of a dropout layer after the feature extraction layer and before the output layer

In detail, there are a total of 10 layers used on the CNN architecture in this research. Consists of three pairs of Convolution and Max-Pool layers, two dense layers, and two dropout layers. Each layer uses ReLU as its activation function, except for the last dense layer that uses the softmax function. Each convolution layer has 32, 50, and 80 filters (kernels), respectively. Each convolution and pool layer have filters of the same size, namely 2x2, and stride = 1. Each input image will be resized to a size of 32x32 with a depth of 3 (RGB) before being incorporated into the CNN architecture.

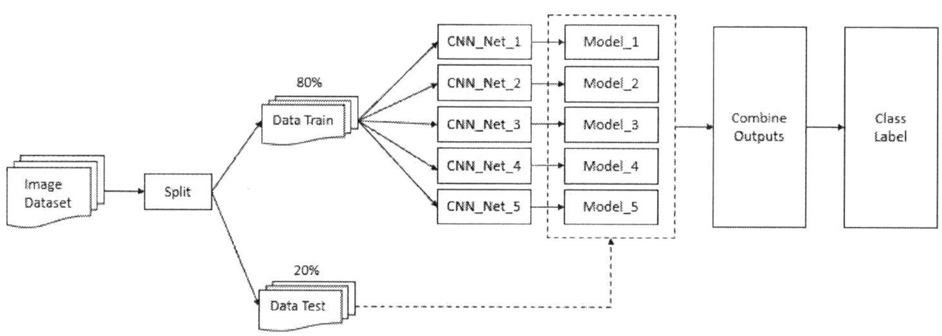

Figure 2. The proposed ensemble CNN method

Apart from adding a dropout layer, the CNN architecture was also combined with the ensemble method. Figure 2 shows the design using the ensemble and CNN methods in classifying batik images. There are two main phases in the design method proposed. The first phase is the training phase. In this phase, the data train will be entered into CNN with the same architecture five times. The result is five different models, with varying values of accuracy. This has happened because, on CNN, the weight of each kernel was initially set at random. As the training process progresses, the filter's weight will become more convergent towards the best value. However, because the initial value is different, each model's best value in the last epoch will also be different. Although the difference is not very significant, it will also affect the classification process of data. And then the second phase is the testing phase. In this phase, the test data will be predicted using the five models that have been created previously. Then, the label of the test data will be determined using voting techniques. All these stages are referred to as the ensemble CNN method.

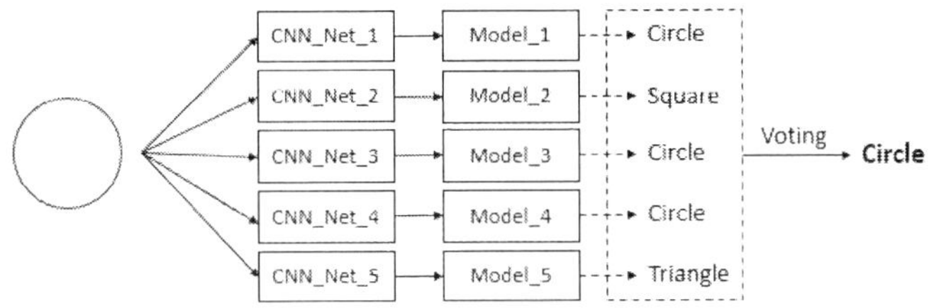

Figure 3. The ensemble (combine output) process on CNN to determine the output label of a data

The ensemble method itself refers to a technique that uses a large number of models (where the exact number of models used is very dependent on the case of classification), then combines the results of the prediction through voting or averaging to improve the accuracy of the classification results. Figure 3 exemplifies how the ensemble method is paired on CNN to combine the output produced by each model. Using the voting technique, the label that is guessed by many models is the label that will be selected as the output label.

3. Result and discussion

There are two types of datasets used in this research. The first dataset is Batik300 (Figure 4). This dataset consists of 300 batik images, which are divided into 50 different classes. Each class consists of 6 images. These six images are the results of the augmentation of an image with a large dimension. All images in this dataset have the same dimensions, i.e., 128 x 128 pixels. The second dataset is Batik41k (Figure 5). This dataset has characteristics similar to Batik300, but with a higher number of classes, namely 351 classes. Another thing that also distinguishes this dataset from Batik300 is the imbalance dataset condition. Batik351 has a different amount of data in each category. This dataset's total batik image is 41,621 data, with the dimensions of each image being the same, namely 400 x 600 pixels. The two datasets were split into two parts, 20% as the test data, and the rest as the data train.

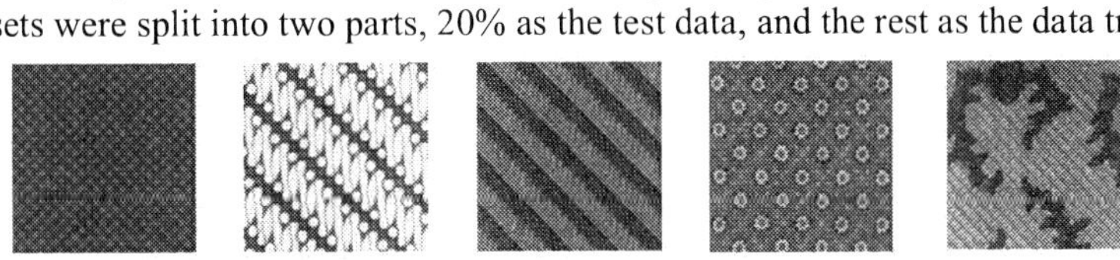

Figure 4. Examples of Image Contained in Batik300 Dataset

Figure 5. Examples of Image Contained in Batik41k Dataset

Three test scenarios will be carried out in this research. The first scenario is to look at adding layer dropouts on CNN architecture in classifying batik images. The purpose of this test is to see how much influence the dropout layer has on handling overfit.

There are two architectures to be tested—one architecture without a layer dropout and another architecture with a layer dropout (Figure 1). Both architectures used the same hyperparameter. Hyperparameter tuning is performed using the Keras Tuner library. The led hyperparameter is the learning rate and optimization function. Of the 5 learning rate scores tested (1e-1, 1e-2, 1e-3, 1e-4, 1e-5), the most optimal results for learning rate is 1e-3. As for optimization function, testing of Adam, SGD, and RMSprop function. Adam gives the best results. The results obtained when testing the two architectures in the Batik300 dataset are shown in Figure 6. Although both architectures can produce reasonably good accuracy values (above 0.8), it is seen that architectures that do not use dropout layers tend to overfit. This is concluded from the observation that the value of loss tends to increase, and the value of accuracy tends to decrease from the 86th epoch. Different things are shown in the results of architectural training using layer dropouts. It is seen that the value of loss continues to fall close to 0 and does not indicate an increase. The same thing happened in the accuracy value plot, which continuously increased to close to 1. From this test, it can be concluded that the addition of the dropout layer proved to be sufficient enough to suppress the tendency to overfit the classification results.

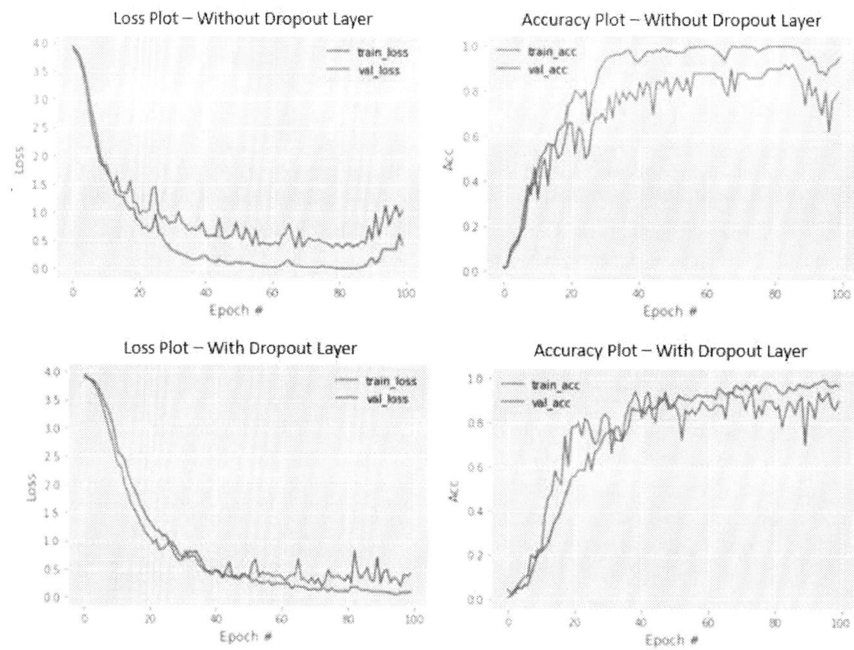

Figure 6. Comparison of loss and accuracy values obtained by models

The second testing scenario conducted in this research is the use of the ensemble method on CNN. The purpose of using this method is to increase the accuracy value of CNN. In this research, the ensemble method was carried out by combining the results of the classification of 5 CNN models trained using the same architecture and dataset. Comparing the proposed method and the previous method is also applied using the same hardware and the same dataset. The results can be seen in Table 1 and Table 2.

Table 1. The accuracy values of the proposed method compared with previous studies for the Batik300 dataset

Method	Accuracy
MTH+SVM [26]	0.92
MTCD+SVM [27]	0.96
CNN	0.98
Ensemble CNN	1

In Table 1, the proposed method (ensemble CNN) can produce an excellent accuracy value (100%) for the Batik300 dataset. It outperformed the results obtained by previous studies in [26] and [27].

Even when compared to traditional CNN methods, the ensemble CNN is still superior. Meanwhile, in Table 2, if we focus on the comparison between the traditional CNN method and the ensemble CNN for the two types of datasets (Batik300 and Batik41k), it can be seen that the ensemble CNN can outperform almost all measurement parameters. Especially for the batik41k dataset, the Ensemble CNN's accuracy value is the same as [27]. But, the use of the Ensemble CNN method was considered to have advantages. For example, at the speed of the extraction of its features, this hypothesis was proven using the third testing scenario.

Table 2. Comparison of result before and after the ensemble method was applied to the CNN architecture

Dataset	Parameter	CNN	Ensemble CNN
Batik300	Accuracy	0.98	1
	Precision	0.95	1
	Recall	0.98	1
	F1-Score	0.96	1
Batik41k	Accuracy	0.99	1
	Precision	0.97	1
	Recall	0.99	1
	F1-Score	0.98	1

In the third test scenario, predictions were made using models that have been formed by the ensemble CNN and MTCD + SVM methods. Because both methods have been tested using the same dataset, and both can get perfect accuracy values (i.e., 100%). Both methods were considered to have the same ability in classifying batik images. Variables that will be compared in this test are the time needed by the two methods in extracting features for testing data (new data). Both methods will be given the same dataset, namely Batik41k. Then, both methods were trained using the same number and variety of train data. After the model was formed, the two methods were asked to predict as many as 8,094 batik images, which were still in the form of raw images. The results obtained in Table 3 show that the ensemble CNN method is superior to the MTCD + SVM method in terms of its feature extraction speed. In the MTCD method, each feature (color, edge, and texture) must be extracted separately. For color and edge features, extraction is done by converting six filters one by one. This process needs to be done for each batik image (be it a training image or test image). Unlike MTCD, the extraction of features on CNN is done by creating a feature map in each layer. The process of creating this feature map can be done very quickly at the testing stage because the kernel weight in each layer has been obtained at the time of training. On the MTCD+SVM method, feature extracted is performed by MTCD. While on CNN, the feature extraction process is done on the input layer until the flatten layer. The total time to classify an image is calculated from the image is inserted into the model until all features have been extracted (histogram feature vector for MTCD and flatten feature map for CNN). Since each kernel's best weight value has been obtained during the training process, then at the time of testing, CNN only needs to perform one convoluted process in each convolutional layer to get the best feature map (only one feed-forward step from the input layer to output layer). This is what causes the testing process of the CNN ensemble to be faster than MTCD+SVM.

Table 3. Comparison of the time needed to extract features in batik41k test data between the previous and proposed method

Method	Time (s)
MTCD+SVM [27]	235.23
Ensemble CNN	9.54

4. Conclusion

This research has presented a method for classifying two batik datasets using dropout, ensemble testing scenarios, and comparing testing times between Ensemble CNN vs. previous research method.

Based on the test results, it can be concluded that the addition of the dropout layer was able to overcome overfitting, and the ensemble technique can improve the accuracy of the previous method, like MTH+SVM or MTCD+SVM. Ensemble CNN can also increase the accuracy, precision, recall, and f-measure value of the CNN method, both for batik300 and batik41k dataset. Although ensemble CNN and MTCD+SVM reach accuracy 1, in the testing process, Ensemble CNN has a faster computing process than MTCD + SVM with a time difference of 225.69 seconds.

References

[1] Bartleson F E 1973 Totem name batik kit *United State US Patent No US3840113A*

[2] Handhayani T 2016 Batik Lasem images classification using voting feature intervals 5 and statistical features selection approach *Proc. - 2016 Int. Semin. Intell. Technol. Its Appl. ISITIA 2016 Recent Trends Intell. Comput. Technol. Sustain Energy* pp 13–6

[3] Suciati N, Kridanto A, Naufal M F, Machmud M and Wicaksono A Y 2015 Fast discrete curvelet transform and HSV color features for batik image clansificotlon *2015 Int. Conf. on Information & Communication Technol. and Systems (ICTS) Surabaya* pp 99–104

[4] Rangkuti A H, Rasjid Z E and Santoso D J 2015 Batik image classification using treeval and treefit as decision tree function in optimizing content based batik image retrieval *Procedia Computer Sci.* **59** pp 577–83

[5] Nurhaida I, Noviyanto A, Manurung R and Arymurthy A M 2015 Automatic Indonesian's batik pattern recognition using SIFT approach *Procedia Computer Sci.* **59** pp 567–76 DOI: 10.1016/j.procs.2015.07.547

[6] Caraka R E and Pardamean T W C B 2018 Batik parang rusak detection using geometric invariant moment *2018 Indonesian Association for Pattern Recognition Int. Conf. (INAPR)* pp 71–4

[7] Aditya C S K, Hani'ah M, Bintana R R and Suciati N 2015 Batik classification using neural network with gray level co-occurence matrix and statistical color feature extraction *2015 Int. Conf. on Information & Communication Technol. and Systems (ICTS)* Surabaya pp 163–8

[8] Sulistianingsih N, Soesanti I and Hartanto R 2018 Classification of batik image using grey level co-occurrence matrix feature extraction and correlation based feature selection *2018 Int. Seminar on Research of Information Technol. and Intelligent Systems (ISRITI) Yogyakarta, Indonesia* pp 492–7

[9] Juwita A R and Solichin A 2018 Batik pattern identification using GLCM and artificial neural network backpropagation *2018 Third Int. Conf. on Informatics and Computing (ICIC) Palembang, Indonesia* pp 1–6

[10] Irawan C, Ardyastiti E N, Setiadi D R I M, Rachmawanto E H and Sari C A 2018 A survey: Effect of the number of GLCM features on classification accuracy of Lasem batik images using K-nearest neighbor *2018 Int. Seminar on Research of Information Technol. and Intelligent Systems, ISRITI 2018* pp 33–8

[11] Minarno A E, Munarko Y, Kurniawardhani A and Bimantoro F 2014 Texture feature extraction using co-occurrence matrices of sub-band image for batik image classification *Information and Communication Technol. (ICoICT)* (New Jersey: IEEE) pp 249–54

[12] Azhar Y, Minarno A E, Munarko Y and Ibrahim Z 2020 Image retrieval based on texton frequency-inverse image frequency *Kinet. Game Technol. Inf. Syst. Comput. Network, Comput. Electron. Control* **5(2)** pp 87–94

[13] Azhar R, Tuwohingide D, Kamudi D, Sarimuddin and Suciati N 2015 Batik image classification using SIFT feature extraction, bag of features and support vector machine *Procedia Computer Sci.* **72** pp 24–30 DOI: http://dx.doi.org/10.1016/j.procs.2015.12.101

[14] Setyawan I, Timotius I K and Kalvin M 2015 Automatic batik motifs classification using various combinations of SIFT features moments and k-Nearest Neighbor *2015 7th Int. Conf. on Information Technol. and Electrical Engineering (ICITEE) Chiang Mai* pp 269–74

[15] Pradnyana G A, Suryantara I K A and Darmawiguna I G M 2018 Impression classification of

Endek (Balinese fabric) image using K-nearest neighbors method *Kinet. Game Technol. Inf. Syst. Comput. Network, Comput. Electron. Control* **3(3)** pp 213–20
[16] Rasyidi M A and Bariyah T 2020 Batik pattern recognition using convolutional neural network *Bull. Electr. Eng. Informatics* **9(4)** pp 1430–7
[17] Wicaksono A Y, Suciati N, Fatichah C, Uchimura K and Koutaki G 2017 Modified convolutional neural network architecture for batik motif image classification *IPTEK J. Sci.* **2(2)** pp 26–30
[18] Prasetyo H and Akardihas B A P 2019 Batik image retrieval using convolutional neural network *Telkomnika (Telecommunication Comput. Electron. Control)* **17(6)** pp 3010–8
[19] Tristanto J, Hendryli J and Herwindiati D E 2018 Classification of batik motifs using convolutional neural networks *SSRN Electron. J.* pp 1–5
[20] Puarungroj W and Boonsirisumpun N 2019 Convolutional neural network models for handwoven fabric motif recognition *2019 Joint Int. Conf. on Digital Arts, Media and Technol. with ECTI Northern Section Conf. on Electrical, Electronics, Computer and Telecommunications Eng. (ECTI DAMT-NCON)* pp 300–3
[21] Handhayani T, Hendryli J and Hiryanto L 2017 Comparison of shallow and deep learning models for classification of Lasem batik patterns *Proc. - 2017 1st Int. Conf. on Informatics and Computational Sciences - ICICoS 2017* pp 11–6
[22] Agastya I M A and Setyanto A 2018 Classification of Indonesian batik using deep learning techniques and data augmentation *2018 3rd Int. Conf. on Information Technol., Information System and Electrical Eng. (ICITISEE) Yogyakarta, Indonesia* pp 27–31
[23] Arsa D M S and Susila A A N H 2019 VGG16 in batik classification based on random forest *2019 Int. Conf. on Information Management and Technol. (ICIMTech) Jakarta/Bali, Indonesia* pp 295–9
[24] Minarno A E, Munarko Y, Kurniawardhani A and Bimantoro F 2015 Classification of texture using multi texton histogram and probabilistic neural network *Int. Conf. on Eng. and Technol. for Sustainable Development* pp 12–22
[25] Minarno A E, Maulani A S, Kurniawardhani A, Bimantoro F and Suciati N 2018 Comparison of methods for batik classification using multi texton histogram *Telkomnika (Telecommunication Comput. Electron. Control* **16(3)** pp 1358–66
[26] Minarno A E, Sumadi F D S, Wibowo H and Munarko Y 2020 Classification of batik patterns using K-Nearest neighbor and support vector machine *Bull. Electr. Eng. Informatics* **9(3)** pp 1260–7
[27] Minarno A E, Azhar Y, Sumadi F D S and Munarko Y 2020 A robust batik image classification using multi texton co-occurrence descriptor and Support Vector Machine *2020 3rd Int. Conf. on Intelligent Autonomous Systems (ICoIAS) Singapore* pp 51–5

IOP Conf. Series: Materials Science and Engineering 1077 (2021) 012054 doi:10.1088/1757-899X/1077/1/012054

The Physarum Algorithm with Adaptive Power Control: A Routing Optimization to Overcome Obstacle Problems in IoT Network

Muh Asnoer Laagu[1,*], and Muhamad Asvial[1]

[1]Department of Electrical Engineering, Faculty of Engineering, Universitas Indonesia, Kampus Baru UI – Depok, 16424, Jakarta, Indonesia

E-mail: *muh.asnoer@ui.ac.id

Abstract. The existence of obstacles on the Internet of Thing (*Io*T) network area is the biggest challenge in the IoT network infrastructure. The obstacle in the IoT network can be a wall or the absence of a router in a certain area. Routing optimization to facing obstacles problems has been done a lot, however, no routing algorithm that can overcome these obstacles. This paper proposed Physarum Routing based Adaptive Power Control (*PR-APC*), the routing optimization algorithm to resolve obstacles problems in the IoT network by controlling and maximizing the available power. This paper compares the routing optimization algorithm Physarum inspired Routing Protocol (*P-iRP*) and Lightweight and Energy-Aware Routing (*LEAR*). As a result, *PR-APC* can minimize the number of hops that are passed when sending data packets from the source node to the final destination node. *PR-APC* takes 3 hops with a delay of 6.4ms, while *P-iRP* takes 9 hops with a delay of 10.55ms, and *LEAR* 8 hops with a delay of 9.19ms. *PR-APC* optimizes the last node before the obstacle to be given strong power and be able to send data packets to the final destination node. *P-iRP* and *LEAR* in their routing decisions use parameter energy saving and will take a detour to avoid obstacles.

1. Introduction

Revolution 4.0 is a new chapter in the dynamics of human life. We entered the trend of automation and data exchange belonging to cyber systems, cloud computing (cloud), Internet of Things (IoT), and cognitive computing. Internet of Things (IoT) technology has been using to improve the quality of human life. Limited resources are one of the obstacles in the IoT network, other things such as restricted devices, energy resources, memory resources, and computing resources of nodes on the local IoT network[1]. Nodes in a local IoT network can be low-power devices and rely on batteries as a resource to carry out all their activities. This resource limitation causes the nodes in the local IoT network to be unable to pass through existing obstacles. The obstacle can be a wall or the absence of a router in a specific area[2].

Various algorithms had been propose in the routing optimization method, consists of conventional methods, mathematical programming methods, and bio-inspired methods[3], such as ant colony[4], fireflies, and *physarum polycephalum*. Taro et al[5], research the phenomena of *physarum* organisms in a physical and mathematical model called the basic *physarum*. It inspired Zhang et al[6] to develop a Physarum-inspired Routing Protocol (*P-iRP*) in a wireless network that utilizes energy residue and geographic location of nodes to create an energy balance in the network. Rosyidi et.al[7] developed the *P-iRP* protocol by adding a tree-level parameter to the gateway so that the routing activity can be calculated as the minimum number of hops required to send data packets to the final destination gateway. This protocol is Lightweight and Energy Aware Routing (*LEAR*). The studies show that both *P-iRP* and *LEAR* have problems when they encounter obstacles[8]. The obstacle could be a wall or the absence of a router in a specific location.

This paper will optimize the *P-iRP* and *LEAR* algorithms to pass-through obstacle problems. We proposed Physarum Routing based Adaptive Power Control (*PR-APC*), that maximizes the power of the last node before facing an obstacle[9]. In particular, the increase in routing performance from previous studies is the basis for the development of routing optimization methods. This process to obtain the required routing efficiency and power distribution, especially when there is an obstacle in the network[8].

2. Related Work

2.1. Routing problem in IoT Network

IoT network architecture, consisting of many IoT node devices as endpoints, such as sensors, actuators, IoT gateway or base station, or a server/cloud platform on the internet, connected via Wi-fi / Ethernet / Cellular. There are two networks in IoT, local and large-scale local IoT networks. The local IoT network is a network of all IoT nodes with gateways. A large-scale local IoT network is a network that can be divide into several clusters. Each cluster can have a node that functions as a cluster head. The cluster head had a significant role as an intra-cluster coordinator and as a point of contact for data before sent to the IoT gateway. Gateway is a destination point that collects data before sent over the internet server or IoT cloud platform. Gateways on IoT networks are usually highly resourceful devices. These devices can be computers with a great processor and memory resources, as well as unlimited energy resources because they connect to a power source. The gateway provides the communication interface required for transmission to the internet, such as via wi-fi, Ethernet, or cellular communications. The nodes in IoT devices are low power devices, usually using batteries to carry out various activities. This resource limitation causes the nodes in the local IoT network didn't have sufficient power to pass-through the obstacles on the local network in a specific area. The barrier/obstacle on the IoT network can be a wall/wall or the absence of a router in the IoT network area

2.2. Physarum Polychepalum

Physarum polycephalum is a single-celled organism that lives in a humid environment, also known as a slime-mold. Biologists have long conduct research into the intelligence and ability of these organisms to find the most effective pathways to food sources. This phenomenon led the researchers to design various optimization methods inspired by the behavior of the *physarum polycephalum* organism. Some variants of the physarum algorithm, such as basic physarum, rapid physarum, and improved physarum. Nagaki et.al[10] investigate the physarum phenomenon when searching for food sources found the fact that physarum will find the shortest route by spreading the network of vessels throughout the maze, then cut off and kill the tissue of the vessels, leaving a network of vessels leading to the food source.

Taro et.al[5][11] describes a mathematical model *physarum polycephalum* that has flux and is determined by the pressure and conductivity between two points.

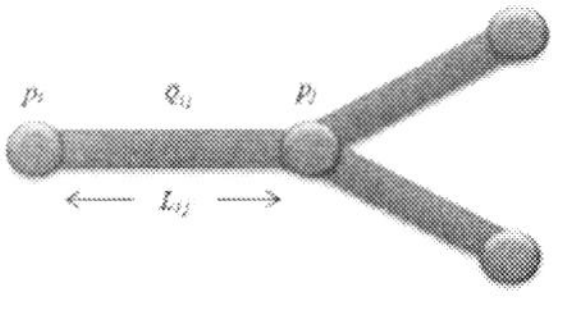

Figure 1. Physical model physarum polychepalum[5]

Q_{ij} is the flux through the vessel from point i to j. The measurement results of the flux Q_{ij} can be known by[11]:

$$Qij = D_{ij}\frac{P_i - P_j}{L_{ij}} \quad (1)$$

D_{ij} is the conductivity of vessel, P is the pressure on the point of the source (i) and destination (j), and L is the vessel length from i to j. Vessel thickness dynamic adaptation can be formulated by assuming D_{ij} will change with the flux Q_{ij} [11]:

$$dD_{ij} = f\left(|Q_{ij}|\right) - rD_{ij} \quad (2)$$

The channel length is assumed to be constant during the adaptation process f, a function of growth $f(0) = 0$, and r is the depreciation rate. Through this equation can be seen that the effect of incoming flux implies that conductivity will shrink and also increase.

2.3. Physarum Inspired Routing Protocol (P-iRP)

Zhang et.al[6] modified the previous *physarum* model by introducing several *P-iRP* routing parameters. D_{ij} is a physical characteristic in the channel vessels replaced by the quality of the link ϕ_{ij}. ϕ_{ij} is a characteristic of a wireless channel. In wireless communication, there is a component *path loss* α, although L_{ij} as the channel length can also be the same as the wireless communication transmission distance. L_{ij} replaced $L_{ij^{\alpha}}$ as signal attenuation in wireless communication.

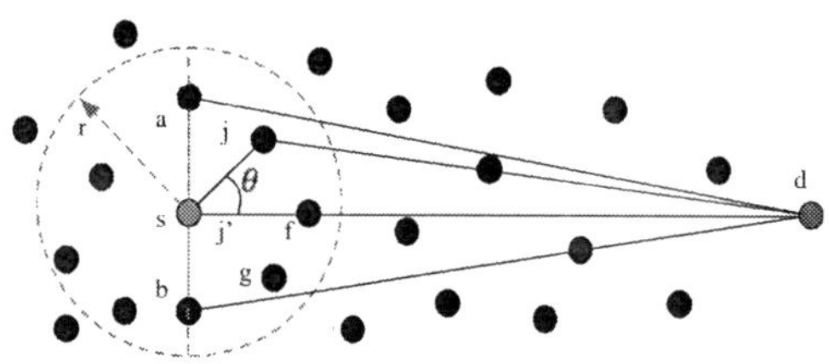

Figure 2. *P-iRP* flux equation[6]

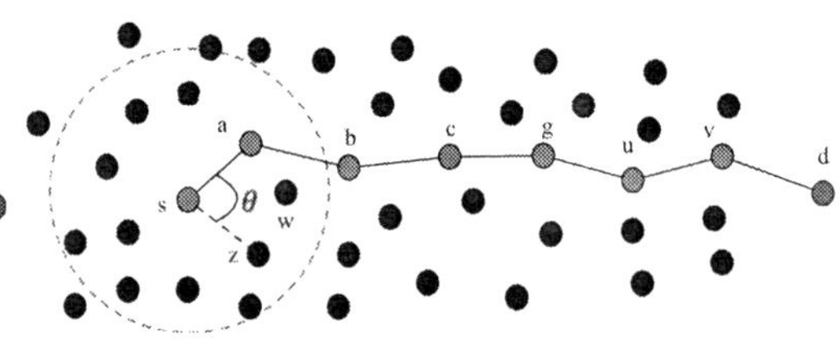

Figure 3. Routing decision process P-iRP[6]

The flow of data packets directed from the source to the node has a deviation angle ϕ_{sjd}, the smallest hop node among the candidate destination nodes, as depicted in the Figure 2. The dashed red circle of radius r represents the range of the source node (s). ϕ_{scd} is the deviation angle between the next-hop candidate and the end node (d), measured from the source node (s).

In *P-iRP*, the sum of next-hop energy residue ER_j and transmission direction Q_{ij} can compensate for the pressure difference ($P_i - P_j$), we get *P-iRP* physarum modification[6]:

$$Q_{ij} = \frac{\phi_{ij} \times [k \times \widehat{ER}_j + (1-k) \times cos\theta_{jid}]}{L_{ij^{\alpha}}} \quad (3)$$

Q_{ij} is the virtual flux over the wireless link ϕ_{ij} represents the link quality, $\widehat{ER}_j$ is the energy residue of the next-hop candidate, $L_{ij^{\alpha}}$ is the signal attenuation factor for distance node i to node j. θ_{jid} is the deviation angle formed from the node i, node j, and node d with the cosine function for normalization, k is the priority constant for energy residual considerations ER_j with the intention of $cos\ \theta_{jid}$. The higher k value, the stronger the energy residue consideration in determining the virtual flux value Q_{ij}. Figure 3 describes the results of the *P-iRP* routing decision. Data packet from source (s) to destination (d) selected path *s-a-b-c-g-u-v-d*. Node *a* selected because it has a smaller energy residue than node *w* and node *z*, although the distance of the three nodes is almost the same.

P-iRP routing process has problems with obstacles, the impact of the energy residual consideration results in the *P-iRP* routing decision to find the closest source node. *P-iRP* routing algorithm tends to consider the selection of the router with the smallest deviation angle to destination. Figure 4 describes the *P-iRP* routing decision to choose the *s-a-b-c-e-f-g-h-i-d* path of 9 hops has the smallest deviation angle compared to *s-j-k-e-f-g-h-i*-d path of 8 hops. In conclusion, *P-iRP* routing mechanism is not efficient for areas where there are obstacles.

2.4. Lightweight and Energy-Aware Routing (LEAR)

Rosyidi et.al[7] make improvements to *P-iRP* algorithm by developing the Lightweight and Energy Aware Routing (*LEAR*) algorithms. *LEAR* still takes the physarum flux equation by adding the *HL* (Hierarchy Level) parameter that is the minimum number of hops. Figure 5 shows the measurement of *HL* value in the network.

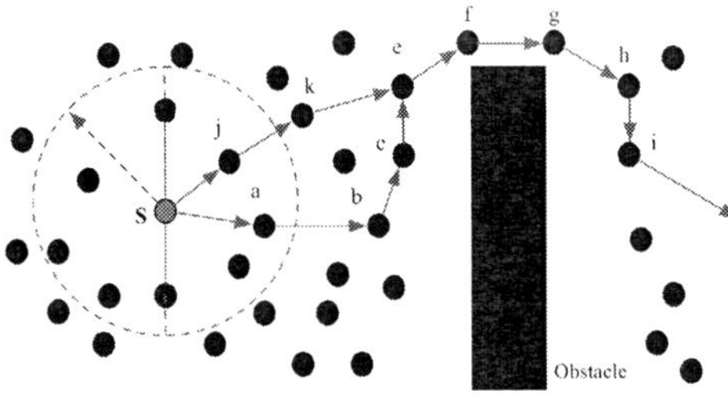

Figure 4. P-iRP routing decision with obstacle area[7]

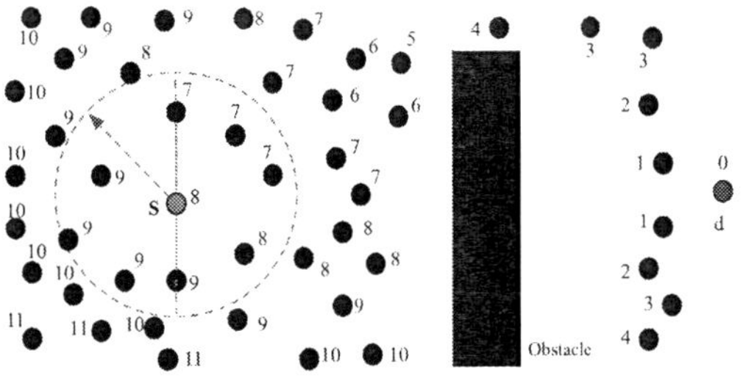

Figure 5. Determination of HL value in LEAR[7]

HL value becomes the reference for determining the routing direction, starting from the gateway (*HL*=0) increasing by one every next hop. The node can assign its *HL* values by the taking value of the smallest neighboring node plus one. Priority factor *k* sets the weights between *ER* and *HL* whose values range from 0 to 1 in the next-hop selection. Signal attenuation parameter L_{ij^α} replaced by a factor *Receive Signal Strength Indicator (RSSI)* easily scalable on real devices. The closer the *RSSI* value to 0, signal will get strong. RSSI value in normal conditions between -30dBm to -70dBm. Link quality parameter (∅) replaced by parameters *LQI* (*Link Quality Indicator*) generally available on transceiver devices. *LQI* parameter used in the routing shown in the following equation[7]:

$$Q_{ij} = \widehat{LQI_{ij}} \times \widehat{RSSI_{ij}} \times [k \times \widehat{ER_j} + (1-k) \times (HL_i - HL_j)] \quad (4)$$

Routing strategies are more likely to prefer larger *RSSI* and *LQI* values for the next-hop candidate. That provides an advantage in the routing process because the signal will get strong or the whatever the distance is closer. Q_{ij} parameter will be the total of virtual flux score and the neighboring nodes with the largest value Q_{ij} will be selected as the next hop. *LEAR* makes the routing method more efficient than *P-iRP*, especially when facing a network that has obstacles. Although it is more efficient when encountering obstacles, *LEAR* has not been able to pass through nodes that are far enough away or have obstacles. Both of these algorithms, *P-iRP* and *LEAR*, will rotate to find the nearest node with power-saving considerations.

3. Method

This paper proposed Physarum Routing based Adaptive Power Control (*PR-APC*) improving *P-iRP* and *LEAR* algorithms when there is an obstacle in the network. *LEAR* and *P-iRP* have the main differences in terms of routing direction strategies. *P-iRP* uses the parameter of deviation angle as the destination, *LEAR* exploits routing based hierarchy that takes advantage of the difference in Hierarchy Level (*HL*) nodes. *PR-APC* will get the information for each node in the network, then determine its routing decisions.

3.1. Model system PR-APC

PR-APC initializes all existing nodes, including the location of the obstacle on the network[2]. Information required is the position of the source node (s) then the location of the next node or normal node (d), node before obstacle also called obstacle node (h), location of the latitude and longitude obstacle, and the position of the final destination node (e). Figure6 describe the model system of *PR-APC*:

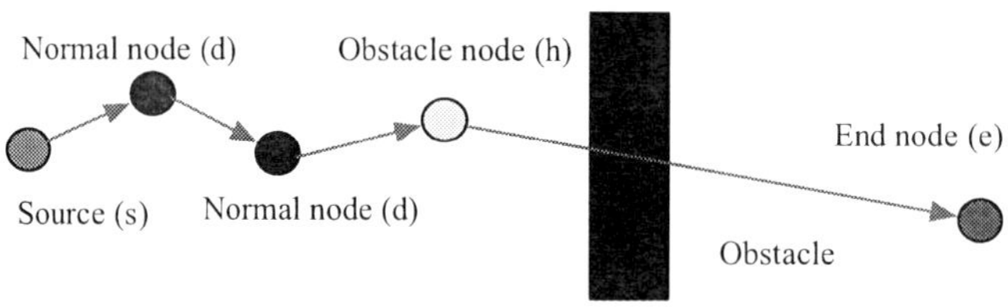

Figure 6. Model system *PR-APC*

The proposed routing algorithm initializes all routing information and requirements. All nodes that are not near the obstacle will be associated as a normal node (d), while those near the barrier will be referred to as the obstacle node (h). The position of nodes on the network is made static and random based on the longitude and latitude. The process of initializing the information for each node becomes easier. After the initialization process, each node will send broadcast information about the identity of that node, making it easier for routing to determine the next hop candidate[12]. Data received will be stored as information for the next-hop candidate.

Q_{ij} value can be updated using an equation (1). We assume for the pressure value, if $p \geq 1$ then the next-hop destination is obstacle node (h), if $0 < p < 1$ then the next-hop destination is normal node (d). The destination end node (e), pressure value p must first go through the obstacle node whose pressure value is greater or equal to 1. The next-hop routing decision can determine by the following equation:

Source node (s) to normal node (d) $$: dD_{ij} = \frac{D_{ij} \times (P_i - P_j)}{L_{ij} \times (P_s - P_d)} - D_{ij} \quad (5)$$

Source node (s) to obstacle node (h) $$: dD_{ij} = \frac{D_{ij} \times (P_i - P_j)}{L_{ij} \times (P_s - P_h)} - D_{ij} \quad (6)$$

Obstacle node (h) to end node (e) $$: dD_{ij} = \frac{D_{ij} \times (P_i - P_j)}{L_{ij} \times (P_h - P_e)} - D_{ij} \quad (7)$$

The location and coordinates of the obstacle node itself are the primary information in *PR-APC* routing decision. Routing direction must first pass through the obstacle node to get strong power and be able to pass through existing obstacles before going to the end node (e). There are three crucial parameters in *PR-APC*[13], conductivity D_{ij}, pressure difference$(P_i - P_j)$ simplified to be $\Delta P_{i,j}$, and flux length $L_{i,j}$. Conductivity D_{ij} is a physical character inherent in the network. P_j received power can be normalized at node j as the initial value D_{ij}. P_j indicates the quality of the link, so the conductivity equation on the wireless network can be seen as follows:[13]:

$$\widehat{D_{ij}} = \frac{P_j - P_{min}}{P_{max} - P_{min}} \quad (8)$$

$\widehat{D_{ij}}$ represents the normalization of the signal received from the link (i,j), P_{min} is the minimum power value that can communicate between nodes, P_{max} is the maximum power between each node. The distance of nodes needs to be adjusted in length for broadcast. To find out the length $L_{i,j}$ we use the projection of the distance between the nodes so that it is easy to calculate. Length $L_{i,j}$ can determine in advance the projection distance between nodes ($L'_j d = L_{sd} - L_{sj}$) as in the equation below[13]:

$$L_{i,j} = \frac{L'_j d}{L_{id}} \quad (9)$$

$L'_j d$ is the projection distance from node j relative to the next node, L_{id} is the distance between node i and node d. In addition to determining the difference of pressure $\Delta P_{i,j}$, it is necessary to pay attention to the remaining energy (E_{res}) and buffer (B_{res}) in the flux flow, as in the following equation[13]:

$$\Delta P_{i,j} = \alpha E_{res} + \beta B_{res}$$
$$0 \leq \alpha \leq 1, 0 \leq \beta \leq 1, \alpha + \beta = 1 \quad (10)$$

Equation (1) can be rewritten in the following equation[13]:

$$Q_{i,j} = \frac{\widehat{D_{ij}}\,(\alpha E_{res} + \beta B_{res})}{L_{i,j}} \quad (11)$$

After all the parameters are fulfilled, the conductivity value can then be calculated using the equation (2) and flux value can use the equation (11).

3.2. Routing Decision

Figure 7 shows *PR-APC* routing decision. *PR-APC* routing decisions are greatly influenced by the distance and location of nodes in network[14].

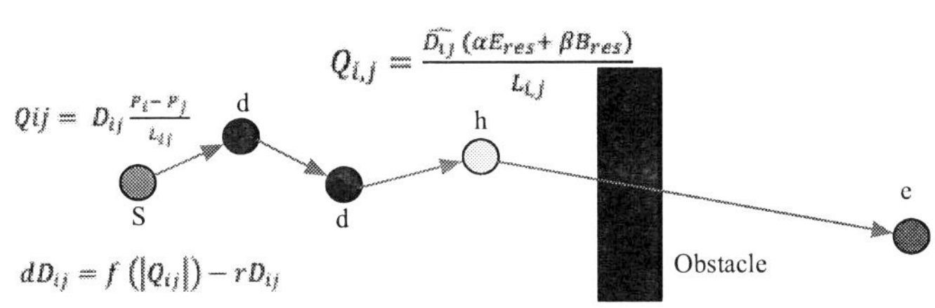

Figure 7. *PR-APC* routing decision

PR-APC solved *P-iRP* and *LEAR* problems in facing obstacles. An explanation of the *PR-APC* routing decision as follows:

- Determine the flux value at the time of initial initialization in the network using the equation (1), conductivity D_{ij} based on equation (2), pressure difference $\Delta P_{i,j}$ according to the equation (10), and length $L_{i,j}$ as in the equation (9).
- Routing direction will get the obstacle node (h) when routing started, so the flux value must be solved by the equation (11).
- Routing decisions will continue until routing finds the destination end node (e) has passed the obstacle.

Figure 8 and figure 9 shows flowchart system and *PR-APC* routing process algorithm.

PR-APC routing algorithm starts with initialization process. In this process, the *PR-APC* routing algorithm will send broadcast route request for all crucial information such as the location of the source node (s), normal node (d), obstacle node (h), end node (e), and obstacle location. Each node will broadcast its information to make node identification easier at routing decisions. After the initialization, PR-APC will start the routing decision algorithm. all data and information from the initialization process save as crucial information. Routing will calculate all node pressure (p) and update value the flux through vessels (Q) using equation (1) and renew conductivity (D) using equation (2). Before determining the next-hop candidate, the *PR-APC* routing algorithm will calculate the pressure $\Delta P_{i,j}$. When $0 \leq \Delta P_{i,j} \leq 1$, the routing decision will direct towards normal node (d), but if $\Delta P_{i,j} \geq 1$, routing decision towards obstacle node (*h*). In the obstacle node (*h*), the node will have more power using equation (10), so that it can make routing decision directly to the end node (*e*).

4. Result

In this experiment, simulations were carried out using Matlab software on a Macbook Pro 2012 intel core i5 laptop with a clock of 2.5 GHz and RAM 16 GHz. The purpose of this simulation is to compare the delay and the number of hops passed by *PR-APC* with other optimization methods, such as *P-iRP* and *LEAR* when facing obstacle problems. The obstacle used in this experiment is perpendicular and is positioned at the predetermined longitude and latitude.

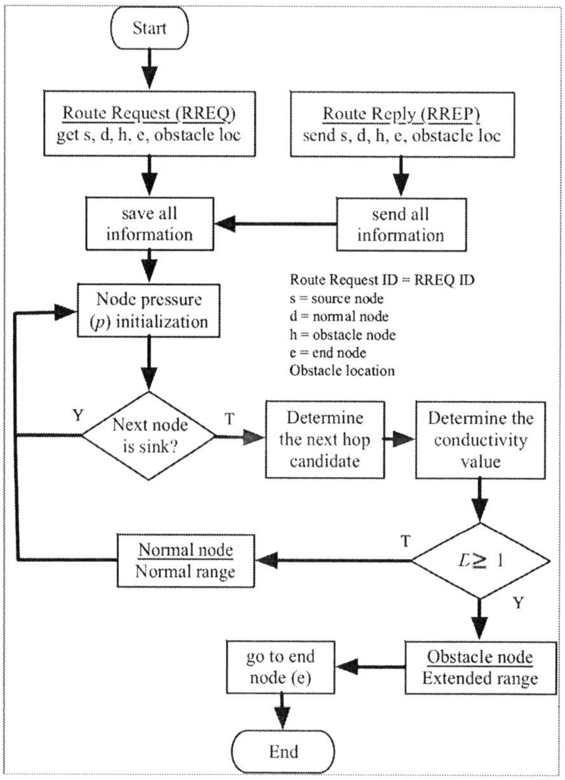

Figure 8. Flow Chart *PR-APC*

```
PR-APC initialization algorithm
send route request_broadcast
while not time out
        for each incoming_reply
            get all data nodes (s, d, h, e)
            get obstacle location
            save s, d, h, e
            save obstacle location
        end
end
if no_incoming_reply
        go to delay before inizialtization
end

PR-APC routing decission algorithm
for each incoming_packet
        get all data nodes (s, d, h, e)
        get obstacle location
        calculate node pressure, eq (1)
        update value Q, eq (2)
        renew eq (3)
        count + 1
until criteria fulfilled
return
        compare pressure difference ΔP_i,j, eq (10)
if 0 ≤ ΔP_i,j ≤ 1                #normal node
        calculate value normal hop, eq (5)
        go to normal node (d)
repeat
else if ΔP_i,j ≥ 1               #obstacle node
        calculate value obstacle hop, eq (6)
        go to obstacle node (h)
        calculate routing to end node (e), eq (7)
        send package to end node (e), eq (11)
end
```

Figure 9. Pseudocode PR-APC routing

Table 1 shows the simulation parameters for *PR-APC* evaluation. The simulated network consists of randomly distributed nodes. The total number of nodes in this simulation is 86 nodes with 1 source node and 1 final destination node.

Table 1. *PR-APC* simulation parameters

Parameter	Value
Number of nodes	86
Distribution of nodes	Random, static
Source node (s) position	Static
Obstacle node (h)position	Static
End node (e) position	Static
Obstacle location	Static and perpendicular

Figure 10 is a simulation overview of the distribution of nodes, location of nodes and obstacle, position of source node, and destination node. The red perpendicular line is the obstacle, the red dot is source node (*s*) dan the green dot is destination end node (*e*). There are 5 nodes near obstacle, which are associated as obstacle node (*d*). Comparison the number of nodes between *PR-APC, P-iRP* and *LEAR* is shown in figure 11.

P-iRP routing algorithm[6] tends to choose the smallest deviation angle in routing decisions so the path that traverse is longer if there are obstacles. Figure 11, describe *P-iRP* goes through nine hops *s-a-b-c-e-f-g-h-i*. Meanwhile, *LEAR*[7] initializes all hops with the Hierarchy Level (*HL*) parameter passing only eight hops *s-j-k-e-f-g-h-i*. *PR-APC* in overcoming obstacle problems will determine which node is closest to the obstacle, then controlling power to the node before sending data packets to the end node destination. *PR-APC* only passes three hops *s-a-b* because it detects node *b* as the obstacle node that is near the obstacle, routing decision can be directed directly to the end node (*e*) destination. *PR-APC* is more efficient in generating routing paths in obstructed area conditions.

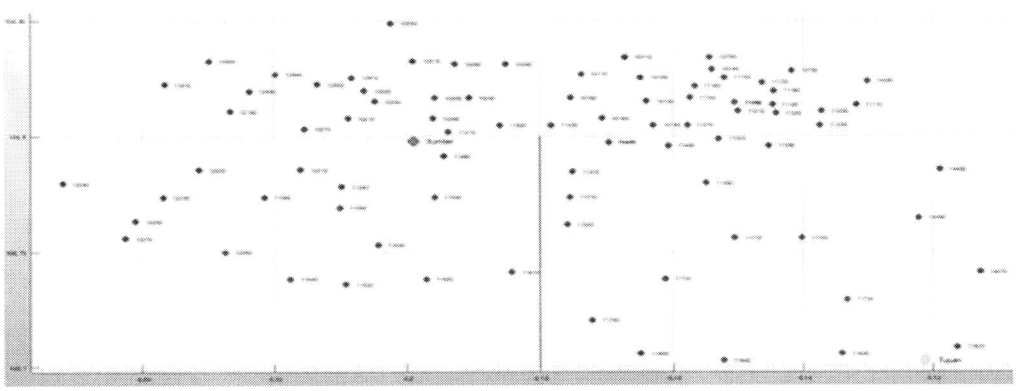

Figure 10. *PR-APC* simulation area

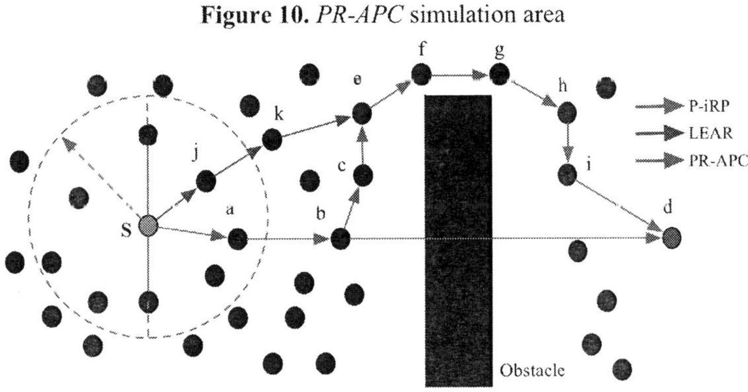

Figure 11. Comparison the number of nodes, PR-APC, LEAR and P-iRP

Table 2 shows the total delay in each routing algorithm. Until iteration 100, *PR-APC* has a smaller delay of 6.46ms with a total of 3 hops passed. *LEAR* has a 9.19ms delay go through 8 hops, and *P-iRP* has a 10.55ms delay with a total of 9 hops to reach the final destination node. *P-iRP* starts the routing path considers the deviation angle. At the time when routing facing an obstacle, the required delay is longer. *LEAR* has consistently made routing routes more efficient when facing obstacles. Routing decisions will choose a longer path due to energy residue considerations. *PR-APC* get all routing information, such as source node position (*s*), obstacle node (*h*), obstacle position, end node position (*e*), resulting routing decisions are faster than *LEAR* or *P-iRP*

Table 2. Total delay in *PR-APC, LEAR* and *P-iRP*

Algorithm	Delay (ms)	Number of hops
PR-APC	6.46	3
LEAR	9.19	8
Pi-RP	10.55	9

Figure 12 and figure 13 shows the number of hops is strongly influenced by the total iteration performed. During total iteration 10 routing through four hops, when the total iteration above 100 the number of hops passes only three hops, the source node (*s*), then the obstacle node (*h*), then go directly to the final destination node (*e*).

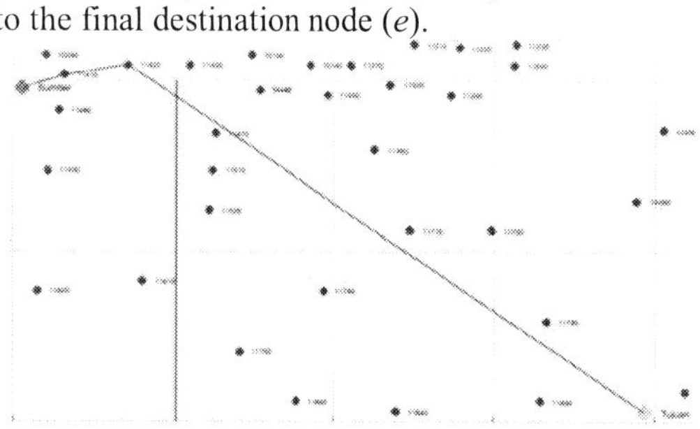

Figure 12. Total iterations 10

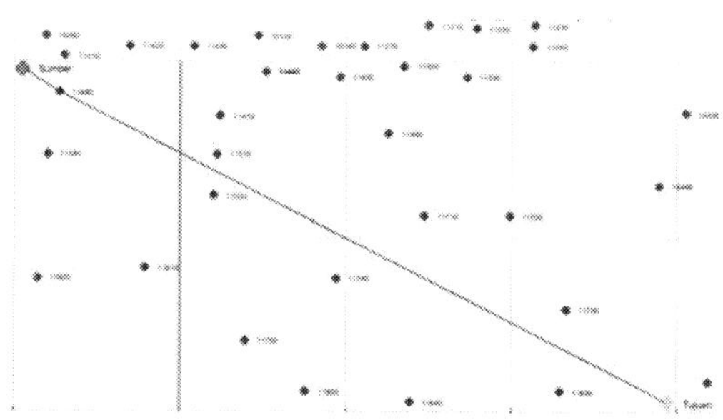

Figure 13. Total iterations 100

PR-APC algorithm will choose the closest distance to the obstacle node (*h*), it can directly forward the data packet to the end node (*e*). The number of iterations is directly proportional to the delay, total iteration affected to delay, when the number of iteration increase, the delay will also increase.

5. Conclusion

In this paper, we have presented the Physarum Routing Algorithm based Adaptive Power Control (*PR-APC*) as our proposed routing mechanism for pass-throgh obstacle in IoT network. *PR-APC* algorithm performs routing optimization for networks that has an obstacle. Number of hops required by *PR-APC* is much more efficient than routing optimization using *P-iRP*[6] and *LEAR*[7]. *PR-APC* routing decision will choose the shortest distance to the node that is closer to the obstacle, power, energy and distance can be calculated to forward the packet to the end node destination. PR-APC can pass-through obstacle by adding more power to the last node before obstacle.

PR-APC in this paper only describes a few performance indicators in simulation result, we only show the total hops efficiency and delay of *PR-APC* compared with to the other algorithm, *P-iRP and LEAR*. For future work, we will add more performance indicator results such as network lifetime, throughput, and end-to-end delay. We will also validate *PR-APC* in the real testbed for investigation of some important aspects and performance indicators in the real implementation. This is important to ensure that the proposed protocol design meets the requirements for routing in the IoT low power network.

References

[1] Marjani M *et al* 2017 Big IoT data analytics: architecture, opportunities, and open research challenges *IEEE Access* vol **5** pp 5247–61

[2] Liu L, Han G, Wang H and Wan J 2017 Obstacle-avoidance minimal exposure path for heterogeneous wireless sensor networks *Ad Hoc Networks* vol **55** pp 50–61

[3] Hamidouche R, Aliouat Z, Gueroui A M, Ari A A A and Louail L 2018 Classical and bio-inspired

mobility in sensor networks for IoT applications *J. Netw. Comput. Appl.* vol **121** no July pp 70–88
[4] Liu Y *et al* 2015 Solving NP-Hard Problems with Physarum Based Ant Colony System *IEEE/ACM Trans. Comput. Biol. Bioinforma.* vol **14** no 1 pp 108–20
[5] Tero A, Kobayashi R and Nakagaki T 2006 Physarum solver: A biologically inspired method of road-network navigation *Phys. A Stat. Mech. its Appl.* vol **363** no 1 pp 115–9
[6] Zhang M, Xu C, Guan J, Zheng R, Wu Q and Zhang H 2013 P-iRP: Physarum-inspired routing protocol for wireless sensor networks *IEEE Veh. Technol. Conf.* pp 1–6
[7] Rosyidi L, Asvial M and Sari R F 2018 Lightweight and energy-aware wireless mesh routing for autonomic internet of things network (accepted) *Int. J. Grid Distrib. Comput.* vol **11** no 12
[8] Guo W and Huang X 2020 PORA: A Physarum-inspired obstacle-avoiding routing algorithm for integrated circuit design *Appl. Math. Model.* vol **78** pp 268–86
[9] Ahmed G and Khan N M 2017 Adaptive power-control based energy-efficient routing in wireless sensor networks *Wirel. Pers. Commun.* vol **94** no 3 pp 1297–329
[10] Nakagaki T, Kobayashi R and Nishiura Y 2004 Obtaining multiple separate food sources: behavioural intelligence in the physarum plasmodium *Proc. of the Royal Society B: Biological Sciences* **271(1554)** pp 2305–10
[11] Tero A, Kobayashi R and Nakagaki T 2007 A mathematical model for adaptive transport network in path finding by true slime mold *J. of Theoretical Biology* vol **244** pp 553–64
[12] Won J and Park H K 2018 An adaptive power-controlled routing protocol for energy limited wireless sensor networks *J. Inf. Commun. Converg. Eng.* vol **16** no 3 pp 135–41
[13] Jiang H, Liu X, Xiao S, Tang C and Chen W 2020 Physarum-inspired autonomous optimized routing protocol for coal mine MANET *Wireless Communications and Mobile Computing* vol **2020** pp 1–14
[14] Gao C, Zhang X, Yue Z and Wei D 2018 An accelerated physarum solver for network optimization *IEEE Trans. Cybern.* vol **50** no 2 pp 765–76

Predicting COVID-19 Confirmed Case in Surabaya using Autoregressive Integrated Moving Average, Bivariate and Multivariate Transfer Function

A Saikhu[1], C V Hudiyanti[2], J L Buliali[3], V Hariadi[4]

Department of Informatics, Institut Teknologi Sepuluh Nopember, Surabaya, Indonesia

E-mail: [1]saikhu@if.its.ac.id, [2]cvairra@gmail.com, [3]joko@if.its.ac.id, [4]victor@if.its.ac.id

Abstract. In March 2020, the first case of Covid-19 was found in Indonesia. The increase of confirmed, suspected, and exposed in Surabaya has also significantly. Some studies show there is a relation among temperature, humidity, suspected, and exposed patients in an area with the number of confirmed COVID-19. Several statistical techniques that can be used to determine this relationship are to analyze and predict it using the ARIMA, bivariate, and multivariate transfer functions. The aim of this study is the performance of three models and determine the best model. The performance on the training data for ARIMA is 0.376, which shows that the accuracy of the model is 37.6%. The bivariate transfer function accuracy is 0.409, and the accuracy of the multivariate transfer function is 0.478. The result performance of ARIMA testing is 0.074, the bivariate transfer function is 0.055, and the multivariate transfer function is 0.108. The multivariate transfer function forecasting model is a technique in this case with the best performance.

1. Introduction

In recent, the Corona-virus disease (COVID-19) has become a serious concerned for public health around the world. Outbreaks of infectious diseases such as COVID-19 can also significantly affect many aspects of each region. March 2, 2020, the first case of COVID-19 was found in Indonesia, until now the number of confirmed, suspected, and exposed is still increasing. Since the beginning of the case, until the last data was obtained in August 2020, Surabaya is the second city with the highest number of COVID-19 confirmed, suspected, and exposed in Indonesia based on government data daily new cases of COVID-19 [1].

COVID-19 confirmed that patients are patients infected with COVID-19 with positive test results through the PCR Swab examination. A suspected is a person with an acute respiratory infection, i.e. fever (> 38°C) or history of fever, accompanied by any of the symptoms or signs of respiratory disease. It has are cough, shortness of breath, sore throat, runny nose, pneumonia mild to severe, and there were no other causes based on the conclusive clinical picture. In a suspected patient, 14 days before symptoms appear, patients have a travel history or live in a country or region that has a report of the occurrence of COVID-19 transmission. Exposed patients are people who have a fever (≥38°C) or a history of fever or respiratory system disorders. In the 14 days before symptoms appeared, exposed patients had a travel history or lived in a country or region that had reports of COVID-19 transmission.

Based on government data, the number of confirmed, suspected, and exposed patients in the Surabaya has also increased significantly every day. Coronavirus transmission can be influenced by many factors, including climatic conditions, such as temperature (temp) and relative humidity (RH) [2].

Several previous research studies on the influence of climatic conditions and the spread of COVID-19 in a region, research by Fabiola is the influence of temperature, evaporation, rainfall, and regional climate on local transmission of the SARS-CoV-2 corona-virus in 31 states and Mexican capitals [3]. The statistical analysis obtained is the Local Transmission Ratio (LTR) calculated by region. Some days of infection effective since regional attacks in each state. The results obtained are the regional beginnings in dry climates appear earlier. It occurs because of lower temperatures than other regions and higher rainfall than tropical climates [3].

Research by Mazhar et al. analyzes the relation among COVID-19 and local climate parameters using existing global climate scales [4]. The results of the average day-time working hours correlate with the total cases of COVID-19 with a coefficient of determination of 0.42, and the average high temperature shows 0.59 and 0.42 with sum case of COVID-19 and case of deaths respectively [4]. Based on several previous research, it can be seen that using a variety of methods can be used to obtain the relation among regional climate variables and COVID-19 transmission in a region. In this research, we will predict the number of COVID-19 transmissions based on regional climate variables, i.e. temp and RH.

Previous research, predicting the transmission of COVID-19 using the LSTM network method in Canada, this research used Long Short Term Memory (LSTM) method which obtained the results that the transmission rate in Canada is a linear trend and in the USA it is exponential growth. One of these is another popular statistical technique and often used is Autoregressive Integrated Moving Model (ARIMA) [5]. Farajzadeh et al. are used the ARIMA model for forecasting monthly rainfall in the northwest from Iran [6]. ARIMA has become a valuable method for Multivariate phenomenon modelling, namely Vector Autoregression (VAR) rainfall forecast [6], Vector transfer function [7], and Spatio-temporal [8].

Based on the results of this research, this research compares the performance of three methods for predicting the number of COVID-19 confirmations, i.e. ARIMA, bivariate, and multivariate transfer functions method function which also in some research on other topics showed good performance in predicting time series data. We use Bivariate and multivariate transfer function methods to evaluate the effectiveness of adding external variables (independent variables) compared to ARIMA, which only uses confirmation cases variable. The dependent variable in this research is the confirmation case, and the independent variable is exposed COVID-19, suspect COVID-19, temp, and RH.

2. Data Set and Processing

2.1. Case Study

The observations daily are from March 23 until August 2, 2020. The number of data is 131 data. The distribution of the confirmed infected cases, suspect, and exposed COVID-19 by the COVID-19 in Surabaya was provided by COVID-19 Response Acceleration Task Force and available online https://covid19.go.id/peta-sebaran.

The climate data used in this research observation were obtained from the Meteorology, Climatology, and Geophysics Agency for the cities of Surabaya and Sidoarjo, Class I Juanda Meteorological Station, WMO-ID: 96935 (http://dataonline.bmkg.go.id/). The data used are daily data. This weather station observes a 30 km^2 area in Surabaya, Sidoarjo, and surrounding. The daily average temperature and humidity were used in our research to forecast confirmed cases.

The definition of these variables is: Confirmed is anyone who has tested positive for COVID-19 by the laboratory, suspected is anyone who has symptoms such as someone positive for COVID-19 but not tested positive for Covid-19 by the laboratory, and exposed is anyone who has a history of contact with probable cases or confirmed cases of COVID-19 [9].

2.2. Data Preprocessing
Figure 1 shows the fluctuation in the number of COVID-19 confirmations. Based on the ACF and PACF correlograms in Figure 2 and Figure 3, it can be identified that the time series data already has a stationary-pattern. It is because the ACF chart has an exponentially decreasing pattern and the PACF chart has a 'cut-off' pattern after the 2nd lag [10].

The identification of stationary patterns can use testing, namely an *augmented dickey-fuller test* (*adf.test*). The test defines an alternative hypothesis that the time series data already has a stationary-pattern. The "adf.test" value is -3.5217 with a lag on order 15 and p-value 0.0432. The results show that the COVID-19 confirmation data has a stationary-pattern. The pre-processing required is to divide the total amount of data into two subsets, namely the training and the testing subset.set. The data will divide by a ratio of 70% for training and 30% for testing. The first subset is consists of 92 records and the second is consists of 39 records.

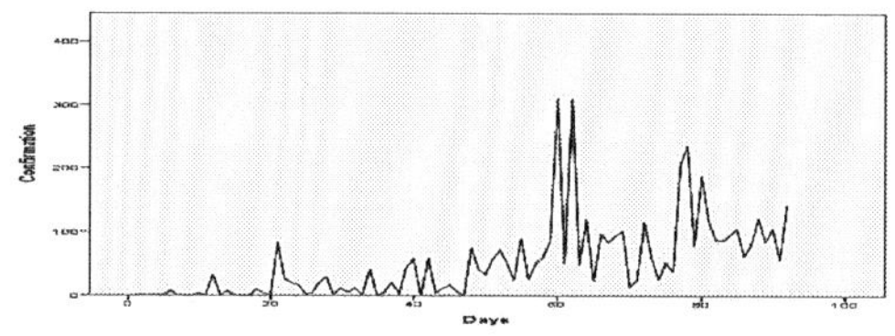

Figure 1. Confirmed case COVID-19 time-series at Surabaya.

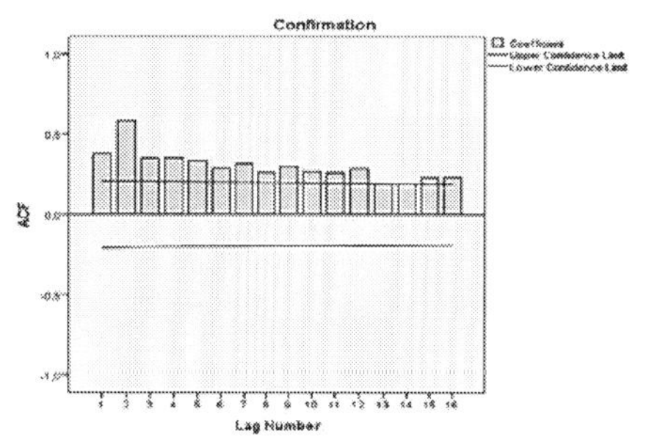

Figure 2. ACF Correlogam of COVID-19 confirmation cases.

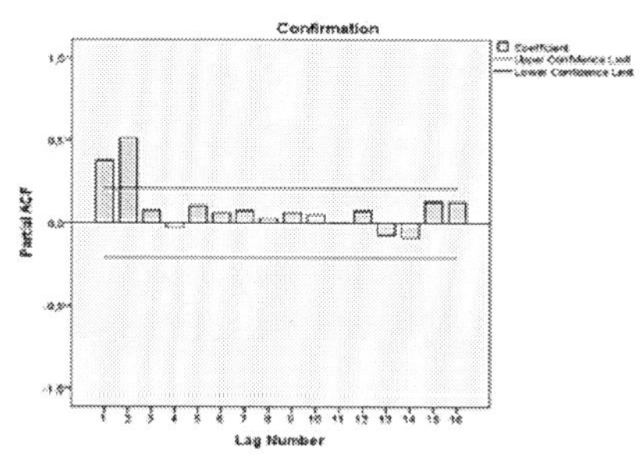

Figure 3. PACF Correlogam of COVID-19 confirmation cases.

3. Forecasting Method

3.1. Autoregressive Integrated Moving Average
One of the statistical methods which are the most on forecasting or prediction research is the ARIMA model. Based on the research of Babu [11], the autoregressive model shows that there is a relation among a random value added by a and values in the past (Z_{t-k}) and value in the present (Z_t). Moving Average model shows that there is a relation among a value in the present (Z_t) and residuals in the past (a_{t-k} with $k = 1, 2, \ldots$). ARIMA(p, d, q) model is a mixture of AR(p) and MA(q) with d differencing order and a non-stationary data pattern. Equation (1) is the form of ARIMA(p, d, q).

$$\emptyset_p(B)(1 - B)^d Z_t = \theta_q(B)a_t \tag{1}$$

IOP Conf. Series: Materials Science and Engineering 1077 (2021) 012055 doi:10.1088/1757-899X/1077/1/012055

where p is AR model order q is MA model order d is difference order and $\emptyset_p(B), \theta_q(B)$ based on equation (2,3).

$$\emptyset_p(B) = \left(1 - \emptyset_1 B - \emptyset_2 B^2 - \cdots - \emptyset_p B^p\right) \quad (2)$$

$$\theta_q(B) = (1 - \theta_1 B \text{ - } \theta_2 B^2 - \cdots - \theta_q B^q) \quad (3)$$

Generalized ARIMA model for seasonal pattern data which is written as ARIMA $(p, d, q)\ (P\ D\ Q)^s$ in equation (4).

$$\emptyset_p(B)\phi_p(B^s)(1 - B)^d(1 - B^s)^D Z_t = \theta_q(B)\Theta_Q(B^s)a_t \quad (4)$$

where s is a period of seasonal and $\phi_P(B), \Theta_Q(B)$ based on equation (5, 6).

$$\phi_P(B) = (1 - \phi_1 B - \phi_2 B^2 - \cdots - \phi_p B^p) \quad (5)$$

$$\Theta_Q(B) = (1 - \Theta_1 B \text{ - } \Theta_2 B^2 - \cdots - \Theta_Q B^q) \quad (6)$$

Flow-chart for the ARIMA model is described in Figure 4.

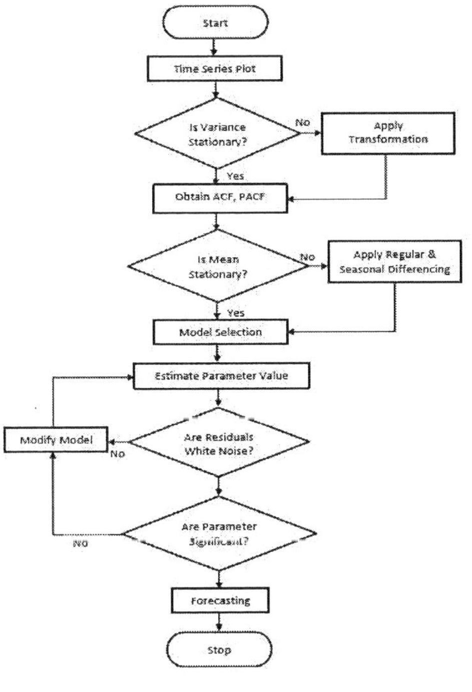

Figure 4. Flow-Chart of ARIMA modelling.

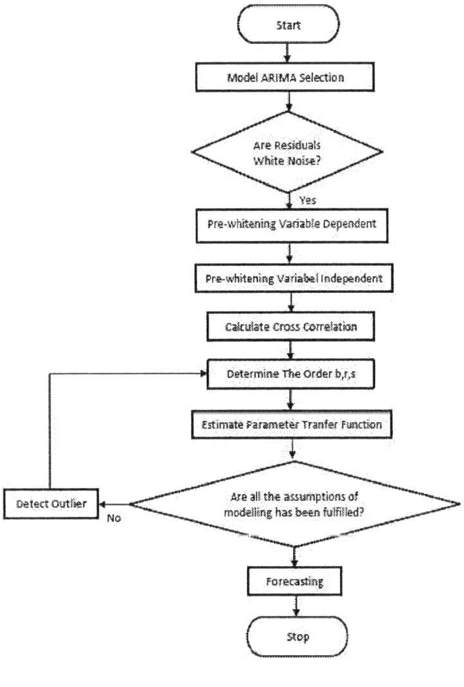

Figure 5. Flow-chart of transfer function modelling.

3.2. Bivariate Transfer Function (BTF)

The transfer function model is not the same method as the ARIMA model. In the time series model, the ARIMA model is a univariate model, but the transfer function is a multivariate model. So it can be concluded that the ARIMA model only connects the circuit with its past. In addition to the past series, the transfer function model also relates the series to other time series as described in Figure 5 and Figure 6. Other time series variables are used as predictor variables in the transfer function model.

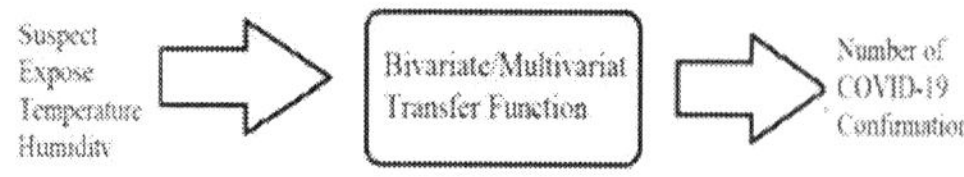

Figure 6. Transfer function modelling.

Transfer function models can be used to model with one output and more than one output systems [6]. The first stage of pre-whitening is the dependent variable pre-accounting to eliminate the existing pattern dependent variable series so that there is only a white noise input series. Second, the independent variable pre-whitening to maintain the functional relationship on the transfer function and noise. The next stage is to detect and measure the power relation and by using the Cross-Correlation Function [12]. input the value of X_t into Z_t. However, the transformation on Z_t does not have to convert Z_t to white.

In the model with one output, only one equation is required to describe the model. It is showed to as a single-input transfer function model, as equation (7).

$$Z_t = C + \frac{\omega_S(B)}{\delta_r(B)} B^b X_t + N_t \quad (7)$$

where $N_t = \frac{\theta(B)}{\Phi(B)} a^t$

$$\omega_S(B) = \omega_1 B + \omega_2 B^2 \ldots + \omega_s B^s$$
$$\delta_S(B) = \delta_1 B + \delta_2 B^2 \ldots + \delta_s B^s$$
$$\Phi_S(B) = \Phi_1 B + \Phi_2 B^2 \ldots + \Phi_s B^s \quad (8)$$

with b is a delay, s is the numerator, and r is the dominator.

3.3. *Multivariate Transfer Function (MTF)*

Like multiple regression models, transfer function models can also contain more than one input variable. In this paper, the single input and multi-input transfer functions are applied to assume stationary for the entire input and output series the general form of a single-input transfer function model in equation (9). We can apply the differencing technique if those variables are non-stationary.

$$Z_t = C + \frac{\omega_{S1}(B)}{\delta_{r1}(B)} B^{b1} X_t + \frac{\omega_{S2}(B)}{\delta_{r2}(B)} B^{b2} X_{2t} + \cdots + \frac{\omega_{Sm}(B)}{\delta_{rm}(B)} B^{bm} X_{mt} + N_t \quad (9)$$

3.4. *Performance Model*

The best model is evaluated using four performance measures, i.e. Root Mean Square Error (RMSE), Mean Absolute Error (MAE), R square (R_{sq}) and Correlation (r) [13]. A good model is to have small RMSE and MAE and have a large R_{sq} and r. The four performance measures are equation (10-13).

$$RMSE = \sqrt{\frac{\sum_{t=1}^{n}(Z_t - \hat{Z}_t)^2}{n}} \quad (10)$$

$$MAE = \frac{\sum_{t=1}^{n}|Z_t - \hat{Z}_t|}{n} \quad (11)$$

$$R_{sq} = 1 - \frac{\sum_{t=1}^{n}(Z_t - \hat{Z}_t)^2}{\sum_{t=1}^{n}(Z_t - \bar{\hat{Z}}_t)^2} \quad (12)$$

$$\text{Correlation} = r = \sqrt{R_{sq}} \quad (13)$$

4. Result and Discussion

4.1. *Autoregressive Integrated Moving Average (ARIMA)*

Based on the flowchart of the ARIMA model (Figure 4), the confirmed COVID-19 data is identified, it has been stationary using time-series plots. The COVID-19 confirmation data is stationary according to figure 1 because the time-series plot shows that the pattern is a fluctuating in average-value.

Secondly, we identified the stationarity time-series based on ACF and PACF pattern. The ACF correlogram is shown in Figure 4. Events have an exponential decrease which is thought to be a non-seasonal model of AR. Figure 2 and Figure 3 shows that the confirmation COVID-19 data is stationary because PACF has been cut off after the 2nd lag. No transformation is needed nor difference.

Confirmed-data is identified as AR(2), ARMA(1,1), MA(2), etc. Based on the ACF and the PACF, several ARIMA models are proposed $(p\ d\ q)(P\ D\ Q)^s$ using training data. Table 1 shows the six significant models in which the best model is the third model as an equation (14):

$$y_t = 6.693 + 0.297y_{t-1} + 0.47y_{t-2} + e_t \tag{14}$$

This model has the least RMSE and the highest R_{sq}. Therefore it was selected as the best model.

The PACF correlogram of the residual model (2 0 0) has a white-noise pattern so that the model is acceptable. White noise means that the partial correlation value of the residuals is not significant; it can be seen from the PACF value not exceeding the UCL-LCL limit, according to figure 7. In figure 8, from the forecasting results, the graph that formed on the forecasting plot does not represent a graph similar to actual testing data; this results in low accuracy values. Table 2 shows that R_{sq}-value is a few.

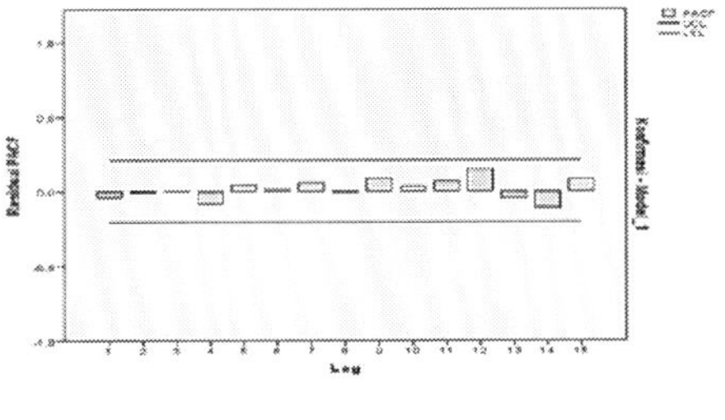

Figure 7. PACF Correlogam of COVID-19 confirmation cases of error of the third model.

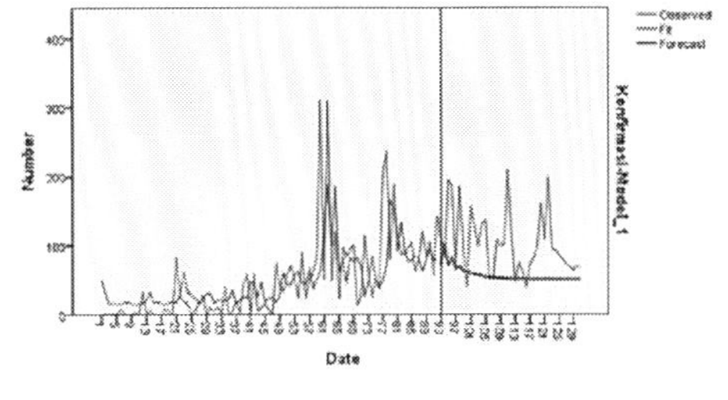

Figure 8. Forecasting plot-based model 1.

Table 1. Performance of ARIMA based on training data.

Performance of ARIMA				
Model	*RMSE*	*MAE*	R_{sq}	*r*
(2 0 0)	49.884	32.499	0.376	0.613
(1 0 2)	51.009	32.856	0.355	0.596
(1 0 1)	51.775	29.887	0.330	0.574
(1 0 0)	58.132	39.693	0.146	0.382

Table 2. Performance of ARIMA based on testing data.

Performance of ARIMA				
Model	*RMSE*	*MAE*	R_{sq}	*r*
$(2\ 0\ 0)(1\ 0\ 0)^6$	61	47.8	0.074	0.272

4.2. Bivariate Transfer Function (BTF)

The prediction of the model is improved by adding one predictor variable. This model is called the bivariate input transfer function. The independent variables used by the predictors are carried out alternately to get the best BTF model.

According to figure 10, the number of exposed does not affect the number of confirmations, so it is not suitable for BTF. Cross-Correlation Function (CCF) from suspected variable to COVID-19 confirmed has a relationship. It is proven that the BTF (suspected and confirmed) model is significant. Partially, the other three variables influence the number of confirmed cases.

The best predictor is the suspected variable. The increase in the performance of BTF is better than ARIMA, so it can be concluded that the BTF model is better. From figure 9, plot "training – testing": among actual and predicted for the BTF (suspected – confirmed) model is better than ARIMA, and then the graph can follow fluctuations in the observed data. Table 3 shows that the suspect variable is the best predictor for Covid-19 Confirmed in the BTF model. While the testing accuracy is still low, as shown in table 4.

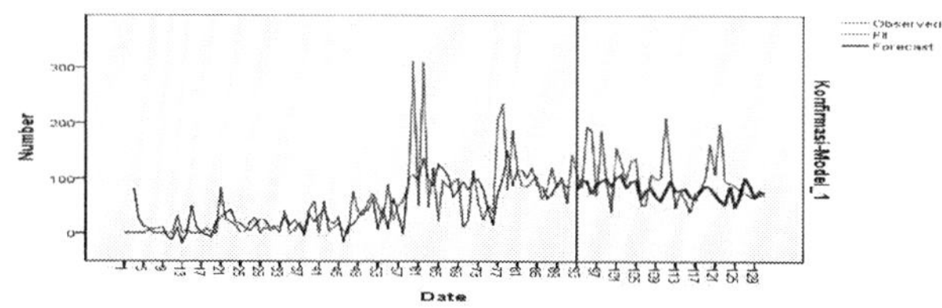

Figure 9. Forecasting plot Bivariate Transfer Function.

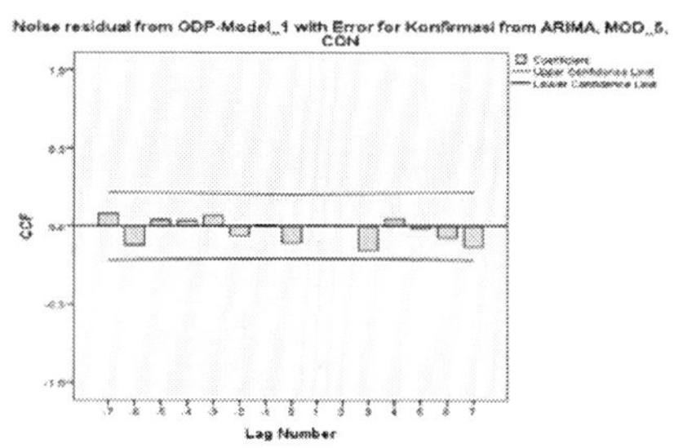

Figure 10. The cross-correlation function of exposed and confirmed.

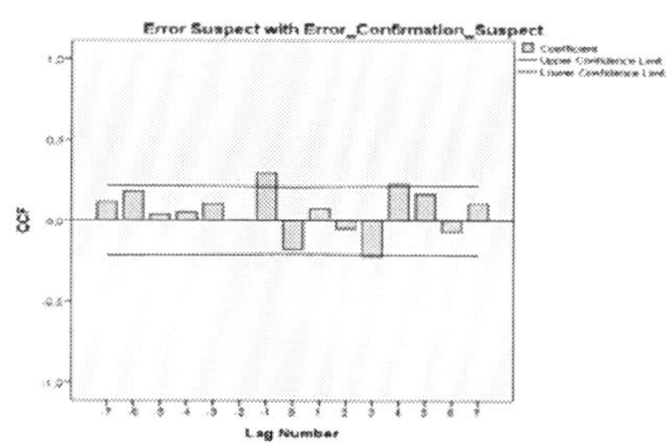

Figure 11. The cross-correlation function of suspected and confirmed.

Table 3. Performance of BTF based on training data.

Performance of BTF				
Input	*RMSE*	*MAE*	R_{sq}	*r*
Expose (0 0 0)	-	-	-	-
Suspect (1 0 1)	50.309	31.562	0.409	0.639
Temp (2 0 0)	49.407	34.491	0.397	0.630
RH (1 0 1)	49.490	31.244	0.388	0.618

4.3. Multivariate Transfer Function (MTF)

The third model is MTF that the predictor variables as input are four variables. The results of the training process for the MTF model are shown in table 5. The exposed variable is not as significant as a predictor based on the BTF model.

Meanwhile, the suspected, temperature and humidity variables are significant as predictors. The training results show that the performance of the MTF model is quite good. In comparison, the performance of the testing process is shown in table 6 with a low R_{sq}-value. Figure 12 shows that the training results are close to the actual value, but the test results still cannot follow the fluctuation of real data.

Table 4. Performance of BTF based on testing data.

Performance of BTF				
Input	*RMSE*	*MAE*	R_{sq}	*r*
Suspect	50	35.59	0.055	0.234

Table 7 is a summary of the performance comparison of the ARIMA, BTF, and MTF models. These results indicate that the MTF has better performance than the other two models.

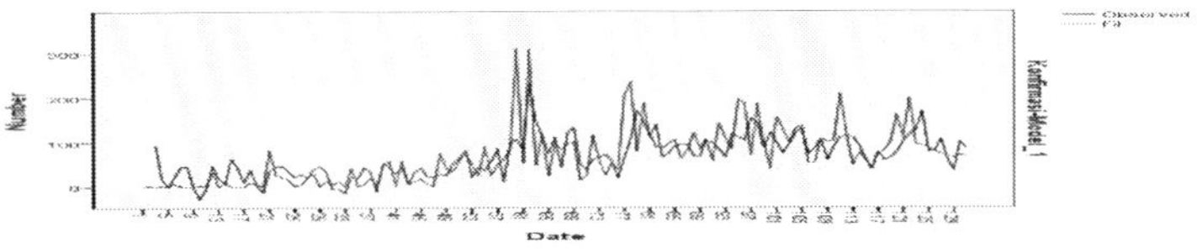

Figure 12. Forecasting plot Multivariate Transfer Function.

Table 5. Performance of MTF based on training data.

Performance of MTF				
Input	*RMSE*	*MAE*	R_{sq}	*r*
Exposed Suspect (0 0 2) Temperature (0 1 0) Humidity (0 2 0)	46.544	32.333	0.478	0.691

Table 6. Performance of MTF based on testing data.

Performance of MTF				
Input	*RMSE*	*MAE*	R_{sq}	*R*
Suspect, Temp, RH	46	34.87	0.108	0.328

Table 7. Performance of 3 models based on testing data.

Performance 3 Model				
	RMSE	*MAE*	R_{sq}	*R*
Training				
ARIMA	49.884	32.499	0.376	0.613
BTF	50.339	31.560	0.409	0.639
MTF	46.544	32.333	0.478	0.691
Testing				
ARIMA	61	47.8	0.074	0.272
BTF	50	35.59	0.055	0.235
MTF	46	34.87	0.108	0.328

5. Conclusion

The ARIMA model has the worst training and testing performance because it cannot follow the actual data pattern. The smaller accuracy value also evidences this than the other two methods, the R_{sq} value of 0.376 for training and 0.074 for testing. BTF model has better accuracy than ARIMA because the relative test results can follow the real data pattern. However, it has a small correlation, the R_{sq} value of 0.409 for training and 0.055 for testing. The best model is MTF because it has better accuracy than the other two models for training and testing processes. The result of training is 0.478, and training is 0.108 for the R_{sq} value obtained. Statistical-based modeling of COVID-19 confirmation cases is recommended to use MTF if external variable observations are available.

Recommendations that can be given for future research are that researchers can add a lot of data and can try using other methods that can be used to predict such as Recurrent Neural Network (RNN) or Long Short-term Memory (LSTM).

Acknowledgements
This research and publication were funded by **Penelitian Dana Departemen,** a grant from Department of Informatics, ITS, Indonesia, the budget year 2020 (No: 850/PKS/ITS/2020).

References

[1] Gugus Tugas Percepatan Penanganan COVID-19 2020 *Peta Sebaran COVID-19* Retrieved from https://covid19.go.id/peta-sebaran-covid19
[2] Wang J, Tang K, Feng K, and Lv W 2020 High temperature and high humidity reduce the transmission of covid-19 *SSRN Electronic. J.* 3551767
[3] Méndez-Arriaga F 2020 The temperature and regional climate effects on communitarian COVID-19 contagion in Mexico throughout phase 1 *Sci. of the Total Environ.* **735** 139560
[4] Iqbal M M, Abid I, Hussain S, Shahzad N, Waqas M S, and Iqbal M J 2020 The effects of regional climatic condition on the spread of COVID-19 at global scale *Sci. of the Total Environ.* **739** 140101
[5] Chimmula V K R and Zhang L 2020 Time series forecasting of COVID-19 transmission in Canada using LSTM networks *Chaos, Solitons and Fractals* **135** 109864
[6] Farajzadeh J, Fakheri Fard A, and Lotfi S 2014 Modeling of monthly rainfall and runoff of Urmia lake basin using "feed-forward neural network" and "time series analysis" model *Water Resour. Ind.* **7–8** pp 38–48
[7] Nugroho A, Subanar, Hartati S and Mustofa K 2014 Vector Autoregression (Var) Model for Rainfall Forecast and Isohyet Mapping in Semarang – Central Java – Indonesia *Int. J. Adv. Comput. Sci. and Appl.* **5** pp 44–9
[8] Fuentes M, Reich B and Lee G 2008 Spatial–temporal mesoscale modeling of rainfall intensity using gage and radar data *Annals of Appl. Stat.* **2** pp 1148–69
[9] World Health Organization 2020 *World Health Organization* Retrieved from https://www.who.int/
[10] Saikhu A, Arifin A Z and Fatichah C 2017 Rainfall Forecasting by Using Autoregressive Integrated Moving Average, Single Input and Multi Input Transfer Function *Int. Conf. on Information, Communication Technol. and System (ICTS)*
[11] Babu C N and Reddy B E 2014 A moving-average filter based hybrid ARIMA–ANN model for forecasting time series data *Applied Soft Computing* **23** pp 27–38
[12] Remple R K and Tischler M B 2006 *Aircraft and Rotorcraft System Identification* (Virginia: American Institute of Aeronautics and Astronautics, Inc.)
[13] Chai T and Draxler R R 2014 Root mean square error (RMSE) or mean absolute error (MAE)?– Arguments against avoiding RMSE in the literature *Geosci. Model Dev.* **7** pp 1247–50

Realization of Ball on Plate Control using PID Controller and Webcam

Dahnur Agus Pahlawan, Agung Surya Wibowo, Erwin Susanto[1]
School of Electrical Engineering, Telkom University

E-mail: [1]erwinelektro@telkomuniversity.ac.id

Abstract. This study aims to apply the PID controller to the ball on plate control realisation to reach a stable position of ball on plate. The ball on plate props adjust the position at the desired coordinates on a flat square plate. The targeted position is set as set point coordinates in the form of X pixels and Y pixels. The input used is in the form of coordinates obtained from the webcam through image processing methods and sent serially to the microcontroller. The output from the microcontroller is fed into the actuator, namely the servo motor. From the experiment results and its implementation, it is shown that ball on plate can move towards to the desired set point. For PWM signal set point, the value referred to centre of plate is 250 and the response of X-Y axes have errors 3.64% and 8%, respectively.

1. Introduction

Ball on plate control instruments are popular props in many basic control system laboratories. They were used for education purposes and showing how a kind control method can be applied. However, to provide these instruments and others is often constrained by cost problem. Building a simple ball and plate control instrument using low cost electronics and mechanics parts is an alternative solution for providing aid learning or trainer module in laboratory.

This research aims to build a simple and inexpensive ball on plate control system for the benefit of providing a control system learning model using PID control. In addition, this research also uses a camera to detect the position of the ball on the plate. Therefore, image processing is needed to apply control strategies.

Ball on Plate control is a tool that adjusts the position of a ball on plate so that it is located at a certain point and stable. Some researches on ball and plate control with any kind control methods can be found in some manuscripts, [1][2] for instances. The control system is designed in the form of a closed loop control whereas the feedback sensor is fed to reference, providing measured output information.

Ball on Plate system is a mechanical system in which a ball is placed on a flat plane that can move freely and can be noised by disturbances from its environment, such as wind or shocks. Mechanical construction of this system can be shown in Figure 1.

In references [1], [2], the ball position data is detected using the touch panel so that it is mechanically contacted. In this research, the ball position data is detected using a web camera so that it is not physically touching and avoiding measurement noise caused by friction between the ball and the plate. They used the MPC (Model Predictive Control) and SMC (Sliding Mode Control) method which required a fairly mathematical model. It led to complex computations and advanced control. In this research, a simple PID control is applied with intuitively adjusted gains so that it has the advantage, once a suitable gain is obtained it can be used for further works with lighter computations.

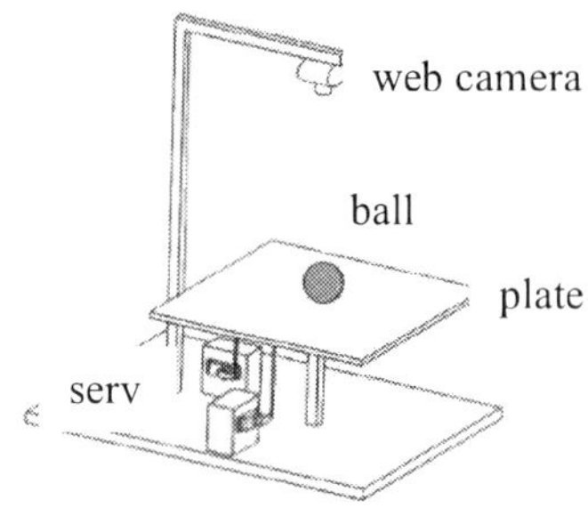

Figure 1. Mechanical Structure of a Ball on Plate

2. Control Method

Different from that was done by [1] they used the touch panel sensor to obtain position data of ball, in this research, the ball on plate system uses a camera as a sensor to determine the position of the ball. The captured image will be processed using image processing and fed to PID controller which is built in Personal Computer. The controller actuates servomotor to regulate slope of plate. This condition led a ball move to set point or targeted position, on the centre of plate for instance. Servomotor is an actuator designed as closed loop system along with some parts such as servo motor, a series of gear, control circuit and potentiometer. Because it is a closed loop system then it can be adjusted to determine the angular position of the motor output shaft.

The servo motor is controlled by providing a pulse width modulation (PWM) signal. PID controller was used to regulate PWM signal as given control signal that determine the rotational angle position of the servo motor shaft.

2.1. PID Control

PID (Proportional-Integral-Derivative) control is a classical controller that often used to determine the control action for system the feedback [3]. The PID controller yields a control action by comparing the error signal which is the difference between the process variable and set point. This controller is represented by the following equation:

$$u(t) = K_p.e(t) + \frac{K_p}{T_i}\int_0^t e(t)dt + K_pT_d\frac{de(t)}{dt} \quad (1)$$

$$u(t) = K_p.e(t) + K_i\int_0^t e(t)dt + K_d\frac{de(t)}{dt} \quad (2)$$

where:

$u(t)$ = PID controller output signal
K_p = proportional constant
T_i = integral time constant
T_d = derivative time constant
K_i = integral constant (Kp/Ti)
K_d = derivative constant (Kp.Td)
$e(t)$ = error signal = reference – plant output
= set point – sensor signal

The figure 2 is a block diagram of the generalized PID control.

2.2. Image Processing

Webcam has promised low cost for detecting motion and position of the objects thus supporting the development of computer vision, including eye movement sensing [4]. The webcam functions as a

IOP Conf. Series: Materials Science and Engineering 1077 (2021) 012056 doi:10.1088/1757-899X/1077/1/012056

sensor that is used to detect the X and Y coordinates of the ball, the direction of movement of the ball on a plate is read as an input signal.

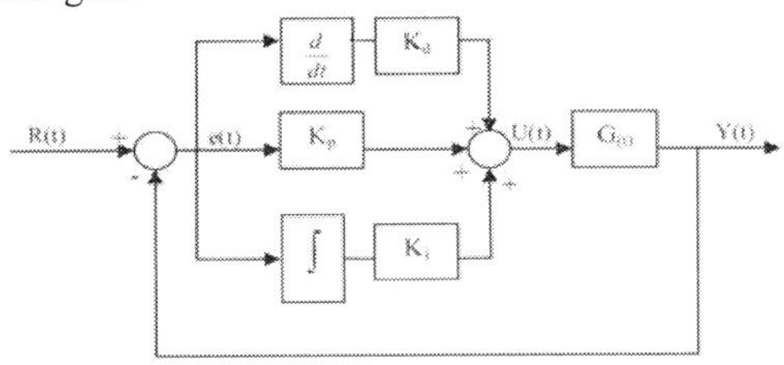

Figure 2. Generalized PID Control

Image processing is a signal processing of a set of image input. The output can be in the form of an image or a set of characteristics or parameters related to the image. The term of digital image processing is generally defined as two-dimensional image processing with a series of real or complex numbers represented by certain bits. Digital images are formed by a collection of image dots called pixels.

Pixel is the smallest part of a digital image. The monitor displays the image by dividing the screen into thousands (even millions) of pixels. These pixels are so close together that they appear to be connected to one another.

The number of bits is used to represent each pixel determines how many colours of the pixel can be displayed. This is commonly referred to colour depth. For example, if the monitor uses 8 bits for each pixel, then each pixel is capable of displaying 256 different colours. Each pixel is represented as one small square. Each pixel has a position coordinate. Row and column index (x, y) in a pixel are represented by integers. Pixel (0,0) is located in the upper left corner of the image, index x moves to the right and index y moves down. Figure 3 shows a top of view from captured web camera.

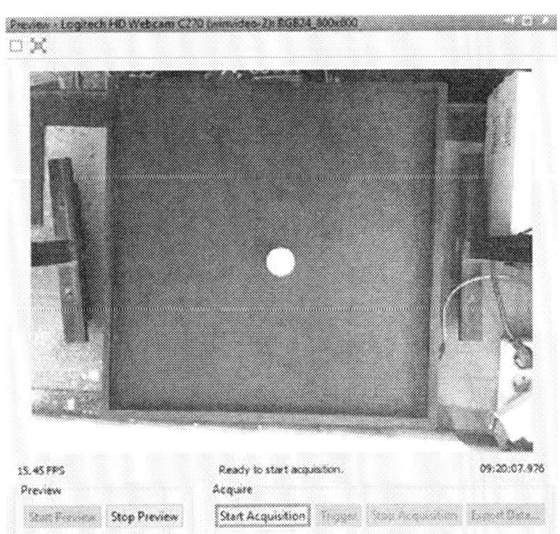

Figure 3. Webcam top view of Ball on Plate

3. System Design

Figure 4 shows the developed ball on plate control system. Computers provide extensive image processing due to the large capacity, software and algorithms that are embedded. For this reason, detection of the position of the ball on the plate uses a camera connected to a computer as a processor.

Moreover, development of computer vision itself began in the early 1970s [5], so the use of cameras and computers in this research is a potential option.

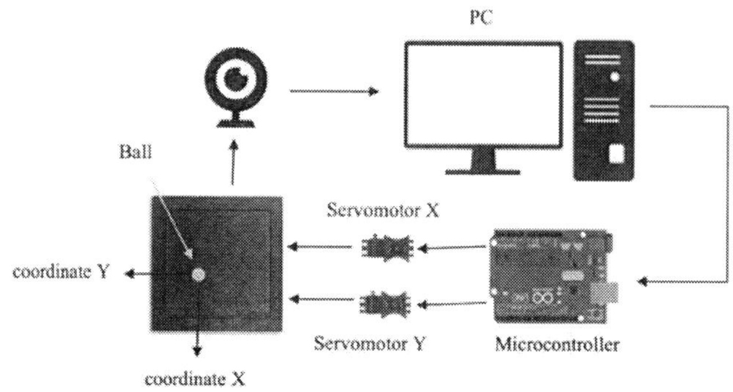

Figure 4. System Design of Ball on Plate Control

Figure 4 also displays entire work process flow. This process is conducted by algorithm and programming using a PC. The results of the programming is processed and sent serially into the Arduino microcontroller. This microcontroller actuates servo motor, and the servo motor drives the plate movement such that the ball move to desired position. The position of the ball is captured by camera to determine whether the ball's position is around desired position. The specifications of designed instruments listed here:

- The plates are 45 cm x 45 cm square.
- The webcam has 720p high quality
- Microcontroller that can control servo motor rotation
- Servo motor can rotate 180 degrees.
- PC is used for image processing and PID control actuating.

Realization of Ball on Plate Control is built to provide the advantage of PID control method in physical implementation. The system output is the ball on the plate in desired position according to user setting. The coordinates of the ball are known after the data from the webcam is processed by the image processing and PID control, forwarded to the microcontroller and sent to the servo motor.

Table 1. PWM Signal measurement of Ball position, based on X-Y coordinates

No	Set Point (PWM signal value)	X-axis Coordinate	Y-axis Coordinate	Error on X-axis	Error on Y-axis	Disturbance	Time
1	250	244	220	6	30	No	110
2	250	249	287	1	37	Yes	310
3	250	274	260	24	10	Yes	780
4	250	233	234	17	16	Yes	1050
5	250	238	242	12	8	Yes	1400
6	250	234	235	16	15	Yes	1810
7	250	243	277	7	27	Yes	2500
8	250	248	204	2	46	Yes	2550
9	250	253	242	3	8	Yes	3000
10	250	247	247	3	3	Yes	3500

4. Results and Analysis

The purpose of this testing is to find out the error between set point and actual position. The test is carried out by giving interference to the ball. The ball is stopped at the specified set point. The error is calculated from the difference between the set point and the obtained coordinates of the ball. Measurement of PWM signal value according to set point and X-Y coordinate is provided in Tabel 1 and Figure 5. PID gain constants are given, K_p= 0.3, K_d = 4.5 and K_i = 0.003.

Disturbance is given by touching the ball randomly with finger when the ball has stopped around the centre of the plate.

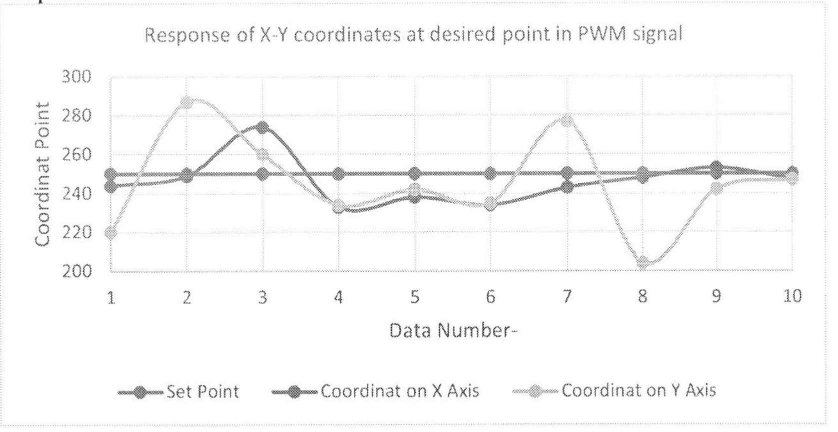

Figure 5. Response of X-Y coordinates at desired point in PWM signal

The measurement results show that the average error of X-Y coordinates are 3.64% and 8%, respectively. The final position of the ball depends on initial ball location processed by web camera image processing. The size of the plate is 45 cm x 45 cm square, it means that if the error between the average distance of the ball to the centre point is 3.64% and 8%, then it will be about 1.64 cm and 3.6 cm. Dynamic system with PID control usually has maximum error of 5%, see [3]. It means that PID can work well to control ball position in x coordinate. However, it still needs to be corrected for ball position in y coordinate which can be done by adjusting the gain control.

5. Conclusion

This research had built a simple and inexpensive ball on plate control system for the benefit of providing a control system learning model using PID control. To verify the result, some points of conclusion are provided here:

1. Based on the measurement of set point and actual position for realization of ball on plate control system, the testing leads to errors on X-Y coordinates: 3.64% on the x-axis and 8% error on the y-axis.
2. To reach around set point position at centre of plate, ball on plate system needs 94.46 seconds.
3. Interface of object detection using web camera and PID control action is handled by personal computer and embedded software.

From the analysis, simulation and experimentation, it can be shown in student learning that PID control can be applied to control of the ball position on the plate in the x and y coordinates.

Acknowledgments

This research was partially funded by the Ministry of Research and Technology, Decree number 25/EI/KPT/2020, contract number 037/PNLT2/PPM/2020. The authors express the appreciation to Directorate of Research and Service Community, Telkom University for technical and financial assistances.

References

[1] Bang H and Lee Y S 2018 Implementation of a ball and plate control system using sliding mode

control *IEEE Access* **6** pp 32401–8
[2] Bang H and Lee Y S 2019 Embedded model predictive control for enhancing tracking performance of a ball and plate system *IEEE Access* **7** pp 39652 –9
[3] Nise N S 2011 *Control Systems Engineering* 6th edition (New Jersey: John Wiley & Sons)
[4] Meng C and Zhao X 2017 Webcam-based eye movement analysis using CNN *IEEE Access* **5** pp 19581–7
[5] Szeliski R 2011 *Computer Vision Algorithm and Applications* (London: Springer-Verlag)

Formal Verification of the Authentication and Voice Communication Protocol Security on Device X Using Scyther Tool

Muhamad Al Fikri[1], Kalamullah Ramli, Dodi Sudiana

National Cyber and Crypto Agency, Universitas Indonesia

E-mail: [1]alfikri92@gmail.com

Abstract. In the current era, the ownership of strategic information and the ability to effectively manage it has become a significant advantage. Reflecting on the experience of attacks on strategic communications in Indonesia, including the tapping of the former President Susilo Bambang Yudhoyono's conversation through the cellular network and President Jokowi's official residence, Indonesia has begun paying more attention to security in this sector. Device X is one of the secret strategic communication tools used in Indonesia. The XYZ Agency initiated the use of this device. As of 2020, there have been 1,284 units of Device X widely used by the army, police officers, and other strategic agencies in Indonesia. In its 5 years of operation, the XYZ Agency has researched the algorithm security used in Device X. However, there has never been a study of thwe security regarding the authentication and communication protocols of this device. This research aims to make a security analysis of voice communication and authentication protocols of Device X. The research was implemented using Scyther Tool as a formal verification approach. The analysis focuses on guaranteeing the confidentiality of information and authentication with four criteria, namely, secrecy, aliveness, synchronization, and agreement. The experimental results demonstrate that the authentication and voice communication protocol of Device X satisfy the secrecy criteria for transmitted confidential information but does not satisfy the criteria of aliveness, synchronization, and agreement on several entities involved in the protocol. Thus, it can be claimed that the authentication and voice communication protocol of Device X is provably secure based on the confidentiality aspect of information but is not secure in terms of authentication.

1. Introduction

From an organizational perspective, information is regarded as a resource that can be vitally utilized for organizational growth and development [1]. In the current era, ownership of strategic information and the ability to effectively manage it remains crucial. An organization is said to be successful if it can continuously update and maintain the authenticity of information to be disseminated to all relevant divisions and integrate it with the latest technology and communication devices. This view is known as the information management perspective. Based on this perspective, dissemination is one of the most vital stages. The purpose of information dissemination is to communicate information from the information maker to the parties in need. Thus, communication is necessary for organizations to conduct their business processes.

More and more experts across disciplines affirm that communication is essential to how an organization is established and designed and to how it survives due to the dynamic, collective nature of continuous interactions among units in the organization itself [2]. Previous research has demonstrated that stakeholders have different ways of responding to these views. This is true in particular in terms of understanding and interpreting the value of communication, which is then used as the basis for implementing a communication system to achieve organizational goals [3]. Different views in determining communication value have resulted in two general communication classifications, namely, strategic and ordinary communication [3].

Strategic communication is defined as communication that is entirely in line with the organization's communication strategy and aims to improve the organization's strategic position. Effective communication will encourage organizations to convey clear, understandable, and correct information adequately and consistently to each communicating entity. Others have defined strategic communication as communication used to formulate how the organization can achieve its primary goals and then support these efforts in specific and consistent ways for all organizational entities [4].

Considering the substance of strategic communication, which is closely related to the direction of organizational goals, generally, strategic communication is conducted at high-level management or in covert operations. Given the high value of ongoing communication, strategic communication often becomes the target of attacks. This is, of course, a threat because if the content of strategic communication is leaked, it can pose a risk to state security. Several incidents regarding attacks on strategic communications in Indonesia include the tapping of the former President Susilo Bambang Yudhoyono's conversations through the cellular network in Cikeas in November 2013 by the Australian embassy, which installed devices at the nearest base station [5] [6]. Furthermore, there are also cases of alleged tapping of President Jokowi's official residence, as conveyed by the secretary general of the Indonesian Democratic Party of Struggle, Tjahjo Kumolo, in February 2014 [7].

In Indonesia, Device X is one of the devices used for strategic communication. Device X is used by the Indonesian National Army (TNI), the Indonesian National Police (POLRI), and strategic civilian agencies. The XYZ Agency defines device X as classified tactical communication equipment provided by the XYZ Agency. The device's designation is set according to request or analysis of the need for cybersecurity in government agencies or in local governments [8]. Device X by the XYZ agency has been used since 2015 in government agencies and local governments. Although it is not the primary secret communication device, the demand for this device is relatively high. Based on data from Directorate A of the XYZ Agency, 1,284 pieces of Device X have been distributed throughout Indonesia [8].

Device X is used for strategic communication needs with confidential content. Information security usually focuses on three main aspects: confidentiality, integrity, and information availability [9] [10] [11]. To provide information security assurance, Device X applies cryptographic techniques, such as encryption and hash functions [9] [12]. Since the use of Device X is associated with strategic information with confidential content, the ability of Device X to provide information security assurance must be ensured. To determine Device X's information security assurance, an analyst can perform security analysis. In the context of secret communication, cryptographic security analysis of communication algorithms and protocols can be used as an approach to determine whether the security features provided by Device X are as expected [13] [14].

Based on in-depth analysis, Device X, which has been distributed throughout Indonesia, has previously undergone a modification process by embedding a proprietary encryption algorithm from the XYZ Agency. The modification was made based on the results of a study conducted by the XYZ Agency regarding the security of the built-in encryption algorithm for Device X. However, there has been no review within 5 years on the operation of the communication protocol security for Device X.

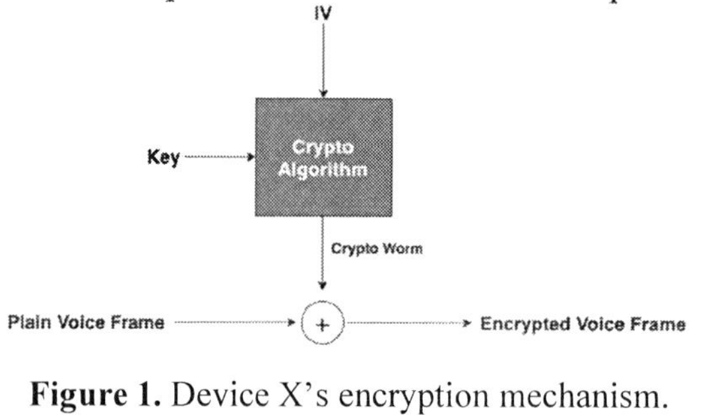

Figure 1. Device X's encryption mechanism.

Cryptographic protocol security analysis usually focuses on cryptographic features used to ensure the security of both the confidentiality and integrity of the information transmitted and the non-denial and authentication criteria of the parties involved in the communication [14]. Cryptographic protocol analysis is generally conducted in two ways: direct observation and formal verification, which includes either the use or non-use of tools [15] [16]. In response to these conditions, in this current study, we are encouraged to perform a security analysis of the communication and authentication protocol of Device X to complete the security study of Device X as one of the secret strategic communication devices in Indonesia.

2. Device X

2.1. Device X Overview

Device X is a type of secret strategic communication tool used in Indonesia. It is a DMR-based handy talky. Device X is used by the Indonesian National Army (TNI), the Indonesian National Police (POLRI), and strategic civilian agencies. The XYZ agency defines device X as a classified tactical communication device. According to a request or analysis of the need for cybersecurity in government agencies or local governments [8], the devices' designation plan was formulated. Device X by the XYZ Agency has been used in government agencies and local governments in Indonesia since 2015. Although Device X is not the primary secret communication tool, this device's need is relatively high. It has been established that as of 2019, based on data from Directorate A in XYZ Agency, 1,284 pieces of Device X have been distributed throughout Indonesia [8].

2.2. General System Overview

Because Device X is used for confidential communication, encryption is used as a security measure. During communication, every voice message is sent in an unencrypted form to the module. Digital voice messages in each frame will be operated using the XoR function with the crypto worm and will generate encrypted voice messages. Crypto worm is an output from a computational process using an algorithm with a key and an initial vector as the input. The encryption process during the sending process (TX) is depicted in Figure 1.

As with the encryption process during the sending process, the decryption during the receiving process is also carried out in the module. In this process, every encrypted voice message is sent to the module. The decryption process is performed to obtain voice messages from the encrypted voice messages received. The decryption process is conducted by performing an XoR operation between encrypted voice messages and the crypto worm. The crypto worm used in the decryption process is the same crypto worm used in the encryption process. Close attention needs to be paid to the initial vector used at the time of this message encryption-decryption process. IV is generated when communication is established. A new 56 bits (7 Bytes) IV will be generated and transmitted every 720 ms (12 voice frames) as depicted in Figure 2.

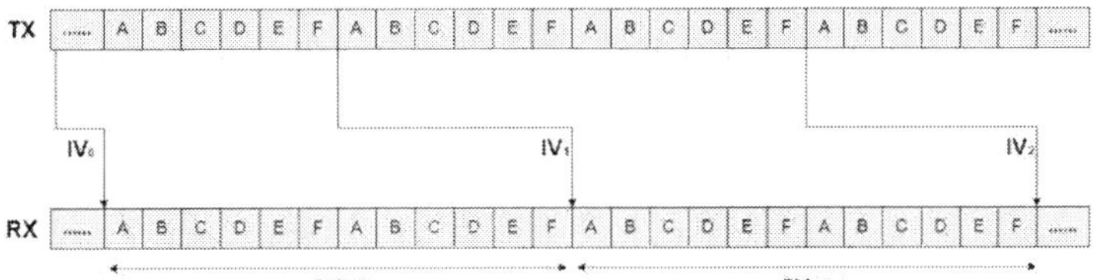

Figure 2. New IV generation and distribution.

3. Scyther Tool

3.1. Scyther Tool Overview

The Scyther Tool is the result of Cas Cremers' research in 2006. During that time, Cas Cremers published his research related to a methodology for formal analysis and verification of a cryptographic protocol in the form of an analysis tool known as the Scyther Tool. This tool can analyze the confidentiality (secrecy) and authentication property in a cryptographic protocol. With the Scyther Tool, the objectivity of a formal protocol verification can be guaranteed.

3.2. Security Claims on Scyther Tool

The claim event is used by the role to model the security properties in question. There are two security properties, including secrecy (confidentiality) and authentication.

1) *Secrecy*. A *Secrecy* claim states that certain information was not disclosed to the attacking party even though the data was communicated over an untrusted network.
2) *Authentication*. *Authentication* exists to ensure the existence of the communicating parties. There are three criteria for authentication, namely, aliveness, synchronization, and agreement. The authentication criteria are hierarchical in that the synchronization criteria can be met only if the aliveness criteria have been met. Furthermore, the agreement criteria will be met only if synchronization and aliveness have been met [17]. The explanation of the authentication criteria on Scyther Tool is as follows:
    a) *Aliveness*: Aliveness is a form of authentication that aims to ensure that the communicating party has performed several events, which indicates that the party is 'alive'.
    b) *Synchronization*: Synchronization requires that the message has been sent or received by the communicating parties. Synchronization only reviews content and sorts the messages. This criterion implies that each entity has sent and received messages. There are two types of synchronization, namely, injective and non-injective synchronization. Non-injective synchronization ensures that the protocol is executed in the absence of an attacker, leaving the protocol vulnerable to replay attacks. Furthermore, injective synchronization prevents replay attacks, meaning the recipient can only communicate once for the same message.
    c) *Agreement*: This form of authentication focuses on the data agreement exchanged between agents, which requires that the content of the message received matches the message sent as determined by the protocol.

4. Research Methodology

In this research, we utilized a literature review and experimental methods for investigation. The literature review analyzed theories and concepts related to DMR, SDR, security requirements and threats to SDR, Device X, authentication and voice communication protocols, and Scyther Tool. Sources used for the literature review in the current study included books, papers, theses, standards, official documentation, and other sources supporting this research. The experimental method was conducted by simplifying the authentication and voice communication protocol before modelling it using a Scyther Tool-appropriate programming language. After that, the model was tested and analyzed. The analysis results were used to make recommendations for the protocol weaknesses.

The research stages are illustrated in the research flow depicted in Figure 3. The explanation of each stage is as follows:

1) Literature Review
   A literature review was conducted of theories, references, and concepts regarding DMR, SDR, security requirements and threats to SDR, Device X, authentication and voice communication protocols, and the Scyther Tool. The primary sources of reference in this literature review included the ETSI TS 102 361-1 Standard, ETSI TS 102 361-2, ETSI TS 102 361-4, and Device X documentation. The result of this process was a general description of Device X authentication and voice communication protocols.
2) Experiment
    a) Protocol Simplification: The next step was protocol simplification based on the general description of Device X authentication and voice communication protocols obtained from the literature review. Simplification was made based on the research focus. The simplification results only included relevant protocol data units (PDUs) to the research

focus. The protocol was described with the assumption that all requests sent by the device during the authentication and communication processes were received without experiencing time outs, misidentification of address, or other errors.

After protocol simplification, the next step included validation. Validation was conducted by scanning using RTL-SDR, DSR # v.1.0.0.1635, and DSD+ software. Validation aims to see whether the PDUs described in the simplification results referring to the ETSI TS 102 361 Part 1, Part 2, and Part 4 standards is true.

b) Modelling: The second experimental stage was modelling. At this stage, the simplified authentication and voice communication protocols of Device X were modelled using Scyther Tool proprietary programming language. Modelling was made by referring to the rules determined in the Scyther Tool. The model was then validated using two approaches. *Component Checklist*: This approach was performed using a checklist table containing information on each component in the modelling, including role, atomic term, event, and claim. *Model Checking* in Scyther Tool: This approach follows the modelling requirements in Scyther Tool. Checks were performed automatically by the Scyther Tool.
c) Testing: Testing was conducted by verifying the protocol security on the Scyther Tool. The verification performed was based on the assumptions included in the Scyther Tool. Testing on the Scyther Tool was performed to prove two things: the possibility of an attack pattern on the protocol and claims of the confidentiality of information and authentication of the parties involved in the communication. The confidentiality aspect was represented by secrecy criteria, whereas authentication aspects were represented by *aliveness*, *synchronization*, and *agreement* criteria.
d) Analysis: The current study included two analyses: the security analysis and the Scyther Tool's output analysis. The security analysis was conducted by considering the security aspects of SDR that also applied to Device X. The Scyther Tool output analysis was performed by examining possible attack patterns and ensuring that each claim was met as expected.

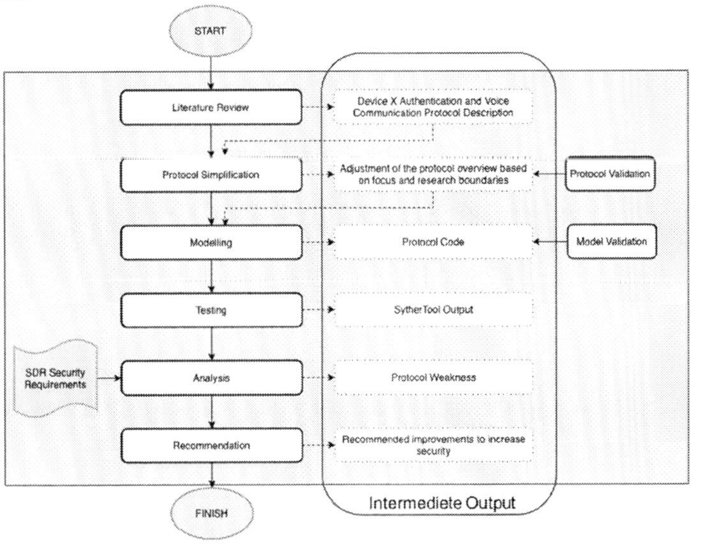

Figure 3. Research flow diagram.

5. System and Simplified Protocol Description

The general illustration of the existing system on Device X is depicted in Figure 4. Two devices, Mobile Station A (MSA) and Mobile Station B (MSB), were used to communicate. Prior to communication, each MSA and MSB was registered to Trunk Station Control Channel (TSCC). After the registration process was completed, communication occurred from MSA to MSB.

For each component, the MSA, MSB, and TSCC were instrumented with a security system as follows:

1) TSCC was instrumented with RC4 and maintained a database that contained information about authorized devices instrumented with each device's authentication key.
2) MSA and MSB were each instrumented with two modules, namely, the default module and the custom module. The default module was instrumented with RC4 and the respective device authentication key, which was set by the factory. Furthermore, the custom module was instrumented with a high-level encryption system used in the communication process.

Based on Figure 4, it can be seen that the device has two types of pre-installed keys: one used for authentication and one used for communication. The pre-installed key used for authentication is specified in the 56-bit long device creation process. Furthermore, the pre-installed key used for communication is determined by the user and entered into a customizable module. There are differences in security systems in the authentication process and device voice communication. The use of more complex algorithms and longer key makes the security system in the communication process more secure than the security system in the authentication process.

The protocol simplification in this research was made according to the research focus to test the security requirements of the aspects of confidentiality of information and authentication. The confidentiality of information focused on confidential information. Thus, PDUs or information that did not contain confidential information was not described. Additionally, authentication focused on the parties or entities that communicate in the protocol. In the simplified protocol, only the main communicating entities at the highest layer were described. These entities included MS or Device X and TSCC. The simplification results are depicted in Figure 5.

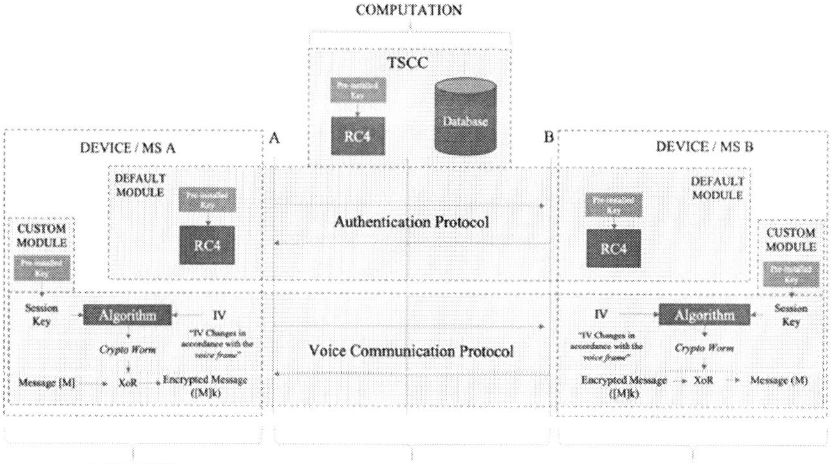

Figure 4. System overview of Device X.

Based on Figure 5, there are eight stages to complete the MSA and MSB authentication process on the TSCC. The necessary PDUs in the authentication process are C_RAND, C_AHOY, C_ACKU, and C_ACKD. Each PDU is a collection of information elements. The following process was used to perform device authentication:

1) MS, who will register to TSCC, sends C_RAND PDU along with CC.
2) After receiving the C_RAND PDU from the registering device, TSCC will perform an authentication process using a challenge-response mechanism that uses known information about the device by sending the C_AHOY PDU along with the CC.
3) When the registering MS receives C_AHOY PDU, the MS will calculate the response to the received challenge. When finished, the MS sends a response with C_ACKU PDU along with CC.

4) After receiving C_ACKU PDU, TSCC will perform calculations with the same information and algorithms, then match the results. If the information received is the same, TSCC will send C_ACKD PDU along with CC, indicating that the device registration was successful.

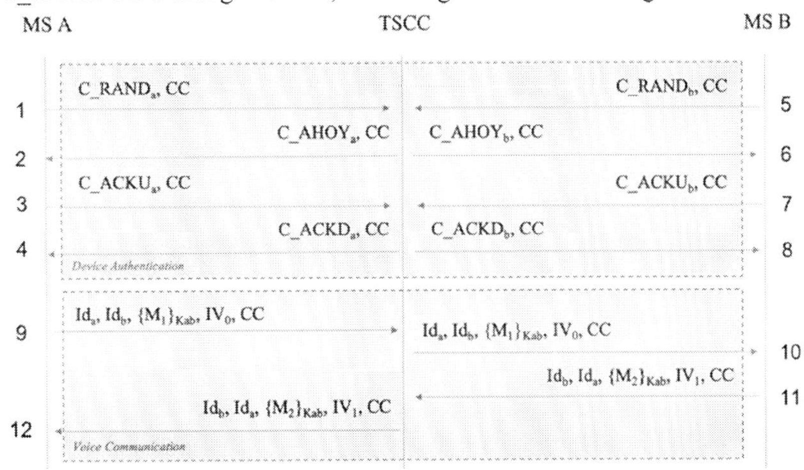

Figure 5. Device X's simplified authentication and voice communication protocols.

The communication process includes some crucial pieces of information, namely, device identifier (Id_a and Id_b), message (M), symmetric key shared between devices A and B (K_{ab}), an Initial Vector (IV) of the voice frame. The Id is the address and destination marker of the communicating device. Messages are sound transmitted via radio waves. Symmetric key and IV are the parameters used as input for the encryption algorithm to secure the message. The following process was used to perform voice communication:

1) MSA sends a series of signals consisting of Id_a as the source address, Id_b as the destination address, $\{M_1\}$ K_{ab} as the M_1 message which is encrypted with K_{ab}, IV_0 as the parameters used to increase the configuration aspect of the encryption result.
2) After receiving the message, MSB will send a series of signals consisting of Id_b as the source address, Id_a as the destination address, $\{M_2\}$ K_{ab} as the M_2 message that is encrypted with K_{ab}, IV_0 as the parameters used to increase the configuration aspect of the encryption result.

To ensure that the PDU contained in the simplified protocol is correct, a simple experiment was conducted using the RTL-SDR and the DSR # v.1.0.0.1635 and DSD + software. The tools used in this experiment were RTL-SDR (DVB-T Dongle with RTL2832U chipset), an antenna with MCX female and MCX male connectors, DSD+ software, Windows computers, and two units of Device X.

Using DSD+, researchers monitored encrypted communication between two devices, namely the X-1 and the X-2. It can be seen that based on monitoring results, the PDU Devices Identifier (Ida and Idb), Color Code (CC), and encrypted messages ([M]Kab) were captured. Although there was no information explaining the initial vector (IV), the IV-sent information was considered valid based on the information in the official documentation from Device X owned by the XYZ Agency. Based on the findings and supporting data above, the authentication and voice communication protocol of Device X resulting from the simplification was considered valid to represent the transmitted PDU.

6. Modelling and Testing

Device X's authentication and voice communication protocols were developed in two separate models to simplify the data presentation. Even though they were manufactured separately, the entities involved in the protocol and the transmitted PDUs in the protocol still followed the configuration depicted in Figure 5.

The modelling of the authentication and voice communication protocols for Device X was performed by compiling the protocol MSC following the Scyther Tool provisions. In the authentication protocol, the two MSA and MSB entities acted as initiators in the device registration process. However, in the voice communication protocol, MSA acted as a communication initiator.

The protocol model's validity was tested using two approaches, namely, the component checklist and model checking. Device X's authentication protocol model was verified on Scyther Tool. As depicted in Figure 6, the model was promptly validated with the output of protocol verification based on the model checking approach. The component checklist was performed by creating a control table according to the protocol specifications. The results demonstrated that the model made with the Scyther Tool's local programming language followed Device X's authentication protocol specifications.

After creating the model, the next step was to test its verification. Verification was done using the *Verify → Verify Protocol* menu. Figure 13 depicts the verification results of Device X's authentication protocol. Of the nine claims, three were declared valid or OK, and six claims were declared as failed or FAIL. Figure 14 illustrates the verification results of Device X's voice communication protocol. Of the 13 claims, 10 were declared valid or OK, and three claims were declared failed or FAIL.

7. Analysis

The analysis was divided into two: an analysis of the Scyther Tool output and an analysis of the relevant SDR security requirements for this research.

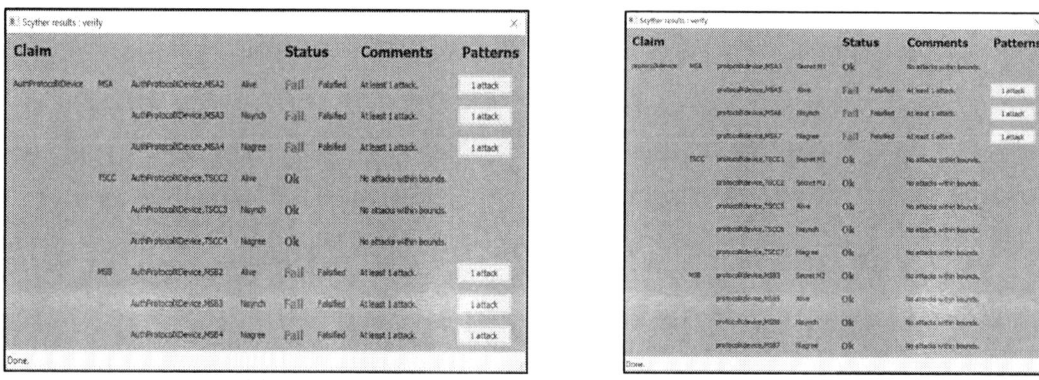

(a) (b)

Figure 6. Device X's authentication protocol (a) and voice communication protocol (b) verification results.

7.1. *Scyther Tool Output Analysis*

The Scyther Tool analysis included two components: the interpretation of test results and analysis of potential attack patterns in the protocol. The interpretation of the test results provides an overview of the protocol's overall condition according to the claims made. Analysis of attack patterns demonstrates attacks that resulted in unfulfilled protocol security claims. Table I and Table II provides a summary of Device X's authentication and voice communication verification results.

Table 1. Recapitulation of device X's authentication protocol verification result

Claim	*Confidentiality*	*Authentication*
OK	*Role MSA: -* *Role TSCC: -* *Role MSB: -*	*Role MSA: -* *Role TSCC: Aliveness, Nisynch, and Niagree* *Role MSB: -*
FAIL	*Role MSA: -* *Role TSCC: -* *Role MSB: -*	*Role MSA: Aliveness, Nisynch, and Niagree* *Role TSCC: -* *Role MSB: Aliveness, Nisynch, and Niagree*

In Table 1, the claim is not fulfilled because there is at least one possible type of attack against the device authentication protocol based on the test results. As the initiator, each MSA and MSB experienced the same failure in fulfilling all security criteria on authentication type claims, namely, aliveness, synchronization, and agreement. The possible attack patterns against these criteria on each device were the same as illustrated in Figure 7 (a).

The first failure was in the aliveness criteria. Referring to the definition, Device X's authentication protocol was stated as guaranteeing the MSA/MSB entity's aliveness if another entity communicating with MSA/MSB, namely TSCC, can be assured the existence of MSA/MSB. Many protocols failed to meet this authentication criterion despite simple attacks. As with most cases, an attacker can mirror attack by returning the message sent by entity A back to him. This is illustrated by the attack patterns that fail MSA and MSB's aliveness criteria as depicted in Figure 7 (a).

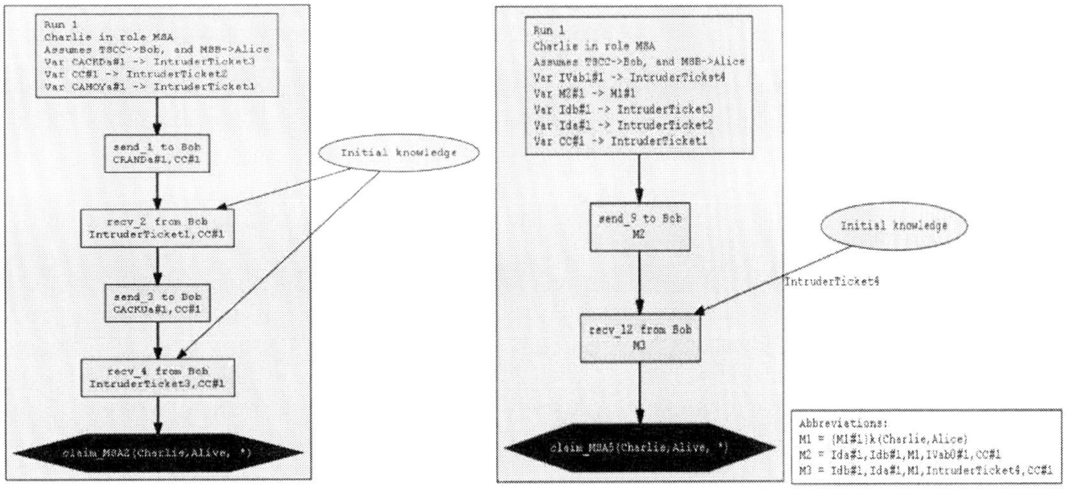

(a) (b)

Figure 7. Attack scenarios that fail the aliveness-type authentication claims by role MSA in the authentication protocol (a) and in the voice communication protocol (b).

The second failure was in the synchronization criteria. The aliveness criterion only requires that some events be executed by the real communication partner entity when the communication takes place without limiting the content of the messages exchanged. Synchronization has higher authentication requirements. Synchronization requires that the communication partner send all messages received and that the communication partner receive the messages sent. This condition conforms to the requirement that the actual message exchange has occurred precisely as specified by the protocol description. However, in Figure 7 (a), the attack pattern proves that there is no guarantee that the message MSA receives is actually from TSCC. In Device X's authentication protocol, some parties can pretend to be communication partners, thereby eliminating the guarantee of the communication partner's correctness.

The third failure was in the agreement criteria. The synchronization criteria ensure that the protocol can behave according to predefined descriptions even in an attacker's presence. The agreement is another criterion of authentication that focuses on the agreement on exchanged between entities. The idea behind the agreement criteria is that after executing the protocol, the parties agree on certain variables' values. The agreement is defined as a criterion that requires that the message's content is following the message sent as determined by the protocol. As a result, after the protocol has been run, the variables' contents will be precise as defined by the protocol. In that sense, there is no possibility of changing the content of the message. If it happens, then it can be known. Under the Device X authentication protocol, there was no guarantee of the contents of the messages sent.

In Table 2, the claim was not fulfilled because there is at least one possible type of attack against the device authentication protocol based on the test results. As the initiator, each MSA experienced the same failure in fulfilling all security criteria on authentication type claims, namely, aliveness, synchronization, and agreement. The possible attack patterns against these criteria on each device are the same as depicted in Figure 7 (b).

Table 2. Recapitulation of device X's voice communication protocol verification result.

Claim	*Confidentiality*	*Authentication*
OK	Role MSA: *M1* Role TSCC: *M1 and M2* Role MSB: *M2*	Role MSA: - Role TSCC: *Aliveness, Nisynch, and Niagree* Role MSB: *Aliveness, Nisynch, and Niagree*
FAIL	Role MSA: - Role TSCC: - Role MSB: -	Role MSA: *Aliveness, Nisynch, and Niagree* Role TSCC: - Role MSB: -

As a communication initiator, MSA failed to fulfil all security criteria on authentication type claims. The cause of failure of Device X's voice communication protocol resembled that of Device X's authentication protocol failure.

The first failure was in the aliveness criteria. Device X voice communication protocol is stated to guarantee the MSA entity's aliveness if another entity communicating with MSA, namely TSCC and MSB, can be assured of the existence of MSA. The mirror attack is illustrated by the attack patterns that fail MSA's aliveness criteria as depicted in Figure 7 (b).

The second failure was in the synchronization criteria. Synchronization requires that the communication partner indeed sends all messages received on Device X's voice communication protocol and that the communication partner indeed receives the messages sent. This conforms to the requirement that the actual message exchange has occurred precisely as specified by the protocol description. However, the attack pattern in Figure 7 (b) proves that there is no guarantee that the message MSA receives is actually from the TSCC. In Device X's authentication protocol, a party can pretend to be a communication partner, thereby eliminating the guarantee of truth from the communication partner.

The third failure was in the agreement criteria. The agreement criterion requires that the content of the message received matches the message sent as determined by the protocol. As a result, after the protocol has been run, the variables' contents will be precisely as defined by the protocol. In that sense, there is no possibility of changing the content of the message. In Device X's voice communication protocol, there is no guarantee for the contents of the message sent as the possibility is illustrated in Figure 7 (b). The absence of such a guarantee invalidates the agreement claim in the communication.

7.2. Security Analysis

Referring to the definition of security requirements from the International Telecommunications Union, security requirements have eight aspects of security requirements as follows:

1) Access control to resources: The system must ensure that actors are prevented from gaining access to information or resources that do not comply with their authority.
2) Robustness: The system must provide the necessary communication services following a specific service level agreement (SLA). For example, the SLA can determine the Quality of Service (QoS) or traffic capacity required. This requirement relates to the system's ability to deal with threats that aim to thwart one or more system services.
3) Protection of confidentiality: The system must provide the ability to ensure the confidentiality of data stored and communicated.
4) Protection of system integrity: The system must guarantee the integrity of the system and its components.
5) Protection of data integrity: The system must guarantee the integrity of the data stored and communicated.
6) Compliance with the regulatory framework: The system must be able to ensure compliance with existing regulations in the area where the system operates.
7) Accountability: The system must ensure that an entity cannot deny responsibility for the actions taken. In this context, accountability is used as a synonym for non-denial.

8) Identity verification: The telecommunications network must provide the ability to establish and verify the identity of any authorized user.

Following research that focuses on secrecy and authentication, the other six security aspects are not discussed. According to references from ITU and the definitions of Scyther Tool's criteria, the relevant aspects of the safety requirements are detailed in points 3, 5, and 8.

The results from testing and analyzing Scyther Tool have demonstrated that the confidentiality aspect of information has been guaranteed both for the authentication protocol and the voice communication of Device X. This aspect of confidentiality can be achieved by applying the encryption of confidential information. In the protocol description, the classified information is denoted as M1 and M2.

Nevertheless, unfortunately, data integrity and identity verification aspects are not fulfilled in several parts of the authentication and voice communication protocols of Device X. This is because the authentication and voice communication protocols of Device X have not guaranteed the following:

1) the presence (aliveness) of several entities in the protocol for its communication partners:
2) that the message received was sent by the communication partner and that the message sent was received by the communication partner;
3) and that the contents of the message received can be guaranteed its integrity.

8. Recommendations

It should be borne in mind that the protocol that is used as the object of this research is run on a radio device that has limitations in terms of hardware capacity for implementation. On the one hand, the hardware limitation is a plausible explanation for any flaws in the system. The flaw in question can be seen in terms of the system (e.g., unstandardized encryption algorithms) or the weakness of the implemented cryptographic protocols. However, from a theoretical perspective, the security aspect, which has been proven to have weaknesses, cannot be ignored. Rather than being seen as an excuse to justify using a less secure system, hardware limitations are better viewed as a challenge to create a lightweight but robust system.

Concerning the results, the weaknesses found can be overcome by using existing cryptographic techniques such as the following:

1) the use of a timestamp as a marker to guarantee the timeliness and uniqueness of a message;
2) the use of a hash function to ensure the authenticity of the message received;
3) and use of a variety of signature techniques, such as blind signatures and public-key mechanisms, to verify the identity of the entities involved in communication.

For hardware limitations, implementation can be done by implementing lightweight algorithms or schemes. In the future, if no algorithm or scheme is considered qualified, this will be an opportunity and a challenge for researchers in the information security field to develop algorithms and lightweight schemes that can better meet security criteria for the specific case.

9. Conclusion

Based on the analysis result, the following can be concluded that Device X's authentication and voice communication protocols have fulfilled the secrecy criteria for the transmitted confidential information but has not fulfilled the criteria of aliveness, synchronization, and agreement on several entities involved in the protocol. Thus, Device X's authentication and voice communication protocols are considered to be provably secure based on the confidentiality aspect, but not based on the authentication aspect.

10. Future Work

Future work that can be developed from this research includes the following:

1) a possible updating of the security review document for Device X according to the research results by the XYZ agency;
2) research to apply cryptographic techniques to the recommendations for improvement in this research to overcome Device X's authentication and voice communication protocols' identified weaknesses;

3) and research that examines more deeply Device X's vulnerability to eavesdropping and cloning.

References

[1] Eroğlu Ş and Çakmak T 2018 Information as an organizational asset: assessment of a public organization's capabilities in Turkey *Information Development* **36(1)** pp 58–77
[2] Cooren F, Kuhn T, Cornelissen J P and Clark T 2011 Communication, organizing and organization: an overview and introduction to the special issue *Organization Studies* **32(9)** pp 1149–70
[3] Johansson C and Ottestig A T 2011 Communication executives in a changing world *J. of Communication Management* **15(2)** pp 144–64
[4] Hirsch P 2000 Book reviews: corporate communication 2nd edition by Paul A. Argenti *Business Communication Quarterly* **63(1) pp** 111–3
[5] BBCIndonesia 2013 *BIN: Australia Menyadap Indonesia Sejak 2007* Retrieved on 1 June 2020 from https://www.bbc.com/indonesia/berita_indonesia/2013/11/131120_bin_sadap_australia
[6] BIN 2013 *Kepala BIN: Evaluasi Sistem Keamanan Komunikasi* Retrieved on 1 June 2020 from http://www.bin.go.id/nasional/detil/255/1/29/11/20%2013/kepala-bin-evaluasi-sistem-keamanankomunikasi
[7] Tempo.co 2014 *4 Kasus Penyadapan Besar di Indonesia* Retrieved on 1 June 2020 from https://nasional.tempo.co/read/556304/4-kasus-penyadapan-besar-di-indonesia
[8] XYZ Agency 2019 *Device X* Unpublished
[9] Menezes A J, van Oorschot P C and Vanstone S A 2018 *Handbook of Applied Cryptography* (Florida: CRC Press)
[10] Shim J 2013 *The International Handbook of Computer Security* (New York: Routledge)
[11] Komperda T and Boudriga N 2013 Telecommunications and network security: communications and network security *Information Security Management Handbook* 6th Edition ed R O'Hanley and J S Tiller (New York: Routledge) chapter 1
[12] Michael K 2012 The basics of information security: understanding the fundamentals of InfoSec in theory and practice *Computers & Security* **31(4)** pp 634–5
[13] ISO 2916 *Information Technology – Security Techniques– Cryptographic Algorithms and Security Mechanisms Conformance Testing* Retrieved from https://www.sis.se/en/produkter/information-technology-office-machines/it-security/isoiec183672016/
[14] Chen Q, Zhang C and Zhang S 2008 Overview of security protocol analysis *Lecture Notes in Computer Science* vol **5111** pp 17–72
[15] Avalle M, Pironti A and Sisto R 2012 Formal verification of security protocol implementations: a survey *Formal Aspects of Computing* **26(1)** pp 99–123
[16] Chadha R, Cheval V, Ciobâcă Ş and Kremer S 2016 Automated verification of equivalence properties of cryptographic protocols *ACM Transactions on Computational Logic* **17(4)** pp 1–32
[17] Lowe G 1997 A hierarchy of authentication specifications *Proc. 10th Computer Security Foundations Workshop* pp 31–43

Usability Evaluation of a Fintech Lending Mobile Application for University Student: A Case Study

Ellen Emanuela*, Ari Widyanti, Gradiyan Budi Pratama

Institut Teknologi Bandung

E-mail: *ellenemanuela12@gmail.com

Abstract. The rapid development of information technology has changed various sector of our life, including economy. A number of startup companies in Indonesia has been utilized technology to deliver financial services to its customer. One of them attempts to answer the needs of university students in Indonesia for an alternative source of funding, who develop a lending mobile application for student. An initial study using interview and secondary data found that the users still find difficulties to use the mobile application. Usability is one approach that can be used to minimize the problem. The purpose of this study is to increase a fintech mobile application's usability by conducting evaluation of several features including, 'creating a new account', 'completing profile registration', 'applying for university tuition lending', and 'applying for goods lending.' Usability evaluation was conducted by testing the effectiveness, efficiency, and satisfaction through a set of remote usability testing involving 20 participants. The results shows that the mobile application is not optimized in terms of effectiveness, efficiency, and satisfaction as yet. Using the Concept Generation Method, 21 recommendations were proposed to increase the usability of the lending mobile application.

1. Introduction

Technology utilization in Indonesia's financial lending industry enables lending activity to be conducted more effectively and efficiently, so called financial technology (fintech). According to the Indonesian Financial Services Authority, fintech lending has the most aggressive growth in the overall fintech category. The rapid growth of fintech lending businesses is influenced by public demand for an alternative source of funding which could go beyond geographic limitation, convoluted administration, and other barriers that exist in the traditional financial institution.

This paper studied a fintech lending mobile application, Cicil, which is developed by PT Cicil Solusi Mitra Teknologi. The apps could be downloaded from Google PlayStore. Cicil provides a lending services for university students in Indonesia, particularly for school tuition and purchases of goods. There are administration barriers for university students in Indonesia to access loan services through the traditional financial institution and this lending service aims to help students with a safer yet convenient alternative source of funding for their needs.

A preliminary study was conducted to examine the experience in using the fintech lending mobile application (from now on called lending mob apps). The investigation includes interview and observation to five respondents as well as documentation of user feedbacks in Google PlayStore. Resemblant problems were identified from those three methods that users found it difficult to operate and access the product. There are confusions during the process of proposing the loan in the lending mobile apps and 80% of the respondent claimed that it discourages them to use the lending mobile application even further.

Usability evaluation can be used as an approach to solving the problems in mobile applications in general. Usability is a crucial factor for software products and has a vital role in the success of software product development [1]. Usability describes a product's ability to reach user expectations and evaluation could be conducted through a set of procedures to collect end-user interaction data with the product [2].

This research aimed to evaluate the interface of a lending mob apps from the view of usability by investigating three parameters namely effectivity, efficiency, and satisfaction [3]. The output of the study is a detailed assessment of the mobile application and identification of room for improvements for the application which can be used as a reference for other lending mob apps to improve the interface and use of the apps.

2. Methods

2.1. *Usability Testing*

The usability testing procedure was conducted remotely via Google Meet (recorded) and several controls have been made. The researcher confirmed all the participant used a mobile phone for the testing (not tablets) with Android Operating System. The testing consisted of three methods: performance measurement, Retrospective Think Aloud (RTA), and questionnaires. Participants were asked to perform four tasks in the mobile application. Performance was measured based on the success rate, total time on task, and errors. The RTA was applied to capture user experience, by asking the participants to reflect their experience of using the application and pointed out the problem they found during the task. They were supported by video recording of their own test session in order to be able to recall the problem and make the most representative justification from it.

Two questionnaires were used to measure user satisfaction that are After Scenario Questionnaire (ASQ, Sauro & Lewis, 2016) and System Usability Scale (SUS, Sauro & Lewis, 2016), with 7-point Likert Scale and 5-point Likert Scale respectively [4]. At the end of every task, participants were asked to fill out the ASQ. The session closed with filling out the SUS after finishing all the tasks, RTA sessions, and post-task questionnaires. Qualitative data collected through the RTA protocols are used to fit up with the questionnaire result for determining the user satisfaction level.

2.2. *Participants*

There are twenty participants involved in the usability testing (10 males, 10 females) [5]. All of the participants were active university students ranging from 17 to 21 years old. Participants were recruited based on the company's target market profile consisting new users and existing users with a balanced proportion. New users were university students who have never had prior experience using the mobile application.

2.3. *Testing scenario*

There are four scenarios/tasks set for the testing procedure. First, participants were asked to create a new account in the mobile application. The second scenario is completing the profile of the user account. Third, the participants were asked to perform a loan request for school tuition and, four, a request for purchasing goods. All the tasks that were set in this research were further divided into subtasks to conduct an in-depth data processing and analysis.

3. Results

3.1. *Effectiveness*

Tasks in this research were divided into several subtask and the success criteria were defined prior the testing for each of them. A task/ subtask with success rate above 70% categorized as effective [5]. There were one ineffective subtask identified in the first task (creating password that met the specifications), two ineffective subtasks in the second task (open profile form easily, find & click the "simpan &

selanjutnya" button), one ineffective subtask in the third task (find & click the "simpan & selanjutnya" button), and no ineffective subtask for the fourth task. Success rate and user group contribution of failure of all task presented in Table 1.

Table 1. List of innefective sub task and its success rate (%)

Total	New User Group	Existing User Group
Task 1		
Creating password that met the specifications (60%)	Find and click "Daftar" easily (50%) Creating password that met the specifications (50%)	-
Task 2		
Open profile form easily (55%)	Open profile form easily (40%)	Find & click the "simpan & selanjutnya" button (50%)
Find & click the "simpan & selanjutnya" button (45%)	Find & click the "simpan & selanjutnya" button (40%)	
Task 3		
Find & click the "simpan & selanjutnya" button (65%)	-	Find & click the "simpan & selanjutnya" button" (50%)

Table 2. Time on task and *t-test* value across different tasks (measured in unit of time: second)

		t-value	Total		New User Group		Existing User Group	
			Mean	*Std. Dev*	*Mean*	*Std. Dev*	*Mean*	*Std. Dev*
Task 1: Create new account								
1.1	Find account registration form	0.00	21.55	17.75	35.1	15.71	8	3.23
1.2	Fill out account registration form	0.64	44.8	10.06	45.9	11.34	43.7	9.07
1.3	Verify and confirm account registration	0.16	60.5	15.66	66.2	16.11	54.8	13.64
Task 2: Completing account profile								
2.1	Find profile form	0.01	53.6	58.66	88.5	67.15	18.7	6.96
2.2	Fill out profile form	0.02	659.5	103.24	710.5	116.31	608.5	56.49
2.3	Verify and confirm profile	0.03	38.9	26.12	51.6	32.44	26.2	5.47
Task 3: University tuition loan request								
3.1	Find university tuition loan page	0.02	12.7	12.88	19.3	15.49	6.1	3.70
3.2	Fill out university tuition loan form	0.01	163.5	61.29	198.3	39.40	128.7	60.72
3.3	Verify and confirm loan request	0.13	24.5	12.56	28.8	12.51	20.2	11.65
Task 4: Goods loan request								
4.1	Find goods loan page	0.02	5.45	4.41	5.9	4.98	5	3.97
4.2	Fill out goods loan form	0.00	154.6	109.54	221.6	121.74	87.6	23.16
4.3	Verify and confirm loan request	0.07	33	7.17	34.7	7.02	31.3	7.27

3.2. *Efficiency*

Efficiency measurement were conducted by looking at the total time on each task and occurrence of error during the task. Subtask were considered inefficient when there is a significant difference in time of New User Group versus Existing User Group. Error on task also indicated inefficiency due to the need of user to revise or re-do the process [6]. There are two inefficient subtasks for task 1 (find account registration form and fill out account registration form); three inefficient subtasks for task 2 (find profile form, fill out profile form, verify and confirm profile); two inefficient subtasks for task 3 (find university tuition loan page, fill out university tuition loan form); two inefficient subtasks for task 4 (find goods loan page, fill out goods loan form). The time on task, *t-test* value and error for task 1, 2, 3, and 4 are presented in table 2 and table 3 respectively.

Table 3. List of error occurrence

Errors in 1st task	Errors in 2nd task	Errors in 3rd task	Errors in 4th task
KA1: Failed to find and click on "daftar" button easily	KB1: Failed to find profile form easily	KC1: Miss to input payment deadline	KD1: Failed to input product URL
KA2: Failed to create qualified password	KB2: try to submit the uncomplete personal data form	KC2: Miss in uploading the proof of bill	KD2: Miss filling the seller note
KA3: Submitting the form without finishing the questions	KB3: Input wrong optional data	KC3: Input wrong optional data	KD3: Proceed request before finishing the loan request form
	KB4: try to submit the uncomplete academic data form	KC4: Failed to request university loan	
	KB5: unaware of optional data in students job field	KC5: Miss filling the payment type	
	KB6: Miss in uploading ID card	KC6: Miss uploading payment step document	
	KB7: Miss in uploading ID card and selfie		
	KB8: Miss in uploading Student ID Card		
	KB9: Miss in registering parents' phone numbers		
	KB10: Try to submit uncomplete parents data form		

Table 4. Results of After Scenario Questionnaire (ASQ)

Task	ASQ Questions	*Mean*					
		New User Group		Existing User Group		Total	
	ASQ1	5		6.1		5.55	
1	ASQ2	5.6	5.17	6.18	6.16	5.85	5.65
	ASQ3	4.9		6.2		5.55	
	ASQ1	4.5		5.9		5.2	
2	ASQ2	4.1	4.33	5.9	6.1	5	5.22
	ASQ3	4.4		6.5		5.45	
	ASQ1	4.7		6.5		5.6	
3	ASQ2	4.9	4.63	6.4	6.43	5.65	5.53
	ASQ3	4.3		6.4		5.35	
	ASQ1	5.1		6.2		5.65	
4	ASQ2	5.2	5.13	5.9	6.07	5.55	5.6
	ASQ3	5.1		6.1		5.6	

3.3. *After Scenario Questionnaire (Post-task Questionnaire)*

User satisfaction in each subtask was calculated using the After Scenario Questionnaire (ASQ). ASQ questionnaire in this study consist of 3 questions:

1. *Secara keseluruhan, saya puas dengan kemudahan penyelesaian tugas pada skenario ini*
2. *Secara keseluruhan, saya puas dengan waktu penyelesaian tugas pada skenario ini*

3. *Secara keseluruhan, saya puas dengan informasi pendukung yang tersedia (pusat bantuan online, pesan, dokumentasi) ketika menyelesaikan tugas ini*

Respond collected in 7-point Likert scale with score 1 as very disagree and score 7 as very agree. Task 2 has the lowest satisfaction among other tasks. The result of the ASQ questionnaire presented in Table 4.

3.4. System Usability Scale (Post-test Questionnaire)

The overall satisfaction measured using the System Usability Scale. In total, the mobile application scored 56.4 out of 100 and compared to previous studies is within the 15-34 percentile [7].

Table 5. System Usability Scale (SUS) result

	SUS Score	Grade	Percentile
New User Group	46.3	F	0-14
Existing User Group	66.5	C	41-59
Total Score	56.4	D	15-34

3.5. Retrospective Think Aloud

By using Pareto Method, the researcher prioritizes problems, which identified by the participant in the RTA session for each scenario. The problems prioritized for the 1st, 2nd, 3rd, and 4th scenario could be observed in Table 6.

Table 6. List of prioritized usability problems

1st Scenario	2nd Scenario	3rd Scenario	4th Scenario
Complaints related to verification process	Complaints related to the profile form features	Complaints related to the university loan form features	Complaint related to the product input process
Complaints related to the initial steps on making the new account	Complaints related to the instruction to fill the profile form	Complaint related to the loan calculation	Complaint related to form instructions and information
Complaints related to form filling activity	Complaints related to the initial steps to complete user profile	Complaint related to diction selection in the university loan form	Complaints related to the goods loan form features
	Complaint related to diction selection in the profile form	Complaint related to loan submission flow	Complaint related to the loan calculation
		Complaint related to form instructions and information.	

3.6. Room for Improvements

Referring to the definition of usability from International Organization for Standardization (ISO), a system/product should be effective, efficient, and satisfy the user therefore improvement shall be cultivated when one or more of the parameters are not reached. This research formulates recommendations based on the scenario from the usability evaluation. There are 21 recommendations proposed based on the result of the evaluation of the Mobile Application listed in Table 7. Recommendations formulated using the concept generation method, which consist of 5 steps [8]:

1. Usability problem definition
2. External exploration (benchmarking, external discussion, etc)
3. Internal exploration (internal brainstorming, etc)
4. Concept exploration (combining internal & external exploration)
5. Recommendation design (2D wireframe)

Table 7. List of improvement recommendations

No	Scenario	Usability Parameters	Related Errors	Recommendation
1	(1st Scenario) Find account registration form	Efficiency, Satisfaction	KA1	Resize and shift "Daftar" button
2	(1st Scenario) Create password	Effectiveness, Efficiency, Satisfaction	KA2	Create realtime feedback for every password criteria with color indicator.
3	(1st Scenario) Submit account registration form	Efficiency	KA3	Differentiate optional and non-optional field using different color and add information about referral code.
4	(1st Scenario) Account verification	Efficiency, Satisfaction	-	Sending OTP code through SMS
5	(2nd Scenario) Find profile form	Effectiveness, Efficiency, Satisfaction	KB1	Re-design profile menu with a clear profile button
6	(2nd, 3rd, 4th Scenario) Click "Simpan & Selanjutnya" button	Effectiveness, Efficiency, Satisfaction	KB2, KB4, KB6-KB22, KB24-KB32, KC1, KC2, KC5, KC6, KD2	Move "Simpan & Selanjutnya" button to the end of the form
7	(2nd, 3rd, 4th Scenario) Upload photos	Efficiency, Satisfaction	KB7, KB8, KB17, KB18,	Redesign upload status bar with clear status and color indicator.
8	(2nd Scenario) Input unregistered faculty	Efficiency, Satisfaction	-	Shift additional faculty button bellow the registered faculty
9	(2nd, 3rd, 4th Scenario) Input date information	Efficiency, Satisfaction	KC1	Use dropdown menu for to input date information
10	(2nd Scenario) Input GPA	Efficiency, Satisfaction	KB23	Create GPA sample
11	(2nd Scenario) Input house type	Efficiency, Satisfaction	-	Create "rumah" option
12	(2nd Scenario) Information about emergency contact person	Efficiency, Satisfaction	-	Provide simpler information with pictures as an illustration.
13	(2nd, 3rd, 4th Scenario) Input optional and non-optional information	Effectiveness, Efficiency, Satisfaction	KB3, KB5, KC3	Differentiate optional and non-optional field using different color
14	(2nd Scenario) Input Guarantor (penjamin) data	Efficiency, Satisfaction	KB33	Create pop-up information about Guarantor (Penjamin)
15	(3rd & 4th Scenario) Initiate loan request	Effectiveness, Efficiency, Satisfaction	KC4, KD1	Change "Simulasi Cicilan" button to"Mulai Hitung Cicilan" button
16				Add loan button in "Cicilan" tab
17	(3rd & 4th Scenario) input down payment	Efficiency, Satisfaction	-	Empty down payment field as default
18	(3rd & 4th Scenario) Loan calculation	Efficiency, Satisfaction	-	Add pop up information about load detail
19	(3rd Scenario) Loan insurance information	Efficiency, Satisfaction	-	Add pop up information about loan insurance
20	(4th Scenario) Input product's URL	Efficiency, Satisfaction	KD1	Create how-to information to input product's URL under URL field.
21	(4th Scenario) create note for seller	Efficiency, Satisfaction	KD2	Remove "Keterangan tambahan produk"

4. Conclusions

Usability testing has been conducted to the Fintech Lending Mobile Application involving three parameters: effectivity, efficiency, and satisfaction. According to the result, the mobile application still has room for improvements in regards with the usability aspect. In total, there are 4 out of 12 subtasks were identified as ineffective and 9 out of 12 subtasks were identified as inefficient category. The overall satisfaction of the mobile application was measured using the System Usability Scale with the score result 56.4 out of 100 points which is considered as low user satisfaction. In order to achieve a better usability for the mobile application, 21 interface recommendations were formulated and had passed the initial validation as proper solutions for the existing usability problems.

It should be taken into consideration that the evaluation sessions for all of the participants was held online and remotely, while it is considered as a good alternative if an in-house usability testing was not possible to conduct [9]. As a control, the researcher makes sure that all the devices used by participants are mobile phones (not tablets) with Android OS. For further research, it is recommended to conduct further usability testing, such as the AB testing for the 21 recommendation that is already formulated in this study.

References

[1] Madan A and Dubey S K 2012 Usability evaluation method: a literature review *Int. J. Eng. Sci. Technol.* vol **4(2)** pp 590–9

[2] Fernandez A, Insfran E and Abrahao S 2011 *Usability Evaluation Methods for the Web: A Systematic Mapping Study* (Amsterdam: Elsevier)

[3] ISO 2016 *System and Software Engineering – System and Software Quality Requirements and Evaluation – Measurement of Quality in Use* (ISO/IEC 25022)

[4] Sauro J and Lewis J R 2016 *Quantifying the User Experience: Practical Statistics for User Research* (Cambridge: Morgan Kaufmann)

[5] Rubin J and Chisnell D 2008 *Handbook of Usability Testing: How to Plan, Design, and Conduct Effective Test* (Indiana: Willey Publishing)

[6] Tullis T and Albert B 2013 *Measuring the User Experience* (Waltam: Elsevier)

[7] Kortum P and Bangor A 2013 Usability ratings for everyday products measured with the System Usability Scale *Int. J. Hum.-Comput. Interact.* vol **29(2)** pp 67–76

[8] Ulrich K T and Eppinger S D 2016 *Product Design and Development* (New York: McGraw-Hill Education)

[9] Schade A 2013 *Nielsen Norman Group* Retrieved from https://www.nngroup.com/articles/remote-usability-tests/

Blockchain-based Model for Health Information Exchange: A Case for Simulated Patient Referrals Using an Electronic Medical Record

Alison O. Perez[1], **Thelma Domingo-Palaoag**[2]

[1,2]College of Information Technology and Computer Science, University of the Cordilleras, Baguio City, Philippines

E-mail: [1]perez.alison@gmail.com, [2]tpalaoag@gmail.com

Abstract. Health information exchange (HIE) has been recognized as an effective mechanism to improve service delivery. The benefits are compounded with the introduction of electronic medical records (EMR). Exchanging information between health facilities remain a challenge due to the absence of an integrated health information system and issues pertaining to data integrity. The emergence of blockchain has provided an innovative solution to data sharing challenges in security, transparency and accessibility. This paper aims to develop a framework for health information exchange between health facilities with the use of EMR and blockchain-based network. The EMR was developed with functionalities such as adding and updating patient profile and health records, referring patient to another facility and monitoring patients referred. These functionalities have been simulated using hypothetical test cases. The simulation has demonstrated blockchain-related functions such as registration of health facility in the network, defining consensus rules and adding patient referral transactions in the network. The description of the proposed framework, role of blockchain and key features of the prototype EMR is summarized in the conclusion part.

1. Introduction

Healthcare data is one of the most valuable shared resources within the medical community. Resolving of medical cases, developing health researches and establishing health program policies are just some of the many areas that healthcare data is heavily utilized. With the development of electronic medical records (EMR), access to the health data have become much faster and its quality has been more enhanced.

Sharing of healthcare data is essential towards improving the quality of care [1]. As the patients moves from one healthcare provider to another, there is need for a unified and integrated health records to come up with proper diagnosis and treatment of medical case. This remains a challenge because the health records are often paper-based where data is stored in the local health facility and not shared with other health facilities. As such, monitoring of compliance of the patient and assuring continuity can be challenging.

Blockchain offers a practical solution to the issues relating to health information exchange [2]. With blockchain, participants (healthcare providers and patients) with appropriate permissions shall have access to the distributed ledger containing secured health data while ensuring that it is safe from tampering. While blockchain can either be public or private, sharing of health information can be best done using a consortium-based setup wherein only official providers and patient are enrolled into the network [3]. Participants in the consortium blockchain can agree on a consensus mechanism when in validating transactions being added into the blockchain.

To address the issues mentioned, this study proposes a health information exchange model for patient referral demonstrated through a prototype electronic medical record. The author has identified the following research questions: 1). What is the framework used to develop health information exchange via patient referral system? 2). How is blockchain used and integrated into the framework? 3). How can the framework be demonstrated on functional electronic medical record prototype? Target users of the system are health care service providers in Philippine public health facilities. Using simulated data sets from three different EMR instances, personal patient data such as name, date of birth and gender and medical notes in SOAP format (subjective, objective, assessment and plan) from the databases of the EMRs and receiving provider for referral cases.

2. Methodology

2.1. Characterization of Health Information Exchange in Patient Referral System

The Philippine Service Delivery Network (SDN) manual of operations has defined steps to refer patients from one health facility to another as shown in Figure 1. A patient visits a health facility for consultation and data such as patient profile and demographics, vital signs, diagnosis and treatment are being recorded. If a referral is needed, the health service provider fills out a referral slip that the patient needs to give to the receiving health facility. Upon visiting and being seen by the receiving health facility, the patient is being sent back to the originating health facility for continuity of care.

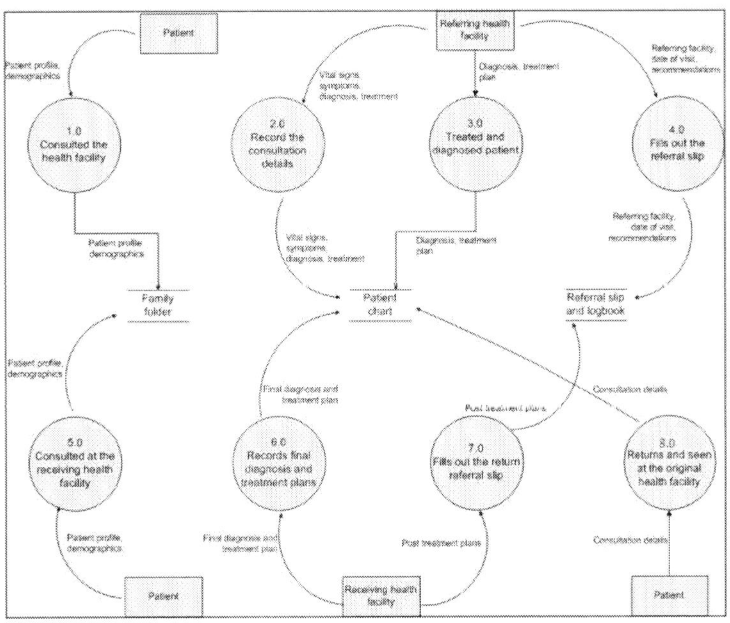

Figure 1. General Steps in Patient Referral

2.2. Data Preparation

During the process of data gathering and requirements analysis, initial and baseline data libraries were collected and entered in the system's database. Among the data collected include list of all departments and sections, schedule and designated personnel for services offered and protocols for referral in and out of their facility. In addition, internationally accepted health data standards such as International Coding of Diseases 10 (ICD-10) for disease classification and Current Procedural Terminologies (CPT) for coding of health services have been included in the system database. The system was hosted on a third-party web-based cloud server and can be made access via the URL https://alisonperez.net/ers. Users were registered centrally into the system and can be designated into more than one health facilities.

2.3. Development of Electronic Medical Record Prototype

The study makes use of an electronic medical record prototype developed by the author called "Electronic Health Facility Information Exchange and Referral System (ERS)" and currently hosted in https://alisonperez.tech/ers. ERS is suitable for a variety of health facilities such as hospitals, health centers and private clinics. ERS was created using PHP 7.0, Javascript and MySQL database. The forms and reports generated by system were derived from the actual forms used in Philippine public health centers. The system will be referred to as ERS EMR from here onwards.

In the development of the ERS EMR, the author makes use of agile software development methodology particularly Extreme Programming (XP). This development methodology was adopted as it requires constant changes in the system such updating of the system's features and fixing software related bugs within limited amount of time. On the other hand, lean development model was used for the the integration of blockchain using the OurSQL network. Lean development model is suitable due to short time required well as the immediate availability of readily-made third party software tool to accomplish this feature [14].

ERS EMR have multiple instances running with each instance representing a health facility. Profile of the patient (i.e. name, birthday, gender, PhilHealth ID, address), profile of the health facility (i.e. name of health facility, facility ID as defined by the Philippines' Department of Health, classification) and consultation details (i.e. consultation date, physical and diagnostic exams, diagnosis, treatment, referred to, reason for referral) are the key data sets collected and processed by ERS EMR. Only consultations that require referrals and its corresponding patient and health facility identifiers will be added by the ERS EMR to the cloud-based blockchain.

3. Literature Review

3.1. Electronic Medical Record

Electronic medical records (EMR) are computerized health information systems that is used to collect, store and a person's medical information such as health history, diagnosis, medicines, tests, allergies, immunizations and treatment plans [4]. Electronic medical records prove to be a relevant tool in transforming healthcare due to benefits such as reducing medical errors, reducing healthcare cost and promoting quality of health services [5].

According to the National Academies of Sciences and Engineering Medicine, an EMR should be able to perform eight core functions such as health information and data (i.e. patients' diagnoses, allergies, lab test results and medications), result management (i.e. accessing current and past test results), order management (i.e. storing prescriptions, test and service legibly), decision support (i.e. using prompts and alerts to improve compliance), electronic communication and connectivity (i.e. efficient and secure communication between healthcare providers and patients), patient support (i.e. providing patient access to records), administrative processes and reporting (i.e. scheduling systems) and reporting and population health (i.e. patient safety and disease surveillance) [6].

3.2. Health Information Exchange

Health Information Exchange (HIE) is a technology that shares clinical and administrative data by various healthcare providers. HIE allows physicians, nurses, other healthcare providers and patient to better track and securely share medical records thereby improving the speed, quality and cost of patient care.

To allow HIE, data standards such as LOINC (for laboratories results) and RxNorm (for medication) and interoperability standards were established. Fast Health Interoperability Resources (FHIR) and Consolidated Clinical Document Architecture (C-CDA). C-CDA was designed to transfer the entire document while FHIR would only transmit piece of data based on the request made by the healthcare provider [7].

The study conducted by Mandel, et. al [8] has explored the integrated of an HIE standard FHIR into an existing health technology platform called Substitutable Medical Applications and Reusable Technologies (SMART). The project SMART on FHIR was designed and developed using clinical data models and application programming interface of FHIR. This allows medical applications to be written once and run unmodified across IT systems.

According to Sadoughi, et. al [9], although HIE can promote patient care quality and reduce cost related to resource utilization, further researches are needed to provide better understanding of this domain and to improve adoption of this technology.

3.3. Blockchain Technology and HIE

Blockchain is a shared, distributed ledger that facilitates the process of recording transactions and tracking assets in a business environment. Rather than having a central administrator, a distributed ledger has a network of replicated databases, synchronized via the Internet and visible to anyone within the network. It is a special database where transactions are added and cannot be deleted. Each block contains information such as hash of the previous block, hash of the block, information of the transaction and other information. As new blocks are added into the blockchain, all nodes should reach an agreement using consensus algorithm [10].

Among the key advantages of blockchain are impossibility to change data of the approved block (immutability), impossibility to change the data of the approved block (cryptographically secured), minimizes failure through decentralization and all transactions are documented and available for all participants (provenance) [11]. This makes the application of blockchain suitable for healthcare data sharing and information exchange.

Fan, et. al [12] proposed MedBloc, a blockchain-based information system that promotes efficient access and retrieval of patient information from EMRs. Medbloc high information security by combining blockchain features such as customized access controls protocols and symmetric cryptography.

In a study conducted by Gordon, et. al. [13] has explored the methods, potentials, risks and barriers of implementing a patient-driven interoperability using blockchain technology. Under this scenario, an individual patient's electronic data is made available to them through standard mechanisms like APIs. The study stated that blockchain could provide high-level framework on how a patient could securely interact with multiple stakeholders using mechanisms such as digital access rules management and data aggregation, availability and liquidity.

Zhang, et. al. [2] has reviewed the application of blockchain technology in sharing clinical data as required by the "Shared Nationwide Interoperability Roadmap" of the Office of the National Coordinator for Health Information Technology (ONC), USA. The study crafted FHIRChain, a blockchain- based decentralized architecture that simulates FHIR data interoperability standard. FHIRChain provides patients with more collaborative clinical environment and has complied with ONC's interoperability requirements of user identifiability and authentication, consistent format and system modularity.

4. Results and Discussion

4.1. Design Framework for Health Information Exchange via Patient Referral

In Figure 2, the patient visits health facility 1 and his health record is recorded in the ERS EMR. The health record is stored in both the local database and the local blockchain of health facility 1. If the patient is referred to health facility 2, the local blockchain instance in health facility 1 is shared and integrated into the local blockchain of health facility 2. Data is decrypted in health facility 2 and merged into its local database. The referral transactions are formatted to make it more readable. Only referred cases intended for health facility 2 will be displayed.

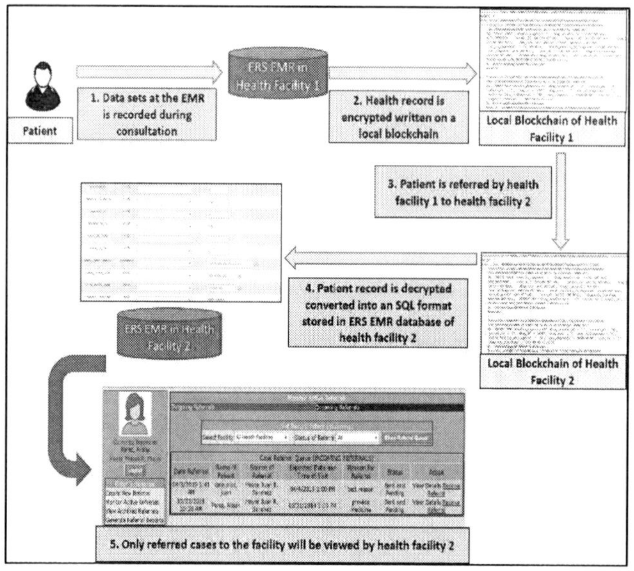

Figure 2. Framework for Health Information Exchange via Patient Referral in ERS EMR

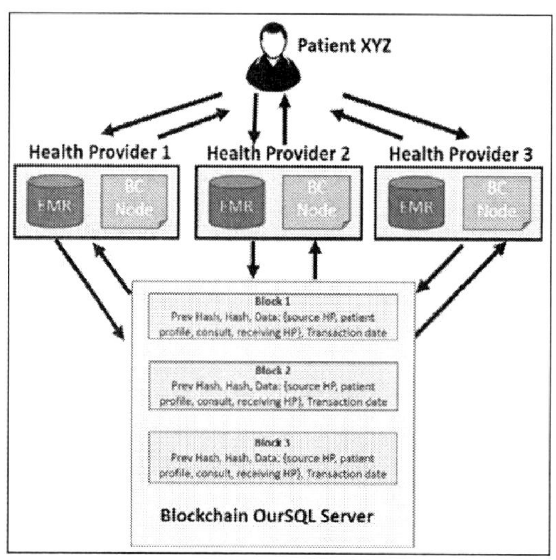

Figure 3. Interaction Between ERS EMR instances and Blockchain server

4.2. Use of OurSQL to Create Blockchain-Based Network of Databases

ERS EMR is binded and connected to OurSQL via MySQL proxy connection [15]. OurSQL (https://oursql.org) is standalone blockchain server that translates MySQL databases into blockchain. Before a transaction is added into the blockchain, OurSQL validates the transaction based on the

IOP Conf. Series: Materials Science and Engineering 1077 (2021) 012059 doi:10.1088/1757-899X/1077/1/012059

consensus rules set in the blockchain. For this study, Proof of Work (PoW) was used as the consensus algorithm. The size of the block is determined by the maximum number of transactions defined in OurSQL configuration file. Figure 3 shows the interaction between ERS EMR instances using and the blockchain server.

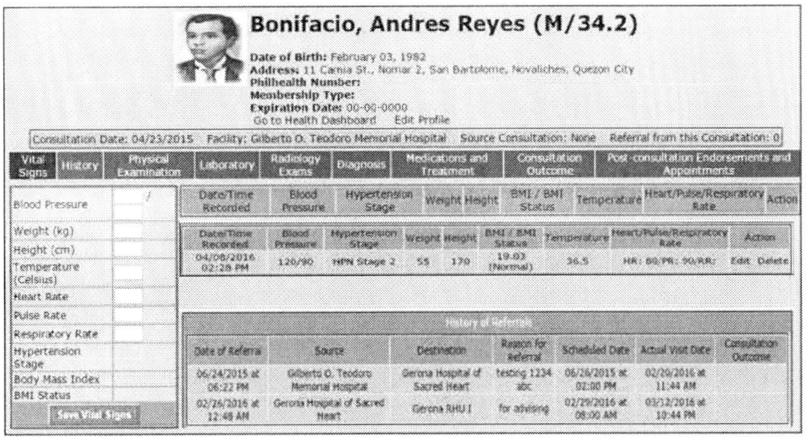

Figure 4. Consultation Recording Form of ERS EMR

Each blockchain transaction is signed using the prime256 ECDSA signature. For the signature to be generated, the public key and the SQL query must be sent to the MySQL proxy attached to the OurSQL server. The MySQL proxy returns a record which contains data to sign (string to sign). Using the server's private key corresponding to the public key used in the previous step, the string will be signed and shall execute a second SQL request where the signature produced is included already.

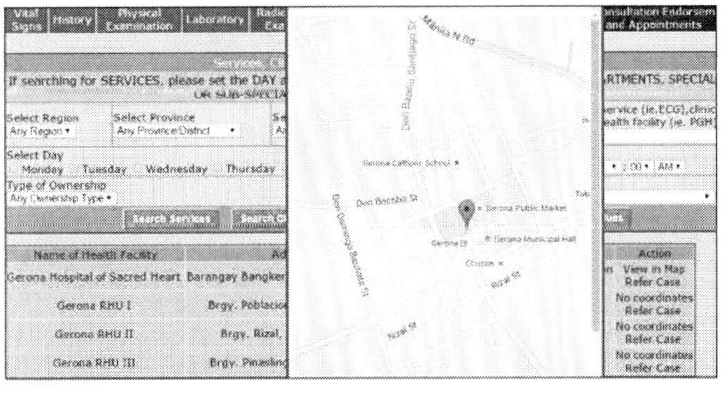

Figure 5. Health facility search feature

Sending the transaction to other nodes (i.e. ERS EMR instances) is also being handled by the OurSQL blockchain server. If a node receives a transaction, it will check if the SQL can be executed against the consensus rules set. Once it is verified for execution, the signed transaction will be decoded using the public key and will be added into the node's own OurSQL instance.

4.3. Utilization of ERS EMR for Health Information Exchange via Patient Referral

Among most relevant features of ERS EMR includes searchable library of health facilities based on location, schedule and services available, a patient and consultation registry, integrated patient dashboard and a case referral component. All transactions would start with searching an existing patient and recording his consultation details in SOAP format (Subjective-Objective-Assessment-Plan). Figure 4 shows the recording form of consultation details at ERS.

If a referral is required for the case, the end user will have to utilization the health facility search feature of ERS and then provide the referral details before sending the case to the receiving health facility. Figure 5 shows health facility search feature while Figure 6 shows the case referral form of ERS.

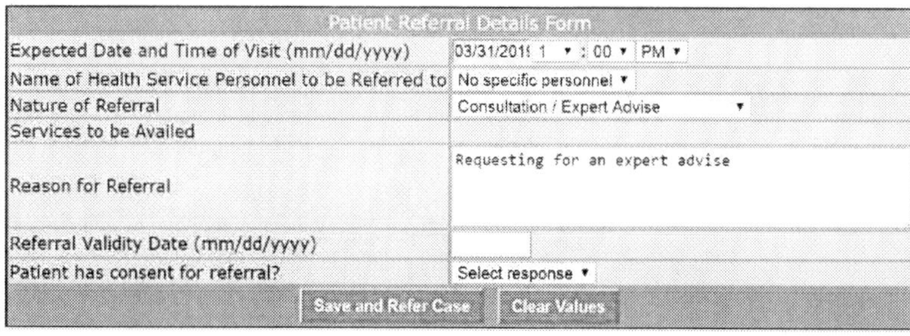

Patient Referral Details Form	
Expected Date and Time of Visit (mm/dd/yyyy)	03/31/201! 1 : 00 PM
Name of Health Service Personnel to be Referred to	No specific personnel
Nature of Referral	Consultation / Expert Advise
Services to be Availed	
Reason for Referral	Requesting for an expert advise
Referral Validity Date (mm/dd/yyyy)	
Patient has consent for referral?	Select response

Save and Refer Case Clear Values

Figure 6. Patient referral form

The receiving health facility shall receive a notification if a referral has been sent. The list of incoming referrals is shown in Figure 7. By clicking the "View Consultation" link, details of patient medical history are going to be visible. The health service provider at the receiving health facility can annotate the results of the consultation done on their facility as shown in Figure 8.

Case Referral Queue (INCOMING REFERRALS)

Date Referred	Name of Patient	Source of Referral	Expected Date and Time of Visit	Reason for Referral	Status	Action
04/18/2016 5:14 PM	Bonifacio, Andres	Gilberto O. Teodoro Memorial Hospital	04/23/2016 9:00 AM	request to conduct a PFT test on the patient	Referral Received (Consult Done)	View Details View Consultation

Figure 7. List of Incoming Referrals

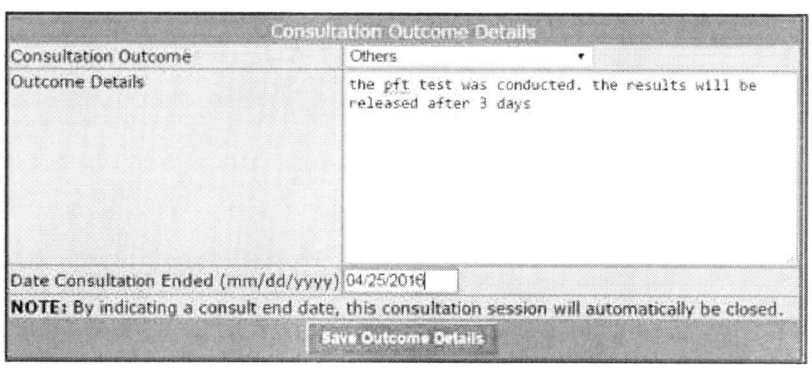

Consultation Outcome Details	
Consultation Outcome	Others
Outcome Details	the pft test was conducted. the results will be released after 3 days
Date Consultation Ended (mm/dd/yyyy)	04/25/2016

NOTE: By indicating a consult end date, this consultation session will automatically be closed.

Save Outcome Details

Figure 8. Updating of Consultation Details

5. Conclusion

Health information exchange via patient referral is demonstrated through a prototype electronic medical record called "Electronic Health Information Exchange and Referral System (ERS)". ERS EMR is based on a framework which features recording of data in local facility, referral of patient case, receiving of patient and monitoring of patient movement. Blockchain is used as a support tool to connect the sending facility to the receiving facility, integrate referral data into a central ledger, secure transaction using the prime256 ECDSA signature and verify the integrity of transaction to the intended receiving facility. Among the key features of ERS EMR is recording of medical information, search engine of participating health facilities and sending and receiving referrals. ERS EMR features were tested using simulated cases.

As a next step, ERS EMR is to be deployed into an actual referral practice. This would aid in validating the perceived benefits of the system and could contribute to its further enhancement.

References

[1] Kuo T *et al* 2017 Blockchain distributed ledger technologies for biomedical and health care applications *J. of the American Medical Informatics Association: JAMIA* vol **24** issue 6 pp 1211–20 DOI: 10.1093/jamia/ocx068

[2] Zhang P *et al* 2018 FHIRChain: applying blockchain to securely and scalably share clinical data *Computational and Structural Biotechnology J.* vol **16** pp 267–78 DOI: 10.1016/j.csbj.2018.07.004

[3] Azaria A *et al* 2016 MedRec: Using blockchain for medical data access and permission management *2nd Int. Conf. on Open and Big Data (OBD)* pp 25–30 DOI: 10.1109/OBD.2016.11

[4] NCI 2020 *Dictionary of Cancer Terms* Retrieved from https://www.cancer.gov/publications/dictionaries/cancer-terms/def/electronic-medical-record

[5] Abdekhoda M 2018 Determinant factors in applying electronic medical records in healthcare *Eastern Mediterranean Health J.* vol **25** no 1 pp 24–33 DOI: 10.26719/emhj.18.007

[6] MD Anderson Cancer Center 2020 *Electronic Health Records Definition and Functionality* Retrieved from https://mdanderson.libguides.com/c.php?g=451500

[7] Ahier B 2015 *FHIR and The Future of Interoperability* Retrieved from https://www.healthcareitnews.com/news/fhir-and-future-interoperability

[8] Mandel J *et al* 2016 SMART on FHIR: a standards-based, interoperable apps platform for electronic health records *J. of the American Medical Informatics Association: JAMIA* vol **23** issue 5 pp 899–908 DOI: 10.1093/jamia/ocv189

[9] Sadoughi F 2018 The impact of health information exchange on healthcare quality and cost-effectiveness. A systematic literature review *Computer Methods and Programs in Biomedicine* vol **161** pp 209–32 DOI: 10.1016/j.cmpb.2018.04.023

[10] Chen J *et al* 2018 A blockchain application for medical information sharing *2018 IEEE Int. Symp. on Innovation and Entrepreneurship (TEMS-ISIE)* DOI: 10.1109/TEMS-ISIE.2018.8478645

[11] Novikov S *et al* 2018 Blockchain and smart contracts in a decentralized health infrastructure *2018 IEEE Int. Conf. Quality Management, Transport and Information Security, Information Technologies (IT&QM&IS)* pp 697–703 DOI: 10.1109/ITMQIS.2018.8524970

[12] Fan K *et al* 2016 MedBlock: efficient and secure medical data sharing via blockchain *J. Med. Syst.* volume **42** issue 8 DOI: 10.1007/s10916-018-0993-7

[13] Gordon W *et al* 2018 Blockchain technology for healthcare: facilitating the transition to patient-driven interoperability *Computational and Structural Biotechnology J.* vol **16** pp 224–30 DOI: 10.1016/j.csbj.2018.06.003

[14] Acodez 2020 *12 Best Software development methodologies with pros and cons* Retrieved from https://acodez.in/12-best-software-development-methodologies-pros-cons/

[15] Gelembjuk R 2020 *OurSQL* Retrieved from https://github.com/gelembjuk/oursql

IOP Conf. Series: Materials Science and Engineering 1077 (2021) 012060 doi:10.1088/1757-899X/1077/1/012060

Developing Facebook Chatbot Based on Deep Learning Using RASA Framework for University Enquiries

Yurio Windiatmoko[1], Ridho Rahmadi[2], Ahmad Fathan Hidayatullah[3]

Department of Informatics, Universitas Islam Indonesia, Yogyakarta, Indonesia

E-mail: [1]yurio.windiatmoko@students.uii.ac.id, [2]ridho.rahmadi@uii.ac.id , [3]fathan@uii.ac.id

Abstract. Smart systems for universities powered by artificial intelligence have been massively developed to help humans in various tasks. The chatbot concept is not something new in today's society which is developing with recent technology. College students or candidates of college students often need actual information like asking for something to customer service, especially during this pandemic, when it is difficult to have an immediate face-to-face meeting. Chatbots are functionally helping in several things such as curriculum information, admission for new students, schedule info for any lecture courses, students grade information, and some adding features for Muslim worships schedule, also weather forecast information. This Chatbot is developed by deep learning models, which was adopted by an artificial intelligence model that replicates human intelligence with some specific training schemes. This kind of deep learning is based on RNN which has some specific memory savings scheme for the deep learning model, specifically this chatbot using LSTM which already integrates by RASA framework. LSTM is also known as Long Short Term Memory which efficiently saves some required memory but will remove some memory that is not needed. This Chatbot uses the Facebook platform because the Facebook users have already reached up to 60.8% of its entire population in Indonesia.

Keywords. Chatbot; Deep Learning; Chatbot Framework; RNN; RASA; LSTM

1. Introduction

Chatbot is the easiest system to access by any users; it is an artificial intelligence technology for communicating through natural language with humans. Chatbot makes it possible to understand human language through natural language processing. Language is a social and cultural reciprocal for a natural protocol. It is time for computer machines to add features to understand humans along with the increased performance of the recent computer machine hardware [1].

College students or candidates of college students often need actual information, e.g., asking for something to customer service, especially during this pandemic, when it is difficult to have an immediate face-to-face meeting. Therefore, a chatbot is needed for answering questions quickly, correctly, and is available 24 hours [2]. Functionally, this Chatbot is helping in several things such as curriculum information, admission for new students, schedule info for any lecture courses, students grade information, and some adding features for Muslim worships schedule, also weather forecast information.

To help solve the problems above, we propose a chatbot using deep learning models. Deep learning consists of high-level abstract modeling algorithms on data using transformation of nonlinear functions arranged by many layers and deeply [3]. Deep learning is also a branch of machine learning, inspired by the neural network of the human brain which is represented by layers on the neural network [4]. The deep learning method used in this case is RNN based called LSTM (Long-Short Term Memory) which has some specific memory savings scheme for the deep learning model. LSTM is able to efficiently save some required memory and also will remove some memory that is not needed. We also use a framework called RASA to build this chatbot application which already integrates LSTM as the deep learning method. The implementation of deep learning on chatbot requires the role of the chatbot framework which is for unifying and pipelining the deep learning model. The sense of a chatbot framework is to implement its conversation flow. RASA works on two main procedures namely RASA NLU and RASA Core. RASA is a library of tools from the Python programming language that is open source for building conversational software. The aim and philosophy of the initiators of the RASA chatbot framework are to create machine learning-based dialogue management that provides ease of use in terms of implementation, as well as bootstrapping, by starting the stages from existing sources to create something more complex and in a more efficient way, even with minimal initial training data [5].

2. Literature Review

There are two main approaches to generate a chatbot response. If the chatbot uses a traditional approach then it is hard-coded rule-based templates and rules that process for responses. However, nowadays many new and interesting approaches allow deep learning approaches to emerging. Neural network models are trained with several data to study the process of generating responses that are grammatically relevant and in terms of user speech intent [6].

These are some previous works related to chatting systems or chatbot for universities. Online Chatting System for college inquiries using a knowledgeable database by Bathe, Malusare, and Kolpe is doing pattern matching to perform information retrieval on chatbots. It has not already used the deep learning approach, so that is still too rigid by rules for users to make inquiries. But, detailed working steps in this research are very good, there such a UML, and various kinds of process diagrams [7]. Erasmus the AI chatbot, Erasmus is a chatbot on Facebook which is used to answer the queries related to the college information. Designing end-to-end systems using cloud services, starting from api.ai (Dialogflow), Mlab (MongoDB cloud), IBM Bluemix (webhook API), scrapper import.io to minimize the scripting code process, but what's happening here, it is all because there are too many diverse cloud services, there is quite a long latency between the services [8]. Then, Eaglebot a Chatbot Based Multi-Tier Question Answering System for Retrieving Answers from Heterogeneous Sources Using BERT. It is a scalable chatbot system using 3 route selection methods with the main framework using Dialogflow, then several completion methods from document retrieval and document reader [9]. Eaglebot which serves to answer various questions commonly asked by students in the university domain. However, the Eaglebot application still has some limitations by Dialogflow chatbot framework which has some request limits, if it wants to unlimit requests you need to subscribe for the member privileged.

Intelligent Chatbot System Based on Entity Extraction Using RASA NLU and Neural Network [10] by Anran Jiao was comparing performance between RASA NLU stack with Neural Network Classifier and Entity Extractor from scratch. This research was told that the RASA NLU method still superior to extract all the entities and classify user speech intent. It's also quite comprehensive research but this research only depends on a free API for response completion and not using its own database. Compared to all those previous works, this research will be more reliable by using deep learning for chatbot natural language understanding and combining free API with it's own database for response completion. So it won't need for pattern matching, not using cloud provider service-based, and not only depends on a free API service so that customization will be easier, scalable and no limit request for user interaction.

Facebook users in Indonesia August 2020. There were 166.500.000 Facebook users in Indonesia in August 2020, which accounted for 60.8% of its entire population. The majority of them were men - 54.6%. People aged 25 to 34 were the largest user group 59.000.000. The highest difference between men and women occurs within people aged 25 to 34, where men lead by 9.000.000. So that this Chatbot uses the Facebook platform as the user interaction medium.

IOP Conf. Series: Materials Science and Engineering 1077 (2021) 012060 doi:10.1088/1757-899X/1077/1/012060

3. Data and Methods

All the processes in this research are following this diagram process flow below.

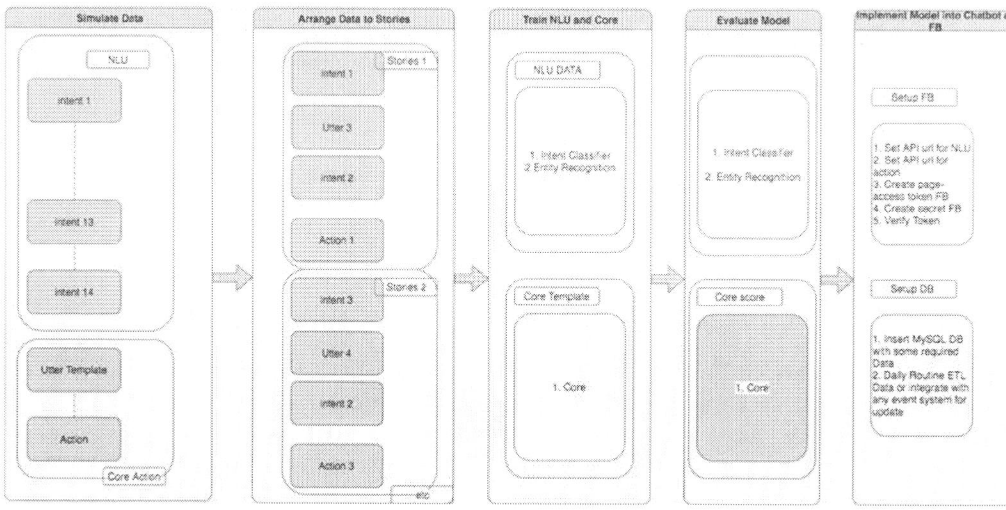

Figure 1. Workflow chart chatbot

Figure 1 describes flow and methods of the research process carried out. Beginning with simulating the data based on common dialogue interaction along with grouping for each sentence into the intent topic, utter chatbot response template, and action. Especially in the intent topic which has been proposed as the featured chatbot also needs defining Entities there. The difference between utter and action here is basically that utter is just a sentence with an expression of the general response, whereas action is utilized when it's response needs some completion based on a database or API, also it needs wrapping-up as the class actions on RASA. At this stage, various data previously seen and considered from general reference schemes or conversation scenarios for each chatbot feature such as communication flow on student questions about the curriculum, admission for new students, schedule for lectures, student grade, Muslim Worships schedule, and weather forecast also adding up with some common dialogue intent like greetings and so on. Then, arrange those intent, utter and action in such a way that it corresponds to the flow of stories and common entities in each conversation sentence. After all the data is already set up, then train those data using RASA Trainer. Check and recheck after training was finished by evaluating the trained model. If the evaluation score obtained has reached the required expectations then just integrate that model with credential setup configuration after the Facebook developer account has been created.

3.1. Framework

Chatbot framework is used for Bootstrapping all the chatbot engine modules. Especially for the Machine Learning Models. Below is the Architecture flow of Rasa framework used.

The architecture or dialogue management flow used in this research approach is as shown in Figure 2 which is the architecture of the chatbot stack framework itself. The first stage of the message is received and forwarded to the interpreter, namely Rasa NLU to extract intents, entities, and other structured information. Both interpreter or tracker is tracking, detecting, and maintain the status of the conversation context through the message notifications it has received. The third policy or policy manager receives context status from the tracker. The fourth policy or policymakers choose which action will be taken next. The five actions or actions are recorded by the tracker. These six actions are executed by sending a message to the user. Seventh, if the action that has been executed is ignored or ignored by the user at a certain time, the process returns to the third step.

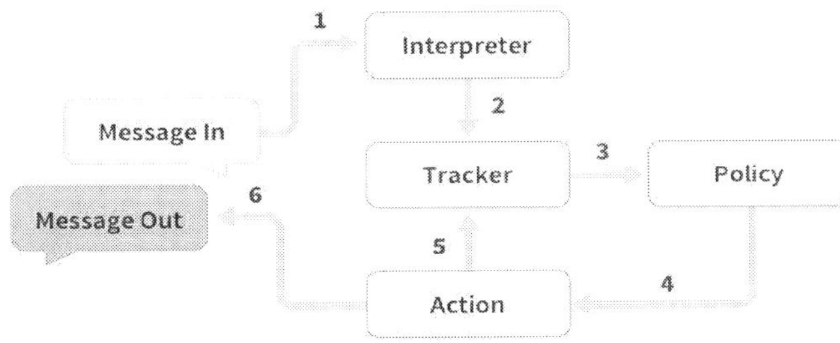

Figure 2. RASA stack framework

3.2. Methods For Training
Figure 3 depicts the intent entity training flow. Our chatbot system has 2 modules and each module is trained separately, those modules are NLU and Dialogue Policy or Dialogue Management. NLU or the interpreter comprises two component models inside. Those are Intent Classifier and Entity Extractor.

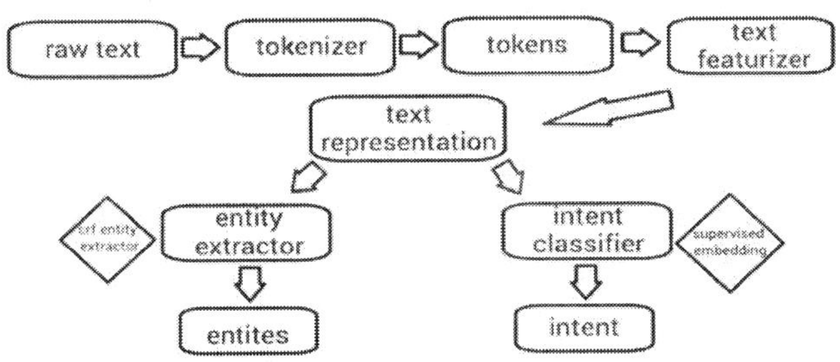

Figure 3. Intent entity training flow

Specifically, this chatbot used Supervised Embeddings as Intent Classifier, which is also known as Embedding Intent Classifier. The Embedding Intent Classifier embeds user inputs and intent labels into the same space. Supervised embeddings are trained by maximizing similarity between them. This algorithm is based on StarSpace [11]. However, in this implementation, the loss function is slightly different and additional hidden layers are added together with dropout. This algorithm also provides similarity rankings of the labels.

The training configuration to build this intent classifier model is 300 for epochs, then for the hidden layers sizes at feed forward neural networks are 256 at h1, 128 at h2, and 14 at output neural network label intent class. Embedding dimension used for training is 20, this is multiple embedding layers inside the model architecture. For example, vector of '__CLS__' token and it's intent is passed on to an embedding layer before they are compared and the loss is calculated. config for batch size is (64, 256) which the initial and final value for batch sizes. Batch size will be linearly increased for each epoch. Learning rate is set as 0.001 for this training config.

Entity Extractor used for this chatbot is crf entity extraction [12] which method is suitable for custom entities extractor. It's component implements a conditional random field (CRF) to do named entity recognition. CRFs can be thought of as an undirected Markov chain where the time steps are words and the states are entity classes. It's features are from the words like capitalization, POS tagging, etc. Those Features are give probabilities to certain entity classes, as are transitions between neighbouring entity tags, the most likely set of tags is then calculated and returned. Training configuration for CRFEntityExtractor has some list of features to use. However, it can be set and chosen by any requirements. Specifically this chatbot uses a list of features in such 'lowercase' to check whether the token is lower case, 'uppercase' to check whether the token is upper case, 'title' to check whether the token starts with an uppercase character and all remaining characters are lowercase, 'digit' to checks whether the token contains just digits, 'prefix5' to take the first five characters of the token, 'prefix2' to take the first two characters of the token, 'suffix5' to take the last five characters of the token, 'suffix3' to take the last three characters of the token, 'suffix2' to take the last two characters of the token, 'suffix1' to take the last character of the token, 'bias' to add an additional "bias" feature to the list of features.

As the featurizer is moving over the tokens in a user message with a sliding window, it defines features for previous tokens, the current token, and the next tokens in the sliding window. so the features are as [before, token, after] the array.

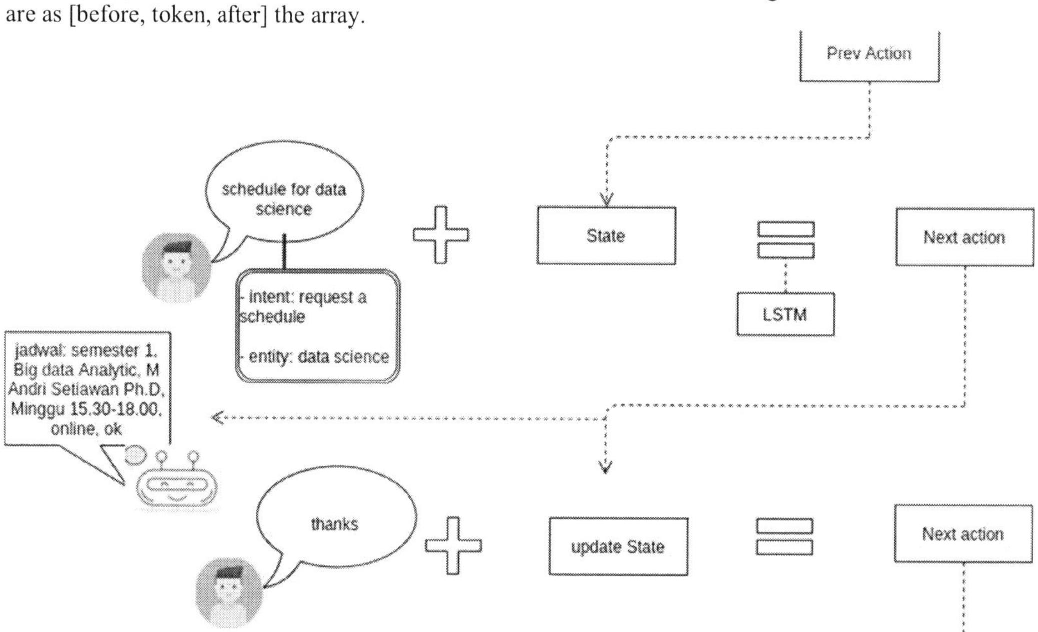

Figure 4. Dialogue management component flow

As for the Dialogue Policy or Dialogue Management, it works by creating training data from the stories and training a model on that data. Stories are like some scenario or scheme from the common topic in any chat flow that we design or we expected. Chat Response works by creating training data from the stories and training a model on that data as described in Figure 4. Dialogue Policie decides which action to take at every step in the conversation. There are different policies to choose from, and it includes multiple policies in a single Agent or chatbot. At every turn, the policy which predicts the next action with the highest confidence will be used.

This chatbot already configured the policies by memoization Policy and neural network Policy. The execution of the policy will be based on the priority; highest priority policy will be executed first and so on. Priority is calculated on the basis of confidence score of each policy, which has a higher score given higher priority. The MemoizationPolicy just memorizes the conversations in training data. It predicts the next action with confidence 1.0 if this exact conversation exists in the training data, otherwise it predicts None with confidence 0.0. A Neural Network Policy implemented in Keras to select the next action. The default architecture is based on an LSTM [13]. It is configured with masking layer (input, 5, 32), lstm units (32) dropout=0.2, dense layer (19) then 100 training epochs.

3.3. Connector Module Platform

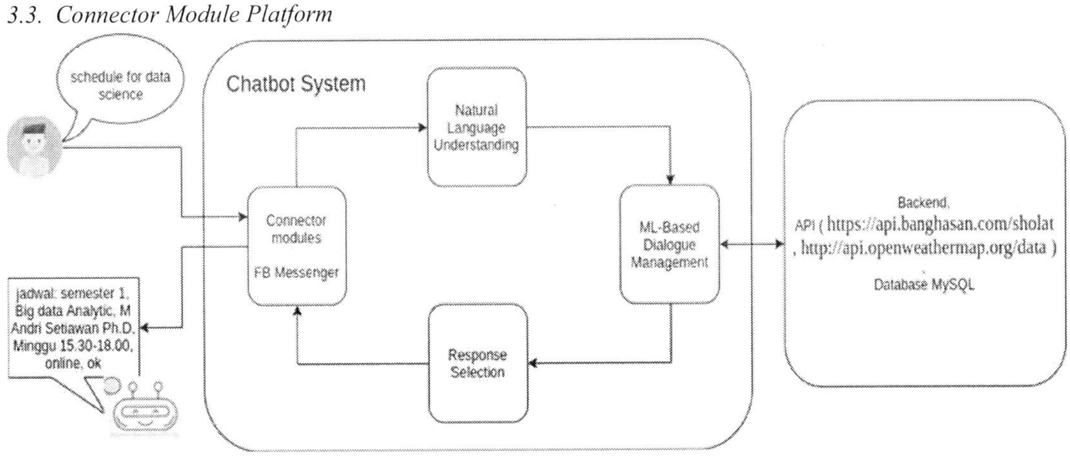

Figure 5. Flow of chat system

Figure 5 illustrates the Facebook Messenger as a connector module platform, first it needs to set up a facebook page and app to get credentials to connect to Facebook Messenger. Once already have all of it, add these to the credentials.yml file. Then set up a Webhook and select at least the messaging and messaging_postback subscriptions. Insert the callback URL which will look like https://<HOST>/webhooks/facebook/webhook. Insert and Verify Token which has to match the verify entry in the credentials.yml file. So then, webhooks allows the chatbot to receive real-time HTTP notifications of changes to specific objects and postback to messaging.

4. Result and Discussion

This chatbot development research that uses deep learning models, it is expected to provide the best performance for chatbots systems through this deep learning model. Furthermore, the representation results of the performance evaluation of the deep learning model on the chatbot, namely intent classification and entity extraction at NLU and Dialogue Policy on the dialogue management chatbot system are measured and evaluated by precision score, recall score and F1-score [14].

Precision may be defined as the probability that an object is relevant given that it is returned by the system or here as the model predicted, while the recall is the probability that a relevant object is returned based on ground truth of each label data. Then, F1-score is needed for seeking a balance between precision and recall.

$$Precision = \frac{True\ Class}{True\ Class + False\ Predicted\ Class} = \frac{True\ Class}{Total\ Predicted\ Class} \tag{1}$$

$$Recall = \frac{True\ Class}{True\ Class + False\ Predicted\ Oth\quad Class} = \frac{True\ Class}{Total\ Actual\ Class} \tag{2}$$

$$F1 \quad = 2 \times \frac{Precision \times Recall}{Precision + Recall} \tag{3}$$

Immediately, that precision shows how precise or accurate the model is out of those predicted Class, how many of them are actual Class. Recall calculates how many of the Actual Class our model captures through labeling it as those Class. Ultimately, F1-score is utilized to find a balance between precision and recall.

Table 1. Metrics evaluation intent classifier 1

	intent agreed	intent requests a schedule	intent goodbye	intent Muslim greeting	mood unhappy	intent only request a schedule	intent weather forecast	intent request a worship schedule
precision	1	1	1	0.75	1	1	1	1
recall	1	1	1	1	1	1	1	1
f1-score	1	1	1	0.85	1	1	1	1
data count	6	100	3	3	3	4	5	11
confused with	-	-	-	-	-	-	-	-

Table 1 shows the metrics evaluation intent for classifier 1. Here what needs to be underlined is the Intent Muslim greeting class where the model can predict accurately to the amount of truth data for that class which the recall score is 1, but the model still predicts the intent Muslim greeting in other intent classes actual which the precision score is 0.75 namely intent greeting class.

Table 2. Metrics evaluation intent classifier 2

	mood happy	intent greeting	intent thanks	intent request grade score	intent refuse	challenge bot	avg	total
precision	1	1	1	1	1	1	0.982	13.75
recall	1	0.85	1	1	1	1	0.98	13.85
f1-score	1	0.92	1	1	1	1	0.98	13.77
data count	4	7	2	10	3	4	11.78	165
confused with	-	'intent Muslim greeting: 1	-	-	-	-	-	-

In table 2, we describe the metrics evaluation for intent classifier 2. While for the intent greeting itself, here is get a 1 precision score where the Intent greeting is sometimes not detected by the model on the contrary intent Muslim greeting is even detected because it uses the same word phrase, which is 'assalamu alaikum' but this is not a problem because the chatbot template utter will answer the same thing as greeting utter.

The table 3 provides a detailed explanation of the precision and recall calculations of the two classes that are often misclassified. It can be seen from table 3 that the precision equals to 1. It shows that our chatbot system is able to detect the intent greeting precisely. As for the intent muslim greeting, the precision is 0.75.

Table 3. Confusion matrix between intent muslim greeting and intent greeting

		predicted			
		intent Muslim greeting	intent greeting	data count	recall
actual	intent Muslim greeting	3	0	3	1
	intent greeting	1	6	7	0.85
	total predicted	4	6	f1-score	
	precision	0.75	1	0.85	0.92

Table 4. Metrics evaluation entity recognition

	name	concentration study program	city	NIM	avg	total
precision	1	1	1	1	1	4
recall	1	1	1	1	1	4
f1-score	1	1	1	1	1	4
data count	10	170	12	3	48.75	195

In table 4, the entity class metrics evaluation table has shown the best performance from the model to be able for extracting the entity labels.

Table 5 illustrates the dialogue policy performance which has shown the best performance of the model in making decisions for the chatbot response. Student conversation flow requesting a lecture schedule from the chatbot has the following process. First, the chat is started with the greetings made by students to the chatbot, and the chatbot detects the greeting intent from users and replies with utter replies to the greeting. Furthermore, students go to the point to ask about their lecture schedule, and Chatbot detects it as an intent to request a schedule from users, then replies with an utter asking about the concentration of the student's study program. Then the student tells the chatbot the concentration of his study program, namely 'fd' or digital forensics, and the chatbot considers this to be the concentration entity of the study program and continues the query to MySQL DB as a completion to get lecture schedule data.

As for the alternative, the process of requesting schedule also may ask directly for a specific study program concentration schedule, for example, 'schedule ds' and the chatbot will detect the intent to request the schedule as well as to detect the concentration entity of the study program 'data science'. And so chatbot continues the query to MySQL DB as a completion to get lecture schedule data.

Figure 6 shows an example of conversation from the student who asks about grade score. This chat flow is about requesting a grade score is almost the same as requesting lecture schedule flow, but only with different intents and entities, also different actions to show a list of grade scores by continuing the completion query to MySQL DB as a completion to get lecture schedule data. There is also a conversation flow of requesting the Muslim worship schedule. Beginning by asking directly for a specific intent about asking the Muslim worship schedule. Then the chatbot asking about a specific location as an entity, after that continuing the completion query directly to API https://api.banghasan.com/sholat/format/json/jadwal/kota/{city}/tanggal/{date.now}[1].

Table 5. Confusion matrix dialogue policy decision

	action schedule list	action listen	utter did it help	utter good bye	utter salam replies	utter say hello	utter cheers	utter im bot	utter happ y	utter asked concentrat ion study program	utter thank you
action schedule list	4	0	0	0	0	0	0	0	0	0	0
action listen	0	27	0	0	0	0	0	0	0	0	0
utter did it help	0	0	2	0	0	0	0	0	0	0	0
utter goodbye	0	0	0	2	0	0	0	0	0	0	0
utter salam replies	0	0	0	0	3	0	0	0	0	0	0
utter say hello	0	0	0	0	0	6	0	0	0	0	0
utter cheers	0	0	0	0	0	0	2	0	0	0	0
utter im bot	0	0	0	0	0	0	0	1	0	0	0
utter happy	0	0	0	0	0	0	0	0	2	0	0
utter asked concentrat ion study program	0	0	0	0	0	0	0	0	0	2	0
utter thank you	0	0	0	0	0	0	0	0	0	0	5

Then a conversation flow of requesting a weather forecast for today. Beginning by asking directly for a specific intent about the weather forecast for today. Then the chatbot asking about a specific location as an entity, after that continuing the completion query directly to API. http://api.openweathermap.org/data/2.5/weather?q={Yogyakarta.id}&APPID={your_app_id}&units=metric[2], then translate the weather forecast description into Indonesian Language.

[1] "Fathimah API." https://api.banghasan.com/. Accessed 16 Sep. 2020.
[2] "Weather API - OpenWeatherMap." https://openweathermap.org/api. Accessed 16 Sep. 2020.

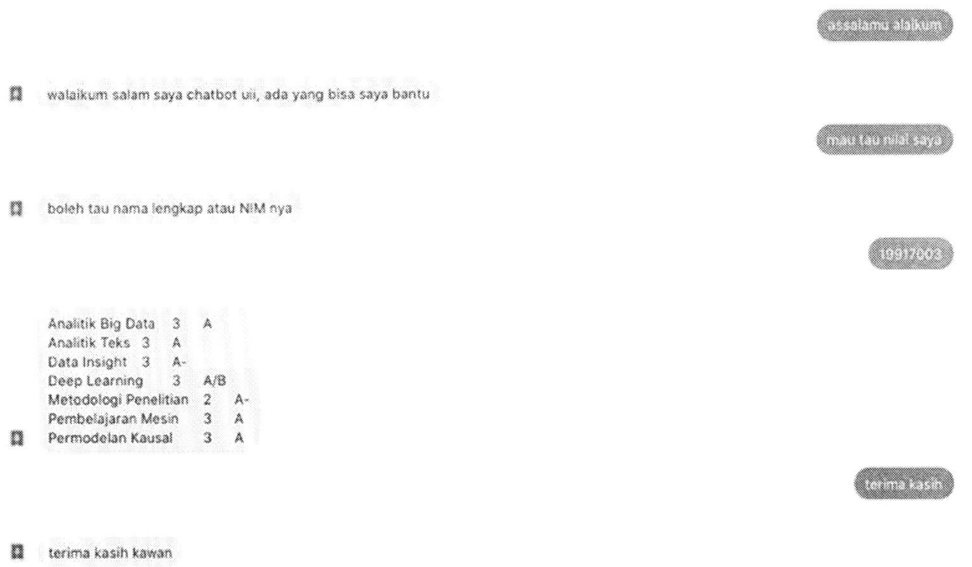

Figure 6. Chat example

5. Conclusion and Future Work

In this work, we have developed a chatbot system to help answer all queries asked by people to the university. Because of the chatbot has not been used operationally on campus, the effectiveness and the ease of use for users cannot be measured yet. So the chatbot performance is only measured through the accuracy of generating sentences that can be read correctly to reply the user. Moreover, the chatbot has a response time from the inference model which is fairly fast within less than a second. In addition, the evaluation results of the chatbot model show fairly good results near the perfect score of precision, recall and F1 which has been explained in the previous section. The scores of these numbers all reach the value of 1, it's just that in the Muslim greeting intent is 0.75 in precision score and 0.85 in f1-score, also in the intent greeting 0.85 in recall score and 0.92 in f1-score.

This research is, however, still in the first stage of development within fairly sufficient simulate data. We also have implemented deep learning models for chatbots, especially by implementing the intent classifier and entity recognition on rasa nlu also dialogue policy on rasa core, then integrating it with Facebook Messenger.

Furthermore, this chatbot needs to be developed by extending the dialogue data and more features for helping the automating task in the universities, then by directing towards the open domain chatbot, for more extensive implementation in deep learning Chatbot, to be the 'chatbot I know everything'. Also the needs for implementing chatbot within its own platform, no limitations then and more customizable.

References

[1] Yoo S Y and Jeong O R 2020 Auto-growing knowledge graph-based intelligent chatbot using BERT *ICIC Express Lett.* vol **14** no 1
[2] Al-fakhri S *et al* 2019 Aplikasi chatbot informasi Kampus Polban menggunakan aplikasi LINE Messenger *Prosiding Industrial Research Workshop and National Seminar* vol **10** no 1
[3] Dadang W 2018 Memahami kecerdasan Buatan berupa deep learning dan machine learning *Prosiding Industrial Research Workshop and National Seminar* vol **10** no 1
[4] Santoso A and Ariyanto G 2018 Implementasi deep learning berbasis keras untuk pengenalan wajah *Emit. Jurnal Teknik Elektro* vol **18** no 1

[5] Bocklisch T, Faulkner J, Pawlowski N and Nichol A 2017 Rasa: open source language understanding and dialogue management *Preprint arXiv:1712.05181*
[6] Csaky R K 2019 Deep Learning Based Chatbot Models *Preprint arXiv:1908.08835*
[7] Prashant B P, Anil M S and Dilip K M 2017 *Online Chatting System for College Enquiry using Knowledgeable Database* undergraduate project (Maharashtra: Savitribai Phule Pune University)
[8] Thakkar J, Raut P, Doshi Y and Parekh K 2018 Erasmus AI chatbot *Int. J. Comput. Sci. Eng.* vol **6** no 10
[9] Rana M 2019 A *Chatbot Based Multi-Tier Question Answering System for Retrieving Answers from Heterogeneous Sources Using BERT* thesis (Georgia: Georgia Southern University)
[10] Jiao A 2020 An Intelligent chatbot system based on entity extraction using RASA NLU and neural network *J. Phys. Conf. Ser.* vol **1487** no 1
[11] Wu L, Fisch A, Chopra S, Adams K, Bordes A and Weston J 2018 StarSpace: embed all the things *32nd AAAI Conf. Artif. Intell. (AAAI 2018)*
[12] Lafferty J and Mccallum A 2001 Conditional random fields probabilistic models *Proc. of the 18th Int. Conf. on Machine Learning 2001 (ICML 2001)*
[13] Hochreiter S and Schmidhuber J 1997 Long short-term memory *Neural Comput.* vol **9** no 8
[14] Goutte C and Gaussier E 2005 A Probabilistic interpretation of precision, recall and F-Score, with implication for evaluation *27th European Conf. on IR Research (ECIR 2005)*

EEG Neurofeedback Training Among Adult with Attention Deficit: A Review Article

Eshrak Alaros[1], Dini Oktarina Dwi Handayani[2], Hamwira Yaacob[3], Muharman Lubis[4]

[12]School of Computer Science & Engineering, Faculty of Innovation & Technology, Taylor's University, Selangor, Malaysia
[3]Computer Science Department, Kulliyyah of Information and Communication Technology, International Islamic University Malaysia, Selangor, Malaysia
[4]Information System, Faculty of Industrial Engineering, Telkom University, Indonesia

E-mail: [1]eshrakabdulqaderali@sd.taylors.edu.my, [2]dinioktarina.dwihandayani@taylors.edu.my, [3]hyaacob@iium.edu.my, [4]muharmanlubis@telkomuniversity.ac.id

Abstract. Neurofeedback is considered a promising therapy for boosting cognitive skills, which 10% of the US medical sector is using as a new training method among patients. Attention is a significant mechanism for learning and is essential for life in general. As the person grows up the level of responsibilities also increases, which needs high concentration in doing normal activities. To succeed in any activity, neurofeedback therapy needs to check brain waves to locate the problem in the brain and train it. This paper aims to review the attention deficit treatment based on neurofeedback using a systematic review method. The results found that the common features are the area of EEG (Frontal, central) in the brain, type of tools and software, time of training and the impact upon the trainer. Neurofeedback showed improvement of brain attention-ability for a longer time as well as reduction in recovery time in case of a disturbance interruption.

1. Introduction

Brain waves are made up of a set of signals divided into five waves, namely delta, theta, alpha, beta, and gamma, which are classified based on frequency [29]. Each wave is linked with specific mental states, with varying values. The classifications of brain waves using electroencephalographic (EEG) separated bands based on frequency found that beta waves are used for focused mental work, attention, and sensory data processing. EEG is a non-invasive neuroimaging technique that has been employed broadly to study many resting states and cognitive processes in the brain [21]. Knowing the elements of the standard EEG is a prerequisite for obtaining expertise in analyzing an abnormal recording and describing the frequency bands and individual waveforms for the waking and sleeping states among adults [23]. Figure 1 shows a typical example of using and utilizing EEG.

In this figure, a standard training (in this figure is a video playing) takes place, using EEG/BCI equipment attached to the head capturing brain signals. Later those captured signals will be processed to figure out patterns. The brain reacts to different stimuli, and the captured brainwaves are usually divided into different frequency rhythms by EEG signals. These band waves are ever-present in different portions of the brain [16]. To capture EEG signals, there is a 10-20 internationally standard placement system [16]. For example, to capture the Emotiv Epoc+ signal, the neuro-sensor headset system is the most used device. The typical electrodes positions over the scalp are shown in figure 2, and there are 16 electrodes [31]. The sampling rate of the band using Emotiv Epoc+ was 128 Hz, and the EEG data are segmented using one second time interval [16].

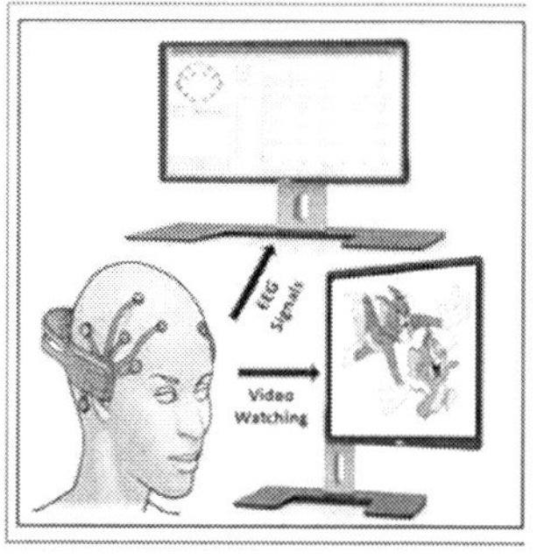

Figure 1. An example of EEG usage (source: [16], [48])

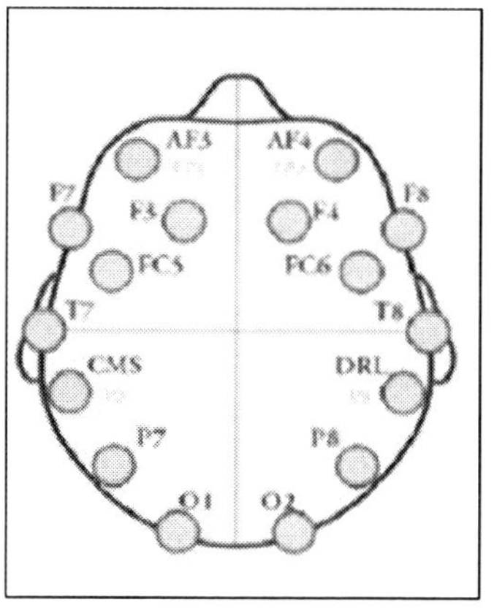

Figure 2. Electrode placement system using Emotiv Epoc+ [16]

2. Attention Study using Neurofeedback

Many approaches and methods have been used to conduct experiments related to attention using EEG. Randomized controlled trials (RCTs) are techniques that employ EEG with those diagnosed with ADHD. The trails involve theta and beta bands training, which target increasing beta band (13–20 Hz) activity while reducing theta band (4–8 Hz) activity. The trails also involve slow cortical potential (SCP) training, which is linked to bidirectional regulation of cortical excitability. Thomas and Vinod [44] investigated the effect of neurofeedback on enhancing the attention and cognitive function of healthy people using games. The researchers experiment took five days, in which they used Neurofeedback-based game played using EEG brain signals related to recognition. They measured the level of attention using entropy values of EEG and compared the cognitive test results. The findings revealed the benefits of utilizing neurofeedback using games in the treatment of attention disorders or improving cognitive skills. Yildirim & Varol [48] used EEG-based games, mainly 'Matching Mind Math' to enhance students' attention. When the student starts the timer, then the average attention is reset, to measure average attention rates. The student maps mathematical equations with relevant answers with a drag and drop button game. The result is then used for further analysis to measure the attention level and its duration. The measurement method used a student's mood to study his/her mediation and attention level.

The frequencies used in detecting and enhancing attention using EEG feature extraction are theta, alpha, and beta frequencies. For instance, Vasiljevic and de Miranda [46] and Thomas and Vinod [45] employed EEG in alpha waves only. They measured the differences in the alpha waves as indicator for increasing in the attention levels. Their findings showed a positive correlation between attention level and game difficulty while some studies focused on in beta waves as in [18, 35, 52]. However; Theta and alpha frequencies get less focused on enhancing attention [17]. Regarding the tools used with EEG, the most used tool is Emotiv Epoc. Another tool, Neurosky/MindWave, is cheaper and easy to use- as it has only one channel for collecting signals. However; different tools with different areas and channels give different results for attention. Meanwhile the prior studies focused on detecting attention by checking the attention level using different stimuli [5]. For instance; Thomas [43] trained participants to enhance their attention level by putting participants under pressure-via exposing them to many questions that they should answer as fast as possible. He found that the attention score increases. In continuing with training for better attention, Ochi et al., [35] employed a game to keep participants' attention all time they used a stricter level of games in each session

The frequency chosen; for EEG during training time; was different for various purposes. EEG-based therapy most of the time rely on increase/ decrease the frequency. Beta frequency rises with healthy adults and reduces with attention deficit group. The arguments among those studies was regarding the length and periodically of the training session; in overall, the arguing to do it with less than one week with 30 minutes per session. To conclude that to produce measurable results in biofeedback therapies, a mean of 30-minute session; three day a week is needed. [50]. All previous work discussed in this study used various stimulation to enhance attention. Sometimes, they employ visual or audio effects, while others put some pressure to motivate participants to increase their concentration. However, most of the studies were observing not interfering, i.e., they try to measure waves related to attention and the various changes occurring during the experiment session but without interfering. On other hand, the theta/beta protocol cannot be used with high expectations because it is not a good index with maturation and it cannot solely be used to describe attention disorder [21]. Thus, it could be said that theta/beta can be an extra diagnostic marker mostly for children but not for adolescents. The theta/beta protocol analysis showed that children could be differentiated from normal people by absolute theta, but this was not the case for adults [24]. This proves the significant role of EEG in diagnosing attention. Some related studies summarized using the different methods without using EEG as a measurement tool to improve the focus. The methodology used no electronic devices whatsoever attached to participants to train their attention. They either used surveys [8], results from playing games [34, 36], or used a questionnaire [2]. Some emphasize on physical activities to improve not only health but also attention [2, 36].

3. Research Gap

The attention of people is influenced by the distraction and the temptations surround them, which the students are among the most affected people due to their interest in social media and video games. Researchers employ EEG-based techniques as they are accurate in reading brain activities-related waves. For the sake of finding threshold values for (alpha, beta, and theta) that tell whether a student has attention issues or his/her attention improved, this study conducted an extensive literature review to students' attention improvement studies. The results have shown that currently, researchers use EEG-based tools and techniques to study the variances in waves emit from the brain, yet, this after employing other methods to train students and enhance their attention. There is inconsistency regarding the most significant wave type and frequency range that describes whether is attention issues or improved. EEG was found to be a quick and objective tool for evaluating attention levels [5]. Yet, EEG works with brain waves that are fairly weak, subject to interference and noise, vary depending on the individual, the subject's physiological status and time [5]. Moreover; a simple action while EEG tool capturing the waves such as fatigue, pain and even closed eyes probably can distort the measurements [26]. Consequently; this brings difficulties to collect brain waves cleared of errors. Due to that investigating techniques employed in prior studies to overcome those challenges; in addition to values that claimed by researchers to be the threshold and can be standardized.

Many studies have designated that different brain activity frequencies in the frontal lobe are highly associated with personality, emotion, attention, rationality, and creativity [39]. EEG signals can be studied in the time, frequency domains, and the most discussed features are waveform, band and band power in the alpha, beta, and theta bands [5]. Researchers have focused on various wave types with multiple justifications. They mainly focused on changes in the wave's strength during the session. For instance; [15] reported that an upsurge in attention marked by an increase in alpha waves and a decrease in theta waves. Moreover, many studies indicate that an increase in beta waves and a decrease in the theta band is a significant sign of good attention/ concentration and memory [53]. Another study reported that alpha amplitudes an increase in frequency, meanwhile means high attention or deep concentration [46]. Sauseng et al., [37] have found that an increase in theta-waves, claiming that means focused attention [5]. Due to no an agreement among researchers regarding using specific wave types (alpha, beta, theta), specific frequency and/or regions of brain to measure people's attention level; many studies focused on power densities in alpha, theta and beta frequencies such as in [5] with attention. Researchers tend to measure attention improvement using several blocks of the games where the first blocks look like training, other blocks to measure attention improvement. Simultaneously, they record waves emit from the brains of the players. Later and using t-test and ANOVA with players' scores to find positive differences. Then after those statistical procedures confirm positive differences in players' ratings, they studied the changes in alpha, beta, and theta. The results presented in table 1, show entirely different and sometimes contradicted findings regarding the changes in waves particularly alpha and theta. Yet, there is no specific range of values that may not use as a trigger for attention problems or improvement. Researchers again have quite a long time to come up with stable and valid EEG-based measurements regarding attention. Studies yet rely a little bit on brain waves reading to confirm attention-related problems, which is most of the time using it as a supplementary method.

Table 1. Monitoring Differences in waves to prove an improvement in attention

Author	Attention improvement signs
(Chiang et al., 2017)	Calculated the relationship between each feature (theta, beta, alpha) and attention
(Zolubak et al., 2018)	An increase in beta waves and a decrease in theta waves
(Sauseng et al., 2007)	Theta waves up and beta down
(Kawatake et al., 2017)	Alpha amplitudes decrease and increase in frequency
(Lansbergen, Van Dongen-Boomsma, Buitelaar, & Slaats-Willemse, 2011)	Increase in theta waves
(Yao et al., 2017)	Alpha power increases
(Zamora Blanón et al., 2016)	Different in alpha and beta declines
(Lim et al., 2019)	High alpha power, beta power and theta power
Author	Attention improvement signs
(Chiang et al., 2017)	Calculated the relationship between each feature (theta, beta, alpha) and attention
(Zolubak et al., 2018)	An increase in beta waves and a decrease in theta waves
(Sauseng et al., 2007)	Theta waves up and beta down
(Kawatake et al., 2017)	Alpha amplitudes decrease and increase in frequency
(Lansbergen, Van Dongen-Boomsma, Buitelaar, & Slaats-Willemse, 2011)	Increase in theta waves
(Yao et al., 2017)	Alpha power increases
(Zamora Blanón et al., 2016)	Different in alpha and beta declines
(Lim et al., 2019)	High alpha power, beta power and theta power

4. The Attention Deficit

Attention issues link mostly to paying less concentration or recalling information in time whenever needed [17]. Attention Deficit (AD) is a condition characterized by the presence of a variety of symptoms that cannot be diagnosed without clinical examination [49]. Some of the symptoms are related to memory, inattention, impulsivity, executive function, trouble with multitasking, minor planning, low frustration tolerance, frequent temper swings, troubleshooting and completion of obligations, and trouble managing stress [38]. These symptoms vary in magnitude and degree from one person to another and need daily activity to test the level of disorder [12].

4.1. Attention Representation

Neuroglia and neurons are two types of cells [29] in the brain, which the function of neuroglia cells is to back up the neurons to supply them with the required oxygen and nutrients [29]. On the other hand, the neurons are in charge of information transmission through nerve impulses, i.e., biochemical and electrical urges. This continuous flow of electrical current in the brain, caused by synaptic excitation of the dendrites in the neurons, produces electrical signals that spread from the encephalic mass to the scalp, which are called brain waves. The first example of a disorder that is similar to AD was inspected by the Scottish physician Sir Alexander Crichton in 1798 [19], wherein he validated the observations of 43 clinic cases of mental illness [19]. Attention deficit affects children and adolescents and may continue until the end of their lives [40]. Boys are more prone to getting attention issues than girls and are usually discovered during the early school years [19, 35]. Individuals may display problems in various ways, and these behaviours interfere with school and home lives [43]. Adults may have trouble managing time, setting goals, being organized, and holding down a job, and they have low self-esteem and self-efficacy [27]. Usually, symptoms of AD disorder may change as a person gets older [45].

4.2. Attention Measurement Tools

A Brain-Computer Interface (BCI) is a medium that allows a human to interconnect with devices using brain signals. Several studies have investigated an EEG and/or BCI as a control system for several tenders [31]. Any individual can have an experience without using physiques [3] as EEG/BCI transfers the command to the device from the brain, rather than the particular way of tensions or physical fitness [22]. EEG-BCIs have been utilized as rehabilitation tools, as assistive technology, and others [32]. Many EEG applications utilize EEG signals. In contrast, other applications use other recording modalities such as magnetoencephalography (MEG), electrocorticography (ECoG), intracortical microelectrode recording of single-neuron action potentials or local field potentials, functional magnetic resonance imaging (fMRI), and functional near-infrared spectroscopy (fNIRS) [28]. EEG works with brain waves of different bands [10, 28, 31]. For instance, when people are exposed to video stimuli, the detected brain waves are gamma and beta wavebands from the frontal lobe [16, 42]. It has been noticed that for different training purposes with different protocols, using different channels can provide different results. For instance, using EEG for attention detection, Chouhan et al. [6] and Kaur, Singh and Roy [16] used alpha and beta wave bands from Fp1, Fp2, F3, and F4 channels. Meanwhile, other channels were used to detect other cognitive skills [44]. Likewise, developing an EEG-based game for attention-deficit treatment aimed to enhance the level of the player's engagement, whether in education for students, work for employees, or medical response for patients [46]. According to Zolubak, Pelc and Kawala-Janik [53], there is a long history of EEG studies with attention, with the first study on resting-state brain oscillations in children with behavioral problems consistent with Attention Deficit Hyperactive Disorder reported in 1938 [20]. The first observations described as the frontcentral slowing in the EEG of affected children, which revealed that a rise in the power expressed slower frequency oscillations (theta band, 4-7 Hertz) over the frontal and central lobe [21]. Consequently, more concentration was given to high theta power, low beta power, and theta to beta band ratio (TBR) [41]. It is worth mentioning that theta power is related to the slowing of brain activity, while low beta power is related to faster frequencies.

5. The Attention Treatment

Using brain training or altering its status are effective means in attention improvement, and the ability to change brain attention to remote tasks is still a topic of concern. Thus, training the brain on a single task may lead to enhancing and elevating its sense of cognition; nevertheless, training the brain on a group of tasks, which could boost and increase attention generally, is still a prospect. The tools may be different, but the goals lead to improving brain cognitive skills. Hence, individuals start with medicine and physical activity (i.e. sport, games, activities) [35]. However, if cognitive skills are not treated, then the researcher moves to another option of using neurofeedback as an alternative method for treating disorders [33, 35].The usage of drugs such as Methylphenidate, Clonidine, Atomoxetine has been approved for the treatment of attention-deficit due to reducing the relevant symptoms. For example, Methylphenidate is a synthetic central nervous system stimulant drug based on the clinical analysis, and it proved to be the most effective treatment available for attention deficit [1]. Another intervention used for AD is cognitive training, which focuses on training the working memory by incorporating adaptive schedules that are hypothesized to strengthen attention neuropsychological processes [4]. In this regard, Furley and Wood [9] concluded that the control of attention by the activated contents in the working memory might not only apply to situations in sport but might generalize to a whole range of other activities. However, they acknowledged that sport could be considered as serving and stimulating attention during physical performance. A study by Han [11] in Taiwan on attention training conducted on 116 college students, where their focus was tested by standardized tests (WMS-III, 1997), found that 15 minutes of walking or jogging while engaging in nature increases their emotions and concentration.

6. Neurofeedback Treatment

Biofeedback or neurofeedback utilizing brain activity has attracted a great deal of attention as self-control therapy for adolescents with developmental disorders, multiple body disorders, and brain disorders. There are seven types of neurofeedback used for the treatment of various kinds of mental and psychological disorders [25]. These types of treatment are usually based on either EEG or magnetic resonance imaging (rtfMRI) in real-time [13]. The First and most commonly used frequency/power is biofeedback, also called surface neurofeedback, which involves using 2-4 surface electrodes. It is used for adjusting the rate or speed of specific brain waves to handle diseases in particular brain locations [25]. Second, the Slow Cortical Potential Neurofeedback (SCPNF) strengthens the course of slow cortical potentials for treating epilepsy, ADD, and migraines [7]. Third, the Low-Energy Neurofeedback System (LENS) generates a mild electromagnetic stimulus to adjust the brain waves of the patient when his/her eyes are closed and motionless [51]. This form of neurofeedback was used for treating traumatic brain injury, ADHD, insomnia, fibromyalgia, syndrome of the restless legs, anxiety, depression, and rage. Fourth, the Hemoencephalographic (HEG) neurofeedback provides data on migraine treated cerebral blood flow [25]. Fifth, the Live Z-score Neurofeedback introduces the continuous comparison of variables of brain electrical activity to a systematic database to provide continuous feedback [25], and it is used to treat insomnia. Sixth, the Low-Resolution Electromagnetic Tomography (LORETA) involves the use of 19 electrodes to monitor phase, power, and coherence [17], and it is used to treat addictions, depression, and obsessive-compulsive disorder. Seventh, Functional Magnetic Resonance Imaging (fMRI) is the most recent type of neurofeedback to regulate brain activity based on the activity feedback from deep subcortical areas in the brain [14].

The primary usage of EEG-BCIs was as a communication tool for disabled people or patients with severe motor impairment [28, 30]. Many EEG-based studies have concentrated on using EEG to re-establish communication and control to patients paralyzed due to chronic neuromuscular disorder [28]. Neurofeedback allows users to improve their dynamic abilities and cognitive skills by employing specific training strategies [44, 46]. Neurofeedback remedial results using neuroplasticity as a therapeutic approach with individuals suffering from AD, Attention-Deficit Hyperactive Disorder (ADHD), stroke, and dementia have been reported in many studies [46]. With children diagnosed with AD, diverse neurofeedback (NF) protocols have been practiced, with the most protuberant difference between EEG frequency-band (theta/beta) training and slow cortical potentials (SCPs) training [10].

Computer interfaces based on brain attention are designed to measure the attention level of the user to facilitate communication [46]. When it comes to computer games, the player's level of attention is used as a means to increase his/her interaction and involvement in the game [46].

Markovska-Simoska and Pop-Jordanova [24] conducted a study to compare children and adults' diagnosed attention highlighting that less concern was given to adults affected by the disorder. The EEG signal used EEG absolute/relative power and theta/beta ratio to assist in treating the AD of both children and adults. EEG was recorded during an eyes-open condition. The findings showed a rise of absolute power in delta and theta frequencies in children, while there were no significant differences in adults compared to healthy subjects. For relative power, the results showed an increase in alpha and beta in adults compared to children. Classification analysis showed that attention disorders in children could be identified using absolute theta and theta-beta ratios from the central lobe of the brain (Cz channel). The authors suggested further research on any significant difference in cortical arousal between children and adults using different methods and features. Another study conducted by [6], using audio and visual stimulation, found that they enhance participants' attention levels. Similarly, the work of [47] targeted how EEG Neurofeedback training can enhance attention and memory, where they used a virtual reality system 3D game. The findings showed that first and second-order attentional control improved, yet SMR and theta training showed no effect. Using different measurements that keep eyes open, [15] employed games to enhance the attention level among 85 children. They found that there is a minor enhancement in two of the six near-transfer measures, particularly: Go-No Go and Focus. In their experiment, they employed 12 channels and used emotiv epic. Outcome assessments include latency (the speed of response of the participant), the number of correct patterns selected, and a statistical measure that provides the frequency of an error following a correct or incorrect answer [33].

7. Conclusion

The article reviewed the available literature on attention deficit based on neurofeedback and it revealed many approaches and different results regarding handling the attention. To come with a conclusion from those studies regarding a specific case, it will be difficult due to variances among those studies. Therefore, this work intends to develop a neurofeedback training system that assists in identifying attention with any case based on prior scholars. EEG-based measurements can be described as the most accurate method to capture waves related to brain activities. However, currently, researchers are in the learning status; they are trying to find the relationship between attention and changes in waves emit from the brain. Yet, there is no confirmation regarding the ranges of frequencies that can reveal attention related issues or improvement. Moreover, researchers have to overcome a lot of errors related to capturing and analysing brain waves.

Acknowledgment
This work is supported by Fundamental Research Grant Scheme (FRGS) funded by the Ministry of Higher Education (Grant code: FRGS/1/2019/ICT02/TAYLOR/03/1)

References
[1] Auvin S 2019 Pharmacological treatment of attention-deficit/hyperactivity disorder in children and adolescents with epilepsy *Revue Neurologique* vol **175(3)** pp 141–3 DOI: 10.1016/j.neurol.2018.11.003
[2] Berrigan F, Turcotte S, Beaudoin S, Gignac C and Bernier P M 2014 Physical activity before class enhances attention during the school day (October) *J. of Physical Activity and Health* vol **11** S130 DOI: 10.13140/2.1.1576.4482
[3] Beveridge R, Wilson S and Coyle D 2016 3D graphics, virtual reality, and motion-onset visual evoked potentials in neurogaming *Progress in Brain Research* 1st edition vol **228** DOI: 10.1016/bs.pbr.2016.06.006
[4] Cardoso-Leite P and Bavelier D 2014 Video game play, attention, and learning *Current Opinion in Neurology* **27(2)** pp 185–91 DOI: 10.1097/wco.0000000000000077

[5] Chiang H S, Hsiao K L and Liu L C 2017 EEG - Based detection model for evaluating and improving learning attention *J. of Medical and Biological Eng.* vol **38** pp 847–56 DOI: DOI: 10.1007/s40846-017-0344-z

[6] Chouhan T, Panse A, Smitha K G and Vinod A P 2016 A comparative study on the effect of audio and visual stimuli for enhancing attention and memory in brain computer interface system *Proc. - 2015 IEEE Int. Conf. on Systems, Man, and Cybernetics, SMC 2015* November pp 3104–39 DOI: 10.1109/SMC.2015.539

[7] Christiansen H, Reh V, Schmidt M H and Rief W 2014 Slow cortical potential neurofeedback and self-management training in outpatient care for children with ADHD : study protocol and first preliminary results of a randomized controlled trial *Frontiers in Human Neuroscience* vol **8** p 943 DOI: 10.3389/fnhum.2014.00943

[8] Fergus T A and Wheless N E 2018 The attention training technique causally reduces self-focus following worry provocation and reduces cognitive anxiety among self-focused individuals *J. of Behavior Therapy and Experimental Psychiatry* vol **61** pp 66–71 DOI: 10.1016/j.jbtep.2018.06.006

[9] Furley P and Wood G 2016 Working memory, attentional control, and expertise in sports: a review of current literature and directions for future research *J. of Applied Research in Memory and Cognition* **5(4)** pp 415–25 DOI: 10.1016/j.jarmac.2016.05.001

[10] Gevensleben H, Moll G H, Rothenberger A, Heinrich H, Duric N S and Trust H F 2014 Neurofeedback in attention-deficit/hyperactivity disorder – different models, different ways of application *Frontiers in Human Neuroscience* **8** p 846 DOI: 10.3389/fnhum.2014.00846

[11] Han K T 2017 The effect of nature and physical activity on emotions and attention while engaging in green exercise *Urban Forestry and Urban Greening* **24** pp 5–13 DOI: 10.1016/j.ufug.2017.03.012

[12] Harris A M, Dux P E, Jones C N and Mattingley J B 2017 Distinct roles of theta and alpha oscillations in the involuntary capture of goal-directed attention *NeuroImage* **152** pp 171–83 DOI: 10.1016/j.neuroimage.2017.03.008

[13] Horki P, Bauernfeind G, Schippinger W, Pichler G and Müller-putz G R 2016 Evaluation of induced and evoked changes in EEG during selective attention to verbal stimuli *J. of Neuroscience Methods* **270** pp 165–76 DOI: 10.1016/j.jneumeth.2016.06.015

[14] Hurt E, Arnold L E and Lofthouse N 2014 Quantitative EEG neurofeedback for the treatment of pediatric attention-deficit/hyperactivity disorder, autism spectrum disorders, learning disorders, and epilepsy *Child and Adolescent Psychiatric Clinics of North America* **23(3)** pp 465–86 DOI: 10.1016/j.chc.2014.02.001

[15] Johnstone S J, Roodenrys S J, Johnson K, Bonfield R and Bennett S J 2017 Game-based combined cognitive and neurofeedback training using Focus Pocus reduces symptom severity in children with diagnosed AD/HD and subclinical AD/HD *Int. J. of Psychophysiology* **116** pp 32–44 DOI: 10.1016/j.ijpsycho.2017.02.015

[16] Kaur B, Singh D and Roy P P 2018 EEG based emotion classification mechanism in BCI *Procedia Computer Sci.* **132** pp 752–8 DOI: 10.1016/j.procs.2018.05.087

[17] Kawatake S, Torii I and Ishii N 2017 Increasing concentration with neurofeedback *Proc. - 4th Int. Conf. on Applied Computing and Information Technol., 3rd Int. Conf. on Computational Sci./Intelligence and Applied Informatics, and 1st Int. Conf. on Big Data, Cloud Computing, Data Sci.* pp 247–52 DOI: 10.1109/ACIT-CSII-BCD.2016.055

[18] Khong A, Jiangnan L, Thomas K P and Vinod A P 2014 BCI based multi-player 3-D game control using EEG for enhancing attention and memory *2014 IEEE Int. Conf. on Systems, Man, and Cybernetics (SMC)* pp 1847–52 DOI: 10.1109/SMC.2014.6974189

[19] La Marca J P 2014 Neurofeedback as an intervention to improve reading achievement in students with attention deficit hyperactivity disorder, inattentive subtype (March) *Neuro Regulation* **3(2)** DOI: 10.15540/nr.3.2.55

[20] Lenartowicz A, Mazaheri A, Jensen O and Loo S K 2017 Aberrant modulation of brain oscillatory

activity and attentional impairment in attention-deficit/hyperactivity disorder *Biological Psychiatry* **1(3)** pp 19–29 DOI: 10.1016/j.bpsc.2017.09.009

[21] Lenartowicz A, Mazaheri A, Jensen O and Loo S K 2018 Review aberrant modulation of brain oscillatory activity and attentional impairment in attention-de fi cit / hyperactivity disorder *Biological Psychiatry: Cognitive Neuroscience and Neuroimaging* **3(1)** pp 19–29 DOI: 10.1016/j.bpsc.2017.09.009

[22] Liarokapis F, Debattista K, Vourvopoulos A, Petridis P and Ene A 2014 Comparing interaction techniques for serious games through brain–computer interfaces: a user perception evaluation study *Entertainment Computing* **5(4)** pp 391–9

[23] Marcuse L V, Fields M C and Yoo J 2016 The normal adult EEG *Rowan's Primer and Mini-Atlas of EEG* **2** pp 39–66 DOI: 10.1016/B978-0-323-35387-8.00002-0

[24] Markovska-Simoska S and Pop-Jordanova N 2017 Quantitative EEG in children and adults with attention deficit hyperactivity disorder: comparison of absolute and relative power spectra and theta/beta ratio *Clinical EEG and Neuroscience* **48(1)** pp 20–32

[25] Marzbani H, Marateb H R and Mansourian M 2016 Neurofeedback: A comprehensive review on system design, methodology and clinical applications *Basic and Clinical Neuroscience* **7(2)** pp 143–58

[26] Mateo J, Torres A M, Sanchez-Morla E M and Santos J L 2015 Eye movement artefact suppression using volterra filter for electroencephalography signals *J. of Medical and Biological Eng* **35(3)** pp 395–405 DOI: 10.1007/s40846-015-0036-5

[27] Mazzone L *et al* 2013 Self-esteem evaluation in children and adolescents suffering from ADHD *Clinical Practice and Epidemiology in Mental Health* **9(1)** pp 96–102 DOI: 10.2174/1745017901309010096

[28] Mcfarland D J and Wolpaw J R 2017 EEG-based brain – computer interfaces *Current Opinion in Biomedical Eng.* **4** pp 194–200 DOI: 10.1016/j.cobme.2017.11.004

[29] Mendes G A V 2017 *Brain-Computer Interface Games based on Consumer-Grade Electroencephalography Devices: Systematic Review and Controlled Experiments* thesis (Natal-RN: Federal University of Rio Grande Do Norte)

[30] Minguillon J, Lopez-Gordo M A and Pelayo F 2016 Detection of attention in multi-talker scenarios: A fuzzy approach *Expert Systems with Applications* **64** pp 261–8 DOI: 10.1016/j.eswa.2016.07.042

[31] Minguillon J, Lopez-Gordo M A and Pelayo F 2017 Biomedical signal processing and control trends in EEG-BCI for daily-life : requirements for artifact removal *Biomedical Signal Processing and Control* **31** pp 407–18 DOI: 10.1016/j.bspc.2016.09.005

[32] Minguillon J, Lopez-Gordo M A and Pelayo F 2017 Trends in EEG-BCI for daily-life: requirements for artifact removal *Biomedical Signal Processing and Control* **31** pp 407–18 DOI: 10.1016/j.bspc.2016.09.005

[33] Mohammadi H S, Pirbabaei E, Sisi M J and Sekhavat Y A 2019 ExerBrain: a comparison of positive and negative reinforcement in attention training using BCI based computer games *2018 2nd National and 1st Int. Digital Games Research Conf.: Trends, Technologies, and Applications (DGRC 2018)* pp 167–71 DOI: 10.1109/DGRC.2018.8712048

[34] Mourik R V, Janssen T W P, Bink M and Gelad K 2016 A randomized controlled trial into the effects of neurofeedback, methylphenidate, and physical activity on EEG power spectra in children with ADHD *J. Child. Psychol. Psychiatr.* **57** pp 633–44 DOI: 10.1111/jcpp.12517

[35] Ochi Y, Laksanasopin T, Kaewkamnerdpong B and Thanasuan K 2017 Neurofeedback game for attention training in adults *BMEiCON 2017 - 10th Biomedical Eng. Int. Conf.* January 2017 pp 1–5 DOI: 10.1109/BMEiCON.2017.8229113

[36] Posner M I, Rothbart M K and Tang Y Y 2015 Enhancing attention through training *Current Opinion in Behavioral Sciences* **4** pp 1–5 DOI: 10.1016/j.cobeha.2014.12.008

[37] Sauseng P, Hoppe J, Klimesch W, Gerloff C and Hummel F C 2007 Dissociation of sustained attention from central executive functions: local activity and interregional connectivity in the

theta range *European J. of Neuroscience* **25(2)** pp 587–93 DOI: 10.1111/j.1460-9568.2006.05286.x

[38] Schönenberg M, Wiedemann E, Schneidt A, Scheeff J, Logemann A, Keune P M and Hautzinger M 2017 Neurofeedback, sham neurofeedback, and cognitive-behavioural group therapy in adults with attention-deficit hyperactivity disorder : a triple-blind, randomised, controlled trial *The Lancet Psychiatry* vol **4** issue 9 pp 673–84 DOI: 10.1016/S2215-0366(17)30291-2

[39] Sirca F, Onorati F, Mainardi L and Russo V 2015 Time-varying spectral analysis of single-channel EEG: application in affective protocol *J. of Medical and Biological Eng* **35(3)** pp 367–74 DOI: 10.1007/s40846-015-0044-5

[40] Smitha K G, Shenjie S, Thomas K P and Vinod A P 2014 Two player EEG-based neurofeedback ball game for attention enhancement *Conf. Proc. - IEEE Int. Conf. on Systems, Man and Cybernetics* January 2014 pp 3150–5 DOI: 10.1109/smc.2014.6974412

[41] Son D V, Blasio F M De, Fogarty J S, Angelidis A, Putman P and Barry R J 2019 Frontal EEG theta/beta ratio during mind wandering episodes *Biological Psychology* **140** pp 19–27 10.1016/j.biopsycho.2018.11.003

[42] Staufenbiel S M, Brouwer A M, Keizer A W and van Wouwe N C 2014 Effect of beta and gamma neurofeedback on memory and intelligence in the elderly *Biological Psychology* **95(1)** pp 74–5 DOI: 10.1016/j.biopsycho.2013.05.020

[43] Thomas K P 2016 A study on the impact of neurofeedback in EEG based attention-driven game *2016 IEEE International Conference on Systems, Man, and Cybernetics (SMC)* pp 320–5 DOI: 10.1109/SMC.2016.7844260

[44] Thomas K P, Vinod A P and Guan C 2013 Design of an online EEG based neurofeedback game for enhancing attention and memory *Proc. of the Annual Int. Conf. of the IEEE Eng. in Medicine and Biology Society - EMBS* pp 433–6 DOI: 10.1109/EMBC.2013.6609529

[45] Thomas K P, Vinod A P and Guan C 2014 Evaluation of EEG features during overt visual attention during neurofeedback game *Conf. Proc. - IEEE Int. Conf. on Systems, Man and Cybernetics* January 2014 pp 1841–6 DOI: 10.1109/smc.2014.6974188

[46] Vasiljevic G A M and de Miranda L C 2019 The influence of graphical elements on user's attention and control on a neurofeedback-based game *Entertainment Computing* **29** pp 10-9 DOI: 10.1016/j.entcom.2018.10.003

[47] Yao L, Chen M L, Sheng X, Mrachacz-kersting N, Zhu X and Farina D 2017 Testing the specificity of EEG neurofeedback training on first- and second-order measures of attention *Proc. of 11th Int. Conf. on Augmented Cognition (AC 2017)* LNCS vol **10284** ed D D Schmorrow C M Fidopiastis (New York: Springer Link) pp 90–8 DOI: 10.1007/978-3-319-58628-1

[48] Yildirim N and Varol A 2013 Developing educational game software which measures attention and meditation with brainwaves *ICEE ICIT 2013 Conf.* pp 325–32 Retrieved from http://asafvarol.com/makaleler/NilayYAsafV.pdf

[49] Yuki K, Bhagia J, Mrazek D and Jensen P S 2016 How does a real-world child psychiatric clinic diagnose and treat attention deficit hyperactivity disorder? *World J. of Psychiatry* **6(1)** 118 DOI: 10.5498/wjp.v6.i1.118

[50] Zamora B D, Muñoz J E, Lopez D S and Henao G O 2016 Influence of a BCI neurofeedback videogame in children with ADHD quantifying the brain activity through an EEG signal processing dedicated toolbox *MindReflector Technologies LLC* **3** pp 1–8

[51] Zandi M Y, Firoozabadi M and Rostami R 2015 Improvement of neurofeedback therapy for improved attention through facilitation of brain activity using local sinusoidal extremely low frequency magnetic field exposure *Clinical EEG and Neuroscience* **46(2)** pp 100–12 DOI: 10.1177/1550059414524403

[52] Zhang S, Wang D, Afzal N, Zhang Y and Wu R 2016 Rhythmic haptic stimuli improve short-term attention *IEEE Transactions on Haptics* **9(3)** pp 437–42 DOI: 10.1109/TOH.2016.2531662

[53] Zolubak M, Pelc M and Kawala-Janik A 2018 Challenges in differentiating between attention disorders based on EEG recordings in neurofeedback therapy *2018 Applications of Electromagnetics in Modern Techniques and Medicine (PTZE 2018)* pp 65–8 DOI: 10.1109/PTZE.2018.8503119

Raspberry Pi Based IoT for Aquaculture Realtime Remote Monitoring System with Self Energy Harvesting

Harnan M. Abdullah[1, a], **Anik M. Hariati**[2], **M. Fakhri**[2], **A. Yuniarti**[2]

[1]Diploma of Information Technology, Universitas Brawijaya
[2]Study Program of Aquaculture, Faculty of Fisheries and Marine Science, Universitas Brawijaya,

E-mail: [a]harnan_malik@ub.ac.id

Abstract. Temperature and pH are two essentials water quality parameters in shrimp culture. Therefore, monitoring these two variables is very important to improve productivity in aquaculture. In this study, a monitoring system was created to monitor the two variables in the pond. For this reason, Raspberry Pi IoT based hardware and software has been made to carry out the monitoring function. Monitoring data is automatically sent to online servers using a cellular internet network. For electrical power needs, remote devices are equipped with solar power plants. The results showed that the system could read the temperature and pH of the water accurately and sent it to an online server. There is a web-based calibration menu that allows users to easily calibrate sensors. Moreover, a user interface is also made to display the sensor readings data that has been stored in the database. In conclusion, the monitoring system is applicable for water quality data logger in the field operation with user friendly operation.

Keywords: Shrimp culture, remote monitoring, real-time, IoT, Raspberry Pi

1. Introduction

Shrimp (Litopenaues vannamei) is one of the main export commodities of Indonesian aquaculture. The productivity of shrimp production is highly dependent on the cultivation process carried out by pond farmers. One factor that is very influential on the production of shrimp culture is pond water quality. There are several variables that affect water quality including pH and temperature. Shrimp will grow well if the water temperature is between 26 and 30 °C [1] with a pH range from 6.5 to 8.5 [2]. Outside the range of values, will greatly affect the shrimp production [2].

Due to the importance of the pH and temperature in shrimp farming in ponds, it is necessary to monitor these variables regularly. Observation by direct measurement is very impractical. On the one hand, the data obtained is not real-time, so it is difficult to do analysis and action if there are variables that are in conditions outside the standard. On the other hand, recording data that is not automatic in a database makes observational data not easy to display or access quickly. Therefore, it is needed a monitoring system of pond water temperature and pH in real time and also monitoring data can be accessed easily at any time.

Data reading in real time can be done using a data logger as shown in [3]. In this work, the historical data is stored and accessed on site. This lack of flexibility to be accessed due to the necessary to visit the logger for accessing the data. It can be overcome by performing monitoring system remotely through communication network. Wahyuni & Wijaya [4] developed remote monitoring based on Xbee wireless network with limited coverage area. This distance limitation between the server and its measurement device can be overcome by utilizing the existing telecommunications network infrastructure such as the cellular telephone network. Kovacs et al. [5] made a remote data logger study for weather monitoring using GPRS GSM network data communication. ATMega64 microcontroller is used as the main component of the data processing processor before being sent to the server. In addition, a PC on site is needed to configure the device. Munandar & Syamsi [6] make Universal Data Logger based on SMS and GPRS Data dial services using GSM networks. The term universal here emphasizes the use of selected communication network choices, namely SMS and GPRS data dial. In both studies, it was not stated that data that has been sent and stored on the server can be accessed online using the internet network.

At present, the world trend of industry has entered the era of fourth generation called Industry 4.0. In further discussion, Lee et al. [7] explains that Industry 4.0 introduces the so-called "Smart Factory", in which Cyber-Physical Systems monitors the factory's physical processes or equipment and makes decentralized decisions. The system is supported by the Internet of Things (IoT) technology, where devices that are used can communicate and cooperate with one another and also with humans in real-time through internet networks [8]. IoT technology is being adapted in agriculture to support increased food security by implementing remote data loggers for agricultural system monitoring and automation [9] as well as monitoring agricultural land.

IoT based monitoring system is applied in aquaculture as well. A raspberry pi-based monitoring system has been studied in [10] with Wi-fi internet connection required for data transfer. On another study, pH and temperature aquarium monitoring system is developed in [11] where the sensed data is transferred to cloud system through Wi-Fi connection as well. Therefore, advanced study needs to be performed to enable it for outdoor and remote application due to limitation of Wi-Fi coverage area. Meanwhile, a system for water quality has been studied in [12] based on IoT with 3G cellular data transfer. However, the system was made only to detect threshold data and not for sending data continuously. On the other hand, there is no local database system so that the disconnection of the data connection makes the system lose history from sensor reading.

In this study, a real time monitoring system for the temperature and pH of shrimp pond water is based on Internet of Things (IoT). Both variables are measured using sensors and the database will be stored both on a remote device and on an online server. Sending data to an online server is done using a cellular network. Thus, this allows the data logger device to be installed in a remote area and easily moved if needed. Because in shrimp ponds there are rarely electricity sources from public power system, the equipment is equipped with solar power plants as a provider of electrical power devices.

2. Materials and Methods

In this session, the proposed methodology will be discussed which starts with a general discussion of the system to the design of each sub-system.

2.1. General System

In general, the system created as shown in Figure 1. In accordance with what is in the picture, there is a sensor module that functions to read the temperature and pH of pond water. Whereas the mini server acts as the main processor that reads the data from the sensor module stores it and then sends the data to the online server. For each sensor reading, it is also necessary to record the reading time so that the Real Time Clock (RTC) module is needed to provide accurate time data. The Wi-Fi router has two functions at the same time, namely providing access to the device directly at the location of the device and also providing a connection to the cellular internet network with the addition of a 4G LTE modem. On the server side, there is a database system that stores sensor reading data that has been sent by a remote device and also provides a web-based application to display the data.

The data flow from the system as shown in Figure 2. Input data from the sensor is read by Raspberry Pi, then calibrated first and then stored from the local database in Raspberry pi. Stored data is marked if it has not been sent. Then a collection of data that has not been sent periodically is sent to an online server and then marked if the data has been sent. On the server side, data will be stored in a database and then displayed in the form of web pages.

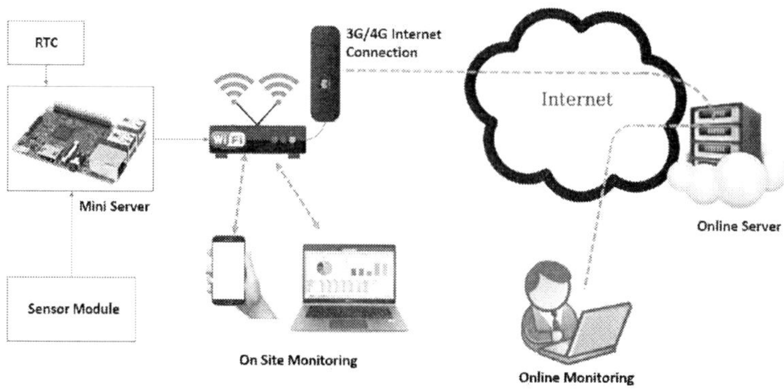

Fig 1. General System of IoT Raspberry Pi Based Monitoring System

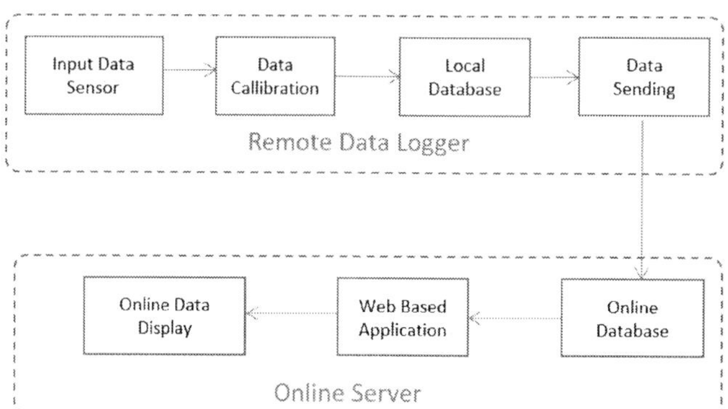

Fig 2. Data Flow from Sensor to Online Server

2.2. IoT Based System Design
The discussion of the design of this system is divided into three parts. The first is related to hardware design, then software design and then solar power system design.

2.2.1. IoT Remote Data Logger Hardware Design. The hardware design is shown in Figure 3. There are two sensors used, the DS18B20 temperature sensor and the liquid pH sensor. Another one is the mini server uses Raspberry Pi 3+. The output of the temperature sensor uses a 1-wire protocol so that data is connected to pin 7 on the Raspberry Pi. While the output of pH sensing is an analogue voltage variable, so an ADC module is needed to be read by Raspberry Pi. For the reason, this study uses the EMS Pi Logger module which is a Raspberry Pi Shield that has ADC (MPC3202) and RTC (DS1370) features. The connection from ADC to Raspberry Pi uses the SPI protocol while RTC uses i2c. To save the operating system of the Raspberry Pi and its applications and databases, a micro SD card with a capacity of 16GB is used.

For connectivity, two devices are needed. The first is the TL-MR3020 Wi-Fi Router and Huawei E3372 3G / 4G modem. Actually, the cellular modem can be connected directly to the USB port of the Raspberry Pi. But not done in this study for reasons of computability and user convenience. Reconfiguration is required if a SIM Card is changed with a different mobile network operator. On the other hand, the use of a Wi-Fi router makes it easy for users to connect to the device to perform maintenance on the site.

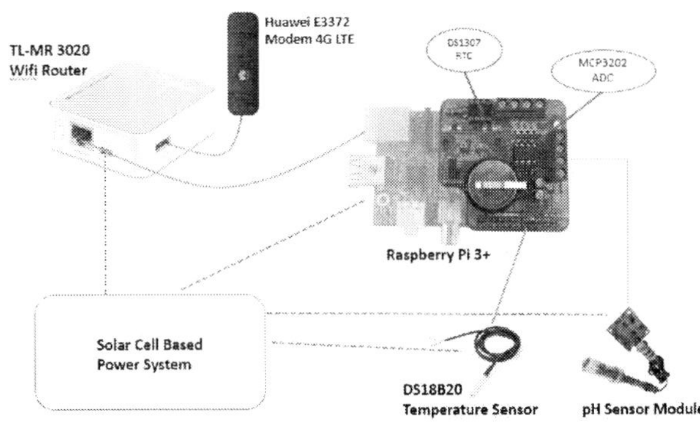

Fig 3. Hardware System Design

2.2.2. IoT Remote Data Logger Database Design. Databases on remote devices have two main functions: the first saves sensor reading data and the second saves system configuration data which includes calibration settings, station identity, and also online server identity for the purpose of sending data over the internet network.

2.3. Acquisition and Transfer Data

Figure 4 is a flowchart for reading sensor data and also sending data to the server. Starting with the sensor reading, the software then calibrates for the correct reading. After that the data will be stored in a database with a sign that the data has not been sent to an online server. The period for reading the sensor data is done once every minute. After that, the data will be sent to the online server every 12 hours. Every data that has been sent will be given a sign that the data has been sent.

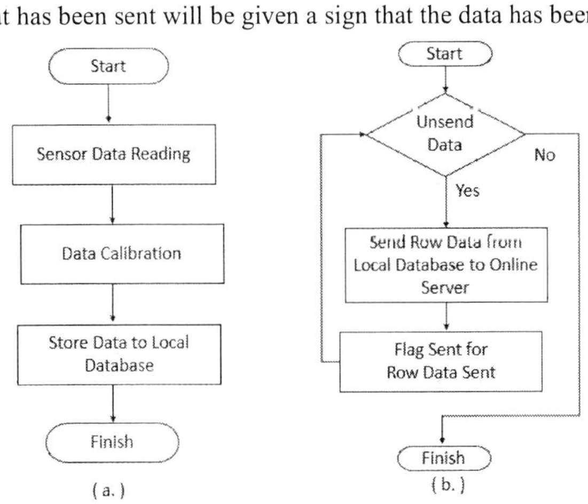

Fig 4. Flowchart for Data Acquisition (a) and Sending Data to Server (b)

2.4. Important Script

The system carried out in this study uses the Python programming base on the Raspberry Pi as a core language other than PHP as the user interface. Therefore, this subsection discusses important scripts in the system, especially those related to data acquisition.

2.4.1. Data Acquisition and Calibration. Acquisition and calibration data are the important part of this system. Sensor readings certainly need to be tested periodically to get accurate readings. This reading adjustment is done in software where the calibration parameters are stored in the database on the Raspberry Pi. In this study, there are two types of sensors, namely a temperature sensor that uses Onewire communication and a pH sensor with a variable voltage output signal. Below is a Python code on the Raspberry Pi to calibrate the output voltage signal reading from the pH sensor read by the Raspberry Pi ADC. As for the temperature sensor, it also uses a similar method.

```
#ConnectDB
db = MySQLdb.connect("localhost","dbUser","dbPassword","db_name")
curs=db.cursor()
#get data, calibration, and store
curs.execute("SELECT StationSensorID,LowOutput,HighOutput, LowValue,
HighValue FROM sensor_detail")
data=curs.fetchall()
vref=5
for row in data:
    StationSensorID = row[0]
        x1 = row[1]
        x2 = row[2]
        y1 = row[3]
        y2 = row[4]
        inADC = 0;
        v=readADC(inADC)
        dataSensor=round(((float(v)-x1)/(x2-x1))*(y2-y1)+y1,2)
        curs.execute("INSERT INTO sensor_data (DateTime, StationSensorID,
value, status,x_value) VALUES (CURRENT_TIMESTAMP,'%s', '%s' ,
0,'%s')"%(StationSensorID,dataSensor,v))
db.commit()
db.close()
```

In the script above, the sensor data is calculated based on the two-point linear equation (x1, y1) and (x2, y2) where the x and y variables are obtained from the measurement results when testing. The value of x is the value of the sensor output reading, while the value of y is the reading of the comparator test. Subset 1 shows the first test with conditioning fluids with a low pH while subset 2 shows the second for liquids conditioned with high pH. Moreover, this method is only suitable for sensors giving a linear reading.

2.4.2. SCrontab Autorun Script. Sensor data acquisition is carried out periodically as well as sending data to online servers. In this study, setting the running period for the script was done using the crontab feature in the Raspbian operating system used. The following is a crontab script that is used to run the data acquisition process (logger.py) and also sends data to an online server (sendingdata.py).

```
pi@raspberrypi:~ $ crontab -l
* * * * *  sudo python /home/pi/logger.py
0 */12 * * *  sudo python /home/pi/sendingdata.py
```

2.5. User Interface Design

To make it easier for users to maintain, especially in terms of sensor calibration, web-based software is created for remote monitoring device as depicted in Figure 5. In general, there are three main types of menus namely menus related to device identity, configuration menus for sensor calibration, and also menus for displaying sensor reading data. Thus, users without programming background or computer configuration can calibrate the sensor if needed. On the server side, a web interface is also made to view the data that has been stored.

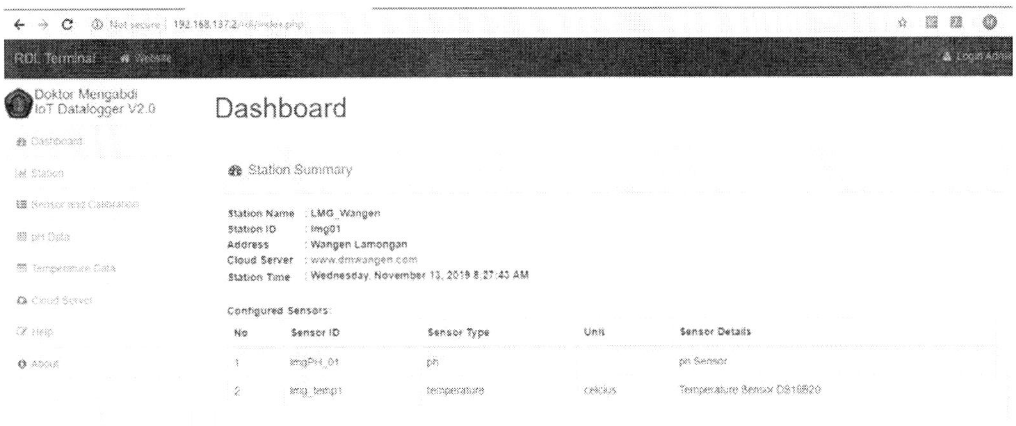

Fig 5. Remote Data Logger Web Interface

2.6. Solar Cell Based Power System Design

The electrical power source in this research uses solar energy. Thus, the system can be installed ponds even though there is no electricity network from the public electricity company. To meet the electrical power requirements of the tool, it is necessary to calculate the power requirements so that the design specifications of the solar panel system can be determined namely the WP solar panel and also the battery capacity [13]. Based on the measurement result, the equipment made requires an electric current of 1A for a voltage source of 5V. Thus, the power needed can be calculated that is 5W. For a day and night, it takes as much as 120-watt hours (Wh) of power.

It is assumed that in a day's radiation, the sun shines optimally to be converted into electricity for 4 hours. So, from the calculations needed 30-watt peak (WP) solar panels to obtain the 120Wh. Battery capacity on the market is usually expressed in units of ampere hour (Ah). If the battery used has a 12V voltage, to store power of 120Wh, a battery with a capacity of 10Ah is needed.

3. Result and Discussion

A remote data logger device that is produced with a temperature sensor and pH sensor is shown in Figure 6. This tool has been tested in Laboratory of Fish Reproduction, Universitas Brawijaya.

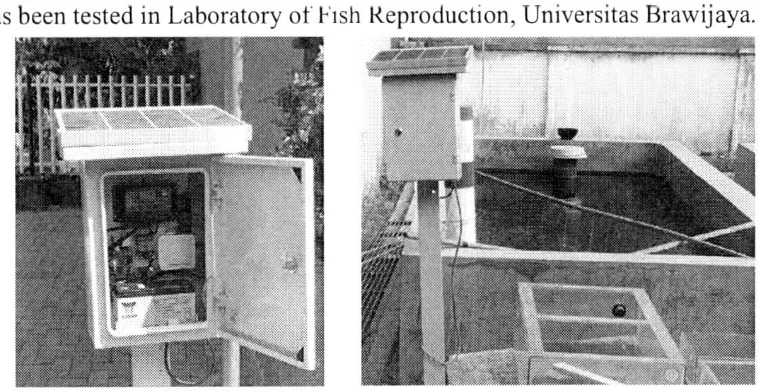

(a.) (b.)

Fig 6. Device Implementation

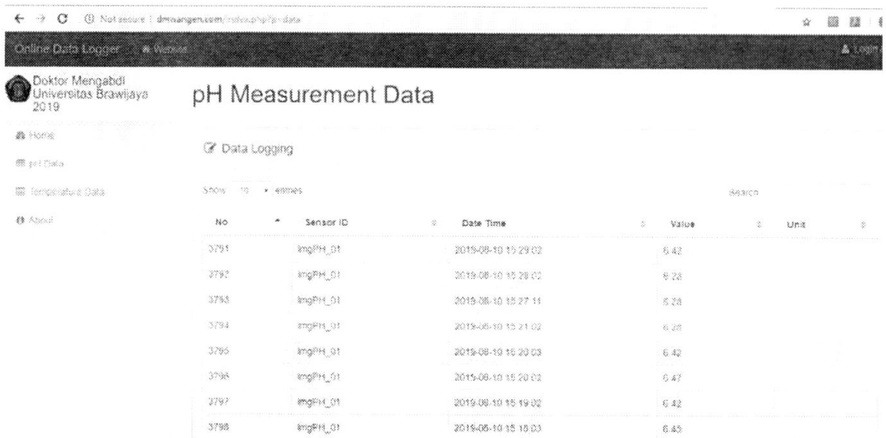

Fig 7. pH Measurement Data

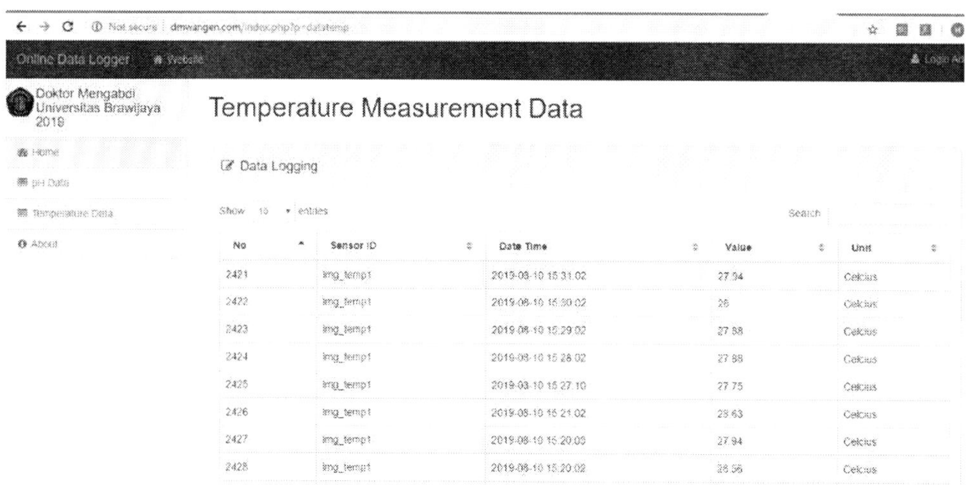

Fig 8. Temperature Measurement Data

The test results showed that the system created could function properly. For this trial, the data reading period was made every minute while the data transmission was set every minute. The results showed that the system automatically sent data to the server and could be accessed online. Figure 7 is a display of temperature data, while the pH data is shown in Figure 8. This monitoring system will be beneficial to optimize the shrimp production because of allowing users to monitor water quality (pH and temperature) in real-time so that they can promptly detect and fix the problem. Uddin et al. [14] suggested that the implementation of a monitoring system based on the internet of things improved the production and survival rate of freshwater shrimp (Macrobrachium rosenbergii) in Bangladesh.

There are some works have been done about water monitoring such as by *Bookingkito Jr.* [15] and Huan *et al.* [16]. However, they did not provide user friendly sensor calibration menu in their work. This item is important for maintaining sensor data accuracy in the field application. In this work, the feature is provided by web-based application. Sensor calibration is done by comparing the sensor reading values with portable measuring devices and then configuring the calibration menu that has been made.

Based on the result, the system has several advantages, namely real time (can be installed in remote areas) and easy to operate. However, one of the weaknesses of this system is that power requirements are still relatively high for the size of remote devices. Therefore, in the future the system can be developed with hardware that does not consume a lot of system power which is made easy to be operated by even an unskilled person.

4. Conclusion

In this study, a remote data recording platform was created to monitor the temperature and pH of shrimp pond. This system has been tested in a laboratory and is ready to be implemented in ponds. The test results showed that monitoring data can be sent properly on an online server and can be accessed using the internet network. This tool can also meet its own energy needs by using its own solar power plants.

Acknowledgments

This research was financially supported by the Institute of Research and Community Services (LPPM), Universitas Brawijaya, Indonesia.

References

[1] Hariati A M, Wiadnya D G R, Tanck M W T, Boon J H and Verdegem M C J 1996 Penaeus monodon (Fabricius) production related to water quality in East Java, Indonesia *Aquaculture Research* **27(4)** pp 255–60

[2] Fakhri M, Budianto B, Yuniarti A and Hariati A M 2015 Variation in water quality at different intensive Whiteleg Shrimp, Litopenaeus vannamei, farms in East Java, Indonesia *Nature Environment and Pollution Technol.* **14(1)** pp 65–70

[3] Lambrou T P, Anastasiou C C, Panayiotou C G and Polycarpou M M 2014 A low-cost sensor network for real-time monitoring and contamination detection in drinking water distribution systems *IEEE Sensors J.* **14(8)** pp 2765–72

[4] Wahyuni R T and Wijaya Y P 2016 Solar panel performance monitoring system in sensor node *Proc. of 2016 Int. Symposium on Electronics and Smart Devices* pp 92–6

[5] Kovacs A, Nicolcioiu A, Arhip J and Cașu G 2014 Design and implementation of a GPRS remote data logger for weather forecasting *Proc. of 2014 10Int. Conf. on Communications* pp 1–4

[6] Munandar D and Syamsi D 2014 Data logger management software design for maintenance and utility in remote *Proc. of The 1st Int. Conf. on Information Technol., Computer, and Electrical Eng* pp 74–8

[7] Lee J, Bagheri B and Kao H 2015 A cyber-physical systems architecture for industry 4.0-based manufacturing systems *Manufacturing Letters* **3** pp 18–23

[8] Shammar E A and Zahary A T 2019 The internet of things (IoT): a survey of techniques, operating systems, and trends *Library Hi Tech.* **38(1)** pp 5–66

[9] Raju K R S R and Varma G H K 2017 Knowledge based real time monitoring system for aquaculture using IoT *Proc. of 2017 The 7th Int. Advance Computing Conf.* pp 318–21

[10] Cordova-Rozas M, Aucapuri-Lecarnaque J and Shiguihara-Juárez P 2019 A cloud monitoring system for aquaculture using IoT *Proc. of 2019 IEEE Sciences and Humanities Int. Research Conf.* pp 1–4

[11] Abinaya T, Ishwarya J and Maheswari M 2019 A novel methodology for monitoring and controlling of water quality in aquaculture using Internet of Things (IoT) *Proc. of Int. Conf. on Computer Communication and Informatics* pp 1–4

[12] Mohanraj I, Ashokumar K and Naren J 2016 Field monitoring and automation using IoT in agriculture domain *Procedia Computer Sci.* **93** pp 931–9

[13] Al-Ali A R, Nabulsi A A, Mukhopadhyay S, Awal M S, Fernandes S and Ailabouni K 2019 IoT-solar energy powered smart farm irrigation system *J. of Electronic Sci. and Technol.* **17(4)**

100017
[14] Uddin M S, Istiaq M F, Rasadin M and Talukder M R 2019 Freshwater shrimp farm monitoring system for Bangladesh based on internet of things *Eng. Reports* **2020** 2:e12184
[15] Bookingkito Jr P B and Llantor O E 2017 Design and implementation of real-time mobile-based water temperature monitoring system *Procedia Computer Sci.* **124** pp 698–705
[16] Huan J, Li H, Wu F and Weijian C 2020 Design of water quality monitoring system for aquaculture ponds based on NB-IoT *Aquacultural Eng.* **90** 102088

Towards Development of a Web-Based System for Analyzing Marine Ecological Carrying Capacity (MECC) for Managing Marine Resources in The Outer Island of Indonesia (Case Study: Nunukan Regency)

A Susandi*, A Wijaya, W S Kuntoro, I Faisal, A Y Pratama, I Nurdin and F G Kertabudi

Bandung Institute of Technology

E-mail: *armi@meteo.itb.ac.id

Abstract. Research on the development of data integration system design and web-based system in the calculation of Marine Ecological Carrying Capacity (MECC) for managing marine and coastal resource management has been carried out. In this study, Nunukan Regency, North Kalimantan was chosen as the study area. The process of integrating data in managing marine and coastal resources in a sustainable manner using the carrying capacity concept approach is carried out by designing an information system development system in processing and displaying the results of MECC calculations. This can make it easier for decision makers to develop the utilization of blue economic potential in accordance with geographic conditions as well as efforts that can be made to increase the productivity of the blue economy in Nunukan Regency, North Kalimantan.

Key Words: Marine Ecological Carrying Capacity (MECC), Nunukan Regency, System Design, Information Systems

1. Introduction

The blue economic potential of Indonesia as an archipelago which reaches 17 thousand trillion rupiah is dominated by the marine capture fisheries sector. However, only 70% of the 12.54 million tonnes of potential marine catch in Indonesia can be realized [1]. This potential has not been able to be utilized and managed optimally because of the many obstacles that must be faced, such as limited access to information, infrastructure, and human resources (HR) capable of managing this potential. On the other hand, as a result of climate change, weather conditions in the sea become erratic (such as strong winds and extreme waves) and endanger fishermen and residents on the coast, as well as impact on weakening economic growth, especially in areas that only depend on coastal and marine resources. . The concept of carrying capacity can be used as a basic principle to explore the use of coastal and marine resources in supporting a sustainable blue economy. This concept directly seeks to demonstrate the relationship between the population as a supported object (carrying object) and the environment as a support (carrier) to ensure sustainability [2]. Theoretically, population carrying capacity evolves into resource carrying capacity, cnvironmental carrying capacity and then into ecological carrying capacity (ECC) [3].

To achieve this, a system that is capable of integrating data based on historical data, monitoring data, and predictive data is needed to see the development of sustainable use of marine and coastal resources. The results of data integration will produce information in the form of a maritime performance index wrapped in an appropriate application, so that decision makers can easily and on target in determining which development of blue economic potential is in accordance with geographic conditions and efforts that can be made to increase productivity the blue economy in the region.

Research on integrating data based on information systems used in managing marine and coastal resources is rarely conducted, especially in Nunukan District, North Kalimantan. Therefore, the challenge of integrating data into an information system to calculate Marine Ecological Carrying Capacity (MECC) is very interesting to do. Based on this explanation, the purpose of this study is to find out how to integrate data in managing marine and coastal resources in a sustainable manner using the carrying capacity concept approach.

2. Study Area

In this research, the area of Nunukan Regency, North Kalimantan was the research location. Astronomically Nunukan Regency is located at 115° 33' 00" to 118° 3' 55" East Longitude and between 3° 15' 00" to 4° 24' 55" North latitude which is the northernmost region of North Kalimantan Province [4].

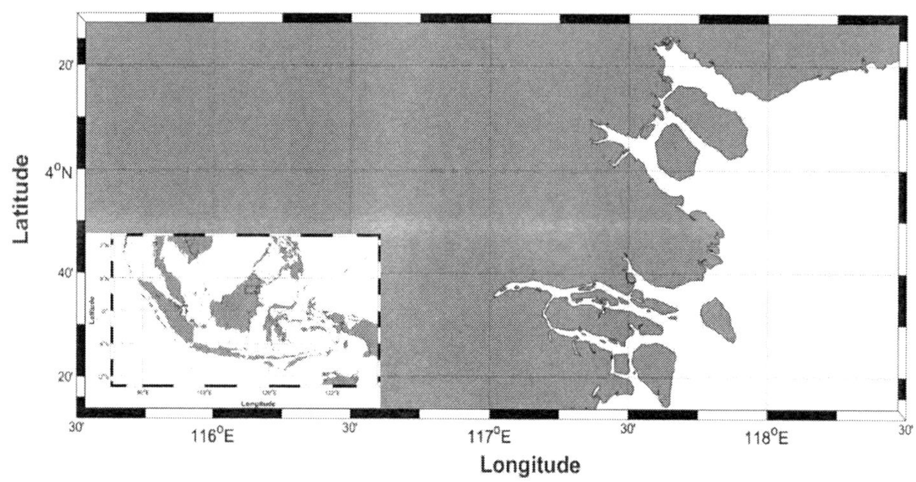

Figure 1. Location of Applied Prototype Test in Nunukan Regency, North Kalimantan

3. Data Acquisition

In the Marine Ecological Carrying Capacity (MECC) calculation, the data for each parameter is obtained from various sources as follows

3.1. Significant Wave Height (SWH)

Simulated significant wave height data were obtained from the University Corporation for Atmospheric Research (USAR). USAR predicts waves using the WAVEWATCH III (WW3) wave model. WW3 is a third-generation wave simulation model developed at the Marine Modeling and Analysis Branch (MMAB) of the Environmental Modeling Center (EMC) of the National Centers for Environmental Prediction (NCEP). WW3 is based on WAVEWATCH I and WAVEWATCH II developed at Delft University of Technology and NASA's Goddard Space Flight Center, respectively [5].

3.2. Current Sea, Sea Surface Temperature (SST), and Sea Water Level (SWL)

The HYbrid Coordinate Oceans Model (HYCOM) is a global ocean circulation model that evolved from the Miami Isopycnic-Coordinate Oceans Model MICOM) developed by Rainer Bleck and colleagues

[6]. HYCOM provides ocean prediction data such as current velocity data, sea surface temperature (SST), and sea level (SWL) which are used as one of the parameters used in calculating carrying capacity.

3.3. Chlorophyll-a
Chlorophyll-a data is obtained from field observation data obtained from the Visible and Infrared Imager / Radiometer Suite (VIIRS) satellite sensors. VIIRS is the successor to the Moderate Resolution Imaging Spectroradiometer (MODIS) developed by the National Aeronautics and Space Administration (NASA) [7].

3.4. VIIRS Boat Detection (VBD)
Vessel monitoring data is obtained from the low light emission of fishing boats captured by VIIRS satellite imagery. The data is processed by the Earth Observation Group by developing a near real time VIIRS boat detection system that produces global mapping called VIIRS Boat Detection (VBD) [8].

3.5. Regional Statistical Data of Nunukan Regency
Regional statistical data used in the calculation of carrying capacity consisting of the number of tourists, fishery production, number of fishery households, population density, Gross Regional Domestic Product (GRDP), and human development index are obtained from reports from the Central Statistics Agency (BPS) for the Nunukan Regency.

4. System Design
In this section will explain how the design of the Marine Ecological Carrying Capacity (MECC) prediction calculation system is shown in Figure 2. The stages of forming the system design consist of three stages, namely data preparing, data processing, and data visualization.

4.1. Data Preparing
The MECC calculation to become an MPI consists of two components, namely the Carrying object component (human activities and socio-economic growth) and the Carrier resistance component (ecological resilience). Human activities include coastal and marine activities that directly suppress ecosystems (Carriers), while socio-economic growth includes elements that represent coastal populations and economies, as well as actions to protect coastal areas. Carrier resilience components consist of elements that maintain or damage the structure and function of coastal and marine ecosystems.

The data required in MECC calculations are obtained by creating an automatic download system for each data source as described in the previous section. The data will be cut automatically according to the study area so that it can represent the actual conditions in the study area.

4.2. Data Preparing
In this section the elements calculated for each component can be adjusted according to local conditions and data availability in the study area. MECC calculations are done using Python software so that calculations can be done automatically. In making the script, follow the steps in calculating MECC which consists of three stages, namely:

4.2.1. Standardizing the value of each indicator (element). In the first stage, Equation (1) and Equation (2) are used to standardize the value of each indicator (element) into the range 0-1. For indicators that are positive, the equation used is as follows:

$$Y_{ij} = \frac{X_{ij} - X_{jmin}}{X_{jmax} - X_{jmin}} \quad (1)$$

Meanwhile, for indicators that are negative, the equation used is as follows:

$$Y_{ij} = \frac{Xj_{max} - X_{ij}}{X_{jmax} - X_{jmin}} \quad (2)$$

where X_{jmax} and X_{jmin} are the maximum and minimum values for the X_j indicator, and Y_{ij} is the normalized value for the X_j indicator.

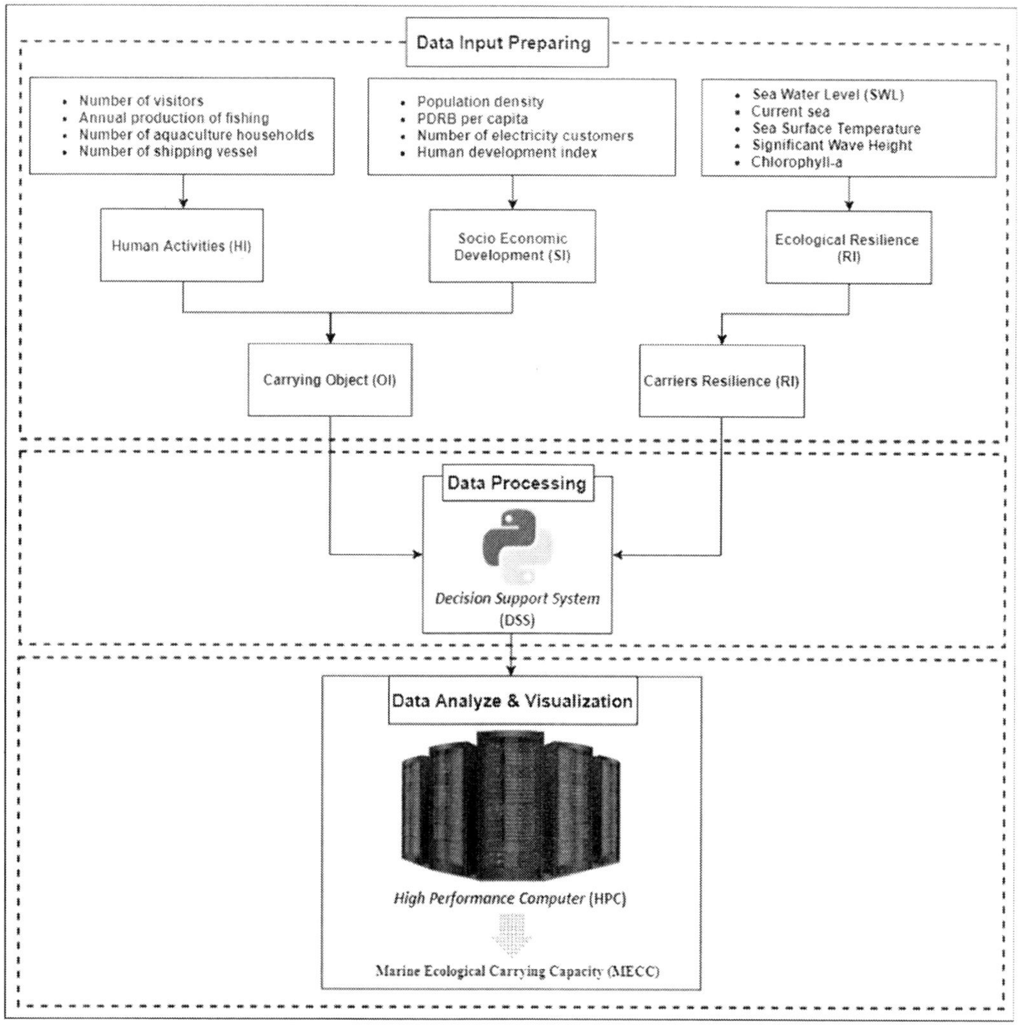

Figure 2. Design of the Marine Ecological Carrying Capacity (MECC) prediction calculation system.

4.2.2. Determining the weight of each indicator (element). The weight determination is based on the effect of each element on changes in environmental conditions in the study area, especially in coastal and marine areas. In this case, Nunukan Regency is the study area in this study.

4.2.3. Calculation of the MECC index (MPI). The principal component analysis (PCA) method is carried out to extract important information from each indicator, namely the human activity index (HI), socioeconomic growth index (SI), and ecosystem resilience index (RI). The results of the contribution of each indicator based on the PCA method are weighted for each indicator.

Based on these two stages, the values for HI, SI, and RI can be calculated by the following equation 3:

$$I_m = \sum_{j=1}^{n} W_j Y_{ij} \tag{3}$$

where I_m is HI, SI, or RI; Wj is the weight for the X_{ij} indicator; and Y_{ij} is the normalized value for X_j indicator.

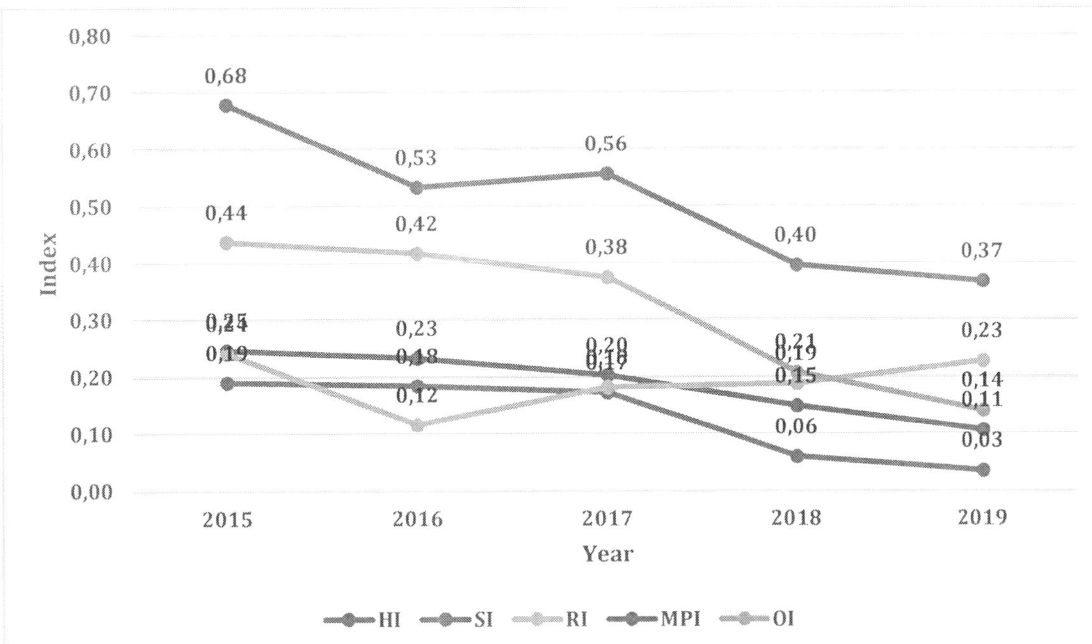

Figure 3. Graph of human activity index (HI), socio-economic development activity index (SI), ecological resilience index (RI), MECC performance index (MPI), and object carrying index (OI)

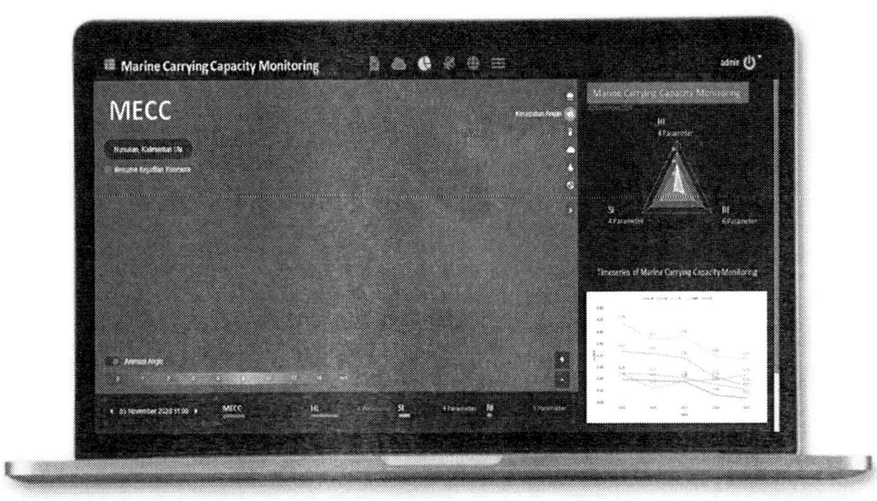

Figure 4. Website-based information system for MECC monitoring

4.3. Data Analysis and Visualization

The calculated data (Figure 3) will be displayed in a website-based information system so that each parameter can be analyzed easily and more interactively (Figure 4). Thus, decision makers can easily and on target in determining which development of blue economic potential is in accordance with geographic conditions as well as efforts that can be made to increase the productivity of the blue economy in Nunukan Regency, North Kalimantan.

5. Conclusions

The development of the MECC calculation system design can be used as the basis for the process of integrating data in managing sustainable marine and coastal resources using the carrying capacity concept approach. The development of a website-based information system makes it easy for decision makers to develop the utilization of blue economic potential in accordance with geographic conditions as well as efforts that can be made to increase the productivity of the blue economy in Nunukan Regency, North Kalimantan.

Acknowledgements

Parts of this research were funded by the Ministry of Research and Technology (Kemenristek) / National Agency for Research and Innovation with number contract 2/AMD/E1/kp.ptnbh/2020. The authors would like to thank for the supports of BRIN during this research.

References

[1] Katadata 2019 *Pengusaha Perikanan Minta KKP Segera Revisi Pembatasan Ukuran Kapal* Retrieved on 4 April 2019 from https://katadata.co.id/berita/2019/01/30/pengusaha-perikanan-minta-kkp-segera-revisi-pembatasan-ukuran-kapal
[2] Ma P, Ye G, Peng X, Liu J, Qi J and Jia S 2017 Development of an index system for evaluation of ecological carrying capacity of marine ecosystems *Ocean Coast. Manag.* **144** pp 23–30
[3] Martire S, Castellani V and Sala S 2015 Carrying capacity assessment of forest resources: Enhancing environmental sustainability in energy production at local scale *Resour. Conserv. Recycl.* **94** pp 11–20
[4] Pemda Kabupaten Nunukan 2018 *Perubahan RPJMD Kabupaten Nunukan Tahun 2016-2021* (Nunukan: Pemda) Retrieved from https://jdihn.go.id/files/838/PERDA%2011-2018.pdf
[5] Tolman H L 2009 *User Manual and System Documentation of WAVEWATCH III TM version 3.14* pp 276–20 Retrieved from https://www.researchgate.net/publication/228750848_User_manual_and_system_documentation_of_WAVEWATCH_III_version_314
[6] Bleck R, Halliwell Jr G R, Wallcraft A J, Carroll S, Kelly K and Rushing K 2002 *HYbrid Coordinate Ocean Model (HYCOM) User's Manual: Details of the Numerical Code* Retrieved from https://www.hycom.org/attachments/063_hycom_users_manual.pdf
[7] Ocean Biology Processing Group 2018 *Suomi-NPP/VIIRS* (NASA Goddard Space Flight Center, Ocean Ecology Laboratory) Retrieved from https://oceancolor.gsfc.nasa.gov/data/viirs-snpp/
[8] EOG VIIRS 2020 *Boat Detection* Retrieved on 15 October 2020 from https://eogdata.mines.edu/vbd/

IOP Conf. Series: Materials Science and Engineering 1077 (2021) 012064 doi:10.1088/1757-899X/1077/1/012064

A Prototype of Portable Digital Forensics Imaging Tools using Raspberry Device

F Yudha[1], E Ramadhani, R M Komaryan

Department of Informatics, Universitas Islam Indonesia, Yogyakarta, Indonesia

E-mail: [1]yudha@uii.ac.id

Abstract. One of the digital forensics activities has the goal to prove a cybercrime. There are several stages in digital forensics when doing an investigation. Each stage has its suitable hardware and software that is used while investigating a case. Standalone forensics hardware is a suitable media in the process of investigation. It can do an acquisition and imaging process at the same time as the investigation process. Nowadays, standalone forensics hardware for imaging devices has a very high price. This paper gives a solution to develop low budget portable imaging forensics devices using a raspberry device. The device enables us to do an acquisition to a hard disk or flash drive. The output has a raw format file type i.e .dd. This device includes a logging file consisting of detailed information related to the digital evidence also hashing to provide file integrity. The result of this paper describes how to construct a prototype low budget portable device for digital forensics acquisition using a raspberry device and how to operate it in GUI. The prototype was successfully created and tested in several scenarios. The performance test of this device has a result of that transfer rate of 1,85 MB/s.

1. Introduction

Digital forensics arose as the implication of computer-related crime. It is the branch of forensics science that deals with the examination and analysis of digital evidence stored on physical(electronic) evidence in the form of personal computers (personal computers-PCs), laptops/notebooks, netbooks, and tablets. These methods needed to prove the crime that happened and connect it with the perpetrators. Based on data from the Internet Crime Complaint Center (IC3), in 2019 there were 467361 complaints in the United States[1]. The development of technology and its applications enforce digital forensics science practice harmonizing the application. The digital forensics investigation becomes complex due to the rapid growth of technology[2].

Process and procedure are the core component of a digital forensics investigation. The common investigation process contains preservation, acquisition, examination, analysis, and reporting. The acquisition process takes an important part in the digital forensics investigation process. It is a process of collecting evidence that correlates with the crime. A mistake in the acquisition process can have an impact on all of the investigation process. One of the techniques used in the acquisition process is imaging. It is the process of creating identical duplicate copies of electronic evidence.

To acquire and analyze digital evidence from cybercrime cases special treatment is required, due to minor errors in the handling of digital evidence that can make digital evidence not legally recognized and automatically cannot serve as valid evidence in court. According to The Indonesian act number 19 of 2016 regarding Information and Electronic Transactions (UU ITE) article 43 paragraph 5 alphabet (J) reads, "Requesting expert assistance needed in investigating criminal acts in the field Information

Technology and Electronic Transactions"[3]. One way to secure digital evidence by copying data from the storage media bitstream image and place it in a safe place. The technique takes bit by bit bits of data from a physical storage medium called imaging, and the result of such cloning is called a disk image.

To perform the disk imaging process, an investigator certainly needs devices and applications which can help the work. Today there are many applications of forensics imaging that can be used, ranging from paid to free. The outline of devices for computer forensics is differentiated by hardware and software, but both in terms of hardware and computer forensics software it is expected to fulfill 5 functions, that are acquisition, validation and discrimination, extraction, reconstruction, and reporting. Reporting from the site forensicswiki.com, here are some applications that can be used for forensics imaging including guymager, ewfacquire, Adepto, aimage, AIR, dcfldd, dd, and many more.

Traditionally, forensics analysis is accomplished by standalone several means to acquire hard drives and external media such as hardware write-blocker, dock attached to a laptop, computer, workstation with the preinstalled operating system. A standalone forensics acquisition and imaging device are very expensive. In this paper, we propose an acquisition system using a raspberry device as an imaging device in the digital forensics investigation. The raspberry device will consist of an operating system and tool. The tool that will be used is a graphical user interface (GUI) based tool. We use raspberry pi because of its low cost and its portability, popularity, and convenience of use.

Based on the fact and brief problem above, this research will propose a prototype portable tool for digital forensics imaging using Raspberry Pi. Raspberry Pi was chosen because of the capability to run it by using micro USB power through any power source with 5V voltage and easy to connect to the network using Wi-Fi. The objective of this paper is to describe how to make a prototype of a forensics acquisition device using a raspberry pi with the GUI interface. We build the system using raspberry pi and python programming language. The result of this paper is the performance of the imaging device. This paper is organized into several sections. Section I is the introduction to explaining the research background. In section II explaining the literature review of the paper. Section III explains the methodology of the paper. Finally, section IV has explained the result of the paper and we conclude the paper in the last section in section V.

2. Literature Review

Digital forensics research that utilizes the Raspberry Pi device has been done before. This research performs digital forensics analysis on the raspberry pi devices in connection with cybersecurity investigation was studied by [4]. They used Raspberry Pi to act as an attacker also as a target. A forensics investigation was performed by acquiring a Raspberry Pi sdcard using Encase Acquisition. This process showed that the researcher uses a traditional forensics method instead of a live forensics method. This research has some information that produces by this research which is the use of Raspberry Pi as an attack platform and Raspberry Pi as a target.

Research conduct by [5] addresses software that was used for digital forensics analysis and examination. Based on that research they found that open source tools might be used for supporting digital forensics investigation processes. The tools produced clear and comprehensive results as closed source tools (paid tools). This research suggests several suggestions that are the development of the comprehensive test, tool publication, creation of a standard, publication of specific procedures for the tool.

Research in the development of open-source digital forensics devices was started by [6] which is called FIREBrick. They used a mini-ITX motherboard and customized operating system to build a digital forensics workstation and used dcfldd as the imaging platform. The result of this research shows that FIREBrick performed acquisition on the SSD devices on 5 GB/min but they were not compared with other storage media devices. This result makes sense because they used full pc features in this device, so they will get better results.

[7] Conducted research to test different ARM processor-based single board computers. ARM-based computer board is used to demonstrate an effective, low cost, and low energy forensics imaging device. They used Raspberry Pi 2 and Odroid XU4 and 3 storage media to perform acquisition tests on the

testbed using a command-line application (DC3DD). This research found that Odroid XU4 takes 2 times faster in acquisition time than Raspberry Pi. Researchers do not create any application to perform forensics imaging easily.

Storage Evaluator and Knowledge Extraction Reader were developed by [8]. It was a triage device that performed data extraction on the storage media setups on Raspberry Pi devices and used a web interface for running the scripts. SEAKER collect data from storage media that was mounted to this device and analyzed the collected data. The result of this research shows that the researcher creates a device and application that perform logical data collection. They do not provide any imaging process in this research.

Based on the research gap in this section, this research will utilize Raspberry Pi devices to create a prototype open-source forensics imaging devices and develop an application interface to make the forensics imaging process better.

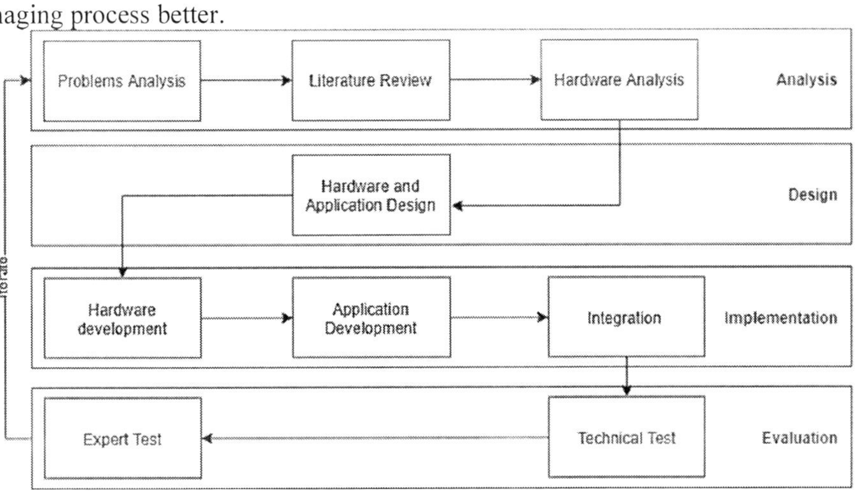

Figure 1. Methodology

3. Method

There are several phases in this paper i.e. analysis, design, implementation, and evaluation. It is divided into 9 stages that are problem analysis, literature review, hardware analysis, hardware and application design, hardware development, application development, integration technical test, and expert test.

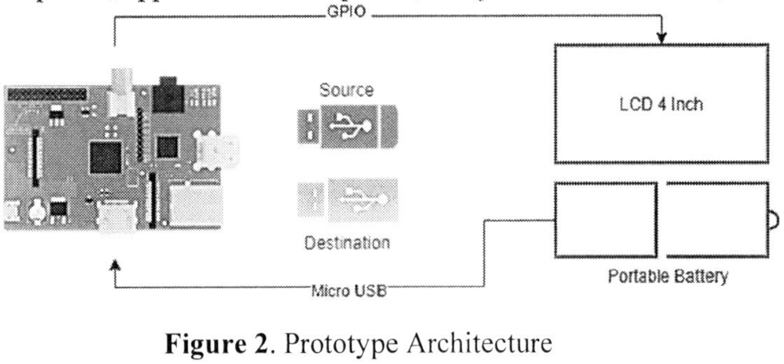

Figure 2. Prototype Architecture

In the analysis phase, preliminary analysis of this research was conducted and the hardware that will be used was chosen based on hardware analysis. It stages to analyze the best hardware for this research by comparing the technical specifications of every single board computer. Design phase we organized the hardware and software that were used to build the system. Then in the design system, a display of the user interface and created pseudo code were made. The implementation section is the process to implement the system related to the design system. The last process was an evaluation phase. It was a process to evaluate the prototype by performing technical and expert testing. Technical testing was evaluated by testing the devices by several scenarios of acquisition and digital forensics experts performed expert testing by evaluating survey instruments.

The prototype used Raspberry Pi 3 Model B+ as a portable device powered by a portable battery and attached an LCD 4 Inch to controlling the devices and application status as shown in **Figure 2**. It consisted of a Raspbian operating system, python shell, python library/module, and GUI based application[9]. Qt Designer was used to making a GUI application for forensics imaging that will be ported into the Raspberry Pi.

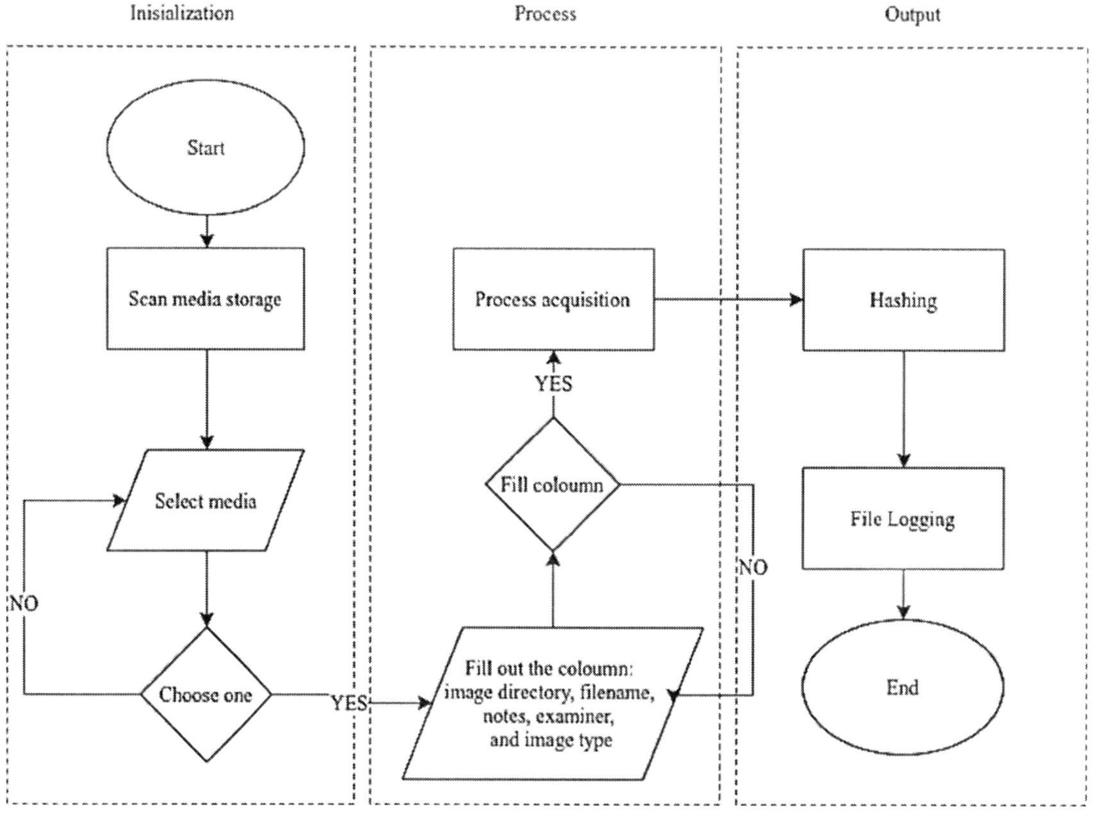

Figure 3. Application Flowchart

This system is divided into three stages i.e. initialization, process, and output. In the initialization stage, the system was able to detect the storage media that was to be acquired. The user chooses which media to process and selects the features in the application such as image directory, filename, notes, examiner, and image type. The main process is scanning, acquisition process, hashing, and generating a log as shown in **Figure 3**. The result of the acquisition process was raw file type. We use MD5 and SHA512 hash to ensure the integrity of the result.

The designing process also designed an application using Qt Designer. The first scenario is the system will present the available storage media in the list form and after choosing the devices that will be acquired the window will be redirected to the pre-process window. Pseudocode is designed by two scenarios i.e. initialization and pre-process. In the first initialization process, three libraries are used such as PyQt5, sys, and os, and the pre-process used libraries such as PyQt5, sys, hashlib, subprocess, logging, time, datetime, and os.

4. Result and Discussion

The prototype was developed with two stages: the first stages was the application development process that develops applications based on Python and Qt5 that port into Raspberry Pi. The second stage was

IOP Conf. Series: Materials Science and Engineering 1077 (2021) 012064 doi:10.1088/1757-899X/1077/1/012064

the performance test of the tools, after the application was ported to the device, the performance of the application tested and monitored. The Tool was tested using 6 scenarios and it was monitored using Cacti. **Figure 4** showed the prototype of a forensics imaging device using Raspberry Pi and was powered by a portable power source.

Application interfaces were designed using the Qt5 library in Python. The application developed following the flowchart in Figure 3. The applications consist of 2 pages that are storage list and acquisition information form with 480 x250 resolution to comply with the display used in this device. Because these devices used 4 inch Raspberry Pi LCD. The device that will be acquired is attached to the devices on the USB port and the application will scan the attached devices.

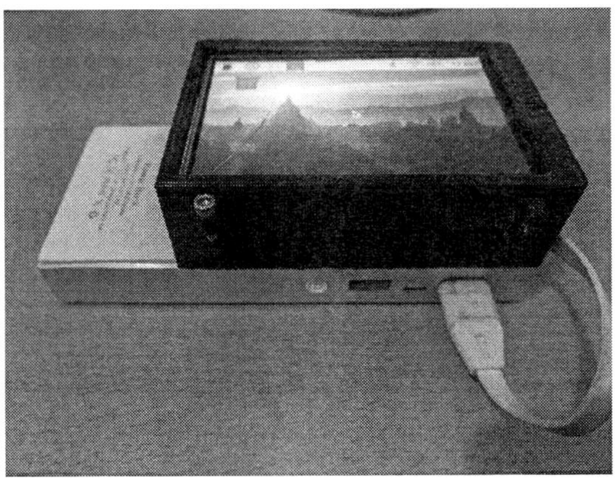

Figure 4. Prototype Forensics Imaging Device

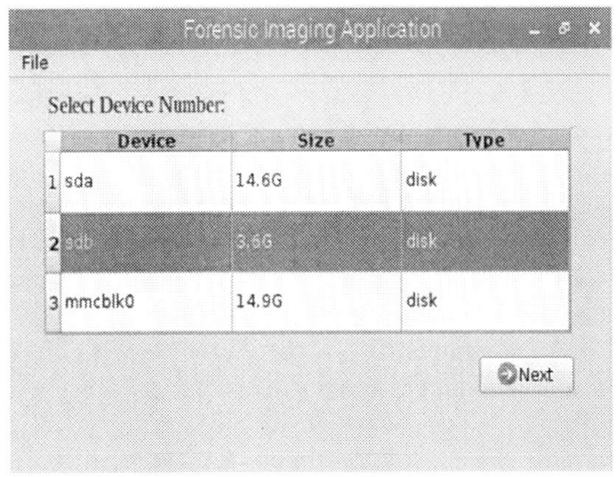

Figure 5. Application Interface – Disk List

Figure 5 depicted the user interface of the GUI application for forensics imaging. The first page shows the list of physical devices that were attached to the device. This is an automated process to detect the disk which is attached to this prototype forensics acquisition device. The device that was selected from the list would be acquired by the highlighted row. Mmcblk0 was a default device that always appeared on the list, which is the primary partition of Raspberry OS.

After the acquisition process was finished, the application will generate a log file that contains information relevant to the investigation process as shown in **Figure** 7. Information's that are generated by the application are as follow:

a. Media information
   Media information showed base information about the device. This information is generated automatically by the application based on physical device information on the fdisk –l command. The information that is generated in this process is Physical source, label, size, number of sectors. and UUID.
b. Acquisition Information
   Acquisition information contains base information about the acquisition process such as result file name, destination directory, note, and examiner name. This information is captured before the acquisition process is started as indicated in **Figure 6**.
c. Time
   This section shows the start time and end time of the acquisition process of the physical device and is displayed in a long time format. It is generated by the system based on the time when the acquisition was launched and finished.
d. Hash Checksum
   After the acquisition process has been finished, the application will generate a hash checksum. There were 2 types of hash checksum which are md5 and sha512. The checksum was generated in 2 forms: source checksum and cloning checksum. The Source was the checksum from the physical device (/dev/sdx) and the cloning checksum is generated from the result of the acquisition process.

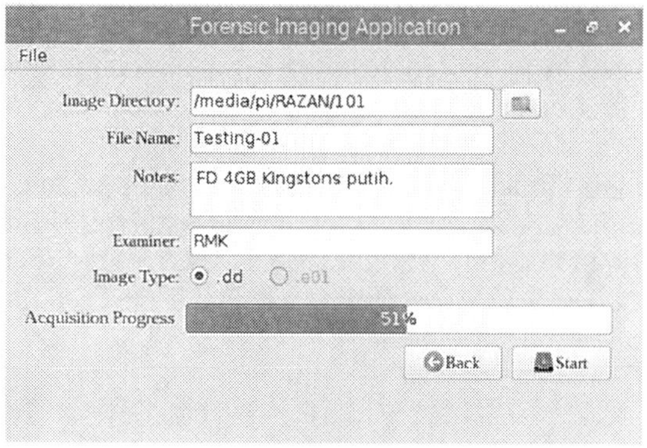

Figure 6. Application Interface – Imaging Information Form

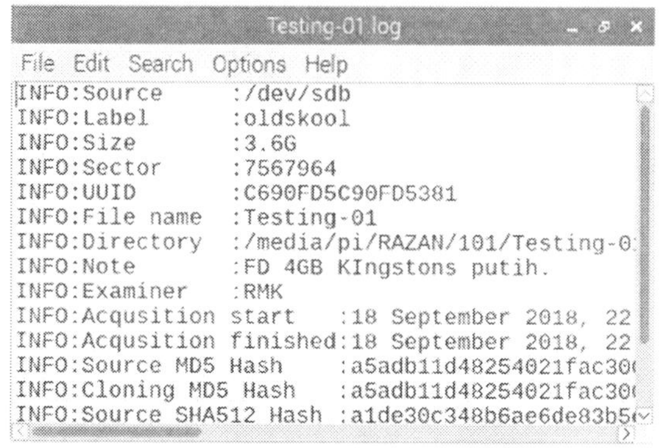

Figure 7. Example of Log File

The prototype was tested using 6 scenarios and 6 samples of various storage devices. There were 3 categories of the sample based on the sample size that are 4 GB, 8 GB, and 16 GB. Table 1 describes the performance of the system from different samples. The objective of this evaluation is to find out if the system opens with several types of samples.

As mentioned before the system was monitored using Cacti to get the memory usage and CPU load in the device. **Figure 8** showed the memory usage of sample 1. Based on the result of Cacti average memory usage for every sample was recorded in Table 1. The data showed that the size of the sample affects the acquisition time. The size of the disk does not affect the transfer speed of the acquisition process. The transfer speed is one of the challenges that must be solved in this research. Transfer speed was calculated by using the formula (1).

$$Transfer\ Speed = \left(\frac{Total\ Bytes\ Acquired}{Total\ Time\ in\ Second}\right) \Big/ 10^{20} \qquad (1)$$

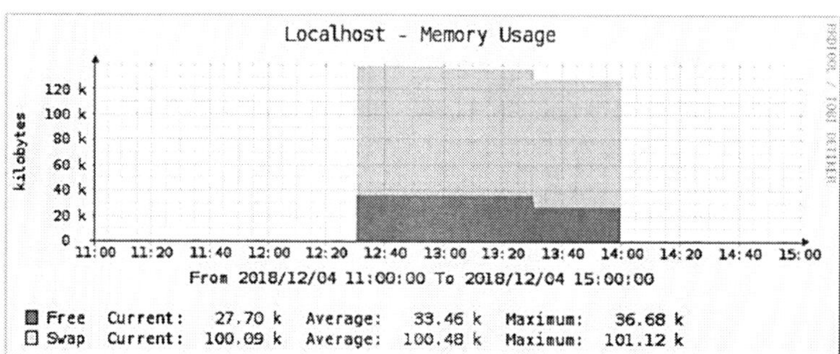

Figure 8 Memory Usage Monitored on Cacti

The transfer speed was spread between 1.69 MB/s to 1.93 MB/s and the average transfer speed was 1.85 MB/s. It is good speed for low-cost digital forensics imaging hardware using Single Board PC. The average total memory used in this prototype tends to be the same in all samples. It took 1048543.65 Kilobytes average of total memory use when the total memory of Raspberry Pi was 1,048,576 Kilobytes. Total memory used means that include memory used by the operating system because in this test Cacti cannot display the individual memory usage of the developed application.

The prototype was also tested on the digital forensics community in the special region of Yogyakarta with 56 respondents who have expertise in the fields of digital forensics. The test was performed by creating a demo video related to the process of acquiring flash drives using these prototypes. There were

3 aspects of the question in the questionnaire that were functionality, user interface, and benefit. It breaks into 14 instruments. The instrument of this test can be shown in **Table 2**. The result was calculated using the Likert formula and the average score of the test was 4.04 in all aspects.

Table 1. Test Result of Samples

No	Sample	Time Taken (Minutes)	Transfer Speed (MB/s)	Average Memory Usage (Kilobytes)
1	Sample 1 - 4 GB	35	1.75	1048543.32
2	Sample 2 - 4 GB	34	1.91	1048543.61
3	Sample 3 - 8 GB	73	1.69	1048542.13
4	Sample 4 - 8 GB	64	1.92	1048545.56
5	Sample 5 - 16 GB	129	1.93	1048543.63
6	Sample 6 - 16 GB	131	1.88	1048543.63
		Average Transfer Speed	**1.85**	**1048543.65**

Table 2. Survey Instruments

No	Aspect	Instrument
1	Benefit	Have the prototype advantages for digital forensics investigator
2		The use of a single-board computer (SBC) specifically Raspberry Pi can decrease the cost of the prototype against the real digital forensics hardware
3		The prototype is easy to carry because it is a small device
4		Will the investigator use this prototype for creating a digital forensic image
5		The imaging result of the prototype has been met with digital forensic investigation methodology
6		The prototype has prospects to used later
7	User Interface	The application interface is interesting to use
8		The application interface is easy to recognize
9		The displayed menu is easy to understand
10		The placement of all the features and application buttons is appropriate
11	Functionality	The autorun feature of the application runs well
12		Auto-scan feature to detect what storage is available is running well
13		All buttons and menus work normally
14		All error handling run well

5. Conclusion

In this paper, this research proposed a prototype portable device for performing digital forensics acquisition by utilizing Raspberry Pi devices with an application developed for this purpose. These tools can be used for supporting digital forensics investigation processes especially in the process of acquisition. The prototype was successfully created and tested, Not only in a scenario-based test but also tested by the digital forensics community. Based on the test result of the scenario-based test, the prototype forensics imaging device can perform the acquisition process on the sample flash drive with 1,85 MB/s speed. Also, responses from the digital forensics community have a good result.

For future work, some improvements are needed, not only in the scope of application development but also in the operating system and hardware development. The application must be improved to cover another disk format such as E01 and AFF4. Besides, Transfer speed needs to be improved for a better and faster acquisition process such as perform acquisition in Raspberry Pi Cluster.

References

[1] FBI 2019 *2019 Internet Crime Report* Retrieved from https://www.fbi.gov/news/stories/2019-internet-crime-report-released-021120

[2] Reedy P 2020 Digital evidence review 2016–2019 *Forensics Sci. Int. Synerg.* **2** pp 489–520

[3] Goverment of Republik Indonesia 2016 *Undang-Undang Republik Indonesia Nomor 19 Tahun 2016 Tentang Perubahan Atas Undang-Undang Nomor 11 Tahun 2008 Tentang Informasi dan Transaksi Elektronik* Retrieved from https://jdih.kominfo.go.id/produk_hukum/view/id/555/t/undangundang+nomor+19+tahun+2016+tanggal+25+november+2016

[4] Feng X, Babatunde O and Liu E 2015 Cyber security investigation for raspberry Pi devices *Int. Ref. J. Eng. Sci.* pp 1–14

[5] Carrier B 2002 *Open Source Digital Forensics Tools: The Legal Argument* Retrieved from https://citeseerx.ist.psu.edu/viewdoc/download?doi=10.1.1.581.6818&rep=rep1&type=pdf

[6] Tobin L and Gladyshev P 2015 Open forensics devices *J. Digit. Forensics, Secur. Law* **10** 1–5

[7] Olson E and Shashidhar N 2016 Low budget forensics drive imaging using arm based single board computers *J. Digit. Forensics, Secur. Law* **11** pp 53–76

[8] Gentry E and Soltys M 2019 SEAKER: A mobile digital forensics triage device *Procedia Computer Sci.* vol **159** pp 1652–61

[9] RASPBERRY PI FOUNDATION 2018 Raspberry Pi 3 Model B+ *Raspberry Pi*

Circuit Breaker in Microservices: State of the Art and Future Prospects

Falahah*, Kridanto Surendro, Wikan Danar Sunindyo

School of Electrical Engineering and Informatics, Institut Teknologi Bandung, Jl. Ganesha 10 Bandung, Indonesia

E-mail: *falahah@telkomuniversity.ac.id

Abstract. Resilience is one of the important issues in the implementation of microservices architecture, and to handle it, various approaches and solutions are available. A circuit breaker is a solution to handle failure in microservices and is now widely implemented both as a library and as a pattern embedded in the service and client modules. Despite the widespread use of circuit breaker, the research on circuit breaker is relatively less than research subject on other microservices area, or in microservices resiliency. This article provides an overview of recent research in circuit breaker, map the research subject, and find the opportunities for future research. The research is conducted using a systematic map study but the results focused on extract the knowledge and find important keywords to build a conceptual overview of the research field circuit breaker. The result shows that there are five categories of research subjects which are concept, strategy, management, implementation, and products. After exploring each category, we can propose the opportunity in enhancing the use of circuit breaker in microservices architecture, by explore the potential implementation of proxy circuit breaker.

1. Introduction

The popularity of microservices architecture kept arise in recent year. Lots of research has been conducted in the field of microservices and it still have some issues need to address on implementation. Some important issues such as managing microservices, monitoring, DevOps culture, integration testing, and fault tolerance [1] are emerging along with the rise of complexity in microservices implementation. The basic of MSA principles that the services can be provided by a different organization and running on different platforms, make resilience become an important issue. The resilience can be viewed from two perspectives, from the client that accessing the services and from the service provider. In the microservices-based system, the client does not interact directly with the service provider, so the responsibility of resiliency that should be a consideration for sides, the client and service provider, usually not as easy as in a monolithic-based system.

The usage of the circuit breaker as a pattern for resilience has been populated since it first mentioned in Martin Fowler's blog [2]. Lots of technology available for supporting the implementation of the circuit breaker in microservices. But there is very little research interest on circuit breaker, as a specific topic in microservices resilient strategy. Most of the research articles mentioned the circuit breaker as a pattern that usually exists in MSA, but not much of them discussed the detailed process on the circuit breaker.

This research aims to find the opportunity for extending existing research on the circuit breaker, by reviewing existing state-of-the-art, classifying, and categorizing existing research on the circuit breaker, exploring the issues and challenges, and finding the opportunities for future research. The rest of the paper will be organized as follows: section 2 will discuss the method we use, section 3 will explain the basic concept of the circuit breaker, section 4 will discuss the research subjects we extracted from recent literature, detailed breakdown and connection between them, and proposed the map of the research subject on the circuit breaker, section 5 will discuss the future research opportunities, and section 6 is the conclusion of our research. To make it easy for writing, we will use some abbreviations according to our topic of discussion as follows: Microservices Architecture (MSA), Circuit Breaker (CB), and Research Questions (RQ).

2. Research Method

The objective of this research is to explore the role of CB in MSA and find the opportunities for enhancing the utilization of circuit breaker on MSA. The motivation of our research is drive by the fact that there is little interest in research that has focused on CB, and most of the research reports mention CB as a component needed to maintain MSA resilience. We conduct the research using a systematic mapping study approach, but we choose to present the result by discussing and summarizing the knowledge extracted from various literature. The steps for doing the research are as follow:

1. Define research questions as the motivation for this study.
2. Define searching strategy that includes source for searching, search criteria or keywords, and inclusion and exclusion criteria.
3. Analysis and extract the data from selected resources.
4. Result, discussion, and summarize the finding from the study.

2.1. Research Questions

We define 3 research questions and backgrounds for motivation. The goal of this research is to find the opportunity of enhancing the use of CB in MSA, based on existing implementation, both concept and the technologies available. From the papers and articles that discuss the MSA, most of them mentioned the circuit breaker as a part of resilience, but only little provides a detailed description of how it works [3]. To achieve our goal, we define some research question as shown on table 1.

Table 1. Research Questions

Research Question	Motivation
RQ1: What is the role of Circuit Breaker in MSA?	Get a basic understanding of how the CB works in MSA and find the internal structure of CB.
RQ2: What is the research subject of the circuit breaker?	Understanding the problem space of CB based on recent research and its result.
RQ3: Are there any opportunities for enhancing the use of circuit breaker in MSA?	Finding the opportunity for enhancing the use of CB in MSA

2.2. Search Strategy

We searched for articles from several databases such as ACM, CiteSeer, IEEE Xplore, Wiley, and Springer Link. We also consider the articles that come from blog and whitepaper, to cover recent issues in CB implementation in practices and industries.

Table 2. Keywords for Searching Criteria

Major terms	Synonym terms
Microservices	Architecture, pattern
Resilience	Performance OR Failure OR failover
Circuit breaker	Communication OR closed OR half-open

Microservices is an emerging architecture style, lots of interesting discussions, and examples on the blog, and some of the sites consist of huge information such as microservices.io and martinfolwer.com.

We also collected some information from GitHub as media for discussion among programmers. In the searching process, we did the four steps as follows: define the major keys, identify alternative words, synonyms, or related terms to major keywords, and verify the contents of articles are matching with the major keyword. We focused on CB in MSA, because the term circuit breaker is commonly used in some field such as electrical circuits, computer networks, and electrical power. Table 2 shows the keywords using in the searching process.

The papers that were found then evaluated using some criteria to find the most important and relevant papers that can help us answer the research questions. We use the criteria as follow:

Inclusion criteria:
1. The article discusses the concept of circuit breaker and its implementation in a microservices architecture.
2. The article shows the example of circuit breaker implementation.
3. The article reports the best practice on circuit breaker technologies.

Exclusion criteria:
1. The article discusses the implementation of circuit breaker outside of microservices, such as in the electricity field or networking system.
2. The article does not describe detailed information about the circuit breaker.
3. The paper was older than five years, to find the most recent improvement in circuit breaker techniques and its implementations.

After implementing the search criteria above, we select 12 papers published between 2015-2020. From the papers we choose, we can identify some important keywords and can extract some important information from the papers.

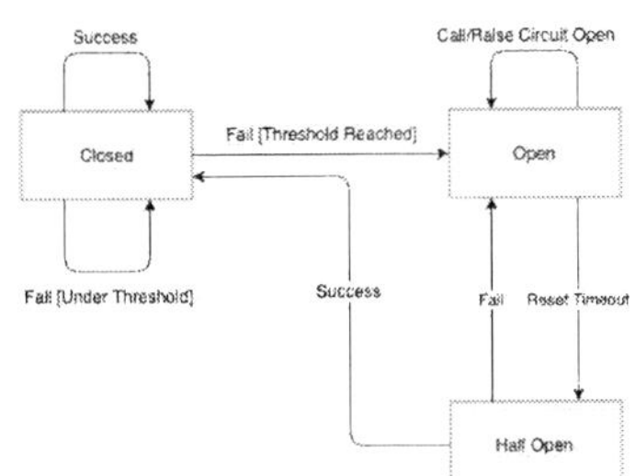

Figure 1. Mechanism in Circuit Breaker [5]

3. Concept and Structure of Circuit Breaker

This section will answer the first research questions: what is the role of CB in MSA? Since it was mentioned in Martin Fowler's blog in 2014 [1] as a technique for preventing the error cascade on the system, some articles and blogs discuss the concept of CB. The idea first mentioned in Nygard book [2] that proposed some pattern to keep the stability of the software, which are: circuit breaker, timeouts, decoupling middleware, handshaking, and test harness.

The benefit of using CB can be mentioned as follow [4]: prevent to access failure component, handling error quickly and gracefully, prevent the caller stuck on failure services, can custom the fallback action, prevent the services for overwhelmed managing request.

Most of the articles have the same perception of implementing the CB concept: a CB is a bridge between the clients to a target service to prevent accessing the service when it failed. As a bridge, CB can have three states, open, close, and half-open. If it closed means that all the requests are forwarding to the service, if it opened means that all the requests are blocked. Figure 1 shows the basic concepts of the circuit breaker.

The three states above can be described as below:
1. Closed: all the request is forwarded to the service targeted, it kept happens until the system detect the failure in certain threshold (such as timeout counters or sum of failure request), the

circuit breaker will become open. The threshold can be defined by the system or pre-set by the user.
2. Open: all the request is blocked by the circuit breaker, and circuit breaker or any others component, send a failure message to the requestor or client, or we can call any failure-handle procedure.
3. Half-open: on this condition -that happens during the services failed and circuit breaker in open state- the circuit breaker can be set half-open, which means that the circuit breaker will try to contact the service to check when it becomes responsive. It can be done periodically by pinging the service or after a certain time. When the circuit breaker received the response from a service, the state can be set closed again and all the requests can be forwarded as normal.

Most previous works use the same concepts of how a circuit breaker works, except there was a proposed idea for modifying the states into two states: open and closed, that implement Markov Chain to detect the fault without a timeout, so the timeout can be set dynamically [6].

As a pattern for microservices, the structure of a circuit breaker can be summarized as below, from several sources as [2], [7], and [8]:

```
[Initialization]
   Parameters (timeout, failure_threshold)
   Initial_state = closed
[the states]
Case state
State=closed
   Begin
   Forward_message (the services)
   If_error
   error_handling
   Set_state= open
   End
State:open
   begin
   error_handling
   Set time_out
   Set state=half-open
   End
State=half-open
   Begin
   Call_services(period)
   If response=true then
        Set state = closed
   End
```

From the structure code above, we can identify the fundamental part of the circuit breaker structure as follow:

1. Parameter, for controlling the circuit breaker status and process.
2. Minimum three-block code for execute the action for each state.
3. Error handling or alternative actions can be written inside the circuit breaker or can be put outside as separated components of the circuit breaker.

The taxonomy of the circuit breaker was proposed by [4] but the author refers to the implementation in IoT. The taxonomy can give us an overview of the general pattern of the circuit breaker. It divides into two segments, structural and behavioral.

The parameters that commonly implemented in a circuit breaker are [4] [9]:

1. Timeout
2. Max Concurrent Request
3. Request Volume threshold
4. Sleep Window
5. Error Percent Threshold

The role and position of circuit breaker compare with others component on microservices architecture implementation:

1. CB as a pattern: Most articles, books, and blogs mentioned CB as a pattern. [9] categorized the CB as a reliability pattern. It was put as a running pattern along with service discovery, security, and observability pattern. [10] put it as generic pattern and then become a decorator. As a pattern, CB structure can be inserted as a block of code when a client calling a service. It can be used for supporting microservices resiliency and prevent failure cascading to other services. The main task of CB is to block the access and send the request to check the availability, but it has no responsibility to repair the failed service. We need to implement a different scenario for handling this.
2. CB as a tactic: CB is also categorized as a tactic that support three case: prevent single dependency, set timeouts, and providing fallbacks [11].
3. CB as a security aspect: the existence of CB can be considered as a security aspect for designing microservices that running on service-mesh architecture, but it can be achieved using the combination of some strategies such as throttling (rate-limiting), load balancing, blue/green deployments, and canary releases [12].

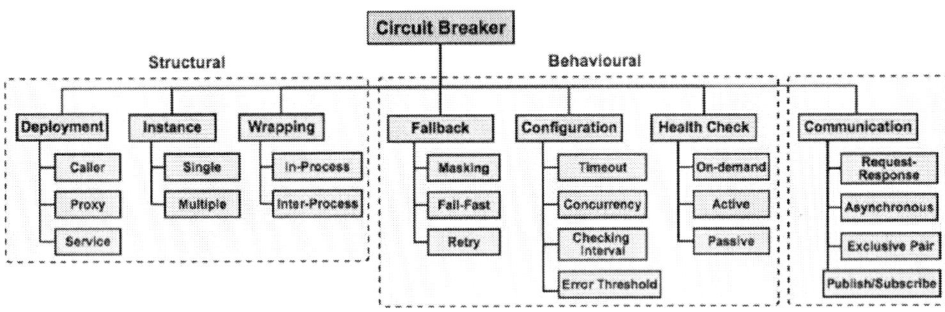

Figure 2. Circuit Breaker Taxonomy [4]

4. Research Subjects in Circuit Breaker

In this section, we will analyze and find the relationship between the keywords and breakdown the main keywords into a specific issue to get the map of the research subject of CB. This section aims to answer the RQ2. what is the research subject of the circuit breaker?

After collecting and reading 23 articles on CB, we use five keywords for identify the research subjects on CB at MSA. The approach of finding the keyword can be defined through answering the sequences of questions as follows:

1. What is the role of CB in MSA? To answer this question, we need to understand what is CB and how it actually works, it leads us to identify the basic concept of CB, what is it? Are there any specific components? How we use it? Is there any parameter or behaviour we need to know? On this section, we discuss the CB only on two elements: the client that request a service, and the service itself.
2. How to implement it? After knowing the basic concept on how the CB works in two elements above, we then expand the exploration that in the MSA, there is many microservices and clients, and usually client and a service is built separately, where and how we insert the CB? In client or in a service or both? We resume this question into the keywords: strategy.
3. In complex MSA, there is possibilities that we have to manage more that CB. Each CB will have specific features on the service under its monitored or supervised. It leads us into the question how to manage the CB if we need to access different services? Do we need to set one CB for each instance of services, and if we have to set one CB for each service, how to control it? We resume this question into a keyword: management.
4. The implementation of CB on MSA will follow the scope of MSA implementation. Since we know that today, the MSA can be exist on web application, mobile application or IoT, then the CB will exist on the same platform of MSA. We use the keyword: implementation.

5. To answer our last question about the opportunities of enhancing existing implementation, we search about the existing product than contain CB and how the product is use. We define the keyword for this issue: product.

After defining 5 main keywords above, we can describe the scope of research subject in CB as follows: Can we explore the possibilities of enhancing the implementation of CB among existing products, technology or concept, so we can find the new approach for managing CB and implement it using certain strategy. Through five keywords above, we extract the relevant keyword from the papers or articles and mapping them into conceptual model as shown in figure 3.

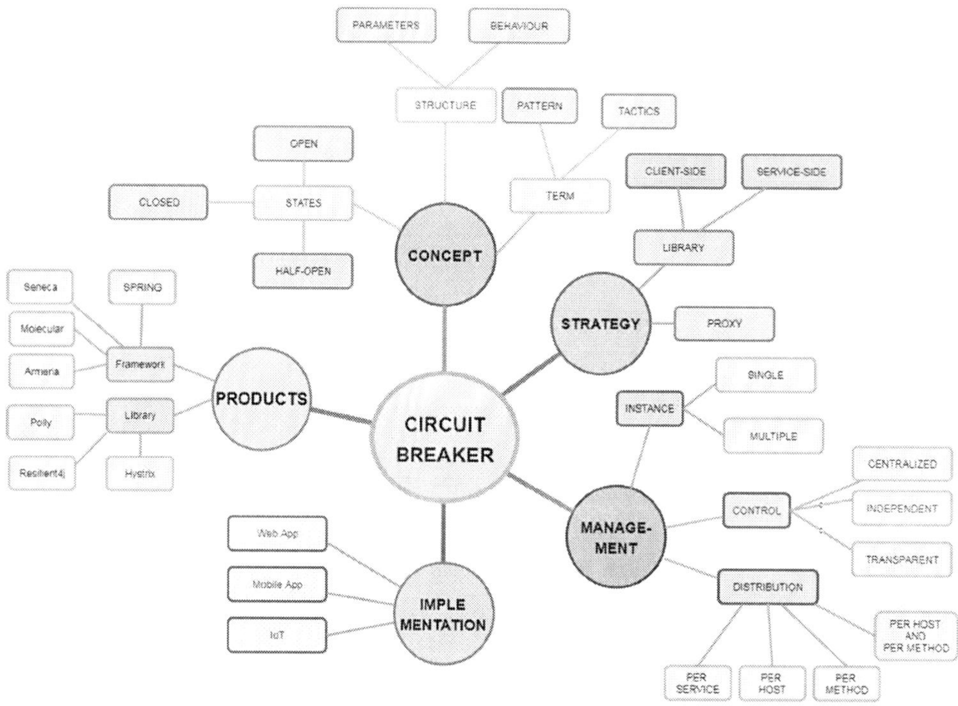

Figure 3. Research Area of Circuit Breaker in Microservices Architecture

4.1. *Strategy*

In this section, we will analyze and find the relationship between the keywords and breakdown the main keywords into a specific issue to get Most of the articles do not describe the specific strategy for implementation. The common strategy for implementing CB is using existing libraries such as Hystrix Netflix, Armeria, Pybreaker, and Polly. By implementing CB as a pattern, the client has the option for calling the service using the CB, by calling the CB library explicitly on the script. The idea for placing the CB as an external component, not just a pattern, called a proxy circuit breaker is explained in [7], [10], and [12]. But since then, there is no real example of how to implement this strategy. The pros and cons about circuit breaker implementation options, according to [7] and [13] can be explained as bellow:

1. Client-side CB, when CB is inserted into the client component, so it provides easy and flexible implementation. But the weakness of this approach is the information about services availability only exist for a local client. If the services are accessed by some clients, each client should implement its CB separately.
2. Server-side CB, when CB is inserted into the service component, so the status of service can be published and read by the clients. The problem with this strategy, usually the service provider does not allow their service modified by the other.

3. Proxy-side, when CB is deployed separated, so we do not have to modify client or service component. To adopt this strategy, we need an additional module or mechanism to arrange the request and manage the CB.

The proxy side can be considered as an alternative strategy to implement CB since it gives us a more flexible way of implementing CB without modifying the client or service. The possibility of implementing proxy side CB can be mapping into a specific issue on another research area as shown in table 3. In the management area, proxy CB allowed us to manage the CB centralized and transparent. The client and the services do not have to be modified and for distributed CB, we can assign CB for each host, method, or both.

Table 3. Keywords for Searching Criteria

strategy	Management									product	
	instance		control			distribution					
	single	multiple	centralized	independent	transparent	per service	per host	per method	per host and per method	library	framework
client	x	x		x			x	x	x	x	x
server	x	x			x	X				x	
proxy			X		x		x	x	x		

4.2. Management

The important issue of implementing CB is how we manage the CB. The complex system of MSA that consists of hundreds of services and with layered services level need the various approach to manage the CB. We have some options for managing the CB such as:

1. Instance: we can create a single instance for CB to manage some instance for one service, or each instance of services need to monitore by one instance of CB
2. Control: we can implement the three options for controlling the CB, which are centrally managed, independently managed, and transparently managed [14]. On centralized CB, the CBs are managed by the specific component that can run in API gateway. All incoming requests from the client should go to this component for checking the service availability. On independent CB, the client should manage the CB itself, or manage by each service. Transparently managed is the condition when CB is managed by an intelligent runtime environment. It is an ideal condition but preparing such an intelligent environment is not easy.
3. Distribution is the issue that considers how CB managed the services. Do we need to assign one CB for each service, one CB for each host or method [15]. Some problems can arise on this condition such as discussed on [16] using Polly library but it has the issue for quorum decision. Armeria, a framework for microservices, provides the option for grouping the circuit breaker using three methods: per host, per method, and per host and per method [17].

4.3. Implementation

The Circuit breaker can be implemented in any area as MSA can be. Some articles show the implementation on IoT [4], Cloud [18], or mobile applications, shows that CB is an important pattern or library that accesses by the client when calling the services.

4.4. Products

Implementation of CB in industries can be as a library such as in Hystrix, or as a component on a complete framework for building microservices-based application.

1. As a library, the CB pattern can be found on Hystrix (Java), Polly (.Net), Resilient4j (Java), Akka (Scala), Rubyist (Go), pybreaker (Python), JRugged, and Javaslang. Apart from those

mentioned earlier, nowadays we can find many alternative libraries for CB by searching database on GitHub [19].

2. CB also usually exist on a framework for building microservices. We can use some popular framework for creating a microservices-based system such as using Spring, Seneca, Armeria, and Molecular, and adding the circuit breaker feature.

5. Future Research Opportunities

In this section, we will answer RQ3: Are there any opportunities for future research in the circuit breaker? As previously mentioned, there is relatively little research on the CB subject compared to other research topics on MSA. In the area of the internal structure of CB, there is a report that proposed to modify the state of open-closed using the Markov chain approach and eliminate half-open state [6]. In [3], the author tries to analyze the behavior of CB, and in [8], there is a formal description of the internal CB structure. The research conduct by [20] and [21] proposed the integrated model for MSA resilience that consist of CB as a component, but the articles do not focus on a specific issue in CB. The idea of future research in CB is building CB as a separate component of services or clients, instead of just attach as a library, is coming from [7] and [10] . It is called as proxy CB. The need for this CB is relevant to the issue of distributed CB ([17], [16], and [14]), and it falls into the area of managing CB. [22] proposed the idea for packaging the CB as well as packaging the microservices and other components in MSA.

The idea for managing CB as a different component and the possibility to manage the CB as managing the services through service discovery, load balancer, and service routing, could become an interesting topic, but rarely addressed in the research subject. To adopt this idea, we need to consider the basic function of CB and set the clear definition, limitation, and task assignment between CB, load balancer, and service discovery.

Existing condition shows that most microsevices-based systems inserted the CB directly as a pattern into a client script may lead to the opportunity to create CB more flexible and intelligent. It opens lots of research opportunities in the future. Some of the opportunities such as:

1. Improving the internal side of how the CB work, such as setting the rule of how CB change the state or identify the basic action need to execute for each state.
2. Existing condition shows that most microservices-based systems inserted the CB directly as a pattern into a client script may lead to the opportunity to create CB more flexible and intelligent. It opens lots of research opportunities in the future. Some of the opportunities such as:
   a. Improve the internal side of how the CB work, such as setting the rule of how CB change the state or identify the basic action need to execute for each state.
   b. Extending the ability of environment where the CB is implemented, by adding the capability how to control and manage the CB from the external component so it can be set dynamically as an aware component for changing in environment, such as automatically adjust the parameters based on services load or based on the previous request, manage the incoming request by periodically checking the queue on a service, or resolve the problem that could exist on distributed CB issue.

6. Conclusion

CB is an important pattern on keeping the resiliency on MSA. The use of CB on MSA becomes a necessary component on building the MS-based system, along with other resilient patterns such as retry, failback or bulkhead. Despite the fact that CB almost exists on each MSA-based system, the research that addresses the CB is relatively less compared with other issues in MSA implementation. We conduct the literature research for understanding the concept of CB and identifying the most important keywords on CB research. It helps us to provide a conceptual overview of the research subject on CB, and we present the map to show the problem space in CB. Based on the map, we explore each subject and find the possibilities for future research directions. We identify that there is a lack of research in CB management, especially the need for managing distributed CB, or manage CB as a separated component from client or services.

References

[1] Ananthasubramanian A 2018 *Challenges in Implementing Microservices DZone* Retrieved on 5 September 2020 from https://dzone.com/articles/challenges-in-implementing-microservices
[2] Fowler M 2014 *Circuit Breaker* Retrieved on 10 September 2020 from https://martinfowler.com/bliki/ CircuitBreaker.html
[3] Mendonca N C, Aderaldo C M, Camara J and Garlan D 2020 Model-based analysis of microservice resiliency patterns *Proc. of IEEE Int. Conf. on Software Architecture (ICSA)*
[4] Aquino G, Queiroz R, Merrett G and Al-Hashimi B 2019 The circuit breaker pattern targeted to future IoT application *Proc. of Service-Oriented Computing: 17th Int. Conf. (ICSOC)*
[5] Goyal S 2019 *Circuit Breaker in Microservices Architecture* Retrieved on 2 September 2020 from https://www.codeprimers.com/circuit-breaker-in-microservice-architecture/
[6] Addeen H H 2019 *A Dynamic Fault Tolerance Model for Microservices Architecture* thesis (South Dakota: South Dakota State University)
[7] Montesi F and Weber J 2016 Circuit Breakers, Discovery, and API Gateways in Microservices *Preprint arXiv:1609.05830*
[8] Vergara S, Gonzalez L and Ruggia R 2020 Towards formalizing microservices architectureal pattern with event-B *Proc. of 2020 Int. Conf. on Software Architecture Companion* pp 71–4
[9] Richardson C 2018 *Microservices Patterns* (South Carolina: Manning)
[10] Montesi F and Weber J 2018 From the decorator pattern to cicruit breakers in microservices *Proc. of 33rd Annual ACM Symp. on Applied Computing* pp 1733–5
[11] Marquez G and Astudillo H 2019 Identifying availability tactics to support security architectural design of microservice-based systems *Proc. of 13th European Conf. on Software Architecture 2019* (New York: ACM) pp 123–9
[12] Chandramouli R and Butcher Z 2020 *Building Secure Microservices-based Application Using Service-Mesh Architecture* (Gaithersburg: NIST, US Department of Commerce)
[13] Chandramouli R 2019 *Security Strategies for Microservices-based Application System* (Gaithersburg: NIST, US Department of Commerce)
[14] Lin T 2018 *Which Strategy You Will Choose to Implement Circuit-Breaker for Microservices Architecture Application, Part 1* Retrieved from https://medium.com/@1000lin/ which-strategy-you-will-choose-to-implement-circuit-breaker-for-microservices-architecture-dcd690f3c90
[15] Scott C 2018 *Designing Resilient System - Part 1* Retrieved on 7 September 2020 from https://engineering.grab.com/designing-resilient-systems-part-1
[16] Reisenberger 2017 *What is Distributed Circuit Breaker?* Retrieved on 1 October 2020 from https://github.com/App-vNext/Polly/issues/287
[17] Yuichi O 2016 *Circuit Breakers for Distributed Services* Retrieved on 5 September 2020 from https://engineering.linecorp.com/en/blog/circuit-breakers-for-distributed-services/
[18] Balalaie A, Heydarnoori A and Jamshidi P 2015 Migrating to cloud-native architectures using microsrevices: an experience report *Proc. of European Conf. on Service-Oriented and Cloud Computing, Advanced in Service-Oriented and Cloud Computing* pp 201–15
[19] Github 2020 *Circuitbreaker* Retrieved from https://github.com/topics/circuitbreaker
[20] Heorhiadi V, Rajagopalan S, Jamjoom H, Reiter M K and Sekar S 2016 Gremlin: systematic resilience testing of microservices *Proc. of 2016 IEEE 36th Int. Conf. on Distributed Computing Systems* pp 57–66
[21] Hoorn A v, Aleti A, Dullmann T F and Pitakrat T 2018 ORCAS: Efficiency resilience benchmarking of microservices architectures *Proc. of 29th IEEE Int. Symp. on Software Reliability Eng. (ISSRE 2018)* pp 146–7
[22] Montesi F and Thrane D 2017 Packaging microservices *Proc. of 17th IFIP Int. Conf. on*

Distributed Applications and Interoperable Systems (DAIS) pp 131–7
[23] Nygard M T 2007 *Release It! Design and Deploy Production-Ready Software (Pragmatic Programmer)* 1st Edition (Pragmatic Bookshelf)
[24] Molchanov H and Zhmaiev A 2018 Circuit breaker in systems based on microservices architecture *Advanced Information Systems* **2** 4 pp 74–7

Exploring Familiarity and Participation in Online Gig Economy Among Indonesians

Nina Mia Aristi[1], **Ahmad R Pratama**[2]
[1]Master Program in Informatics. Universitas Islam Indonesia, Sleman, DIY, Indonesia
[2]Department of Informatics, Universitas Islam Indonesia, Sleman, DIY, Indonesia

E-mail : [1]nina.aristi@students.uii.ac.id, [2]ahmad.rafie@uii.ac.id

Abstract. Online gig economy has been rapidly growing and shown a promising prospect in Indonesia, but little is known about familiarity and participation of Indonesians in this new business model. This study tries to fill that gap in literature by exploring familiarity and participation in four different types of online gig economy, i.e. transportation-based service (TS), professional service (PS), asset-sharing (AS), and handmade goods, household, and miscellaneous services (HGHM) among Indonesians based on an online survey of 385 participants. The primary data were analyzed by using descriptive statistics, multiple correspondence analysis (MCA) and logistic regression models to show that most Indonesians are TS gig users whereas PS is the most preferred category by Indonesian gig workers. This study found some differences in characteristics of gig users and gig workers in different categories of online gig economy. Most notably, it suggests the existence of sexual, generational, geographical, and socioeconomic gaps among Indonesians when it comes to their participation in different types of online gig economy. Despite the high participation rate in the online gig economy, the familiarity rate in both the concept and the term "online gig economy" is relatively low. The findings from this study can shed light and serve as a steppingstone for further research on the online gig economy as a new business model and dig deeper into its potential in Indonesia

1. Introduction

Technological advancement over the time has constantly changed various aspects of human lives, including in economics and job sectors. Recently, many firms have changed their operations and started to look at freelance services as an alternative to their labor needs [1]. The Internet, acting as the bridge between individuals across the globe has also been used as a meeting point between employers and job seekers. Now, it also allows a new business model to grow by connecting freelancers as service providers and firms, organizations, or even other individuals looking for various types of freelance services as their users. This new business model is also known as the *online gig economy,* where the two parties are only bound in either a short-term contract of employment or on a gig or task basis.

The surge of Internet users in Indonesia, specifically that 10,12% increase in 2018 [2], has given a broad opportunity to many Indonesians to be part of that online gig economy, either as gig users or as gig workers. Most notably is in the case of the transportation-based service type of online gig economy that is driven by two major companies, an Indonesia-based Gojek and a Malaysia turned Singapore-based Grab. Both companies have been providing ride-hailing service with motorcycle or car options since 2010 [3] and 2014 in Indonesia [4]. The high popularity of food-based e-commerce at 27.85% [5], especially by small medium enterprises that have no official food delivery services have also contributed to the rising popularity of transportation-based online gig economy. Even more so during the COVID-

19 pandemic that shows a surge in online shopping activities by 320% [6] transportation-based online gig economy in the form of package couriers or delivery service.

Apart from the transportation-based one, 20.83% e-commerce in Indonesia in 2019 provided various services ranging from professional services (PS) to asset sharing (AS) as per Indonesian Census Bureau [5]. Nonetheless, most platforms supporting asset-based or professional service online gig economy, such as Airbnb in the case of AS, or Fiverr and Upwork in the case of PS are either not Indonesian based companies, not operating in Indonesia officially, or not supporting Bahasa Indonesia as one of their official languages. Likewise, information about these types of online gig economy is rather limited. The facts above can serve as another barrier to entry for some Indonesians that can prevent them from joining the other types of online gig economy other than the TS category.

Therefore, this study aims to explore familiarity and participation of Indonesians in various types of online gig economy, including transportation-based, asset-sharing, professional service, and handmade goods as well as to investigate demographic factors associated with different types of online gig economy in Indonesia. Specifically, it will try to answer these research questions: a) what are the familiarity and participation rates of Indonesians in different types of online gig economy? and b) what are the differences in characteristics of Indonesian gig users and gig workers in different types of online gig economy? The findings from this study can shed light on the online gig economy that is shown to be a more preferable business model by the younger generations worldwide. It can also serve as a steppingstone for further research on the online gig economy as a new business model and dig deeper into its potential in Indonesia

2. Literature Review

Online gig economy can be defined as a business model where two parties (i.e., gig users and gig workers) agree to have a temporary or short-term arrangement to do one or more projects without any permanent and fixed employment relationship. In some cases, gig users and gig workers can meet on any social media platform, but in some other cases a dedicated third-party platform is also available. Oftentimes, this dedicated third party platform gives a better experience for both gig workers and gig users as they set some ground rules and establish a sense of security, trust, and safety net between gig workers and gig users who are likely to be strangers to each other. As per MasterCard's classification, there are at least four different types of online gig economy [7]. First, asset sharing services (AS) that facilitates short-term peer-to-peer rentals of property with Airbnb as the most notable platform. Second, transportation-based services (TS) that includes ridesharing, carpooling, food delivery, or package couriers with some famous names like Uber, Lyft, Gojek, or Grab as the most notable platforms. Third, professional services (PS) that connects freelancers with a specialized skill set (e.g., graphic design, programming, copywriting) directly with their firms or individual clients, facilitated by some platforms like Fiverr or Upwork. And fourth, handmade goods, household, and miscellaneous services (HGHM) where freelancers can offer some homemade crafts or on-demand services for household-related tasks to anyone who are interested in buying their products or using their service on platforms like Etsy or Airtasker.

According to surveys [8], about 30 million Americans rely on the online gig economy activities for their living by 2019. It is reported that more females are working in the online gig economy sectors than they are in the traditional sectors despite there being more male gig workers than the female counterparts. Also, most of them are millennials with some college degree. The same case is true in the UK where more than half of the gig workers are between the range of 18 and 34 years of age, far more than the 10% aged at 55 years old or more [9]. The most popular online gig economy type in the UK is TS, with 42% gig workers identified as package couriers, 37% ride sharing drivers, and 21% food delivery service. The second most popular online gig economy type is the professional service (PS) on the freelance marketplace at 37% [9]. It is also confirmed by MasterCard that transaction-wise, TS online gig economy had the highest gross volume for that same year [7]. In terms of income, gig workers tend to treat gig work as a secondary source of income rather than their primary or even more so, their only source of income. No more than 10% of gig workers in the UK made their gig work as their primary source of income [9].

TS online gig economy has also been thriving in Indonesia as it is driven by two major companies: Gojek and Grab. In 2019, there were 29,2 million active users of Gojek, which is an Indonesian company, in Indonesia [10]. Its competitor, Grab, Malaysia turned Singapore based company that operates mostly within the Southeast Asia region had 187 active users in 330 cities in 8 different countries across Asia [11]. The sheer number of Gojek and Grab users in Indonesia shows how familiar most Indonesians are and how high the participation rate of Indonesian is specific to the TS type of online gig economy. Unfortunately, apart from one study in 2019 profiling Indonesian PS gig workers that mostly fall within the creative and multimedia category with an average monthly salary of IDR 3.4 million [12], there is little to no research on the familiarity and participation rates of Indonesians in other types of online gig economy. This study will try to fill that gap by providing a firsthand analysis from the primary data collected specifically to answer the research questions about familiarity with and participation in different types of online gig economy in Indonesia.

3. Methodology

Data collection was conducted using online surveys on social media, specifically Twitter, Facebook, and WhatsApp in the first half of 2020. A total of 385 people completed the survey consisting of 40.26% males (n=155) and 59.74% females (n=230) aged between 15 and 60 years old. Participants also reported their educational attainment, their monthly income, place of origin, place of residence, and their familiarity with both the concept and the term "online gig economy". Participants were then asked about their participation in any of the four online gig economy categories as per MasterCard's classification as follows: Asset Sharing Services (AS), Transportation-Based Services (TS), Professional Services (PS), and Handmade Goods, Household, and Miscellaneous Services (HGHM). Table 1 summarizes the characteristics of participants in this study.

Data analysis was done with descriptive statistics and multiple correspondence analysis (MCA) in R 3.6.2 and RStudio 1.2.5 with the help of FactoMineR [13] and factoextra packages [14] as well as logistic regression models in STATA 15.

4. Results and Discussion

Table 1 provides information about familiarity with and participation in four types of online gig economy among all participants. Unsurprisingly, TS had the highest participation rate of all categories at 93.8%, almost twice that of PS at 57.7% as the second highest while AS and HGHM shared quite similar participation rates at 47.3% and 46% respectively. Most participation in all categories are as gig users at 90% or higher except for PS that falls way behind at 73%. Interestingly, participation in PS as gig workers not only was the highest in all categories at 38.7% or almost twice that of HGHM at 20.9% at the second place, but also had the highest ratio between participation as gig workers and as gig users with slightly more than 1:2, far more than HGHM at roughly 2:9, AS at around 1:7, and TS at approximately 1:11. In other words, most Indonesians involved in the online gig economy participate as gig users in TS and as gig workers in PS category.

Another analysis with MCA by using participation in each category of online gig economy as the active variables and demographic information as the supplementary variables was conducted to see the closeness between variables in the dataset. Based on Figure 1, participation in TS is located at the barycenter of the chart, indicating most Indonesians as TS users, which is also confirmed by Table 1. Participations in the other three categories of PS, AS, and HGHM are closely located with people who are from or live in Java, including Jakarta and who work as freelancers. On the other edge of the spectrum are those who are both unfamiliar with the online gig economy and not participating in any type of online gig economy, not even TS. These people are indicated as those who are from and live outside Java.

Table 1. Familiarity with and participation in online gig economy among all participants (n=385)

Variable	Category	PS		AS		TS		HGHM	
		freq	%	freq	%	freq	%	freq	%
Familiarity	Familiar with both concept and term	40	18.0	34	18.7	53	14.7	28	15.8
	Familiar only with the concept	151	68.0	119	65.4	233	64.5	115	65.0
	Unfamiliar	31	14.0	29	15.9	75	20.8	34	19.2
Participation†	As gig users	162	73.0	168	92.3	349	96.7	160	90.4
	As gig workers	86	38.7	25	13.7	34	9.4	37	20.9
Total participants		**222**	**57.7**	**182**	**47.3**	**361**	**93.8**	**177**	**46.0**

Note: † total do not add up to 100% since some gig workers also indicated some experience as gig users.

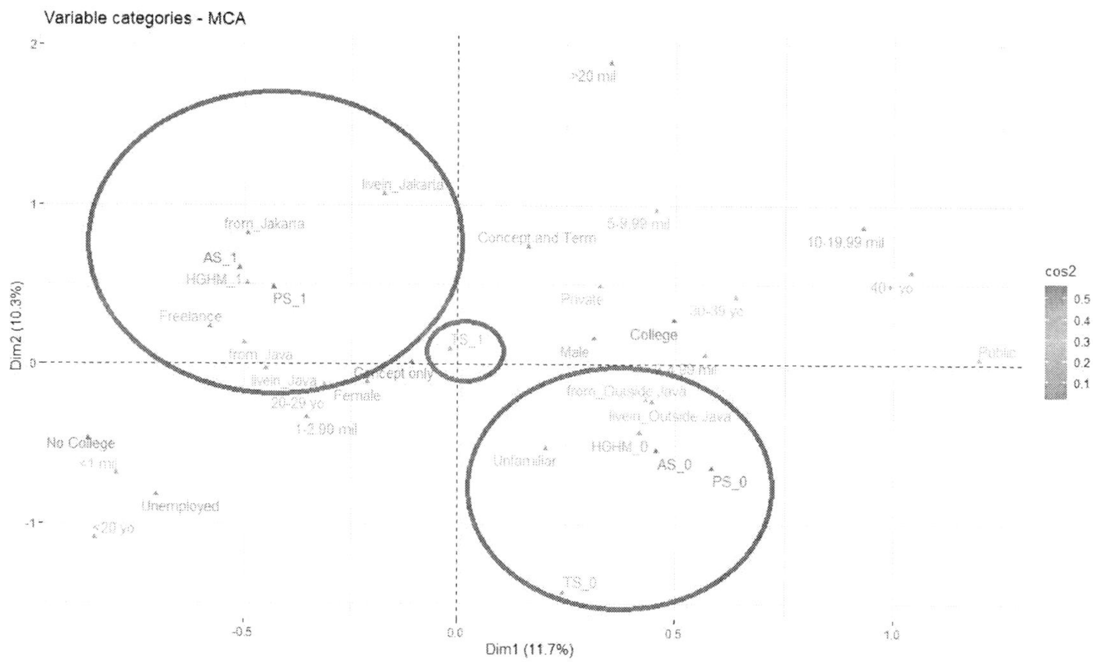

Figure 1. MCA of participation in four types of online gig economy

Considering the huge discrepancy between gig users and gig workers in all categories, it is necessary to split this information accordingly. Table 2 summarizes the demographic information about gig users, whereas Table 3 contains the same demographic information for gig workers. Based on the two tables, some differences in participation in the online gig economy across groups of people can be seen. For example, females have the highest participation rate as gig users in HGHM than in any other categories. Likewise, female participation rates as gig workers in HGHM and PS are much higher than in AS and TS that have more male participation instead. Still among the gig workers, those living in Java tend to have the highest participation rate in PS than in any other categories. There are more differences across groups of people and across different types of online gig economy that might warrant further analysis, especially a multivariate one.

Table 2. Participation in the online gig economy as gig users (n = 385)

Variable	Category	PS		AS		TS		HGHM	
		freq	%	freq	%	freq	%	freq	%
Sex	Male	64	39.5	63	37.5	137	39.3	53	33.1
	Female	98	60.5	105	62.5	212	60.7	107	66.9
Age	≤19	7	4.3	7	4.2	21	6.0	8	5.0
	20-29	107	66.0	109	64.9	209	59.9	99	61.9
	30-39	35	21.6	38	22.6	86	24.6	38	23.8
	40+	13	8.0	14	8.3	33	9.5	15	9.4
Origin	Jakarta	14	8.6	20	11.9	31	8.9	17	10.6
	Sulawesi	17	10.5	13	7.7	31	8.9	14	8.8
	Sumatera	13	8.0	14	8.3	34	9.7	12	7.5
	Jawa	61	37.7	73	43.5	136	39.0	62	38.8
	Kalimantan	30	18.5	30	17.9	72	20.6	36	22.5
	Others	27	16.7	18	10.7	45	12.9	19	11.9
Residence	Jakarta	26	16.0	27	16.1	43	12.3	21	13.1
	Sulawesi	8	4.9	8	4.8	21	6.0	11	6.9
	Sumatera	7	4.3	6	3.6	27	7.7	9	5.6
	Jawa	64	39.5	75	44.6	147	42.1	60	37.5
	Kalimantan	37	22.8	34	20.2	79	22.6	42	26.3
	Others	20	12.3	18	10.7	32	9.2	17	10.6
Occupation	Public Sector	18	11.1	18	10.7	53	15.2	19	11.9
	Private Sector	67	41.4	67	39.9	141	40.4	62	38.8
	Freelance	28	17.3	35	20.8	56	16.0	31	19.4
	Unemployed	49	30.2	48	28.6	99	28.4	48	30.0
Monthly Income (IDR)	< 1 million	41	25.3	43	25.6	74	21.2	45	28.1
	1 - 2,99 million	45	27.8	43	25.6	89	25.5	40	25.0
	3 - 4,99 million	36	22.2	38	22.6	101	28.9	34	21.3
	5 - 9,99 million	30	18.5	33	19.6	66	18.9	29	18.1
	10 - 19,99 million	6	3.7	8	4.8	14	4.0	8	5.0
	> 20 million	4	2.5	3	1.8	5	1.4	4	2.5
Education	Middle School	2	1.2	0	0.0	4	1.1	0	0.0
	High School	67	41.4	74	44.0	121	34.7	70	43.8
	Vocational School	10	6.2	10	6.0	38	10.9	8	5.0
	Bachelor's Degree	63	38.9	64	38.1	147	42.1	61	38.1
	Master's Degree	18	11.1	19	11.3	35	10.0	20	12.5
	Doctoral Degree	2	1.2	1	0.6	4	1.1	1	0.6
All gig users		**162**	**42.1**	**168**	**43.6**	**349**	**90.6**	**160**	**41.6**

To confirm whether the differences in characteristics of gig users and gig workers in all four types of online gig economy were statistically significant, logistic regression models were used by developing two models, one for gig users and another for gig workers in all four categories. All four models for the gig users turned out to be statistically significant whereas three out of four models for the gig workers turned out to be not statistically significant, indicating no significant differences across different groups of people who are involved in the online gig economy as gig workers apart from the PS category. The results of logistic regression analysis are summarized in Table 4.

Table 3. Participation in the online gig economy as gig workers (n = 182)

Variable	Category	PS		AS		TS		HGHM	
		freq	%	freq	%	freq	%	freq	%
Sex	Male	34	39.5	14	56.0	20	58.8	13	35.1
	Female	52	60.5	11	44.0	14	41.2	24	64.9
Age	≤19	8	9.3	3	12.0	2	5.9	5	13.5
	20-29	56	65.1	10	40.0	21	61.8	18	48.6
	30-39	17	19.8	7	28.0	7	20.6	11	29.7
	40+	5	5.8	5	20.0	4	11.8	3	8.1
Origin	Jakarta	10	11.6	0	0.0	3	8.8	0	0.0
	Sulawesi	4	4.7	2	8.0	3	8.8	2	5.4
	Sumatera	6	7.0	3	12.0	5	14.7	4	10.8
	Jawa	45	52.3	12	48.0	14	41.2	18	48.6
	Kalimantan	10	11.6	5	20.0	6	17.6	10	27.0
	Others	11	12.8	3	12.0	3	8.8	3	8.1
Residence	Jakarta	7	8.1	0	0.0	2	5.9	2	5.4
	Sulawesi	2	2.3	2	8.0	2	5.9	1	2.7
	Sumatera	6	7.0	2	8.0	3	8.8	3	8.1
	Jawa	50	58.1	11	44.0	18	52.9	16	43.2
	Kalimantan	13	15.1	8	32.0	8	23.5	12	32.4
	Others	8	9.3	2	8.0	1	2.9	3	8.1
Occupation	Public Sector	5	5.8	5	20.0	3	8.8	4	10.8
	Private Sector	33	38.4	11	44.0	16	47.1	14	37.8
	Freelance	34	39.5	4	16.0	8	23.5	11	29.7
	Unemployed	14	16.3	5	20.0	7	20.6	8	21.6
Monthly Income (IDR)	< 1 million	16	18.6	6	24.0	8	23.5	9	24.3
	1 - 2,99 million	21	24.4	3	12.0	12	35.3	11	29.7
	3 - 4,99 million	23	26.7	6	24.0	5	14.7	7	18.9
	5 - 9,99 million	22	25.6	8	32.0	7	20.6	10	27.0
	10 - 19,99 million	3	3.5	1	4.0	2	5.9	0	0.0
	> 20 million	1	1.2	1	4.0	0	0.0	0	0.0
Education	Middle School	2	2.3	1	4.0	2	5.9	1	2.7
	High School	27	31.4	6	24.0	12	35.3	13	35.1
	Vocational School	8	9.3	1	4.0	4	11.8	3	8.1
	Bachelor's Degree	42	48.8	10	40.0	13	38.2	16	43.2
	Master's Degree	7	8.1	7	28.0	3	8.8	4	10.8
	Doctoral Degree	0	0.0	0	0.0	0	0.0	0	0.0
All gig workers		**86**	**22.3**	**25**	**6.5**	**34**	**8.8**	**37**	**9.6**

Table 4. Logistic regression estimates of participation in online gig economy per category

Variable		TS User	PS User	PS Worker	AS User	HGHM User
Sex	Female	**2.206***	1.010	1.219	1.276	**1.662***
		(.844)	(.224)	(.334)	(.285)	**(.374)**
Age	Teenagers	.6133	**.349***	**3.465***	**.315***	**.442†**
		(.386)	**(.170)**	**(1.864)**	**(.155)**	**(.209)**
Location	In Jakarta	3.167	**2.537****	**.350***	**2.125***	1.774
		(3.375)	**(.916)**	**.164**	**(.775)**	(.640)
	Not in Java	1.637	.835	**.561***	1.222	**.604***
		(.639)	(.196)	**.162**	(.287)	**(.144)**
Occupation	Freelancer	**.454†**	1.119	**8.835*****	**1.713†**	1.646
		(.211)	(.351)	**(3.348)**	**(.536)**	(.518)
	Private Sector	.780	1.303	**2.106***	1.219	1.313
		.368	(.338)	**(.738)**	(.319)	(.345)
Education	College Degree	.868	**.553***	.951	**.441****	**.455****
		(.366)	**(.141)**	(.307)	**(.115)**	**(.118)**
Monthly Income	Less than IDR 3 million	**.213****	1.291	**.501***	.899	.988
		(.103)	(.325)	**(.159)**	(.228)	(.250)
Constant		17.788***	.854	.118*	.872	.908
		(10.551)	(.286)	(.052)	(.295)	(.308)
Model $\chi 2$		24.92**	17.78*	55.18***	26.35***	21.36**
Pseudo R2 (McFadden)		0.104	0.034	0.135	0.050	0.041
df		8	8	8	8	8
Observation		385	385	385	385	385

Note: Numbers reported are odds ratios with standard errors between parentheses.
Signif. codes: '***' 0.001 '**' 0.01 '*' 0.05 '†' 0.1

As shown in Table 4, females are more likely to be gig users than males in TS and HGHM categories, whereas no significant difference between males and females in any other online gig economy categories, either as gig workers or gig users. One plausible explanation is that females are less comfortable driving on the roads their own, and thus relying more on TS as their main source of transportation. As for HGHM, it might have something to do with the patriarchal culture in Indonesia meaning many married females who are housewives, and thus be in charge of their household and domestic affairs. In terms of age, one interesting but not really surprising finding is that teenagers are much more likely to be a PS gig-worker than the older people. It further confirms that the online gig economy, or specifically PS in the case of Indonesia, is loved and more preferred means of living by the younger generations. While it is true that younger people tend to have more exposure on the Internet, including how to use it to earn money than the older people, it might also have something to do with difficulties in getting a more conventional job for these younger people. On the other hand, older generations are more likely to be a gig user in AS and HGHM categories, which is understandable since teenagers are less likely to need some accommodation either for work or leisure activities, just like they are less likely to need help with household chores than older people who are either a house owner or a house renter.

The findings also suggest some geographical gap between people living in Java, including Jakarta, and others who live in any other part of the country as it is shown that Jakartans are more likely to be a gig user in PS and AS whereas those living outside Java are less likely to be a gig user in HGHM or a gig worker in PS. This phenomenon can be attributed to either digital divide between Java and outside Java, especially in terms of access and infrastructure needed as the main driver or gig-economy. It can also be attributed to an urban lifestyle of Jakartans that makes them more likely to be a gig user on a more regular basis than people living in other parts of Java island. Interestingly, most PS workers tend to live in Java but outside of Jakarta. It could be the case that the urban lifestyle doesn't really fit these people, so they are opting for a more suburban lifestyle as long as the infrastructure is adequate. Regarding their occupations, as expected most PS workers identified themselves as freelancers with some other indicating that they have another job in the private sector. In terms of educational attainment, it is interesting to note that those without college degrees are more likely to be a gig user in PS, AS, and HGHM categories. Most probably, these people are college students in their early twenties. Finally, those with a low-income background who earn less than IDR 3 million (roughly USD 200) per month are less likely to be TS users and PS workers. It further confirms that TS users are more likely to be from the middle class and that PS workers can earn a decent amount of money to be considered as a primary source of income for some.

5. Conclusion

This study shows that participation rates in different types of online gig economy among Indonesians are quite high, especially in TS category, which is the most popular type of online gig economy where most Indonesians are participating, mostly as gig users. Apart from PS that has the lowest participation rate at 73%, all other types of online gig economy have participation rates at above 90%. However, familiarity rates are a different story. Apparently, most Indonesians are still not familiar with the term "online gig economy" despite being familiar with the concept itself. Interestingly, despite its status as the most popular type of online gig economy, TS has the lowest familiarity rate among all other types of online gig economy among its participants. It turns out that some Indonesians have been participating in TS, mostly as users, without realizing that it is a form of online gig economy.

This study also found some significant differences in characteristics of Indonesian gig users in all four categories and gig workers in PS category. HGHM users tend to be middle aged or older females without college degree living in Java, while TS and PS users tend to be females from middle class economy or higher, PS users tend to be young adults living in Jakarta. As for the PS gig workers, they are likely to be younger people living in suburban Java who use their gig work as either their main source of living or as a side for their main job in the private sector and they tend to earn more than IDR 5 millions per month.

The findings from this study, despite all of the limitations, especially in terms of geographical distribution of sample, can help fill the gap in the literature about familiarity and participation in different types of online gig economy among Indonesians by providing a first-hand analysis from the primary data collected. This study also confirms that the online gig economy is more popular among the younger generations and that it has tremendous potential to be a new business model that can be a primary source of income for many. Further research can build upon this either by focusing on a specific type of online gig economy to dig deeper into its potential, or by investigating factors determining the accessibility and acceptability of various types of online gig economy among different groups of people across the country.

References

[1] Signes A T 2017 *The 'Online gig economy': Employee, Self-employed or the Need for a Special Employment Regulation?* (New York: SAGEPUB) pp 1– 3
[2] TIM APJII 2019 *Buletin APJII 2019* vol **40** Retrieved from

https://apjii.or.id/downfile/file/BULETINAPJIIEDISI40Mei2019.pdf
[3] Gojek 2020 *Gojek: About Us* Retrieved from https://www.gojek.com/about/
[4] e27 2014 *GrabTaxi launches beta trial in Jakarta* Retrieved from https://e27.co/grabtaxi-launches-beta-trial-in-jakarta-20140606/
[5] Sub Direktorat Statistik Komunikasi dan Teknologi Informasi 2019 *Statistik E-Commerce 2019* ed T Sujono *et al* (Jakarta: Badan Pusat Statistik Indonesia) pp 10–5
[6] Sub Direktorat Pengembangan Model Statistik 2020 *Tinjauan Big Data Terhadap Dampak COVID-19* (Jakarta: Badan Pusat Statistik Indonesia) pp 23–4
[7] Mastercard and Kaiser Associates 2019 *The Global Online Gig Economy: Capitalizing on a ~$500B Opportunity* Retrieved from https://newsroom.mastercard.com/wp-content/uploads/2019/05/Gig-Economy-White-Paper-May-2019.pdf
[8] MetLife 2019 *The Online Gig Economy: Opportunities, Challenges, and Employer Strategies* (New York: MetLife Services and Solution)
[9] Lepanjuuri K, Wishart R and Comick P 2018 *The Characteristics of Those in the Online Gig Economy* Retrieved from https://www.voced.edu.au/content/ngv%3A79237#
[10] Iwamoto K 2019 *Gojek aims to grow non-Indonesia users to 50% of operations* Retrieved from https://asia.nikkei.com/Business/Startups/Gojek-aims-to-grow-non-Indonesia-users-to-50-of-operations
[11] CNBC 2020 *CNBC Disruptor 50* Retrieved from https://www.cnbc.com/2020/06/16/grab-disruptor-50.html
[12] Faisal A L F, Sucahyo Y G, Ruldeviyani Y and Gandhi A 2019 Discovering Indonesian digital workers in online gig economy platforms *2019 Int. Conf. on Information and Communications Technol.*
[13] Sebastien L, Josse J and Husson F 2008 FactoMiner: a package for multivariate analysis *J. of Statistical Software* vol **25** no 1 p 1–18
[14] Kassambara A and Mundt F 2020 *Factoextra: Extract and Visualize the Results of Multivariate Data Analyses* Retrieved from https://CRAN.R-project.org/package=factoextra

Applying PC Algorithm and GES to Three Clinical Data Sets: Heart Disease, Diabetes, and Hepatitis

Nurdi Afrianto, Yopi Azzani, Yuan Sa'adati, Nurhaeka Tou, Putri Mentari Endraswari, Yohani Setiya Rafika Nur, Nur Annisa, Rifai Nur Widyanara, Ridho Rahmadi

Master Program in Informatics, Universitas Islam Indonesia, Yogyakarta, Indonesia

E-mail: [nurdi.afrianto, yopi.azzani, yuan.saadati, nurhaeka.tou, putri.endraswari, yohani.nur, nur.annisa, rifai.widyanara]@students.uii.ac.id

E-mail: ridho.rahmadi@uii.ac.id

Abstract. The goal of many sciences, including those related to the clinical domain, is to discover the generative model, that is, to understand how variables in the data take on their values. This goal cannot be addressed directly using approaches such as machine learning and deep learning, as such methods focus more on the association between input and output variables. In this paper, we aim to show to the readers an alternative approach, which can be a more appropriate method to target such aforesaid research goal. This approach is called causal modeling. We will first begin with some application examples of machine learning and deep learning on clinical data, and then show our applications of causal modeling to three clinical real-world data sets. This paper is projected to be a concise guideline for researchers to causal modeling, as well as to choose suitable approaches for problems of interest.

1. Introduction

Machine learning (ML) and deep learning (DL) are advances in Artificial Intelligence, and have long been applied in the clinical domain [1][2][3]. Those two concepts aim to develop (mostly automatic) systems which can learn models from data. The typical learning methods are generally of two paradigms: supervised learning and unsupervised learning. The supervised learning focuses on tasks, e.g., prediction and classification. Examples of this paradigm are Naïve Bayes classifier, support vector machine (SVM), and convolution neural network. The unsupervised learning, on the other hand, focuses on different tasks, e.g., clustering and components analysis. Examples of this paradigm are K-Means, independent component analysis, and Autoencoders.

Both ML and DL, however, mainly target the associations between input variables X and output variables Y, that is, $P(y|x)$. Here both X and Y are observations on domains of interest. The learning process here is designed to project output Y based on the given input X. A practical example is to learn whether or not an individual has diabetes (output), based on his/her recent clinical examination (input) gathered in a hospital. Such a task is called classification. Despite the accuracy of such classification models, the conducted learning process results in trained model parameters which are numerous and difficult to explain [4]. In some DL models, for example, the number of parameters reaches thousands or even millions [5][6][7]. The lack of explainability of ML and DL models has been discussed in [4].

The goal of many sciences, including those related to the clinical domain, is to discover generative models. Such models explain how variables in the data take on their values [8]. This goal, however, cannot be addressed directly using ML and DL models, which mostly focus on the association between input and output variables $P(Y|X)$. As an alternative, a method called causal modeling or causal discovery can be used to address the aforesaid generative models. Basically, a causal model learns how variable Y (effect) changes its value solely based on another variable X (cause). Often X is obtained from an intervention, e.g., a medical therapy; here we denote $do(x)$ to distinguish a result of an intervention from a merely observation x. Unlike $P(y|x)$ in ML and DL problems, here we are interested in to see the mechanisms entailed by $P(y|do(x), z)$, that is, "what will be y, if I do $X = x$, and observe some z". Understanding such mechanisms helps us to estimate the model which generates the data we have. In clinical practice, obtaining a causal model can lead us to better develop medication or therapy, because we could focus more on the variables which have more impact on the disease or the disorder. Technically, the causal methods can be divided into two approaches, namely the constraint-based approach and the score-based approach [9]. The constraint based approach uses conditional independence tests and orientation rules to obtain a causal model. The score based approach uses scores, often for model refinements, which are the building blocks of the final causal model.

In this study, therefore, we aim to show causal modeling applications in estimating the generative models of some clinical data, as an alternative method to those of ML and DL, which focus more on the association between input and output variables. In particular, we will apply causal modeling on three clinical data sets, namely, hepatitis [10], diabetes mellitus [11], and heart attack/disease [12]. These three diseases are of the five main causes of death worldwide [13], and thus understanding the causal mechanisms underlying each of the diseases is very important. Especially in recent years, we are faced with the challenges of developing more efficient and effective medications and treatments. We hope that the outcome of this study could become both scientific and practical references in the related fields. Technically, we will employ a score-based method called GES (Greedy Equivalent Search)[14] and a constraint-based method called PC algorithm [15]. We are aware that these two causal methods are not the current state-of-the-art. These methods, however, generally reflect the foundation and simplicity of constraint-based procedure (PC algorithm) and score-based procedure. The recent causal methods, on the other hand, are based on these methods with additional steps and assumptions, or even with more complicated algorithms. As the main purpose of the present study is to introduce causal modeling as an alternative approach, we select simpler methods such as PC algorithm and GES.

This paper is organized as follows. Section 1 describes the background and motivation of the study. Section 2 describes the previous studies which are relevant. Section 3 explains the data and detailed methods. Section 4 discusses the results, and Section 5 concludes the study.

2. Previous Studies

There have been numerous clinical studies on heart disease, which applying machine learning or deep learning. For example, to predict the outcome of patients given some clinical measurements. For example, a study by Rufai and Umar [16] attempted to predict heart disease status from 303 patients using a neural network model. Rajesh et al. [17] reviewed the perfomance of classification methods such as Naïve Bayes classifier and decision trees for predicting heart disease from 300 patients. A research conducted by Sisodia and Sisodia [18] attempted to predict diabetes using decision tree, Naive Bayes classifier, and SVM. The study was conducted on the Pima Indians Diabetes Database. Kumar and Vignaswari [19] administered a comparison study of machine learning applications (logistic regression, random forest, C4.5, and multilayer

IOP Conf. Series: Materials Science and Engineering 1077 (2021) 012067 doi:10.1088/1757-899X/1077/1/012067

perceptron) to hepatitis prediction, based on 155 patients. There have been some causal studies on clinical or psychological cases, e.g., Alzheimer's disease, chronic fatigue syndrome, and chronic kidney disease [20][21][22]. In this present study we aim to apply causal modeling to clinical data similar to the studies described in the beginning, i.e., heart disease, hepatitis, and diabetes.

3. Data and methods

We use R [23] to compute the analyses. In particular we use package `pcalg` [24] to apply the PC algorithm and GES to the following data sets. For each data set, we deliberately select some variables which are continuous and of our interest. As our main purpose of the present study is to introduce causal modeling as an alternative method, we will conduct simple applications rather than comparisons. Here, we will apply PC algorithms to the Heart disease and the Diabetes data sets, and GES to the Hepatitis data set.

3.0.1. Heart disease data

The data set is obtained from `kaggle.com`, containing observations on 303 patients who have heart attack possibility. Each patient is characterized by 14 attributes related to their health conditions. In this study, we select in particular variables of interest, that is, resting blood pressure (trestbps), serum cholesterol (chol), maximum heart rate achieved (thalach), and ST depression induced by exercise relative to rest looks at stress of heart during exercise unhealthy heart will stress more (oldpeak).

3.0.2. Diabetes data

The Pima Indians Diabetes is originally from the National Institute of Diabetes and Digestive and Kidney Diseases, and obtained from `kaggle.com`. The data set comprises observations on 768 females. Originally this data set contains nine continuous attributes and one outcome. In this study particularly we select variables of interest, namely Glucose, Blood Pressure, Insulin, and BMI.

3.0.3. Hepatitis

The data set is obtained from `kaggle.com`, containing observations on 142 patients. Originally the data comprises 20 variables. In this study we select some variables of interest, that is, Age, Albumin, Alkaline Phospate, and Protime.

3.1. Methods

3.1.1. PC-Algorithm

PC algorithm [15] seeks causal relations by testing conditional independence among variables. It has two steps. In the first step, PC algorithm starts with a complete graph and repeatedly eliminates edges connecting two variables if those variables are independent based on conditional independence test, resulting in a skeleton. In the second step, the skeleton is directed, resulting in a causal model represented by a partially completed directed acyclic graph (CPDAG). For more detail of the first step, see the Algorithm1. Technically, this step tests for conditional independence among pairs of variable across conditioning set that increases in size $(d + 1)$. When X and Y are independent conditional on Z, that is, $I(X, Y|Z) = \texttt{TRUE}$, we eliminate the edge between X and Y and save Z in the corresponding separating set. Once the skeleton is estimated, it is oriented with the following rules. For each triple of variables (X, Z, Y) such that X and Z are adjacent, Z and Y are adjacent, and X and Y are not adjacent, orient the edges $X - Z - Y$ as $X \rightarrow Z \leftarrow Y$, if Z was not in the set conditioning on which X and Y became independent and the edge between them was accordingly eliminated. We call such a triple of variables a *v-structure*. For each triple of variables such that $X \rightarrow Z - Y$, and X and Y are not adjacent, orient the edge $Z - Y$ as $Z \rightarrow Y$. The pseudocode of PC algorithm is adopted from [25].

Algorithm 1 PC Algorithm

Input Dataset D with a set of variables $\mathbf{V}$ and significant level α
Output The undirected graph G with a set of edges $\mathbf{E}$
Assume all nodes are connected initially
Let depth $d = 0$

repeat
  for *each ordered pair of adjacent vertices* X *and* Y *in* G **do**
    if $|adj(X, G) \setminus \{Y\}| \geq d$ **then**
      for *each subset* $Z \subset adj(X, G) \setminus \{Y\}$ *and* $|Z| = d$ **do**
        Test $I(X, Y|Z)$
        if $I(X, Y|Z)$ **then**
          Remove edge between X and Y
          Save Z as the separating set of (X, Y)
          Update G and $\mathbf{E}$
          break
        end if
      end for
    end if
  end for
  Let $d = d + 1$
until $|adj(X, G) \setminus \{Y\}| <$ *for every pair of adjacent vertices in* G

3.1.2. GES

The Greedy Equivalence Search [14] (GES; see Algorithm 2) is a score-based causal method, consisting of forward equivalence search (FES; see Algorithm 3) and backward equivalence search (BES; see algorithm 4). GES repeatedly adds causal relations via FES until no more improvement on the score (typically Bayesian Information Criterion) and repeatedly deletes causal relations via BES until no more improvement on the score. The pseudocode of GES is adopted from [26].

Algorithm 2 Add Operation (Forward phase)

Input A dataset D and a CPDAG G
Output A CPDAG H
Let M be a set of CPDAGs obtainable from G by **adding** one edge
Let K be a CPDAG from M with the highest score $S(D, K)$

if $S(D, G) < S(D, K)$ **then**
  $H \leftarrow K$
else
  $H \leftarrow G$
end if
return H

Algorithm 3 Delete Operation (Backward phase)

Input A dataset D and a CPDAG G
Output A CPDAG H
Let M be a set of CPDAGs obtainable from G by **deleting** one edge
Let K be a CPDAG from M with the highest score $S(D, K)$

if $S(D, G) < S(D, K)$ **then**
    $H \leftarrow K$
else
    $H \leftarrow G$
end if
return H

Algorithm 4 GES

Input A dataset D
Output A CPDAG G

while $S(D, Add(D, G)) > S(D, G)$ **do**
    $G \leftarrow Add(D, G)$
end while
while $S(D, Del(D, G)) > S(D, G)$ **do**
    $G \leftarrow Del(D, G)$
end while
return G

3.2. Source code

We provide the source code used in all computation in https://github.com/nurdi1995. As and example, the following source code is used for computing heart disease data set using PC algorithm.

```
continuous_data <- data.frame(data$trestbps , data$chol ,
                              data$thalach , data$oldpeak)

suffStat <- list(C=cor(y), n = nrow(continuous_data))
pc.fit <- pc(suffStat , indepTest=gaussCItest ,
     p = ncol(continuous_data), alpha=0.05 , verbose = TRUE)
```

4. Results and Discussion

4.1. Heart disease

We applied PC algorithm to the heart disease data set. We assume the variables are normally distributed and use the Gaussian conditional independence test with $\alpha = 0.05\%$.

Figure 1 shows the CPDAG estimated by PC algorithm. The causal relations between chol and trestbps indicates cholesterol factor will have an effect on blood pressure where the body is at rest. The research that has been done by [27] mentioned that increased serum cholesterol levels are able to influence blood pressure, at least during sympathetic stimulation. We also find that oldpeak, which indicated an myocardial ischemia (condition where blood flow is obstructed by plaques), are influenced by thalach and trestbps. In studies [28] [29], it is explained that a consistent ST depression elevation hypertension affects left ventricular hypertrophy on the ECG is still within normal limits and an increase in the maximum heart rate will affect ST depression during the treadmill process. In general, the causal relations estimated here are corroborated by previous studies.

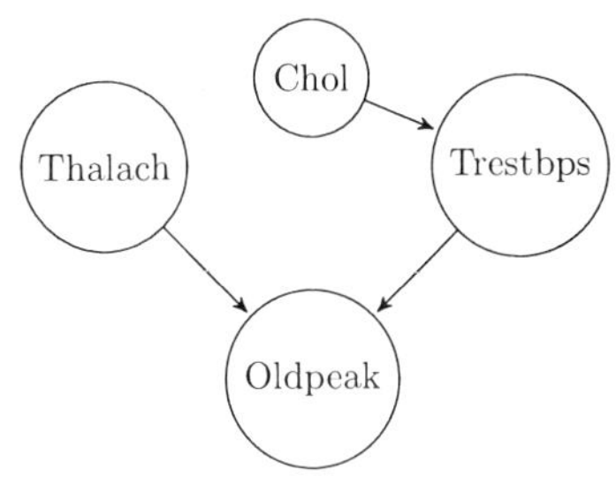

Figure 1. The CPDAG estimated by PC Algorithm on Heart disease data set.

4.2. Diabetes
We used PC algorithms to estimate causal relations from the Diabetes data. Figure 2 shows the result. We found several causal relations here: from insulin and blood pressure to BMI, from insulin to BMI, and from insulin to glucose. Studies by Dewi, et al. [30] found that a positive and significant relationship between BMI and glucose. Moreover, the studies in [31] [32] have found an association between insulin and BMI. Overall, mean systolic and diastolic blood pressures increased significantly and linearly across BMI levels. We found gradient increases in blood pressure with higher BMI levels. The fact that this gradient exists in a fully adjusted analysis suggests that BMI can cause an immediate effect on blood pressure, independent of other clinical risk factors. We can see that our findings are in general, accordance with those of some previous studies [33].

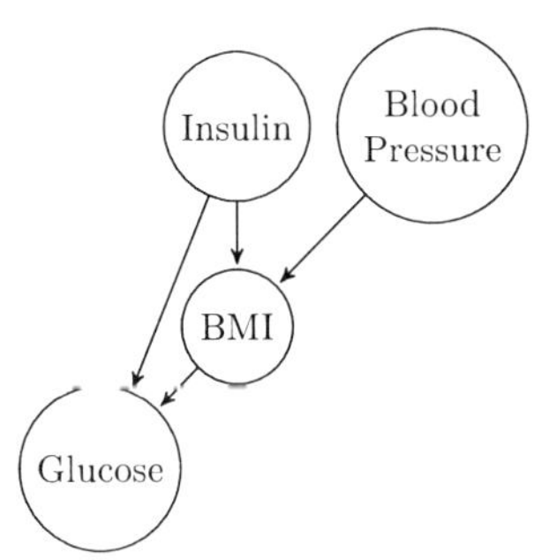

Figure 2. The CPDAG estimated by PC algorithm on Diabetes data set.

4.3. Hepatitis
Figure 3 shows the result of applying GES to the Hepatitis data. We found that age and alkaline phosphatase variables causes the albumin, and albumin causes prothrombin time. Several previous studies are in favor of these findings. The studies in [34] found that a decrease in albumin levels in the blood is significantly associated with aging, i.e., the older a person is, the lower his/her level of albumin in the blood. Moreover, the studies in [35] shows that the Prothrombin time and albumin are positively related to the incidence of hepatitis. Again, in general our causal findings in the Hepatitis data conform those of the previous studies.

5. Conclusion & Future works
ML and DL typically resolves problems by modeling the association between input and output variables. In some domains, however, the main goal is to understand the model generating data. Such case cannot be addressed directly by ML and DL. The aim of the present study is to

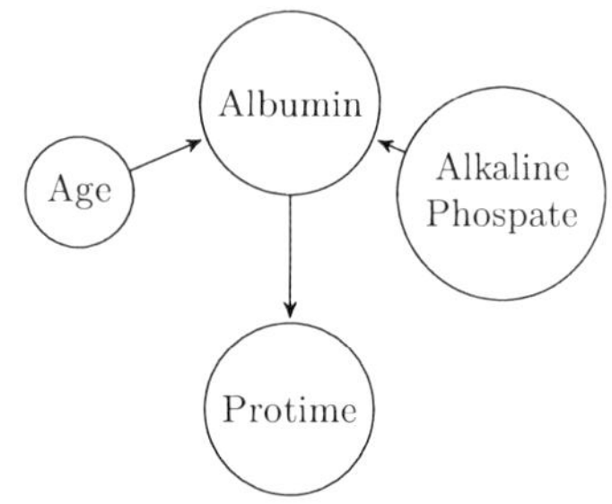

Figure 3. The CPDAG estimated by GES on Hepatitis dataset.

show causal modeling as an approach to such problem. Causal modeling attempts to identify causal relations among variables, which constitute the generative model. Understanding such a model is important, e.g., when developing a new clinical treatment for a disease, we need to understand which symptoms are the main causes of the disease, in order to design the best intervention. We showed that our results of applying causal methods on three clinical data sets. The results are, in general, in accordance with those of relevant previous studies. Our present study, however, does not cover practical problems such as missing values or non-normal data distribution. Future work should attempt to address those problems.

6. References

[1] Kumar P S and Pranavi S 2017 Performance analysis of machine learning algorithms on diabetes dataset using big data analytics *Int. Conf. on Infocom Technologies and Unmanned Systems* pp 509–13
[2] Fatima M and Pasha M 2017 Survey of achine learning algorithms for disease diagnostic *J. of Intelligent Learning Systems and Applications* vol **09** no 01 pp 1–16
[3] Bakator M and Radosav D 2018 Deep learning and medical diagnosis: a review of literature *Multimodal Technologies and Interaction* vol **2** no 3 pp 1–12
[4] Pearl J 2019 The seven tools of causal inference, with reflections on machine learning *Commun. ACM* vol **62** no 3 pp 54–60
[5] Bzdok D, Nichols T E and Smith S M 2019 Towards algorithmic analytics for large-scale datasets *Nat Mach Intell* vol **1** no 7 pp 296–306
[6] Gupta S, Zhang W and Wang F 2016 Model accuracy and runtime tradeoff in distributed deep learning: a systematic study *IEEE 16th Int. Conf. on Data Mining (ICDM)* pp 171–80
[7] Sejnowski T J 2020 The unreasonable effectiveness of deep learning in artificial intelligence *preprint 2002.04806 [q-bio.NC]*
[8] Spirtes P 2010 Introduction to causal inference *J. Mach. Learn. Res* vol **11** no 5 pp 1644–62
[9] Rahmadi R, Groot P, Heins M, Knoop H and Heskes T 2017 Causality on cross-sectional data: stable specification search in constrained structural equation modeling *Appl. Soft Comput.* vol **52** pp 687–98
[10] HariniR 2020 *Hepatitis* Retrieved on 13 Sep. 2020 from https://www.kaggle.com/harinir/hepatitis
[11] Devisangeetha 2017 *Which Factor Causes Diabetes* Retrieved on 13 Sep. 2020 from https://www.kaggle.com/devisangeetha/which-factor-causes-diabetes
[12] Bhat N 2020 *Health Care: Data Set on Heart Attack Possibility* Retrieved on 13 Sep. 2020 from https://www.kaggle.com/nareshbhat/health-care-data-set-on-heart-attack-possibility.
[13] Onggo I T and Wijayanti D 2011 *5 Penyakit Utama Pencabut Nyawa: Jantung - Stroke - Diabetes Mellitus - Kanker - Hepatitis B and C* (Yogyakarta: Mitra Buku)
[14] Chickering D M and Meek C 2015 Selective Greedy Equivalence Search: Finding Optimal Bayesian Networks Using a Polynomial Number of Score Evaluations *preprint 1506.02113 [cs.LG]*
[15] Spirtes P, Glymour C and Scheines R 2000 *Causation, Prediction, and Search, Second Edition* (Cambridge: MIT Press)
[16] Rufai A U S and Umar M 2018 Using artificial neural networks to diagnose heart disease *IJCAI* vol **182** no 19 pp 1–6
[17] Rajesh N, Maneesha T, Hafeez S and Krishna H 2018 Prediction of heart disease using machine learning algorithms *Int. J. of Eng. and Technol.* vol **7** no 2.32 pp 363–6

[18] Sisodia D and Sisodia D S 2018 Prediction of diabetes using classification algorithms *Procedia Comput. Sci.* vol **132** pp 1578–85
[19] Kumar N K and Vigneswari D 2019 Hepatitis-infectious disease prediction using classification algorithms *J. Adv. Pharm. Technol. Res.* vol **12** no 8 pp 3720–5
[20] Rahmadi R, Groot P and Heskes T 2018 Stable specification search in structural equation model with latent variables *preprint 1805.09527 [stat.ML]*
[21] Rahmadi R, Groot P and Heskes T 2017 Causality on cross-sectional data: stable specification search in constrained structural equation modeling *Applied Soft Computing* vol **52** pp 687–98
[22] Rahmadi R *et al* 2018 Causality on longitudinal data: stable specification search in constrained structural equation modeling *Stat. Methods Med. Res.* vol **27** no 12 pp 3814–34
[23] Douglas A C & D L A, Roos D and Mancini F 2018*Quantitative Geography: The Basics* (California: Sage Publication)
[24] R Development Core Team 2020 *The Comprehensive R Archive Network* Retrieved from https://cran.r-project.org
[25] Le T D, Hoang T, Li J,Liu L, Liu H and Hu S 2019 A Fast PC Algorithm for High Dimensional Causal Discovery with Multi-Core PCs *IEEE/ACM Trans. Comput. Biol. Bioinform.* vol **16** no 5 pp 1483–95
[26] Coumans V 2017 *Causal Discovery Algorithms and Real World Systems* (Nijmegen: Radboud University)
[27] Alonso-Barba J I, delaOssa L, Gámez J A and Puerta J M 2013 Scaling up the greedy equivalence search algorithm by constraining the search space of equivalence classes *Int. J. Approx. Reason.* vol **54** no 4 pp 429–51
[28] Rampengan S H 2015 *Kegawatdaruratan Jantung* (Jakarta: Fakultas Kedokteran Universitas Indonesia)
[29] Radi B, Arson I, Sarvasti D, Tadjoedin Y and Tjahjono C 2016 *Pedoman Uji Latih Jantung: Prosedur dan Interprestasi* (Jakarta: PERKI)
[30] Dewi D Y, Subawa A A N and Mahartini N N 2019 Hubungan indeks massa tubuh terhadap kadar gula darah puasa pengunjung lapangan Niti Mandala Renon bulan Juli tahun 2018 *Intisari Sains Medis.* vol **10** no 3 pp 711–4
[31] Jawad S M A, Al-Zubaide B A and Al-Shalan N W K 2019 The relationship between insulin hormone and obesity in type 2 diabetic patients *Drug Invention Today.* vol **11** pp 2589–92
[32] Al-Goblan A S, Al-Alfi M A and Khan M Z 2014 Mechanism linking diabetes mellitus and obesity *Diabetes Metab. Syndr. Obes.* vol **7** pp 587–91
[33] Landi F *et al* 2018 Body mass index is strongly associated with hypertension: Results from the longevity check-up 7+ study *Nutrients* vol **10** no 12 pp 1–12
[34] Gom I *et al* 2007 Relationship between serum albumin level and aging in community-dwelling self-supported elderly population *J. Nutr. Sci. Vitaminol.* vol **53** no 1 pp 37–42
[35] Tungadi N 2017 *Hubungan Nilai Prothrombin Time dan Albumin dengan Staging Pasien Sirosis Hepatitis Di RSU DR. Wahidin Sudirohusodo* (Makasar: Universitas Hasanudin)

Vehicle Classification Based on CCTV Video Recording Using Histogram of Oriented Gradients, Local Binary Patterns, and Hierarchical Multi-SVM

D A Navastara[1], M F Maulana[2], N Suciati[3], and S T Sarena[4]

[1,2,3]Department of Informatics, Faculty of Intelligent Electrical and Informatics Technology, Institut Teknologi Sepuluh Nopember, Surabaya, Indonesia
[4]Marine Electrical Engineering, Ship Building Institute of Polytechnic Surabaya, Indonesia

E-mail: [1]dini_navastara@if.its.ac.id, [2]farhanm.official@gmail.com, [3]nanik@if.its.ac.id, [4]sryang.tera@ppns.ac.id

Abstract. Vehicle classification is a crucial element in a dynamic traffic management system. The system is needed since the number of vehicles gradually increase every year. There are many methods for vehicle classification, one of those is the use of CCTV camera. As it is multifunction and cheaper than other types of camera, the government tend to use the CCTV camera as monitoring equipment of vehicles on the road. Hence, an accurate vehicle classification system from CCTV records is needed for traffic management system as well as other systems. In this study, the method contains four steps: object detection using Gaussian Mixture Model, features extraction using Histogram of Oriented Gradients and Local Binary Patterns, model training and classification using Hierarchical Multi-SVM. The experimental results show that our proposed method works well with the accuracy of 80.28%, precision of 94.27%, recall of 73.79%, and F-measure of 82.76%.

1. Introduction

The level of density of road users in Indonesia can be categorized as crowded. The number of motorized vehicles in Indonesia always increases from year to year. The increase in the number of motor vehicles is motivated by the increasing number of Indonesia's population, which results in an increase in transportation needs in the daily life of the community. The last census in 2010 set Indonesia's population as 237,641,326 people, with an average increase of 1.54% per year [1]. As for the number of motorized vehicles, at the last calculation, there were 13,480,973 passenger car vehicles, 2,420,917 bus cars, 6,611,028 freight cars, and 98,881,267 motorbikes [2]. Data from the Central Statistics Agency shows an increase in the number of vehicles each year. Meanwhile, the growth of new roads cannot keep pace with the rate of increase in the number of vehicles each year. Therefore, we need a solution from the field of information technology in order to be able to analyze conditions on the highway.

There have been many research topics that prioritize the application and application of technology in transportation systems. One topic that is currently hot is the classification of vehicle types from CCTV recordings installed on the side of the road. This classification is very useful in daily life, such as helping to recognize the majority of vehicle types in certain areas, and enforcement of road regulations such as prohibiting road passes for certain types of vehicles. Some existing studies use CCTV recordings to detect vehicles, but only limited between "vehicles" and "non-vehicles" [3] [4]. Other studies have been able to classify more than one type of vehicle but are still weak against changes in vehicle direction [5] and tolerance to lighting [6] [7].

In this study, the process of classification of motorized vehicles in the daytime with sunny conditions. This classification process uses four stages. The first stage, object detection, is the stage for separating objects and backgrounds using the Gaussian Mixture Model, Median Filter, dilation, and erosion. The second stage, feature extraction is the stage for obtaining motor vehicle features using Histograms of Oriented Gradients and Local Binary Patterns. The third stage, the training model is the stage for creating a classification model with data from the feature extraction stage using the Support Vector Machine. The fourth stage, classification, uses Hierarchical Multi-SVM, a variation of how to implement the Support Vector Machine, to classify between types of vehicles.

Figure 1. Examples of CCTV recording frames.

2. Our Approach

2.1. Dataset
The data used in the form of two CCTV video footage on one of the highway sections with a size of 640 × 360 pixels and has an RGB image channel taken from the Youtube video sharing site [16]. Video 1 is 81 seconds long, with a total of 2030 frames. Video 2 is 55 seconds long, with a total of 1640 frames. Examples of recording frames used are shown in Figure 1, where the left frame is an example frame in Video 1, and the right frame is an example frame in Video 2.

From both videos, the object detection process will be carried out to obtain training data and testing data. The training data used are positive samples and negative samples on the results of the detection stage of randomly selected objects. Positive samples are motorized bus_truck, car, minivan, and motorcycle types. Whereas negative samples are non-motorized imagery, where non-motorized imagery can be in the form of streets, trees, buildings, and alike. Negative samples can also be in the form of images that have more than one motorized vehicle in them. Because the data used does not have a ground truth, the process of grouping training data into positive samples and negative samples is done manually. The results of the detection phase of objects not selected as training data will be used as testing data.

Positive and negative samples for dimensions 64x64 pixels amounted to 3840 samples, consisting of 960 bus_truck positive samples, 960 positive car samples, 960 positive minivan samples, and 960 negative samples. While positive and negative samples for the dimensions of 32x64 pixels amounted to 720 samples, consisting of 360 positive motorcycle samples and 360 negative samples. The amount above is determined according to the ratio of 8:2 to class with the smallest number of samples.

2.2. System Design
The processes involved in implementing this system include the detection of motorized vehicle objects, the feature extraction stage, the training model stage, and the classification stage. The flowchart of the whole system process can be seen in Figure 2.

2.2.1. Object Detection. The first stage is the detection of vehicle objects. This stage uses the Gaussian Mixture Model (GMM) to separate vehicle objects from the background. The use of GMM was chosen because of its ability to distinguish lighting changes [8]. The results of GMM in the form of binary images that still have some noise that has the potential to interfere with the detection results. For this reason, several filtering and morphological operations such as median filter, dilation, and erosion are performed to get better detection results. The median filter is useful for removing salt and pepper noise [9], and dilation and erosion are used to refine the image by connecting the fragments and eliminating small holes in the image [9]. The next step is cropping and resizing to get object segments as vehicle candidates based on GMM result image as shown in Figure 3. Because the feature extraction stage uses Histograms of Oriented Gradients (HOG) that depend on image size, and non-uniform cropping results, resizing is needed so that the size becomes uniform.

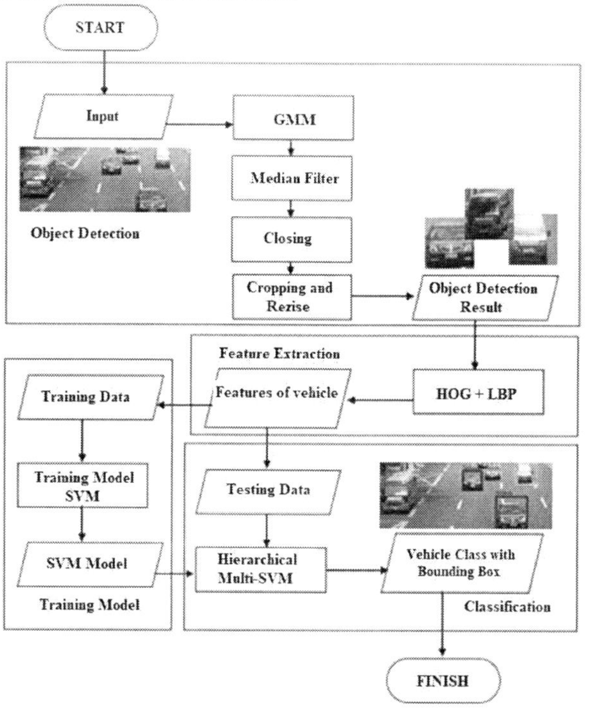

Figure 2. System Design

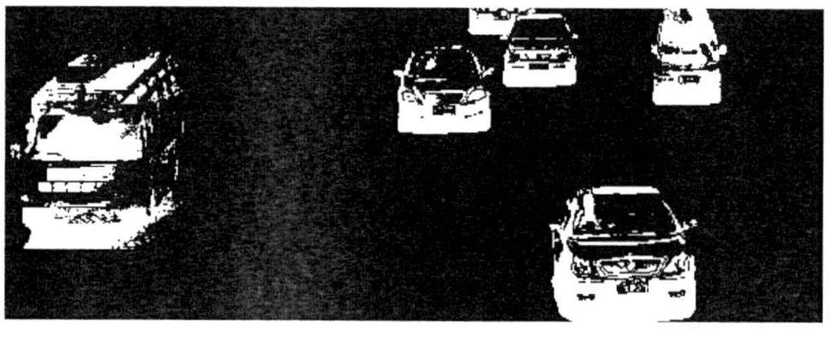

Figure 3. Original Image (top), result image of Gaussian Mixture Model (bottom)

The sizes used in the implementation of this stage are 64x64 pixels and 32x64 pixels, to adjust to the image of 4-wheeled vehicle segments that tend to be square and the image of 2-wheeled vehicle segments which tend to be rectangular in portrait orientation. From this stage, there will be some areas that have potential as vehicle objects. Figure 4 shows a positive sample resulting from this stage, and Figure 5 for a negative sample.

2.2.2. Feature Extraction. The feature extraction that used in this paper are Histograms of Oriented Gradients (HOG) and Local Binary Patterns (LBP). HOG is one method feature extraction commonly

IOP Conf. Series: Materials Science and Engineering 1077 (2021) 012068 doi:10.1088/1757-899X/1077/1/012068

used [10] - [12]. It counts the number of occurrences of the gradient orientation of a particular part of an image. This method is calculated from the grid of uniformly spaced cells and uses normalization of overlapping local contrast to improve accuracy [13]. Whereas LBP is one of the image feature extraction for classification. LBP calculates the pixel value of the image by thresholding neighbourhood pixels and storing the result as a binary number.

At this stage, the best HOG configuration will be searched by giving a variation of values in pixel configuration per cell and cell per block of HOG. The results of this stage are features that are ready to be used for the training model stage. Visualization of the results of the extraction of HOG and LBP features can be seen in Table 1.

(a) (b)

Figure 4. Example cropping positive samples. (a) 64x64 pixels, (b) 32x64 pixels

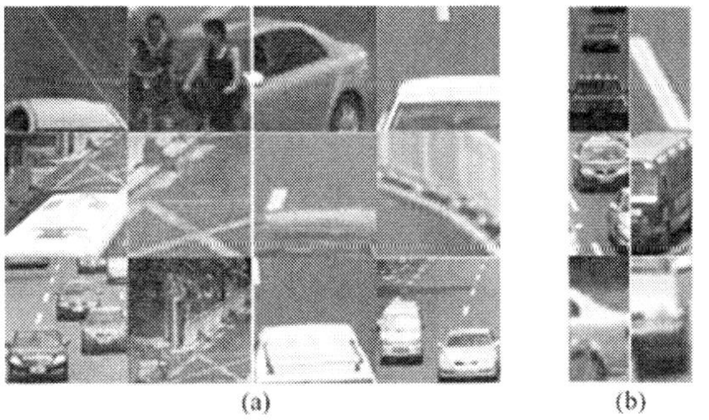

(a) (b)

Figure 5. Example cropping negative samples. (a) 64x64 pixels, (b) 32x64 pixels

2.2.3. Training Model. The training model stage is the stage for creating a classification model based on the features of the training data. This stage uses the Support Vector Machine (SVM) with a one-versus-rest configuration to produce a classification model that will be used at a later stage [14]. At this stage, the results of feature extraction will be classified into two classification models, namely for the effects of feature extraction from 64x64 pixel data and for feature extraction results from 32x64 pixel data. The 64x64 pixel classification model has four classes, namely bus_truck, car, minivan, and other_n. As for the classification model, 32x64 pixels have two classes, namely motorcycle, and other_m.

2.2.4. Classification. The last stage, the classification stage is the stage for classifying vehicle objects into prediction classes according to the classification model produced at the training model stage. Members of vehicle object prediction classes include Bus_Truck, Car, Minivan, and Motorcycle. This stage uses Hierarchical Multi-SVM, where this method is used to overcome heavy traffic conditions with vehicles that are side by side [15]. This method works on the principle that the prediction results of vehicle types do not have to be on the first try. At this stage, two SVM hierarchies are used, namely first-order SVM and second-order SVM. The results of this stage are vehicle objects that have been labeled with classes according to the predicted results. The flowchart of classification process using Hierarchical Multi-SVM can be seen in Figure 6.

Table 1. Visualization of HOG and LBP results

Class	Sample	HOG	LBP
Bus/Truck		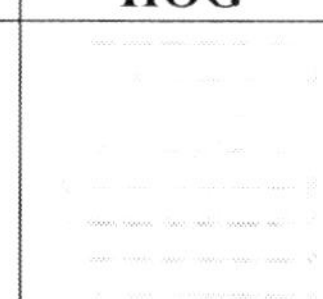	
Car			
Minivan		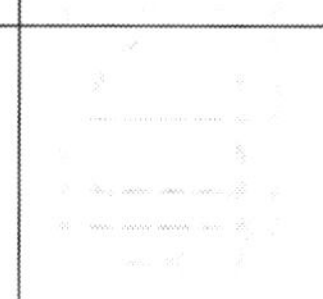	
Motor Cycle		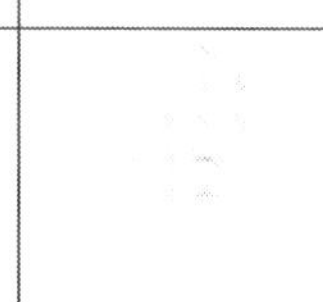	
Negative 64x64 Piksel		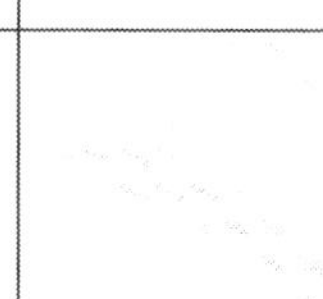	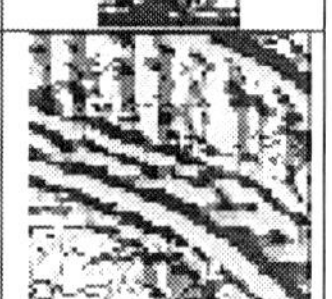
Negative 32x64 Piksel		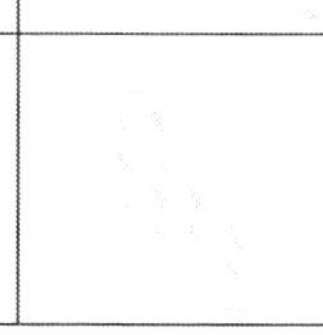	

3. Result and Discussion

In the experiment, some scenarios have been applied to obtain the best model of our proposed method. The initial values of parameters should be set, such as the LBP type used is LBP uniform 3x3 pixels. In hierarchical multi-SVM, we set the kernel as RBF and the regularization value, C=1.

3.1. Pixel Cell HOG Test per Cell

In this scenario, a feature extraction trial is performed using the Histogram of Oriented Gradients (HOG). In this scenario, the pixel value per cell is used as an independent variable, and cells per block are fixed variables. The pixel values per cell used are 4x4, 8x8, and 16x16. The cell value per block as a fixed variable is 2x2, according to the HOG standard configuration, trials using training data obtained from object detection results. Then the accuracy calculation process is done using 10-fold Cross-Validation. Table of trial results can be seen in Table 2.

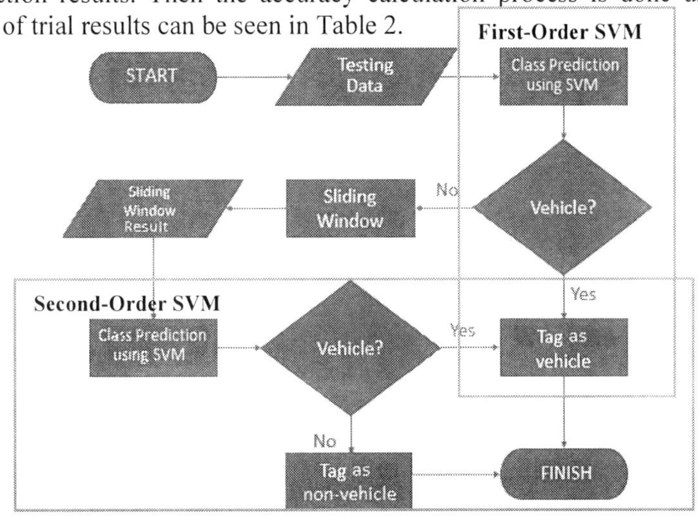

Figure 6. flowchart of classification process using Hierarchical Multi-SVM

Table 2. HOG pixel per cell trial results

n-Fold	Accuracy of Pixel per Cell (%)					
	64x64 Pixels			32x64 Pixels		
	4x4	8x8	16x16	4x4	8x8	16x16
1	83.33	92.71	92.19	95.83	95.83	98.61
2	90.36	91.41	91.41	80.56	79.17	100
3	80.47	87.76	93.49	98.61	100	94.44
4	89.32	92.19	94.27	100	100	100
5	92.97	94.01	96.61	91.67	97.22	97.22
6	89.84	91.93	92.19	86.11	95.83	94.44
7	87.50	90.36	90.62	76.39	94.44	98.61
8	89.32	90.62	87.24	95.83	94.44	98.61
9	91.41	91.93	89.32	94.44	98.61	98.61
10	83.07	83.85	82.29	98.61	98.61	98.61
Average	**87.76**	**90.68**	**90.96**	**91.81**	**95.42**	**97.92**

3.2. Trial HOG Cells per Block
In this scenario, a feature extraction trial is performed using the Histogram of Oriented Gradients. In this scenario, the value of cells per block is used as an independent variable, and pixels per cell are fixed variables. Cell values per block used are 1x1, 2x2, and 4x4. The selection of cell values per block is based on factors on the dimensions of the object detection results, namely 32x64 and 64x64. Then the accuracy calculation process is done using 10-fold Cross-Validation. The table of trial results can be seen in Table 3.

3.3. Comparative Trials of HOG, LBP and HOG + LBP
In this scenario, a comparison of the results of the trial feature extraction of vehicles using Histogram of Oriented Gradients, Local Binary Patterns, and a combination of Histogram of Oriented Gradients and Local Binary Patterns. Trials using training data obtained from object detection results. Then the accuracy calculation process is done using 10-fold Cross-Validation. Table of trial results can be seen in Table 4.

3.4. Discussion

From the first and second test scenarios, it is found that the best accuracy value for the extraction of HOG features is the configuration of 16x16 pixels per cell and 2x2 cell per block, with values of 90.96% for dimensions of 64x64 pixels and 97.92% for dimensions of 32x64 pixels. However, after the third test scenario, it was found that the accuracy value for the 64x64 pixel dimension decreased after combined with LBP to 90.68%. In contrast, the configuration of 8x8 pixels per cell has increased accuracy from 90.68% to 90.99%. For this reason, it was concluded that HOG + LBP with an arrangement of 8x8 pixels per cell is the most optimal configuration for extracting object features with dimensions of 64x64 pixels.

Table 3. HOG cell test results per block

n-Fold	Accuracy of Cell per Block (%)					
	64x64 Pixels			32x64 Pixels		
	1x1	**2x2**	**4x4**	**1x1**	**2x2**	**4x4**
1	87.76	92.71	94.27	95.83	95.83	95.83
2	92.19	91.41	91.41	77.78	79.17	87.50
3	86.20	87.76	86.98	98.61	100	100
4	90.88	92.19	93.49	98.61	100	100
5	92.71	94.01	91.93	98.61	97.22	100
6	88.28	91.93	91.67	90.28	95.83	97.22
7	88.28	90.36	87.76	91.67	94.44	84.72
8	90.36	90.62	89.06	98.61	94.44	97.22
9	92.97	91.93	91.67	98.61	98.61	97.22
10	83.30	83.85	83.85	98.61	98.61	100
Average	**89.30**	**90.68**	**90.21**	**94.72**	**95.42**	**95.97**

Table 4. Test results of feature extraction comparisons

Feature Extraction	Accuracy Average (%)	
	64x64 Pixels	**32x64 Pixels**
LBP	78.80	94.86
HOG 4x4 pixel per cell	87.76	91.81
HOG 8x8 pixel per cell	90.68	95.42
HOG 16x16 pixel per cell	90.96	97.92
HOG 4x4 pixel per cell + LBP	87.53	94.72
HOG 8x8 pixel per cell + LBP	**90.99**	**98.33**
HOG 16x16 pixel per cell + LBP	90.68	97.92

Table 5. System performance

Video	Accuracy (%)	Precision (%)	Recall (%)	F-measure (%)
Video 1	82.60	92.14	74.46	82.36
Video 2	77.96	96.39	73.11	83.15
Average	**80.28**	**94.27**	**73.79**	**82.76**

Whereas for object dimensions of 32x64 pixels, the best accuracy values occur after the third test scenario, where the HOG + LBP configuration of 8x8 pixels per cell and HOG + LBP 4x4 cells per block both have an accuracy of 98.33%. So that the configuration used is the same as the configuration for the 64x64 pixel dimension, HOG + LBP with 8x8 pixel configuration is chosen for the extraction of 32x64 pixel object dimensions. Based on Table 5, the system performance shows the proposed method works quite well with the accuracy, precision, recall, and F-measure of 80.28%, 94.27%, 73.79%, and 82.76%, respectively.

Examples of differences in detection results between Hierarchical Multi-SVM and standard SVM in heavy traffic can be seen in Figure 7, with the top frame using standard SVM, and the bottom frame using Hierarchical Multi-SVM. In the picture, it can be seen that there are two adjoining cars that are unable to be classified by standard SVM but can be sorted by Hierarchical Multi-SVM. Figure 8 shows the example of vehicle type misclassification. The car (red circle) misclassified as minivan because the characteristic of that car similar to minivan.

Figure 7. Differences in the results of SVM detection (top) and Hierarchical Multi-SVM in heavy traffic (bottom)

Figure 8. Example of vehicle type misclassification

4. Conclusion

Vehicle classification systems can be implemented using the Gaussian Mixture Model for object detection, Histogram of Oriented Gradients and Local Binary Patterns for feature extraction, and

Hierarchical Multi-SVM for classification. The best configuration for this system is with a distribution value of 8x8 pixels per cell for HOG, 2x2 cells per block for HOG, and LBP uniform 3x3 pixels. Based on the experimental results, our proposed method gives quite good performance with accuracy, precision, recall, and F-measure of 80.28%, 94.27%, 73.79%, 82.76%, respectively.

In the future, this system can be developed by extracting highways and tracking vehicles to improve vehicle detection results, as well as improving work efficiency so that the operation can do real-time classification.

Acknowledgment

This study was funded by Directorate of Research and Community Service (Direktorat Riset dan Pengabdian Masyarakat, DRPM) Institut Teknologi Sepuluh Nopember (ITS) Surabaya with the grant number of 1665/PKS/ITS/2020. The source code is provided in a repository at https://github.com/mumulmaulana/VehicleClassification.

References

[1] BPS 2010 *Jumlah dan Distribusi Penduduk* Retrieved on 16 April 2018 from http://sp2010.bps.go.id/index.php/site/topik?kid=1&kategori=Jumlah-dan-Distribusi-Penduduk

[2] BPS 2016 *Perkembangan Jumlah Kendaraan Bermotor Menurut Jenis 1949-2018* Retrieved on 16 May 2018 from http://www.bps.go.id/linkTableDinamis/view/id/1133

[3] Saiful A 2016 *Deteksi Mobil dari Citra CCTV Menggunakan Fuzzy Morphology dan Gradient Boosting* thesis (Surabaya: Teknik Informatika ITS)

[4] Prasetya I 2016 *Deteksi Kendaraan Bermotor dari Citra CCTV Menggunakan Gray Level Co-Ocurrence Matrix dan Random Forest* (Surabaya: Teknik Informatika ITS)

[5] Dong Z, Wu Y, Pei M and Jia Y 2015 Vehicle type classification using a semisupervised convolutional neural network *IEEE Trans. Intell. Transp. Syst.* **16** no 4 pp 2247–56

[6] Hsieh J, Yu S and Chen Y 2006 Automatic traffic surveillance system for vehicle tracking and classification *IEEE Trans. Intell. Transp. Syst.* **7** no 2 pp 175–87

[7] Xue X and Ling H 2011 Robust visual tracking and vehicle classification via sparse representation *IEEE Transactions on Pattern Analysis and Machine Intelligences* **33** no 11 pp 2259–72

[8] Stauffer C and Grimson W 1999 Adaptive background mixture models for real-time tracking *IEEE Computer Society Conf. on Computer Vision and Pattern Recognition* pp 246–52

[9] Gonzales R C and Woods R E 2008 *Digital Image Processing* 3rd edition (New Jersey: Pearson Prentice Hall)

[10] Dalal N and Triggs B 2005 Histograms of oriented gradients for human detection *IEEE Computer Society Conf. on Computer Vision and Pattern Recognition* pp 886–93

[11] Yan G, Yu M, Yu Y and Fan L 2016 Real-time vehicle detection using histograms of oriented gradients and AdaBoost classification *Optik – Int. J. for Light and Electron Optics* no 127 pp 7941–51

[12] Wang X, Han T X and Yan S 2009 An HOG-LBP human detector with partial occlusion handling *IEEE 12th Int. Conf. on Computer Vision (ICCV)* pp 32–9

[13] Navastara D A, Putra W Y M and Fatichah C 2020 Drowsiness detection based on facial landmark and uniform local binary pattern *J. of Physics: Conf. Series* vol **1529** no 5 (Bristol: IOP Publishing)

[14] Fatichah C, Alam S P and Navastara D A 2019 Optical flow feature based for fire detection on video data *1st Int. Conf. on Artificial Intelligence and Data Sciences (AiDAS)* (New York: IEEE) pp 100–5

[15] Fu H, Ma H, Liu Y and Lu D 2016 A vehicle classification system based on hierarchical multi-SVMs in crowded traffic scenes *Neurocomputing* vol **211** pp 182–90

[16] Supercircuits 2014 *Alibi ALI-IPU3030RV IP Camera Highway Surveillance* Retrieved on 30 October 2020 from https://www.youtube.com/watch?v=PJ5xXXcfuTc

IOP Conf. Series: Materials Science and Engineering 1077 (2021) 012069 doi:10.1088/1757-899X/1077/1/012069

Performance of DICOM Data De-identification Process in a Single Board Computer

K Y E Aryanto[1], A A G Y Paramartha

Computer Science, Department of Informatics, Faculty of Engineering and Vocational, Universitas Pendidikan Ganesha, Singaraja, Bali, 81116, Indonesia

E-mail: [1]yota.ernanda@undiksha.ac.id

Abstract. This work focused on testing the Raspberry Pi device's performance with the MIRC CTP application, developed by RSNA. The application could de-identify medical image data. The medical images tested in this research were using the DICOM standard. The results have shown that the average data processing time for 100 DICOM files in CT modality was roughly 0.5 seconds per images. Meanwhile, the tests obtained an average of three images per seconds when more extensive data with average size of 132 KB were transferred, using 1000, 5000, and 10000 files, respectively. It was shown that the installation of the RSNA CTP in the Raspberry Pi device is not complicated. Furthermore, the transfer results demonstrated that the device could be used in the medical image data de-identification, specifically in image data sharing

1. Introduction

Medical data sharing has become very common and widely conducted by various health institutions in recent years due to its benefit in advancing research or improving healthcare services [1]. In research, data sharing is recognized to be assuring scientific validity [2]. Meanwhile, sharing data is of great potential to improve public health by providing more objective evidence regarding patient safety and effectiveness [3].

Sharing such data raises awareness of what type of data or information could also be transferred and the medical data itself. In medical images, sensitive data can be embedded inside the images. The data could be in the form of text metadata in the file header or burnt in as pixels within the images. Thus, efforts to remove or modify such sensitive information should be made.

Removing original information that is considered sensitive, contained in the medical image data is known as data de-identification. It is mainly recognized in two types, namely anonymization and pseudonymization. The first one removes the intended data entirely, while the latter is done by replacing the targetted data with some codes generated by the related party. Both processes could deliver anonymity to the data so that the data owner's identification will be difficult or even impossible to do.

Many kinds of research had been done to provide attempts to de-identifying medical image data. Many applications were developed as well to supports the effort. One of the widely used application is the Medical Imaging Resource Center - Clinical Trial Processor (MIRC CTP) [4], developed by the Radiology Society of North America (RSNA) [5]. It is one of the applications that can de-identify metadata of the medical imaging file, in the format of Digital Imaging and Communication in Medicine (DICOM) standard [6], with a very high success rate [7] even when it was set in its default settings. The de-identification process was facilitated in a service that could be called through its pipeline configuration. MIRC CTP was explored in various research or projects, as a single processor application or bigger systems [8]–[11]. The application or the system was reported to be performed well when running under a personal computer or higher specification.

In recent years, computers in small sizes had been massively developed. One such device that has gained its popularity is Raspberry Pi [12]. It has been implemented in various medical fields research, resulting in a good performance of handling sophisticated processes. An assessment of the feasibility of running MIRC on the Raspberry Pi device was done by Pereira et al. [13]. However, the MIRC's performance in de-identifying DICOM data was not described since the work was mainly focused on delivering the requirements of a teaching file system. The single-board computer is practically cheap. The latest version, Raspberry Pi 4, are in the market for as low as US$ 35. It comes in the form of a credit card-sized computer [14]. However, the Raspberry Pi offered various features close to a standard personal computer's ability, and it is easy to use. Therefore, it should be able to process the data de-identification as well. In this work, an observation of the MIRC CTP application's ability installed in the Raspberry Pi device to perform the de-identification processes was conducted.

2. Materials and Method

2.1. MIRC CTP

The RSNA MIRC was published in the early 2000s as a software tool for transferring, processing, sending, and storing medical images for the radiology community. It was developed using Java code. The early version of MIRC required the installation of Apache Tomcat as the HTTP web server environment. In a later development, it was simplified, and therefore the requirement of Apache Tomcat is no longer remains. Furthermore, with the increasing acceptance and implementation of medical image data sharing, the MIRC was split into two separate applications: MIRC Teaching File System (TFS) and MIRC Clinical Trials Processor (CTP).

The MIRC CTP application is a standalone and open-source application developed by researchers and health institutions to exchange images and documents from clinical trial image data [15]. This application has several main features that are claimed to make it easier for users to process the exchanged files in clinical trials or research, including medical image data. Apart from being easy to install on computer devices, several processing features were developed to perform several parallel processes in each pipeline with different settings for each. Therefore, processing and data transfer can be carried out simultaneously through several available protocols [4]. Data transfer can be done using several protocols, including HTTP, secure HTTP, and DICOM.

MIRC CTP was developed based on the Java platform so that this application can be installed under a variety of supporting operating systems such as Microsoft Windows and Linux. Thus, this application should also run under an operating system that is a derivative of such an extensive operating system such as Raspbian [12] or Windows 10 IoT Core [16] as an operating system installed on a Raspberry Pi device. Furthermore, MIRC CTP is an open-source application so that it can be easily modified and integrated according to institutional user needs.

MIRC CTP, apart from being used to transfer data, is also used to process medical image data using the DICOM standard, which consists of a header and a scanned image itself. There is metadata in the DICOM file header containing all data related to medical images in the file, including patient data, patient families, related medical personnel, health institutions, time and image capture devices, and others. This part is an essential concern in the data sharing process because applicable regulations, such as the GDPR in the European Union (EU) [17] and HIPAA in the United States (US) [18], have indeed obliged data collectors to maintain the confidentiality of the data and do not share it with other parties.

One of the MIRC CTP application features is de-identification, a process to remove or encode patient data in medical files so that they do not refer directly to the patient concerned. In this case, the data contained in the tags in the DICOM file's header are deleted or modified in such away. Various functions have been provided to do this, such as deleting, modifying, saving, encoding, and so forth.

2.2. DICOM Standard

DICOM is one of the standards used globally for medical images and other related information [6]. This standard was developed to support the interoperability of various systems that generally involve retrieving, storing, displaying, transmitting, processing, or receiving medical image data. This standard itself is widely used because of several main benefits such as free standards that are not specific to a particular vendor, making it easier for medical images to be stored and operated in or exchanged between systems, and their development is still being actively carried out to meet technological developments and medical imaging needs.

The DICOM standard medical image file structure consists of two main elements, namely the header and the image data set. The header section contains data or information relating to the file itself. Meanwhile, the image data set part contains several data elements related to the stored image. These data elements include various metadata containing important data or information related to the image contained in them. The data or information can be in the form of data associated with the identity of patients, institutions, medical personnel, studies, series, modalities, image formats, pixel attributes, and others.

There are two types of elements or attributes, in a DICOM file, including standard data elements and private data elements. These elements can generally be written as two groups of numbers (xxxx, yyyy) where xxxx and yyyy are in hexadecimal number format, respectively. The first four numbers (xxxx) are generally referred to as the group number, while the rest is the element number. Standard data elements will have even group numbers, while private data elements will have odd group numbers.

Images stored using the DICOM standard can come from a variety of modalities. There are more than 50 types of modalities that can be stored in the DICOM standard, which include Computed Radiography (CR), Computed Tomography (CT), Magnetic Resonance (MR), Positron emission tomography (PET / PT), X-Ray Angiography (XA), and Ultrasound (US). The resulting file size varies considerably between each of these modalities. CT, MR, and the US scanned files are on average in the range of 100 to 600 kilobytes (KB), while images of modalities such as Mammography (MG) and CR can reach file sizes in the range of 27 to 30 megabytes (MB) per image [19].

This standard has been widely implemented in radiology, which allows the process of digitizing workflows to replace scans with final physical outputs. Furthermore, all the modalities used can be retrieved, stored, and distributed in a single system that utilizes existing network technologies, including internet technology. DICOM has made it possible to develop applications that connect all lines and update both doctors' and patients' communication process.

2.3. Method

This study focuses on the data de-identification process carried out using the MIRC CTP application, installed on the Raspberry Pi 4 device. The data used is medical image data using the DICOM standard. Data is public data obtained through the Kaggle.com website. There are no restrictions regarding file size; however, the size used is in the range of 132-512 KiloBytes (KB). This research was conducted in 2 stages, namely: (1) installation of the MIRC CTP application on the Raspberry Pi 4 device, followed by setting up the device and testing to ensure that the "anonymizer" feature can function properly; and (2) Testing the "anonymizer" feature in de-identifying several medical image data. The numbers used in this work are 1000, 5000, and 10000 medical image files, respectively. Observations were made of the time required to carry out the de-identification process, the de-identification process's success, as seen from the absence of quarantined files. The de-identification scheme used is to change the DICOM file header tag's value, which is "de-identified".

Installation was done by first looking at the needs of the CTP application itself. After the module or library of the application has been successfully installed, the CTP application is installed into the device according to the operating system's installation mode. In this study, the Ubuntu operating system was used in this work. The tuning is done by adding a processing feature known as an "anonymizer". The processes conducted in this work can be seen in figure 1.

IOP Conf. Series: Materials Science and Engineering 1077 (2021) 012069 doi:10.1088/1757-899X/1077/1/012069

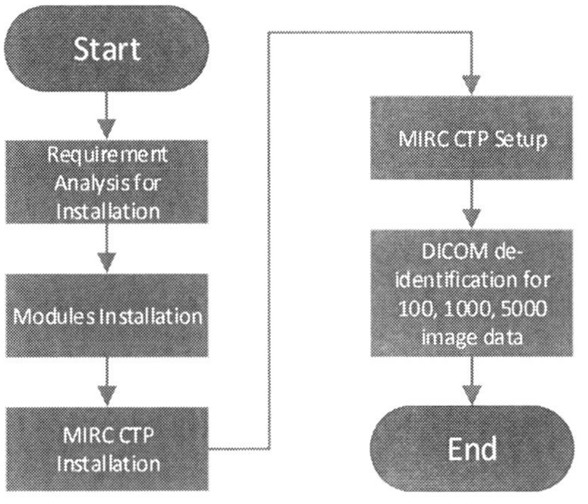

Figure 1. The process flow of the research

3. Results and Discussion

The implementation of this study follows the method described in the previous section. The method implemented in this research is divided into 4 (four) steps, including: (1) problem analysis and installation, (2) setting up limited scope of devices, applications, and infrastructure, (3) data de-identification trials, and (4) analysis of results.

3.1. System Installation

The MIRC CTP application can be downloaded via the website http://mirc.rsna.org. The installation process requires a GUI-based display, where accessing the downloaded files will result in the process of extracting files to the CTP folder. Due to the focus of the process that does not involve many features of the operating system itself, the Ubuntu Server operating system is used in this study. Thus, the extraction process can be carried out on another machine with a GUI-based window display and then transferred to the Raspberry Pi device.

This application is based on the Java programming language. Therefore, it requires the Java running environment (JRE) package on the operating system used. JRE can be provided by installing the appropriate application package. In running the application for the first time, a problem was encountered regarding the Java Imageio library version stored in the imageio folder. However, the library is only used if there is the processing of the image portion of the DICOM file, while in this work, it solely focuses on the header file. Thus, the library can be removed, and after that, the application can run as usual. As for running the application, the file that must be executed is "Runner.jar" with the appropriate command. The system then can be controlled through a website provided by the MIRC CTP itself.

3.2. System Settings

All of the pipelines and containing services should be determined before the application is executed. The setting can be configured following the format stated on the documentation page of the MIRC CTP. In this study, only a de-identification service was required. Therefore, the only setup that should be prepared is where the "DicomAnonymizer" was selected.

There are two ways of configuring the setup of the MIRC CTP. First, the configuration can be done using the CTP DICOM Anonymizer setup format, as mentioned in the documentation page of MIRC CTP. The step should be applied to the "config.xml" file. The example of a simple setup used in this work can be seen in figure 2.

The other way, and the more straightforward way of setting up the required services in MIRC CTP, is through the MIRC CTP launcher program's configuration tab. However, the application required a GUI to implement this option, which is not supported by Ubuntu Server by default. The "trick" is to set it up using another machine with a GUI-based window and transfer the XML file to the Raspberry Pi 4

IOP Conf. Series: Materials Science and Engineering 1077 (2021) 012069 doi:10.1088/1757-899X/1077/1/012069

device's installation folder. The steps of configuring pipelines and services using the GUI-based launcher of the MIRC CTP are shown in figure 3.

```
<Configuration>
    <Server
        port="1080"/>
    <Pipeline
        name="testpipeline"
        root="Anonymize">
        <DicomAnonymizer
            class="org.rsna.ctp.stdstages.DicomAnonymizer"
            lookupTable="scripts/LookupTable.properties"
            name="DicomAnonymizer"
            quarantine="quarantines/DicomAnonymizer"
            root="roots/DicomAnonymizer"
            script="scripts/DicomAnonymizer.script"/>
    </Pipeline>
</Configuration>
```

Figure 2. The configuration of de-identification service in MIRC CTP

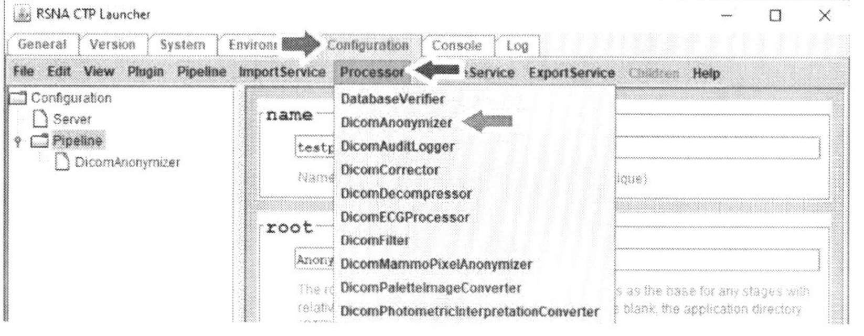

Figure 3. The configuration of de-identification service in MIRC CTP using the launcher program

3.3. De-identification Process

After the required settings were all configured, 100 for functional testing, followed by 1000, 5000, and 1000 data was conducted. The transfer of 100 data was done using a set of DICOM data with the size of each image was approximately 170 KB. From the test, it was found that the overall process required to de-identify 100 DICOM files were nearly 1 MegaByte per second (MBps) or approximately 7,84 Megabits per second (Mbps). It was equal to 2 images per second was obtained, as seen in table 1.

Table 1. Results of de-identification processing of 100 DICOM data.

Feature	Size	Unit
Total Data	100	Images
Average image size	512	KB
Total image size	51,2	MB
Total Time	51	Seconds
Average	7,84	Mbps

The results shown above were roughly half of the transfer process's speed obtained in work by Aryanto et al. [20]. The work was using 25,000 images through a complete process of receiving, processing, and storing images. However, the work was done using an intranet scheme, which means the higher specification of devices and networks were used.

The processing of three datasets was done using images that were available freely on the internet. The DICOM file's average size was 132 KB. The results showed that an average of four images per second was processed through the provided service. There was a slight drop in processing the DICOM images when more images were transferred. The results of the data processing of the three datasets are shown in the table 2.

Table 2. Results of de-identification processing of 1000, 5000, and 10000 DICOM data.

Feature	Test #1	Test #2	Test #3
Total Data (images)	1000	5000	10000
Average of image size (KB)	132	132	132
Total image size (KB)	135,141,662	676,816,698	1,353,633,396
Total Time (seconds)	238	1357	3038
Average (Images/seconds)	4	3	3

The table 2 shows that the amount of data between test #1, test #2, and test #3 did not significantly differ. The results indicate consistency in data processing using one modality type, as seen from the file sizes. The results were comparable in the number of images processed per second compared to the previous other work. However, the difference in image size or modalities may influence the processing speed as well.

The decrease of the average processing time may be caused by the amount of queue processing in the pipeline that was also significantly more than the test of 100 images data. More data to process means that more resources were required. Therefore, with the considerable amount of data in the queue list, it is expected to have a more significant drop of time necessary for the data processing.

4. Conclusion

A data processing regarding the de-identification of DICOM image data were conducted. The system consisted of a Raspberry Pi 4 device implemented with MIRC CTP to perform the intended task. Although the process of more data means consumed more time, the differences were not significantly high. The result shows that the Raspberry Pi can deliver de-identification strategies of DICOM images data and be used in DICOM image data sharing with the advantage of its portability.

Acknowledgments

The authors would like to thank the Study Program of Computer Science, the Department of Informatics, Universitas Pendidikan Ganesha, to conduct this research. The authors would also like to thank Ann for her editorial help in preparing the manuscript.

Funding

This work was supported by Universitas Pendidikan Ganesha under the DIPA BLU funding, grant number SP DIPA – 023.17.2.677530/2020 Revision 1 (March 16, 2020).

References

[1] Lo B 2015 Sharing clinical trial data: maximizing benefits, minimizing risk *JAMA* vol **313** no 8 pp 793–4 DOI: 10.1001/jama.2015.292

[2] Freymann J B, Kirby J S, Perry J H, Clunie D A and Jaffe C C 2012 Image data sharing for biomedical research—meeting HIPAA requirements for de-identification *J. Digit. Imaging* vol **25** no 1 pp 14–24 DOI: 10.1007/s10278-011-9422-x

[3] Lo B, Coetzee T and Demets D 2015 *Committee on Strategies for Responsible Sharing of Clinical Trial Data, Board on Health Sciences Policy, Institute of Medicine. Sharing clinical trial data: Maximizing Benefits, Minimizing Risk* (Washington: National Academies Press)
[4] RSNA 2019 *MIRC CTP* Retrieved on 15 October 2020 from https://mircwiki.rsna.org/index.php?title=MIRC_CTP
[5] RSNA 2020 *Radiological Society of North America* Retrieved on 15 October 2020 from https://www.rsna.org/
[6] National Electrical Manufacturers Association (NEMA) 2020 *DICOM Standard* Retrieved on 15 October 2020 from https://www.dicomstandard.org/
[7] Aryanto K Y E, Oudkerk M and van Ooijen P M A 2015 Free DICOM de-identification tools in clinical research: functioning and safety of patient privacy *Eur. Radiol.* vol **25** no 12 pp 3685–95 DOI: 10.1007/s00330-015-3794-0
[8] Clark K *et al* 2013 The Cancer Imaging Archive (TCIA): maintaining and operating a public information repository *J. Digit. Imaging* vol **26** no 6 pp 1045–57
[9] Erickson B J, Fajnwaks P, Langer S G and Perry J 2014 Multisite image data collection and management using the RSNA image sharing network *Transl. Oncol.* vol **7** no 1 pp 36–9 DOI: 10.1593/tlo.13799
[10] Chatterjee A R *et al* 2017 Image sharing in radiology—a primer *Acad. Radiol.* vol **24** no 3 pp 286–94 DOI: 10.1016/j.acra.2016.12.002
[11] Caturla S I *et al* 2018 Integrating information in radiology: Teaching File System (TFS) implementation in a third-level hospital using MIRC *ECR 2018*
[12] Raspberry Pi Foundation 2020 *Raspbian* Retrieved on 27 October 2020 from https://www.raspbian.org/
[13] Pereira A, Atri M, Rogalla P, Huynh T and M E O'Malley 2015 Assessment of feasibility of running RSNA's MIRC on a raspberry pi: a cost-effective solution for teaching files in radiology *Int. J. Comput. Assist. Radiol. Surg.* vol **10** no 11 pp 1793–801 DOI: 10.1007/s11548-014-1143-4
[14] Raspberry Pi Foundation 2020 *Raspberry Pi 4 Model B – Raspberry Pi* Retrieved on 15 October 2020 from https://www.raspberrypi.org/products/raspberry-pi-4-model-b/
[15] RSNA 2020 *Imaging research tools* Retrieved on 15 October 2020 from https://www.rsna.org/research/imaging-research-tools
[16] Microsoft 2020 *Windows for Internet of Things* Retrieved on 15 October 2020 from https://developer.microsoft.com/en-us/windows/iot/
[17] EC Directorate-General for Communication 2018 *Data Protection in the EU* Retrieved on 15 October 2020 from https://ec.europa.eu/info/law/law-topic/data-protection/data-protection-eu_en
[18] US Department of Health & Human Services 2020 *Health Information Privacy* Retrieved on 15 October 2020 from https://www.hhs.gov/hipaa/index.html
[19] Dandu R 2008 Storage media for computers in radiology *Indian J. Radiol. Imaging* vol **18** no 4 p 287 DOI: 10.4103/0971-3026.43838
[20] Aryanto K Y E, Broekema A, Oudkerk M and van Ooijen P M A 2012 Implementation of an anonymization tool for clinical trials using a clinical trial processor integrated with an existing trial patient data information system *Eur. Radiol.* vol **22** no 1 pp 144–51 DOI: 10.1007/s00330-011-2235-y

Scrum Adoption Challenges in Higher Education in Indonesia: Case Study of Board of Information System, Universitas Islam Indonesia

Mukhammad Andri Setiawan[1] and Sujono[2]

[1,2]Universitas Islam Indonesia, Yogyakarta, INDONESIA

E-mail: [1]andri@uii.ac.id, [2]sujono@students.uii.ac.id

Abstract. The level of success in project development information systems or software in many countries is still low. A study of 110 information system projects in Indonesia found that most information system projects are problematics, and 18% were canceled. Even though it has been around for a while, Scrum as an approach is relatively new in Indonesia's software development. Many efforts to adopt agile in various organizations are not running well. This research aims to identify the challenges in agile adoption in a higher education organization in Indonesia using the Scrum Adoption Challenges Detection Model (SACDM) theory. The study found that even though Scrum is performing well, there are still some works needed. Especially in improving the experiences and transforming the organizational culture and structure to make the scrum adoption more effective.

Keywords: Scrum, Agile Adoption, SCADM.

1. Introduction

The degree of success in project development of information systems or software in many countries is still low. [1] explain that e-government projects in developing countries, only 15% are considered to be successful, while 35% experienced a total failure, and 50% had a partial collapse. Hence, it is necessary to identify successful software development tasks for others to learn. [2] in Chaos Report conducted surveys from 1994 to 2015 on information system projects across the USA and Europe. This survey explained that information system projects' success rate is still between 16% to 31%. Figure 1 shows the success rate of the development that is not stable. Sometimes it went up, but at many times also went down.

A study of a software project in South Africa in 2013[3] explained that 34% of software development projects are considered successful, but at 66% is considered a failure and challenge. By 32% and by 34%, software development projects fail to face challenges due to a software development project not completed on time or does not fit the budget estimates. Meanwhile, [4], in a study of 110 information system projects in Indonesia, 27% of information system projects were within budget and finished on time, up to 55% of projects are problematic, and 18% of project owners cancelled the development. The results showed that the success rate of information system projects in Indonesia is still low as also experienced by other countries. To overcome this issue, many organizations started to look into different kinds of software development approaches, and agile is one of them.

IOP Conf. Series: Materials Science and Engineering 1077 (2021) 012070 doi:10.1088/1757-899X/1077/1/012070

Agile is a relatively new approach in software development [5]. Many leading organizations have adopted Agile in all or some of their projects [6]. Agile is becoming increasingly popular that many organizations seeking to take Agile software development [7]. The need for a better, faster, and cost-effective software solution, and at the same time, also provide a level of employee satisfaction high drove this popularity [5].

Fig. 1. Trend of Information Systems Project Success Rate [2]

A study that takes a sample of data from 1,002 projects in various industries and countries has conducted tests on the effects of the use of Agile in the two-dimensional organization of project success: efficiency and overall satisfaction of stakeholders toward organizational objectives. The research findings indicate that Agile methods positively impact both the dimensions of the project's success [8]. As stated in the Chaos Report conducted by the Standish Group in 2015, other results identified factors that create success: the Agile process and small projects. Small projects that use Agile methods have a failure rate of 4% only [2].

Chaos Report provided reports and showed that Agile projects have nearly four times the success rate compared to the Waterfall. In comparison, the Waterfall project has three times the failure rate compared to Agile [2]. Many studies show the attempts to adopt Agile in various organizations or companies in several countries are not running smoothly. Those studies found a wide range of challenges that affect success in adopting Agile. For example, in Kosovo's financial institutions, the challenges are the lack of support from management and the skills needed to implement the methodology, the allocated budget, and the number of willing to jump into Agile mode [9].

In research with case studies of projects in Brazil's multidisciplinary teams, they explained that direct communications from stakeholders to team members had caused constant disruption in the development process. Other than an excessive number of informal communications, the centralization of information on a particular member represents created challenges in adopting agile [10].

One of the most popular methodologies used in agile is Scrum [11]. The Scrum itself can be characterized as a self-management team that brings decision-making authority to the level of operational problems and uncertainties. However, even though Scrum is the most popular agile approach, research on an eCommerce company in the United States showed challenges in agile adoption with Scrum, such as lacking permission from the board of company [12]. Meanwhile, in the study on government entities in the United Arab Emirates, the challenges of Scrum adoption are missing roles such as lack of Master Agile, the absence of the pilot project, and no employees were experienced enough with agile methods. The entities were also facing working pressure, and it was not very easy to persuade the top management to invest in the new methodology. Lastly, government bureaucracy and extensive documentation requirements hindered the adoption of Scrum [5].

What about Indonesia? To the best of our knowledge, no study was conducted yet on how an organization outside technology company/organization in Indonesia used Scrum. Based on the discussions above, we found that different organizations adopted Scrum in different variations. This study aims to find a model of agile adoption in a higher education institution in Indonesia, with a case study in the Board of Information Systems, Universitas Islam Indonesia. The board of information systems is the IT unit at the university. As part of the university, it is naturally adhering to a higher education industry culture different from software industry culture. However, in recent years, the higher education industry faces an arduous effort to comply due to regulations, internal and external governance compliances, and digitalization. These things have forced the Board of Information System at Universitas Islam Indonesia to adopt and develop information systems faster than it used to be. The board has been experimenting with the use of the Scrum approach since 2018 to develop software faster and robust. A robust system is necessary since almost 30,000 users will access it.

The remainder of this paper is organized as follows. Section 2 provides a background on achieving success in the software development projects and challenges that affect the success in Scrum adoption. We describe the research methodology in section 3. Subsequently, we present our findings in section 4, with detailed discussions in Section 5. In section 5, we explain the limitations of our study. Lastly, we conclude this study in section 6.

2. Methodology

We conducted this research to answer a research question i.e. what are the challenges faced in the agile methodology adoption in Higher Education institutions? This study used quantitative research based on surveys and interviews to examine agile adoption challenges in universities. We used the Scrum Adoption Challenges Model Detection Model (SACDM) [13], which is a model adapted from the theory of Diffusion of Innovation (DOI) to measure the adoption of Scrum. Figure 2 shows the SACDM conceptual framework that consists of four constructs: individual, team factors, organizational factors, and technological factors. We used nineteen independent variables to develop the four constructs.

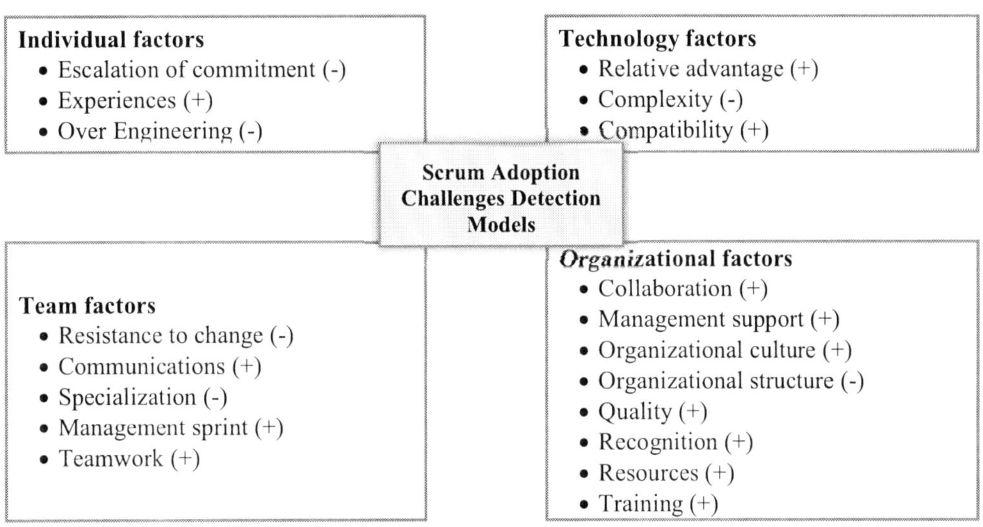

Fig. 2. SACDM conceptual framework [13]

2.1. Respondents

Participants who participated in our research came from different roles in the organization. Those are Scrum Master, Product Owner, Team Member (Development and Operational), and Management team of the Board of Information System, Universitas Islam Indonesia. The number of respondents is 41 out of 50 people who work for the Board of Information System when the research was conducted (82.9% males and 17.1% females) with the details as follows: 21 people from Developer team – *Dev* (51.2%), nine Operational team member – *Ops* (22 %), six Product Owner – *PO* (14.6%), and five people (12.2%) as Scrum Master – *SM_Mg*. Most respondents are generation Y (millennials) amounted to 90.2%, and 9.8% are generation X. From the educational background point of view, most respondents have a bachelor's degree and consist of 70.7% of the respondents, master's degree 12.2 %, doctoral degree 2.4%, polytechnic degree 4.9%, and 9.8% are high-school graduate. We collected the questionnaires from our respondents from December 2019 until January 2020.

2.2. Data analysis

We analyze the data using quantitative data based on completed questionnaires from respondents and validate the test using test validity and analyze the results by calculating each variable's mean of each construct.

2.3. Test Validity

We performed test validity using the bivariate correlation between the respective indicator scores and the constructs' total score. We obtained significant values from the results of the test validity results that indicate the questions are valid. Table 1 presents detailed results.

3. Results

We calculated the means from each variable from each construct, i.e., individual factors (3 variables), team factors (5 variables), organizational factors (8 variables), and technological factors (3 variables). Table 2 shows the means calculation for each variable in the individual factor. Variable is measured in likert scale, where 1 indicate the unlikeliness or strong disagreement and 5 indicate the likeliness or strong agreement.

Of the three variables on individual factors, the largest means average is Over Engineering variable with its score 3.83, and the lowest average is experience. This reflects the fact that Scrum is only applied in the organization in recent years. The board of information system, Universitas Islam Indonesia, started implementing Scrum in mid-2017. Hence most team members felt that they are not confident yet with the experience variable. Since the organization has only been applying recently, it often changes team composition and also the team tasks that contribute to the high score of the Over Engineering variable.

Table 3 shows that all variables in the team factor have average scores above 4.00. The highest average value in the communication variable with a score of 4.43 and indicates that the team members can communicate well and can exchange ideas with other team members easily. The teams also established common goals and collaborated intensively to achieve the goals. Team members felt that working intensively in a team is much more preferable than working alone. With regard to the specialization, respondents expressed that senior team members who have higher capability are helpful and collaborate easily with junior team members. In Sprint Management, respondents have understood its function well, which is to monitor and manage the sprint's progress and prevent the lack of timely feedback. In the change-resistant factor, most respondents see the benefit of using Scrum compared to the waterfall methodology that was previously used extensively and did not felt that Scrum hinders the works.

Table 4 shows only two out of 8 variables are above 4. Only training and collaboration variables scored above 4. The organizational structure scored the lowest with a score of 3.59. The organizational structure variable focused on how to support the organizational structure that enables the effective use of an agile method. The medium score means that most respondents felt that the organization's current structure has not yet accommodated the Scrum method well.

The second-lowest mean score is organizational culture. This variable explains the involvement of team members in the decision makings in the organization. Many still felt that most decisions were only decided by a few people, such as management (Scrum Masters) or product owners, rather than involving all team members.

Table 1. Test Validity in Agile Adoption Questionnaire

No.	Variables	Number of questions	Indicator	Pearson Correlation
1	Escalation of commitment	2	FIN_A1_KomitMg	0.793 **
			FIN_A2_KomitMg	0.845 **
2	Over engineering	2	FIN_B1_RekaTim	0.788 **
			FIN_B2_RekaTim	0.802 **
3	Experiences	3	FIN_C1_P.P Software	0.804 **
			FIN_C2_P.Agile	0.827 **
			FIN_C3_P.BSI	0.853 **
4	Communication	2	FTim_A1_Comm	0.924 **
			FTim_A2_Comm	0.925 **
5	Teamwork	2	FTim_B1_KerTim	0,795 **
			FTim_B2_KerTim	0.866 **
6	Specialization	1	FTim_C_Spesi	1
7	Sprint Management	2	FTim_D1_MnSprint	0.915 **
			FTim_D2_MnSprint	0.932 **
8	Change Resistant	3	FTim_E1_Resis	0.799 **
			FTim_E2_Resis	0.859 **
			FTim_E3_Resis	0.835 **
9	Training	2	FOrg_A1_Pelat	0.785 **
			FOrg_A2_Pelat	0.839 **
10	Recognition	2	FOrg_B1_Rekog	0.801 **
			FOrg_B2_Rekog	0.755 **
11	Quality	4	FOrg_C1_Kual	0.767 **
			FOrg_C2_Kual	0.741 **
			FOrg_C3_Kual	0.777 **
			FOrg_C4_Kual	0.756 **
12	Resource	2	FOrg_D1_SD	0.858 **
			FOrg_D2_SD	0.820 **
13	Collaboration	3	FOrg_E1_Kolab	0.898 **
			FOrg_E2_Kolab	0.773 **
			FOrg_E3_Kolab	0.901 **
14	Management Support	6	FOrg_F1_DukMg	0.431 **
			FOrg_F2_DukMg	0.574 **
			FOrg_F3_DukMg	0,812 **
			FOrg_F4_DukMg	0.823 **
			FOrg_F5_DukMg	0.793 **
			FOrg_F6_DukMg	0,645 **
15	Organizational culture	2	FOrg_G1_BudOrg	0.822 **
			FOrg_G1_BudOrg	0.825 **
16	Organizational structure	1	FOrg_H_StrOrg	1
17	Relative advantage	5	FTek_A1_KeunRel	0,729 **
			FTek_A2_KeunRel	0.839 **
			FTek_A3_KeunRel	0.775 **
			FTek_A4_KeunRel	0.804 **
			FTek_A5_KeunRel	0.671 **
18	Complexity	2	FTek_B1_Kompl	0.895 **
			FTek_B2_Kompl	0.894 **
19	Compatibility	3	FTek_C1_Kompatb	0.682 **
			FTek_C2_Kompatb	0.850 **
			FTek_C3_Kompatb	0,885 **

** Correlation is significant at the 0:01 level (2-tailed).

Table 2. Means in individual factor

No.	Variables	Indicator	Mean					Mean per Varia-bles
			Group				Total (N: 41)	
			Dev (N: 21)	Ops (N: 9)	PO (N: 6)	SM_Mg (N: 5)		
1	Escalation of commit-ment	FIN_A1_KomitMg	2,71	3.44	3.50	3.00	3.02	3.33
		FIN_A2_KomitMg	3.43	3.89	4.00	3.60	3.63	
2	Over engi-neering	FIN_B1_RekaTim	3.52	3.67	3.50	4.20	3.63	3.83
		FIN_B2_RekaTim	3.90	3.89	4.17	4.60	4.02	
3	Experi-ences	FIN_C1_P.P Software	3.48	2.33	4.00	4.60	3.44	2.90
		FIN_C2_P.Agile	2,14	2.33	3.33	3.00	2.46	
		FIN_C3_P.BSI	2.24	2.56	4.67	3.40	2.80	

Table 3. Means in the team factor

No.	Variables	Indicator	Mean					Mean per Varia-bles
			Group				Total (N: 41)	
			Dev (N: 21)	Ops (N: 9)	PO (N: 6)	SM_Mg (N: 5)		
1	Communi-cation	FTim_A1_Comm	4.29	4.56	4.33	4.80	4.41	4.43
		FTim_A2_Comm	4.29	4.56	4.50	4.80	4.44	
2	Teamwork	FTim_B1_KerTim	4.19	4.44	4.17	4.80	4.32	4.27
		FTim_B2_KerTim	4.14	4.44	4.17	4.20	4.22	
3	Specializa-tion	FTim_C_Spesi	3.95	4.22	3.83	4.40	4,05	4,05
4	Sprint Manage-ment	FTim_D1_MnSprint	4.19	3.89	4.50	4.40	4.20	4.21
		FTim_D2_MnSprint	4.19	3.78	4.50	4.80	4.22	
5	Change Re-sistant	FTim_E1_Resis	4,05	4.22	4.00	5.00	4.20	4.01
		FTim_E2_Resis	4.00	4.33	4.00	4.60	4.15	
		FTim_E3_Resis	3.76	3.56	3.00	4.40	3.68	

Table 5 shows that the means of all variables are below 4. This condition means that most respondents still felt that agile has not yet reached a state that can help them complete tasks quicker and easier or more effectively. This situation indicates that the Scrum methodology is still perceived as a complicated approach.

4. Discussion

This research has identified the challenges faced in adopting Agile using the SACDM approach. The score of each question in the assessment questionnaires is from 1.00 to 5.00. We grouped the level of adoption into four levels. The average means value for the group of Level 3 (L3) is from 1.00 to 2.00, Group Level 2 (L2) is from 2.01 to 3.00, Group Level 1 (L1) is from 3.01 to 4.00, and Group Level 0 (L0) is from 4.01 to 5.00. Level 0 means the best adoption possible that adhere the Scrum guidelines, while Level 3 means the lowest compliance from Scrum guidelines.

Figure 3 shows how in overall the Scrum adoption happened in the board of information system, Universitas Islam Indonesia. We discussed the details below.

5. Individual Factors

Our findings on individual factors found that experience is something that needs to be addressed. To address this issue, the team member should work more on various assignments, projects, and teams during a specific time as prescribed by [13]. A high level of experience of the project management software teams and coaches who teach Agile technical practices can improve the experience part [14]. This finding is similar to the research on Government Entities in the United Arab Emirates that one challenge is that there are no employees who are experienced with agile methods [5].

Table 4. Means in organizational factors

No.	Variables	Indicator	Mean Group Dev (N: 21)	Ops (N: 9)	PO (N: 6)	SM_ Mg (N: 5)	Total (N: 41)	Mean per Variables
1	Training	FOrg_A1_Pelat	4.14	4.22	4.33	4.60	4.24	4.11
		FOrg_A2_Pelat	3.86	4.00	4.00	4.40	3.98	
2	Recognition	FOrg_B1_Rekog	4.19	3.89	4.33	4.40	4.17	3.86
		FOrg_B2_Rekog	3.57	3.22	3.67	3.80	3.54	
3	Quality	FOrg_C1_Kual	4.00	3.56	3.83	4.20	3.90	3.95
		FOrg_C2_Kual	3.95	3.67	3.67	4.00	3.85	
		FOrg_C3_Kual	3.95	3.89	3.83	4.00	3.93	
		FOrg_C4_Kual	4.14	3.67	4.17	4.60	4.10	
4	Resource	FOrg_D1_SD	3.86	3.56	3.83	4.20	3.83	3.88
		FOrg_D2_SD	4,05	3.44	3.50	4.80	3.93	
5	Collaboration	FOrg_E1_Kolab	4.48	4.33	4.33	4.80	4.46	4.41
		FOrg_E2_Kolab	4.29	4.11	4.50	4.60	4.32	
		FOrg_E3_Kolab	4.38	4.22	4.50	5.00	4.44	
6	Management Support	FOrg_F1_DukMg	4.33	4.00	4.17	5.00	4.32	3.93
		FOrg_F2_DukMg	4.19	4.22	3.67	4.60	4.17	
		FOrg_F3_DukMg	3.43	3.78	3.83	4.40	3.68	
		FOrg_F4_DukMg	3.43	3.78	4.00	4.60	3.73	
		FOrg_F5_DukMg	3.67	4.11	3.83	4.40	3.88	
		FOrg_F6_DukMg	3.81	3.78	3.33	4.20	3.78	
7	Organizational culture	FOrg_G1_BudOrg	3.71	3.56	3.33	4.00	3.66	3.65
		FOrg_G2_BudOrg	3.52	3.56	3.67	4.20	3.63	
8	Organizational structure	FOrg_H_StrOrg	3.48	3.56	3.50	4.20	3.59	3.59

Table 5. Means in technology factors

No.	Variables	Indicator	Mean Group Dev (N: 21)	Ops (N: 9)	PO (N: 6)	SM_ Mg (N: 5)	Total (N: 41)	Mean per Variables
1	Relative advantage	FTek_A1_KeunRel	3.71	4.11	3.83	3.60	3.80	3.89
		FTek_A2_KeunRel	3.95	4.11	3.83	4.20	4.00	
		FTek_A3_KeunRel	3.90	3.78	3.83	4.00	3.88	
		FTek_A4_KeunRel	3.76	3.78	3.83	3.80	3.78	
		FTek_A5_KeunRel	4.00	3.89	3.83	4.20	3.98	
2	Complexity	FTek_B1_Kompl	3.52	3.44	3.00	3.80	3.46	3.60
		FTek_B2_Kompl	3.76	3.56	3.50	4.20	3.73	
3	Compatibility	FTek_C1_Kompatb	3.48	3.00	3.00	3.40	3.29	3.53
		FTek_C2_Kompatb	3.67	3.56	3.33	4.40	3.68	
		FTek_C3_Kompatb	3.57	3.67	3.17	4.40	3.63	

Meanwhile, in the commitment variable's escalation, a notification is necessary to prevent management from continuing assigning resources to projects that show signs of failure. The Scrum team should notify the issue sooner to limit the waste of resources [13]. Over-engineering should also be avoided to limit additional software features and functions that the customer must not require [15].

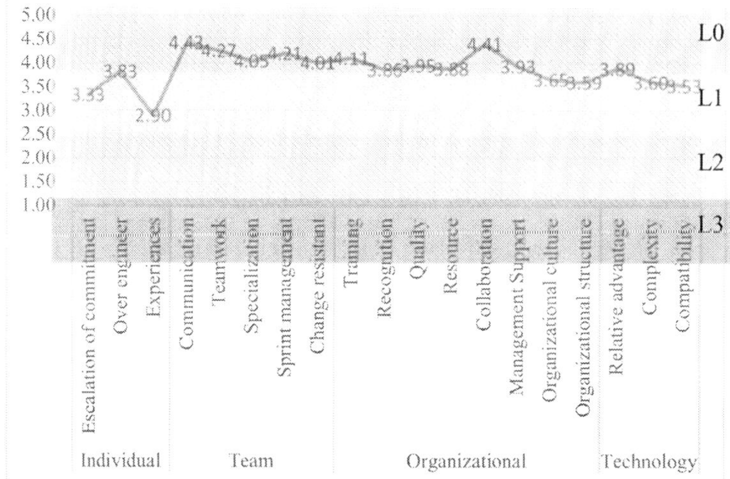

Fig. 3. Challenges in Agile Adoption

5.1. *Team factors*

This construct shows the best performance, with average means score greater than 4.0. Communication is one of the main factors for the success of the product [10]. Communication and collaboration of people in the organization greatly affect the quality of Scrum. Development team members are cross-functional experts who have considerable experience in all technical roles. On top of that, they also require considerable social skills to facilitate communication and collaboration with other team members [16].

To improve the teamwork, the board of information system, Universitas Islam Indonesia, extensively used agile software support such as Jira and Confluence to document the work. The use of the software helps to overcome the recurring problems in teamwork. To improve the work's effectiveness, the organization prepares each team member to specialize in something, e.g., UI/UX engineer, back-end engineer, front end engineer, site reliability engineer, and security engineer. The specialization is achieved through extensive training and weekly TechTalk updates. Sprint management is also managed well to monitor and manage the sprint [13].

5.2. *Organizational factors*

Organization factors are still big homework for the board of information systems, as most of them scored less than 4. More training is needed to improve both individual and group [13]. Lack of training can trigger problems such as ignorance about the importance of meeting in Scrum. As collaboration is one of the values mentioned in the Agile Alliance's Manifesto, this organization needs to collaborate with more stakeholders to lessen the problems encountered in adopting agile [17]. [18] explain that customers play a significant role in software projects' success with their role as customer collaboration, customer satisfaction, and customer commitment.

Another organizational factor that is also important is to match the remuneration, the rewards, and the benefits with the productivity levels of workers [13]. The team member will be pleased more if the acknowledgment of their contribution is recognized. This, in turn, can improve the quality of the developed software, so it meets business and user expectations. Organizational behavior and culture, such as an open environment and collegial (less hierarchical structure), also play an important part in enabling innovation adoption. To improve the organization's culture, the organization should also focus on human aspects [19] so that a conducive situation can emerge.

5.3. Technological factors
The overall average means score in technological factors is less than 4. This indicates that the organization needs to pay more attention to these variables. The relative advantages of Scrum have not been felt yet to its maximum potential. Scrum should be able to contribute positively to each individual and the organization itself [14]. Due to unfamiliarity since the organization's Scrum adoption is still relatively new, the Scrum is perceived to be complex and slightly incompatible with the existing culture.

6. Conclusion
This research has shown how SACDM model can help understand agile adoption challenges in higher education in Indonesia with specific case studies in the Board of Information System, Universitas Islam Indonesia. From the discussion part, the communication problem is addressed well with the help of Scrum. It helps overcoming the recurring problems in teamwork. Overall, the Scrum approach helps to improve the software development cultures in the organization. The Board of Information System, Universitas Islam Indonesia, seems to adopt the approach well. However, some issues persist, especially in improving the experience variable by having more variety in the projects and providing more training. This study itself is without limitation. The questionnaires are only distributed in one organization. Thus, biased answers are unavoidable. A further study is needed in a different scenario, such as in start-ups, where Scrum adoption is far more prevalent.

References
[1] Heeks R 2003 Summary for policymakers *Climate Change 2013 - The Physical Science Basis* **14** pp 1–30 DOI: 10.1017/CBO9781107415324.004
[2] The Standish Group 2015 *CHAOS REPORT 2015* Retrieved on 2020/02/10 from https://www.standishgroup.com/sample_research_files/CHAOSReport2015-Final.pdf
[3] Joseph N 2014 *A Predictive Model for Information Technology Project Success* (Johannesburg: Diss. University of Johannesburg)
[4] Apriyanto R D and Putro H P 2018 Tingkat kegagalan dan keberhasilan proyek sistem informasi di Indonesia *Seminar Nasional Teknologi Informasi Dan Komunikasi 2018 (SENTIKA 2018)*
[5] Hajjdiab H and Taleb A S 2011 Adopting agile software development: issues and challenges *Int. J. of Managing Value and Supply Chains (IJMVSC)* **2(3)** pp 1–10
[6] Smith G and Sidky A 2009 *Becoming agile:--in an imperfect world* (New York: Manning)
[7] Benefield G 2008 Rolling out agile in a large enterprise *Proc. of the 41st Annual Hawaii Int. Conf. on System Sciences (HICSS 2008)* (New Jersey: IEEE)
[8] Serrador P and Pinto J K 2015 Does Agile work?—A quantitative analysis of agile project success *Int. J. of Project Management* **33(5)** pp 1040–51
[9] Hajrizi E and Bytyci F 2015 Agile software development process at financial institution in Kosovo *IFAC-PapersOnLine* **48(24)** pp 153–6
[10] dos Santos L S *et al* 2018 A scrum-based process to distributed projects in multidisciplinary teams: a case study *Proc. of the 13th Int. Conf. on Global Software Eng.*
[11] Rubin K S 2012 *Essential Scrum: A practical guide to the most popular Agile process* (Boston: Addison-Wesley)
[12] Atlas A 2009 Accidental adoption: The story of scrum at amazon.com *2009 Agile Conf.* (New Jersey: IEEE)
[13] Hanslo R and Mnkandla E 2018 Scrum adoption challenges detection model: SACDM *2018 Federated Conf. on Computer Sci. and Information Systems (FedCSIS)* (New Jersey: IEEE)
[14] Senapathi M and Srinivasan A 2012 Understanding post-adoptive agile usage: An exploratory cross-case analysis *J. of Systems and Software* **85(6)** pp 1255–68
[15] Santos R *et al* 2011 Agile technical management of industrial contracts: scrum development of ground segment software at the european space agency *Int. Conf. on Agile Software Development* (Berlin, Heidelberg: Springer)
[16] Sarpiri M N and Gandomani T J 2017 The impact of team climate on failure and success of scrum

adoption *Int. J. of Software Eng. & Applications (IJSEA)* vol **8** no 3 pp 33–8
[17] Beck K *et al* 2006 *Manifesto for Agile Software Development* Retrieved from https://agilemanifesto.org/
[18] Vithana V N 2015 Scrum requirements engineering practices and challenges in offshore software development *Int. J. of Computer Applications* **116(22)** pp 43–9
[19] Gandomani T J and Nafchi M Z 2016 Agile transition and adoption human-related challenges and issues: a grounded theory approach *Computers in Human Behavior* **62** pp 257–66

Penalty Strategy in The Fitness Function of Grey Wolf Optimizer for Minimum Spanning Tree Problem

Z Zukhri[1], I V Paputungan[2], N S Handayani[3] and V Satriadi[4]

[1,2]Department of Informatics, Universitas Islam Indonesia
[3]Master Program in Informatics, Universitas Islam Indonesia
[4]Undergraduate Program in Informatics, Universitas Islam Indonesia

E-mail: [1]zainudin@uii.ac.id, [2]irving@uii.ac.id, [3]nadya.satya007@gmail.com, [4]vebri.satriadi@uii.ac.id

Abstract. The Grey Wolf Optimizer (GWO) is a relatively new population-based optimizer. Various optimization problems have been solved using GWO. This paper presents an experiment on how to apply GWO to solve the minimum spanning tree (MST) problem. MST is normally solved using revision strategy when formulating the fitness in other population-based algorithms. Another strategy, called Penalty strategy, is used in the experiment. Existing dataset for MST problem is tested using GWO. The experiment showed that implementation of penalty strategy in the fitness function of GWO can find a solution with almost 96% accuracy.

1. Introduction

Many real-world problems in the public services area such as minimizing cables of the power electric network, minimizing optical fibre of the communication networks, and minimizing pipes of the water supply networks, can be modelled as minimum spanning tree (MST) problems. In MST, the solution is modelled as a tree that consists of feasible connected edges and vertices where the total length of edges is minimum [1]. The feasible edge is the connector between two edges that are available in the graph. Meanwhile, an infeasible edge is a virtual connector between two edges and not available in the graph. MST is not only able to minimize the use of material to build the network, but also reduce the material waste. Minimum material will consequently save the budget for material procurement. For example, when developing an electrical network, longer cable will definitely need more money. However, at the same time, it makes more power loss [1]. In 2015, there was a power loss of almost 300 MVA within the Java-Bali power transmission line captured by Indonesian State Electricity Company [2]. The alternative solution to reduce this situation is by minimizing the energy transmission line. It can be done by minimizing the length of installation materials.

This paper describes how GWO finds the solution for MST problem. Similar to other population-based optimizer, the solution is normally generated in two different types of process in GWO. First is a random process which usually is done at the initial step, and a process that involves certain operators which usually is performed after initial step. However, both may produce infeasible solution. To handle such issue, two strategies, namely the revision and penalty strategy, can be applied [3]. Revision strategy means the optimizer is designed to avoid infeasible solutions. It works by repeating the process to generate feasible solution or transforming the infeasible solution into feasible one. Whereas in penalty strategy, infeasible solution is managed and carefully handled. When optimizer produces infeasible solutions during the process, the fitness value of these solutions will be reduced in order to lower the probabilities to be selected as the final solution.

GWO is a bio-inspired algorithm that adopts how grey wolf pack hunting process [5]. How the herd finds its prey is modelled as computation to find a solution. It starts from the gathering of wolves at a certain point and will eventually move closer to the target using certain rules until the prey animal can be finally caught. Since the computational process is aimed to find a solution, the positions of the wolf pack must be decoded for each iteration as a set of solution candidates. The quality of each solution in this set is represented by a fitness function. By comparing the fitness value from all solutions, the best one is then produced.

Prior to decoding the position of the wolf pack, it should be noted that candidate solutions of MST are not always obtained merely based on the generated position. It is because a position either randomly generated or another movement of wolf pack. In other words, wolf pack position is occasionally not representing a tree or it is a tree but contains infeasible edge. Penalty strategy is proposed to handle this condition. This strategy is much simpler than revision strategy to find MST solution. The revision strategy, though, will normally add significant computational steps. The proposed fitness function is formulated by reducing the value when the decoding result is not a tree. Since minimization is the ultimate goal of MST, if such goal is used as the objective function, we then need to reverse this assumption in penalty strategy, i.e., penalty value must be escalated. When the decoding result is not a tree, the fitness value must be added by a certain term or it must be multiplied by a certain factor. How the term formulation must be added by the fitness value or the factor formulation must be multiplied by the fitness value will be presented in this paper.

This paper is presented in the following structure. Literature review on related works and infeasible solution in using GWO to solve MST problem is presented in section 2. Section 3 presents the methods of the research. The results and discussions are presented in section 4. Lastly, the conclusions and subsequent work of our research are presented.

2. Literature Review

This section presents several works on MST problem and also discusses what kind of infeasible MST solution that will be handled.

2.1. *Related Works*

MST problem is an interesting classical model that is able to solve real world problems while evolving the model. Currently, it is not only about the power network, the optical fibre network, or the water supply network, but also the design of logistics networks in inter Island Ocean [6] or even MST has been implemented as the standard protocol for wireless Internet access points [7], etc.

Several studies have been performed to solve MST problem. The revision strategy is applied in the placement of phasor measurement units [8]. Genetic Algorithm (GA) was used to solve the MST. The researcher called the chromosome that generates an infeasible solution as an unobservable chromosome. When an unobservable chromosome is generated, they repeat the chromosome generation procedure until a feasible solution is obtained. The revision strategy is also applied in solving MST using evolutionary algorithm [9][10]. Such work shows that only the feasible solutions are inspected. The evolutionary algorithm can only produce less than 70% feasible solutions or almost a third of the total chromosomes would be ignored. Zukhri used a simplified Genetic Algorithm to solve the MST problem [10], however the study implemented a revision strategy. All in all, it is difficult to find studies that proposed penalty strategy to solve MST.

2.2. *Infeasible solution in using GWO to solve MST problem*

As mentioned in the introduction, applying GWO to solve MST problem might produce infeasible solution. It can be caused by either the position of the wolf pack which does not represent the tree or the presence of an infeasible edge in the tree. Pseudocode of GWO in Figure 1 shows how the infeasible solution is generated from the position of a wolf. Such pseudocode is adopted from how the grey wolf pack hunts, see Figure 2 [5].

At the initial point during hunting, the wolves from the gathering point will form a formation to randomly chase (Figure 2.A) and surround the target animal (Figure 2.B–2.E). Likewise, when moving to intimidate prey, wolves will make certain adjustments to each other in response to the movement of the prey. Changing the wolf pack's position will transform the MST solution. Since the updated wolf pack positions can be anywhere in the area around the prey and some conditions must be met to become an MST solution, it is possible that the transformed MST solution will not meet the requirements. It is called an infeasible solution.

Initialize the grey wolf population X_i *(i = 1, 2, ..., n)* *)
Initialize a, A, and C
Calculate the fitness of each search agent
X_α *= the best search agent*
X_β *= the second best search agent*
X_δ *= the third best search agent*
while *(t < Max number of iterations)*
    for *each search agent*
        Update the position of the current search agent by equation *)
    end for
    Update a, A, and C
    Calculate the fitness of all search agents
    Update X_α, X_β, *and* X_δ
    t=t+1
end while
return X_α

Figure 1. The pseudocode of GWO [5]

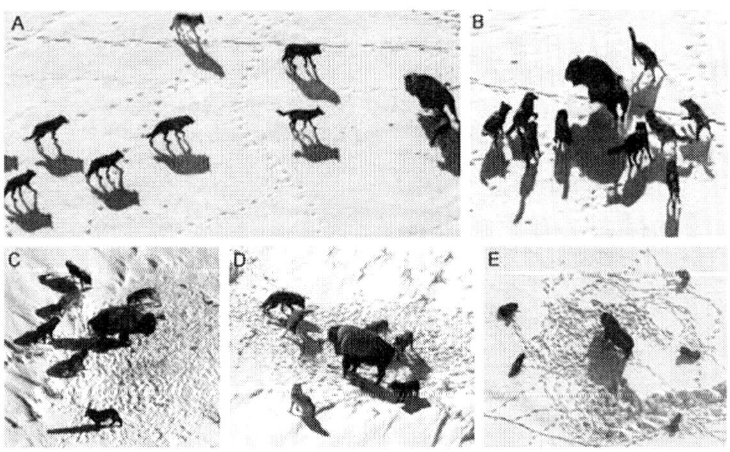

Figure 2. Natural events inspire GWO [5]

In Figure 1, statements that can generate an infeasible solution are considered as the initialization statement on the first line. The agent position then updates statement on a nested loop (the starred lines). Both starred statements are potential to generate random wolf positions. Although the number of edges is five, such edges will not surely form the MST solution since it is not a tree or a tree that contains an infeasible edge.

3. Methods

This section shows the customization part of GWO to solve MST problem with two alternative fitness functions containing 1) modelling MST problem, 2) decoding the position of the wolf pack, 3) fitness function, 4) dataset, 5) measurement of the solution quality, and 6) hypothesis testing.

3.1. Modelling MST problem to be solved by GWO
MST problem is shown in a graph that consists of m edges and n vertices where (n-1) edges must be selected from m to form a tree connection of n vertices. Virtual edges that are not available in the graph are ignored in the proposed method. The selected edges in this method are always available in the graph. Hence, an infeasible solution is only generated by circular edges.

3.2. Decoding the position of the wolf pack as a solution candidate
According to the proposed model, each wolf position in GWO must be decoded into a solution candidate, both feasible and infeasible solution. Wolf position is limited to the interval 0 to 100 in the coordinate system. This value will be used to select one of the available edges in the sorted edges, from 1 to m. This selection will be made (n-1) times expecting that n vertices in the graph will be connected. Consequently, the edges that have already been selected will not be on the set of sorted edges for the next selection. Thus, for each iteration, the number of edges in the set of sorted edges will be decreased by one. Illustration of the decoding wolf position to the solution candidate is shown in Figure 3.

The sequence of available edges in the graph (E)	e_1	e_2	e_3	.	.	.	.	e_{m-2}	e_{m-1}	e_m

Wolf Coordinate	x_1	x_2	x_3	.	.	.	.	x_{n-1}
Selected edges (E')	e'_1	e'_2	e'_3	.	.	.	.	e'_{n-1}

where
e_i is the ith edge in E, $1 < i < m$
x_j is the jth coordinate of the pack
e'_j is the jth edge in the solution candidate, $1 < j < n - 1$
At j^{th} iteration of the edge selection, the number of available edge will decrease as m_j

Figure 3. Illustration of selecting (n-1) edges in the solution candidate of m edges in the graph of the proposed model

The index of the selected edge is selected in the range 1 to m, hence the index of selected edge, j, can be obtained based on equation (1) and (2).

$$e'_j = e_{index} \tag{1}$$

$$index = floor\left(\frac{x_i}{100} m_j\right) + 1 \tag{2}$$

In order to make the index less than or equal to m_j, the maximum value of x_i must be 100. Therefore, if the index is greater than m_j since x_i equal to 100, the index must be equal to m_j as shown in equation (3).

$$index = m_j, for\ x_i = 100 \tag{3}$$

3.3. Fitness function
How to formulate the fitness function is very important in this work. The objective function is set as fitness function. Since the proposed method is penalty strategy, there is a part of this function that penalizes the infeasible solution. Thus, the proposed fitness function will mainly consist of the objective and the penalty function part. As MST is minimizing the total distance, penalty function must be added to the fitness function value. Additional or multiplication form which is equivalent to the infeasibility level of the solution candidates is proposed. A higher level makes a bigger penalty value and vice versa. The lowest level makes the penalty value = 0. It is easy to measure the infeasibility level of the solution candidates. It is indicated by the circular connector of several selected edges. The circular connector will separate one or more vertices. Therefore, the penalty value must be proportional to the number of unconnected vertices.

To formulate the penalty value that is proportional to the infeasibility of the solution, we do not make an experiment with penalty functions mentioned by Zukhri for generic constrained optimization problems [4]. We propose the penalty function as simple as possible. The first formula is equal to the multiplication of unconnected vertices number (n_c) and the maximum weight of the edge in the graph (w_i), where i in range between 1 and number of edge (n) and w_j is the weight of e_j' in Figure 3. The second one is equal to the multiplication of unconnected vertices number and the square of the maximum weight of the edge. We also do not use the MST objective function that is a function defined on the matrix that is usually applied in operational research. We prefer to define the MST objective function based on the MST definition as mention in the introduction of this paper [1] since it is a function defined on the array as equation (4) that more simple, less complicated. Finally, we get two alternatives of fitness function we proposed as the sum of the objective function & penalty function as equation (5) and equation (6). We needed to simulate both alternatives in the experiments to get one of them whose better performance.

$$f_{obj} = min \sum_{j=1}^{n-1} w_j \quad (4)$$

$$f_{fitness_1} = f_{obj} + n_c . max(w_i) \quad (5)$$

$$f_{fitness_2} = f_{obj} + n_c . (max(w_i))^2 \quad (6)$$

where i is the index of edge in the graph and j is the index of edge in the solution.

3.4. Dataset

The experiments to test the proposed method are dataset from previous research as shown in Figure 4 [11], Figure 5 [11], and Figure 6 [12]. In order to make the graph model read easily, the model is not-to-scale. The minimum solutions for the three datasets are 16, 13, and 202 [11, 12], respectively.

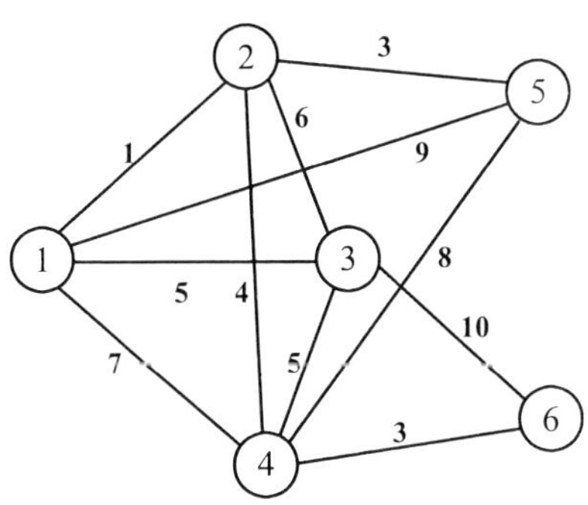

Figure 4. Data [11].

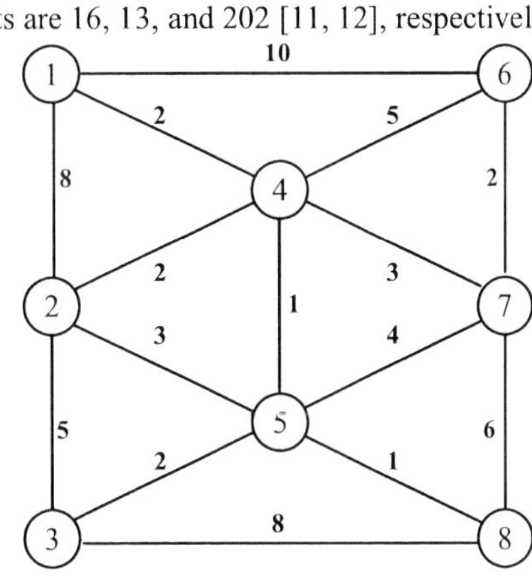

Figure 5. Data 2 [11].

3.5. Measurement of the solution quality

We have discussed that a fitness function is the measurement of the solution quality in the computation. But it is only used to compare the solution with each other & it cannot be used to measure how good the solution is compared by the known solution. For this matter, we use the concept of accuracy to measure the quality of the solution. The concept should be applied for all dataset and support to the determined hypotheses testing. The best solution is equal to the known solution from the previous researches. We will call it as the minimum solution. Oppositely, there is the worst solution. It is the maximum value of feasible solution found in the computation. This value is called the maximum solution. Hence, the accuracy is defined as the proportion of the difference between the maximum solution and the solution found by the optimizer compared to the difference between the maximum solution and the minimum solution as shown in equation (7).

$$accuracy = \frac{max\ solution - solution}{max\ solution - min\ solution} . 100\% \quad (7)$$

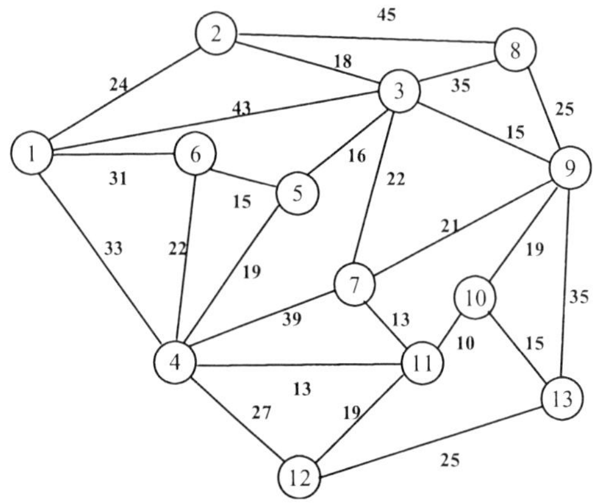

Figure 6. Data 3 [12].

3.6. Hypothesis testing

The hypothesis testing in this study is carried out based on the statistical testing data, i.e. the solution quality found by GWO with $fitness_1$ is assumed to equal the quality found by GWO with $fitness_2$ as in equation (8) and equation (9).

$$H_0: \mu_1 = \mu_2 \tag{8}$$

$$H_1: \mu_1 \neq \mu_2 \tag{9}$$

We use the two-sample t-test. The null hypothesis (H_0) fails to be rejected if the test value calculated based on statistical data is at the critical t interval on both sides and vice versa if the statistical value is outside the critical t interval on both sides, the alternative hypothesis (H_1) will be accepted.

The experiment is performed using an open source program at https://github.com/7ossam81/EvoloPy with inserting the proposed MST fitness functions into benchmarks.py file and adding dataset as input text.

4. Results and Discussions

Different fitness function does not make a significant difference to the GWO algorithm. The penalty function shows that the second fitness function will require a longer computation time, but it is not that significant. Therefore, we can skip the computational time. To shorten the experiment time, we ran the simulation software on 2 different machines. Each data will be subjected to 349 trials for performance testing. The number of trials is based on the Isaac and Michael Tables [13] and the optimizer parameter adjusts to the number of vertices in the graph. The parameters of each testing, the minimum known solution, and the maximum solution of each data were found in the test as shown in Table 1. In testing, the optimizer with both penalty functions can find the minimum solution. The results of two-sample t-test at the level of significance = 1% are shown in Table 2.

Table 1. Dataset and parameter testing.

	Data 1	Data 2	Data 3
Population Size	6	10	45
Iteration	50	200	3000
Max Solution	37	37	287
Min Solution	16	13	202

The experiments showed that GWO with two alternatives penalty functions can find solutions to the MST problem in all datasets. Based on the accuracy means in the t-sample t-test table, GWO with $fitness_1$ showed slightly better results than another. But the results of the two-sample t-test indicated that the difference in mean value is not significant enough. It is indicated by the t-critical value in the interval of ± t critical two-tail value interval or H_0 is accepted. However, we have to consider the computation time that we don't need to test. It is clear that the quadratic form in the penalty function will slightly increase the time complexity of the algorithm. Of course, the multiplication operator will make the fitness function simpler than the squares operator. Finally, it can be concluded that penalty strategy in GWO for solving the MST problem can be formulated by increasing the value of objective function slightly more than the possible maximum solution. We do not much more value than the maximum solution.

Table 2. Two-sample t-test results.

Accuracy	Data 1		Data 2		Data 3	
	Fitness 1	*Fitness 2*	*Fitness 1*	*Fitness 2*	*Fitness 1*	*Fitness 2*
Mean	98.376	98.717	96.538	97.576	96.572	96.760
Variance	18.142	13.598	38.565	29.380	7.379	8.658
Observations	349	349	349	349	349	349
Pooled Variance	15.870		33.973		8.019	
Hypothesized Mean Difference	0		0		0	
df	696		696		696	
t Stat	-1.131		-2.354		-0.881	
P(T<=t) one-tail	0.129		0.009		0.189	
t Critical one-tail	2.332		2.332		2.332	
P(T<=t) two-tail	0.258		0.019		0.379	
t Critical two-tail	2.583		2.583		2.583	

Meanwhile, the value of p on the second test shows that the accuracy value of the experiment is not homogeneous ($p \leq 0.05$). Although the accuracy value in the second test is homogeneous ($p> 0.05$) we can say that GWO produces a non-homogeneous solution. It is possible since we cannot give specific parameter values to simulate wolf pack behaviour. We can only set the common parameter of the algorithm, i.e., population size and number of iterations. It is possible that a more homogeneous solution value would be generated if we could set the specific algorithm parameter. GWO needs to be developed by observing other behaviour of wolf pack to insert a certain specific parameter for this algorithm in order to explore more searching space.

5. Conclusions

The penalty strategy can be applied at GWO to solve MST problems with almost 96% accuracy regardless of the computation time. We do not need to penalize all the infeasible solutions to this problem. The coding of the problem solutions in GWO for solving this problem as agent positions can be designed to prevent the formation of unavailable edges in the graph. Penalties are only applied to solution candidates generated by circular edges. It will reduce the number of infeasible solutions. Based on the tested penalty functions, we only need a penalty formula that gives slightly greater than the maximum possible solution. It is simpler than penalizing the infeasible solution with much more value than the maximum solution. In the future, GWO can be developed by observing wolf pack behaviour in more depth so that there are certain parameters besides population size and number of iterations. That way, the behaviour of the wolf pack can be changed and provide the possibility for exploration of search space.

Acknowledgments

This work is part of the project sponsored by the Department of Informatics, Universitas Islam Indonesia, Yogyakarta, Indonesia, 2020.

References

[1] Sedgewick R and Wayne K 2018 *Minimum Spanning Tree* Retrieved on October 2020 from https://algs4.cs.princeton.edu/43mst/

[2] Pamungkas M A P, Priharto D and Putranto H 2019 Study of technical and non-technical factors in energy consumption on 20 kV distribution networks *Frontier Energy System and Power Eng.* vol **1** no 2 (Malang: Electrical Engineering, Universitas Negeri Malang) pp 1–6

[3] Sulistiawati I B and Ashari M I 2016 Artificial bee colony algorithm for optimal power flow on transient stability of Java-Bali 500 KV *Proc. of Second Int. Conf. on Electrical Systems, Technol. and Information 2015* (Singapore: Springer Science+Business Media) p 247

[4] Zukhri Z 2014 *Algoritma Genetika, Metode Komputasi Evolusioner untuk Menyelesaikan Masalah Optimasi* (Yogyakarta: Penerbit Andi)

[5] Mirjalili S, Mirjalili S M and Lewis A 2014 Grey wolf optimizer *Advances in Eng. Software* vol **69** (Amsterdam: Elsevier) pp 46–61

[6] Kismanti T S 2016 *Penentuan Jaringan Logistik pada Transportasi Laut Menggunakan Fuzzy C-Means dan Minimum Spanning Tree Berbasis Hybrid Genetic Algorithm* (Surabaya: Institut Teknologi Sepuluh Nopember Surabaya)

[7] Cisco Systems Inc 2017 *Cisco IOS Configuration Guide for Autonomous Cisco Aironet Access Points - Releases 15.3(3) JE and Later* (Cisco Systems Inc)

[8] Devi M M and Geethanjali M 2020 Hybrid of genetic algorithm and minimum spanning tree method for optimal PMU placements *Measurement* vol **154** (Amsterdam: Elsevier) pp 1–12

[9] Bolton C C, Rey C, Cossio S R, Rodríguez C, Gatica F and Parada V 2016 Automatically produced algorithms for the generalized minimum spanning tree problem *Scientific Programming* vol **2016** (London: Hindawi) pp 1–11

[10] Zukhri Z 2017 Penyelesaian masalah pohon rentang minimum dengan algoritme genetika yang disederhanakan suatu adopsi operator dari sistem kekebalan buatan *Seminar Nasional Inovasi dan Aplikasi Teknologi di Industri* (Malang: Institut Teknologi Nasional) p A.39.1

[11] Hidayat T 2000 Algoritma genetik untuk pemecahan persoalan minimum spanning tree *Seminar Nasional Aplikasi Sistem Cerdas dalam Rekayasa dan Bisnis* (Yogyakarta: Universitas Islam Indonesia) p 53

[12] Setiawan S and Wibowo A M 1999 *Struktur Data Graf* Retrieved on October 2020 from http://aren.cs.ui.ac.id/sda/archive/1998/handout/handout20.html

[13] Sugiyono 2018 *Statistik Nonparametris untuk Penelitian 2nd edition* (Bandung: Alfabeta)

AUTHOR INDEX